Summability Theory and Its Applications

Summability Theory and Its Applications

Feyzi Başar
İnönü University, Turkey

CRC Press
Taylor & Francis Group
Boca Raton London New York

CRC Press is an imprint of the
Taylor & Francis Group, an **informa** business
A CHAPMAN & HALL BOOK

Second edition published 2022
by CRC Press
6000 Broken Sound Parkway NW, Suite 300, Boca Raton, FL 33487-2742

and by CRC Press
4 Park Square, Milton Park, Abingdon, Oxon, OX14 4RN

First edition published by Bentham Books 2012

CRC Press is an imprint of Taylor & Francis Group, LLC

ISBN: 978-1-032-27536-9 (hbk)
ISBN: 978-1-032-25949-9 (pbk)
ISBN: 978-1-003-29415-3 (ebk)

DOI: 10.1201/9781003294153

Typeset in Latin Modern
by KnowledgeWorks Global Ltd.

This study is dedicated to the memory of respected Professor Emeritus Ekrem Öztürk who was the supervisor of the author during his master's and doctoral degree research.

Feyzi Başar

Contents

Foreword

This book is actually timely and is intended for graduate and research students who have an interest in sequence spaces, summability methods and their applications. The author of the book, Professor Feyzi Başar, is one of the renowned researchers in this field, whose conscientiousness is reflected in the organization the contents of the book.

Professor Başar has successfully tried to capture the spirit of this emerging and fascinating discipline in his book and present the width and depth of the topics intelligently. His survey starting from the very basic definitions highlights the progress and developments of the subject in a well-organized manner in order to motivate the readers.

This book covers many interesting studies on sequence spaces, e.g., topological properties, matrix transformations, matrix domains of triangles, spectrum, core and fuzzy study. So, it would attract researchers from various fields.

Two other books on this subject are also worth mentioning here: One by A. Wilansky (*Summability through Functional Analysis*, North-Holland Mathematics Studies 85, Amsterdam-New York-Oxford, 1984) and the other by J. Boos (*Classical and Modern Methods in Summability*, Oxford University Press Inc., New York, 2000). Chapters 5–8 of this book make it different from the other two books on this topic, which cover the most recent work on sequence spaces. The presentation by Professor Başar is very simple and straightforward.

This book also provides the basic tools for researchers using directly or indirectly the notion of sequences and series and their convergence problems using modern summability methods. Professor Başar has given emphasis on the use of soft analysis, which makes the subject matter easily comprehensible. This book is a Bible of modern summability methods.

Professor M. Mursaleen
Aligarh Muslim University

Preface to Second Edition

This second edition is a corrected, revised and reprinted version of our original textbook. Besides, we also introduce the domain of four-dimensional binomial matrix in the spaces of bounded, convergent in the Pringsheim's sense, both convergent in the Pringsheim's sense and bounded, and regularly convergent double sequences, in Chapter 7.

In Chapter 9, after giving the concepts related to the absolute summability of sequences and series, it is investigated some properties of absolute summability methods. Additionally, some inclusion and summability factors theorems are given together with the general absolute inclusion theorem involving a pair of triangles.

We are particularly grateful to Professor Bilâl Altay, Department of Mathematical Education, İnönü University, Malatya, Türkiye, who have carefully read the revised form of the textbook and a number of suggestions for corrections. We express our thanks and gratitude to Professor Malkowsky, Department of Mathematics, State University in Novi Pazar, Serbia, who provided us the proof of Part (b) of Theorem 2.5.6 at our request. Finally, we wish also to thank Professor Serkan Demiriz, Department of Mathematics, Tokat Gaziosmanpaşa University, Tokat, Türkiye, for sending the required TEX document related to the domains of four-dimensional matrices in some spaces of double sequences. I must express my thanks and gratitude to my dear colleague Prof. Dr. Rifat Çolak, Department of Mathematics, Faculty of Science, Fırat University, Elâzığ, Türkiye, who was the editor of the first edition of the book. Finally, I would like to express my sincere thanks to Professor Callum Fraser, Editor of Mathematics Books of CRC Press/Taylor & Francis Group, who decided to include the second edition of the book among the CRC Press/Taylor & Francis Group publications without delay and thanks to Mansi Kabra from the Editorial Department, who quickly handled the bureaucratic procedures related to the printing process.

ORIGINAL PREFACE TO FIRST EDITION

This book is intended for graduate students and researchers with a (special) interest in spaces of single and double sequences, matrix transformations and matrix domains. Besides Preface and Index, the book consists of eight chapters and is organized as follows:

The book *Infinite Matrices and Sequence Spaces* of Cooke is fundamental for referring to the theory of infinite matrices. So, we introduce the required definitions and topics related to infinite matrices in Chapter 1.

In Chapter 2, certain normed and paranormed sequence spaces are studied, and the α-, β-, γ- and continuous duals of the spaces ℓ_∞, c, c_0 and ℓ_p of all bounded, convergent, null and absolutely p-summable sequences are determined together with some other sequence spaces isomorphic to them. Additionally, a table of the α-, β- and γ-duals of certain normed sequence spaces is given.

In Chapter 3, the matrix transformations in sequence spaces are studied and the characterizations of the classes of Schur, Kojima and Toeplitz matrices together with their versions for the series-to-sequence, sequence-to-series and series-to-series matrix transformations are given.

Chapter 4 is devoted to the domains of some particular summability matrices, with a special emphasize on the Cesàro, difference, m^{th}-order difference, Euler, Riesz, Cesàro and weighted mean sequence spaces, and other spaces derived in this way. Also, the Schauder bases of those spaces, their α-, β- and γ-duals, and the characterizations of some classes of matrix transformations are given.

In Chapter 5, the spectrum and the fine spectrum of the Cesàro operator C_1, the difference operator $\Delta^{(1)}$, the generalized difference operator $B(r,s)$ and the operator generated by the triple band matrix $B(r,s,t)$ acting on the sequence spaces c_0, c, ℓ_p and bv_p with respect to Goldberg's classification are determined, where $1 \leq p < \infty$.

In Chapter 6, the Knopp core, σ-core, $\mathcal{I}$-core and $F_{\mathcal{B}}$-core of a sequence are studied. Also, a short survey for the results related to the core of a sequence is given.

In Chapter 7, the fundamental results on double sequences and related topics are given. In particular, the concept of convergence of double series in the Pringsheim's sense is defined, certain spaces of double sequences are introduced, and their α- and β-duals are determined. Additionally, some classes of four-dimensional matrices are characterized.

Chapter 8 is devoted to the sequences of fuzzy numbers. After presenting the fundamental facts concerning convergent sequences of fuzzy numbers, some results on statistical convergence of sequences of fuzzy numbers and related results are given. Also the α-, β- and γ-duals of the classical sets $\ell_\infty(F)$, $c(F)$, $c_0(F)$ and $\ell_p(F)$ of all bounded, convergent, null and absolutely p-summable sequences of fuzzy numbers are determined, and the classes $(\mu(F) : \ell_\infty(F))$, $(c_0(F) : c(F))$, $(c_0(F) : c_0(F))$, $(c(F) : c(F); p)$, $(\ell_p(F) : c(F))$, $(\ell_p(F) : c_0(F))$ and $(\ell_\infty(F) : c_0(F))$ of infinite matrices of fuzzy numbers are characterized, where $\mu \in \{\ell_\infty, c, c_0, \ell_p\}$. Finally, the quasilinearity of the classical sets of sequences of fuzzy numbers is investigated.

Feyzi Başar
April 2011, İstanbul

ACKNOWLEDGEMENTS

During the preparation of the present text, a number of friends, colleagues and students have helped and made useful comments and suggestions. The author wishes to express his warmest thanks to Professor Mohammad Mursaleen and Professor Syed Abdul Mohiuddine, Department of Mathematics, Aligarh Muslim University, Aligarh, India, for their valuable help on the α-, β-, γ- and continuous duals of certain sequence spaces in Section 2.5. We also thank Professor Mikâil Et, Department of Mathematics, Fırat University, Elâzığ, Türkiye, for his valuable help related to difference sequence spaces, especially on the difference sequence spaces of order m, in Section 4.3. We benefited a lot from Professor Celal Çakan's knowledge on the core of a sequence in Chapter 6. So, the author thank Professor Çakan, Department of Mathematical Education, İnönü University, Malatya, Türkiye, for his valuable help. We also express our sincerest thanks to Professor Maria Zeltser, Department of Mathematics, Tallinn University, Narva mnt. 25, Tallinn, Estonia, who kindly sent us personal copies of valuable document concerning spaces of double sequences. We are also especially indebted to Professor Salih Aytar, Department of Mathematics, Süleyman Demirel University, Isparta, Türkiye, for his valuable help on preparing Sections 7.1–7.3. The author also thank to Dr. Özer Talo, who is among his best graduate level students, for his careful reading and helpful corrections concerning Chapter 8. He would also like to thank Professor Bilâl Altay, Department of Mathematical Education, İnönü University, Malatya, Türkiye, who is his best student and collaborator, read in detail the whole manuscript, added required changes, corrected the errors in the TEX files and suggested numerous improvements. The author is particularly grateful to Professor Eberhard Malkowsky, Department of Mathematics, Fatih University, Büyükçekmece/İstanbul, Türkiye, for his careful reading and making several valuable suggestions and corrections, preparing the main TEX file and helping in correcting both the language and mathematical contents across the book.

The author expresses his gratitude to Sara Moqeet and Sarah A. Khan, Assistant Managers, e-books Publications Department, Bentham Science Publishers, for their kindly cooperation and also sincerely thanks Professor Mohamed Bakari, Department of American Culture and Literature, Fatih University, Büyükçekmece/ İstanbul, Türkiye, for his valuable help. Finally, he would like to extend his sincerest thanks to Professor Rifat Çolak, Department of Mathematics, Fırat University, Elâzığ, Türkiye, for his careful redaction of the whole of the text and several suggestions, which improved the presentation and readability of the book.

Feyzi Başar
April 2011, İstanbul

List of Tables

List of Abbreviations and Symbols

A' : transpose of a matrix A
$\overline{A}$: conjugate of a matrix A
$\overline{A'}$: conjugate transpose of a matrix A
$|A|$: determinant of a matrix A
A^k : k^{th} column of the matrix $A = (a_{nk})$, i.e., $A^k = (a_{nk})_{n=0}^{\infty}$ for all $k \in \mathbb{N}$
A_n : n^{th} row of the matrix $A = (a_{nk})$, i.e., $A_n = (a_{nk})_{k=0}^{\infty}$ for all $n \in \mathbb{N}$
A^{-1} : inverse of a matrix A
$\mathfrak{F}$: associative field of matrices
$x^{[n]}$: n^{th} section of a sequence $x = (x_k)$
$\{(Ax)_n\}$: A-transform of a sequence x
$\mathcal{A}x$: $\{(Ax)_n^i\}_{i,n=0}^{\infty}$
$A^{(-1)}$: right inverse of a matrix A
${}^{(-1)}A$: left inverse of a matrix A
A^0 : right zero-divisor of a matrix A
${}^0A$: left zero-divisor of a matrix A
$\mathrm{tr}(A)$: trace of a matrix A
$\chi(A)$: characteristic of a matrix A
C_1 : Cesàro mean of order 1
$C^{(m)}$: Cesàro mean of order m
E^r : Euler (Euler-Knopp) mean of order r
E_1 : original Euler matrix
T^r : Taylor matrix
R^t : Riesz mean generated by the sequence of coefficients $t = (t_k)$
N^t : Nörlund mean generated by the sequence of coefficients $t = (t_k)$
$H(\mu)$: Hausdorff matrix associated with the sequence (μ_n)
GWM : generalized weighted means
$|A|$: absolute A summability
$|C_1|$: absolute Cesàro summability of order 1
$|C_1|_k$: k^{th} order $|C_1|$ summability
$|C^{(m)}|$: absolute Cesàro summability of order m
$|C^{(m)}|_k$: k^{th} order $\left|C^{(m)}\right|$ summability
$|N^t|$: absolute Nörlund summability
$|N^t|_k$: k^{th} order absolute Nörlund summability
$|R^t|$: absolute Riesz summability

$\|R^t\|_k$	: k^{th} order absolute Riesz summability
PSM	: pair of summability methods
ω	: space of all sequences with complex entries
ϕ	: space of all finitely non-zero sequences
ℓ_∞	: space of bounded sequences
f	: space of almost convergent sequences
f_0	: space of almost null sequences
c	: space of convergent sequences
c_0	: space of null sequences
ℓ_p	: space of absolutely p-summable sequences
bs	: space of bounded series
fs	: space of almost convergent series
fs_0	: space of series almost converging to zero
cs	: space of convergent series
cs_0	: space of series converging to zero
bv	: space of sequences of bounded variation
bv_p	: space of sequences of p-bounded variation
ω_0^p	: space of sequences that are strongly summable to zero of index $p \geq 1$ by the Cesàro method of order 1
ω^p	: space of sequences that are strongly summable of index $p \geq 1$ by the Cesàro method of order 1
ω_∞^p	: space of sequences that are strongly bounded of index $p \geq 1$ by the Cesàro method of order 1
λ^α	: α-dual of a sequence space λ
λ^β	: β-dual of a sequence space λ
λ^γ	: γ-dual of a sequence space λ
λ^*	: continuous dual of a sequence space λ
λ^f	: f-dual of a sequence space λ
$\lambda^{r\alpha}$	: $r\alpha$-dual of a sequence space λ
$\lambda^{r\beta}$	: $r\beta$-dual of a sequence space λ
$\lambda^{r\gamma}$	: $r\gamma$-dual of a sequence space λ
λ^{rN}	: rN-dual of a sequence space λ
θ	: zero vector in a linear space X
$\ell_\infty(p)$	: space of sequences (x_k) such that $\sup_{k\in\mathbb{N}} \|x_k\|^{p_k} < \infty$
$c(p)$	: space of sequences (x_k) such that $\|x_k - l\|^{p_k} \to 0$, as $k \to \infty$
$c_0(p)$	: space of sequences (x_k) such that $\|x_k\|^{p_k} \to 0$, as $k \to \infty$
$\ell(p)$	: space of sequences (x_k) such that $\sum_k \|x_k\|^{p_k} < \infty$
$(a_k * b_k)$	: Cauchy product of sequences (a_k) and (b_k)
$(a_{kl} * b_{kl})$	: Cauchy product of double sequences (a_{kl}) and (b_{kl})
$\sum_{i,j=0}^{m,n}$	: $\sum_{i=0}^{m}\sum_{j=0}^{n}$
$\sum_{i,j}$	: $\sum_{i=0}^{\infty}\sum_{j=0}^{\infty}$
$[A,p]_0$	: set of sequences which are strongly summable to zero by $A = (a_{nk})_{n,k\in\mathbb{N}}$
$[A,p]$	: set of sequences which are strongly summable by $A = (a_{nk})_{n,k\in\mathbb{N}}$
$[A,p]_\infty$	: set of sequences which are strongly bounded by $A = (a_{nk})_{n,k\in\mathbb{N}}$
$L(\mathbb{R})$	: set of all fuzzy numbers on $\mathbb{R}$

W : set of all closed bounded intervals A with endpoints $\underline{A}$ and $\overline{A}$
$[u]_\alpha$: α-level set of $u \in L(\mathbb{R})$
$supp(u)$: set of real numbers t such that $u(t) > 0$
$u \not\sim v$: neither $u \preceq v$ nor $u \preceq v$
BSFN : a bounded sequence of fuzzy numbers
$\delta(K)$: natural density of a set K
Λ_x : set of all statistical limit points of the sequence $x = (x_k) \in \omega(F)$
Γ_x : set of all statistical cluster points of the sequence $x = (x_k) \in \omega(F)$
$x_k \sim \infty$: $x = (x_k)$ is definitely divergent
$\mathcal{A}$: sequence of infinite matrices $A^i = \{a_{nk}(i)\}$
$\mathcal{B}x$: $\{(Bx)^i_m\}_{i,m=0}^{\infty}$
$F_{\mathcal{B}}$: space of all $F_{\mathcal{B}}$-convergent sequences
$F_{0\mathcal{B}}$: space of all $F_{\mathcal{B}}$-null sequences
st : space of all statistically convergent sequences
$\|\mathcal{A}\|$: $\sup_{n,i\in\mathbb{N}} \sum_k |a_{nk}(i)|$
$core\{x\}$: core of a bounded sequence $x = (x_k)$
$\mathcal{K}$-$core(x)$: Knopp core of a sequence $x = (x_k)$
K_n : set $\{k \in K : k \leq n\}$
$\delta(K)$: natural density of K
Λ_x : set of all statistical limit points of a sequence $x = (x_k) \in \omega(F)$
Γ_x : set of all statistical cluster points of a sequence $x = (x_k) \in \omega(F)$
$\overline{A'_x}$: set of statistical lower bounds of a sequence $x = (x_k) \in \omega(F)$
$\mathcal{I}(\Gamma_x)$: set of all $\mathcal{I}$-cluster points of x
$\omega(F)$: set of all sequences of fuzzy numbers
$\ell_\infty(F)$: set of bounded sequences of fuzzy numbers
$c(F)$: set of convergent sequences of fuzzy numbers
$c_0(F)$: set of null sequences of fuzzy numbers
$\ell_p(F)$: set of absolutely p-summable sequences of fuzzy numbers
$bs(F)$: set of bounded series of fuzzy numbers
$cs(F)$: set of convergent series of fuzzy numbers
$cs_0(F)$: set of series of fuzzy numbers converging to zero
$F_{\mathcal{I}}(b)$: set of all $\mathcal{I}$-convergent and bounded sequences
D_∞ : Hausdorff metric on the set $\ell_\infty(F)$
D_p : Hausdorff metric on the set $\ell_p(F)$
$F(\mathfrak{X})$: set of fuzzy sets such that α-level sets $[u]_\alpha = \{t \in \mathfrak{X} : u(t) \geq \alpha\}$ are non-empty for every $\alpha \in (0, 1]$
$(c : c)$: class of conservative matrices
$(c : c; p)$: class of Toeplitz (regular) matrices
$(\lambda : \mu)_\varrho$: class of ϱ-multiplicative matrices
$(cs : c; p)$: class of γ-matrices
$(\ell_\infty : c)$: class of Schur (coercive) matrices
$rad(X)$: radical of an algebra X
δ_{ij} : Kronecker delta which is $= 1$ if $i = j$ and $= 0$ if $i \neq j$
$e^{(k)}$: sequences whose only non-zero term is a 1 in k^{th} place for each $k \in \mathbb{N}$
$\emptyset$: empty set

$(AB)_{ij}$	:	i^{th} row and j^{th} column entry of the matrix product AB
I	:	unit matrix
$D(T)$	:	domain of a linear operator T
$R(T)$	:	range of a linear operator T
$Ker(T)$	:	kernel or null space of a linear operator T
$r_\sigma(T)$	:	spectral radius of an operator $T \in B(X)$
T^*	:	adjoint of a bounded linear operator T
$B(X)$	:	unit ball in a normed space X
$S(X)$	:	unit sphere in a normed space X
$\mathcal{L}(X:Y)$	:	set of linear operators from a space X into a space Y
$\mathcal{B}(X:Y)$	:	set of bounded linear operators from a space X into a space Y
$B(X)$	:	set of bounded linear operators on a space X
$\sigma(T,X)$	:	spectrum of a linear operator T on a space X
$\rho(T,X)$	:	resolvent set of a linear operator T on a space X
$\sigma_p(T,X)$	:	point (discrete) spectrum of a linear operator T on a space X
$\sigma_c(T,X)$	:	continuous spectrum of a linear operator T on a space X
$\sigma_r(T,X)$	:	residual spectrum of a linear operator T on a space X
$\sigma_{ap}(T,X)$	:	approximate point spectrum of a linear operator T on a space X
$\sigma_\delta(T,X)$	:	defect spectrum of a linear operator T on a space X
$\sigma_{co}(T,X)$	:	compression spectrum of a linear operator T on a space X
$\mathbb{N}$	:	set of natural numbers, i.e., $\mathbb{N} = \{0, 1, 2, \ldots\}$
$\mathbb{N}_r$	:	set of integers which are greater than or equal to $r \in \mathbb{N}$
$\mathbb{Q}$	:	set of rational numbers
$\mathbb{R}^+$	:	set of non-negative real numbers
$\mathbb{R}$	:	set of real numbers, the real field
$\mathbb{C}$	:	set of complex numbers, the complex field
$Re[z]$	:	real part of $z \in \mathbb{C}$
$Im[z]$	:	imaginer part of $z \in \mathbb{C}$
$\mathbb{R}^n$	:	n-dimensional Euclidean space
$\mathbb{C}^n$	:	n-dimensional complex Euclidean space
$\mathbb{N}_1^2$	:	set of all pairs of positive integers
$\mathbb{N}^2$	:	set of all pairs of non-negative integers
$\mathcal{F}$	:	collection of all finite subsets of $\mathbb{N}$
$\{0,1\}^{\mathbb{N}\times\mathbb{N}}$	:	set of all double sequences consisting of 0's and 1's
$[a]$	:	integer part of a number a
D_x	:	derived set, the set of limit points, of a sequence x

Author

Dr. Feyzi Başar has been Professor Emeritus since 2016, at İnönü University, Malatya, Turkey. He has published three e-books for graduate students and researchers and more than 160 scientific papers in the field of summability theory, sequence spaces, FK-spaces, Schauder bases, dual spaces, matrix transformations, the spectrum of certain linear operators represented by a triangle matrix over some sequence spaces, the α-, β- and γ-duals and some topological properties of the domains of some two- and four-dimensional triangles in certain spaces of single and double sequences, sets of the sequences of fuzzy numbers and multiplicative calculus. As of April 2021, 5746 citations to Professor Başar's research and publications were made (see https://scholar.google.com.tr/citations?hl=tr&user= uhWs28gAAAAJ). He has guided 17 MA and 10 Ph.D. students, and served as a referee for 142 international scientific journals. He is a reviewer of Mathematical Reviews since 2007 and Zentralblatt MATH, and a member of editorial boards of 21 scientific journals. He is also a member of scientific committees of 17 mathematics conferences, delivered talks at 14 different universities as an invited speaker and participated in more than 70 mathematics symposiums with papers.

CHAPTER 1

Infinite Matrices

The book *Infinite Matrices and Sequence Spaces* of Cooke is fundamental for referring to the theory of infinite matrices. So, we introduce the necessary definitions and topics related to infinite matrices in the present chapter.

Keywords: Infinite matrix, addition, product and scalar multiplication of infinite matrices, unit matrix, left and right inverse of a matrix, transpose of a matrix, lower and upper semi-matrix, orthogonal and unitary matrices, semi-continuous and continuous matrices, left and right zero-divisors of a matrix.

1.1 PRELIMINARIES

Following Cooke [111], we start by giving a short survey related to infinite matrices. An infinite matrix $A = (a_{ij})$ of complex numbers is a double sequence of complex numbers defined by a function A from the set $\mathbb{N} \times \mathbb{N}$ into the complex field $\mathbb{C}$, where $\mathbb{N}$ denotes the set of natural numbers, i.e., $\mathbb{N} = \{0, 1, 2, \ldots\}$. The complex number a_{ij} denotes the value of the function at $(i, j) \in \mathbb{N} \times \mathbb{N}$ and is called the entry of the matrix in the ith row and j^{th} column. We wish to emphasize, shortly, on the differences between finite and infinite matrix theories.

The *addition* and *scalar multiplication* of the infinite matrices $A = (a_{ij})$ and $B = (b_{ij})$ are defined by $A + B = (a_{ij} + b_{ij})$ and $\alpha A = (\alpha a_{ij})$, where α is a scalar. The *product* AB of the infinite matrices $A = (a_{ij})$ and $B = (b_{ij})$ is defined by

$$(AB)_{ij} := \sum_k a_{ik} b_{kj} \quad \text{for all} \quad i, j \in \mathbb{N} \tag{1.1.1}$$

provided the series on the right-hand side of (1.1.1) converge for all $i, j \in \mathbb{N}$, where $(AB)_{ij}$ denotes the entry of the matrix AB in the ith row and j^{th} column. For simplicity in notation, here and in what follows, the summation without limits runs from 0 to ∞. Since the series on the right-hand side of (1.1.1) may diverge for some, or all, values of i, j; the product AB of the infinite matrices A and B may not exist.

A large number of theorems had been established for finite n-square matrices. However, the corresponding theorems for infinite matrices are rarely obtained from them by letting n tend to infinity.

DOI: 10.1201/9781003294153-1

1.1.1 Some Problems Involving the Use of Infinite Matrices

(a) Consider the following system of an infinite number of linear equations in infinitely many unknowns $x_0, x_1, x_2, \ldots$

$$\sum_k a_{ik}x_k = y_i \quad \text{for all} \quad i \in \mathbb{N}. \tag{1.1.2}$$

If we construct an infinite matrix $A = (a_{ik})$ with the coefficients a_{ik} of the unknowns x_k and denote the vectors of unknowns and constants by X and Y, then (1.1.2) can be expressed in matrix form as follows:

$$AX := Y \tag{1.1.3}$$

Define the infinite matrix B such that the pth row is zero except b_{pq} which is 1. Then, the pth row of BC is the same as the qth row of C. Similarly, if the elements of the qth column of C are zero except c_{pq} which is 1, then the qth column of BC is the same as the pth column of B. Hence, $IA = AI = A$, where $I = (\delta_{ij})$ is called *unit matrix* and is defined for all $i, j \in \mathbb{N}$ by

$$\delta_{ij} := \begin{cases} 1 & , \quad i = j, \\ 0 & , \quad i \neq j. \end{cases} \tag{1.1.4}$$

Let us suppose that *the left inverse* ${}^{(-1)}A$ of a matrix A exists, such that ${}^{(-1)}AA = I$. Then, under certain conditions, if we multiply both sides of (1.1.3) on the left by ${}^{(-1)}A$, we obtain the solution of (1.1.3) as $X = {}^{(-1)}AY$.

(b) A very important application of infinite matrices is used in the theory of summability of divergent sequences and series which is considered in Chapters 2–7.

A simple example of this is the *Cesàro mean of order 1* which is the well-known method of summability and is defined by the matrix $C_1 = (c_{nk})$, as follows:

$$c_{nk} := \begin{cases} \frac{1}{n+1} & , \quad 0 \leq k \leq n, \\ 0 & , \quad k > n. \end{cases} \tag{1.1.5}$$

The C_1-transform of a sequence $x = (x_k)$ is the sequence $y = (y_n)$ defined by

$$y_n := (C_1x)_n = \frac{1}{n+1}\sum_{k=0}^{n} x_k \quad \text{for all} \quad n \in \mathbb{N}.$$

One can easily see that the C_1-transform of the sequence $x = \{(-1)^k\}$, which is obviously bounded but divergent, is the null sequence

$$y = (y_n) := \left\{\frac{1 + (-1)^n}{2(n+1)}\right\}.$$

(c) A further important application of infinite matrices is the Heisenberg-Dirac theory of quantum mechanics. Here, two basic problems consist in solving two linear equations in infinite matrices:

(i) For a given matrix A, the *quantization equation* which is in the form $AX - XA = I$.

(ii) For a given matrix A and a diagonal matrix D, the equation is in the form $AX - XD = 0$.

1.2 SOME DEFINITIONS

The matrix obtained from an infinite matrix A by interchanging rows and columns is called the *transpose* of $A = (a_{ij})$ and is denoted by $A' = (a'_{ij})$ so that $a'_{ij} = a_{ji}$ for all $i, j \in \mathbb{N}$.

The *zero matrix* 0 is the matrix whose entries are all equal to zero. Thus, it is obvious that $A0 = 0A = 0$. But, $AB = 0$ does not imply $A = 0$ or $B = 0$. Indeed, this does not follow even for the simplest types of matrices, such as diagonal matrices. To see this, define the diagonal matrices $D = diag(d_0, d_1, d_2, \ldots)$ and $E = diag(e_0, e_1, e_2, \ldots)$ by $d_i = 0$ when $i \in \{1, 3, 5, \ldots\}$ and $d_i \neq 0$ when $i \in \{0, 2, 4, \ldots\}$; and $e_i = 0$ when $i \in \{0, 2, 4, \ldots\}$ and $e_i \neq 0$ when $i \in \{1, 3, 5, \ldots\}$. Then, it is immediate that $DE = diag(d_0e_0, d_1e_1, d_2e_2, \ldots) = 0$ although neither D nor E is the zero matrix.

The diagonal matrix αI is called a *scalar matrix* whose leading diagonal entries are all equal to α.

If every row of a matrix A contains only a finite number of non-zero elements, then A is said to be *row-finite*. If the same is true with respect to every column, then A is said to be *column-finite*. Thus, if $A = (a_{ij})$ is row-finite, then $a_{ij} = 0$ for $j \geq q_i$, where q_i is a function of i. If $a_{ij} = 0$ for $j \geq q$, then A is called a *row-bounded matrix*, where q is independent of i. Similarly, if $a_{ij} = 0$ for $i \geq r$, then A is called a *column-bounded matrix*, where r is independent of j.

If $a_{ij} = 0$ for $j > i$, $A = (a_{ij})$ is called a *lower semi-matrix* or *lower triangular matrix* . The matrix C_1 of arithmetic mean defined by (1.1.5) is an example of a lower semi-matrix. Similarly if $a_{ij} = 0$ for $j < i$, $A = (a_{ij})$ is called an *upper semi-matrix* or *upper triangular matrix*.

A matrix $A = (a_{ij})$ is said to be *symmetric* if $A' = A$ and *skew-symmetric* if $A' = -A$, that is $a_{ij} = a_{ji}$ and $a_{ij} = -a_{ji}$ for all $i, j \in \mathbb{N}$, respectively.

The *conjugate* $\overline{A}$ of a complex matrix $A = (a_{ij})$ is the matrix $\overline{A} = (\overline{a}_{ij})$, where $\overline{a}_{ij}$ is the conjugate of the complex number a_{ij} in the usual sense.

A matrix $A = (a_{ij})$ is said to be *Hermitian* if $\overline{A}' = A$ and *skew-Hermitian* if $\overline{A}' = -A$, i.e., $\overline{a}_{ij} = a_{ji}$ and $\overline{a}_{ij} = -a_{ji}$ for all $i, j \in \mathbb{N}$, respectively.

From the above definitions, it is easy to see the following relations:

$$\begin{array}{llllll}
\text{(i) } \overline{(\overline{A})} & = A, & \text{(ii) } (A')' & = A, & \text{(iii) } \overline{(\overline{A}')'} & = A, \\
\text{(iv) } \overline{AB} & = \overline{A}\,\overline{B}, & \text{(v) } (AB)' & = B'A', & \text{(vi) } \overline{(AB)}' & = \overline{B}'\,\overline{A}'.
\end{array}$$

The results given by (i)–(iii) are trivial. The result (iv) follows from the fact that the series $\sum_k \alpha_k$ and $\sum_k \overline{\alpha}_k$ with complex terms are both convergent or both divergent. To prove (v), it is not hard to see by taking into account the definitions of matrix product and the transpose of a matrix that

$$(AB)'_{ij} := (AB)_{ji} = \sum_k a_{jk}b_{ki} = \sum_k b'_{ik}a'_{kj} = (B'A')_{ij}$$

holds for all $i, j \in \mathbb{N}$. The result (vi) follows by combining the facts (iv) and (v).

If $AB = I$, then B is called a *right inverse* of A which is denoted by $A^{(-1)}$, and A is called a *left inverse* of B which is denoted by ${}^{(-1)}B$.

If A and B are both different from 0, and if $AB = 0$, then B is called a *right zero-divisor* of A, and A is called a *left zero-divisor* of B which are denoted by A^0 and ${}^0B$, respectively.

When the sum converges $\sum_i a_{ii}$ is called the *trace of the matrix* A and is denoted by $tr(A)$. This concept is used in some applications of the quantum theory of radiation.

1.3 SOME CHARACTERISTIC PROPERTIES OF INFINITE MATRICES

Since the product of two diagonal matrices $D = diag(d_0, d_1, d_2, \ldots)$ and $E = diag(e_0, e_1, e_2, \ldots)$ is the diagonal matrix

$$DE = diag(d_0e_0, d_1e_1, d_2e_2, \ldots) = diag(e_0d_0, e_1d_1, e_2d_2, \ldots) = ED,$$

multiplication is commutative for diagonal matrices.

But in general, the products of matrices are not commutative, since $\sum_k a_{ik}b_{kj}$ is not equal to $\sum_k b_{ik}a_{kj}$ for every i and j, even assuming that both series converge for every i and j, in general. In fact, AB may not exist when BA exists. For example, if $b_{ij} = 0$ when $j \in \mathbb{N}_1$, then $BA = (b_{i0}a_{0j})$ which exists for arbitrary b_{i0} and A; but $(AB)_{ij} = \sum_k a_{ik}b_{kj} = 0$ when $j \in \mathbb{N}_1$ and $(AB)_{i0} = \sum_k a_{ik}b_{k0}$, so that AB does not exist if the last sum diverges.

The sum of two matrices always exists and is commutative and associative, i.e.,

$$A + B := B + A \quad \text{and} \quad A + (B + C) := (A + B) + C.$$

The *distributive property*

$$A(B + C) := AB + AC, \quad (A + B)C := AC + BC$$

holds provided that if AB and AC exist, then also $A(B + C)$ exists and is equal to $AB + AC$. But, $A(B + C)$ may exist when AB and AC do not exist. For example, this holds if $a_{ij} = 1$ for all $i, j \in \mathbb{N}$, $b_{ij} = d_i + 1$, $c_{ij} = d_i - 1$ for all $j \in \mathbb{N}$ with $\sum_i d_i$ converges.

It is not hard to see from the above note on diagonal matrices that the multiplication of finite number of diagonal matrices is associative. The similar result also holds for the lower semi-matrices.

But, the multiplication of infinite matrices is not associative, in general. For example, if $a_{ij} = c_{ij} = 1$ for all $i, j \in \mathbb{N}$, and if

$$\sum_j \left(\sum_i b_{ij} \right) \neq \sum_i \left(\sum_j b_{ij} \right) \tag{1.3.1}$$

then, $(AB)C \neq A(BC)$, where $\{(AB)C\}_{ij}$ and $\{A(BC)\}_{ij}$ are given in the left and right-hand sides of (1.3.1).

As an example of a matrix satisfying (1.3.1), we can define the matrix $B = (b_{ij})_{i,j\in\mathbb{N}}$ by

$$b_{ij} := \begin{cases} 0 & , \quad i = j = 0 \\ 2^{-i} & , \quad j = 0, i \in \mathbb{N}_1, \\ -2^{-j} & , \quad i = 0, j \in \mathbb{N}_1, \\ \frac{(i-j)}{2^{i+j}} \frac{(i+j-1)!}{i!j!} & , \quad i, j \in \mathbb{N}_1. \end{cases}$$

Then, it is easily seen that

$$\sum_j \left(\sum_i b_{ij} \right) = b_{00} + \sum_{i=1}^{\infty} b_{i0} = 1 \text{ and } \sum_i \left(\sum_j b_{ij} \right) = b_{00} + \sum_{j=1}^{\infty} b_{0j} = -1.$$

Consider the set S of infinite matrices such that

(a) S contains the scalar matrices.

(b) Every finite product of the elements of S exists and is associative.

(c) S is closed under the finite sum and finite matrix product.

Then, the set S is called an *associative field $\mathfrak{F}$ of matrices.*

Diagonal matrices, row-finite and column-finite matrices form associative fields.

The positive integral powers of a matrix A may be defined by induction as

$$A^2 := A \cdot A, \quad A^3 := A^2 \cdot A, \ldots, A^n := A^{n-1} \cdot A.$$

A non-zero matrix A is said to be *idempotent*, if $A^2 = A$.

If p is the least positive integer such that $A^p = 0$, then the non-zero matrix A is said to be *nilpotent with index p*.

It is trivial that I is idempotent. Besides, the matrix $A = (a_{nk})_{n,k\in\mathbb{N}}$ whose only non-zero element a_{13} is nilpotent with index 2.

A is called *orthogonal matrix*, if $A'A = AA' = I$.

U is called *unitary matrix*, if $\overline{U}'U = U\overline{U}' = I$.

If A and B are orthogonal, $AB(AB)'$ and $(AB)'AB$ exist and are associative, then by taking into account the hypothesis $A'A = AA' = I$ and $B'B = BB' = I$ it is easy to see that

$$\begin{aligned} AB(AB)' &= ABB'A' = AA' = I \\ (AB)'AB &= B'A'AB = B'B = I, \end{aligned}$$

that is, AB is also orthogonal.

1.4 SOME SPECIAL INFINITE MATRICES

(i) Let P be the matrix obtained from unit matrix I by changing its pth and q^{th} rows, where $p, q \in \mathbb{N}$. Then, PA can be found from A by changing its pth and q^{th} rows. The matrix P is called a *permutator*. Permutators for columns can be derived in a similar way, with AP instead of PA. Indeed, if P is the matrix obtained from unit matrix I by changing its pth and q^{th} columns, then AP can be obtained from A by changing its pth and q^{th} columns.

(ii) Consider the matrix $E = (e_{ij})$ defined by

$$e_{ij} := \begin{cases} 1 & , \quad i = j = 0 \text{ or } j = i + 1, i \in \mathbb{N}_1, \\ 0 & , \quad \text{otherwise.} \end{cases}$$

Then, the matrices EA and AE' correspond to the matrix A without its second row and second column, respectively. In fact, if $EA = (f_{ij})_{i,j \in \mathbb{N}}$ and $AE' = (g_{ij})_{i,j \in \mathbb{N}}$, then

$$f_{ij} = \sum_k e_{ik} a_{kj} := \begin{cases} a_{0j} & , \quad i = 0, \\ a_{i+1,j} & , \quad i \in \mathbb{N}_1 \end{cases}$$

and

$$g_{ij} = \sum_k a_{ik} e'_{kj} := \begin{cases} a_{i0} & , \quad j = 0, \\ a_{i,j+1} & , \quad j \in \mathbb{N}_1. \end{cases}$$

A matrix such as E is called a *selector*. More generally, if E is the matrix derived from the unit matrix by omitting its pth, q^{th}, r^{th}, etc. rows, then EA is the matrix derived from the matrix A by omitting its pth, q^{th}, r^{th}, etc. rows. Therefore, we can formulate this, as follows: Let $f(i)$ be an increasing function such that $f(i) \in \mathbb{N}$ for all $i \in \mathbb{N}$. Consider the matrix $E = (e_{ij})$ defined by

$$e_{i,f(i)} := \begin{cases} 1 & , \quad i \in \mathbb{N}, \\ 0 & , \quad \text{otherwise.} \end{cases}$$

Then, E is called a *row-selector*. In this situation, EA is obtained from A by omitting the rows whose suffixes are missed by $f(i)$.

(iii) F is called *combinator*, if its only non-zero off-diagonal element is f_{mn}. If all the diagonal elements of F are equal to 1 and $f_{mn} = r$, then FA is obtained from A by adding r times of the nth row to the mth row. Then, the entries of the matrix $FA = (g_{ij})$ are given by

$$g_{ij} := \sum_k f_{ik} a_{kj} := \begin{cases} a_{ij} & , \quad i \neq m, \\ a_{m,j} + r a_{nj} & , \quad i = m. \end{cases}$$

The concept of the combinator can be generalized. The matrix F derived by replacing the rth row of the unit matrix by the numbers $c_0, c_1, c_2, \ldots$ such that $\sum_k c_k$

absolutely converges is called a *row combinator*. Therefore, all the rows of FA are the same as in A and the rth row is the combination of rows

$$\sum_i c_i a_{ij} \text{ for all } j \in \mathbb{N}. \tag{1.4.1}$$

Define $c = [c_0, c_1, c_2, \ldots]$ and $e = [e_0, e_1, e_2, \ldots]$, where e_j denotes the sum of (1.4.1). Then, we write in matrix notation $cA = e$. Thus, the row combinator F has a row c replacing the rth row of I and FA replaces the rth row of A by cA. Combinators are used for producing a zero row in A when there is a linear homogenous relation between the rows, say $\sum_i c_i a_{ij} = 0$ for all $j \in \mathbb{N}$.

Similarly, the matrix F derived by replacing the rth column of the unit matrix by the numbers $c_0, c_1, c_2, \ldots$ such that $\sum_k c_k$ absolutely converges is called a *column combinator*. Therefore, all the columns of AF are the same as in A and the rth column is the combination of columns $\sum_j a_{ij} c_j$ for all $i \in \mathbb{N}$.

1.5 THE STRUCTURE OF AN INFINITE MATRIX

Define the matrix $A + \lambda I$ whose leading diagonal elements are obtained from the leading elements of A adding λ and all of the off-diagonal elements are the same as in A, i.e.,

$$(A + \lambda I)_{ij} := \begin{cases} a_{ij} & , \quad i \neq j, \\ a_{ij} + \lambda & , \quad i = j, \end{cases} \text{ for all } i, j \in \mathbb{N}.$$

Since *positive integral powers of a matrix* A are defined, a polynomial $f(A)$ of A has a meaning in a self-associative field.

In the theory of finite matrices, the characteristic roots play an important role. The characteristic roots of an n-square matrix A are the roots of the *characteristic equation*

$$p(\lambda) := |\lambda I - A| = \begin{vmatrix} \lambda - a_{11} & -a_{12} & -a_{13} & \cdots & -a_{1n} \\ -a_{21} & \lambda - a_{22} & -a_{23} & \cdots & -a_{2n} \\ -a_{31} & -a_{32} & \lambda - a_{33} & \cdots & -a_{3n} \\ \vdots & \vdots & \vdots & \ddots & \vdots \\ -a_{n1} & -a_{n2} & -a_{n3} & \cdots & \lambda - a_{nn} \end{vmatrix} = 0,$$

where $p(\lambda)$ is called the *characteristic polynomial* of the matrix A. It is well-known by the Cayley-Hamilton theorem that every n-square matrix A satisfies its characteristic equation $p(A) = 0$.

We also have a classical theorem on the structure of a finite matrix A that when the characteristic roots of A are distinct, A can be expressed in a simple way in terms of these characteristic roots and certain idempotent matrices associated with A. This result also holds in the case of multiple characteristic roots.

It is natural to expect the extension of the above results to infinite matrices. The first and perhaps most serious difficulty which is encountered in attempting to extend these structural theorems to general infinite matrices is that of obtaining what corresponds to the characteristic roots of a finite matrix. We can define the *characteristic root of an infinite matrix* A as any scalar λ for which $Ax = \lambda x$, where $x = (x_0, x_1, x_2, \ldots)$ is a non-zero column vector. This leads us to the discussion of the spectrum of a linear operator represented by an infinite matrix and is examined in Chapter 5.

If A is a lower semi-matrix, since $|\lambda I - A| = 0$ when $\lambda_i = a_{ii}$ for all $i \in \mathbb{N}$, then the characteristic roots of A are its leading diagonal entries.

1.6 THE EXPONENTIAL FUNCTION OF A LOWER-SEMI MATRIX

We have seen that a definite meaning can be assigned to a polynomial in an infinite matrix A. For suitably restricted classes of infinite matrices, this may be extended to more general functions than polynomials, e.g., the exponential function.

It is well-known that the exponential function $f(z) = e^z$ has no zero in the complex plane. The exponential function exists for every lower semi-matrix, all its values are lower semi-matrices and it misses all the nilpotent values in the field of lower semi-matrices.

(i) Since lower semi-matrices form a field, the exponential function may be defined by

$$f(A) := I + \frac{A}{1!} + \frac{A^2}{2!} + \cdots + \frac{A^n}{n!} + \cdots ,$$

where

$$\{f(A)\}_{ij} = \delta_{ij} + a_{ij} + \frac{1}{2!}\sum_{k=j}^{i} a_{ik}a_{kj} + \cdots + \frac{1}{n!}\sum_{k} a_{ik} \sum_{l} a_{kl} \cdots \sum_{q} a_{pq}a_{qj} + \cdots .$$

Let b be the maximum modulus of the elements a_{kj} in the first i rows of A. Then,

$$|\{f(A)\}_{ij}| \leq 1 + \frac{b}{1!} + \frac{ib^2}{2!} + \cdots + \frac{i^k b^{k+1}}{(k+1)!} + \cdots \leq e^{ib}.$$

Hence, $f(A)$ has a definite meaning in the field of lower semi-matrices.

(ii) In the Cauchy product of $f(A)$, $f(B)$, the terms that are homogenous and of the nth degree in A and B are

$$\frac{A^n}{n!} + \frac{A^{n-1}B}{1!(n-1)!} + \frac{A^{n-2}B^2}{2!(n-2)!} + \cdots + \frac{B^n}{n!},$$

which is equal to $(A + B)^n/n!$ in an associative field when A and B commute.

(iii) Since, $f(A + B) = f(A)f(B)$ if A and B are commutative by (ii), and $f(0) = I$ by definition, one can derive from here by putting $B = -A$ that $f(0) = f(A)f(-A) = I$. This shows that $f(A)$ does not vanish for any A.

1.7 SEMI-CONTINUOUS AND CONTINUOUS MATRICES

It is often convenient to use the operator obtained by replacing the natural number n in the infinite matrix $A = (a_{nk})$ by a continuous non-negative variable, say x. Then, it is usually written $a_k(x)$ instead of a_{xk} and A is called as a *semi-continuous matrix*. Besides, both natural numbers n and k are sometimes replaced by continuous non-negative variables, say x and y, and then it is also usually written $a(x, y)$ instead of $a_{x,y}$ and A is called as a *continuous matrix*. It is obvious that a continuous matrix seems to a function of two non-negative variables and is employed in the quantum theory and in abstract statistics.

The usual matrix $A = (a_{nk})$ is sometimes called as a *dis-continuous matrix*. Throughout the text, we employ unless stated otherwise the term *matrix* instead of *dis-continuous matrix*.

1.8 INVERSES OF INFINITE MATRICES

In this section, we give some knowledge on the inverses of infinite matrices.

1.8.1 Inverses of Lower Semi-Matrices

In Section 1.2, the concepts of right and left inverses of an infinite matrix were defined. Therefore, the left inverse of the infinite matrix A is a solution in X of the linear matrix equation

$$AX = I. \tag{1.8.1}$$

Similarly, the right inverse of the infinite matrix A is a solution in X of the linear matrix equation

$$XA = I. \tag{1.8.2}$$

It is obvious that the linear matrix equations in (1.8.1) and (1.8.2) consist of an infinite number of unknowns and an infinite number of linear equations. To solve them, it is important to find the left inverse and right inverse of the coefficients matrix A.

Now, we may give the following result on the lower semi-matrices:

Proposition 1.8.1. (cf. Cooke [111, (2.1, I); p. 19]) *The lower semi-matrix A has no right inverse if $a_{ii} = 0$ for one or more values of $i \in \mathbb{N}$. But, if $a_{ii} \neq 0$ holds for all $i \in \mathbb{N}$, then A has a unique right inverse which is a lower semi-matrix and whose leading diagonal elements are $1/a_{ii}$.*

Proof. Since A is a lower semi-matrix, the matrix equation in (1.8.1) can be expressed in the coordinate form, as follows

$$\sum_{k=0}^{i} a_{ik}x_{kj} := \delta_{ij} \text{ for all } i, j \in \mathbb{N}. \tag{1.8.3}$$

When $i = 0$, we obtain $a_{00}x_{00} = 1$, $a_{00}x_{0j} = 0$ if $j \in \mathbb{N}_1$. If $a_{00} = 0$, the first equation has no solution in x_{00} so that in this case the solution vector X does not exist. If $a_{00} \neq 0$, then $x_{00} = 1/a_{00}$ and $x_{0j} = 0$ for $j \in \mathbb{N}_1$.

When $i = 1$, we have

$$a_{10}x_{00} + a_{11}x_{10} = 0, \quad a_{11}x_{11} = 1, \quad a_{11}x_{1j} = 0 \tag{1.8.4}$$

for $j \geq 2$. If $a_{11} = 0$, the second equation has no solution in x_{11} so that in this case the solution vector X does not exist. If $a_{11} \neq 0$, then $x_{11} = 1/a_{11}$. Therefore, the first equation in (1.8.4) yields $x_{10} = -a_{10}/a_{11}^2$ and the third equation gives that $x_{1j} = 0$ for $j \geq 2$, when $a_{11} \neq 0$.

Proceeding in this way, we have a direct row by row determination of X which is unique. □

Example 1.8.2. (cf. Cooke [111, p. 20]) *The Euler semi-matrix* $A = (a_{ij})$ *defined by*

$$a_{ij} := \begin{cases} \frac{(-1)^{j-1}(i-1)!}{(i-j)!(j-1)!} & , \quad 1 \leq j \leq i, \\ 0 & , \quad j > i \end{cases}$$

is its own inverse, both left and right.

Indeed one can easily observe that for $j < i$,

$$\begin{aligned} \sum_{k=j}^{i} a_{ik}a_{kj} &= (-1)^{j-1}\frac{(i-1)!}{(j-1)!}\sum_{k=j}^{i}\frac{(-1)^{k-1}(k-1)!}{(i-k)!(k-1)!(k-j)!} \\ &= \frac{(i-1)!}{(i-j)!(j-1)!}\left[1 - \frac{i-j}{1!} + \frac{(i-j)(i-j-1)}{2!} - \cdots + (-1)^{i-j}\right] \\ &= \frac{(i-1)!}{(i-j)!(j-1)!}(1-1)^{i-j} = 0 \end{aligned}$$

and for $j > i$, *the sum* $\sum_{k=j}^{i} a_{ik}a_{kj}$ *is obviously* 0 *and for* $j = i$, *the sum* $\sum_{k=j}^{i} a_{ik}a_{kj}$ *consists of the single term for which* $k = j = i$ *and is equal to* 1 *which means that the inverse of* A *is itself.*

Proposition 1.8.3. (cf. Cooke [111, (2.1, II); p. 20]) *If* A *has a zero row (or column),* A *has no right (or left) inverse.*

Proof. Since the proof is similar for the case of a zero column, we consider only the case of a zero row. Suppose that $a_{nk} = 0$ for a fixed n and for every $k \in \mathbb{N}$. Then, $\sum_k a_{nk}x_{kj} = \delta_{nj}$ reduces to the contradiction $0 = 1$ when $j = n$, so that A^{-1} does not exist. □

Proposition 1.8.4. (cf. Cooke [111, (2.1, III); p. 21]) *If the elements of a row (or column) in* A *are the same multiple of the corresponding elements of another row (or column),* A *has no right (or left) inverse.*

Proof. Since the proof is similar for the case of proportional columns, we consider only the case of proportional rows. Suppose that $a_{i_1,k} = ca_{i_2,k}$ for fixed i_1, i_2 and for every $k \in \mathbb{N}$. Then, we have $\delta_{i_1,j} = c\delta_{i_2,j}$ which leads to the contradiction $1 = 0$ when $j = i_1$. □

If A^{-1} is a solution of linear matrix equation (1.8.1), then $A^{-1} + A^0$ is also a solution and every solution is in this form. Further, in the same associative field, there is no matrix A has both a right inverse $A^{(-1)}$ and a left zero-divisor ${}^0A$ or both a left inverse ${}^{(-1)}A$ and right zero-divisor A^0.

The relation between left and right inverses is given by the following theorem:

Theorem 1.8.5. (cf. Cooke [111, (2.2, I); p. 21])

(i) *If $A^{(-1)}$ is unique and $(AA^{(-1)})A$ is associative, then $A^{(-1)}$ is also a left inverse of A and the only one for which ${}^{(-1)}AAA^{(-1)}$ is associative.*

(ii) *If A has both a right inverse $A^{(-1)}$ and a left inverse ${}^{(-1)}A$, and ${}^{(-1)}AAA^{(-1)}$ is associative, then ${}^{(-1)}A = A^{(-1)}$, and A has no other two-sided inverse for which ${}^{(-1)}AAA^{(-1)}$ is associative.*

Proof. (i) If Ω is a solution of $AX = B$, then all other solutions are of the form $\Omega + A^0$. In particular, the solutions of $AX = A$ are all of the form $I + A^0$. If $A^{(-1)}$ is unique, there is no A^0, since $A^{(-1)} + A^0$ would be another right inverse; and hence $X = I$ is the only solution of $AX = A$. Now, if $(AA^{(-1)})A$ is associative, $A(A^{(-1)}A) = (AA^{(-1)})A = A$ so that $A^{(-1)}A = I$, and hence, $A^{(-1)}$ is also a left inverse of A.

Moreover, if B is a left inverse of A such that $BAA^{(-1)}$ is associative we have $B = B(AA^{(-1)}) = (BA)A^{(-1)} = A^{(-1)}$. In a similar way, the second part can be established.

(ii) Since ${}^{(-1)}AA = I$ and $AA^{(-1)} = I$ by the hypothesis, we have

$$\left({}^{(-1)}AA\right)A^{(-1)} = A^{(-1)} \quad \text{and} \quad {}^{(-1)}A\left(AA^{(-1)}\right) ={}^{(-1)}A$$

so that when ${}^{(-1)}AAA^{(-1)}$ is associative, it follows that ${}^{(-1)}A = A^{(-1)}$.

Fixing $A^{(-1)}$, every ${}^{(-1)}A$ such that ${}^{(-1)}AAA^{(-1)}$ is associative is equal to this fixed $A^{(-1)}$ which means that there is one ${}^{(-1)}A$ only in this associative field. Similarly, the proof can be given for $A^{(-1)}$ and this step completes the proof. □

Following Cooke [111], we summarize the results concerning the inverse of an infinite matrix.

(a) Let $A = (a_{nk})$ be a lower semi-matrix for which $a_{ii} \neq 0$ holds for every $i \in \mathbb{N}$, so that by Proposition 1.8.1, A has a unique right inverse X. Then, X is also a left inverse of A and is the only two-sided inverse of A.

For $AX = I$, i.e., $AXA = A$ or $A(XA - I) = 0$. But since $X = A^{(-1)}$ is unique, there is no A^0. Hence, $XA - I = 0$ so that X is also a left inverse of A. Since A and X are lower semi-matrices, the associative property is valid, above. However, it does not follow that X is the unique left inverse of A. For example, if the non-zero elements of A are $a_{i,i-1}$ and a_{ii} for every $i \in \mathbb{N}$, then the matrix equation $XA = I$ reduces to

$$x_{i,j+1}a_{j+1,j} + x_{ij}a_{jj} = \delta_{ij} \tag{1.8.5}$$

which by fixing i and varying j yields the ith row of X when the first column of X is arbitrarily given. Thus, this matrix has only one right inverse but has an infinity of left inverse.

(b) It follows by Theorem 1.8.5 that if, in a suitable associative field $\mathfrak{F}$, A has a unique $A^{(-1)}$ which is a two-sided inverse of A and is unique in $\mathfrak{F}$. In $\mathfrak{F}$, if A has both a right inverse $A^{(-1)}$ and a left inverse ${}^{(-1)}A$, then $A^{(-1)}$ and ${}^{(-1)}A$ coincide and form the unique two-sided inverse of A in $\mathfrak{F}$.

Hence, in the above example in (a), one solution of (1.8.5) must be the unique $A^{(-1)}$ but the other infinity of solutions cannot belong to a field $\mathfrak{F}$ in which ${}^{(-1)}AAA^{(-1)}$ is associative. In fact, the other infinity of solutions are all of the form $A^{(-1)} +^0 A$ and as we have already seen ${}^0AAA^{(-1)}$ is not associative.

(c) If the conditions of Theorem 1.8.5 are not satisfied, it may happen that A has no $A^{(-1)}$ but has an infinity of ${}^{(-1)}A$ or no ${}^{(-1)}A$ but has an infinity of $A^{(-1)}$.

Consider the selector $S = (s_{ij})$ defined by

$$s_{ij} := \begin{cases} 1 & , \quad i = j = 0, \\ 1 & , \quad j = i + 1, \ i \in \mathbb{N}_1, \\ 0 & , \quad \text{otherwise.} \end{cases}$$

Since the second column consists entirely of zero elements, S has no left inverse by Proposition 1.8.3. But, S has infinitely many right inverses.

Proposition (1.8.1) gives $s_{i,i+1}x_{i+1,j} = \delta_{ij}$, $(i \in \mathbb{N}_1)$, $s_{00}x_{0j} = \delta_{1j}$, whence

$$x_{ij} := \begin{cases} 1/s_{00} & , \quad i = j = 0, \\ 1/s_{i,i+1} = 1 & , \quad j = i - 1, \ i \in \mathbb{N}_1, \\ 0 & , \quad i = 1, \ j \in \mathbb{N}_1 \text{ or } i \in \mathbb{N}_1, \ j \neq i. \end{cases}$$

Thus, x_{2j} is left arbitrary for every $j \in \mathbb{N}$. If x_{2j} is taken to be 0 for every $j \in \mathbb{N}$, we get S' as one right inverse of S; hence, all right inverses of S are given by $S' + S^0$, where the S^0 are the right zero-divisors of S.

Similarly, S' has infinitely many left inverses but no right inverse, and all the left inverses of S' are given by $S +^0 S'$, where the ${}^0S'$ are the left zero-divisors of S'.

(d) A matrix may have a unique left and a unique right inverse, and no left or right zero-divisors, i.e., a matrix may have a unique two-sided inverse, not merely in an associative field $\mathfrak{F}$ but have no other inverses outside $\mathfrak{F}$.

Thus, consider the permutator $P = (p_{ij})$ with $p_{ij} = 0$ except when $j = r_i$, r_i running through all natural numbers and different for different values of i; and $p_{i,r_i} = 1$. Then, it is obvious that $PP' = I$ and $P'P = I$, where $P' = (p'_{ij})$ with $p'_{r_i,i} = 1$ and $p'_{ki} = 0$, $(k \neq r_i)$. Moreover, P has no zero-divisors, either left or right. Also, the associative law for ${}^{(-1)}PPP^{(-1)}$ holds automatically, since the sums involved are finite.

To sum up, we see from the above notes that an infinite matrix A may have

(i) an unique left and right inverse, i.e., ${}^{(-1)}A$ and $A^{(-1)}$ both exist such that ${}^{(-1)}A = A^{(-1)}$, and there is no ${}^{0}A$ and no A^0.

(ii) an unique two-sided inverse in a field $\mathfrak{F}$ in which ${}^{(-1)}AAA^{(-1)}$ is associative and infinitely many ${}^{(-1)}A$ or $A^{(-1)}$ not belonging to $\mathfrak{F}$.

(iii) no inverse, either left or right.

(iv) no left inverse but infinitely many right inverses.

(v) no right inverse but infinitely many left inverses.

But Theorem 1.8.6 shows that if A is restricted to a given associative field, the possibilities are reduced in number.

Theorem 1.8.6. (cf. Cooke [111, (2.2, II); p. 24]) *If A is a matrix of a given associative field $\mathfrak{F}$, then either*

(i) *A has a unique two-sided inverse in $\mathfrak{F}$.*

or

(ii) *A has no inverse in $\mathfrak{F}$ on one side and no inverse in $\mathfrak{F}$ or infinitely many inverses in $\mathfrak{F}$ on the other side.*

Proof. The zero matrix or the diagonal matrix one at least of whose diagonal element is 0 shows that A may have no inverse whatever on either side. Also, as we have seen, A cannot have both a right inverse $A^{(-1)}$ in $\mathfrak{F}$ and a left zero-divisor ${}^{0}A$ in $\mathfrak{F}$. Hence, when $A^{(-1)}$ exists in $\mathfrak{F}$, A has either one left inverse ${}^{(-1)}A$ in $\mathfrak{F}$ or none. If A has both a right and a left inverse, each in $\mathfrak{F}$, then $({}^{(-1)}AA)A^{(-1)} = A^{(-1)}$ and ${}^{(-1)}A(AA^{(-1)}) ={}^{(-1)} A$, which shows that ${}^{(-1)}A = A^{(-1)}$.

In this case, both ${}^{(-1)}A$ and $A^{(-1)}$ are unique in $\mathfrak{F}$, since, by the preceding argument, every ${}^{(-1)}A$ is equal to any fixed $A^{(-1)}$. Finally, we show that when A has no right inverse in $\mathfrak{F}$ but has a left inverse in $\mathfrak{F}$, then A has infinitely many left inverses in $\mathfrak{F}$. By hypothesis, $AX = I$ has no solution in $\mathfrak{F}$ and $YA = I$ has a solution, say Y in $\mathfrak{F}$. It follows for every X in $\mathfrak{F}$ that $AX = I + Z$ with $Z \in \mathfrak{F} \setminus \{0\}$. In particular, $AY = I + Z$; hence from $YA = I$ and the associative law we have $A = A + ZA$, i.e., $ZA = 0$. Consequently, $Y + cZ$ are all different left inverses of A in $\mathfrak{F}$ for different values of any scalar c. □

Theorem 1.8.7. (cf. Cooke [111, (2.2, III); p. 24]) *If A and B have right inverses such that $ABB^{(-1)}A^{(-1)}$ is associative, then $B^{(-1)}A^{(-1)}$ is a right inverse of AB. If there is a left inverse of A and a right inverse of AB such that $[{}^{(-1)}AAB(AB)^{(-1)}]A$ is associative, then $(AB)^{(-1)}A$ is a right inverse of B.*

Proof. From the hypotheses, we have

$$(AB)B^{(-1)}A^{(-1)} = A\left(BB^{(-1)}\right)A^{(-1)} = AA^{(-1)} = I,$$

which proves the first statement.

For the second statement, multiply $AB(AB)^{(-1)} = I$ by ${}^{(-1)}A$ on the left and by A on the right, then $B[(AB)^{(-1)}A] = I$, so that $(AB)^{(-1)}A$ is a right inverse of B, as desired. □

Corollary 1.8.8. (cf. Cooke [111, Corollary; p. 24]) *If A and B have left inverses such that ${}^{(-1)}B^{(-1)}AAB$ is associative, then ${}^{(-1)}B^{(-1)}A$ is a left inverse of AB.*

Proof. It follows that when A and B belong to the same associative field, and $A^{(-1)}$ and $B^{(-1)}$ are unique right inverses in $\mathfrak{F}$, then $B^{(-1)}A^{(-1)}$ is the unique right inverse of AB in $\mathfrak{F}$. In this case, $A^{(-1)}$ and $B^{(-1)}$ are also left inverses, Theorem 1.8.5, and so it follows from $(AB)X = I$ that $X =^{(-1)} B^{(-1)}A$. □

Theorem 1.8.9. (cf. Cooke [111, (2.2, IV); p. 25]) *If $A^{(-1)}$ and $(I+BA^{(-1)})^{(-1)}$ exist, and $(A+B)A^{(-1)}(I+BA^{(-1)})^{(-1)}$ is associative, then $A^{(-1)}(I+BA^{(-1)})^{(-1)}$ is a right inverse of $A+B$. If ${}^{(-1)}A$ and ${}^{(-1)}(I+{}^{(-1)}AB)$ exist, and ${}^{(-1)}(I+{}^{(-1)}AB)^{(-1)}A(A+B)$ is associative, then ${}^{(-1)}(I+{}^{(-1)}AB)^{(-1)}A$ is a left inverse of $A+B$. In these statements, A and B may be interchanged.*

Proof. From the hypotheses, we have

$$\left[(A+B)A^{(-1)}\right]\left(I+BA^{(-1)}\right)^{(-1)} = \left(I+BA^{(-1)}\right)\left(I+BA^{(-1)}\right)^{(-1)} = I,$$
$${}^{(-1)}\left(I+{}^{(-1)}AB\right)\left[{}^{(-1)}A(A+B)\right] = {}^{(-1)}\left(I+{}^{(-1)}AB\right)\left(I+{}^{(-1)}AB\right) = I$$

which proves the result. □

CHAPTER 2

Normed and Paranormed Sequence Spaces

In this chapter, certain normed and paranormed sequence spaces are studied, and the α-, β-, γ- and continuous duals of the spaces ℓ_∞, c, c_0 and ℓ_p of all bounded, convergent, null and absolutely p-summable sequences are determined together with some other sequence spaces isomorphic to them. Additionally, a table of the α-, β- and γ-duals of certain normed sequence spaces is given.

Keywords: Metric sequence spaces, normed sequence spaces, K-space, FK-space, BK-space, AK-, AB-, AD- and KB-properties, Schauder basis, paranormed sequence spaces, α-, β-, γ- and continuous duals of a sequence space.

2.1 LINEAR SEQUENCE SPACES

We denote the set of all sequences with complex terms by ω. It is a routine verification that ω is a linear space with respect to the co-ordinatewise addition and scalar multiplication of sequences, which are defined, as usual, by

$$x + y = (x_k) + (y_k) := (x_k + y_k) \text{ and } \alpha x = \alpha(x_k) := (\alpha x_k),$$

respectively, where $x = (x_k)$, $y = (y_k) \in \omega$ and $\alpha \in \mathbb{C}$. By a *sequence space*, we understand a linear subspace of the space ω.

Let $e = (e_k)$ and $e^{(n)} = (e_k^{(n)})$, $(n \in \mathbb{N})$ be the sequences with $e^{(k)} = 1$ for all $k \in \mathbb{N}$, and $e_k^{(n)} = \delta_{nk}$ for all $k, n \in \mathbb{N}$, where δ_{nk} denotes the Kronecker delta defined by (1.1.4).

We write ℓ_∞, c, c_0 and ℓ_p for the classical sequence spaces of all bounded, convergent, null and absolutely p-summable sequences of complex numbers, respectively, where $0 < p < \infty$. By ϕ, we mean the space of all finitely non-zero sequences. Also by bs and cs, we denote the spaces of all bounded and convergent series, respectively. Here, bv_1 and bv are the spaces of all sequences of bounded variation i.e., consisting of all sequences (x_k) such that $(x_k - x_{k-1})$ and $(x_k - x_{k+1})$ in ℓ_1, respectively, and bv_0 is the intersection of the spaces bv and c_0. ω_0^p, ω^p and ω_∞^p are the spaces of sequences that are strongly summable to zero, summable and bounded of index $p \geq 1$ by the Cesàro method of order 1.

DOI: 10.1201/9781003294153-2

We assume throughout, unless stated otherwise, that $1 \leq p < \infty$ and q is the conjugate number of p, i.e., $q = \infty$ if $p = 1$ and $q = p/(p-1)$ for $1 < p < \infty$, and use the convention that any term with a negative subscript is equal to naught.

Let λ denote any of the sets ℓ_∞, c, c_0, ℓ_p, bs, cs, bv_1, bv, ω_∞^p, ω^p and ω_0^p. It is a routine verification that λ is a linear space with respect to the co-ordinatewise addition and scalar multiplication of sequences.

2.2 METRIC SEQUENCE SPACES

Prior to giving examples of metric sequence spaces, we note the definition of the concepts of a metric and a metric space.

Definition 2.2.1. *Let X be a non-empty set and d be a distance function from $X \times X$ to the set $\mathbb{R}^+$ of non-negative real numbers. Then, the pair (X, d) is called a metric space and d is a metric for X, if the following metric axioms are satisfied for all elements $x, y, z \in X$:*

(M.1) $d(x, y) = 0$ *if and only if* $x = y$.

(M.2) $d(x, y) = d(y, x)$ *(the symmetry property).*

(M.3) $d(x, z) \leq d(x, y) + d(y, z)$ *(the triangle inequality).*

Now, we can present some examples of metric sequence spaces:

2.2.1 The Space ω

By ω, we mean the set of all sequences with complex terms, i.e., $\omega := \{x = (x_k) : x_k \in \mathbb{C} \text{ for all } k \in \mathbb{N}\}$. The most popular metric on the space ω is defined by

$$d_\omega(x, y) := \sum_k \frac{|x_k - y_k|}{2^k(1 + |x_k - y_k|)}; \quad x = (x_k), \ y = (y_k) \in \omega.$$

It is obvious that another metric for the space ω can be derived from d_ω replacing 2^{-k} by μ_k such that $\sum_k \mu_k$ is a convergent series with positive terms.

2.2.2 The Space ℓ_∞

The space ℓ_∞ of bounded sequences is defined by

$$\ell_\infty := \left\{ x = (x_k) \in \omega : \sup_{k \in \mathbb{N}} |x_k| < \infty \right\}.$$

The natural metric on the space ℓ_∞ is defined by

$$d_\infty(x, y) := \sup_{k \in \mathbb{N}} |x_k - y_k|; \quad x = (x_k), \ y = (y_k) \in \ell_\infty.$$

2.2.3 The Spaces f and f_0

The spaces f and f_0 of almost convergent and almost null sequences are defined by

$$f := \left\{x = (x_k) \in \ell_\infty : \lim_{m\to\infty} \sum_{k=0}^{m} \frac{x_{n+k}}{m+1} = \alpha \text{ uniformly in } n \text{ for some } \alpha \in \mathbb{C}\right\},$$

$$f_0 := \left\{x = (x_k) \in \ell_\infty : \lim_{m\to\infty} \sum_{k=0}^{m} \frac{x_{n+k}}{m+1} = 0 \text{ uniformly in } n\right\}.$$

We give a short survey on the concept of *almost convergence*. The *shift operator* P is defined on ω by $P_n(x) = x_{n+1}$ for all $n \in \mathbb{N}$. A Banach limit L is defined on ℓ_∞, as a non-negative linear functional, such that $L(Px) = L(x)$ and $L(e) = 1$. A sequence $x = (x_k) \in \ell_\infty$ is said to be almost convergent to the generalized limit α if all Banach limits of x are coincide and are equal to α [246], and Is denoted by $f - \lim x_k = \alpha$. Let P^i be the composition of P with itself i times and write for a sequence $x = (x_k)$

$$t_{mn}(x) := \frac{1}{m+1} \sum_{i=0}^{m} P_n^i(x) \quad \text{for all} \quad m, n \in \mathbb{N}.$$

Lorentz [246] proved that $f - \lim x_k = \alpha$ if and only if $\lim_{m\to\infty} t_{mn}(x) = \alpha$, uniformly in n. It is well-known that a convergent sequence is almost convergent such that its ordinary and generalized limits are equal. For more detail on the Banach limit, the reader may refer to Çolak and Çakar [130], and Das [140].

The metric d_∞ is also a metric for the spaces f and f_0.

2.2.4 The Spaces c and c_0

The spaces c and c_0 of convergent and null sequences are given by

$$c := \left\{x = (x_k) \in \omega : \lim_{k\to\infty} |x_k - l| = 0 \text{ for some } l \in \mathbb{C}\right\},$$

$$c_0 := \left\{x = (x_k) \in \omega : \lim_{k\to\infty} x_k = 0\right\}.$$

The metric d_∞ is also a metric for the spaces c and c_0. It is trivial that since the concepts supremum and maximum are equivalent on the space c_0, the metric d_∞ is reduced to the metric d_0 defined by

$$d_0(x, y) := \max_{k\in\mathbb{N}} |x_k - y_k|; \quad x = (x_k), \ y = (y_k) \in c_0$$

on the space c_0.

2.2.5 The Space ℓ_p

The space ℓ_p of absolutely p-summable sequences is defined as

$$\ell_p := \left\{x = (x_k) \in \omega : \sum_k |x_k|^p < \infty\right\}, \quad (0 < p < \infty).$$

In the case $1 < p < \infty$, the metric d_p on the space ℓ_p is given by

$$d_p(x,y) := \left(\sum_k |x_k - y_k|^p\right)^{1/p}; \quad x = (x_k),\ y = (y_k) \in \ell_p.$$

Also, in the case $0 \leq p \leq 1$, the metric $\widetilde{d_p}$ on the space ℓ_p is given by

$$\widetilde{d_p}(x,y) := \sum_k |x_k - y_k|^p; \quad x = (x_k),\ y = (y_k) \in \ell_p.$$

2.2.6 The Space bs

The space bs of bounded series is defined by

$$bs := \left\{x = (x_k) \in \omega : \sup_{n\in\mathbb{N}} \left|\sum_{k=0}^{n} x_k\right| < \infty\right\}.$$

The natural metric on the space bs is defined by

$$d(x,y) := \sup_{n\in\mathbb{N}} \left|\sum_{k=0}^{n} (x_k - y_k)\right|; \quad x = (x_k),\ y = (y_k) \in bs. \tag{2.2.1}$$

2.2.7 The Spaces cs and cs_0

The space cs of convergent series and the space cs_0 of the series converging to zero are defined, as follows:

$$\begin{aligned} cs &:= \left\{x = (x_k) \in \omega : \lim_{n\to\infty} \left|\sum_{k=0}^{n} x_k - l\right| = 0 \text{ for some } l \in \mathbb{C}\right\}, \\ cs_0 &:= \left\{x = (x_k) \in \omega : \lim_{n\to\infty} \left|\sum_{k=0}^{n} x_k\right| = 0\right\}. \end{aligned}$$

The relation d defined by (2.2.1) is the natural metric on the spaces cs and cs_0.

2.2.8 The Space bv_1

The space bv_1 of sequences of bounded variation is defined by

$$bv_1 := \left\{x = (x_k) \in \omega : \sum_k |x_k - x_{k-1}| < \infty\right\}.$$

Define the difference sequence $\Delta^{(1)}u = \{(\Delta^{(1)}u)_k\}$ by $(\Delta^{(1)}u)_k = u_k - u_{k-1}$ for all $k \in \mathbb{N}$ with $u_{-1} = 0$. The natural metric on the space bv_1 is defined by

$$d(x,y) := \sum_k |\Delta^{(1)}(x_k - y_k)|; \quad x = (x_k),\ y = (y_k) \in bv_1. \tag{2.2.2}$$

2.2.9 The Spaces ω_0^p, ω^p and ω_∞^p

The spaces ω_0^p, ω^p and ω_∞^p of sequences that are strongly summable to zero, summable and bounded of index $p \geq 1$ by the Cesàro method of order 1 are defined by

$$\begin{aligned} \omega_0^p &:= \left\{ x = (x_k) \in \omega : \lim_{n\to\infty} \frac{1}{n+1} \sum_{k=0}^{n} |x_k|^p = 0 \right\}, \\ \omega^p &:= \left\{ x = (x_k) \in \omega : \lim_{n\to\infty} \frac{1}{n+1} \sum_{k=0}^{n} |x_k - l|^p = 0 \text{ for some } l \in \mathbb{C} \right\}, \\ \omega_\infty^p &:= \left\{ x = (x_k) \in \omega : \sup_{n\in\mathbb{N}} \frac{1}{n+1} \sum_{k=0}^{n} |x_k|^p < \infty \right\}. \end{aligned}$$

The metric $\widehat{d}_p$ on the spaces ω_0^p, ω^p and ω_∞^p is defined by

$$\widehat{d}_p(x,y) := \begin{cases} \sup_{n\in\mathbb{N}} \frac{1}{n+1} \sum_{k=0}^{n} |x_k - y_k|^p &, \quad 0 < p \leq 1, \\ \sup_{n\in\mathbb{N}} \left(\frac{1}{n+1} \sum_{k=0}^{n} |x_k - y_k|^p \right)^{1/p} &, \quad 1 < p < \infty. \end{cases}$$

As an easy consequence of Hölder's inequality that the inclusion $\omega^q \subset \omega^p$ holds when $0 < p < q$.

We now give the definition of a Fréchet space.

Definition 2.2.2. *Let X be a real or complex linear space and d be a metric on X. Then, (X, d) is said to be a linear metric space, if the algebraic operations addition $(+)$ and scalar multiplication $(\cdot)$ on X are continuous. A complete linear metric space is said to be a Fréchet space.*

One can show that the metric space ω of all sequences with complex terms forms a Fréchet space with respect to the metric d_ω.

2.3 NORMED SEQUENCE SPACES

We begin with giving the definitions of a norm and a p-norm.

Definition 2.3.1. *Let X be a real or complex linear space and $\|\cdot\|$ be a function from X to the set $\mathbb{R}^+$ of non-negative real numbers. Then, the pair $(X, \|\cdot\|)$ is called a normed space and $\|\cdot\|$ is a norm for X, if the following norm axioms are satisfied for all elements $x, y \in X$ and for all scalars α:*

(N.1) $\|x\| = 0$ *if and only if* $x = \theta$.

(N.2) $\|\alpha x\| = |\alpha|\|x\|$ *(the absolute homogenity property).*

(N.3) $\|x + y\| \leq \|x\| + \|y\|$ *(the triangle inequality).*

Definition 2.3.2. *Let X be a real or complex linear space, $\|\cdot\|$ be a function from X to the set $\mathbb{R}^+$ of non-negative real numbers and $p > 0$. Then, the pair $(X, \|\cdot\|)$ is called a p-normed space and $\|\cdot\|$ is a p-norm for X, if the following p-norm axioms are satisfied for all elements $x, y \in X$ and for all scalars α:*

(P.1) $\|x\| = 0$ *if and only if* $x = \theta$.

(P.2) $\|\alpha x\| = |\alpha|^p \|x\|$.

(P.3) $\|x + y\| \leq \|x\| + \|y\|$.

Now, we can give the examples of normed and p-normed sequence spaces. First, consider the following relations, which are derived from the corresponding metrics given in Section 2.2 by putting, as usual, $\|x\| = d(x, \theta)$:

$$\begin{aligned}
\|x\|_\infty &:= \sup_{k\in\mathbb{N}} |x_k|, \quad \|x\|_0 := \max_{k\in\mathbb{N}} |x_k|, \\
\|x\|_p &:= \begin{cases} \sum_k |x_k|^p &, \ 0 < p \leq 1, \\ \left(\sum_k |x_k|^p\right)^{1/p} &, \ 1 < p < \infty, \end{cases} \\
\|x\|_{bv} &:= |x_0| + \sum_{k=1}^{\infty} |x_k - x_{k+1}|, \quad \|x\|_{bv_1} := \sum_k |x_k - x_{k-1}|, \\
\|x\|_{cs} &:= \|x\|_{bs} = \sup_{k\in\mathbb{N}} \left| \sum_{j=0}^{k} x_j \right|, \qquad (2.3.1) \\
\|x\widehat{\|}_p &:= \begin{cases} \sup_{n\in\mathbb{N}} \frac{1}{n+1} \sum_{k=0}^{n} |x_k|^p &, \ 0 < p \leq 1, \\ \sup_{n\in\mathbb{N}} \left(\frac{1}{n+1} \sum_{k=0}^{n} |x_k|^p\right)^{1/p} &, \ 1 < p < \infty. \end{cases}
\end{aligned}$$

It is not hard to show that ℓ_∞ and c are Banach spaces endowed with the norm $\|\cdot\|_\infty$. Additionally, c_0 and bv are Banach spaces with the norms $\|\cdot\|_0$ and $\|\cdot\|_{bv}$. The spaces bs and cs are Banach spaces endowed with the norm $\|\cdot\|_{bs}$. In the case $1 < p < \infty$, the spaces ℓ_p and ω_0^p, ω^p, ω_∞^p are Banach spaces with the norms $\|\cdot\|_p$ and $\|\cdot\widehat{\|}_p$, respectively. In the case $0 < p \leq 1$, the spaces ℓ_p and ω_0^p, ω^p, ω_∞^p are complete p-normed spaces with the p-norms $\|\cdot\|_p$ and $\|\cdot\widehat{\|}_p$, respectively.

Consider *Jensen's inequality*

$$\left(\sum_{k=0}^{n} |x_k|^r\right)^{1/r} < \left(\sum_{k=0}^{n} |x_k|^s\right)^{1/s},$$

where $r > s > 0$ and $x_0, x_1, \ldots, x_n \in \mathbb{C}$. Jensen's inequality gives the inclusion $\ell_r \supset \ell_p$ for $r > p$.

A sequence space λ with a linear topology is called a *K-space* provided each of the maps $p_i : \lambda \to \mathbb{C}$ defined by $p_i(x) = x_i$ is continuous for all $i \in \mathbb{N}$. A K-space λ is called an *FK-space* provided λ is a complete linear metric space. An FK-space whose topology is normable is called a *BK-space*.

Given an FK-space $\lambda \supset \phi$, we denote the nth *section of a sequence* $x = (x_k) \in \lambda$ by $x^{[n]} = \sum_{k=0}^{n} x_k e^{(k)}$, and we say that x has the property

$$\begin{gathered}
AK \text{ if } x^{[n]} \to x \text{ as } n \to \infty \text{ (Abschnittskonvergenz)}, \\
AB \text{ if } \left(x^{[n]}\right) \text{ is bounded (Abschnittsbeschränktheit)}, \\
AD \text{ if } x \in \overline{\phi} \text{ (abschnittsdicht)}, \\
KB \text{ if the set } \{x_k e^{(k)}\} \text{ is bounded in } \lambda \text{ (koordinatenweise beschränkt)}.
\end{gathered}$$

If one of these properties holds for every $x \in \lambda$, then we say that the space λ has that property (cf. [194]). It is trivial that AK implies AD and AK iff $AB + AD$. For example, c_0 and ℓ_p are AK-spaces and, c and ℓ_∞ are not AD-spaces.

Now, we give the following definition concerning with the perfectness, normality and monotonicity of a sequence space (see [211, p. 48]):

Definition 2.3.3. *Let λ be a sequence space. Then,*

(i) *λ is called perfect if $\lambda = \lambda^{\alpha\alpha}$, where $\lambda^{\alpha\alpha}$ is the α-dual of the space λ^α.*

(ii) *λ is called solid or normal if and only if*

$$\widetilde{\lambda} := \{(u_k) \in \omega : \exists (x_k) \in \lambda \text{ such that } |u_k| \leq |x_k| \text{ for all } k \in \mathbb{N}\} \subset \lambda.$$

(iii) *For a sequence J of $\mathbb{N}$ and a sequence space λ, we define λ_J by*

$$\lambda_J := \{x = (x_i) : \text{there is a } y = (y_i) \in \lambda \text{ with } x_i = y_{n_i} \text{ for all } n_i \in J\}$$

and call λ_J the J-stepspace or J-sectional subspace of λ. If $x_J \in \lambda_J$, then the canonical preimage of x_J is the sequence $\bar{x}_J$ which agrees with x_J on the indices in J and is zero elsewhere. Then, λ is called monotone provided λ contains the canonical preimages of all its stepspaces.

If a normed sequence space λ contains a sequence (b_n) with the property that for every $x \in \lambda$ there is a unique sequence of scalars (α_n) such that

$$\lim_{n\to\infty} \|x - (\alpha_0 b_0 + \alpha_1 b_1 + \cdots + \alpha_n b_n)\| = 0$$

then, (b_n) is called a *Schauder basis* (or briefly *basis*) for λ. The series $\sum_k \alpha_k b_k$ which has the sum x is then called the expansion of x with respect to (b_n) and written as $x = \sum_k \alpha_k b_k$. The concepts of Schauder and algebraic bases coincide for finite-dimensional spaces. Nevertheless, there are linear spaces without a Schauder basis.

Example 2.3.4. *The sequence $\{e^{(k)}\}_{k\in\mathbb{N}}$ is a Schauder basis for the space ω under its natural paranorm g_ω* (cf. [285, Example 23, p. 147]) *defined by*

$$g_\omega(x) := \sum_k \frac{|x_k|}{2^k(1 + |x_k|)}, \quad x = (x_k) \in \omega$$

and any $x = (x_k) \in \omega$ has a unique representation of the form $\lim_{n\to\infty} x^{[n]} = x$.

Example 2.3.5. *The sequence $\{e, e^{(0)}, e^{(1)}, \dots\}_{k\in\mathbb{N}}$ is a Schauder basis for the space c and any $x = (x_k) \in c$ has a unique representation of the form*

$$x = \alpha e + \sum_k (x - \alpha) e^{(k)}, \quad \text{where } \alpha = \lim_{k\to\infty} x_k. \tag{2.3.2}$$

It is obvious that the inclusions $\ell_p \subset c_0 \subset c \subset f \subset \ell_\infty$ hold. It follows from

$$|x_k| \leq \sup_{k\in\mathbb{N}} |x_k| = \|x\|_\infty \quad \text{and} \quad |x_k| \leq \left(\sum_k |x_k|^p\right)^{1/p} = \|x\|_p$$

that $\lim_{n\to\infty} x^{(n)} = x$ in any of the spaces ℓ_∞, c, c_0 and ℓ_p with $p \geq 1$ implies $\lim_{n\to\infty} x_k^{(n)} = x_k$ for all $k \in \mathbb{N}$. Thus, norm convergence in the classical spaces ℓ_∞, c, c_0 and ℓ_p with $p \geq 1$ implies co-ordinatewise convergence. Since convergence in (ω, d_ω) and co-ordinatewise convergence are equivalent, Banach spaces ℓ_∞, c, c_0 and ℓ_p with $p \geq 1$ have the property that the metrics obtained by their norms are stronger than the metrics of ω on them. Hence, the spaces ℓ_∞, c, c_0 and ℓ_p with $p > 1$ are BK-spaces with their natural norms.

Lemma 2.3.6. *The space bs is norm isomorphic to the space ℓ_∞, i.e., $bs \cong \ell_\infty$.*

Proof. To prove this, we should show the existence of a norm preserving linear bijection between the spaces bs and ℓ_∞.

Consider the transformation T defined from bs to ℓ_∞ by $x \mapsto y = Tx = \left(\sum_{j=0}^{k} x_j\right)_{k\in\mathbb{N}}$. The linearity of T is obvious. Further, it is trivial that $x = \theta$ whenever $Tx = \theta$ and hence, T is injective.

Let $y = (y_k) \in \ell_\infty$ and define the sequence $x = (x_k)$ by $x_k := y_k - y_{k-1}$ for all $k \in \mathbb{N}$ with $y_{-1} \equiv 0$. Then, we observe that

$$\|x\|_{bs} = \sup_{k\in\mathbb{N}} \left|\sum_{j=0}^{k} x_j\right| = \sup_{k\in\mathbb{N}} \left|\sum_{j=0}^{k} (y_j - y_{j-1})\right| = \sup_{k\in\mathbb{N}} |y_k| = \|y\|_\infty < \infty.$$

Thus, we observe that $x \in bs$. Consequently, T is surjective and is norm preserving. Hence, T is a linear bijection, which therefore says us that the spaces bs and ℓ_∞ are norm isomorphic, as desired. □

2.4 PARANORMED SEQUENCE SPACES

Firstly, we give the definition of the concept of paranorm.

Definition 2.4.1. *Let X be a real or complex linear space, g be a function from X to the set $\mathbb{R}$ of real numbers. Then, the pair (X, g) is called a paranormed space and g is a paranorm for X, if the following axioms are satisfied for all elements $x, y \in X$ and for all scalars α:*

(PN.1) $g(x) = 0$ *if* $x = \theta$.

(PN.2) $g(-x) = g(x)$.

(PN.3) $g(x + y) \leq g(x) + g(y)$.

(PN.4) *If (α_n) is a sequence of scalars with $\alpha_n \to \alpha$ as $n \to \infty$ and $x_n, x \in X$ for all $n \in \mathbb{N}$ with $x_n \to x$ as $n \to \infty$, then $\alpha_n x_n \to \alpha x$ as $n \to \infty$, in the sense that $g(\alpha_n x_n - \alpha x) \to 0$ as $n \to \infty$.*

Assume here and after that $p = (p_k)$ is a bounded sequence of strictly positive real numbers, so that $0 < p_k \leq \sup p_k = H < \infty$ and $M = \max\{1, H\}$. We shall assume throughout that $p_k^{-1} + (p_k')^{-1} = 1$ provided $1 < p_k$ for all $k \in \mathbb{N}$. Now, we may give the Maddox's spaces $\ell_\infty(p)$, $c(p)$, $c_0(p)$, $\ell(p)$, $\omega_\infty(p)$, $\omega(p)$ and $\omega_0(p)$, and $bs(p)$, $cs(p)$, $cs_0(p)$ as well-known examples of the paranormed sequence spaces:

2.4.1 The Spaces $\ell_\infty(p)$, $c(p)$ and $c_0(p)$

The spaces $\ell_\infty(p)$, $c(p)$ and $c_0(p)$ were defined by Maddox [252] (see also Simons [364]) as follows:

$$\begin{aligned} \ell_\infty(p) &:= \left\{x = (x_k) \in \omega : \sup_{k \in \mathbb{N}} |x_k|^{p_k} < \infty\right\}, \\ c(p) &:= \left\{x = (x_k) \in \omega : \lim_{k \to \infty} |x_k - l|^{p_k} = 0 \text{ for some } l \in \mathbb{C}\right\}, \\ c_0(p) &:= \left\{x = (x_k) \in \omega : \lim_{k \to \infty} |x_k|^{p_k} = 0\right\}. \end{aligned}$$

It is a routine verification that each of the sets $\ell_\infty(p)$, $c(p)$ and $c_0(p)$ forms a linear space with respect to the co-ordinatewise addition and scalar multiplication of sequences. Define the function g_1 on the spaces $\ell_\infty(p)$, $c(p)$ or $c_0(p)$ by

$$g_1(x) := \sup_{k \in \mathbb{N}} |x_k|^{p_k/M}.$$

Then, $c_0(p)$ and $c(p)$ are complete paranormed spaces paranormed by g_1 if $p \in \ell_\infty$; (cf. [253, Theorem 6]). It is known from [252] that the inclusion $c_0(p) \subset c_0(q)$ holds if and only if $\liminf q_k/p_k > 0$. $\ell_\infty(p)$ is also complete paranormed space by g_1 if and only if $\inf p_k > 0$.

2.4.2 The Space $\ell(p)$

The space $\ell(p)$ was defined by Maddox [252] (see also Nakano [298]) as follows:

$$\ell(p) := \left\{x = (x_k) \in \omega : \sum_k |x_k|^{p_k} < \infty\right\}, \quad (0 < p_k < \infty).$$

One can show that the set $\ell(p)$ is a linear space with respect to the co-ordinatewise addition and scalar multiplication of sequences. Maddox [252] showed that the inclusion $\ell(p) \subset \ell(q)$ holds when $0 < p_k \leq q_k$ for all $k \in \mathbb{N}$. We note here that no restriction such as boundedness has to be places on the sequences (p_k) and (q_k) for the validity of the inclusion. Define the function g_2 on the space $\ell(p)$ by

$$g_2(x) := \left(\sum_k |x_k|^{p_k}\right)^{1/M}.$$

$\ell(p)$ is complete paranormed space paranormed by g_2 and $\{e^{(k)}\}_{k \in \mathbb{N}}$ is a basis in the space $\ell(p)$.

2.4.3 The Spaces $\omega_\infty(p)$, $\omega(p)$ and $\omega_0(p)$

The spaces $\omega_\infty(p)$, $\omega(p)$ and $\omega_0(p)$ are defined (cf. [252]) by

$$\begin{aligned}
\omega_\infty(p) &:= \left\{x=(x_k)\in\omega : \sup_{n\in\mathbb{N}} \frac{1}{n+1}\sum_{k=0}^{n}|x_k|^{p_k}<\infty\right\},\\
\omega(p) &:= \left\{x=(x_k)\in\omega : \lim_{n\to\infty} \frac{1}{n+1}\sum_{k=0}^{n}|x_k-l|^{p_k}=0 \text{ for some } l\in\mathbb{C}\right\},\\
\omega_0(p) &:= \left\{x=(x_k)\in\omega : \lim_{n\to\infty} \frac{1}{n+1}\sum_{k=0}^{n}|x_k|^{p_k}=0\right\}.
\end{aligned}$$

It is known that the sets $\omega_\infty(p)$, $\omega(p)$ and $\omega_0(p)$ are the linear spaces with respect to the co-ordinatewise addition and scalar multiplication of sequences. Maddox proved in [253] that the inclusion $\omega_0(p)\subset\omega_0(q)$ holds if and only if $\omega_0(p/q)\subset\omega_0(e)$, where p/q denotes the sequence (p_k/q_k). Define the function g_3 on the spaces $\omega_\infty(p)$, $\omega(p)$ and $\omega_0(p)$ by

$$g_3(x) := \sup_{n\in\mathbb{N}}\left(\frac{1}{n+1}\sum_{k=0}^{n}|x_k|^{p_k}\right)^{1/M}.$$

$\omega(p)$ is a complete paranormed space with g_3 (cf. [253, Theorem 6]) and $\omega_\infty(p)$ is paranormed with g_3 if and only if $0<\inf\ p_k\le\sup\ p_k<\infty$ (cf. [251, Theorem 2]).

2.4.4 The Spaces $bs(p)$, $cs(p)$ and $cs_0(p)$

The spaces $bs(p)$, $cs(p)$ and $cs_0(p)$ are defined by

$$\begin{aligned}
bs(p) &:= \left\{x=(x_k)\in\omega : \sup_{k\in\mathbb{N}}\left|\sum_{j=0}^{k}x_j\right|^{p_k}<\infty\right\},\\
cs(p) &:= \left\{x=(x_k)\in\omega : \lim_{k\to\infty}\left|\sum_{j=0}^{k}x_j-l\right|^{p_k}=0 \text{ for some } l\in\mathbb{C}\right\},\\
cs_0(p) &:= \left\{x=(x_k)\in\omega : \lim_{k\to\infty}\left|\sum_{j=0}^{k}x_j\right|^{p_k}=0\right\}.
\end{aligned}$$

It is not hard to establish that each of the sets $bs(p)$, $cs(p)$ and $cs_0(p)$ forms a linear space with respect to the co-ordinatewise addition and scalar multiplication of sequences.

Define the function g_4 on the spaces $bs(p)$, $cs(p)$ and $cs_0(p)$ by

$$g_4(x) := \sup_{k\in\mathbb{N}}\left|\sum_{j=0}^{k}x_j\right|^{p_k/M}.$$

$bs(p)$, $cs(p)$ and $cs_0(p)$ are the complete paranormed spaces by g_4 if and only if $\inf p_k>0$.

In general, $\ell_\infty(p)$, $c(p)$, $c_0(p)$, $\ell(p)$, $\omega_\infty(p)$, $\omega(p)$, $\omega_0(p)$, $bs(p)$, $cs(p)$ and $cs_0(p)$ are not normed spaces. If we choose $p_k = p$ for all $k \in \mathbb{N}$, then the paranormed spaces $\ell_\infty(p)$, $c(p)$, $c_0(p)$, $\ell(p)$, $\omega_\infty(p)$, $\omega(p)$, $\omega_0(p)$, $bs(p)$, $cs(p)$ and $cs_0(p)$ are reduced to Banach spaces ℓ_∞, c, c_0, ℓ_p, ω_∞^p, ω^p, ω_0^p, bs, cs and cs_0, respectively.

2.5 THE DUAL SPACES OF A SEQUENCE SPACE

In this section, we focus on the α-, β-, γ- and continuous duals of the classical sequence spaces. For the sequence spaces λ and μ, the set $S(\lambda,\mu)$ defined by

$$S(\lambda,\mu) := \{z = (z_k) \in \omega : xz = (x_k z_k) \in \mu \text{ for all } x = (x_k) \in \lambda\} \quad (2.5.1)$$

is called the *multiplier space* of λ and μ. One can easily observe for a sequence space ν that the inclusions

$$S(\lambda,\mu) \subset S(\nu,\mu) \text{ if } \nu \subset \lambda \text{ and } S(\lambda,\mu) \subset S(\lambda,\nu) \text{ if } \mu \subset \nu \quad (2.5.2)$$

hold. With the notation of (2.5.1), the *α-, β- and γ-duals of a sequence space* λ, which are denoted by λ^α, λ^β and λ^γ, respectively, are defined by

$$\lambda^\alpha := S(\lambda,\ell_1), \quad \lambda^\beta := S(\lambda, cs) \quad \text{and} \quad \lambda^\gamma := S(\lambda, bs).$$

The α-dual, β-dual and γ-dual are also referred to as *Köthe-Toeplitz dual*, *generalized Köthe-Toeplitz dual* and *Garling dual*, respectively. The *continuous dual* λ^* of a sequence space λ is defined as the set of all bounded linear functionals on the space λ.

Now, we give the following lemma and theorem which shows the connection of the β- and continuous duals of FK-spaces:

Lemma 2.5.1. (cf. [269, Remark 1.16, p. 152]) *Let $X \supset \phi$ be an FK-space and $a = (a_k)$ be a given sequence. If the series $\sum_k a_k x_k$ converges for every $x = (x_k) \in X$, then the linear functional $f_a : X \to \mathbb{C}$ defined by $f_a(x) = \sum_k a_k x_k$ is continuous for all $x \in X$.*

Proof. The linear functional $f_{a,n} = \sum_{k=0}^n a_k P_k : X \to \mathbb{C}$ is continuous as a finite linear combination of the continuous co-ordinates P_k on the FK-space X. Since $f_a(x) = \lim_{n\to\infty} f_{a,n}(x)$ for every $x \in X$, the linear functional f_a is continuous by the Banach-Steinhaus theorem. □

Theorem 2.5.2. [396, Theorem 7.2.9, p. 107] *Let $X \supset \phi$ be an FK-space.*

(a) *Then there is a linear one-to-one map $\widehat{\ }: X^\beta \to X^*$.*

(b) *If X has AK, then the map $\widehat{\ }: X^\beta \to X^*$ is an isomorphism.*

Proof. We define the map $\widehat{\ }$ for all $a \in X^\beta$ by $\widehat{a} : X \to \mathbb{C}$, where $\widehat{a}(x) = \sum_k a_k x_k$ for all $x \in X$.

(a) First we show $\widehat{\ }: X^\beta \to X^*$. Let $a = (a_k) \in X^\beta$ be given. Then the series $\sum_k a_k x_k$ converge for all $x = (x_k) \in X$, hence $\widehat{a}(x)$ is defined for all $x \in X$. Obviously,

$\widehat{a}$ is linear, and $\widehat{a} \in X^*$ by Lemma 2.5.1. Therefore, we have $\widehat{a} \in X^*$ for all $a \in X^\beta$, and so $\widehat{\ }: X^\beta \to X^*$.

Now, we show that $\widehat{\ }$ is linear. Let $a = (a_k), b = (b_k) \in X^\beta$ and $\lambda \in \mathbb{C}$ be given. Then, it follows for all $x = (x_k) \in X$

$$\begin{aligned} (\widehat{\lambda a + b})(x) &= \sum_k (\lambda a_k + b_k)x_k = \lambda \sum_k a_k x_k + \sum_k b_k x_k \\ &= \lambda \widehat{a}(x) + \widehat{b}(x) = (\lambda \widehat{a} + \widehat{b})(x), \end{aligned}$$

that is, $\widehat{\lambda a + b} = \lambda \widehat{a} + \widehat{b}$. This shows that $\widehat{\ }$ is linear.

Finally, we show that $\widehat{\ }$ is one-to-one. We assume $\widehat{a} = \widehat{0}$. Then, we have

$$\widehat{a}(x) = \sum_k a_k x_k = 0 \quad \text{for all} \quad x \in X,$$

in particular, for $e^{(j)} \in \phi \subset X$, $(j \in \mathbb{N})$ $\widehat{a}(e^{(j)}) = a_j = 0$, and so $a = (a_j)_{j \in \mathbb{N}} = 0$. This shows that $\widehat{\ }$ is one-to-one.

(b) We show that $\widehat{\ }: X^\beta \to X^*$ is onto. Let $f \in X^*$ be given. We put $a_k = f(e^{(k)})$ for $k \in \mathbb{N}$ and $a = (a_k)_{k \in \mathbb{N}}$. Let $x = (x_k) \in X$ be given. Since X has AK, we have $x = \sum_k x_k e^{(k)}$ and $f \in X^*$ implies

$$f(x) = f\left(\sum_k a_k x_k\right) = \sum_k x_k f(e^{(k)}) = \sum_k a_k x_k = \widehat{a}(x).$$

Thus, $\sum_k a_k x_k$ converges for all $x \in X$, i.e., $a \in X^\beta$, and $\widehat{a} = f$. Thus, we have shown that $\widehat{\ }: X^\beta \to X^*$ is onto. □

Now, we determine the α-, β- and γ-duals of the classical sequence spaces, and the continuous duals of the spaces ℓ_p, c_0 and c, where $1 < p < \infty$.

Theorem 2.5.3. *The following statements hold:*

(i) $\ell_\infty^\alpha = c^\alpha = c_0^\alpha = \ell_1$.

(ii) $\ell_1^\alpha = \ell_\infty$.

Proof. (i) Since one can prove in the similar way that $c_0^\alpha = c^\alpha = \ell_1$, we only show that $\ell_\infty^\alpha = \ell_1$.

Let $a = (a_k) \in \ell_1$ and $x = (x_k) \in \ell_\infty$. Then, $\|x\|_\infty = \sup_{k \in \mathbb{N}} |x_k| < \infty$. Therefore,

$$\sum_k |a_k x_k| \leq \left(\sup_{k \in \mathbb{N}} |x_k|\right) \sum_k |a_k| = \|x\|_\infty \left(\sum_k |a_k|\right) < \infty,$$

since $a \in \ell_1$ and this implies that $a \in \ell_\infty^\alpha$, and hence the inclusion $\ell_1 \subseteq \ell_\infty^\alpha$ holds.

Conversely, let $a = (a_k) \in \ell_\infty^\alpha$. Then, $f(x) = \sum_k a_k x_k$ is a continuous linear functional on c_0. Let $m \in \mathbb{N}$ be fixed. Define the sequence $x^{(m)} = \sum_{k=0}^m \operatorname{sgn}(a_k) e^{(k)}$, where $\operatorname{sgn}(z)$ of $z \in \mathbb{C}$ is defined by (cf. [258, p. 109])

$$\operatorname{sgn}(z) := \begin{cases} |z|/z &, \quad z \in \mathbb{C} \setminus \{0\}, \\ 1 &, \quad z = 0. \end{cases}$$

Then, $x^{(m)} \in c_0$ and $\|x^{(m)}\| \leq 1$, and

$$\sum_{k=0}^{n} |a_k| = \left|\sum_k a_k x^{(m)}\right| = \left|f\left(x^{(m)}\right)\right| \leq \|f\|\|x^{(m)}\| \leq \|f\| < \infty.$$

Since m is fixed, it follows that $\sum_k |a_k| < \infty$, i.e., $a \in \ell_1$. That is to say that the inclusion $\ell_\infty^\alpha \subseteq \ell_1$ also holds.

Therefore, the inclusions $\ell_1 \subseteq \ell_\infty^\alpha$ and $\ell_\infty^\alpha \subseteq \ell_1$ give that $\ell_\infty^\alpha = \ell_1$.

(ii) Let $a = (a_k) \in \ell_\infty$ and $x = (x_k) \in \ell_1$. Then, $\sum_k |x_k| < \infty$. Now,

$$\sum_k |a_k x_k| \leq \|a\|_\infty \left(\sum_k |x_k|\right) < \infty$$

and hence $a \in \ell_1^\alpha$. Therefore, we have the inclusion $\ell_\infty \subseteq \ell_1^\alpha$.

Conversely, suppose that $a = (a_k) \in \ell_1^\alpha$. Therefore,

$$\sum_k |a_k x_k| < \infty \quad \text{for all} \quad x = (x_k) \in \ell_1.$$

If $a \notin \ell_\infty$, then there exists a strictly increasing sequence (n_i) with $|a_{n_i}| > i^3$ for all $i \in \mathbb{N}_1$. If we define $x = (x_n)$ by

$$x_n := \begin{cases} i^{-2} & , \quad n = n_i, \\ 0 & , \quad n \neq n_i, \end{cases} \quad (i \in \mathbb{N}_1)$$

then $x \in \ell_1$ but $\sum_k |a_k x_k| = \infty$. Thus, $a \notin \ell_1^\alpha$, which is a contradiction. This yields the inclusion $\ell_1^\alpha \subseteq \ell_\infty$.

Therefore, by combining the inclusions $\ell_\infty \subseteq \ell_1^\alpha$ and $\ell_1^\alpha \subseteq \ell_\infty$ we obtain the desired result $\ell_1^\alpha = \ell_\infty$. □

Theorem 2.5.4. *Let $1 < p, q < \infty$ with $p^{-1} + q^{-1} = 1$. Then, $\ell_p^\alpha = \ell_q$.*

Proof. Suppose that $1 < p, q < \infty$ with $p^{-1} + q^{-1} = 1$ and $a = (a_k) \in \ell_p^\alpha \setminus \ell_q$. Then, $\sum_k |a_k x_k| < \infty$ for all $x = (x_k) \in \ell_p$. Then, we can find an increasing sequence (n_k) of positive integers such that $|a_{n_k}|^q > k^2$ for all $k \in \mathbb{N}_1$. Define the sequence $x = (x_n)$ by $x_n := \begin{cases} 1/a_{n_k} & , \quad n = n_k, \\ 0 & , \quad n \neq n_k, \end{cases}$ where $k \in \mathbb{N}_1$. Then, $x \in \ell_p$ but

$$\sum_k |a_k x_k| = \sum_k |a_{n_k}||a_{n_k}|^{-1} = 1 + 1 + \cdots = \infty,$$

which is a contradiction. Therefore, $a \in \ell_q$ and we have $\ell_p^\alpha \subseteq \ell_q$.

Conversely, suppose that $x = (x_k) \in \ell_p$ and $a = (a_k) \in \ell_q$. Then, by Hölder's inequality

$$\sum_k |a_k x_k| \leq \left(\sum_k |a_k|^q\right)^{1/q} \left(\sum_k |x_k|^p\right)^{1/p} = \|a\|_q \|x\|_p < \infty$$

which gives that $a \in \ell_p^\alpha$. Hence, the inclusion $\ell_q \subseteq \ell_p^\alpha$ holds.

Therefore, by combining the inclusions $\ell_p^\alpha \subseteq \ell_q$ and $\ell_q \subseteq \ell_p^\alpha$ we obtain that $\ell_p^\alpha = \ell_q$. □

Now, we give the following lemma concerning with the relation between the α-, β- and γ-duals of a solid sequence space:

Lemma 2.5.5. (cf. [94, Theorem 7.1.10 (b), p. 343]) *If X is solid, then $X^{\alpha} = X^{\beta} = X^{\gamma}$.*

Although, one can directly derive the β- and γ-duals of the spaces ℓ_∞, c_0, ℓ_p, ω and ϕ which are solid, by combining the results of Theorems 2.5.3 and 2.5.4 and Remark 2.5.10 with Lemma 2.5.5, we give their detailed proof.

Theorem 2.5.6. [395, Examples 6.2–6.5, pp. 91–92] *The following statements hold:*

(a) *$\ell_p^{\beta} = \ell_q$ for all p with $1 \le p < \infty$ and $\ell_\infty^{\beta} = c^{\beta} = c_0^{\beta} = \ell_1$.*

(b) *$\ell_p^{\beta} = \ell_\infty$ for all p with $0 < p \le 1$.*

(c) *Let $1 < p < \infty$. Then the continuous dual ℓ_p^* of ℓ_p is norm isomorphic to ℓ_q; this means $f \in \ell_p^*$ if and only if $f(x) = \sum_k a_k x_k$, $(x \in \ell_p)$, for some sequence $a = (a_k)_{k\in\mathbb{N}} \in \ell_q$, and $\|f\| = \|a\|_q$.*

(d) *The continuous dual c_0^* of c_0 is norm isomorphic to ℓ_1.*

(e) *We have $f \in c^*$ if and only if*

$$f(x) = \chi_f \lim_{k\to\infty} x_k + \sum_k a_k x_k, \quad x = (x_k) \in c \tag{2.5.3}$$

with $a = (a_k) = \left\{f\left(e^{(k)}\right)\right\}_{k\in\mathbb{N}} \in \ell_1$, where

$$\chi_f = f(e) - \sum_k f(e^{(k)}), \tag{2.5.4}$$

$$\|f\| = |\chi_f| + \|a\|_1. \tag{2.5.5}$$

Proof. (a/i) First let $p = 1$. Then we show $\ell_\infty \subset \ell_1^{\beta}$. Let $a = (a_k) \in \ell_\infty$ be given. Then we have for all $x = (x_k) \in \ell_1$

$$\sum_k |a_k x_k| \le \sup_{k\in\mathbb{N}} |a_k| \sum_k |x_k| = \|a\|_\infty \|x\|_1 < \infty, \tag{2.5.6}$$

hence $a \in \ell_1^{\beta}$. Thus, we have shown that

$$\ell_\infty \subset \ell_1^{\beta}. \tag{2.5.7}$$

Conversely, let $a = (a_k) \in \ell_1^{\beta}$ be given. Then, we have $\hat{a} \in \ell_1^*$ by Part (a) of Theorem 2.5.2. It follows for $x = \operatorname{sgn}(a_k)e^{(k)}$ for all $k \in \mathbb{N}$ that $ax \in \ell_1$, $\|x\|_1 < 1$ and

$$|\hat{a}(x)| = |a_k| \le \|\hat{a}\| \cdot \|x\|_1 \le \|\hat{a}\| < \infty, \tag{2.5.8}$$

for all $k \in \mathbb{N}$, that is, $sup_{k\in\mathbb{N}}|a_k| < \infty$ by (2.5.8), hence $a \in \ell_\infty$. Thus, we have shown that

$$\ell_1^{\beta} \subset \ell_\infty. \tag{2.5.9}$$

By combining the inclusion relations (2.5.7) and (2.5.9) we have $\ell_1^{\beta} = \ell_\infty$.

(a/ii) Now let $1 < p < \infty$ and $a = (a_k) \in \ell_q$ be given. Then, we have by Hölder's inequality for all $x = (x_k) \in \ell_p$

$$\sum_k |a_k x_k| \leq \left(\sum_k |a_k|^q\right)^{1/q} \left(\sum_k |x_k|^p\right)^{1/p} = \|a\|_q \|x\|_p < \infty. \tag{2.5.10}$$

Hence, $a = (a_k) \in \ell_p^\beta$ by (2.5.10) which gives the inclusion

$$\ell_q \subset \ell_p^\beta. \tag{2.5.11}$$

Conversely, let $a = (a_k) \in \ell_p^\beta$. Then, we have $\widehat{a} \in \ell_p^*$ by Part (a) of Theorem 2.5.2. We may assume $a \neq \theta$. Then, there is $k_0 \in \mathbb{N}$ with $a_{k_0} \neq 0$. Let $m \in \mathbb{N}$ with $m \geq k_0$ be given, and

$$x^{(m)} = \left(\sum_{k=0}^m |a_k|^q\right)^{-1/p} \sum_{k=0}^m \operatorname{sgn}(a_k)|a_k|^{q-1}e^{(k)}.$$

Then, we have $x^{(m)} \in \ell_p$,

$$\left(\left\|x^{(m)}\right\|_p\right)^p = \left(\sum_{k=0}^m |a_k|^q\right)^{-1} \sum_{k=0}^m |a_k|^{pq-p} = \left(\sum_{k=0}^m |a_k|^q\right)^{-1} \sum_{k=0}^m |a_k|^q = 1$$

and

$$\begin{aligned} \left|\widehat{a}(x^{(m)})\right| = \left|\sum_{k=0}^m a_k x_k^{(m)}\right| &= \left(\sum_{k=0}^m |a_k|^q\right)^{-1/p} \sum_{k=0}^m |a_k|^q = \left(\sum_{k=0}^m |a_k|^q\right)^{1/q} \\ &\leq \|\widehat{a}\| \|x^{(m)}\|_p = \|\widehat{a}\| < \infty. \end{aligned}$$

Since m was arbitrary, it follows that

$$\left(\sum_k |a_k|^q\right)^{1/q} \leq \|\widehat{a}\| < \infty. \tag{2.5.12}$$

Thus, we have shown by (2.5.12) that

$$\ell_p^\beta \subset \ell_q. \tag{2.5.13}$$

By combining the inclusion relations (2.5.11) and (2.5.13) we have $\ell_p^\beta = \ell_q$.

(a/iii) Now we show $\ell_\infty^\beta = c^\beta = c_0^\beta = \ell_1$.

Let $a = (a_k) \in \ell_1$ be given. Then, we have for all $x = (x_k) \in \ell_\infty$ as in (2.5.6), interchanging the roles of a and x,

$$\sum_k |a_k x_k| \leq \|a\|_1 \|x\|_\infty < \infty, \tag{2.5.14}$$

hence $a \in \ell_\infty^\beta$ by (2.5.14). Thus, we obtain the inclusion

$$\ell_1 \subset \ell_\infty^\beta. \tag{2.5.15}$$

Conversely, let $a = (a_k) \in c_0^\beta$. Then, we have $\hat{a} \in c_0^*$ by Part (a) of Theorem 2.5.2. Let $m \in \mathbb{N}$ be given. Then, it follows for the sequence $x^{(m)} = \sum_{k=0}^m \operatorname{sgn}(a_k)e^{(k)}$ that $x^{(m)} \in c_0$ with $\|x^{(m)}\|_\infty \leq 1$ and

$$\left|\hat{a}(x^{(m)})\right| = \sum_k |a_k| \leq \|\hat{a}\| \cdot \|x^{(m)}\|_\infty \leq \|\hat{a}\| < \infty.$$

Since m was arbitrary, it follows that

$$\sum_k |a_k| \leq \|\hat{a}\| < \infty. \tag{2.5.16}$$

Thus, $a \in \ell_1$ by (2.5.16) and we have the inclusion

$$c_0^\beta \subset \ell_1. \tag{2.5.17}$$

Now, by considering the inclusion relations (2.5.15) and (2.5.17) it follows that $\ell_\infty^\beta = \ell_1$. Additionally, by the inclusions $c_0 \subset c \subset \ell_\infty$ that $\ell_\infty^\beta \subset c^\beta \subset c_0^\beta \subset \ell_1 \subset \ell_\infty^\beta$, we have $\ell_\infty^\beta = c^\beta = c_0^\beta = \ell_1$.

This completes the proof of Part (a).

(b) First, we observe that $\ell_p \subset \ell_1$ implies $\ell_\infty = \ell_1^\beta \subset \ell_\infty$. Now, we assume $a = (a_k)_{k\in\mathbb{N}_1} \notin \ell_\infty$. Then, there exists a subsequence $(a_{k(j)})_{j\in\mathbb{N}_1}$ such that $a_{k(j)} \geq 2^j$ for all $j \in \mathbb{N}_1$. We define the sequence $x = (x_k)_{k\in\mathbb{N}_1}$ by

$$x_k := \begin{cases} \frac{1}{j^{2/p}} & , \quad k = k(j), \\ 0 & , \quad k \neq k(j) \end{cases}$$

for all $j \in \mathbb{N}_1$. Then, clearly $x \in \ell_p$, but

$$\sum_{k=1}^\infty a_k x_k = \sum_{j=1}^\infty \frac{2^j}{j^{2/p}} = \infty,$$

hence $a \notin \ell_p^\beta$. For this, the reader can refer to [236, Lemma 2 (i)] or [364, Theorem 7].

(c), (d) Since the spaces ℓ_p for $1 < p < \infty$ and c_0 have AK, it follows from Part (a) and Theorem 2.5.2 that ℓ_p^* is isomorphic to ℓ_q and c_0^* is isomorphic to ℓ_1.

If $f \in \ell_1$, then it follows from (2.5.6) and (2.5.8) that $\|f\| = \|\hat{a}\| = \|a\|_\infty$.

If $f \in \ell_p$ for $1 < p < \infty$, then it follows from (2.5.10) and (2.5.12) that $\|f\| = \|\hat{a}\| = \|a\|_q$.

If $f \in c_0$, then it follows from (2.5.14) and (2.5.16) that $\|f\| = \|\hat{a}\| = \|a\|_1$.

This completes the proof of Parts (c) and (d).

(e/i) First we show that if $f \in c^*$, then f has the representation given in (2.5.3) with χ_f defined by (2.5.4). First, we observe that since $f \in c^*$, we have $f|_{c_0} \in c_0^*$ for the restriction of f on c_0, hence

$$a = \left\{f|_{c_0}(e^{(k)})\right\}_{k\in\mathbb{N}} = \left\{f(e^{(k)})\right\}_{k\in\mathbb{N}} \in \ell_1$$

by Part (b). This implies that the series $\sum_k f(e^{(k)})$ converges and so χ_f in (2.5.4) is defined.

Now, let $x = (x_k) \in c$ be given. Then, we have by (2.3.2) in Example 2.3.5, the continuity of f, the fact that $a = (a_k) \in \ell_1 = c^\beta$ by Part (a) and (2.5.4)

$$\begin{aligned} f(x) &= f\left(\alpha e + \sum_k (x_k - \alpha)e^{(k)}\right) = \alpha f(e) + \sum_k (x_k - \alpha) f(e^{(k)}) \\ &= \alpha \left[f(e) - \sum_k f(e^{(k)}) \right] + \sum_k x_k f(e^{(k)}) = \alpha \chi_f - \sum_k a_k x_k, \end{aligned}$$

where $\alpha = \lim_{k\to\infty} x_k$. Thus, we have shown the representation of f in (2.5.3) with χ_f from (2.5.4).

(e/ii) Now, we assume that if f has the representation of f in (2.5.3) with χ_f from (2.5.4), then $f \in c^*$. Since $a = (a_k) \in \ell_1$ and $\ell_1 = c^*$, obviously f is defined on c and linear. Let $x = (x_k) \in c$ be given. Since $|\lim_{k\to\infty} x_k| \leq \|x\|_\infty$, we have

$$\begin{aligned} |f(x)| &= \left| \chi_f \lim_{k\to\infty} x_k + \sum_k a_k x_k \right| \\ &\leq |\chi_f| \|x\|_\infty + \sum_k |a_k x_k| \leq (|\chi_f| + \|a\|_1) \|x\|_\infty. \end{aligned} \tag{2.5.18}$$

Hence, the linear functional f is continuous by (2.5.18). Thus, we have shown that if f has the representation of f in (2.5.3) with χ_f from (2.5.4), then $f \in c^*$.

(e/iii) Finally, we show that if $f \in c^*$, then the norm of f satisfies (2.5.5). Let $f \in c^*$ be given. Then, f can be represented as in (2.5.3) with χ_f from (2.5.4) by Part (e/i) of the proof, and then it follows from (2.5.18) in Part (e/ii) of the proof that

$$\|f\| \leq |\chi_f| + \|a\|_1. \tag{2.5.19}$$

Define the sequence $x^{(m)} = \left\{x_k^{(m)}\right\}$ for every $m \in \mathbb{N}$ by $x_k^{(m)} := \begin{cases} \operatorname{sgn}(a_k), 0 \leq k \leq m, \\ \operatorname{sgn}(\chi_f), k > m. \end{cases}$ Then, we have

$$x^{(m)} \in c, \quad \lim_{k\to\infty} x_k^{(m)} = \operatorname{sgn}(\chi_f), \quad \|x^{(m)}\|_\infty \leq 1,$$

and we obtain from (2.5.3)

$$\begin{aligned} \left| f(x^{(m)}) \right| &= \left| |\chi_f| + \sum_{k=0}^{m} |a_k| + \sum_{k=m+1}^{\infty} \operatorname{sgn}(\chi_f) \cdot a_k \right| \\ &\leq \|f\| \cdot \left\|x^{(m)}\right\|_\infty \leq \|f\|. \end{aligned} \tag{2.5.20}$$

It follows from $a = (a_k) \in \ell_1$ that

$$\lim_{m\to\infty} \sum_{k=m+1}^{\infty} \operatorname{sgn}(\chi_f) a_k = 0,$$

hence we obtain from (2.5.20) by letting $m \to \infty$ that

$$|\chi_f| + \sum_k |a_k| = |\chi_f| + \|a\|_1 \leq \|f\|. \tag{2.5.21}$$

Now, (2.5.19) and (2.5.21) imply (2.5.5). Thus, we have shown that if $f \in c^*$, then the norm of f satisfies (2.5.5).

This completes the proof of the theorem. □

We remark that the continuous dual of ℓ_∞ is not given by a sequence space.

Now, we give the following theorem without proof related to the γ-duals of the classical sequence spaces since one can prove each part on the same lines as α- and β-duals:

Theorem 2.5.7. *The following statements hold:*

(i) $\ell_\infty^\gamma = c^\gamma = c_0^\gamma = \ell_1$.

(ii) $\ell_1^\gamma = \ell_\infty$.

(iii) $\ell_p^\gamma = \ell_q$.

Now, we give the following easy result about the α-, β- and γ-duals of the spaces f_0 and f:

Corollary 2.5.8. (see [94, (d) of Theorem 7.1.11] and [74, Proposition 4.5]) *The α-, β- and γ-duals of the spaces f_0 and f are the space ℓ_1.*

Proof. Combining the inclusions $c_0 \subset f_0 \subset \ell_\infty$ and $c \subset f \subset \ell_\infty$ with the facts stated in Part (i) of Theorem 2.5.3, Part (a) of Theorem 2.5.6 and Part (i) of Theorem 2.5.7 with (2.5.2), one can immediately see that $f_0^\xi = f^\xi = \ell_1$, where ξ denotes any of the symbols α, β and γ. This is what we wished to prove. □

Theorem 2.5.9. [264, Example 3.11] $cs^\alpha = bv_1^\alpha = bv_0^\alpha = bs^\alpha = \ell_1$.

Proof. We prove the case $cs^\alpha = \ell_1$ and the rest can be obtained similarly.

Let $x = (x_k) \in cs$ and $a = (a_k) \in \ell_1$. Then, $\sum_k |a_k x_k| \leq \|x\|_{cs} \left(\sum_k |a_k|\right) < \infty$, where $\|x\|_{cs}$ is defined by (2.3.1). Therefore, $a \in cs^\alpha$ which gives $\ell_1 \subseteq cs^\alpha$.

Conversely, suppose that $a = (a_k) \in cs^\alpha \setminus \ell_1$. Then, to every natural number i we can find an odd n_i with $n_i < n_{i+1}$ and $\sum_{k=n_i+1}^{n_{i+1}} |a_k| > 2^i$ for all $i \in \mathbb{N}$. Define $x = (x_k)$ by

$$x_k := \begin{cases} (-1)^k 2^{-i/2} & , \quad n_i < k \leq n_{i+1}, \\ 0 & , \quad \text{otherwise}, \end{cases}$$

where $i \in \mathbb{N}$. Then, $(x_k) \in cs$ but $\sum_k |a_k x_k| = \infty$. This contradicts that $a \in cs^\alpha$, and so $a = (a_k)$ must be in ℓ_1 which gives the inclusion $cs^\alpha \subseteq \ell_1$.

Hence, $cs^\alpha = \ell_1$. □

Remark 2.5.10. *Note that $\omega^\alpha = \phi$ and $\phi^\alpha = \omega$.*

Theorem 2.5.11. *The following statements hold:*

(i) $bv_0^\beta = bs$, (cf. [396, Part (ii) of Theorem 7.3.5]).

(ii) $bv^\beta = cs$, (cf. [396, Part (iii) of Theorem 7.3.5]).

(iii) $cs^\beta = bv_1$, (cf. [396, Part (v) of Theorem 7.3.5]).

(iv) $bs^\beta = bv_0$, (cf. [396, Part (vi) of Theorem 7.3.5]).

Proof. Since the other parts can be similarly proved, we consider only Part (iii).

Let $a = (a_k) \in cs^\beta$ and $z = (z_k) \in c_0$. Define the sequence $x = (x_k) \in cs$ by $x_k := z_k - z_{k+1}$ for all $k \in \mathbb{N}$. Therefore, $\sum_k a_k x_k$ converges, but

$$\sum_{k=0}^{n}(z_k - z_{k+1})a_k = \sum_{k=0}^{n-1} z_k(a_k - a_{k-1}) - z_{n+1}a_n$$

and the inclusion $cs \supset \ell_1$ yields that $(a_k) \in cs^\beta \subset \ell_1^\beta = \ell_\infty$ which implies that

$$\sum_k (z_k - z_{k+1})a_k = \sum_k z_k(a_k - a_{k-1}).$$

Hence, $(a_k - a_{k-1}) \in c_0^\beta = c_0^\alpha = \ell_1$, i.e., $a \in bv_1$. Therefore, $cs^\beta \subseteq bv_1$.

Conversely, let $a = (a_k) \in bv_1$. Then, $(a_k - a_{k-1}) \in \ell_1$. Further, if $x = (x_k) \in cs$, the sequence (z_n) defined by $z_n := \sum_{k=0}^n x_k$ for all $n \in \mathbb{N}$ is an element of the space c. Since $c^\alpha = \ell_1$, the series $\sum_k z_k(a_k - a_{k+1})$ is absolutely convergent. Also, we have

$$\left|\sum_{k=m}^{n}(z_k - z_{k-1})a_k\right| \le \left|\sum_{k=m}^{n-1} z_k(a_k - a_{k+1})\right| + |z_n a_n - z_{m-1}a_m|. \quad (2.5.22)$$

As $(z_n) \in c$ and $(a_k) \in bv_1 \subset c$, the right-hand side of the inequality (2.5.22) converges to zero as $m, n \to \infty$. Hence, the series $\sum_k (z_k - z_{k-1})a_k$ or $\sum_k a_k x_k$ converges and so, $bv_1 \subseteq cs^\beta$. Thus, $cs^\beta = bv_1$. □

Theorem 2.5.12. *The following statements hold:*

(i) $bs^\gamma = cs^\gamma = bv_1$, (cf. [396, Parts (v) and (iii) of Theorem 7.3.5]).

(ii) $bv_1^\gamma = bv_0^\gamma = bs$, (cf. [396, Parts (iv) and (ii) of Theorem 7.3.5]).

Proof. We prove only the fact that $cs^\gamma = bv_1$, the rest can be proved along similar lines.

By Theorem 2.5.11, we have $bv_1 \subset cs^\beta$ and since $cs^\beta \subset cs^\gamma$, so $bv_1 \subset cs^\gamma$. We need to show that $cs^\gamma \subset bv_1$. Let $a = (a_n) \in cs^\gamma$ and $z = (z_n) \in c_0$. Then, for the sequence $(t_n) \in cs$ defined by $t_n = z_n - z_{n+1}$ for all $n \in \mathbb{N}$, we can find a constant $K > 0$ such that $|\sum_{k=0}^n a_k t_k| \le K$ for all $n \in \mathbb{N}$. Since $(z_n) \in c_0$ and $(a_n) \in cs^\gamma \subset \ell_\infty$, there exists a constant M such that $|a_n z_n| \le M$ for all $n \in \mathbb{N}$. Now,

$$\left|\sum_{k=0}^{n} z_k(a_k - a_{k-1})\right| \le \left|\sum_{k=0}^{n+1}(z_k - z_{k+1})a_k\right| + |z_{n+2}a_{n+1}| \le K + M.$$

Hence, $(a_k - a_{k-1}) \in c_0^\gamma = \ell_1$ by Part (i) of Theorem 2.5.7, i.e., $(a_k) \in bv_1$. Therefore, since the inclusion $cs^\gamma \subset bv_1$ holds, we conclude that $cs^\gamma = bv_1$. □

TABLE 2.1 The α-, β- and γ-duals of certain sequence spaces

λ	λ^{α}	λ^{β}	λ^{γ}	Refer to:
ω	ϕ	ϕ	ϕ	[94, Theorem 7.1.11 (c), p. 344]
ϕ	ω	ω	ω	[94, Part (c) of Theorem 7.1.11, p. 344]
ℓ_1	ℓ_∞	ℓ_∞	ℓ_∞	[94, Part (c) of Theorem 7.1.11, p. 344]
ℓ_p $(p>1)$	ℓ_q	ℓ_q	ℓ_q	[94, Part (c) of Theorem 7.1.11, p. 344]
ℓ_∞	ℓ_1	ℓ_1	ℓ_1	[94, Part (c) of Theorem 7.1.11, p. 344]
f	ℓ_1	ℓ_1	ℓ_1	[94, Part (d) of Theorem 7.1.11, p. 344], [74]
f_0	ℓ_1	ℓ_1	ℓ_1	[94, Part (d) of Theorem 7.1.11, p. 344], [74]
c	ℓ_1	ℓ_1	ℓ_1	[94, Part (d) of Theorem 7.1.11, p. 344]
c_0	ℓ_1	ℓ_1	ℓ_1	[94, Part (d) of Theorem 7.1.11, p. 344]
bs	ℓ_1	bv_0	bv_1	[94, Theorem 7.1.12, p. 347]
cs	bv_1	bv_1	bv_1	[94, Theorem 7.1.12, p. 347]
bv_1	ℓ_1	cs	bs	[94, Theorem 7.1.12, p. 347]
bv_0	ℓ_1	bs	bs	[94, Theorem 7.1.12, p. 347]
fs		bv_0	bv	[74, Theorem 4.8]

Table 2.1 collects the results on the α-, β- and γ-duals of certain sequence spaces.

Finally, we state and prove the following theorem that yields the continuous dual of the space cs given by Coşkun and Başar [113]:

Theorem 2.5.13. *The continuous dual cs^* of cs is norm isomorphic to bv.*

Proof. Since $\left\{e^{(k)}\right\}$ is a basis for cs, every $x=(x_k)\in cs$ has a unique representation $x=\sum_k x_k e^{(k)}$. Consider any $f\in cs^*$. Then, since f is linear and bounded, we have $f(x)=\sum_k a_k x_k$, where $a_k=f(e^{(k)})$ for all $k\in\mathbb{N}$. Since $cs^{\beta}=bv$, the convergence of the series $\sum_k a_k x_k$ for every $x=(x_k)\in cs$ implies that $a=(a_k)\in bv$.

Now, if we take $x=(x_k)\in cs$ and $a=(a_k)\in bv$, then the series $\sum_k a_k x_k$ converges, and so we can obtain a corresponding bounded linear functional on cs. That is, we may define f on cs by $f(x)=\sum_k a_k x_k$. Then, the linearity of f is clear and the boundedness also follows from the fact that

$$\begin{aligned} |f(x)| &= \left|\sum_k a_k x_k\right| \\ &= \left|\lim_{n\to\infty}\sum_{k=0}^{n} a_k x_k\right| = \left|\lim_{n\to\infty}\left[\sum_{k=0}^{n} x_k a_{n+1} - \sum_{k=0}^{n}\sum_{j=0}^{k} x_j(a_{k+1}-a_k)\right]\right| \\ &\le \|x\|_{cs}\left(\lim_{n\to\infty}|a_{n+1}| - \lim_{n\to\infty}\sum_{k=0}^{n}|a_{k+1}-a_k|\right) = \|x\|_{cs}\|a\|_{bv}. \end{aligned}$$

Hence, by taking the supremum over all x of norm 1, we get $\|f\|\le\|a\|_{bv}$ for each $f\in cs^*$. Define the mapping T from cs^* to bv by

$$\begin{aligned} T\ :\ cs^* &\longrightarrow bv \\ f &\longmapsto T(f)=(a_0,a_1,a_2,\ldots);\ \ a_k=f\left(e^{(k)}\right)\ \text{ for all }\ k\in\mathbb{N}. \end{aligned}$$

We have just seen that T maps cs^* onto bv and $\|f\| \leq \|T(f)\|$, and it is easy to verify that T is linear and one-to-one.

Now, consider

$$\|T(f)\| = \|a\| = \lim_{n\to\infty} \sum_{k=0}^{n} \left| f\left(e^{(k+1)} - e^{(k)}\right) \right| + \lim_{m\to\infty} \left| f\left(e^{(m)}\right) \right| \qquad (2.5.23)$$

and choose the scalars α_k's with $|\alpha_k| = 1$ such that

$$\left| f\left(e^{(k+1)} - e^{(k)}\right) \right| = \alpha_k f\left(e^{(k+1)} - e^{(k)}\right) \text{ for all } k \in \mathbb{N}.$$

Then, $\sum_{k=0}^{n} \left| f\left(e^{(k+1)} - e^{(k)}\right) \right| = f(u)$; where

$$\begin{aligned} u = (u_n) &= \left\{ \sum_{k=0}^{n} \alpha_k \left(e^{(k+1)} - e^{(k)}\right) \right\} \\ &= \left\{ \sum_{k=0}^{n-1} (\alpha_k - \alpha_{k+1}) e^{(k+1)} + \alpha_n e^{(n+1)} - \alpha_0 e^{(0)} \right\} \\ &= (-\alpha_0, \alpha_0 - \alpha_1, \ldots, \alpha_{n-1} - \alpha_n, \alpha_n, 0, \ldots). \end{aligned}$$

It is clear that $u = (u_n) \in cs$ and $\|u\|_{cs} = 1$. Therefore, by (2.5.23)

$$\|T(f)\| = \|a\| = \lim_{n\to\infty} f(u) + \lim_{m\to\infty} \left| f\left(e^{(m)}\right) \right| \leq \|f\| + \|f\| = 2\|f\|.$$

Hence, $\|f\| \leq \|T(f)\| \leq 2\|f\|$ for $f \in cs^*$, which means that cs^* is congruent to bv. Further, observe for $f \in cs^*$ with $\{f(e^{(n)})\} \in bv_0$ that $\|T(f)\| = \|f\|$. Since the map f_a, defined on cs by $f_a(x) = \sum_k a_k x_k$ for $a = (a_k) \in bv$, is in cs^* and $T(f_a) = a$, and $bv = bv_0 \oplus lhull(e)$, one can observe for $e \in bv$ that $\|T(f_e)\| = \|f_e\| = 1$. In this respect we have the continuous dual of cs as bv.

This completes the proof. □

In the next chapter, we redetermine the β-dual of the space cs using Theorem 3.3.6.

CHAPTER 3

Matrix Transformations in Sequence Spaces

In this chapter, the matrix transformations in sequence spaces are studied and the characterizations of the classes of Schur, Kojima and Toeplitz matrices together with their versions for the series-to-sequence, sequence-to-series and series-to-series matrix transformations are given.

Keywords: Matrix transformations between sequence spaces, ordinary, absolute and strong summability, Kojima-Schur theorem, Silverman-Toeplitz theorem, Schur's theorem, algebra, normed and Banach algebra, characteristic of a matrix, sequence-to-sequence, series-to-sequence, series-to-series and sequence-to-series matrix transformations, dual summability methods, arithmetics, Cesàro, Euler, Riesz and Nörlund means, Taylor, A^r, Hausdorff, Borel and Abel matrices.

3.1 INTRODUCTION

In some cases, the most general linear operator between two sequence spaces is given by an infinite matrix. So, the theory of *matrix transformations* has always been of great interest in the study of sequence spaces. The study of the general theory of matrix transformations was motivated by special results in the summability theory.

Suppose $A = (a_{nk})$ is an infinite matrix of real or complex numbers a_{nk}, where $k, n \in \mathbb{N}$ and $x = (x_k) \in \omega$. Then, we obtain the sequence Ax, the *A-transform of* x,

DOI: 10.1201/9781003294153-3

by the usual matrix product

$$Ax := \begin{pmatrix} a_{00} & a_{01} & a_{02} & \cdots & a_{0k} & \cdots \\ a_{10} & a_{11} & a_{12} & \cdots & a_{1k} & \cdots \\ a_{20} & a_{21} & a_{22} & \cdots & a_{2k} & \cdots \\ \vdots & \vdots & \vdots & \cdots & \vdots & \cdots \\ a_{n0} & a_{n1} & a_{n2} & \cdots & a_{nk} & \cdots \\ \vdots & \vdots & \vdots & \cdots & \vdots & \cdots \end{pmatrix} \begin{pmatrix} x_0 \\ x_1 \\ x_2 \\ \vdots \\ x_k \\ \vdots \end{pmatrix}$$

$$:= \begin{pmatrix} a_{00}x_0 + a_{01}x_1 + a_{02}x_2 + \cdots + a_{0k}x_k + \cdots \\ a_{10}x_0 + a_{11}x_1 + a_{12}x_2 + \cdots + a_{1k}x_k + \cdots \\ a_{20}x_0 + a_{21}x_1 + a_{22}x_2 + \cdots + a_{2k}x_k + \cdots \\ \vdots \\ a_{n0}x_0 + a_{n1}x_1 + a_{n2}x_2 + \cdots + a_{nk}x_k + \cdots \\ \vdots \end{pmatrix}.$$

Hence, in this way, we transform the sequence x into the sequence $Ax = \{(Ax)_n\}$ with

$$(Ax)_n = \sum_k a_{nk}x_k, \ (n \in \mathbb{N}), \tag{3.1.1}$$

provided the series on the right-hand side of (3.1.1) converges for each $n \in \mathbb{N}$.

Let λ and μ be any two sequence spaces. If Ax exists and is in μ for every sequence $x = (x_k) \in \lambda$, then we say that A defines a matrix mapping from λ into μ, and we denote it by writing $A : \lambda \to \mu$. By $(\lambda : \mu)$, we denote the class of all matrices A such that $A : \lambda \to \mu$. Thus, $A \in (\lambda : \mu)$ if and only if the series on the right-hand side of (3.1.1) converges for each $n \in \mathbb{N}$ and every $x \in \lambda$, and we have $Ax = \{(Ax)_n\}_{n\in\mathbb{N}} \in \mu$ for all $x \in \lambda$.

Now, we may give the following lemma due to Wilansky [396, p. 57] which is needed in the text:

Lemma 3.1.1. *The matrix mappings between the BK-spaces are continuous.*

3.2 INTRODUCTION TO SUMMABILITY

In this subsection, following Malkowsky [260], we give some lines about ordinary, absolute and strong summability.

The theory of sequence spaces is the fundamental of summability. Summability is a wide field of mathematics, mainly in analysis and functional analysis, and has many applications, for instance in numerical analysis to speed up the rate of convergence in operator theory, the theory of orthogonal series and approximation theory.

3.2.1 Summability

This subsection serves as a motivation of what follows. The classical *summability* theory deals with the generalization of the convergence of sequences or series of real or complex numbers. The idea is to assign a limit of some sort to divergent sequences or series by considering a transform of a sequence or series rather than the original sequence or series.

One can ask why we employ the special transformations represented by infinite matrices instead of general linear operators? The answer to this question is, in many cases, the most general linear operators between two sequence spaces are given by an infinite matrix.

Let $A = (a_{nk})_{n,k\in\mathbb{N}}$ be an infinite matrix of complex numbers throughout. There are three kinds of summability: Ordinary, absolute and strong summability, as mentioned below:

(a) *Ordinary Summability.* A sequence $x = (x_k)_{k\in\mathbb{N}} \in \omega$ is said to be summable A to a complex number α if the A-limit of x is α, i.e.,

$$\lim_{n\to\infty}(Ax)_n = \alpha,$$

and is denoted by $x \to \alpha(A)$. The matrix A defines a summability method A, or a matrix transformation by (3.1.1).

Let $s = (s_n)$ be the sequence of partial sums of the series $\sum_{k=0}^{\infty} x_k$. If $(As)_n \to \alpha$, as $n \to \infty$, then we say that the series $\sum_{k=0}^{\infty} x_k$ is A-summable to α.

Example 3.2.1. *The Cesàro transform of a sequence $x = (x_k) \in \omega$ is given by $C_1x = \{(C_1x)_n\}_{n\in\mathbb{N}}$, where the Cesàro method C_1 of order 1 is given by (1.1.5). Consider $x = (x_k) = \{(-1)^k\}_{k\in\mathbb{N}} \in \ell_\infty \setminus c$. Then, since*

$$0 \le |(C_1x)_n| \le \frac{1}{n+1}$$

for all $n \in \mathbb{N}$, $(C_1x)_n \to 0$ as $n \to \infty$. This means that the divergent sequence $x := \{(-1)^k\}_{k\in\mathbb{N}}$ is C_1-summable to zero.

Let us consider the divergent series $\sum_{k=0}^{\infty}(-1)^k$ with $s_n := \sum_{k=0}^{n}(-1)^k := [1-(-1)^{n+1}]/2$ for all $n \in \mathbb{N}$. Since $(C_1s)_n := \frac{1}{2} + \frac{1+(-1)^n}{4(n+1)}$ for all $n \in \mathbb{N}$, it is immediate that $(C_1s)_n \to 1/2$, as $n \to \infty$. Therefore, we conclude that the series $\sum_{k=0}^{\infty}(-1)^k$ is C_1-summable to $1/2$.

Proposition 3.2.2. *Every convergent sequence $x = (x_k)$ is C_1-summable and $\lim_{k\to\infty} x_k = \lim_{k\to\infty}(C_1x)_k$.*

Proof. Suppose that $x = (x_k) \in c$ with $x_k \to l$, as $k \to \infty$. Then, there is a constant $M \ge |l|$ such that $|x_k| \le M$ for all $k \in \mathbb{N}$ and given $\varepsilon > 0$, there is $k_0 \in \mathbb{N}$ such that $|x_k - l| < \varepsilon$ for all $k > k_0$. Let us choose a positive integer $K \ge k_0$ such that $K > 2Mk_0/\varepsilon$. Then, since for all $k \ge K$

$$\begin{aligned} |(C_1x)_k - l| &:= \left|\frac{(x_0 - l) + \cdots + (x_k - l)}{k+1}\right| \\ &\le \frac{|x_0 - l| + \cdots + |x_{k_0} - l|}{k+1} + \frac{|x_{k_0+1} - l| + \cdots + |x_k - l|}{k+1} \\ &\le \frac{2M(k_0 + l)}{k+1} + \frac{(k-k_0)\varepsilon}{k+1} < 2\varepsilon, \end{aligned}$$

$(C_1x)_k \to l$, as $k \to \infty$.

This step completes the proof. □

(b) *Absolute Summability.*

Let $0 < p < \infty$. A sequence $x = (x_k)_{k\in\mathbb{N}}$ is said to be absolutely summable with index p to a complex number s if the series on the right-hand side of (3.1.1) converge for all $n \in \mathbb{N}$ and $\sum_n |(Ax)_n|^p = s$. This is denoted by $x \to s|A|^p$.

Now, we may give the following proposition which gives the relation between ordinary and absolute summability:

Proposition 3.2.3. [260, Remark 1.3, p. 4] *Every $|A|^p$ summable sequence is summable A to 0. However, A summability does not imply $|A|^p$-summability for any p, in general.*

Proof. Let $p < 1$ and the sequence $x = (x_k)_{k\in\mathbb{N}}$ be $|A|^p$ summable to s. Then, obviously $\sum_n |(Ax)_n|^p = s$ implies $(Ax)_n \to 0$, as $n \to \infty$.

For the second part, we define the sequence $x = (x_k)$ by

$$x_k := \frac{k+1}{\log(k+3)} - \frac{k}{\log(k+2)} \quad \text{for all } k \in \mathbb{N}. \tag{3.2.1}$$

Then, we have

$$\begin{aligned}\lim_{n\to\infty}(C_1x)_n &= \lim_{n\to\infty}\frac{1}{n+1}\sum_{k=0}^{n}\left[\frac{k+1}{\log(k+3)} - \frac{k}{\log(k+2)}\right]\\ &= \lim_{n\to\infty}\frac{n+1}{(n+1)\log(n+3)} = \lim_{n\to\infty}\frac{1}{\log(n+3)} = 0,\end{aligned}$$

but for any $p < 1$,

$$\sum_n |(C_1x)_n|^p = \sum_n \frac{1}{[\log(n+3)]^p} = \infty.$$

This shows that although the sequence $x = (x_k)$, defined by (3.2.1), is C_1-summable to zero but not $|C_1|^p$-summable for any $p < 1$. □

(c) *Strong Summability.*

Following Maddox [252], we give the definition of strong summability.

Let $A = (a_{nk})_{n,k\in\mathbb{N}}$ be an infinite matrix over the complex field $\mathbb{C}$ and $p = (p_k)$ also be a sequence of strictly positive numbers. Then, a sequence $x = (x_k)_{k\in\mathbb{N}} \in \omega$ is said to be strongly summable by A to l, if $\sum_k a_{nk}|x_k - l|^{p_k}$ exists for each $n \in \mathbb{N}$ and tends to zero as $n \to \infty$, this is denoted by $x_k \to l[A,p]$. If $\sum_k a_{nk}|x_k|^{p_k} = O(1)$, then we say that $x = (x_k)$ is strongly bounded by A and write $x_k = O(1)[A,p]$.

In case $p_k = p$ for all $k \in \mathbb{N}$, strong summability and boundedness are reduced to the strong summability and boundedness with index p. The sets of sequences which are strongly summable to zero, strongly summable and strongly bounded by A are denoted by $[A,p]_0$, $[A,p]$ and $[A,p]_\infty$, respectively. Some of the well-known sequence spaces are derived by specializing A. For example, in the case $A = I$ we have $[I,p]_0 = c_0(p)$, $[I,p] = c(p)$ and $[I,p]_\infty = \ell_\infty(p)$; in the case $A = C_1$ we have $[C_1,p]_0 = \omega_0(p)$, $[C_1,p] = \omega(p)$ and $[C_1,p]_\infty = \omega_\infty(p)$.

Note that there is no simple relation between strong summability and absolute or ordinary summability, in general.

3.3 CHARACTERIZATIONS OF SOME MATRIX CLASSES

In this section, we give the characterizations of some matrix classes.

We begin with stating the necessary and sufficient condition on an infinite matrix belonging to the class $(\ell_\infty : \ell_\infty)$.

Theorem 3.3.1. *$A = (a_{nk}) \in (\ell_\infty : \ell_\infty)$ if and only if*

$$\sup_{n\in\mathbb{N}} \sum_k |a_{nk}| < \infty. \tag{3.3.1}$$

Proof. Suppose that the condition (3.3.1) holds and $x = (x_k) \in \ell_\infty$. In this situation, since $(a_{nk})_{k\in\mathbb{N}} \in \ell_\infty^\beta = \ell_1$ for every fixed $n \in \mathbb{N}$, the A-transform of x exists. Taking into account the hypothesis, one can easily observe that

$$\sup_{n\in\mathbb{N}} |(Ax)_n| = \sup_{n\in\mathbb{N}} \left| \sum_k a_{nk}x_k \right| \leq \|x\|_\infty \left(\sup_{n\in\mathbb{N}} \sum_k |a_{nk}| \right) < \infty,$$

which leads us to the fact that $Ax \in \ell_\infty$, as desired.

Conversely, suppose that $A \in (\ell_\infty : \ell_\infty)$. Put $Ax = \{(Ax)_n\}$ and observe that (A_n) is a sequence of bounded linear operators on ℓ_∞ such that $\sup_{n\in\mathbb{N}} |(Ax)_n| < \infty$. Now, the result follows from an application of Banach-Steinhaus theorem.

This completes the proof. □

Theorem 3.3.2. (cf. [237] and [300, p. 245]) *$A = (a_{nk}) \in (\ell_\infty(p) : \ell_\infty)$ if and only if*

$$\sup_{n\in\mathbb{N}} \sum_k |a_{nk}| N^{1/p_k} < \infty \quad \textit{for all integers } N > 1. \tag{3.3.2}$$

Proof. Suppose that $A = (a_{nk}) \in (\ell_\infty(p) : \ell_\infty)$, but there is an integer $N > 1$ such that $\sup_{n\in\mathbb{N}} \sum_k |a_{nk}| N^{1/p_k} = \infty$. Define the matrix $B = (b_{nk})$ by $b_{nk} := a_{nk}N^{p_k}$ for all $k, n \in \mathbb{N}$. Then, the matrix $B \notin (\ell_\infty : \ell_\infty)$ and so there is an $x = (x_k) \in \ell_\infty$ with $\|x\| = 1$ such that $\left(\sum_k a_{nk}N^{1/p_k}x_k\right)$ does not belong to the space ℓ_∞. Hence, although $y = (y_k) := (N^{1/p_k}x_k) \in \ell_\infty(p)$, the sequence $\{(Ay)_n\} \notin \ell_\infty$ which contradicts the fact that $A \in (\ell_\infty(p) : \ell_\infty)$. This shows the necessity of the condition (3.3.2).

Conversely, assume that condition (3.3.2) holds and $x = (x_k) \in \ell_\infty(p)$. Let $N > \max\{1, \sup_{k\in\mathbb{N}} |x_k|^{p_k}\}$. Then,

$$\sup_{n\in\mathbb{N}} |(Ax)_n| \leq \sup_{n\in\mathbb{N}} \sum_k |a_{nk}| N^{1/p_k} < \infty,$$

which leads us to the fact that $Ax \in \ell_\infty$, as desired.

This completes the proof. □

We state and prove the Kojima-Schur theorem, which gives the necessary and sufficient conditions on an infinite matrix that maps the space c into itself. A matrix satisfying the conditions of the Kojima-Schur theorem is called a *conservative matrix* or *convergence preserving matrix*.

Theorem 3.3.3. [Kojima-Schur] $A = (a_{nk}) \in (c : c)$ *if and only if (3.3.1) holds, and there exist* $\alpha_k, \alpha \in \mathbb{C}$ *such that*

$$\lim_{n\to\infty} a_{nk} = \alpha_k \quad \textit{for each} \quad k \in \mathbb{N}, \tag{3.3.3}$$

$$\lim_{n\to\infty} \sum_k a_{nk} = \alpha. \tag{3.3.4}$$

Proof. Suppose that the conditions (3.3.1), (3.3.3) and (3.3.4) hold and $x = (x_k) \in c$ with $x_k \to l$ as $k \to \infty$. Then, since $(a_{nk})_{k\in\mathbb{N}} \in c^\beta = \ell_1$ for each $n \in \mathbb{N}$, the A-transform of x exists. In this situation, the equality

$$\sum_k a_{nk}x_k = \sum_k a_{nk}(x_k - l) + l\sum_k a_{nk} \tag{3.3.5}$$

holds for each $n \in \mathbb{N}$. In (3.3.5), since the first term on the right-hand side tends to $\sum_k \alpha_k(x_k - l)$ by (3.3.3) and the second term on the right-hand side tends to $l\alpha$ by (3.3.4) as $n \to \infty$, we have

$$\lim_{n\to\infty} \sum_k a_{nk}x_k = \sum_k \alpha_k(x_k - l) + l\alpha.$$

Hence, $Ax \in c$, that is the conditions are sufficient.

Conversely, suppose that $A \in (c : c)$. Then, Ax exists for every $x \in c$. The necessity of the conditions (3.3.3) and (3.3.4) is immediate by taking $x = e^{(k)}$ and $x = e$, respectively. Since $c \subset \ell_\infty$, the necessity of the condition (3.3.1) is obtained from Theorem 3.3.1.

This completes the proof. □

We state and prove the Silverman-Toeplitz theorem which gives the necessary and sufficient conditions on an infinite matrix that maps the space c into itself, leaving the limit of each convergent sequence invariant. We should note here that Silverman proved the sufficiency of the conditions, which is the easy part of the proof. A matrix satisfying the conditions of the Silverman-Toeplitz theorem is called a *Toeplitz matrix* or *regular matrix*. By $(c : c; p)$, we denote the class of Toeplitz matrices. Now, we may give the corollaries characterizing the classes of $(c : c; p)$ and $(c_0 : c_0)$.

Corollary 3.3.4. [Silverman-Toeplitz theorem] $A = (a_{nk}) \in (c : c; p)$ *if and only if (3.3.1) holds, and (3.3.3), (3.3.4) also hold with* $\alpha_k = 0$ *for all* $k \in \mathbb{N}$, $\alpha = 1$, *respectively.*

Corollary 3.3.5. $A = (a_{nk}) \in (c_0 : c_0)$ *if and only if (3.3.1) holds and (3.3.3) also holds with* $\alpha_k = 0$ *for all* $k \in \mathbb{N}$.

As an application of matrix transformations, we give the following theorem on the β-dual of the space cs proved with the technique due to Altay and Başar (cf. [21, 23, 26, 27, 28, 29, 30, 31]):

Theorem 3.3.6. *β-dual of the space cs is the space bv.*

Proof. Let $x = (x_k) \in cs$ and $y = (y_k)$ be the sequence of partial sums of the series $\sum_k x_k$. Then, we derive the following equality by applying the Abel's partial summation to the nth-partial sums of the series $\sum_k a_k x_k$:

$$\sum_{k=0}^{n} a_k x_k = \sum_{k=0}^{n-1} (a_k - a_{k+1}) y_k + a_n y_n = (By)_n \tag{3.3.6}$$

for all $n \in \mathbb{N}$ with the matrix $B = (b_{nk})$ defined by

$$b_{nk} := \begin{cases} a_k - a_{k+1} & , \quad 0 \leq k \leq n-1, \\ a_n & , \quad k = n, \\ 0 & , \quad k > n, \end{cases}$$

for all $k, n \in \mathbb{N}$. Therefore, one can see from (3.3.6) that

$$ax = (a_k x_k) \in cs \quad \text{whenever} \quad x \in cs \quad \text{if and only if} \quad B \in (c : c).$$

This means that $a = (a_k) \in cs^\beta$ if and only if $a = (a_k) \in bv$, by Theorem 3.3.3 which completes the proof. □

Prior to stating and proving Schur's theorem we give the following lemma which is needed:

Lemma 3.3.7. *If* $\sum_k |b_{nk}| < \infty$ *for each* $n \in \mathbb{N}$ *and* $\sum_k |b_{nk}| \to 0$, *as* $n \to \infty$ *then* $\sum_k |b_{nk}|$ *is uniformly convergent in* $n \in \mathbb{N}$.

Proof. $\sum_k |b_{nk}| \to 0$ as $n \to \infty$ implies that $\sum_k |b_{nk}| < \varepsilon$ for $n \geq N(\varepsilon)$. Since $\sum_k |b_{nk}| < \infty$, $0 \leq n \leq N(\varepsilon)$, there exists $m = M(n, \varepsilon)$ such that $\sum_{k \geq M} |b_{nk}| < \varepsilon$. Hence, there exists $M = M(\varepsilon) \geq 1$ such that $\sum_{k \geq M} |b_{nk}| < \varepsilon$ for all $n \in \mathbb{N}$, which means that $\sum_k |b_{nk}|$ is uniformly convergent in $n \in \mathbb{N}$. □

A matrix transforming every bounded sequence to a convergent sequence is called a *coercive matrix*. Now, we can give Schur's theorem characterizing the class of coercive matrices:

Theorem 3.3.8. [258, Theorem 10, pp. 223–225] $A = (a_{nk}) \in (\ell_\infty : c)$ *if and only if (3.3.3) holds, and*

$$\sum_k |a_{nk}| \quad \text{converges uniformly in} \quad n \in \mathbb{N}. \tag{3.3.7}$$

Proof. Suppose that the conditions (3.3.3) and (3.3.7) hold, and $x = (x_k) \in \ell_\infty$. Then, $\sum_k a_{nk} x_k$ is absolutely and uniformly convergent in $n \in \mathbb{N}$. Hence, $\sum_k a_{nk} x_k \to \sum_k \alpha_k x_k$, as $n \to \infty$ which leads us to the fact that $A \in (\ell_\infty : c)$.

Conversely, suppose that $A = (a_{nk}) \in (\ell_\infty : c)$. Then, Ax exists for every $x = (x_k) \in \ell_\infty$. The necessity of the condition (3.3.3) is immediate on taking $x = e^{(k)}$ for each fixed $k \in \mathbb{N}$.

Define $b_{nk} := a_{nk} - \alpha_k$ for all $k, n \in \mathbb{N}$. Since $\sum_k |\alpha_k| < \infty$, $(\sum_k b_{nk} x_k)$ converges whenever $x = (x_k) \in \ell_\infty$. We are going to show that this implies

$$\lim_{n\to\infty} \sum_k |b_{nk}| = 0. \tag{3.3.8}$$

Then, by Lemma 3.3.7 we obtain that $\sum_k |b_{nk}|$ converges uniformly in $n \in \mathbb{N}$. Suppose to the contrary that $\sum_k |b_{nk}|$ does not tend to 0, as $n \to \infty$. Then, it follows that $\sum_k |b_{nk}| \to l > 0$ as $n \to \infty$ through some subsequence of the positive integers. Also, we have $b_{mk} \to 0$ as $m \to \infty$ for each $k \in \mathbb{N}$. Hence, we may determine $m(1)$ such that

$$\left|\sum_k |b_{m(1),k}| - l\right| < \frac{l}{10} \quad \text{and} \quad |b_{m(1),1}| < \frac{l}{10}.$$

Since $\sum_k |b_{m(1),k}| < \infty$ we may choose $k(2) > 1$ such that

$$\sum_{k=k(2)+1}^{\infty} |b_{m(1),k}| < \frac{l}{10}.$$

It follows that

$$\left|\sum_{k=2}^{k(2)} |b_{m(1),k}| - l\right| < \frac{3l}{10}.$$

To contract notation we are going to write

$$\sum_{k=p}^{q} |b_{mk}| = B(m,p,q).$$

Now, we choose $m(2) > m(1)$ such that $|B(m(2),1,\infty) - l| < l/10$ and $B(m(2),1,k(2)) < l/10$. Then, choose $k(3) > k(2)$ such that $B(m(2),k(3)+1,\infty) < l/10$. It follows that $|B(m(2),k(2)+1,k(3)) - l| < 3l/10$. Continuing in this way and find $m(1) < m(2) < \cdots$, $1 = k(1) < k(2) < \cdots$ such that

$$\left.\begin{array}{rcl} B(m(r),1,k(r)) & < & \frac{l}{10} \\ B(m(r),k(r+1)+1,\infty) & < & \frac{l}{10} \\ |B(m(r),k(r)+1,k(r+1)) - c| & < & \frac{3l}{10} \end{array}\right\}. \tag{3.3.9}$$

Let us define $x = (x_k) \in \ell_\infty$ such that $\|x\|_\infty = 1$ by

$$x_k := \begin{cases} 0 & , \quad k = 1, \\ (-1)^r \operatorname{sgn}(b_{m(r),k}) & , \quad k(r) < k \le k(r+1), \end{cases} \tag{3.3.10}$$

for $r \in \mathbb{N}_1$. Then, write $\sum_k b_{m(r),k} x_k$ as $\sum_1 + \sum_2 + \sum_3$, where $\sum_1$ is over $1 \le k \le k(r)$, $\sum_2$ over $k(r) \le k \le k(r+1)$ and $\sum_3$ over $k > k(r+1)$. It follows immediately from (3.3.9) with the sequence x given by (3.3.10) that

$$\left| \sum_k b_{m(r),k} x_k - (-1)^r l \right| < \frac{l}{2}.$$

Consequently, it is clear that the sequence $Bx = \left(\sum_k b_{nk} x_k\right)$ is not a Cauchy sequence and so is not convergent. Thus, we have proved that Bx is not convergent for $x \in \ell_\infty$ which contradicts the fact that $A \in (\ell_\infty : c)$. Hence, (3.3.8) must be hold. Now, it follows by Lemma 3.3.7 that $\sum_k |b_{nk}|$ converges uniformly in $n \in \mathbb{N}$. Therefore, $\sum_k |a_{nk}| = \sum_k |b_{nk} + \alpha_k|$ converges uniformly in $n \in \mathbb{N}$.

This completes the proof. □

Now, we define some concepts from the theory of Banach algebras.

Definition 3.3.9. (see Maddox [258, p. 153]) *An algebra X is a linear space together with an internal operation of multiplication of elements of X, such that*

$$xy \in X, \quad x(yz) = (xy)z, \quad x(y+z) = xy + xz, \quad (x+y)z = xz + yz$$

and

$$\alpha(xy) = x(\alpha y) \quad \text{for all scalars} \quad \alpha.$$

In some algebras, there exists a non-zero element u such that $ux = xu = x$ for all $x \in X$. If such an u exists, it is obviously unique and is then called the *identity of the algebra*. An algebra is called *commutative* if it has the property that $xy := yx$, for all $x, y \in X$.

Definition 3.3.10. (see Maddox [258, p. 153]) *A normed algebra is an algebra which is normed as a linear space and has the submultiplicative property of the norm, i.e., $\|xy\| \le \|x\|\|y\|$ for all $x, y \in X$.*

Definition 3.3.11. (see Maddox [258, p. 153]) *A Banach algebra is a complete normed algebra, i.e., an algebra which is a Banach space.*

It is known that (cf. Nanda [300, p. 13]) the class $(\ell_\infty : \ell_\infty)$ is a Banach algebra with respect to the usual matrix product and the norm defined by

$$\|A\| := \sup_{n \in \mathbb{N}} \sum_k |a_{nk}| < \infty. \tag{3.3.11}$$

The class $(c : c)$ of conservative matrices is also a Banach algebra with respect to the usual matrix product and the norm defined by (3.3.11). It is trivial that $(c : c)$ is a subalgebra of the Banach algebra $(\ell_\infty : \ell_\infty)$. Since a_{nk} and $\sum_k a_{nk}$ are all linear functionals of unit norm, $(c : c)$ is a closed subalgebra of the Banach algebra $(\ell_\infty : \ell_\infty)$. In addition to this, it is known by Wilansky and Zeller [397] that $(c : c)$ contains its inverse, that is $A \in (c : c)$ and $A^{-1} \in (\ell_\infty : \ell_\infty)$ imply that $A^{-1} \in (c : c)$.

Definition 3.3.12. (see Maddox [258, p. 160]) *Let X be an algebra and S be the set of all scalar homomorphisms on X. Then, the radical $rad(X)$ of X is defined by*

$$rad(X) := \cap\{Ker(f) : f \in S\}.$$

A Banach algebra is said to be *semisimple* if the radical consists of zero vector. Now, we give the result related with the semisimplicity of the classes $(\ell_\infty : \ell_\infty)$ and $(c : c)$, which are proved by Wilansky and Zeller [398]:

Theorem 3.3.13. (cf. Nanda [300, p. 13]) *The classes $(\ell_\infty : \ell_\infty)$ and $(c : c)$ are semisimple.*

Proof. Suppose that $A = (a_{nk}) \in (\ell_\infty : \ell_\infty) \setminus \{\theta\}$ and $x \in c$ with $Ax \neq 0$. Then, for example if we have $a_{nk} \neq 0$ for some $k, n \in \mathbb{N}$ we can take $x := e^{(k)}$. Define $B \in (c : c)$ as follows: B consists entirely of zeros except for a single column and $BAx = x$. Then, the $(I - BA)$-transform of x is θ, hence it is not injective and has no inverse. This means that $A \notin rad[(\ell_\infty : \ell_\infty)]$ and $A \notin rad[(c : c)]$. Hence, the classes $(\ell_\infty : \ell_\infty)$ and $(c : c)$ are semisimple. □

The characteristic $\chi(A)$ of a matrix $A = (a_{nk}) \in (c : c)$ is defined by

$$\chi(A) := \lim_{n\to\infty} \sum_k a_{nk} - \sum_k \left(\lim_{n\to\infty} a_{nk}\right)$$

which is a multiplicative linear functional. The numbers $\lim_{n\to\infty} a_{nk}$ for each fixed $k \in \mathbb{N}$ and $\lim_{n\to\infty} \sum_k a_{nk}$ are called as the *characteristic numbers of A*. A matrix A is called *coregular* if $\chi(A) \neq 0$ and is called *conull* if $\chi(A) = 0$

The Silverman-Toeplitz theorem yields for a regular matrix A that $\chi(A) = 1$ which leads us to the fact that Toeplitz matrices form a subset of coregular matrices. One can easily see for a Schur matrix A that $\chi(A) = 0$ which says us that coercive matrices form a subset of conull matrices.

Now, we may give the *Steinhaus theorem* whose original proof can be found in Cooke [111, pp. 75–76]:

Theorem 3.3.14. [Steinhaus] *There is no regular matrix which transforms every bounded sequence into a convergent sequence.*

Proof. Suppose that the classes $(c : c; p)$ and $(\ell_\infty : c)$ are not disjoint. Then, there exists at least one infinite matrix A such that $A \in (c : c; p) \cap (\ell_\infty : c)$. Therefore, one can easily see that $\chi(A) = 1$ since A is regular and $\chi(A) = 0$ since A is coercive, which is a contradiction. Hence, a matrix cannot be both regular and coercive.

This completes the proof. □

A matrix that transforms a sequence to a sequence is called a *sequence-to-sequence method* or a *sequence method.* A method that transforms a series to a sequence is called a *series-to-sequence method* or a *series method.* A matrix $A = (a_{nk}) \in (cs : c)$ is called a *β-matrix.* Similarly, series-to-series and sequence-to-series methods can be defined.

In the light of Öztürk [313], we give the characterization of the classes $(cs : c)$ and $(cs : cs)$ of series-to-sequence and series-to-series conservative matrices:

Theorem 3.3.15. [396, Example 8.4.5B] $A = (a_{nk}) \in (cs : c)$ *if and only if (3.3.3) holds and*

$$\sup_{n\in\mathbb{N}} \sum_k |\Delta a_{nk}| < \infty, \tag{3.3.12}$$

where $\Delta a_{nk} := a_{nk} - a_{n,k+1}$ *for all* $k, n \in \mathbb{N}$.

Proof. Let $x = (x_k) \in cs$ with $\sum_k x_k = s$ and $y_k = \sum_{j=0}^k x_j$ for all $k \in \mathbb{N}$, and define the matrix $B = (b_{nk})$ by $b_{nk} := \Delta a_{nk}$ for all $k, n \in \mathbb{N}$.

Suppose that $A \in (cs : c)$. Then, Ax exists for every $x \in cs$ and is in c. Since this also holds for $x = e^{(k)} \in cs$, the necessity of (3.3.3) is obvious.

Consider the following relation obtained from m^{th}-partial sums of the series $\sum_k a_{nk}x_k$ by applying Abel's partial summation:

$$\begin{aligned}\sum_{k=0}^{m} a_{nk}x_k &= \sum_{k=0}^{m-1} \Delta a_{nk}y_k + a_{nm}y_m \\ &= \sum_{k=0}^{m-1} \Delta a_{nk}(y_k - s) + s\sum_{k=0}^{m-1} \Delta a_{nk} + a_{nm}y_m \\ &= \sum_{k=0}^{m-1} \Delta a_{nk}(y_k - s) + s(a_{n0} - a_{nm}) + a_{nm}y_m \quad \text{for all } m, n \in \mathbb{N}.\end{aligned} \tag{3.3.13}$$

Therefore, taking into account the fact that $(a_{nk})_{k\in\mathbb{N}} \in cs^\beta = bv \subset \ell_\infty$ for every fixed $n \in \mathbb{N}$ we derive by passing to limit in (3.3.13) as $m \to \infty$ that

$$(Ax)_n = \sum_k a_{nk}x_k = \sum_k b_{nk}(y_k - s) + sa_{n0} \quad \text{for all } n \in \mathbb{N}. \tag{3.3.14}$$

Since $\lim_{n\to\infty} (Ax)_n$ exists and $\lim_{n\to\infty} a_{n0} = \alpha_0$, we see by letting $n \to \infty$ in (3.3.14) that $\lim_{n\to\infty} \sum_k b_{nk}(y_k - s)$ also exists. This yields the fact that $B \in (c_0 : c)$, because $y - se \in c_0$ if and only if $x \in cs$. Hence, the matrix $B = (b_{nk})$ satisfies the condition (3.3.1), which is equivalent to the condition (3.3.12), i.e., the condition (3.3.12) is necessary.

Conversely, suppose that conditions (3.3.3) and (3.3.12) hold. First, (3.3.12) implies $A_n = (a_{nk})_{k\in\mathbb{N}} \in bv = cs^\beta$ for each $n \in \mathbb{N}$, hence Ax exists for every $x \in cs$. Also, (3.3.12) and (3.3.3) imply by Corollary 3.3.5 that $B = (b_{nk}) \in (c_0 : c_0)$. Thus, it follows from (3.3.14) that $\sum_k a_{nk}x_k \to \alpha_0 s$, as $n \to \infty$, i.e., $A = (a_{nk}) \in (cs : c)$.

This completes the proof. □

A β-matrix $A = (a_{nk})$ such that $\sum_k a_{nk}u_k \to \alpha$, as $n \to \infty$, whenever $\sum_k u_k = \alpha$ is called as a *γ-matrix*. The class of all γ-matrices is denoted by $(cs : c; p)$. Since the proof is similar to the proof of Theorem 3.3.15, we give the characterization of the class $(cs : c; p)$ of γ-matrices without proof:

Corollary 3.3.16. $A = (a_{nk}) \in (cs : c; p)$ *if and only if (3.3.12) holds and (3.3.3) also holds with* $\alpha_k = 1$ *for all* $k \in \mathbb{N}$.

Theorem 3.3.17. [396, Example 8.4.6B] $A = (a_{nk}) \in (cs : cs)$ *if and only if*

$$\sup_{n\in\mathbb{N}} \sum_k \left| \sum_{j=0}^{n} \Delta a_{jk} \right| < \infty, \tag{3.3.15}$$

$$\sum_n a_{nk} = \alpha_k \text{ for each } k \in \mathbb{N}. \tag{3.3.16}$$

Proof. Let $x = (x_k) \in cs$ and define the matrix $C = (c_{nk})$ by $c_{nk} := \sum_{j=0}^{n} a_{jk}$ for all $k, n \in \mathbb{N}$.

Suppose that $A \in (cs : cs)$. Then, Ax exists for every $x \in cs$ and is in cs. This yields for $x := e^{(k)} \in cs$ that the condition (3.3.16) is necessary.

It is clear that the following equality

$$\sum_{j=0}^{n}\sum_{k=0}^{m} a_{jk}x_k := \sum_{k=0}^{m}\sum_{j=0}^{n} a_{jk}x_k := \sum_{k=0}^{m} c_{nk}x_k \tag{3.3.17}$$

derived from n^{th}, m^{th}-partial sums of the double series $\sum_j \sum_k a_{jk}x_k$ holds for all $m, n \in \mathbb{N}$. Therefore, by letting $m \to \infty$ in (3.3.17) we have

$$\sum_{j=0}^{n}(Ax)_j = (Cx)_n \quad \text{for all} \quad n \in \mathbb{N}. \tag{3.3.18}$$

Then, since $Ax \in cs$ by the hypothesis $\lim_{n\to\infty} \sum_{j=0}^{n}(Ax)_j$ exists, on the left-hand side of (3.3.18), which leads us to the consequence that $C = (c_{nk}) \in (cs : c)$. Therefore, condition (3.3.12) of Theorem 3.3.15 is satisfied by the matrix $C = (c_{nk})$ which is equivalent to condition (3.3.15).

Conversely, suppose that conditions (3.3.15) and (3.3.16) hold, which imply the existence of the A-transform of $x \in cs$. Then, since (3.3.18) also holds, the matrix C satisfies the conditions of Theorem 3.3.15. Hence, $\lim_{n\to\infty}(Cx)_n$ exists which says by (3.3.18) that $Ax \in cs$, as was desired.

This step completes the proof. □

Corollary 3.3.18. $A = (a_{nk}) \in (cs : cs; p)$ *if and only if (3.3.15) holds and (3.3.16) also holds with* $\alpha_k = 1$ *for all* $k \in \mathbb{N}$.

Finally, we characterize the class $(c : cs)$ of sequence-to-series conservative matrices:

Theorem 3.3.19. [396, Example 8.4.6A] $A = (a_{nk}) \in (c : cs)$ *if and only if (3.3.16) holds and*

$$\sup_{n\in\mathbb{N}} \sum_k \left| \sum_{j=0}^{n} a_{jk} \right| < \infty, \tag{3.3.19}$$

$$\sum_n \sum_k a_{nk} \text{ is convergent.} \tag{3.3.20}$$

Proof. Let $x = (x_k) \in c$ and define the matrix $C = (c_{nk})$ as in the proof of Theorem 3.3.17.

Suppose that $A \in (c : cs)$. Then, Ax exists for every $x \in c$ and is in cs. This yields for $x = e^{(k)} \in c$ and $x = e \in c$, which give the necessity of the conditions (3.3.16) and (3.3.20), respectively.

It is clear that we have the relation (3.3.18), derived by the same way used in the proof of Theorem 3.3.17. Then, since $Ax \in cs$, that is the series $\sum_j (Ax)_j$ converges by the hypothesis $\lim_{n\to\infty} \sum_{j=0}^{n} (Ax)_j$ exists, on the left-hand side of (3.3.18), which leads us to the consequence that $C = (c_{nk}) \in (c : c)$. Therefore, the condition (3.3.1) of Kojima-Schur theorem is satisfied by the matrix $C = (c_{nk})$, which is equivalent to the condition (3.3.19).

Conversely, suppose that the conditions (3.3.16), (3.3.19) and (3.3.20) hold, which imply the existence of the A-transform of $x \in c$. Then, since (3.3.18) also holds the matrix C satisfies the conditions of Theorem 3.3.3. Hence, $\lim_{n\to\infty} (Cx)_n$ exists which says by (3.3.18) that $Ax \in cs$, as was desired.

This step completes the proof. □

Corollary 3.3.20. $A = (a_{nk}) \in (c : cs; p)$ *if and only if (3.3.19), (3.3.16) with* $\alpha_k = 0$ *hold and* $\sum_n \sum_k a_{nk} = 1$.

3.4 DUAL SUMMABILITY METHODS

In this section, following Başar [64], we give a short survey related to dual summability methods and finally mention a generalization of this concept.

3.4.1 Dual Summability Methods Dependent on a Stieltjes Integral

In this subsection, following Lorentz [246], we summarize the historical knowledge about the dual summability methods dependent on a Stieltjes integral.

Consider the function $\sigma(z)$ defined by the Stieltjes integral of a real-valued function $a(z, t)$ of real variables z and t with respect to a real function $s(t)$ of real variable t as

$$\sigma(z) := \int_0^{+\infty} a(z, t)ds(t), \tag{3.4.1}$$

which must be meaningful for all $z \geq 0$. If $\lim_{z\to+\infty} \sigma(z) := \sigma$ exists as a finite limit, then $\sigma(z)$ is called the generalized limit of the function $s(t)$ obtained by the method (3.4.1). In order to exist the improper integral on the right-hand side of (3.4.1), the Stieltjes integral

$$\int_0^{c} a(z, t)ds(t) \tag{3.4.2}$$

must exist for all finite $c > 0$, firstly. At this stage, (3.4.1) is the limit of (3.4.2), as $c \to +\infty$. There are two ways for guaranteeing the existence of the integral (3.4.2):

(I) $a(z,t)$ is a continuous function with respect to the variable t for every fixed $z \geq 0$ and $s(t)$ is of bounded variation on all intervals $(0,c)$.

(II) Conversely, $s(t)$ is continuous and $a(z,t)$ is of bounded variation on all intervals $(0,c)$.

The necessary and sufficient condition in order to the integral (3.4.1) exists for the continuous function $s(t)$ such that $\lim_{t\to+\infty} s(t)$ to be finite is the function $a(z,t)$ is of bounded variation on the interval $0 \leq t < +\infty$ for every fixed $z \geq 0$. At this stage, one can apply integration by parts to (3.4.1) and then (3.4.1) is reduced to the transformation

$$\sigma'(z) := \int_0^{+\infty} s(t)d_t b(z,t) \tag{3.4.3}$$

by depending on suitable choice of the bounded variation function $b(z,t)$ on the interval $0 \leq t < +\infty$. The methods given by (3.4.1) and (3.4.3) which are reduced each to other one by a formal integration by parts are called *dual summability methods.*

If one seeks a transformation in the form

$$\sigma_n := \sum_k a_{nk}x_k \quad \text{for all} \quad n \in \mathbb{N}, \tag{3.4.4}$$

which contains as a special case of the transformation of the sequence (y_k) with $y_k := \sum_{j=0}^{k} x_j$ for all $k \in \mathbb{N}$, of course the limitation method (3.4.1) is found, under the hypothesis (I).

Let us consider the step function

$$s(t) := \begin{cases} y_k & , \quad k < t \leq k+1, \\ 0 & , \quad t = 0, \end{cases} \tag{3.4.5}$$

for all $k \in \mathbb{N}$. It is obvious that the function $s(t)$ defined by (3.4.5) has the jumps x_k at the points $t = k$ and elsewhere is partially constant. So,

$$\sigma(z) = \int_0^{+\infty} a(z,t)ds(t) = \sum_k a(z,k)x_k$$

holds for the function, which corresponds to the transformation (3.4.4) replacing n by the new variable z. However, the substituting of the function $s(t)$, defined by (3.4.5) under the hypothesis (II), in (3.4.1) does not lead to the desired result. Because, the existence of the integral (3.4.1) is guaranteed only for the continuous functions $s(t)$.

3.4.2 Relation Between the Dual Summability Methods

The limit $\lim_{t\to\infty} a(z,t) = \alpha(z)$ exists for the convergence preserving method (3.4.1). One can say for $\alpha(z)$ that if the limit $\lim_{z\to\infty} \sigma(z)$ exists for the bounded function $s(t)$, which tends to infinity as $t \to \infty$, then

$$\lim_{t\to\infty} a(z,t) = 0, \quad (z \geq 0). \tag{3.4.6}$$

Let us suppose that $s(0) = 0$ for the function $s(t)$ of bounded variation in (3.4.1). Additionally, if (3.4.6) also holds, then

$$\sigma'(z) := -\int_0^{+\infty} s(t)d_t a(z,t) \tag{3.4.7}$$

is obtained by a formal integration by parts. Relation (3.4.7) is the same type method of (3.4.3) and is called the dual method of the method (3.4.1).

Define the function $s(t)$ as

$$s(t) := \begin{cases} 0 & , \quad t \leq t_0, \\ 1 & , \quad t > t_0, \end{cases}$$

and replace the variable z by the integer number variable n and suppose the function $a(n,t)$ is interpolated for all n and write a_{nk} instead of $a(n,k)$. Therefore, we derive the series transformation

$$A: \qquad \sigma_n := \sum_k a_{nk} x_k \ \text{ for all } \ n \in \mathbb{N}$$

from (3.4.1). Then, the dual transformation is given by

$$B: \qquad \sigma'_n := \sum_k b_{nk} y_k \ \text{ for all } \ n \in \mathbb{N},$$

where $b_{nk} := a_{nk} - a_{n,k+1}$ for all $k, n \in \mathbb{N}$.

Throughout the text, we apply the infinite matrices $A = (a_{nk})$ and $B = (b_{nk})$ to the sequences $x = (x_k)$ and $y = (y_k)$, respectively.

3.4.3 Usual Dual Summability Methods

Let λ be a sequence space and consider the *summation matrix* $S = (s_{nk})$, the γ-matrix corresponding to the ordinary convergence, defined by

$$s_{nk} := \begin{cases} 1 & , \quad 0 \leq k \leq n, \\ 0 & , \quad k > n, \end{cases} \tag{3.4.8}$$

for all $k, n \in \mathbb{N}$. Then, the space of the series λ_S whose sequences of partial sums are in the space λ is defined by

$$\lambda_S := \left\{ x = (x_k) \in \omega : \left(\sum_{j=0}^{k} x_j \right) \in \lambda \right\}.$$

Define the sequence $y = (y_k)$ via the sequence $x = (x_k) \in \omega$ by

$$y_k := (Sx)_k := \sum_{j=0}^{k} x_j \ \text{ for all } k \in \mathbb{N}. \tag{3.4.9}$$

Then, since $\lambda_S \cong \lambda$ it is obvious that "*$x \in \lambda_S$ if and only if $y \in \lambda$*".

Let us suppose that the infinite matrices $A = (a_{nk})$ and $B = (b_{nk})$ map the sequences $x = (x_k)$ and $y = (y_k)$ which are connected with the relation (3.4.9) to the sequences $u = (u_n)$ and $v = (v_n)$, respectively, i.e.,

$$u_n := (Ax)_n = \sum_k a_{nk} x_k \text{ for all } n \in \mathbb{N}, \tag{3.4.10}$$

$$v_n := (By)_n = \sum_k b_{nk} y_k \text{ for all } n \in \mathbb{N}. \tag{3.4.11}$$

It is clear here that the method B is applied to the S-transform $y = (y_k) := \{(Sx)_k\}$ of the sequence $x = (x_k)$, while the method A is directly applied to the terms of the sequence $x = (x_k)$. So, the methods A and B are essentially different.

Let us assume that the matrix product BS exists, which is a much weaker assumption than the conditions on the matrix B belonging to any matrix class, in general. We shall say in this situation that the methods A and B in (3.4.10) and (3.4.11) are *dual summability methods* if u_n reduced to v_n (or v_n reduced to u_n) under the application of formal summation by parts. This leads us to the fact that BS exists and is equal to A and $Ax = (BS)x = B(Sx) = By$ formally holds, if one side exists. This statement is equivalent to the following relation between the entries of the matrices $A = (a_{nk})$ and $B = (b_{nk})$:

$$a_{nk} := \sum_{j=k}^{\infty} b_{nj} \quad \text{or} \quad b_{nk} := a_{nk} - a_{n,k+1} \text{ for all } k, n \in \mathbb{N}. \tag{3.4.12}$$

Now, we may give a short analysis on the dual summability methods. One can see that v_n reduced to u_n, as follows:

$$v_n = \sum_k b_{nk} y_k = \sum_k b_{nk} \left(\sum_{j=0}^{k} x_j \right) = \sum_j \sum_{k=j}^{\infty} b_{nk} x_j = u_n.$$

But, the order of summation may not be reversed and thus, the methods A and B are not necessarily equivalent.

The partial sums of the series on the right-hand side of (3.4.10) and (3.4.11) are connected with the relation

$$\sum_{k:=0}^{m} a_{nk} x_k := \sum_{k=0}^{m-1} (a_{nk} - a_{n,k+1}) y_k + a_{nm} y_m \text{ for all } m, n \in \mathbb{N}. \tag{3.4.13}$$

Hence if, for a given $n \in \mathbb{N}$, one of the series on the right-hand side of (3.4.10) and (3.4.11) converges, then the other converges if and only if

$$\lim_{m \to \infty} a_{nm} y_m = z_n \tag{3.4.14}$$

for every fixed $n \in \mathbb{N}$. If (3.4.14) holds, then we have from (3.4.13) by letting $m \to \infty$ that

$$u_n := v_n + z_n \text{ for all } n \in \mathbb{N}. \tag{3.4.15}$$

Hence, if (y_n) is summable by one of the methods A and B, then it is summable by the other one if and only if (3.4.14) holds and

$$\lim_{n\to\infty} z_n := \alpha. \tag{3.4.16}$$

Hence, the limits of (u_n) and (v_n) differ by α. Therefore, the A- and B-limits of any sequence summable by one of them agree if and only if B summability implies that (3.4.16) holds with $\alpha = 0$. A similar result holds with A and B interchanged. It follows by the validity of (3.4.16) with $\alpha \neq 0$ that the methods A and B are inconsistent, and conversely.

This type pair of summability methods was generally used by several authors to characterize the class of series-to-sequence matrix transformations, for example the reader may refer to Lorentz and Zeller [248], Kuttner [232], Öztürk [312], Başar and Öztürk [75], Orhan and Öztürk [310], Başar and Çolak [70, 69], Savaş [346], Başar [60] and Başar and Altay [65].

3.5 SOME EXAMPLES OF TOEPLITZ MATRICES

Now, we give some examples of particular summability matrices satisfying the Toeplitz conditions.

3.5.1 Arithmetic Means

The corresponding Toeplitz matrix $C_1 = (c_{nk})$ to the *arithmetic means* is defined by (1.1.5). The inverse matrix $C_1^{-1} = (d_{nk})$ of the matrix $C_1 = (c_{nk})$ is obtained by a routine calculation as

$$d_{nk} := \begin{cases} (-1)^{n-k}(k+1) & , \quad n-1 \leq k \leq n, \\ 0 & , \quad 0 \leq k \leq n-2 \ \text{ or } \ k > n, \end{cases}$$

for all $k, n \in \mathbb{N}$.

3.5.2 Cesàro Means

Let $r > -1$ and define A_n^r by

$$A_n^r := \begin{cases} \frac{(r+1)(r+2)\cdots(r+n)}{n!} & , \quad n \in \mathbb{N}_1, \\ 1 & , \quad n = 0. \end{cases}$$

Then, the *Cesàro means of order* r is defined by the matrix $C_r = (c_{nk}^r)$ as

$$c_{nk}^r := \begin{cases} \frac{A_{n-k}^{r-1}}{A_n^r} & , \quad 0 \leq k \leq n, \\ 0 & , \quad k > n, \end{cases}$$

for all $k, n \in \mathbb{N}$. When $r \geq 0$, $A_n^{r-1} \geq 0$ and since

$$\frac{1}{(1-z)^{r+1}} := \sum_n A_n^r z^n, \quad (|z| < 1)$$

one can easily see that

$$\sum_{k=0}^{n} A_{n-k}^{s} A_{k}^{r} = A_{n}^{s+r+1},$$

which leads us to the fact that $\sum_k |a_{nk}| = \sum_k a_{nk} = 1$ for all $n \in \mathbb{N}$. It is also known by elementary analysis that

$$\lim_{n\to\infty} \frac{A_n^r}{n^r} = \frac{1}{\Gamma(r+1)},$$

where $\Gamma(x)$ denotes the gamma function. It is trivial that C_r is reduced to the unit matrix I and to the arithmetic means C_1 in the cases $r = 0$ and $r = 1$, respectively. It is not hard to show that the Cesàro matrix of order r is a Toeplitz matrix when $r \geq 0$.

3.5.3 Euler Means

Let $0 < r < 1$ and $\binom{n}{k} := n!/[k!(n-k)!]$ for all $k, n \in \mathbb{N}$. Then, the *Euler means* E^r *of order* r is defined by the matrix $E^r = (e_{nk}^r)$, where

$$e_{nk}^{r} := \begin{cases} \binom{n}{k}(1-r)^{n-k}r^k & , \quad 0 \leq k \leq n, \\ 0 & , \quad k > n, \end{cases}$$

for all $k, n \in \mathbb{N}$. The Euler matrix $E^r = (e_{nk}^r)$ of order r is a Toeplitz matrix.

The original Euler mean $E_1 := E^{1/2}$ of order 1 was given by L. Euler in 1755 and was given by the matrix $E_1 = (a_{nk})$, where

$$a_{nk} := \begin{cases} \binom{n}{k}2^{-n} & , \quad 0 \leq k \leq n, \\ 0 & , \quad k > n, \end{cases}$$

for all $k, n \in \mathbb{N}$ whose generalization E_q of order q was defined by the matrix $E_q := (b_{nk}^q)$, where

$$b_{nk}^{q} := \begin{cases} \binom{n}{k}(q+1)^{-n}q^{n-k} & , \quad 0 \leq k \leq n, \\ 0 & , \quad k > n, \end{cases}$$

for all $k, n \in \mathbb{N}$. It is clear that E^r corresponds to E_q for $r = (q+1)^{-1}$. Much of the work on the Euler means of order r was done by Knopp (cf. [222, 223]). So, some authors refer to E^r as the *Euler-Knopp matrix*. It is known that

$$E^r E^s = (e_{nk}^r)(e_{nk}^s) := (e_{nk}^{rs}) = E^{rs}$$

and E^r is invertible such that $(E^r)^{-1} := E^{1/r}$ with $r \neq 0$ (see [319, pp. 58–59]).

3.5.4 Taylor Matrices

Let $r \in \mathbb{C}\setminus\{0\}$. Then, the *Taylor transform* $T(r)$ is defined by the matrix $T^r = (t^r_{nk})$, where

$$t^r_{nk} := \begin{cases} \binom{k}{n}(1-r)^{n+1}r^{k-n} & , \quad k \geq n, \\ 0 & , \quad 0 \leq k < n, \end{cases}$$

for all $k, n \in \mathbb{N}$. In case $r = 0$, it is immediate that $T^0 = (t^0_{nk}) = (\delta_{nk}) := I$. The Taylor method was introduced by Hardy and Littlewood [199] in 1916. Following Powell and Shah [319, pp. 61–62], we note a few lines on the Taylor transform: In working with the Taylor method, the fact is frequently used that

$$\frac{1}{(1-z)^{n+1}} := \sum_{k=n}^{\infty} \binom{k}{n} z^{k-n} \text{ if } |z| < 1$$

and additionally $\sum_{k=n}^{\infty}\binom{k}{n} z^{k-n}$ converges only if $|z| < 1$. The Taylor transform $T(r)$ is regular if and only if $0 \leq r < 1$. The product of the matrix T^r with the matrix T^s is the transpose of the matrix $(1-r)(1-s)E^{(1-r)(1-s)}$. It is known that $T(r)$ is invertible, and the inverse of the matrix T^r is the matrix $(T^r)^{-1} := T^{-r/(1-r)}$, provided $r \neq 1$.

3.5.5 Riesz Means

Let $t = (t_k)$ be a sequence of non-negative real numbers with $t_0 > 0$ and write

$$T_n := \sum_{k=0}^{n} t_k \text{ for all } n \in \mathbb{N}.$$

Then, the *Riesz means with respect to the sequence* $t = (t_k)$ is defined by the matrix $R^t = (r^t_{nk})$ which is given by

$$r^t_{nk} := \begin{cases} t_k/T_n & , \quad 0 \leq k \leq n, \\ 0 & , \quad k > n, \end{cases}$$

for all $k, n \in \mathbb{N}$. It is known that the Riesz matrix R^t is a Toeplitz matrix if and only if $T_n \to \infty$, as $n \to \infty$. It is obvious that the Riesz matrix R^t is reduced in the case $t = e$ to the matrix C_1 of arithmetic means. The inverse matrix $S^t = (s^t_{nk})$ of the matrix $R^t = (r^t_{nk})$ is given by

$$s^t_{nk} := \begin{cases} \frac{(-1)^{n-k}T_k}{t_n} & , \quad n-1 \leq k \leq n, \\ 0 & , \quad 0 \leq k \leq n-2 \text{ or } k > n, \end{cases}$$

for all $k, n \in \mathbb{N}$.

3.5.6 Nörlund Means

Let $t = (t_k)$ and T_n be as in the Riesz means. Then the *Nörlund means with respect to the sequence* $t = (t_k)$ is defined by the matrix $N^t = (a^t_{nk})$ which is given by

$$a^t_{nk} := \begin{cases} t_{n-k}/T_n &, \quad 0 \le k \le n, \\ 0 &, \quad k > n, \end{cases}$$

for all $k, n \in \mathbb{N}$. It is known that the Nörlund matrix N^t is a Toeplitz matrix if and only if $t_n/T_n \to 0$, as $n \to \infty$. One can easily observe that $(t_n/T_n) \in c$ whenever the Nörlund matrix N^t is conservative. It is also obvious that the Nörlund matrix N^t is reduced in the case $t = e$ to the matrix C_1 of arithmetic means.

In order to obtain the inverse of the N^t transform (cf. Peyerimhoff [317, p. 17]), we write formally $t(z) := \sum_n t_n z^n$, $k(z) := 1/t(z) := \sum_n k_n z^n$. The inverse P^t of N^t is then given by $(P^t)_{nj} := k_{n-j}T_j$ for $j \le n$. For $t_n := A^{r-1}_n$ for all $n \in \mathbb{N}$, the method N^t is reduced to the Cesàro method C_r of order $r > -1$.

3.5.7 A^r Matrices

Let $0 < r < 1$. Then the new class $A^r = (a^r_{nk})$ of Toeplitz matrices, introduced by Başar [61], is given by

$$a^r_{nk} := \begin{cases} \frac{1+r^k}{n+1} &, \quad 0 \le k \le n, \\ 0 &, \quad k > n, \end{cases}$$

for all $k, n \in \mathbb{N}$. A straightforward calculation shows that the inverse matrix $B^r = (b^r_{nk})$ of the matrix $A^r = (a^r_{nk})$ is given by

$$b^r_{nk} := \begin{cases} \frac{(-1)^{n-k}(k+1)}{1+r^n} &, \quad n-1 \le k \le n, \\ 0 &, \quad 0 \le k \le n-2 \ \text{ or } \ k > n, \end{cases}$$

for all $k, n \in \mathbb{N}$.

3.5.8 Hausdorff Matrices

Let $\mu = (\mu_n) \in \omega$, $M = (m_{nk})$ be the diagonal matrix with $m_{nn} = \mu_n$ for $n \in \mathbb{N}$, and $D = (d_{nk})$ be the triangle with $d_{nk} := (-1)^k\binom{n}{k}$ for $0 \le k \le n$, $(n \in \mathbb{N})$. Then, the matrix $H := H(\mu) := DMD$ is called the *Hausdorff matrix* associated with the sequence (μ_n). Therefore, the matrix $H = (h_{nk})$ is given by

$$h_{nk} := \begin{cases} \sum_{j=k}^{n}(-1)^{j+k}\binom{n}{j}\binom{j}{k}\mu_j &, \quad 0 \le k \le n, \\ 0 &, \quad k > n, \end{cases}$$

for all $k, n \in \mathbb{N}$.

The Cesàro matrix C_r of order $r > -1$ is the Hausdorff matrix associated with the sequence $\mu = (\mu_n)$, where $\mu_n := 1/A^r_n = 1/\binom{n+r}{n}$ for all $n \in \mathbb{N}$. The *Hölder matrix* H_r of order $r > -1$ is the Hausdorff matrix associated with the sequence $\mu = (\mu_n)$, where $\mu_n := (n+1)^{-r}$ for all $n \in \mathbb{N}$. There is no explicit formula is known for the entries of the matrices H_r, in general. The Euler matrix E_r of order $r > 0$ is the Hausdorff matrix associated with the sequence $\mu = (\mu_n)$, where $\mu_n := (r+1)^{-n}$ for all $n \in \mathbb{N}$.

3.5.9 Borel Matrix

The semi-continuous *Borel matrix* $A = \{a_k(t)\}$ is defined by

$$a_k(t) := \frac{e^{-t}t^k}{k!}, \quad (k \in \mathbb{N},\ t > 0).$$

Since, $\lim_{t\to\infty} a_k(t) = 0$ for each $k \in \mathbb{N}$ and $\sum_k |a_k(t)| = \sum_k a_k(t) = 1$ whence $\sup_{t>0} \sum_k |a_k(t)| < \infty$, the Borel matrix is regular. This shows the essential property of the Borel matrix is that it maps convergent sequences into convergent functions, leaving the limit unchanged.

3.5.10 Abel Matrix (cf. Peyerimhoff [317, p. 24])

A sequence (s_k) is called limitable by the *Abel method* A to l if $\sum_k x^k s_k$ exists for $|x| < 1$, and if

$$\lim_{x\to 1-0} \frac{\sum_k x^k s_k}{\sum_k x^k} = \lim_{x\to 1-0} (1-x) \sum_k x^k s_k = l.$$

This shows that the Abel method A is regular.

It is natural that one can derive the corresponding γ-matrices of the examples of Toeplitz matrices, mentioned in the chapter above, by using the relation between the terms of a series and its sequence of partial sums.

Matrix Domains in Sequence Spaces

This chapter is devoted to the domains of some particular summability matrices, with a special emphasize on the Cesàro, difference, m^{th}-order difference, Euler, Riesz and weighted mean sequence spaces, and other spaces derived in this way. Also, the Schauder bases of those spaces, their α-, β- and γ-duals, and the characterizations of some matrix transformations are given.

Keywords: Domain of an infinite matrix, Cesàro, difference, Euler, Riesz, generalized difference and weighted mean sequence spaces and concerning dual methods, space of p-bounded variation sequences, Schauder bases, α-, β- and γ-duals of a matrix domain, characterization of the matrix transformations related to the matrix domains, paranormed difference sequence spaces and moduli, Orlicz functions.

4.1 PRELIMINARIES, BACKGROUND AND NOTATIONS

The *domain* λ_A of an infinite matrix A in a sequence space λ is defined by

$$\lambda_A := \Big\{x = (x_k) \in \omega : Ax \in \lambda\Big\}, \tag{4.1.1}$$

which is a sequence space. If A is triangle, then one can easily observe that the sequence spaces λ_A and λ are linearly isomorphic, i.e., $\lambda_A \cong \lambda$. If λ is a sequence space, then the continuous dual λ_A^* of the space λ_A is given (cf. [396, Theorem 4.4.2, p. 66]) by $\lambda_A^* := \{f : f = g \circ A,\ g \in \lambda^*\}$.

Let us give the definition of some triangle summability matrices, which are needed in the text. We write $\mathcal{U}$ for the set of all sequences $u = (u_k)$ such that $u_k \neq 0$ for all $k \in \mathbb{N}$. For $u \in \mathcal{U}$, let $1/u := (1/u_k)$. Let $u, v, z \in \mathcal{U}$ and define the difference matrix $\Delta^{(1)} = (\Delta_{nk}^{(1)})$, the generalized difference matrix $B(r,s) = \{b_{nk}(r,s)\}$, the generalized weighted means or factorable matrix $G(u,v) = (g_{nk})$, $A_u^r = (a_{nk}^r)$, the difference matrix $\Delta^{(m)} = (\Delta_{nk}^{(m)})$ of order m, the summation matrix $S^{(m)} = (s_{nk}^{(m)})$ of

DOI: 10.1201/9781003294153-4

order m, the Cesàro matrix $C^{(m)} = (c_{nk}^{(m)})$ of order m and $A^z = (a_{nk}^z)$ by

$$\Delta_{nk}^{(1)} := \begin{cases} (-1)^{n-k} & , \quad n-1 \le k \le n, \\ 0 & , \quad 0 \le k < n-1 \text{ or } k > n, \end{cases}$$

$$b_{nk}(r,s) := \begin{cases} r & , \quad k = n, \\ s & , \quad k = n-1, \qquad\qquad (r, s \in \mathbb{R} \setminus \{0\}) \\ 0 & , \quad 0 \le k < n-1 \text{ or } k > n, \end{cases}$$

$$g_{nk} := \begin{cases} u_n v_k & , \quad 0 \le k \le n, \\ 0 & , \quad k > n, \end{cases} \qquad a_{nk}^r := \begin{cases} \frac{1+r^k}{n+1} u_k & , \quad 0 \le k \le n, \\ 0 & , \quad k > n, \end{cases}$$

$$\Delta_{nk}^{(m)} := \begin{cases} (-1)^{n-k}\binom{m}{n-k} & , \quad \max\{0, n-m\} \le k \le n, \\ 0 & , \quad 0 \le k < \max\{0, n-m\} \text{ or } k > n, \end{cases}$$

$$s_{nk}^{(m)} := \begin{cases} \binom{m+n-k-1}{n-k} & , \quad 0 \le k \le n, \\ 0 & , \quad k > n, \end{cases} \qquad c_{nk}^{(m)} := \begin{cases} \frac{\binom{m+n-k-1}{n-k}}{\binom{m+n}{n}} & , \quad 0 \le k \le n, \\ 0 & , \quad k > n, \end{cases}$$

$$a_{nk}^z := \begin{cases} (-1)^{n-k} z_k & , \quad n-1 \le k \le n, \\ 0 & , \quad 0 \le k < n-1 \text{ or } k > n, \end{cases}$$

for all $k, n \in \mathbb{N}$; where u_n depends only on n and v_k only on k. For the concerning literature about the domain λ_A of an infinite matrix A in a sequence space λ, Table 4.1 may be useful:

The approach constructing a new sequence space by means of the matrix domain of a particular limitation method has recently been employed by Wang [393], Ng and Lee [304], Malkowsky [263], Altay and Başar [20, 21, 23, 26, 31, 27, 28, 29, 30], Malkowsky and Savaş [272], Başarır [79, 80], Başarır and Kayıkçı[84], Başarır and Öztürk [85], Kara and Başarır [212], Kara et al. [213], Aydın and Başar [42, 43, 44, 45, 46], Başar et al. [67], Şengönül and Başar [373], Altay [20], Polat and Başar [318] and, Malkowsky et al. [267]. In Table 4.1, Δ, Δ^2 and Δ^m are the transposes of the matrices $\Delta^{(1)}$, $\Delta^{(2)}$ and $\Delta^{(m)}$, respectively, and $B^{(m)}$ is the composition of the matrix $B(r,s)$ with itself m times. Additionally, $c_0(u,p)$ and $c(u,p)$ are the spaces consisting of the sequences $x = (x_k)$ such that $ux = (u_k x_k)$ is in the spaces $c_0(p)$ and $c(p)$ for $u \in \mathcal{U}$, respectively, and studied by Başarır [79]. Finally, the new technique for deducing certain topological properties, for example AB-, KB-, AD-properties, solidity and monotonicity, etc., and determining the β- and γ-duals of the domain of a triangle matrix in a sequence space is given by Altay and Başar [29].

Although in most cases, the new sequence space λ_A generated by the summability matrix A from a sequence space λ is the expansion or the contraction of the original space λ, it may be observed in some cases that those spaces overlap. Indeed, one can easily see that the inclusion $\lambda_S \subset \lambda$ strictly holds for $\lambda \in \{\ell_\infty, c, c_0\}$. As this, one can deduce that the inclusion $\lambda \subset \lambda_{\Delta^{(1)}}$ also strictly holds for $\lambda \in \{\ell_\infty, c, c_0, \ell_p\}$, where $0 < p < \infty$. However, if we define $\lambda = c_0 \oplus span\{z\}$ with $z := \{(-1)^k\}$, i.e., $x \in \lambda$ if and only if $x := s + \alpha z$ for some $s \in c_0$ and some $\alpha \in \mathbb{C}$, and consider the matrix A with the rows A_n defined by $A_n := (-1)^n e^{(n)}$ for all $n \in \mathbb{N}$, we have $Ae := z \in \lambda$ but

TABLE 4.1 The domains of some triangle matrices in certain sequence spaces

λ	A	λ_A	refer to:
$\ell_p,\ \ell_\infty$	N_q	$X_{a(p)},\ X_{a(\infty)}$	[393]
$\ell_p,\ (1<p\leq\infty)$	C_1	$X_p,\ X_\infty$	[304]
$X_p,\ (1<p\leq\infty)$	Δ^m	$C_p(\Delta^m),\ C_\infty(\Delta^m)$	[156]
$c_0,\ c,\ \ell_\infty$	R^q	$(\overline{N},q)_0,\ (\overline{N},q),\ (\overline{N},q)_\infty$	[263]
$c_0,\ c,\ \ell_\infty$	Δ	$c_0(\Delta),\ c(\Delta),\ \ell_\infty(\Delta)$	[217]
$c_0,\ c,\ \ell_\infty$	Δ^2	$c_0(\Delta^2),\ c(\Delta^2),\ \ell_\infty(\Delta^2)$	[155]
$c_0,\ c,\ \ell_\infty$	$u\Delta^2$	$c_0(u;\Delta^2),\ c(u;\Delta^2),\ \ell_\infty(u;\Delta^2)$	[283]
$\ell_p,\ c_0,\ c,\ \ell_\infty$	$B(r,s)$	$\widehat{\ell_p},\ \widehat{c_0},\ \widehat{c},\ \widehat{\ell_\infty}$	[220]
$f,\ f_0$	$B(r,s)$	$\widehat{f},\ \widehat{f_0}$	[74]
$\ell_p,\ c_0,\ c,\ \ell_\infty$	$G(a,b)$	$\ell_p(F),\ c_0(F),\ c(F),\ \ell_\infty(F)$	[76]
$c_0,\ c,\ \ell_p$	$G(u,v)$	$Z(u,v;c_0),\ Z(u,v;c),\ Z(u,v;\ell_p)$	[272]
$\ell_p,\ (1<p\leq\infty)$	$C^{(m)}$	$\ell_p\big(C^{(m)}\big),\ \ell_\infty\big(C^{(m)}\big)$	[338]
$c_0,\ c,\ \ell_p,\ (0<p\leq 1)$	$C^{(m)}$	$c_0\big(C^{(m)}\big),\ c\big(C^{(m)}\big),\ \ell_p\big(C^{(m)}\big)$	[339]
$c_0,\ c$	C_1	$\widetilde{c_0},\ \widetilde{c}$	[373]
$c_0,\ c$	E^r	$e_0^r,\ e_c^r$	[23]
$e_0^t,\ e_c^t,\ e_\infty^t$	$B^{(m)}$	$e_0^t(B^{(m)}),\ e_c^t(B^{(m)}),\ e_\infty^t(B^{(m)})$	[212]
$c_0,\ c$	$G(u,v)$	$(c_0)_{G(u,v)},\ c_{G(u,v)}$	[330]
$c_0,\ c$	A^r	$a_0^r,\ a_c^r$	[42]
$\ell_p,\ (1<p\leq\infty)$	A^r	$a_p^r,\ a_\infty^r$	[45]
$\ell_p,\ (1<p\leq\infty)$	E^r	$e_p^r,\ e_\infty^r$	[31, 289]
$a_0^r,\ a_c^r$	$\Delta^{(1)}$	$a_0^r(\Delta),\ a_c^r(\Delta)$	[44]
$\ell_p,\ (1<p<\infty)$	$G(u,v)$	ℓ_A^p	[216]
$\ell_p,\ (1<p<\infty)$	$\Delta^{(1)}$	bv_p	[66, 133]
$\ell_p,\ (0<p\leq 1)$	$\Delta^{(1)}$	bv_p	[30]
$c_0,\ c,\ \ell_\infty$	Δ^m	$c_0(\Delta^m),\ c(\Delta^m),\ \ell_\infty(\Delta^m)$	[166, 131]
$\ell_p,\ (1<p<\infty)$	$\Delta^{(m)}$	$\ell_p(\Delta^{(m)})$	[20]
$c_0,\ c,\ \ell_\infty$	$\Delta^{(m)}$	$c_0(\Delta^{(m)}),\ c(\Delta^{(m)}),\ \ell_\infty(\Delta^{(m)})$	[268]
$e_0^r,\ e_c^r$	$\Delta^{(m)}$	$e_0^r(\Delta^{(m)}),\ e_c^r(\Delta^{(m)})$	[318]
$\ell_\infty(p)$	S	$bs(p)$	[60, 65]
$\ell(p)$	A_u^r	$a^r(u,p)$	[46]
$\ell(p)$	S	$\overline{\ell(p)}$	[105]
$c_0(p),\ c(p),\ \ell_\infty(p)$	Δ	$\Delta c_0(p),\ \Delta c(p),\ \Delta\ell_\infty(p)$	[8]
$c_0(p),\ c(p),\ \ell_\infty(p)$	$u\Delta$	$c_0(u,\Delta,p),\ c(u,\Delta,p),\ \ell_\infty(u,\Delta,p)$	[40]
$c_0(p),\ c(p),\ \ell_\infty(p)$	$u\Delta^2$	$c_0(u,\Delta^2,p),\ c(u,\Delta^2,p),\ \ell_\infty(u,\Delta^2,p)$	[86]
$c_0(p),\ c(p),\ \ell_\infty(p)$	$G(u,v)$	$c_0(u,v;p),\ c(u,v;p),\ \ell_\infty(u,v;p)$	[27]
$\ell(p)$	$G(u,v)$	$\ell(u,v;p)$	[28]
$\ell(p),\ \ell_\infty(p)$	A^z	$bv(z,p),\ bv_\infty(z,p)$	[67]
$c_0(u,p),\ c(u,p)$	A^r	$a_0^r(u,p),\ a_c^r(u,p)$	[43]
$\ell(p)$	R^t	$r^t(p)$	[21]
$c_0(p),\ c(p),\ \ell_\infty(p)$	R^t	$r_0^t(p),\ r_c^t(p),\ r_\infty^t(p)$	[26]
$r^q(p)$	Δ	$r^q(p,\Delta)$	[80, 85]
$r^q(p)$	B^m	$r^q(p,B^m)$	[84]
$\ell(p)$	E^r	$e^r(p)$	[213]
$c_0(p),\ c(p),\ \ell_\infty(p)$	Δ^m	$\Delta^m c_0(p),\ \Delta^m c(p),\ \Delta^m\ell_\infty(p)$	[165]
$c_0(p),\ c(p),\ \ell_\infty(p)$	$u\Delta^{(m)}$	$\Delta_u^{(m)}c_0(p),\ \Delta_u^{(m)}c(p),\ \Delta_u^{(m)}\ell_\infty(p)$	[267]

$Az := e \notin \lambda$ which lead us to the consequences that $z \in \lambda \setminus \lambda_A$ and $e \in \lambda_A \setminus \lambda$. That is to say that the sequence spaces λ_A and λ overlap but neither contains the other.

Prior to giving the theorem on the Schauder basis of the matrix domain of a triangle in a linear metric space with a Schauder basis, we record the following result concerning the unique inverse of a triangle:

Proposition 4.1.1. ([396, 1.4.8, p. 9] and [111, Remark 22 (a), p. 22]) *Every triangle U has a unique inverse V which also is a triangle and $x := U(Vx) := V(Ux)$ for all $x \in \omega$.*

Throughout this chapter, let U always denote a triangle and V be its inverse. The reader may refer to [270] for the general results on the Schauder bases of matrix domains of triangles in FK spaces and the β–duals of such matrix domains in FK spaces with AK, and in ℓ_∞ and c, and the characterizations of matrix transformations on such spaces. The following general result technically reduces the determination of a Schauder basis of the matrix domain of a triangle in a linear metric space with a Schauder basis to the determination of the inverse of the triangle:

Theorem 4.1.2. [207, Theorem 2.3] *If $\{b(n)\}_{n\in\mathbb{N}}$ is a Schauder basis of the linear metric sequence space (X, d), then $\{V(b^{(n)})\}_{n\in\mathbb{N}}$ is a basis of $Z = X_U$ with the metric d_U defined by $d_U(z, \tilde{z}) := d(Uz, U\tilde{z})$ for all $z, \tilde{z} \in Z$.*

Proof. It is clear that d_U is a metric on $Z = X_U$. We write $c^{(n)} := Vb^{(n)}$ for $n \in \mathbb{N}$. First, we note that $c^{(n)} \in Z$ for all $n \in \mathbb{N}$, since $Uc^{(n)} := U(Vb^{(n)}) := b^{(n)}$ by Proposition 4.1.1, and $b^{(n)} \in X$.

Let $z \in Z$ be given. Then $x = Uz \in X$ and there is a unique sequence $(\lambda_n)_{n\in\mathbb{N}}$ of scalars such that $x^{\langle m\rangle} = \sum_{n=0}^{m} \lambda_n b^{(n)} \to x$ as $m \to \infty$. We put $x^{\langle m\rangle} = \sum_{n=0}^{m} \lambda_n c^{(n)}$ for $m \in \mathbb{N}$. Then, it follows that

$$U(x^{\langle m\rangle}) = \sum_{n=0}^{m} \lambda_n U(c^{(n)}) = \sum_{n=0}^{m} \lambda_n b^{(n)} = x^{\langle m\rangle}$$

for all $m \in \mathbb{N}$, hence

$$\lim_{m\to\infty} d_U(z^{\langle m\rangle}, z) = \lim_{m\to\infty} d(Uz^{\langle m\rangle}, Uz) \sum_{n=0}^{m} \lambda_n U(c^{(n)}) = \lim_{m\to\infty} d(x^{\langle m\rangle}, x) = 0,$$

as desired. □

Since $X = (X_U)_V$ by Proposition 4.1.1, an application of Theorem 4.1.2 leads the following:

Remark 4.1.3. *The matrix domain X_U of a linear metric sequence space has a basis if and only if X has a basis.*

Now, we give the Schauder bases of FK-spaces with AK.

Corollary 4.1.4. [207, Corollary 2.5] *Let X be an FK-space with AK and the sequences $c^{(n)} = \left\{c_k^{(n)}\right\}$ and $c^{(-1)} = \left\{c_k^{(-1)}\right\}$ be defined by*

$$c_k^{(n)} := \begin{cases} 0 & , \quad 0 \leq k \leq n-1, \\ v_{kn} & , \quad k \geq n, \end{cases} \qquad \text{and } c_k^{(-1)} := \sum_{j=0}^{k} v_{kj}$$

for all $k \in \mathbb{N}$. Then, the following statements hold:

(a) *Every sequence $z = (z_n)_{n\in\mathbb{N}} \in Z = X_U$ has a unique representation*

$$z = \sum_n (Uz)_n c^{(n)}. \tag{4.1.2}$$

(b) *Every sequence $v = (v_n)_{n\in\mathbb{N}} \in Y = X_U \oplus e$ has a unique representation*

$$v := le + \sum_n \{U(v - le)\}_n, \tag{4.1.3}$$

where $l \in \mathbb{C}$ is the uniquely determined such that $v := z + le$ for $z \in Z = X_U$.

(c) *Every sequence $t = (t_n)_{n\in\mathbb{N}} \in T = (X \oplus e)_U$ has a unique representation*

$$t := lc^{(-1)} + \sum_n [(Ut)_n - l]c^{(n)}, \tag{4.1.4}$$

where $l \in \mathbb{C}$ is the uniquely determined such that $Ut - le \in X$.

Proof. First, we note that $c^{(n)} = Ve^{(n)}$ for all $n \in \mathbb{N}$ and $c^{(-1)} = Ve$, hence the sequences $\{c^{(n)}\}_{n\in\mathbb{N}}$ and $\{c^{(n)}\}_{n\in\mathbb{N}_{-1}}$ are bases of Z and T, respectively, by Theorem 4.1.2.

(a) Let $z = (z_n)_{n\in\mathbb{N}} \in Z$ be given. Then, $x = Uz \in X$ and (4.1.2) follows if we take $\lambda_n = (Uz)_n$ for all $n \in \mathbb{N}$ in the proof of Theorem 4.1.2.

(b) Let $v = (v_n)_{n\in\mathbb{N}} \in Y = X_U \oplus e$ be given. Then, there are a uniquely determined sequence $z \in Z$ and $\lambda \in \mathbb{C}$ such that $v := z + le$ and we have $z = \sum_n (Uz)_n c^{(n)}$ by Part (a). It follows that $v := le + z := le + \sum_n \{U(v - le)\}_n c^{(n)}$.

(c) Let $t = (t_n)_{n\in\mathbb{N}} \in T$. Then, $u = Ut \in X \oplus e$ and there are uniquely determined $x \in X$ and $l \in \mathbb{C}$ such that $u := x + le$. We put $z := t - lc^{(-1)}$. Then, $z \in Z$, since $Uz = U(t - lc^{(-1)}) = U(t - lc^{(-1)}) = Ut - lUc^{(-1)} = u - le = x \in X$, and so we have $z = \sum_n (Uz)_n c^{(n)} = \sum_n [(Ut)_n - l]c^{(n)}$ by Part (a). Now, (4.1.4) is an immediate consequence, since $t = z + lc^{(-1)}$. □

Now, we can emphasize the concerning duality relations on matrix domains of triangles in certain sequence spaces.

4.2 CESÀRO SEQUENCE SPACES AND CONCERNING DUALITY RELATION

Let λ denotes any of the classical sequence spaces ℓ_∞, c, c_0 and ℓ_p. Then, the domain λ_{C_1} is called as a *Cesàro sequence space.* The Cesàro sequence spaces $X_\infty := (\ell_\infty)_{C_1}$,

$X_p := (\ell_p)_{C_1}$ were introduced by Ng and Lee [304], and $\widetilde{c} := (c)_{C_1}$, $\widetilde{c_0} := (c_0)_{C_1}$ have been recently studied by Şengönül and Başar [373].

Define the sequence $y = (y_k)$ by the C_1-transform of a sequence $x = (x_k)$, i.e.,

$$y_k := (C_1 x)_k = \frac{1}{k+1}\sum_{j=0}^{k} x_j \quad \text{for all } k \in \mathbb{N}. \tag{4.2.1}$$

Then, since $\lambda_{C_1} \cong \lambda$ it is trivial that the two-sided implication "$x \in \lambda_{C_1}$ *if and only if* $y \in \lambda$" holds. In this situation, the duality relation between the entries of the matrices $A = (a_{nk})$ and $B = (b_{nk})$ is

$$a_{nk} := \sum_{j=k}^{\infty} \frac{1}{j+1} b_{nj} \quad \text{or} \quad b_{nk} := (k+1)(a_{nk} - a_{n,k+1}) \tag{4.2.2}$$

for all $k, n \in \mathbb{N}$. The pair of such type matrices was called as *dual matrices of the new type* by Başar [61] and used by Savaş [347], Başar [61, 62, 63], and Şengönül and Başar [373, 374].

4.2.1 The Cesàro Sequence Spaces of Non-absolute Type

Following Şengönül and Başar [373, 374], we emphasize the Cesàro sequence spaces $\widetilde{c_0}$ and $\widetilde{c}$ of non-absolute type. In [304], Ng and Lee defined the Cesàro sequence spaces X_p and X_∞ of non-absolute type as the sets of the sequences, whose C_1-transforms are in ℓ_p and ℓ_∞, respectively, and computed their β-duals; where $1 < p < \infty$. Following Ng and Lee [304], Şengönül and Başar [373] have introduced the Cesàro sequence spaces $\widetilde{c_0}$ and $\widetilde{c}$ of non-absolute type, as the set of all sequences such that their C_1-transforms are in the spaces c_0 and c, respectively, i.e.,

$$\begin{aligned} \widetilde{c_0} &:= \left\{ x = (x_k) \in \omega : \lim_{n\to\infty} \frac{1}{n+1}\sum_{k=0}^{n} x_k = 0 \right\}, \\ \widetilde{c} &:= \left\{ x := (x_k) \in \omega : \lim_{n\to\infty} \frac{1}{n+1}\sum_{k=0}^{n} x_k \text{ exists} \right\}. \end{aligned}$$

With the notation of (4.1.1), we may redefine the spaces X_∞, X_p, $\widetilde{c_0}$ and $\widetilde{c}$ as follows:

$$X_\infty := (\ell_\infty)_{C_1}, \quad X_p := (\ell_p)_{C_1}, \quad \widetilde{c_0} := (c_0)_{C_1} \quad \text{and} \quad \widetilde{c} := (c)_{C_1}. \tag{4.2.3}$$

It is trivial that $X_p \subset \widetilde{c_0} \subset \widetilde{c} \subset X_\infty$. Now, we may begin with the following theorem:

Theorem 4.2.1. [373, Theorem 2.1] *The sets X_∞, X_p, $\widetilde{c_0}$ and $\widetilde{c}$ are the linear spaces with the co-ordinatewise addition and scalar multiplication, which are BK-spaces endowed with the norm $\|x\|_{X_\infty} := \|x\|_{\widetilde{c_0}} := \|x\|_{\widetilde{c}} := \|C_1 x\|_\infty$ and $\|x\|_{X_p} := \|C_1 x\|_p$.*

Proof. The first part of the theorem is a routine verification and so we omit details. Furthermore, since (4.2.3) holds and ℓ_∞, ℓ_p, c_0 and c are BK-spaces with respect to their natural norms (see Maddox [258, pp. 217–218]), and the matrix $C_1 = (c_{nk})$ is normal(triangle), Theorem 4.3.2 of Wilansky [396, p. 61] gives the fact that the spaces X_∞, X_p, $\widetilde{c_0}$ and $\widetilde{c}$ are BK-spaces. □

One can easily check that the absolute property does not hold on the spaces $\widetilde{c}_0$ and $\widetilde{c}$, since $\|x\|_{X_\infty} \neq \||x|\|_{X_\infty}$, $\|x\|_{X_p} \neq \||x|\|_{X_p}$, $\|x\|_{\widetilde{c}_0} \neq \||x|\|_{\widetilde{c}_0}$ and $\|x\|_{\widetilde{c}} \neq \||x|\|_{\widetilde{c}}$ for at least one sequence in the spaces X_∞, X_p, $\widetilde{c}_0$ and $\widetilde{c}$; where $|x| = (|x_k|)$. This shows that X_∞, X_p, $\widetilde{c}_0$ and $\widetilde{c}$ are the sequence spaces of non-absolute type. Since the sequence spaces λ_A and λ are norm isomorphic whenever A is a triangle, $X_\infty \cong \ell_\infty$, $X_p \cong \ell_p$, $\widetilde{c}_0 \cong c_0$ and $\widetilde{c} \cong c$.

Now, we may give the theorems on the inclusion relations concerning the spaces $\widetilde{c}_0$ and $\widetilde{c}$, and the AD property of the sequence space $\widetilde{c}_0$.

Theorem 4.2.2. [373, Theorem 2.3] *The inclusions $c_0 \subset \widetilde{c}_0$ and $c \subset \widetilde{c}$ strictly hold. But, the spaces $\widetilde{c}_0$ and ℓ_∞ (of course, $\widetilde{c}$ and ℓ_∞) are overlap.*

Proof. To prove the validity of the inclusion $c_0 \subset \widetilde{c}_0$, let us take any $y \in c_0$. Then, bearing in mind the regularity of the method C_1 we immediately observe that $C_1 y \in c_0$, which means that $y \in \widetilde{c}_0$. Hence, the inclusion $c_0 \subset \widetilde{c}_0$ holds. Furthermore, let us consider the sequence $v = (v_k)$ defined by $v_k := (-1)^k$ for all $k \in \mathbb{N}$. Then, since

$$(C_1 v)_n \quad := \quad \frac{1+(-1)^n}{2(n+1)} \quad \text{for all} \quad n \in \mathbb{N},$$

$C_1 v \in c_0$, which implies that v is in $\widetilde{c}_0$ but not in c_0. This shows that the inclusion $c_0 \subset \widetilde{c}_0$ is strict.

On the other hand, if we choose the sequences $x = (x_k) := \{(-1)^{k-1}\sqrt{k}\}_{k=1}^{\infty}$ and $y = e := (1,1,1,\ldots)$, then it is easily seen that $x \in \widetilde{c}_0 \setminus \ell_\infty$ and $y \in \ell_\infty \setminus \widetilde{c}_0$. It is also trivial that all null sequences are in the spaces both $\widetilde{c}_0$ and ℓ_∞. This shows that the spaces $\widetilde{c}_0$ and ℓ_∞ are overlap.

One can see by analogy that the strict inclusion $c \subset \widetilde{c}$ also holds and the spaces $\widetilde{c}$ and ℓ_∞ are overlap. So, we omit details.

This completes the proof. □

Theorem 4.2.3. [373, Theorem 2.4] *$\widetilde{c}_0$ has AD property.*

Proof. For this, we prove that the set ϕ, the space of all finitely non-zero sequences, is dense in $\widetilde{c}_0$.

Note that if $f \in \widetilde{c}_0^*$, then $f(x) := g(C_1 x)$ for some $g \in c_0^*$. Since c_0 has AK property and $c_0^* \cong \ell_1$, $f(x) := \sum_j t_j (C_1 x)_j$ for some $t = (t_j) \in \ell_1$. For any $f \in \widetilde{c}_0^*$ and $e^{(k)} \in \phi$ for each $k \in \mathbb{N}$, we have

$$f(e^{(k)}) = \sum_j t_j \left(C_1 e^{(k)}\right)_j = \left(C_1' t\right)_k;$$

where C_1' denotes the transpose of the matrix C_1. Hence, from Hahn-Banach theorem, $\phi \subset \widetilde{c}_0$ is dense in $\widetilde{c}_0$ if and only if for $t \in \ell_1$, $C_1' t = \theta$ implies $t = \theta$.

By solving the homogenous system of linear equations

$$\left.\begin{array}{rcccccccccl} t_0 & + & \frac{1}{2}t_1 & + & \frac{1}{3}t_2 & + & \frac{1}{4}t_3 & + & \cdots & = & 0 \\ & & \frac{1}{2}t_1 & + & \frac{1}{3}t_2 & + & \frac{1}{4}t_3 & + & \cdots & = & 0 \\ & & & & \frac{1}{3}t_2 & + & \frac{1}{4}t_3 & + & \cdots & = & 0 \\ & & & & & & & & & \vdots & \end{array}\right\}$$

we obtain that $t = \theta$, which completes the proof. □

Because of the isomorphism T, defined on the space c_0 or c by $Tx = C_1x$ is onto, Theorem 4.1.2 implies that the inverse image of the basis of those spaces c_0 and c are the basis of the new spaces $\widetilde{c}_0$ and $\widetilde{c}$, respectively. Therefore, we have the following:

Theorem 4.2.4. [373, Theorem 2.5] *Let $\alpha_k = (C_1x)_k$ for all $k \in \mathbb{N}$. Define the sequence $b^{(k)} := \{b_n^{(k)}\}_{n\in\mathbb{N}}$ of the elements of the space $\widetilde{c}_0$ by*

$$b_n^{(k)} := \begin{cases} (-1)^{k-n}(k+1) & , \quad k \leq n \leq k+1, \\ 0 & , \quad 0 \leq n < k \text{ or } n > k+1, \end{cases}$$

for every fixed $k \in \mathbb{N}$. Then, the following statements hold:

(a) *The sequence $\{b^{(k)}\}_{k\in\mathbb{N}}$ is a basis for the spaces $\widetilde{c}_0$ and X_p, and any x in $\widetilde{c}_0$ or in X_p has a unique representation of the form $x := \sum_k \alpha_k b^{(k)}$.*

(b) *The set $\left\{e, b^{(k)}\right\}$ is a basis for the space $\widetilde{c}$ and any $x \in \widetilde{c}$ has a unique representation of the form $x := lb + \sum_k(\alpha_k - l)b^{(k)}$, where $l := \lim_{k\to\infty}(C_1x)_k$.*

4.2.2 The α-, β- and γ-Duals of the Spaces $\widetilde{c}_0$ and $\widetilde{c}$

We state and prove the theorems determining the α-, β- and γ-duals of the sequence spaces $\widetilde{c}_0$ and $\widetilde{c}$ of non-absolute type.

We shall begin with to quote a lemma, due to Stieglitz and Tietz [368], which is needed in proving Theorem 4.2.6, below.

Lemma 4.2.5. *$A = (a_{nk}) \in (c_0 : \ell_1) = (c : \ell_1)$ if and only if*

$$\sup_{K,N\in\mathcal{F}} \left|\sum_{n\in N}\sum_{k\in K} a_{nk}\right| < \infty.$$

Here and in what follows, we denote the collection of all finite subsets of $\mathbb{N}$ by $\mathcal{F}$.

Theorem 4.2.6. [373, Theorem 3.3] *The α-dual of the spaces $\widetilde{c}_0$ and $\widetilde{c}$ is the set*

$$d_1 := \left\{a = (a_k) \in \omega : \sup_{K,N\in\mathcal{F}} \left|\sum_{n\in N}\sum_{k\in K}(-1)^{n-k}(k+1)a_n\right| < \infty\right\}.$$

Proof. Let $a = (a_n) \in \omega$ and define the matrix $B = (b_{nk})$ by

$$b_{nk} := \begin{cases} (-1)^{n-k}(k+1)a_n & , \quad n-1 \leq k \leq n, \\ 0 & , \quad 0 \leq k < n-1 \text{ or } k > n, \end{cases}$$

for all $k, n \in \mathbb{N}$. Bearing in mind the relation (4.2.1), we immediately derive that

$$a_n x_n := \sum_{k=n-1}^{n} (-1)^{n-k}(k+1)a_n y_k = (By)_n \tag{4.2.4}$$

for all $n \in \mathbb{N}$. We therefore observe by (4.2.4) that $ax = (a_n x_n) \in \ell_1$ whenever $x = (x_n) \in \widetilde{c}_0$ or $\widetilde{c}$ if and only if $By = \{(By)_n\} \in \ell_1$ whenever $y = (y_k) \in c_0$ or c. Then, we derive by Lemma 4.2.5 that

$$\sup_{K,N \in \mathcal{F}} \left| \sum_{n \in N} \sum_{k \in K} (-1)^{n-k}(k+1)a_n \right| < \infty,$$

which yields the desired result that $\widetilde{c}_0^{\alpha} = \widetilde{c}^{\alpha} = d_1$. □

Theorem 4.2.7. [373, Theorem 3.4] *Define the sets d_2 and d_3 by*

$$\begin{aligned} d_2 &:= \left\{ a = (a_k) \in \omega : \sum_k (k+1) |\Delta a_k| < \infty \right\}, \\ d_3 &:= \{ a = (a_k) \in \omega : ((k+1)a_k) \in \ell_\infty \}, \end{aligned}$$

where $\Delta a_k := a_k - a_{k+1}$ for all $k \in \mathbb{N}$. Then, $\widetilde{c}^{\beta} := d_2 \cap cs$ and $\widetilde{c}_0^{\beta} := d_2 \cap d_3$.

Proof. Because, the proof can also be obtained for the space $\widetilde{c}_0$ in the similar way, we give the proof only for the space $\widetilde{c}$.

Consider the equality

$$\sum_{k=0}^{n} a_k x_k := \sum_{k=0}^{n-1} (k+1)\Delta a_k y_k + (n+1)a_n y_n = (Ty)_n, \tag{4.2.5}$$

for all $n \in \mathbb{N}$, where $T = (t_{nk})$ is defined by

$$t_{nk} := \begin{cases} (k+1)\Delta a_k & , \quad 0 \leq k \leq n-1, \\ (n+1)a_n & , \quad k = n, \\ 0 & , \quad k > n, \end{cases} \tag{4.2.6}$$

for all $k, n \in \mathbb{N}$. Thus, we deduce from Kojima-Schur theorem with (4.2.5) that $ax = (a_k x_k) \in cs$ whenever $x = (x_k) \in \widetilde{c}$ if and only if $Ty = \{(Ty)_n\} \in c$ whenever $y = (y_k) \in c$. It is obvious that the columns of that matrix T, defined by (4.2.6), are in the space c. Therefore, we derive the consequences from (3.3.1) and (3.3.4) that

$$\sum_k (k+1) |\Delta a_k| < \infty, \tag{4.2.7}$$

$$a = (a_k) \in cs, \tag{4.2.8}$$

respectively. The consequences (4.2.7) and (4.2.8) show that $\widetilde{c}^{\beta} = d_2 \cap cs$. □

Theorem 4.2.8. [373, Theorem 3.5] *The γ-dual of the spaces $\widetilde{c}_0$ and $\widetilde{c}$ is the set $d_2 \cap d_3$.*

Proof. This is similar to the proof of Theorem 4.2.7 and so, we leave the detail to the reader. □

4.2.3 The Characterization of Some Matrix Mappings Related to the Space $\widetilde{c}$

We characterize the matrix mappings from $\widetilde{c}$ into any given sequence space via the concept of the dual summability methods of the new type.

Theorem 4.2.9. [373, Theorem 4.1] *Let $A = (a_{nk})$ and $B = (b_{nk})$ be the dual matrices of the new type and μ be any given sequence space. Then, $A = (a_{nk}) \in (\widetilde{c} : \mu)$ if and only if $B \in (c : \mu)$ and*

$$(na_{nk})_{n\in\mathbb{N}} \in c_0 \text{ for each } k \in \mathbb{N}. \tag{4.2.9}$$

Proof. Suppose that $A = (a_{nk})$ and $B = (b_{nk})$ are dual matrices of the new type, that is to say that (4.2.2) holds, μ is any given sequence space and take account that the spaces $\widetilde{c}$ and c are norm isomorphic.

Let $A = (a_{nk}) \in (\widetilde{c} : \mu)$ and take any $y = (y_k) \in c$. Then, BC_1 exists and $(a_{nk})_{k\in\mathbb{N}} \in d_2 \cap cs$ which yields that $(b_{nk})_{k\in\mathbb{N}} \in \ell_1$ for each $n \in \mathbb{N}$. Hence, By exists for all $y \in c$ and thus letting $m \to \infty$ in the equality

$$\sum_{k=0}^{m} b_{nk}y_k := \sum_{k=0}^{m}\sum_{j=k}^{m} \frac{1}{j+1} b_{nj}x_k \text{ for all } m, n \in \mathbb{N}, \tag{4.2.10}$$

we have by (4.2.2) that $By = Ax$, which leads us to the consequence $B \in (c : \mu)$.

Conversely, let $B = (b_{nk}) \in (c : \mu)$ and (4.2.9) holds, and take any $x = (x_k) \in \widetilde{c}$. Then, we have $(b_{nk})_{k\in\mathbb{N}} \in \ell_1$, which gives together with (4.2.9) that $(a_{nk})_{k\in\mathbb{N}} \in \widetilde{c}^{\beta}$ for each $n \in \mathbb{N}$. Hence, Ax exists. Therefore, we obtain from the equality

$$\sum_{k=0}^{m} a_{nk}x_k := \sum_{k=0}^{m-1} b_{nk}y_k + (m+1)a_{nm}y_m \text{ for all } m, n \in \mathbb{N},$$

as $m \to \infty$ that $Ax = By$ and this shows that $A \in (\widetilde{c} : \mu)$.

This completes the proof. □

Because of the two-sided implication does not hold, we remark the reader from now on that the condition (4.2.9) of Theorem 4.2.9 cannot be dropped and is missed in Theorem 3.1 of Başar [66].

By changing the roles of the spaces $\widetilde{c}$ and c with μ, we have:

Theorem 4.2.10. [373, Theorem 4.2] *Suppose that the elements of the infinite matrices $D = (d_{nk})$ and $E = (e_{nk})$ are connected with the relation*

$$e_{nk} := \frac{1}{n+1}\sum_{j=0}^{n} d_{jk} \text{ for all } k, n \in \mathbb{N}, \tag{4.2.11}$$

and μ be any given sequence space. Then, $D \in (\mu : \widetilde{c})$ if and only if $E \in (\mu : c)$.

Proof. Let $x = (x_k) \in \mu$ and consider the following equality with (4.2.11)

$$\frac{1}{i+1}\sum_{j=0}^{i}\sum_{k=0}^{n} d_{jk}x_k := \sum_{k=0}^{n} e_{ik}x_k \quad \text{for all} \quad i, n \in \mathbb{N},$$

which yields as $n \to \infty$ that

$$\frac{1}{i+1}\sum_{j=0}^{i} (Dx)_j := (Ex)_i \quad \text{for all} \quad i \in \mathbb{N}. \tag{4.2.12}$$

Now, we immediately observe by (4.2.12) that $Dx \in \widetilde{c}$ whenever $x \in \mu$ if and only if $Ex \in c$ whenever $x \in \mu$.

This step completes the proof. □

It is to be noted that Theorems 4.2.9 and 4.2.10 have several consequences depending on the choice of the sequence space μ. Therefore, by Theorems 4.2.9 and 4.2.10, the necessary and sufficient conditions on the matrices A and D in the classes $(\widetilde{c} : \mu)$ and $(\mu : \widetilde{c})$ can be derived by replacing the entries of A and D by those of the entries of $B := A(C_1)^{-1}$ and $E := C_1 D$, respectively; where the necessary and sufficient conditions on the matrices B and E are read from the concerning results in the existing literature.

4.3 DIFFERENCE SEQUENCE SPACES AND CONCERNING DUALITY RELATION

Let λ denotes any of the classical sequence spaces ℓ_∞, c or c_0. Then, $\lambda(\Delta)$ consisting of the sequences $x = (x_k)$ such that $\Delta x = (x_k - x_{k+1}) \in \lambda$ is called the *difference sequence space*, which was introduced by Kızmaz [217]. Kızmaz [217] proved that $\lambda(\Delta)$ is a Banach space with the norm $\|x\|_\Delta := |x_1| + \|\Delta x\|_\infty$, where $x = (x_k) \in \lambda(\Delta)$ and the inclusion relation $\lambda \subset \lambda(\Delta)$ strictly holds. He also determined the α-, β- and γ-duals of the difference spaces and characterized the classes $(\lambda(\Delta) : \mu)$ and $(\mu : \lambda(\Delta))$ of infinite matrices, where $\lambda, \mu \in \{\ell_\infty, c\}$. Following Kızmaz [217], Sarıgöl [343] extended the difference spaces $\lambda(\Delta)$ to the spaces $\lambda(\Delta_r)$ defined by

$$\lambda(\Delta_r) := \{x = (x_k) \in \omega : \Delta_r x := \{k^r (x_k - x_{k+1})\} \in \lambda \ \text{ for } r < 1\}$$

and computed the α-, β-, γ-duals of the space $\lambda(\Delta_r)$, where $\lambda \in \{\ell_\infty, c, c_0\}$. It is easily seen that $\lambda(\Delta_r) \subset \lambda(\Delta)$, if $0 < r < 1$ and $\lambda(\Delta) \subset \lambda(\Delta_r)$, if $r < 0$.

In the same year, Ahmad and Mursaleen [8] extended these spaces to $\lambda(p, \Delta)$ and studied the related problems. Malkowsky [261] determined the Köthe-Toeplitz duals of the sets $\ell_\infty(p, \Delta)$ and $c_0(p, \Delta)$, and gave new proofs of the characterization of the matrix transformations considered in [8]. In 1993, Choudhary and Mishra [106] studied some properties of the sequence space $c_0(\Delta_r)$, for $r \geq 1$. In the same year, Mishra [278] gave a characterization of BK-spaces, which contain subspace isomorphic to $sc_0(\Delta)$ in terms of matrix maps and sufficient condition for a matrix map from $s\ell_\infty(\Delta)$ into a BK-space to be a compact operator. He showed that any matrix from $s\ell_\infty(\Delta)$ into a BK-space which does not contain any subspace isomorphic to $s\ell_\infty(\Delta)$ is compact, where

$$s\lambda(\Delta) := \{x := (x_k) \in \omega : (\Delta x_k) \in \lambda,\ x_1 = 0 \text{ for } \lambda = \ell_\infty \text{ or } c_0\}.$$

In 1996, Mursaleen et al. [291] defined and studied the sequence space

$$\ell_\infty(p, \Delta_r) := \{x = (x_k) \in \omega : \Delta_r x \in \ell_\infty(p)\}, \quad (r > 0).$$

Gnanaseelan and Srivastava [184] defined and studied the spaces $\lambda(u, \Delta)$ for a sequence $u = (u_k)$ of non-complex numbers such that

(i) $\dfrac{|u_k|}{|u_{k+1}|} = 1 + O(1/k)$ for each $k \in \mathbb{N}_1$.

(ii) $k^{-1}|u_k| \sum_{i=0}^{k} |u_i|^{-1} = O(1)$.

(iii) $\left(k\left|u_k^{-1}\right|\right)$ is a sequence of positive numbers increasing monotonically to infinity.

In the same year, Malkowsky [262] defined the spaces $\lambda(u, \Delta)$ for an arbitrary fixed sequence $u = (u_k)$ without any restrictions on u. He proved that the sequence spaces $\lambda(u, \Delta)$ are BK-spaces endowed with the norm defined by

$$\|x\| := \sup_{k \in \mathbb{N}} |u_{k-1}(x_{k-1} - x_k)| \quad \text{with} \quad u_0 := x_0 := 1.$$

Later, Gaur and Mursaleen [181] extended the space $S_r(\Delta)$ to the space $S_r(p, \Delta)$, where

$$S_r(p, \Delta) := \{x = (x_k) \in \omega : (k^r |\Delta x_k|) \in c_0(p)\}, \quad (r \geq 1)$$

and characterized the matrix classes $(S_r(p, \Delta) : \ell_\infty)$ and $(S_r(p, \Delta) : \ell_1)$. Malkowsky et al. [266], and independently, Asma and Çolak [40] extended the space $\lambda(u, \Delta)$ to the space $\lambda(p, u, \Delta)$ and gave Köthe-Toeplitz duals of these spaces, for $\lambda = \ell_\infty$, c or c_0. Recently, Malkowsky and Mursaleen [265] characterized the matrix classes $(\Delta\lambda : \mu)$ and $(\Delta\lambda : \Delta\mu)$ for $\lambda \in \{c_0(p), c(p), \ell_\infty(p)\}$ and $\mu \in \{c_0(q), c(q), \ell_\infty(q)\}$.

Recently, the difference spaces bv_p consisting of the sequences $x = (x_k)$ such that $(x_k - x_{k-1}) \in \ell_p$ have been studied in the case $0 < p \leq 1$ by Altay and Başar [30], and in the case $1 \leq p \leq \infty$ by Başar and Altay [66], Malkowsky et al. [271] and Çolak et al. [133]. Additionally, the reader may refer to Çolak [125], Çolak and Et [132] and Çolak et al. [127] for some different approaches.

Let us define the sequence $y = (y_k)$ by the $\Delta^{(1)}$-transform of a sequence $x = (x_k) \in \omega$, i.e.,

$$y_k := \left(\Delta^{(1)} x\right)_k := \begin{cases} x_0 &, \quad k = 0, \\ x_k - x_{k-1} &, \quad k \in \mathbb{N}_1. \end{cases} \tag{4.3.1}$$

Then, since $bv_p \cong \ell_p$ it is trivial that the two-sided implication "*$x \in bv_p$ if and only if $y \in \ell_p$*" holds. In this situation, the duality relation between the entries of the matrices $A = (a_{nk})$ and $B = (b_{nk})$ is

$$a_{nk} := \sum_{j=k}^{k+1} (-1)^{j-k} b_{nj} \quad \text{or} \quad b_{nk} := \sum_{j=k}^{\infty} a_{nj} \tag{4.3.2}$$

for all $k, n \in \mathbb{N}$. The pair of such type matrices was used by Başar and Altay [66] and Başar et al. [67].

We write for brevity that

$$\widetilde{a}_{nk} := \sum_{j=k}^{\infty} a_{nj} \quad \text{and} \quad a(n,k) := \sum_{j=0}^{n} a_{jk},$$

for all $n, k \in \mathbb{N}$. We also use the similar notations with other letters.

We introduce the space bv_p of sequences of p-bounded variation and derive some related results. Furthermore, we construct the basis and determine the α-, β- and γ-duals of the space bv_p. Besides, we essentially characterize the matrix classes $(bv_p : \ell_\infty)$, $(bv_\infty : \ell_p)$ and $(bv_p : \ell_1)$, and also derive the characterization of some other classes by means of a suitable relation.

4.3.1 The Space bv_p of Sequences of p-Bounded Variation

Although, in the existing literature, the matrix domain λ_Δ is called *the difference sequence space* whenever λ is a normed or paranormed sequence space, in the case $\lambda = \ell_p$ we prefer calling this difference sequence space as *the space of sequences of p-bounded variation* and denote it by bv_p instead of the usual notation $\ell_p(\Delta)$. We treat slightly more different than Kızmaz [217] and the other authors following him, and employ the technique obtaining a new sequence space by the matrix domain of a triangle limitation method. Following this way, we introduce the sequence space bv_p as the set of all sequences such that their $\Delta^{(1)}$-transforms are in the space ℓ_p, i.e.,

$$\begin{aligned} bv_p &:= \left\{ x = (x_k) \in \omega : \sum_k |x_k - x_{k-1}|^p < \infty \right\}, \quad (0 < p < \infty), \\ bv_\infty &:= \left\{ x = (x_k) \in \omega : \sup_{k \in \mathbb{N}} |x_k - x_{k-1}| < \infty \right\}. \end{aligned}$$

With the notation of (4.1.1), we can redefine the spaces bv_p and bv_∞ by

$$bv_p := (\ell_p)_{\Delta^{(1)}}, \ (0 < p < \infty) \ \text{ and } \ bv_\infty := (\ell_\infty)_{\Delta^{(1)}}.$$

Now, we may begin with the following theorem:

Theorem 4.3.1. (cf. [66, Theorem 2.1] and [30, Theorem 3.1]) *The set bv_p is a linear space with the co-ordinatewise addition and scalar multiplication, which is the BK-space endowed with the norm $\|x\|_{bv_p} = \left\|\Delta^{(1)}x\right\|_p$ if $1 < p \leq \infty$ and is the p-normed space with the p-norm $\|x\|_{bv_p} = \sum_k |x_k - x_{k-1}|^p$ if $0 < p \leq 1$.*

Proof. Since the proof can be obtained in the similar way used in proving Theorem 4.2.1, we omit details.

It is immediate by the fact $bv_p \cong \ell_p$ that bv_p is a complete p-normed linear space with the p-norm $\|x\|_{bv_p} := \sum_k |x_k - x_{k-1}|^p$, in the case $0 < p \leq 1$. □

One can easily see that bv_p is a sequence space of non-absolute type. Since $\Delta^{(1)}$ is a triangle, it is trivial that $bv_p \cong \ell_p$; where $1 < p \leq \infty$. One may expect the similar result for the space bv_p as observed for the space ℓ_p and ask the natural question: *Isn't the space bv_p a Hilbert space with $p \neq 2$?* The answer is positive and is given by the following theorem:

Theorem 4.3.2. [66, Theorem 2.3] *Except the case $p = 2$, the space bv_p is not an inner product space, hence not a Hilbert space for $1 < p < \infty$.*

Proof. We prove that the space bv_2 is the only Hilbert space among the bv_p spaces for $1 < p < \infty$, firstly. Since the space bv_2 is BK-space endowed with the norm $\|x\|_{bv_2} = \|\Delta^{(1)}x\|_2$ by Theorem 4.3.1 and its norm can be obtained from an inner product, that is the equality $\|x\|_{bv_2} = \sqrt{\langle \Delta^{(1)}x, \Delta^{(1)}x \rangle}$ holds, the space bv_2 is a Hilbert space.

Let us now consider the sequences $u = (u_k)$ and $e^{(0)}$ given by

$$u_k := \begin{cases} 1 & , \quad k = 0, \\ 2 & , \quad k \in \mathbb{N}_1, \end{cases} \qquad \text{and} \quad e^{(0)} = (1, 0, 0, \ldots, 0, \ldots).$$

Then, we see for $p \neq 2$ that

$$\left\| u + e^{(0)} \right\|_{bv_p}^2 + \left\| u - e^{(0)} \right\|_{bv_p}^2 = 8 \neq 4(2^{2/p}) = 2\left(\|u\|_{bv_p}^2 + \left\| e^{(0)} \right\|_{bv_p}^2 \right).$$

This shows that the norm on the space bv_p does not satisfy the parallelogram equality, which means that the norm cannot be obtained from an inner product. Hence, the space bv_p with $p \neq 2$ is a Banach space, which is not a Hilbert space.

This completes the proof. □

We wish to derive some inclusion relations concerning with the space bv_p.

Theorem 4.3.3. (cf. [66, Theorem 2.4] and [30, Theorem 3.3]) *The inclusion $\ell_p \subset bv_p$ strictly holds for $0 < p \leq \infty$.*

Proof. To prove the validity of the inclusion $\ell_p \subset bv_p$ for $0 < p \leq \infty$, it suffices to show the existence of a number $K > 0$ such that $\|x\|_{bv_p} \leq K\|x\|_p$ for every $x \in \ell_p$.

Let $x = (x_k) \in \ell_p$ and $0 < p \leq \infty$. Then, we, respectively, obtain in the cases of $0 < p \leq 1$, $1 < p < \infty$ and $p = \infty$ that

$$\sum_k |x_k - x_{k-1}|^p \leq 2 \sum_k |x_k|^p < \infty,$$

$$
\begin{aligned}
\sum_k |x_k - x_{k-1}|^p &\leq \sum_k 2^{p-1}\left(|x_k|^p + |x_{k-1}|^p\right) \\
&\leq 2^{p-1}\left(\sum_k |x_k|^p + \sum_k |x_{k-1}|^p\right), \\
\sup_{k\in\mathbb{N}} |x_k - x_{k-1}| &\leq 2\sup_{k\in\mathbb{N}} |x_k|
\end{aligned}
$$

which together yield us as expected that $\|x\|_{bv_p} \leq 2\|x\|_p$. Besides, since $e = (1,1,\dots) \in bv_p \setminus \ell_p$ and $(k) \in bv_\infty \setminus \ell_\infty$, the inclusion $\ell_p \subset bv_p$ is strict for $0 < p \leq \infty$.

This completes the proof. □

Theorem 4.3.4. [66, Theorem 2.5] *Neither of the spaces bv_p and ℓ_∞ includes the other one, where $1 < p < \infty$.*

Proof. Let us consider the sequences $u = (u_k)$ and $x = (x_k)$ defined by $u_k := \sum_{j=0}^{k} 1/(j+1)$ and $x_k := (-1)^k$ for all $k \in \mathbb{N}$, respectively. Then, $u \in bv_p \setminus \ell_\infty$, since $\Delta^{(1)}u = \{1/(k+1)\} \in \ell_p$. Nevertheless, $x \in \ell_\infty \setminus bv_p$. Hence, the sequence spaces bv_p and ℓ_∞ overlap but neither contains the other.

This completes the proof. □

Now, we can give the corresponding theorem to Theorem 2.5 of [66] with the space c instead of the space ℓ_∞.

Theorem 4.3.5. [15, Theorem 2.2] *Neither of the spaces bv_p and c includes the other one, while the inclusion $bv_1 \subset c$ strictly holds.*

Proof. Let $1 < p < \infty$. Since $\ell_p \subset c$ is well known and the inclusion $bv_p \supset \ell_p$ strictly holds by Theorem 4.3.3, we have $\ell_p \subset bv_p \cap c$. Let us consider the sequences $u = (u_k)$ and $x = (x_k)$ defined by

$$
u_k := \sum_{j=0}^{k} \frac{1}{j+1} \quad \text{and} \quad x_{2k} := \frac{1}{\sqrt[p]{k+1}}, \quad x_{2k+1} := \frac{1}{\sqrt[p]{k+1}} + \frac{1}{\sqrt[p]{k+2}}
$$

for all $k \in \mathbb{N}$, respectively. Then, we have $\Delta^{(1)}u = \{1/(k+1)\} \in \ell_p$, which gives that u is in $bv_p \setminus c$. Nevertheless, x is in $c \setminus bv_p$. Hence, the sequence spaces bv_p and c overlap but neither contains the other.

Since the second part of theorem is trivial, we omit details.

This completes the proof. □

Theorem 4.3.6. [66, Theorem 2.6] *If $1 < p < s$, then $bv_p \subset bv_s$.*

Proof. Suppose that $1 < p < s$ and $x \in bv_p$. With the notation of (4.3.1), $bv_p \cong \ell_p$ implies that $y \in \ell_p$. Then, the well-known inclusion $\ell_p \subset \ell_s$ yields the fact that $y \in \ell_s$. This means that $x \in bv_s$ and hence, the inclusion $bv_p \subset bv_s$ holds, as asserted. □

Because of the isomorphism T, defined on the space bv_p by $Tx = \Delta^{(1)}x$ is onto, by Theorem 4.1.2, the inverse image of the basis of those space ℓ_p is the basis of the space bv_p with $1 \leq p < \infty$. Therefore, we have the following:

Theorem 4.3.7. [66, Theorem 3.1] *Let $1 < p < \infty$ and $\alpha_k = \left(\Delta^{(1)}x\right)_k$ for all $k \in \mathbb{N}$. Define the sequence $b^{(k)} = \{b_n^{(k)}\}_{n\in\mathbb{N}}$ of the elements of the space bv_p for every fixed $k \in \mathbb{N}$ by*

$$b_n^{(k)} := \begin{cases} 0 \ , & n < k, \\ 1 \ , & n \geq k. \end{cases} \tag{4.3.3}$$

Then, the sequence $\{b^{(k)}\}_{k\in\mathbb{N}}$ is a basis for the space bv_p and any $x \in bv_p$ has a unique representation of the form

$$x := \sum_k \alpha_k b^{(k)}. \tag{4.3.4}$$

4.3.2 The Dual Spaces of the Space bv_p

We state and prove the theorems determining the continuous, α-, β- and γ-duals of the space bv_p. Because of the case $p = 1$ may be proved in the similar way and found in the literature, we omit the proof of that case and consider only the case $1 < p \leq \infty$ in the proof of Theorems 4.3.15 and 4.3.16, respectively.

Theorem 4.3.8. *Define the sets D_∞ and D_q by*

$$\begin{aligned} D_\infty &:= \left\{ x = (x_k) \in \omega : \sup_{k\in\mathbb{N}} \left| \sum_{j=k}^{\infty} x_j \right| < \infty \right\}, \\ D_q &:= \left\{ x = (x_k) \in \omega : \sum_k \left| \sum_{j=k}^{\infty} x_j \right|^q < \infty \right\}, \ (1 < q < \infty). \end{aligned}$$

Then, the sets D_∞ and D_q are Banach spaces with the norms

$$\|a\|_{D_\infty} := \sup_{k\in\mathbb{N}} \left| \sum_{j=k}^{\infty} a_j \right|, \tag{4.3.5}$$

$$\|a\|_{D_q} := \left(\sum_k \left| \sum_{j=k}^{\infty} a_j \right|^q \right)^{1/q}. \tag{4.3.6}$$

Additionally,

(i) *D_∞ is isometrically isomorphic to bv_1^*,* [202, Theorem 3.3].

(ii) *D_q is isometrically isomorphic to $(bv_p)^*$,* [15, Theorem 2.3].

Proof. Since it is a routine verification to show that D_∞ and D_q are Banach spaces with the norms defined by (4.3.5) and (4.3.6), we omit details.

To prove the second part of the theorem, we should show the existence of a linear and norm preserving bijection from the spaces bv_1^* and $(bv_p)^*$ to the spaces D_∞ and D_q, respectively.

Define the operator T from bv_1^* to D_∞ by

$$Tf := \left\{f(e^{(0)}), f(e^{(1)}), f(e^{(2)}), \ldots\right\}.$$

It is obvious that T is linear and since $Tf := \theta$ implies $f = \theta$, T is injective. Let us take any $\widetilde{g} := (g_0, g_1, g_2, \ldots)$ in D_∞ and define $f : bv_1 \to \mathbb{C}$ by $f(x) = \sum_k \left(\Delta^{(1)}x\right)_k \sum_{j=k}^{\infty} g_j$. Then, the linearity of f is clear and since

$$\begin{aligned} |f(x)| &= \left|\sum_k \left(\Delta^{(1)}x\right)_k \sum_{j=k}^{\infty} g_j\right| \leq \sum_k \left|\left(\Delta^{(1)}x\right)_k\right| \left|\sum_{j=k}^{\infty} g_j\right| \\ &\leq \sum_k \left|\left(\Delta^{(1)}x\right)_k\right| \sup_{k\in\mathbb{N}} \left|\sum_{j=k}^{\infty} g_j\right| = \|x\|_{bv_1} \|\widetilde{g}\|_{D_\infty}, \end{aligned}$$

f is continuous such that $Tf = \widetilde{g}$. This means that T is surjective.

Finally, we show that T is norm preserving. Since,

$$\begin{aligned} |f(x)| &= \left|f\left(\sum_k \left(\Delta^{(1)}x\right)_k \sum_{j=k}^{\infty} e^{(j)}\right)\right| = \left|\sum_k \left(\Delta^{(1)}x\right)_k \sum_{j=k}^{\infty} f\left(e^{(j)}\right)\right| \\ &\leq \sum_k \left|\left(\Delta^{(1)}x\right)_k\right| \left|\sum_{j=k}^{\infty} f\left(e^{(j)}\right)\right| = \|x\|_{bv_1} \|Tf\|_{D_\infty}, \end{aligned}$$

we conclude that

$$\|f\| \leq \|Tf\|_{D_\infty}. \tag{4.3.7}$$

On the other hand, one can easily see that

$$\left|\sum_{j=k}^{\infty} f\left(e^{(j)}\right)\right| = \left|f\left(b^{(k)}\right)\right| \leq \|f\| \left\|\left(b^{(k)}\right)\right\|_{bv_1} = \|f\|,$$

which leads us for all $k \in \mathbb{N}$ that $\left|\sum_{j=k}^{\infty} f\left(e^{(j)}\right)\right| \leq \|f\|$. Hence, we have

$$\|Tf\|_{D_\infty} \leq \|f\|. \tag{4.3.8}$$

Therefore, we observe by combining (4.3.7) and (4.3.8) that T is norm preserving. Hence, bv_1^* is congruent to the space D_∞.

Let $1 < p < \infty$ and define the transformation $U : (bv_p)^* \to D_q$ by $f \mapsto Uf = a = (a_0, a_1, a_2, \ldots)$. The linearity of U is obvious. Additionally, since $Uf = \theta = (0, 0, 0, \ldots)$ trivially implies $f = \theta = (0, 0, 0, \ldots)$, U is injective.

Let $a = (a_k) \in D_q$. Consider the functional f defined on the space bv_p such that $f(x) = \sum_k \left(\Delta^{(1)}x\right)_k \sum_{j=k}^{\infty} a_j$. Then, f is linear and since

$$\begin{aligned} |f(x)| &\leq \sum_k \left| \left(\Delta^{(1)}x\right)_k \sum_{j=k}^{\infty} a_j \right| \\ &\leq \left[\sum_k \left| \left(\Delta^{(1)}x\right)_k \right|^p \right]^{1/p} \left(\sum_k \left| \sum_{j=k}^{\infty} a_j \right|^q \right)^{1/q} \\ &= \|x\|_{bv_p} \left(\sum_k \left| \sum_{j=k}^{\infty} a_j \right|^q \right)^{1/q} \end{aligned}$$

which yields us that

$$\|f\| \leq \left(\sum_k \left| \sum_{j=k}^{\infty} a_j \right|^q \right)^{1/q} = \|a\|_{D_q}, \tag{4.3.9}$$

f is bounded, i.e., $f \in (bv_p)^*$. Therefore, $U^{-1}a = f \in (bv_p)^*$ whenever $a \in D_q$, which means that U is surjective.

Let us take any $x = (x_k) \in bv_p$ and $f \in (bv_p)^*$. Then, we have by applying f to (4.3.4) that

$$f(x) = \sum_k \left(\Delta^{(1)}x\right)_k f\left(b^{(k)}\right) = \sum_k \left(\Delta^{(1)}x\right)_k \sum_{j=k}^{\infty} f_j,$$

where $f_j = f\left(e^{(j)}\right)$ for each $j \in \mathbb{N}$. Thus, we observe that

$$\|f\| \leq \left(\sum_k \left| \sum_{j=k}^{\infty} f_j \right|^q \right)^{1/q}, \tag{4.3.10}$$

since

$$|f(x)| \leq \|x\|_{bv_p} \left(\sum_k \left| \sum_{j=k}^{\infty} f_j \right|^q \right)^{1/q}.$$

Let us define the sequence $x^{(n)} = \{x_k^{(n)}\} \in bv_p$ by

$$\left(\Delta^{(1)}x\right)_k^{(n)} := \begin{cases} \left| \sum_{j=k}^{n} f_j \right|^{q-1} \operatorname{sgn}\left(\sum_{j=k}^{n} f_j \right) &, \quad 0 \leq k \leq n, \\ 0 &, \quad k > n. \end{cases}$$

Then, it is clear that

$$\left| f\left(x^{(n)}\right) \right| = \sum_{k=0}^{n} \left| \sum_{j=k}^{n} f_j \right|^q \leq \|f\| \|x^{(n)}\|_{bv_p} = \|f\| \left(\sum_{k=0}^{n} \left| \sum_{j=k}^{n} f_j \right|^q \right)^{1/p},$$

which leads us to the consequence that

$$\|f\| \geq \left(\sum_k \left|\sum_{j=k}^{\infty} f_j\right|^q\right)^{1/q}. \tag{4.3.11}$$

Therefore, we have by combining the results (4.3.10) and (4.3.11) that

$$\|f\| = \left(\sum_k \left|\sum_{j=k}^{\infty} f_j\right|^q\right)^{1/q},$$

which is the norm on D_q, i.e., U is norm preserving. Hence, $(bv_p)^*$ is isometrically isomorphic to D_q.

This step completes the proof. □

Theorem 4.3.9. [15, Theorem 2.4] *Although the sequence space bv_p has not AD-property for $p = 1$, it has AD-property for $1 < p < \infty$.*

Proof. Note that if $f \in (bv_p)^*$, then $f(x) = g(\Delta^{(1)}x)$ for some $g \in \ell_p^*$. Since ℓ_p has AK-property and $\ell_p^* \cong \ell_q$,

$$f(x) = \sum_k t_k \left(\Delta^{(1)}x\right)_k$$

for some $t = (t_k) \in \ell_q$. For any $f \in (bv_p)^*$ and $e^{(k)} \in \phi$, we have

$$f(e^{(k)}) = \sum_j t_j \left(\Delta^{(1)}e^{(k)}\right)_j = t_k - t_{k+1} \quad \text{for all} \quad k \in \mathbb{N}.$$

Hence, an application of Hahn-Banach theorem yields the fact that $\phi \subset bv_p$ is dense in bv_p if and only if $(t_k - t_{k+1}) = \theta$ for $t \in \ell_q$ implies $t = \theta$.

For $p = 1$, consider the sequence $e \in \ell_1^\beta = \ell_\infty$. Since $\Delta' e = \theta$, bv_1 has not AD-property; where Δ' denotes the transpose of the matrix $\Delta^{(1)}$.

Since the null space of the operator Δ' on ℓ_q is $\{\theta\}$, bv_p has AD-property. Therefore, the proof of theorem is completed. □

Subsequent to defining the concept of *f-duality of a sequence space*, we give the f-dual of the space bv_p. The f-dual λ^f of a sequence space λ is defined by

$$\lambda^f := \left\{\{f(e^{(k)})\} : f \in \lambda^*\right\}.$$

Lemma 4.3.10. [396, p. 108] *Let λ be an FK-space which contains ϕ. Then, λ has AD iff $\lambda^f := \lambda^*$.*

Thus, as an easy consequence of combining Theorem 4.3.8 or Theorem 4.3.9 with Lemma 4.3.10, we have

Corollary 4.3.11. [15, Corollary 2.6] *Let* $1 < p < \infty$. *Then,*

$$(bv_p)^f := (bv_p)^* := \left\{ a = (a_k) \in \omega : \sum_k \left| \sum_{j=k}^{\infty} a_j \right|^q < \infty \right\}.$$

We shall begin with to quote the lemmas, due to Stieglitz and Tietz [368], which are needed in proving Theorems 4.3.15-4.3.17, below.

Lemma 4.3.12. $A = (a_{nk}) \in (\ell_p : \ell_1)$ *if and only if*

$$\sup_{K \in \mathcal{F}} \sum_k \left| \sum_{n \in K} a_{nk} \right|^q < \infty, \quad (1 < p \leq \infty).$$

Lemma 4.3.13. $A = (a_{nk}) \in (\ell_p : c)$ *if and only if (3.3.3) holds and*

(i) *In the case* $0 < p \leq 1$,

$$\sup_{k,n \in \mathbb{N}} |a_{nk}|^p < \infty, \tag{4.3.12}$$

$$\sup_{k,n \in \mathbb{N}} |a_{nk} - \alpha_k|^p < \infty. \tag{4.3.13}$$

(ii) *In the case* $1 < p < \infty$,

$$\sup_{n \in \mathbb{N}} \sum_k |a_{nk}|^q < \infty. \tag{4.3.14}$$

Lemma 4.3.14. $A = (a_{nk}) \in (\ell_p : \ell_\infty)$ *if and only if*

(i) *(4.3.12) holds, in the case* $0 < p \leq 1$.

(ii) *(4.3.14) holds, in the case* $1 < p < \infty$.

Theorem 4.3.15. [66, Theorem 4.1] *Define the set* a_q *by*

$$a_q := \left\{ a = (a_k) \in \omega : \sup_{K \in \mathcal{F}} \sum_k \left| \sum_{n \in K} a_n \right|^q < \infty \right\}.$$

Then, $(bv_p)^\alpha = a_q$; *where* $1 < p \leq \infty$.

Proof. Let us define the matrix B whose rows are the product of the rows of the matrix Δ^{-1} and the sequence $a = (a_n)$, that is to say that $B_n := (\Delta^{-1})_n a$, where B_n and $(\Delta^{-1})_n$ denote the sequences in the nth rows of the matrices B and the inverse Δ^{-1} of $\Delta^{(1)}$, respectively. Bearing in mind the relation (4.3.1), we easily obtain that

$$a_n x_n := \sum_{k=0}^{n} a_n y_k = (By)_n \text{ for all } n \in \mathbb{N}. \tag{4.3.15}$$

We therefore observe by (4.3.15) that $ax = (a_n x_n) \in \ell_1$ whenever $x = (x_n) \in bv_p$ if and only if $By := \{(By)_n\} \in \ell_1$ whenever $y = (y_k) \in \ell_p$. Then, we derive by Lemma 4.3.12 that

$$\sup_{K \in \mathcal{F}} \sum_k \left| \sum_{n \in K} a_n \right|^q < \infty,$$

which yields the expected result that $(bv_p)^\alpha = a_q$. □

Theorem 4.3.16. (cf. [30, Theorem 3.6] and [66, Theorem 4.2]) *Define the sets* d_q *and* d *by*

$$\begin{aligned} d_q &:= \left\{ a = (a_k) \in \omega : \sup_{n \in \mathbb{N}} \sum_k \left| \sum_{j=k}^{n} a_j \right|^q < \infty \right\}, \\ d &:= \left\{ a = (a_k) \in \omega : \lim_{n \to \infty} \sum_k \left| \sum_{j=k}^{n} a_j \right| = \sum_k \left| \sum_{j=k}^{\infty} a_j \right| < \infty \right\}. \end{aligned}$$

Then, $(bv_p)^\beta := \begin{cases} cs & , \quad 0 < p \le 1, \\ d_q & , \quad 1 < p < \infty, \\ d & , \quad p = \infty. \end{cases}$

Proof. Let $x = (x_k) \in bv_p$ with $1 < p < \infty$ and $a = (a_k) \in \omega$. Consider the equality

$$\sum_{k=0}^{n} a_k x_k := \sum_{k=0}^{n} a_k \sum_{j=0}^{k} y_j := \sum_{k=0}^{n} \sum_{j=k}^{n} a_j y_k := (By)_n, \tag{4.3.16}$$

where $B = (b_{nk})$ defined by

$$b_{nk} := \begin{cases} \sum_{j=k}^{n} a_j & , \quad 0 \le k \le n, \\ 0 & , \quad k > n, \end{cases}$$

for all $k, n \in \mathbb{N}$. Thus, we deduce from Part (ii) of Lemma 4.3.13 with (4.3.16) that $ax = (a_k x_k) \in cs$ whenever $x = (x_k) \in bv_p$ if and only if $By \in c$ whenever $y = (y_k) \in \ell_p$. Thus, $(a_k) \in cs$ and $(a_k) \in d_q$ by (3.3.3) and (4.3.14), respectively. Nevertheless the inclusion $d_q \subset cs$ holds, thus one can see that $(a_k) \in d_q$, which gives us $(bv_p)^\beta = d_q$.

It is obvious that one can also prove the cases $0 < p \le 1$ and $p = \infty$ by using the technique in proving the case $1 < p < \infty$ with Part (i) of Lemma 4.3.13 and Schur's theorem, respectively, instead of Part (ii) of Lemma 4.3.13. So, we leave the detailed proof of these cases to the reader. □

Now, we can give the following theorem, determining the γ-dual of the space bv_p, without proof which is proved in the similar way used in the proof of Theorem 4.3.16 with Lemma 4.3.14 instead of Lemma 4.3.13:

Theorem 4.3.17. (cf. [30, Theorem 3.7] and [66, Theorem 4.3]) *Define the set* d_1 *by*

$$d_1 := \left\{ a = (a_k) \in \omega : \sup_{k,n\in\mathbb{N}} \sum_k \left| \sum_{j=k}^{n} a_j \right| < \infty \right\}.$$

Then, $(bv_p)^{\gamma} := \begin{cases} d_1 & , \quad 0 < p \leq 1, \\ d_q & , \quad 1 < p \leq \infty. \end{cases}$

Before giving our corollary on the monotonicity of the space bv_p, we give the following lemma concerning with the perfectness, normality and monotonicity of a sequence space (see [211, p. 52]):

Lemma 4.3.18. *Let* λ *be a sequence space. Then, the following statements hold:*

(i) λ *is monotone if and only if* $m_0\lambda \subset \lambda$, *where* m_0 *is the span of the set of all sequences of zeros and ones and* $m_0\lambda = \{ax = (a_k x_k) : a \in m_0,\ x \in \lambda\}$.

(ii) λ *is perfect* $\Rightarrow$ λ *is normal* $\Rightarrow$ λ *is monotone.*

(iii) λ *is normal* $\Rightarrow \lambda^{\alpha} = \lambda^{\gamma}$.

(iv) λ *is monotone* $\Rightarrow \lambda^{\alpha} = \lambda^{\beta}$.

Combining Lemma 4.3.18 and Theorems 4.3.15–4.3.17, we have the following:

Corollary 4.3.19. [66, Corollary 4.1] *The space* bv_p *is not monotone and so it is neither normal nor perfect.*

4.3.3 Certain Matrix Mappings Related to the Sequence Space bv_p

In this subsection, we desire to characterize the matrix mappings from the sequence space bv_p to some known sequence spaces and to itself. We directly prove the theorems characterizing the classes $(bv_p : \ell_\infty)$, $(bv_\infty : \ell_p)$ and $(bv_p : \ell_1)$, and derive some other characterizations from them by using the suitable relation between the concerning matrix classes.

Theorem 4.3.20. [66, Theorem 5.1] *Let* $1 < p < \infty$. *Then,* $A = (a_{nk}) \in (bv_p : \ell_\infty)$ *if and only if*

$$\sup_{m\in\mathbb{N}} \sum_k \left| \sum_{j=k}^{m} a_{nj} \right|^q < \infty \quad \text{for each } n \in \mathbb{N}, \tag{4.3.17}$$

$$\sup_{n\in\mathbb{N}} \sum_k |\widetilde{a}_{nk}|^q < \infty \tag{4.3.18}$$

where $\widetilde{a}_{nk} := \sum_{j=k}^{\infty} a_{nj}$.

Proof. Let $A = (a_{nk}) \in (bv_p : \ell_\infty)$ and $1 < p < \infty$. Then, Ax exists and is in ℓ_∞ for all $x \in bv_p$. This leads us to the fact that $(a_{nk})_{k\in\mathbb{N}} \in d_q$ for all $n \in \mathbb{N}$ which shows the necessity of (4.3.17).

Suppose that the sequences $x = (x_k)$ and $y = (y_k)$ are connected with the relation (4.3.1). Then, since $x_k = \sum_{j=0}^{k} \left(\Delta^{(1)}x\right)_j$ holds for every $k \in \mathbb{N}$, we obtain the equality

$$\sum_k a_{nk}x_k = \sum_k a_{nk}\sum_{j=0}^{k}\left(\Delta^{(1)}x\right)_j = \sum_j\sum_{k=j}^{\infty} a_{nk}\left(\Delta^{(1)}x\right)_j = \sum_j \widetilde{a}_{nj}y_j \tag{4.3.19}$$

for all $n \in \mathbb{N}$. Since bv_p and ℓ_∞ are BK-spaces, there exists some real constant $K > 0$ by Lemma 3.1.1 such that $\|Ax\|_\infty \leq K\|x\|_{bv_p}$, for all $x \in bv_p$. Therefore, by the Hölder's inequality we derive from (4.3.19) that

$$\begin{aligned}\frac{\|Ax\|_\infty}{\|y\|_p} = \sup_{n\in\mathbb{N}}\frac{\left|\sum_k \widetilde{a}_{nk}y_k\right|}{\|y\|_p} &\leq \sup_{n\in\mathbb{N}}\frac{\left(\sum_k|\widetilde{a}_{nk}|^q\right)^{1/q}\left(\sum_k|y_k|^p\right)^{1/p}}{\|y\|_p} \\ &= \sup_{n\in\mathbb{N}}\left(\sum_k|\widetilde{a}_{nk}|^q\right)^{1/q} < \infty\end{aligned}$$

for all $x \in bv_p$ and this proves the necessity of (4.3.18).

Conversely, suppose the conditions (4.3.17) and (4.3.18) hold, and take any $x = (x_k) \in bv_p$. Then, the sequence $(a_{nk})_{k\in\mathbb{N}} \in d_q$ for all $n \in \mathbb{N}$ and this implies the existence of A-transform of x. Therefore, taking into account the fact $y = (y_k) \in \ell_p$ by the isomorphism $bv_p \cong \ell_p$, we again obtain by applying the Hölder's inequality to (4.3.19) that

$$\|Ax\|_\infty = \sup_{n\in\mathbb{N}}\left|\sum_k \widetilde{a}_{nk}y_k\right| \leq \sup_{n\in\mathbb{N}}\left(\sum_k|\widetilde{a}_{nk}|^q\right)^{1/q}\left(\sum_k|y_k|^p\right)^{1/p} < \infty,$$

which means that $A \in (bv_p : \ell_\infty)$.

This step concludes the proof. □

We give a lemma concerning the characterization of the class $(\ell_\infty : \ell_p)$, which is needed in proving the next theorem and is due to Stieglitz and Tietz [368].

Lemma 4.3.21. $A = (a_{nk}) \in (\ell_\infty : \ell_p)$ *if and only if*

$$\sup_{F\in\mathcal{F}}\sum_n\left|\sum_{k\in F}a_{nk}\right|^p < \infty, \quad (1 < p < \infty).$$

Theorem 4.3.22. [66, Theorem 5.2] $A = (a_{nk}) \in (bv_\infty : \ell_p)$ *if and only if*

$$\lim_{m\to\infty}\sum_k\left|\sum_{j=k}^{m}a_{nj}\right| = \sum_k|\widetilde{a}_{nk}| \text{ for each } n \in \mathbb{N}, \tag{4.3.20}$$

$$\sup_{F\in\mathcal{F}}\sum_n\left|\sum_{k\in F}\widetilde{a}_{nk}\right|^p < \infty, \quad (1 < p < \infty), \tag{4.3.21}$$

$$\sup_{n\in\mathbb{N}}\sum_k|\widetilde{a}_{nk}| < \infty, \quad (p = \infty) \tag{4.3.22}$$

where $\widetilde{a}_{nk} = \sum_{j=k}^{\infty}a_{nj}$.

Proof. Let $A = (a_{nk}) \in (bv_\infty : \ell_p)$ and $1 < p \le \infty$. Then, Ax exists and is in ℓ_p for every $x \in bv_\infty$. This leads us to the necessity of the conditions (4.3.21) and (4.3.22) with the sequences x_F and $x = (x_k)$, which are defined by

$$x_F := \sum_{k \in F} b^{(k)} \text{ for each } F \in \mathcal{F} \text{ and } x_k = k+1 \text{ for all } k \in \mathbb{N},$$

respectively, where $b^{(k)}$ is defined by (4.3.3). The necessity of the condition (4.3.20) in the cases both $1 < p < \infty$ and $p = \infty$ is immediate, since $(a_{nk})_{k \in \mathbb{N}} \in d$ for all $n \in \mathbb{N}$.

Conversely, suppose the conditions (4.3.20) and (4.3.21) hold, and take any $x \in bv_\infty$. Then, Ax exists. Since $x \in bv_\infty$ if and only if $y \in \ell_\infty$, we therefore have from (4.3.19) by Lemma 4.3.21 that $Ax \in \ell_p$ whenever $By \in \ell_p$, where $b_{nk} = \widetilde{a}_{nk}$ for all $k, n \in \mathbb{N}$. This proves the sufficiency of the conditions (4.3.20) and (4.3.21).

Let us finally suppose that the conditions (4.3.20) and (4.3.22) hold, and take any $x \in bv_\infty$. Then, Ax exists and again by (4.3.19) one can easily see that

$$\|Ax\|_\infty := \sup_{n \in \mathbb{N}} \left| \sum_k \widetilde{a}_{nk} y_k \right| \le \|y\|_\infty \left(\sup_{n \in \mathbb{N}} \sum_k |\widetilde{a}_{nk}| \right) < \infty,$$

which shows the sufficiency of (4.3.20) and (4.3.22).

This completes the proof. □

Now, we may give the theorem characterizing the class $(bv_p : \ell_1)$, whose proof is similar to the proof of Theorem 4.3.20, and so we omit details.

Theorem 4.3.23. [66, Theorem 5.3] $A = (a_{nk}) \in (bv_p : \ell_1)$ *if and only if (4.3.17) holds, and*

$$\sup_{F \in \mathcal{F}} \sum_k \left| \sum_{n \in F} \widetilde{a}_{nk} \right|^q < \infty, \quad (1 < p < \infty). \tag{4.3.23}$$

Lemma 4.3.24. [66, Lemma 5.3] *Let λ, μ be any two sequence spaces, A be an infinite matrix and B a triangle matrix. Then, $A \in (\lambda : \mu_B)$ if and only if $BA \in (\lambda : \mu)$.*

Proof. Let B be a triangle matrix, A be an infinite matrix and $x = (x_k) \in \omega_A$. Then, we have by Theorem 1.1.4 of Wilansky [396, p. 8] that $B(Ax) = (BA)x$. This leads us to the desired consequence that $Ax \in \mu_B$ whenever $x \in \lambda$ if and only if $(BA)x \in \mu$ whenever $x \in \lambda$, which is what we wished to prove. □

4.4 DOMAIN OF GENERALIZED DIFFERENCE MATRIX $B(R, S)$

In this section, following Kirişçi and Başar [220] and Başar and Kirişçi [74], we concentrate on the domain of generalized difference matrix $B(r, s)$ in the classical sequence spaces ℓ_∞, c, c_0 and ℓ_p, and in the space f of almost convergent sequences.

4.4.1 Domain of Generalized Difference Matrix $B(r,s)$ in the Classical Sequence Spaces

In this subsection, we introduce the sequence spaces $\widehat{\ell}_\infty$, $\widehat{c}$, $\widehat{c}_0$ and $\widehat{\ell}_p$, and determine the β- and γ-duals of the spaces and give the Schauder bases for the spaces $\widehat{c}$, $\widehat{c}_0$ and $\widehat{\ell}_p$, and examine some topological properties of the spaces $\widehat{c}_0$, $\widehat{\ell}_1$ and $\widehat{\ell}_p$. Additionally, we characterize the classes $(\widehat{\lambda}_1 : \lambda_2)$ and $(\widehat{\lambda}_1 : \widehat{\lambda}_2)$ of infinite matrices, where $\lambda_1 \in \{\ell_\infty, c, c_0, \ell_p, \ell_1\}$ and $\lambda_2 \in \{\ell_\infty, c, c_0, \ell_1\}$.

We define the sequence spaces $\widehat{\ell}_\infty$, $\widehat{c}$, $\widehat{c}_0$ and $\widehat{\ell}_p$ as the set of all sequences whose $B(r,s)$-transforms are in the spaces ℓ_∞, c, c_0 and ℓ_p, respectively, i.e.,

$$\begin{aligned}
\widehat{\ell}_\infty &:= \left\{x = (x_k) \in \omega : \sup_{k\in\mathbb{N}} |sx_{k-1} + rx_k| < \infty\right\}, \\
\widehat{c} &:= \left\{x = (x_k) \in \omega : \exists l \in \mathbb{C} \ni \lim_{k\to\infty} |sx_{k-1} + rx_k - l| = 0\right\}, \\
\widehat{c}_0 &:= \left\{x = (x_k) \in \omega : \lim_{k\to\infty} |sx_{k-1} + rx_k| = 0\right\}, \\
\widehat{\ell}_p &:= \left\{x = (x_k) \in \omega : \sum_k |sx_{k-1} + rx_k|^p < \infty\right\}.
\end{aligned}$$

With the notation of (4.1.1), we can redefine the spaces $\widehat{\ell}_\infty$, $\widehat{c}$, $\widehat{c}_0$ and $\widehat{\ell}_p$ by

$$\widehat{\ell}_\infty := (\ell_\infty)_{B(r,s)}, \quad \widehat{c} := c_{B(r,s)}, \quad \widehat{c}_0 := (c_0)_{B(r,s)}, \quad \widehat{\ell}_p := (\ell_p)_{B(r,s)}.$$

Define the sequence $y = (y_k)$ by the $B(r,s)$-transform of a sequence $x = (x_k)$, i.e.,

$$y_k := sx_{k-1} + rx_k \quad \text{for all} \quad k \in \mathbb{N}. \tag{4.4.1}$$

Since the spaces λ and $\lambda_{B(r,s)}$ are norm isomorphic one can easily observe that $x = (x_k) \in \lambda_{B(r,s)}$ if and only if $y = (y_k) \in \lambda$, where the sequences $x = (x_k)$ and $y = (y_k)$ are connected with the relation (4.4.1).

Prior to quoting the lemmas which are needed for deriving some consequences given in Corollary 4.4.3, we give an inclusion theorem related to these new spaces.

Theorem 4.4.1. [220, Theorem 3.1] *Let $\lambda \in \{\ell_\infty, c, c_0, \ell_p\}$ and $B = B(r,s)$. Then, the following statements hold:*

(i) $\lambda = \lambda_B$, *if* $|s/r| < 1$.

(ii) $\lambda \subset \lambda_B$ *is strict, if* $|s/r| \geq 1$.

Proof. Let $\lambda \in \{\ell_\infty, c, c_0, \ell_1\}$ and $B = B(r,s)$. Since the matrix B satisfies the conditions:

$$\begin{aligned}
&\sup_{n\in\mathbb{N}} \sum_k |b_{nk}| = |r| + |s|, \\
&\lim_{n\to\infty} b_{nk} = 0, \\
&\lim_{n\to\infty} \sum_k b_{nk} = r + s, \\
&\sup_{k\in\mathbb{N}} \sum_n |b_{nk}| = |r| + |s|,
\end{aligned}$$

$B \in (\lambda : \lambda)$. For any sequence $x \in \lambda$, $Bx \in \lambda$, hence $x \in \lambda_B$. This shows that $\lambda \subset \lambda_B$.

(i) Let $|s/r| < 1$. Since the inverse matrix $B^{-1} := (b_{nk}^{-1})$ of the matrix B also satisfies the conditions:

$$\sup_{n\in\mathbb{N}} \sum_k \left|b_{nk}^{-1}\right| = \left|\frac{1}{r}\right| \sum_k \left|\frac{s}{r}\right|^k < \infty,$$

$$\lim_{n\to\infty} b_{nk}^{-1} = \lim_{n\to\infty} \frac{1}{r}\left(\frac{-s}{r}\right)^n = 0,$$

$$\lim_{n\to\infty} \sum_k b_{nk}^{-1} = \lim_{n\to\infty} \sum_{k=0}^{n} \left(\frac{-s}{r}\right)^k \text{ exists,}$$

$$\sup_{n\in\mathbb{N}} \sum_k \left|b_{nk}^{-1}\right| = \left|\frac{1}{r}\right| \sum_k \left|\frac{s}{r}\right|^k < \infty,$$

$B^{-1} \in (\lambda : \lambda)$. Therefore, if $x \in \lambda_B$ then $y = Bx \in \lambda$ and $x := B^{-1}y \in \lambda$. Thus the opposite inclusion $\lambda_B \subset \lambda$ is just proved. This completes the proof of Part (i).

(ii) Let us consider the sequences $u^1 := \{(-s/r)^n/r\}$, $u^2 := (n/r)$, $u^3 := \{(-1)^n(n+1)\}$ and $u^4 := \{[1+(-1)^n]/2\}$.

If $|s/r| > 1$, then $Bu^1 := e^{(0)} = (1,0,0,\ldots) \in \lambda$. Hence, $u^1 \in \lambda_B \setminus \lambda$.

Suppose that $|s/r| = 1$.

(a) If $\lambda \in \{c_0, \ell_p\}$, then $u^1 \in \lambda_B \setminus \lambda$.

(b) Let $\lambda \in \{\ell_\infty, c\}$. If $s = -r$, then $Bu^2 = e \in \lambda$. Hence, $u^2 \in \lambda_B \setminus \lambda$. If $s = r$, then $Bu^3 = \{r(-1)^n\} \in \ell_\infty$, $Bu^4 = (r,r,r,\ldots) \in c$. Hence, $u^3 \in (\ell_\infty)_B \setminus \ell_\infty$ and $u^4 \in c_B \setminus c$.

This step completes the proof. □

We read the following useful results from Stieglitz and Tietz [368]:

$$\sup_{n\in\mathbb{N}} \sum_k |a_{nk}|^q < \infty, \tag{4.4.2}$$

$$\sup_{k,n\in\mathbb{N}} |a_{nk}| < \infty, \tag{4.4.3}$$

$$\lim_{n\to\infty} a_{nk} = \alpha_k \quad \text{for each} \quad k \in \mathbb{N}, \tag{4.4.4}$$

$$\lim_{n\to\infty} \sum_k |a_{nk}| = \sum_k |\alpha_k|, \tag{4.4.5}$$

$$\lim_{n\to\infty} \sum_k a_{nk} = \alpha. \tag{4.4.6}$$

Lemma 4.4.2. *The necessary and sufficient conditions for $A \in (\lambda : \mu)$ when $\lambda \in \{\ell_\infty, c, c_0, \ell_1, \ell_p\}$ and $\mu \in \{\ell_\infty, c\}$ can be read from Table 4.2:*

TABLE 4.2 The characterization of the class $(\lambda_1 : \lambda_2)$ with $\lambda_1 \in \{\ell_\infty, c, c_0, \ell_p, \ell_1\}$ and $\lambda_2 \in \{\ell_\infty, c\}$, where

From / To	ℓ_∞	c	c_0	ℓ_p	ℓ_1
ℓ_∞	**1.**	**1.**	**1.**	**2.**	**3.**
c	**4.**	**5.**	**6.**	**7.**	**8.**

1. *(4.4.2) with $q = 1$.*
2. *(4.4.2).*
3. *(4.4.3).*
4. *(4.4.4) and (4.4.5).*
5. *(4.4.2) with $q = 1$, (4.4.4) and (4.4.6).*
6. *(4.4.2) with $q = 1$ and (4.4.4).*
7. *(4.4.2) and (4.4.4).*
8. *(4.4.3) and (4.4.4).*

Basic Lemma. [29, Theorem 3.1] *Let $C = (c_{nk})$ be defined via a sequence $a = (a_k) \in \omega$ and the inverse matrix $V = (v_{nk})$ of the triangle matrix $U = (u_{nk})$ by*

$$c_{nk} := \begin{cases} \sum_{j=k}^{n} a_j v_{jk} & , \quad 0 \le k \le n, \\ 0 & , \quad k > n, \end{cases}$$

for all $k, n \in \mathbb{N}$. Then, the following statements hold:

$$\begin{aligned} \lambda_U^{\gamma} &:= \{a = (a_k) \in \omega : C \in (\lambda : \ell_{\infty})\}, \\ \lambda_U^{\beta} &:= \{a = (a_k) \in \omega : C \in (\lambda : c)\}. \end{aligned}$$

Combining Lemma 4.4.2 with Basic Lemma, we have the following:

Corollary 4.4.3. [220, Corollary 3.4] *Define the sets $d_1(r,s)$, $d_2(r,s)$, $d_3(r,s)$, $d_4(r,s)$ and $d_5(r,s)$ by*

$$\begin{aligned} d_1(r,s) &:= \left\{a = (a_k) \in \omega : \sup_{n \in \mathbb{N}} \sum_{k=0}^{n} |T_k^n|^q < \infty \right\}, \\ d_2(r,s) &:= \left\{a = (a_k) \in \omega : \lim_{n \to \infty} \sum_{j=k}^{n} \left(\frac{-s}{r}\right)^{j-k} a_j \text{ exists} \right\}, \\ d_3(r,s) &:= \left\{a = (a_k) \in \omega : \lim_{n \to \infty} \sum_{k=0}^{n} |T_k^n| = \sum_{k=0}^{\infty} \left|\lim_{n \to \infty} T_k^n \right| \right\}, \\ d_4(r,s) &:= \left\{a = (a_k) \in \omega : \lim_{n \to \infty} \sum_{k=0}^{n} \left[\frac{1 - (\frac{-s}{r})^{k+1}}{1 + \frac{s}{r}}\right] a_k \text{ exists} \right\}, \\ d_5(r,s) &:= \left\{a = (a_k) \in \omega : \sup_{k,n \in \mathbb{N}} \left|\sum_{j=k}^{n} \left(\frac{-s}{r}\right)^{j-k} a_j \right| < \infty \right\}, \end{aligned}$$

where $T_k^n = \sum_{j=k}^{n} \left(\frac{-s}{r}\right)^{j-k} \frac{a_j}{r}$. Then, the following statements hold:

(i) $\widehat{\ell_{\infty}^{\gamma}} := \widehat{c}^{\gamma} := \widehat{c_0^{\gamma}} := d_1(r,s)$ *with $q = 1$.*

(ii) $\widehat{\ell_p^{\gamma}} := d_1(r,s)$.

(iii) $\widehat{\ell_1^{\gamma}} := d_5(r,s)$.

(iv) $\widehat{\ell_{\infty}^{\beta}} := d_2(r,s) \cap d_3(r,s)$.

(v) $\widehat{c}^{\beta} := d_1(r,s) \cap d_2(r,s) \cap d_4(r,s)$ *with $q = 1$.*

(vi) $\widehat{c}_0^{\beta} := d_1(r,s) \cap d_2(r,s)$ *with* $q = 1$.

(vii) $\widehat{\ell}_p^{\beta} := d_1(r,s) \cap d_2(r,s)$.

(viii) $\widehat{\ell}_1^{\beta} := d_2(r,s) \cap d_5(r,s)$.

Since, it is known that the matrix domain λ_A of a normed sequence space λ has a basis if and only if λ has a basis whenever $A = (a_{nk})$ is a triangle (cf. [206, Remark 2.4]), we have the following:

Corollary 4.4.4. [220, Corollary 3.5] *Define the sequences* $z = (z_n)$ *and* $b^{(k)}(r,s) = \{b_n^{(k)}(r,s)\}_{n\in\mathbb{N}}$ *for every fixed* $k \in \mathbb{N}$ *by*

$$z_n := \frac{1}{r}\sum_{k=0}^{n}\left(\frac{-s}{r}\right)^k \quad \text{and} \quad b_n^{(k)}(r,s) := \begin{cases} 0 & , \quad n < k, \\ \frac{1}{r}(\frac{-s}{r})^n & , \quad n \geq k. \end{cases}$$

Then, the following statements hold:

(a) *The sequence* $\{b^{(k)}(r,s)\}_{k\in\mathbb{N}}$ *is a basis for the spaces* $\widehat{c}_0$ *and* $\widehat{\ell}_p$, *and any* x *in* $\widehat{c}_0$ *or in* $\widehat{\ell}_p$ *has a unique representation of the form*

$$x := \sum_k \alpha_k(r) b^{(k)}(r,s),$$

where $\alpha_k(r) := \{B(r,s)x\}_k$ *for all* $k \in \mathbb{N}$.

(b) *The set* $\left\{z, b^{(k)}(r,s)\right\}$ *is a basis for the space* $\widehat{c}$, *and any* x *in* $\widehat{c}$ *has a unique representation of the form*

$$x := lz + \sum_k [\alpha_k(r) - l] b^{(k)}(r,s),$$

where $l := \lim_{k\to\infty} \{B(r,s)x\}_k$.

Lemma 4.4.5. [29, Theorem 2.1] *Let* λ, μ *be* BK*-spaces and* $C_\mu^U = (c_{nk})$ *be defined via the sequence* $\alpha = (\alpha_k) \in \mu$ *and the triangle matrix* $U = (u_{nk})$ *by*

$$c_{nk} := \sum_{j=k}^{n} \alpha_j u_{nj} v_{jk}$$

for all $k, n \in \mathbb{N}$. *Then, the domain of the matrix* U *in the sequence space* λ *has the property*

(i) KB *if and only if* $C_{\ell_1}^U \in (\lambda : \lambda)$.

(ii) AB *if and only if* $C_{bv_0}^U \in (\lambda : \lambda)$.

(iii) *monotone if and only if* $C_{m_0}^U \in (\lambda : \lambda)$.

(iv) *solid if and only if* $C_{\ell_\infty}^U \in (\lambda : \lambda)$.

From Lemma 4.4.5, we have

Corollary 4.4.6. [220, Corollary 3.7] *If $|s/r| = 1$, then $\widehat{\ell_1}$ has KB- and AB-properties.*

Lemma 4.4.7. [29, Theorem 2.2] *Let λ be BK-space which has AK-property, U be a triangle matrix and $\lambda_U \supset \phi$. Then, the sequence space λ_U has AD-property if and only if the fact $tU = \theta$ for $t \in \lambda^\beta$ implies the fact $t = \theta$.*

Since c_0 and ℓ_p have AK-property, we can employ Lemma 4.4.7 for the matrix $U = B(r,s)$. Then, we have

Corollary 4.4.8. [220, Corollary 3.9] *$\widehat{c_0}$ and $\widehat{\ell_p}$ $(p > 1)$ have AD-property if and only if $|s/r| \le 1$.*

4.4.2 Some Matrix Transformations Related to the Sequence Spaces $\widehat{\ell_\infty}$, $\widehat{c}$, $\widehat{c_0}$ and $\widehat{\ell_1}$

In this subsection, we characterize some classes of infinite matrices related to the new sequence spaces $\widehat{\ell_\infty}$, $\widehat{c}$, $\widehat{c_0}$ and $\widehat{\ell_1}$.

Theorem 4.4.9. [220, Theorem 4.1] *Let λ be an FK-space, U be a triangle, V be its inverse and μ be arbitrary subset of ω. Then, we have $A = (a_{nk}) \in (\lambda_U : \mu)$ if and only if*

$$C^{(n)} = (c_{mk}^{(n)}) \in (\lambda : c) \quad \text{for all} \quad n \in \mathbb{N}, \tag{4.4.7}$$
$$C = (c_{nk}) \in (\lambda : \mu), \tag{4.4.8}$$

where $c_{mk}^{(n)} := \begin{cases} \sum_{j=k}^{m} a_{nj}v_{jk} & , \quad 0 \le k \le m, \\ 0 & , \quad k > m \end{cases}$ *and* $c_{nk} := \sum_{j=k}^{\infty} a_{nj}v_{jk}$ *for all* $k, m, n \in \mathbb{N}$.

Proof. Let $A = (a_{nk}) \in (\lambda_U : \mu)$ and take $x \in \lambda_U$. Then, we obtain the equality

$$\begin{aligned} \sum_{k=0}^{m} a_{nk}x_k &= \sum_{k=0}^{m} a_{nk}\left(\sum_{j=0}^{k} v_{kj}y_j\right) \\ &= \sum_{k=0}^{m}\left(\sum_{j=k}^{m} a_{nj}v_{jk}\right) y_k = \sum_{k=0}^{m} c_{nk}^{(n)} y_k \end{aligned} \tag{4.4.9}$$

for all $m, n \in \mathbb{N}$. Since Ax exists, $C^{(n)}$ must belong to the class $(\lambda : c)$. Letting $m \to \infty$ in the equality (4.4.9), we have $Ax := Cy$. Since $Ax \in \mu$, then $Cy \in \mu$, i.e., $C \in (\lambda : \mu)$.

Conversely, let (4.4.7) and (4.4.8) hold and take $x \in \lambda_U$. Then, we have $(c_{nk})_{k\in\mathbb{N}} \in \lambda^\beta$, which gives together with (4.4.7) that $(a_{nk})_{k\in\mathbb{N}} \in \lambda_U^\beta$ for all $n \in \mathbb{N}$. Hence, Ax exists. Therefore, we obtain from the equality (4.4.9) as $m \to \infty$ that $Ax := Cy$ and this shows that $A \in (\lambda_U : \mu)$. □

Now, we list the following conditions:

$$\sup_{m\in\mathbb{N}}\sum_{k=0}^{m}\left|\frac{1}{r}\sum_{j=k}^{m}\left(\frac{-s}{r}\right)^{j-k}a_{nj}\right|^{q}<\infty \tag{4.4.10}$$

$$\lim_{m\to\infty}\frac{1}{r}\sum_{j=k}^{m}\left(\frac{-s}{r}\right)^{j-k}a_{nj}=c_{nk} \tag{4.4.11}$$

$$\lim_{m\to\infty}\sum_{k=0}^{m}\left|\frac{1}{r}\sum_{j=k}^{m}\left(\frac{-s}{r}\right)^{j-k}a_{nj}\right|=\sum_{k}|c_{nk}| \text{ for each } n\in\mathbb{N} \tag{4.4.12}$$

$$\lim_{m\to\infty}\sum_{k=0}^{m}\frac{1}{r}\left[\frac{1-(\frac{-s}{r})^{k+1}}{1+\frac{s}{r}}\right]a_{nk}=\alpha_n \text{ for each } n\in\mathbb{N} \tag{4.4.13}$$

$$\sup_{k,m\in\mathbb{N}}\left|\frac{1}{r}\sum_{j=k}^{m}\left(\frac{-s}{r}\right)^{j-k}a_{nj}\right|<\infty \tag{4.4.14}$$

$$\sup_{n\in\mathbb{N}}\sum_{k}|c_{nk}|^{q}<\infty \tag{4.4.15}$$

$$\lim_{n\to\infty}c_{nk}=\beta_k \tag{4.4.16}$$

$$\lim_{n\to\infty}\sum_{k}|c_{nk}|=\sum_{k}|\beta_k| \tag{4.4.17}$$

$$\lim_{n\to\infty}\sum_{k}c_{nk}=\beta \tag{4.4.18}$$

$$\sup_{k,n\in\mathbb{N}}|c_{nk}|<\infty \tag{4.4.19}$$

$$\sup_{k\in\mathbb{N}}\sum_{n}|c_{nk}|<\infty \tag{4.4.20}$$

$$\lim_{n\to\infty}\sum_{k}c_{nk}=0 \tag{4.4.21}$$

$$\sup_{N,K\in\mathcal{F}}\left|\sum_{n\in N}\sum_{k\in K}c_{nk}\right|<\infty \tag{4.4.22}$$

$$\sup_{N\in\mathcal{F}}\sum_{k}\left|\sum_{n\in N}c_{nk}\right|^{q}<\infty \tag{4.4.23}$$

We have from Theorem 4.4.9:

Corollary 4.4.10. [220, Corollary 4.2] *The necessary and sufficient conditions for $A\in(\lambda:\mu)$ when $\lambda\in\{\widehat{\ell_\infty},\widehat{c},\widehat{c_0},\widehat{\ell_p},\widehat{\ell_1}\}$ and $\mu\in\{\ell_\infty,c,c_0,\ell_1\}$ can be read from Table 4.3:*

TABLE 4.3 The characterization of the class $(\widehat{\lambda}:\mu)$ with $\lambda\in\{\ell_\infty,c,c_0,\ell_p,\ell_1\}$ and $\mu\in\{\ell_\infty,c,c_0,\ell_1\}$, where

From / To	$\widehat{\ell_\infty}$	$\widehat{c}$	$\widehat{c_0}$	$\widehat{\ell_p}$	$\widehat{\ell_1}$
ℓ_∞	**1.**	**2.**	**3.**	**4.**	**5.**
c	**6.**	**7.**	**8.**	**9.**	**10.**
c_0	**11.**	**12.**	**13.**	**14.**	**15.**
ℓ_1	**16.**	**17.**	**18.**	**19.**	**20.**

1. *(4.4.11), (4.4.12) and (4.4.15) with* $q = 1$.
2. *(4.4.11), (4.4.13) and (4.4.10), (4.4.15) with* $q = 1$.
3. *(4.4.11) and (4.4.10), (4.4.15) with* $q = 1$.
4. *(4.4.10), (4.4.11) and (4.4.15).*
5. *(4.4.11), (4.4.14) and (4.4.19).*
6. *(4.4.11), (4.4.12), (4.4.16) and (4.4.17).*
7. *(4.4.11), (4.4.13), (4.4.16), (4.4.18) and (4.4.10), (4.4.15) with* $q = 1$.
8. *(4.4.11), (4.4.16) and (4.4.10), (4.4.15) with* $q = 1$.
9. *(4.4.10), (4.4.11), (4.4.15) and (4.4.16).*
10. *(4.4.11), (4.4.14), (4.4.16) and (4.4.19).*
11. *(4.4.11), (4.4.12) and (4.4.21).*
12. *(4.4.11), (4.4.13), (4.4.16) with* $\beta_k = 0$ *and (4.4.18) with* $\beta = 0$, *and (4.4.10), (4.4.15) with* $q = 1$.
13. *(4.4.11), (4.4.16) with* $\beta_k = 0$ *and (4.4.10), (4.4.15) with* $q = 1$.
14. *(4.4.10), (4.4.11), (4.4.15) and (4.4.16) with* $\beta_k = 0$.
15. *(4.4.11), (4.4.14), (4.4.16) with* $\beta_k = 0$ *and (4.4.19).*
16. *(4.4.11), (4.4.12) and (4.4.22).*
17. *(4.4.10) with* $q = 1$, *(4.4.11), (4.4.13) and (4.4.22).*
18. *(4.4.10) with* $q = 1$, *(4.4.11) and (4.4.22).*
19. *(4.4.10), (4.4.11) and (4.4.23).*
20. *(4.4.11), (4.4.14) and (4.4.20).*

We should finally note that, if a_{nk} is replaced by $ra_{nk} + sa_{n-1,k}$ for all $k, n \in \mathbb{N}$ in Corollary 4.4.10, then one can derive the characterization of the class $(\widehat{\lambda} : \widehat{\mu})$ from Lemma 4.3.24 with $B = B(r, s)$.

4.4.3 Domain of Generalized Difference Matrix $B(r,s)$ in the Spaces f_0 and f

In this subsection, we introduce the sequence spaces $\widehat{f_0}$ and $\widehat{f}$ as the domain of generalized difference matrix $B(r, s)$ in the spaces f_0 and f of almost null and almost convergent sequences, and determine the β- and γ-duals of the spaces f, fs and $\widehat{f}$. Furthermore, we prove two basic results on the space f and characterize the classes $(\widehat{f} : \mu)$ and $(\mu : \widehat{f})$ of infinite matrices, and also give the characterizations of some other classes as an application of those main results, where μ is any given sequence space.

By fs, we denote the space of all almost convergent series, i.e.,

$$fs := \left\{ x = (x_k) \in \omega : \exists \alpha \in \mathbb{C} \ni \lim_{m\to\infty} \sum_{k=0}^{m} \sum_{j=0}^{n+k} \frac{x_j}{m+1} = \alpha \text{ uniformly in } n \right\}.$$

It is immediate that the inclusion relations $c \subset f \subset \ell_\infty$ hold. Consider the sequences $x = (x_k)$ and $u = (u_k)$ defined for all $k \in \mathbb{N}$ by

$$x_k := \frac{1 - (-1)^k}{2} \quad \text{and} \quad u := (0, \ldots, 0, 1, \ldots, 1, 0, \ldots, 0, 1, \ldots, 1, \ldots),$$

where the blocks 0's are increasing by factors of 100 and the blocks of 1's are increasing by factors of 10 (cf. Miller and Orhan [277]). Then, the sequence $x = (x_k)$ is not convergent but it is almost convergent to the generalized limit $1/2$. Also, the sequence u is not almost convergent but it is bounded. This shows that the inclusions $c \subset f$ and $f \subset \ell_\infty$ are strict.

Now, following Mursaleen [286], we give the first theorem on the space f:

Theorem 4.4.11. (cf. [94, Example 7.3.2 (b)] and [74, Theorem 3.1]) *f_0 and f endowed with the norm $\|\cdot\|_\infty$ are BK-spaces.*

Proof. It is known from Ahmad et al. [10] that the space V_σ of σ-convergent sequences is a Banach space endowed with the norm

$$\|x\|_\sigma := \sup_{m,n\in\mathbb{N}} \frac{\left|x_n + x_{\sigma(n)} + x_{\sigma^2(n)} + \cdots + x_{\sigma^m(n)}\right|}{m+1}.$$

So, as a particular case, f is a Banach space endowed with the norm

$$\|x\|_f := \sup_{m,n\in\mathbb{N}} |t_{mn}(x)|.$$

The inclusion relations $c \subset f \subset \ell_\infty$ give that there exist positive reals a and b with $a < b$ such that $a\|x\|_\infty \leq \|x\|_f \leq b\|x\|_\infty$. That is to say that two norms $\|\cdot\|_\infty$ and $\|\cdot\|_f$ are equivalent. Since c and ℓ_∞ endowed with the norm $\|\cdot\|_\infty$ are BK-spaces, it follows that f endowed with the norm $\|\cdot\|_\infty$ is also BK-space.

Of course, one can show by the similar way that f_0 endowed with the norm $\|\cdot\|_\infty$ is a BK-space. So, we omit details.

This completes the proof. □

Our second basic result concerning the space f is about its non-separability:

Theorem 4.4.12. [74, Theorem 3.2] *f is a non-separable closed subspace of ℓ_∞.*

Proof. Define the set X by

$$X := \{x = (x_k) \in \omega : x_k \in \{0,1\} \text{ for every } k \in \mathbb{N}\} \subset \ell_\infty$$

and let (x_k) is given as follows;

$$x_k := \begin{cases} 1 \ , & k = n^2, \ n \in \mathbb{N}, \\ 0 \ , & k \neq n^2, \ n \in \mathbb{N}. \end{cases}$$

Then, it is obvious that X is uncountable and $f - \lim x_k = 0$.

Let $x \in X$ and $\varepsilon > 0$ be given. Then, we have for all positive integers k_0 with $1/\sqrt{k_0} < \varepsilon/2$ that

$$t_{mn}(x) = \frac{1}{m+1}\sum_{j=0}^{m} x_{n+j} \leq \frac{1}{m+1}(\sqrt{m+1}+1) \leq \frac{2}{\sqrt{k_0}} < \varepsilon$$

for each positive integer m with $m \geq k_0$ and all $n \in \mathbb{N}$. Hence, $t_{mn}(x) \to 0$ as $m \to \infty$ uniformly in n. This means that $x \in f$, which gives the inclusion $X \subset f$. Therefore, we have shown that f contains an uncountable set X of sequences of 0's and 1's. Also, since $\|x - y\|_\infty = 1$ for all $x, y \in X$ with $x \neq y$, the space $(f, \|\cdot\|_\infty)$ cannot be separable.

It is well-known that f is a proper subset of the space ℓ_∞. f is an FK–space and the FK–topology is stronger than the $\|\cdot\|_\infty$-topology. However, since the natural seminorms generating the FK–topology are obviously $\|\cdot\|_\infty$–continuous, the FK–topology of f is generated by $\|\cdot\|_\infty$. Hence, f is closed in the space ℓ_∞.

This step completes the proof. □

As a direct consequence of Theorem 4.4.12, we have

Corollary 4.4.13. [74, Corollary 3.3] *The Banach space f has no Schauder basis.*

4.4.4 The Sequence Spaces $\widehat{f_0}$ and $\widehat{f}$ Derived by the Domain of the Matrix $B(r, s)$

In this subsection, we define the sequence spaces $\widehat{f_0}$ and $\widehat{f}$, and give an isomorphism between the spaces $\widehat{f_0}$, $\widehat{f}$ and f_0, f, respectively. Finally, we determine the β- and γ-duals of the spaces fs and $\widehat{f}$.

We introduce the sequence spaces $\widehat{f_0}$ and $\widehat{f}$ as the set of all sequences, whose $B(r, s)$-transforms are in the spaces f_0 and f, respectively, i.e.,

$$\widehat{f_0} := \left\{ x = (x_k) \in \omega : \lim_{m \to \infty} \sum_{j=0}^{m} \frac{s x_{k-1+j} + r x_{k+j}}{m+1} = 0 \text{ uniformly in } k \right\},$$

$$\widehat{f} := \left\{ x = (x_k) \in \omega : \exists \alpha \in \mathbb{C} \ni \lim_{m \to \infty} \sum_{j=0}^{m} \frac{s x_{k-1+j} + r x_{k+j}}{m+1} = \alpha \text{ uniformly in } k \right\}.$$

With the notation of (4.1.1), we can redefine the spaces $\widehat{f_0}$ and $\widehat{f}$ by

$$\widehat{f_0} := (f_0)_{B(r,s)} \qquad \text{and} \qquad \widehat{f} := (f)_{B(r,s)}.$$

Now, we may give the following theorem concerning the isomorphism between the spaces $\widehat{f_0}$, $\widehat{f}$ and f_0, f, respectively:

Theorem 4.4.14. [74, Theorem 4.1] *The sequence spaces $\widehat{f_0}$ and $\widehat{f}$ are norm isomorphic to the spaces f_0 and f, respectively; i.e., $\widehat{f_0} \cong f_0$ and $\widehat{f} \cong f$.*

Proof. Since the fact $\widehat{f_0} \cong f_0$ can be similarly proved, we consider only the case $\widehat{f} \cong f$. To prove this, we should show the existence of a linear bijection between the spaces $\widehat{f}$ and f, which preserves the norms. Consider the transformation T defined, with the notation of (4.4.1), from $\widehat{f}$ to f by $x \mapsto y = Tx = B(r, s)x$. The linearity of T is clear. Further, it is trivial that $x = \theta$ whenever $Tx = \theta$ and hence T is injective.

Let us take any $y = (y_k) \in f$ and consider the sequence $x = (x_k)$ using the inverse $B^{-1}(r, s)$ defined by

$$x_k := \{B^{-1}(r, s)y\}_k = \sum_{j=0}^{k} \frac{1}{r} \left(\frac{-s}{r} \right)^j y_{k-j}, \tag{4.4.24}$$

for all $k \in \mathbb{N}$. Then, we have

$$\begin{aligned} sx_{k-1} + rx_k &= \frac{s}{r}\sum_{j=0}^{k-1}\left(\frac{-s}{r}\right)^j y_{k-1-j} + \sum_{j=0}^{k}\left(\frac{-s}{r}\right)^j y_{k-j} \\ &= \sum_{j=0}^{k-1}\left(\frac{-s}{r}\right)^j \left(\frac{s}{r}y_{k-1-j} + y_{k-j}\right) + \left(\frac{-s}{r}\right)^k y_0 = y_k, \end{aligned}$$

for all $k \in \mathbb{N}$, which leads us to the fact that

$$\begin{aligned} \lim_{m\to\infty}\frac{1}{m+1}\sum_{j=0}^{m}(sx_{k-1+j} + rx_{k+j}) &= \lim_{m\to\infty}\frac{1}{m+1}\sum_{j=0}^{m} y_{k+j} \text{ uniformly in } k \\ &= f - \lim y_k. \end{aligned}$$

This means that $x = (x_k) \in \widehat{f}$. Consequently, we see from here that T is surjective.

It is obvious that

$$\|x\|_{\widehat{f}} = \|B(r,s)x\|_f = \|y\|_\infty.$$

This shows that the transformation T preserves the norms.

Hence, T is a norm preserving linear bijection, which therefore says that the spaces $\widehat{f}$ and f are norm isomorphic, as desired. □

Since, it is known by Theorem 4.1.2 that the matrix domain λ_A of a normed sequence space λ has a basis if and only if λ has a basis whenever $A = (a_{nk})$ is a triangle and the space f has no Schauder basis by Corollary 4.4.13, we have

Corollary 4.4.15. [74, Corollary 4.2] *The space $\widehat{f}$ has no Schauder basis.*

We give the lemmas which are needed for deriving some consequences given in Theorem 4.4.21 and Theorem 4.4.22.

Lemma 4.4.16. [211, p. 52, Exercise 2.5 (i)] *Let λ, μ be the sequence spaces and $\xi \in \{\alpha, \beta, \gamma\}$. If $\lambda \subset \mu$, then $\mu^\xi \subset \lambda^\xi$.*

We have from Theorem 2.5.3, Theorem 2.5.6 and Theorem 2.5.7:

Lemma 4.4.17. $c_0^\xi = c^\xi = \ell_\infty^\xi = \ell_1$ *with* $\xi \in \{\alpha, \beta, \gamma\}$.

Lemma 4.4.18. $A = (a_{nk}) \in (f : \ell_\infty)$ *if and only if*

$$\sup_{n\in\mathbb{N}}\sum_k |a_{nk}| < \infty. \tag{4.4.25}$$

Lemma 4.4.19. (cf. Sıddıqi [363]) $A = (a_{nk}) \in (f : c)$ *if and only if (4.4.25) holds, and there are $\alpha_k, \alpha \in \mathbb{C}$ such that*

$$\lim_{n\to\infty} a_{nk} = \alpha_k \text{ for each } k \in \mathbb{N}, \tag{4.4.26}$$

$$\lim_{n\to\infty}\sum_k a_{nk} = \alpha, \tag{4.4.27}$$

$$\lim_{n\to\infty}\sum_k |\Delta(a_{nk} - \alpha_k)| = 0. \tag{4.4.28}$$

Theorem 4.4.20. [74, Theorem 4.8] *The following statements hold:*

(a) *The γ-dual of the space fs is the space bv.*

(b) *The β-dual of the space fs is the space bv_0.*

Proof. (a) Take any $a = (a_k) \in \omega$ and consider the equality

$$\sum_{k=0}^{n} a_k x_k = \sum_{k=0}^{n-1} (a_k - a_{k+1}) y_k + a_n y_n = (Dy)_n \text{ for all } n \in \mathbb{N}, \tag{4.4.29}$$

where $D = (d_{nk})$ is defined by

$$d_{nk} := \begin{cases} a_k - a_{k+1} & , \quad 0 \leq k < n, \\ a_n & , \quad k = n, \\ 0 & , \quad k > n, \end{cases} \tag{4.4.30}$$

for all $k, n \in \mathbb{N}$. Thus, we deduce from (4.4.29) that $ax = (a_k x_k) \in bs$ whenever $x = (x_k) \in fs$ if and only if $Dy \in \ell_\infty$ whenever $y = (y_k) \in f$, where $D = (d_{nk})$ is defined by (4.4.30). Therefore, $D \in (f : \ell_\infty)$ and we obtain from Lemma 4.4.18 that $(fs)^\gamma = bv$.

(b) This can be obtained by analogy to the proof of Part (a) of Theorem 4.4.20 with Lemma 4.4.19 instead of Lemma 4.4.18. So, we omit details. □

Theorem 4.4.21. [74, Theorem 4.9] *The γ-dual of the space $\widehat{f}$ is the set $d_1(r,s)$, where*

$$d_1(r,s) := \left\{ a = (a_k) \in \omega : \sup_{n \in \mathbb{N}} \sum_{k=0}^{n} \left| \frac{1}{r} \sum_{j=k}^{n} \left(\frac{-s}{r} \right)^{j-k} a_j \right| < \infty \right\}.$$

Proof. Take any $a = (a_k) \in \omega$ and consider the equality obtained with (4.4.24) that

$$\sum_{k=0}^{n} a_k x_k = \sum_{k=0}^{n} \left[\sum_{j=k}^{n} \frac{1}{r} \left(\frac{-s}{r} \right)^{j-k} a_j \right] y_k = (Ey)_n \text{ for all } n \in \mathbb{N}, \tag{4.4.31}$$

where $E = \{e_{nk}(r,s)\}$ is defined by

$$e_{nk}(r,s) := \begin{cases} \sum_{j=k}^{n} \frac{1}{r} \left(\frac{-s}{r} \right)^{j-k} a_j & , \quad 0 \leq k \leq n, \\ 0 & , \quad k > n, \end{cases} \tag{4.4.32}$$

for all $k, n \in \mathbb{N}$. Thus, we deduce from (4.4.31) that $ax = (a_k x_k) \in bs$ whenever $x = (x_k) \in \widehat{f}$ if and only if $Ey \in \ell_\infty$ whenever $y = (y_k) \in f$, where $E = \{e_{nk}(r,s)\}$ is defined by (4.4.32). Therefore, we obtain from Lemma 4.4.18 that $(\widehat{f})^\gamma = d_1(r,s)$.

This completes the proof. □

Theorem 4.4.22. [74, Theorem 4.10] *Define the sets $d_2(r,s)$, $d_3(r,s)$, $d_4(r,s)$ and $d_5(r,s)$ by*

$$
\begin{aligned}
d_2(r,s) &:= \left\{a=(a_k)\in\omega : \lim_{n\to\infty}\sum_{j=k}^{n}\left(\frac{-s}{r}\right)^{j-k} a_j \text{ exists}\right\},\\
d_3(r,s) &:= \left\{a=(a_k)\in\omega : \lim_{n\to\infty}\sum_{k=0}^{n}\left[\frac{1-\left(\frac{-s}{r}\right)^{k+1}}{1+\frac{s}{r}}\right] a_k \text{ exists}\right\},\\
d_4(r,s) &:= \left\{a=(a_k)\in\omega : \lim_{n\to\infty}\sum_{k=0}^{n}\left|\sum_{i=n}^{\infty}\left(\frac{-s}{r}\right)^{i-k} a_i\right| = 0\right\},\\
d_5(r,s) &:= \left\{a=(a_k)\in\omega : \lim_{n\to\infty}\sum_{k=n+1}^{\infty}\left|\frac{1}{r}a_k-\frac{r+s}{rs}\sum_{i=k+1}^{\infty}\left(\frac{-s}{r}\right)^{i-k} a_i\right| = 0\right\}.
\end{aligned}
$$

Then, $(\widehat{f})^{\beta} := \bigcap_{i=1}^{5} d_i(r,s)$.

Proof. Let us take any $a=(a_k)\in\omega$ and again consider equality (4.4.31) obtained with (4.4.24). Then, one can easily see from (4.4.31) that $ax=(a_kx_k)\in cs$ whenever $x=(x_k)\in\widehat{f}$ if and only if $Ey\in c$ whenever $y=(y_k)\in f$. Therefore, we derive from Lemma 4.4.19 that $ax=(a_kx_k)\in cs$ whenever $x=(x_k)\in\widehat{f}$ if and only if $a=(a_k)\in\bigcap_{i=1}^{5} d_i(r,s)$. This means that $(\widehat{f})^{\beta}=\bigcap_{i=1}^{5} d_i(r,s)$ which concludes the proof. □

4.4.5 Some Matrix Transformations Related to the Sequence Space $\widehat{f}$

In this subsection, we characterize the matrix transformations from $\widehat{f}$ into any given sequence space μ and from a given sequence space μ into $\widehat{f}$. Since $\lambda_{B(r,s)}\cong\lambda$ for any sequence space λ, it is trivial that the equivalence "$x\in\lambda_{B(r,s)}$ if and only if $y\in\lambda$" holds. The reader may refer to Çolak and Çakar [130], for Banach limits and related matrix transformations.

We shall write throughout for brevity that

$$
\begin{aligned}
\widetilde{a}_{nk} &:= \tfrac{1}{r}\textstyle\sum_{j=k}^{\infty}\left(\frac{-s}{r}\right)^{j-k} a_{nj}, & \widehat{a}_{nk} &:= s a_{n-1,k}+r a_{nk}\\
a(n,k) &:= \textstyle\sum_{j=0}^{n} a_{jk}, & a(n,k,m) &:= \tfrac{1}{m+1}\textstyle\sum_{j=0}^{m} a_{n+j,k}\\
\Delta a_{nk} &:= a_{nk}-a_{n,k+1},
\end{aligned}
$$

for all $k,m,n\in\mathbb{N}$. Now, we may give the following theorem:

Theorem 4.4.23. [74, Theorem 5.1] *Suppose that the entries of the infinite matrices $A=(a_{nk})$ and $F=(f_{nk})$ are connected with the relation*

$$f_{nk} := \widetilde{a}_{nk} \tag{4.4.33}$$

for all $k,n\in\mathbb{N}$ and μ be any given sequence space. Then, $A\in(\widehat{f}:\mu)$ if and only if $(a_{nk})_{k\in\mathbb{N}}\in(\widehat{f})^{\beta}$ for all $n\in\mathbb{N}$ and $F\in(f:\mu)$.

Proof. Let μ be any given sequence space. Suppose that (4.4.33) holds between the entries of $A = (a_{nk})$ and $F = (f_{nk})$, and take into account that the spaces $\widehat{f}$ and f are norm isomorphic.

Let $A \in (\widehat{f} : \mu)$ and take any $y = (y_k) \in f$. Then, $FB(r, s)$ exists and $(a_{nk})_{k\in\mathbb{N}} \in \bigcap_{i=1}^{5} d_i(r, s)$, which yields that $(f_{nk})_{k\in\mathbb{N}} \in \ell_1$ for each $n \in \mathbb{N}$. Hence, Fy exists and thus

$$\sum_k f_{nk}y_k = \sum_k a_{nk}x_k \quad \text{for all} \quad n \in \mathbb{N},$$

we have by (4.4.33) that $Fy = Ax$, which leads us to consequence $F \in (f : \mu)$.

Conversely, let $(a_{nk})_{k\in\mathbb{N}} \in (\widehat{f})^{\beta}$ for each $n \in \mathbb{N}$ and $F \in (f : \mu)$ holds, and take any $x = (x_k) \in \widehat{f}$. Then, Ax exists. Therefore, we obtain from the equality

$$\sum_{k=0}^{m} a_{nk}x_k = \sum_{k=0}^{m} \left[\sum_{j=k}^{m} \frac{1}{r} \left(\frac{-s}{r} \right)^{j-k} a_{nj} \right] y_k \quad \text{for all} \quad n \in \mathbb{N},$$

as $m \to \infty$ that $Ax = Fy$ and this shows that $A \in (\widehat{f} : \mu)$.

This completes the proof. □

By changing the roles of the spaces $\widehat{f}$ and f with μ, we have

Theorem 4.4.24. [74, Theorem 5.2] *Suppose that the entries of the infinite matrices $A = (a_{nk})$ and $G = (g_{nk})$ are connected with the relation $g_{nk} = \widehat{a}_{nk}$ for all $k, n \in \mathbb{N}$ and μ be any given sequence space. Then, $A \in (\mu : \widehat{f})$ if and only if $G \in (\mu : f)$.*

Proof. Let $x = (x_k) \in \mu$ and consider the following equality

$$\begin{aligned} \{B(r, s)(Ax)\}_n &= s(Ax)_{n-1} + r(Ax)_n \\ &= s\sum_k a_{n-1,k}x_k + r\sum_k a_{nk}x_k \\ &= \sum_k (sa_{n-1,k} + ra_{nk})x_k = (Gx)_n \quad \text{for all} \quad n \in \mathbb{N}, \end{aligned}$$

which yields by letting to the generalized limit that $Ax \in \widehat{f}$ if and only if $Gx \in f$.

This step completes the proof. □

Now, we list the following conditions:

$$\lim_{n\to\infty} a_{nk} = \alpha_k \quad \text{for each fixed } k \in \mathbb{N} \tag{4.4.34}$$

$$\lim_{n\to\infty} \sum_k a_{nk} = \alpha \tag{4.4.35}$$

$$\lim_{n\to\infty} \sum_k |\Delta(a_{nk} - \alpha_k)| = 0 \tag{4.4.36}$$

$$\sup_{n\in\mathbb{N}} \sum_k |\Delta a_{nk}| < \infty \tag{4.4.37}$$

$$\lim_{k\to\infty} a_{nk} = 0 \text{ for each fixed } n \in \mathbb{N} \tag{4.4.38}$$

$$\lim_{n\to\infty} \sum_k \left|\Delta^2 a_{nk}\right| = \alpha \tag{4.4.39}$$

$$f - \lim a_{nk} = \alpha_k \text{ exists for each fixed } k \in \mathbb{N} \tag{4.4.40}$$

$$\lim_{m\to\infty} \sum_k |a(n,k,m) - \alpha_k| = 0 \text{ uniformly in } n \tag{4.4.41}$$

$$f - \lim \sum_k a_{nk} = \alpha \tag{4.4.42}$$

$$\lim_{m\to\infty} \sum_k |\Delta[a(n,k,m) - \alpha_k]| = 0 \text{ uniformly in } n \tag{4.4.43}$$

$$\lim_{q\to\infty} \sum_k \frac{1}{q+1} \left|\sum_{i=0}^{q} \Delta[a(n+i,k) - \alpha_k]\right| = 0 \text{ uniformly in } n \tag{4.4.44}$$

$$\sup_{n\in\mathbb{N}} \sum_k |\Delta a(n,k)| < \infty \tag{4.4.45}$$

$$f - \lim a(n,k) = \alpha_k \text{ exists for each fixed } k \in \mathbb{N} \tag{4.4.46}$$

$$\lim_{q\to\infty} \sum_k \frac{1}{q+1} \left|\sum_{i=0}^{q} \Delta^2[a(n+i,k) - \alpha_k]\right| = 0 \text{ uniformly in } n \tag{4.4.47}$$

$$\sup_{n\in\mathbb{N}} \sum_k |a(n,k)| < \infty \tag{4.4.48}$$

$$\sum_n a_{nk} = \alpha_k \text{ for each fixed } k \in \mathbb{N} \tag{4.4.49}$$

$$\sum_n \sum_k a_{nk} = \alpha \tag{4.4.50}$$

$$\lim_{n\to\infty} \sum_k |\Delta[a(n,k) - \alpha_k]| = 0 \tag{4.4.51}$$

Of course, Theorems 4.4.23 and 4.4.24 have several consequences depending on the choice of the sequence space μ. Prior to giving some results as an application of this idea, we give the following basic lemma, which is the collection of the characterizations of matrix transformations related to almost convergence:

Lemma 4.4.25. *Let $A = (a_{nk})$ be an infinite matrix. Then, the following statements hold:*

(i) $A = (a_{nk}) \in (\ell_\infty : f)$ *if and only if (4.4.25), (4.4.40) and (4.4.41) hold* (cf. Duran [154]).

(ii) $A = (a_{nk}) \in (f : f)$ *if and only if (4.4.25), (4.4.40), (4.4.42) and (4.4.43) hold* (cf. Duran [154]).

(iii) $A = (a_{nk}) \in (fs : \ell_\infty)$ *if and only if (4.4.37) and (4.4.38) hold.*

(iv) $A = (a_{nk}) \in (fs : c)$ *if and only if (4.4.34), (4.4.37), (4.4.38) and (4.4.39) hold* (cf. Öztürk [312]).

(v) $A = (a_{nk}) \in (c : f)$ *if and only if (4.4.25), (4.4.40) and (4.4.42) hold* (cf. King [219]).

(vi) $A = (a_{nk}) \in (bs : f)$ *if and only if (4.4.37), (4.4.38), (4.4.40) and (4.4.44) hold* (cf. Başar and Solak [78]).

(vii) $A = (a_{nk}) \in (fs : f)$ *if and only if (4.4.38), (4.4.40), (4.4.43) and (4.4.44) hold* (cf. Başar [59]).

(viii) $A = (a_{nk}) \in (cs : f)$ *if and only if (4.4.37) and (4.4.40) hold* (cf. Başar and Çolak [70]).

(ix) $A = (a_{nk}) \in (bs : fs)$ *if and only if (4.4.38), (4.4.44), (4.4.45) and (4.4.46) hold* (cf. Başar and Solak [78]).

(x) $A = (a_{nk}) \in (fs : fs)$ *if and only if (4.4.44), (4.4.45),(4.4.46) and (4.4.47) hold* (cf. Başar [59]).

(xi) $A = (a_{nk}) \in (cs : fs)$ *if and only if (4.4.45) and (4.4.46) hold* (cf. Başar and Çolak [70]).

(xii) $A = (a_{nk}) \in (f : cs)$ *if and only if (4.4.48), (4.4.49), (4.4.50) and (4.4.51) hold* (cf. Başar [58]).

Now, we can give the following results:

Corollary 4.4.26. [74, Corollary 5.4] *The following statements hold:*

(i) $A = (a_{nk}) \in (\widehat{f} : \ell_\infty)$ *if and only if* $(a_{nk})_{k\in\mathbb{N}} \in (\widehat{f})^\beta$ *for each* $n \in \mathbb{N}$ *and (4.4.25) holds with* $\widetilde{a}_{nk}$ *instead of* a_{nk}.

(ii) $A = (a_{nk}) \in (\widehat{f} : c)$ *if and only if* $(a_{nk})_{k\in\mathbb{N}} \in (\widehat{f})^\beta$ *for all* $n \in \mathbb{N}$ *and (4.4.25), (4.4.34), (4.4.35) and (4.4.36) hold with* $\widetilde{a}_{nk}$ *instead of* a_{nk}.

(iii) $A = (a_{nk}) \in (\widehat{f} : bs)$ *if and only if* $(a_{nk})_{k\in\mathbb{N}} \in (\widehat{f})^\beta$ *for all* $n \in \mathbb{N}$ *and (4.4.48) holds.*

(iv) $A = (a_{nk}) \in (\widehat{f} : cs)$ *if and only if* $(a_{nk})_{k\in\mathbb{N}} \in (\widehat{f})^\beta$ *for each* $n \in \mathbb{N}$ *and (4.4.48), (4.4.49), (4.4.50) and (4.4.51) hold with* $\widetilde{a}_{nk}$ *instead of* a_{nk}.

Corollary 4.4.27. [74, Corollary 5.5] *The following statements hold:*

(i) $A = (a_{nk}) \in (\ell_\infty : \widehat{f})$ *if and only if (4.4.25), (4.4.40) and (4.4.41) hold with* $\widehat{a}_{nk}$ *instead of* a_{nk}.

(ii) $A = (a_{nk}) \in (f : \widehat{f})$ *if and only if (4.4.25), (4.4.40), (4.4.42) and (4.4.43) hold with* $\widehat{a}_{nk}$ *instead of* a_{nk}.

(iii) $A = (a_{nk}) \in (c : \widehat{f})$ *if and only if (4.4.25), (4.4.40) and (4.4.42) hold with* $\widehat{a}_{nk}$ *instead of* a_{nk}.

Corollary 4.4.28. [74, Corollary 5.6] *The following statements hold:*

(i) $A = (a_{nk}) \in (bs : \widehat{f})$ *if and only if (4.4.37), (4.4.38), (4.4.40) and (4.4.44) hold with* $\widehat{a}_{nk}$ *instead of* a_{nk}.

(ii) $A = (a_{nk}) \in (fs : \widehat{f})$ *if and only if (4.4.38), (4.4.40), (4.4.43) and (4.4.44) hold with* $\widehat{a}_{nk}$ *instead of* a_{nk}.

(iii) $A = (a_{nk}) \in (cs : \widehat{f})$ *if and only if (4.4.37) and (4.4.40) hold with* $\widehat{a}_{nk}$ *instead of* a_{nk}.

Corollary 4.4.29. [74, Corollary 5.7] *Let* $\widehat{fs}$ *denotes the domain of the generalized difference matrix* $B(r,s)$ *in the sequence space* fs. *Then, the following statements hold:*

(i) $A = (a_{nk}) \in (bs : \widehat{fs})$ *if and only if (4.4.38), (4.4.44), (4.4.45) and (4.4.46) hold with* $\widehat{a}_{nk}$ *instead of* a_{nk}.

(ii) $A = (a_{nk}) \in (fs : \widehat{fs})$ *if and only if (4.4.44), (4.4.45), (4.4.46) and (4.4.47) hold with* $\widehat{a}_{nk}$ *instead of* a_{nk}.

(iii) $A = (a_{nk}) \in (cs : \widehat{fs})$ *if and only if (4.4.45) and (4.4.46) hold with* $\widehat{a}_{nk}$ *instead of* a_{nk}.

4.5 SPACES OF DIFFERENCE SEQUENCES OF ORDER m

In this section, we emphasize the spaces of difference sequences of order m. Throughout this section, we suppose that m is an arbitrary positive integer and $\binom{-1}{-1} = 1$, although in some documents it is assumed that $\binom{r}{k} = 0$ whenever $k < 0$. Then, the operators Δ^m, $(\Delta^m)^{-1} = S^m : \omega \to \omega$ are defined by

$$
\begin{aligned}
(\Delta^m x)_k &= \{(\Delta \circ \Delta^{m-1})x\}_k \text{ for all } k \in \mathbb{N}_1 \text{ and } m \in \mathbb{N}_2, \\
\{(\Delta^m)^{-1}x\}_k &= (S^m x)_k = \{(S \circ S^{m-1})x\}_k \text{ for all } k \in \mathbb{N}_1 \text{ and } m \in \mathbb{N}_2, \\
(\Delta x)_k &= x_k - x_{k+1} \text{ for all } k \in \mathbb{N}, \\
\left(\Delta^{-1} x\right)_k &= (Sx)_k = \sum_{j=k}^{\infty} x_j \text{ for all } k \in \mathbb{N}, \\
\left(\Delta^{-2} x\right)_k &= \{(\Delta^2)^{-1}x\}_k = \sum_{j=k}^{\infty} (j-k+1)x_j \text{ for all } k \in \mathbb{N}_1, \\
\Delta^m \circ S^m &= S^m \circ \Delta^m = I \text{ for all } m \in \mathbb{N}_1,
\end{aligned}
$$

where I denotes the identity on the space ω. For any subset λ of ω, we write

$$\begin{aligned}
\lambda(\Delta^m) &= \{x = (x_k) \in \omega : \Delta^m x \in \lambda\}, \\
\lambda(p, \Delta^m) &= \{x = (x_k) \in \omega : \Delta^m x \in \lambda(p)\}, \\
\lambda(u, \Delta^m) &= \left(u^{-1} * \lambda\right)(\Delta^m) \\
&= \{x = (x_k) \in \omega : \{u_k(\Delta^m x)_k\}_{k\in\mathbb{N}} \in \lambda\}, \\
\lambda(p, u, \Delta^m) &= \left(u^{-1} * \lambda(p)\right)(\Delta^m) \\
&= \{x = (x_k) \in \omega : \{u_k(\Delta^m x)_k\}_{k\in\mathbb{N}} \in \lambda(p)\}.
\end{aligned}$$

In particular, for $m = 1$ and $m = 2$, we get the sets $\lambda(\Delta)$, $\lambda(\Delta^2)$, $\lambda(p, \Delta)$, $\lambda(p, \Delta^2)$, $\lambda(u, \Delta)$, $\lambda(u, \Delta^2)$, $\lambda(p, u, \Delta)$ and $\lambda(p, u, \Delta^2)$, respectively. For $u = e$, we have

$$\lambda(u, \Delta) = \lambda(\Delta),\ \lambda\left(u, \Delta^2\right) = \lambda\left(\Delta^2\right),\ \lambda(p, u, \Delta) = \lambda(p, \Delta),\ \lambda(p, u, \Delta^2) = \lambda\left(p, \Delta^2\right).$$

Let m be a positive integer, then we can find positive constants M_1 and M_2 such that

$$M_1 k^m \leq \binom{m+k}{k} \leq M_2 k^m \quad \text{for all} \quad k \in \mathbb{N} \tag{4.5.1}$$

and

$$\sum_{k=0}^{n} \binom{n+m-k-1}{m-1} = \binom{n+m}{m} = \binom{n+m}{n} \quad \text{for all} \quad n \in \mathbb{N}. \tag{4.5.2}$$

In 1993, Et [155] defined the sets

$$\lambda\left(\Delta^2\right) := \left\{x = (x_k) \in \omega : \left\{(\Delta^2 x)_k\right\} \in \lambda\right\}.$$

He obtained Köthe-Toeplitz duals of these sets and characterized the matrix class $(\lambda : \mu(\Delta^2))$, where $\lambda, \mu \in \{\ell_\infty, c, c_0\}$. Extending these spaces Mursaleen [283] defined the difference sequence spaces

$$\lambda\left(u, \Delta^2\right) := \left\{x = (x_k) \in \omega : \left\{u_k(\Delta^2 x)_k\right\} \in \lambda\right\}$$

and studied some properties of those spaces. Recently, Bektaş [86] and, independently, Ahmad et al. [9] extended these spaces to

$$\lambda\left(p, u, \Delta^2\right) := \left\{x = (x_k) \in \omega : \left\{u_k(\Delta^2 x)_k\right\} \in \left(u^{-1} * \lambda\right)\left(\Delta^2\right)\right\},$$

determined α-, β-duals of those spaces and characterized some matrix maps which are more general than the results of Mursaleen [283] and include them as a special case $p_n = C$, a constant, for all $n \in \mathbb{N}$.

Recently, Et and Çolak [131, 166], Et and Nuray [168], and Et [157] extended the spaces $\lambda(\Delta)$ and $\lambda(\Delta^2)$ to the spaces $\lambda(\Delta^m)$. Let λ be any sequence space and define

$$\lambda(\Delta^m) := \{x = (x_k) \in \omega : \{(\Delta^m x)_k\} \in \lambda\},$$

where $m \in \mathbb{N}$ and $(\Delta^m x)_k = \{(\Delta \circ \Delta^{m-1})x\}_k$ for all $k \in \mathbb{N}$.

The sequence spaces $\lambda(\Delta^m)$ contain some unbounded sequences for $m \in \mathbb{N}_1$, for example let $x = (k^m)$, then $(k^m) \in \ell_\infty(\Delta^m) \setminus \ell_\infty$. Now, following Et and Nuray [168] we give the next two theorems related to m^{th} order difference spaces:

Theorem 4.5.1. *If λ is a linear space, then $\lambda(\Delta^m)$ is also a linear space.*

Theorem 4.5.2. *If $\mu \subset \lambda$, then the following statements hold:*

(i) $\mu(\Delta^m) \subset \lambda(\Delta^m)$.

(ii) $\lambda(\Delta^{m-1}) \subset \lambda(\Delta^m)$.

Theorem 4.5.3. [157] *Let E be a vector space and let $A \subset E$. If A is a convex set, then $A(\Delta^m)$ is a convex set in $E(\Delta^m)$.*

Proof. Let $x, y \in A(\Delta^m)$. Then, $\Delta^m x, \Delta^m y \in A$. Since Δ^m is linear, we have

$$\alpha\Delta^m x + (1-\alpha)\Delta^m y = \Delta^m[x + (1-\alpha)y], \quad (0 \leq \alpha \leq 1).$$

Since A is convex $\alpha\Delta^m x + (1-\alpha)\Delta^m y \in A$ and so $x + (1-\alpha)y \in A(\Delta^m)$. □

Theorem 4.5.4. [168] *If λ is a BK-space normed by $\|\cdot\|$, then $\lambda(\Delta^m)$ is also a BK-space normed by*

$$\|x\|_\Delta = \sum_{i=1}^{m} |x_i| + \|\Delta^m x\|. \tag{4.5.3}$$

Proof. It is a routine verification that $\lambda(\Delta^m)$ is a normed space normed by (4.5.3). Now, we show that $\lambda(\Delta^m)$ is complete. Let (x^s) be a Cauchy sequence in $\lambda(\Delta^m)$, where $x^s = (x_1^s, x_2^s, \ldots) \in \lambda(\Delta^m)$. Then, $\|x_k^s - x_k^t\|_\Delta \to 0$, as $s, t \to \infty$. Hence,

$$\lim_{s,t\to\infty} \left[\sum_{i=1}^{m} |x_i^s - x_i^t| + \|\Delta^m (x_k^s - x_k^t)\|\right] = 0.$$

Therefore, $(x_i^1, x_i^2, \ldots)$, $(i \leq m)$ and $\{(\Delta^m x)_k^1, (\Delta^m x)_k^2, \ldots\}$ are Cauchy sequences in $\mathbb{C}$ and λ, respectively. Since $\mathbb{C}$ and λ are complete, they are convergent. Suppose that $x_i^s \to x_i$, $(i \leq m)$ in $\mathbb{C}$ and $\{\Delta^m(x^s)\} \to (y_k)$ in λ, as $s \to \infty$. Let $y_k = (\Delta^m x)_k$ so that

$$x_k = \sum_{v=1}^{k-m} (-1)^m \binom{k-v-1}{m-1} y_v = \sum_{v=1}^{k} (-1)^m \binom{k+m-v-1}{m-1} y_{v-m}, \tag{4.5.4}$$

$y_{1-m} = y_{2-m} = \cdots = y_0 = 0$, for sufficiently large k, for instance $k > 2m$. Then, $\Delta^m x^s = \left\{(\Delta^m x)_k^1, (\Delta^m x)_k^2, \ldots\right\}$ converges to $\{(\Delta^m x)_k\}$ in λ. Hence, $\|x^s - x\|_\Delta \to 0$ as $s \to \infty$. Since $(x^s - x)$, $(x^s) \in \lambda(\Delta^m)$ and the space $\lambda(\Delta^m)$ are linear we have $x = x^s - (x^s - x) \in \lambda(\Delta^m)$. Hence, $\lambda(\Delta^m)$ is complete. Since $\lambda(\Delta^m)$ is a Banach space with continuous coordinates, i.e., $\|x^n - x\|_\Delta \to 0$ implies for each $k \in \mathbb{N}$ that $|x_k^n - x_k| \to 0$, as $n \to \infty$, it is a BK-space. □

Corollary 4.5.5. *$\ell_\infty(\Delta^m)$, $c(\Delta^m)$ and $c_0(\Delta^m)$ are BK-spaces endowed with the norm*

$$\|x\|_\Delta = \sum_{i=1}^{m} |x_i| + \|\Delta^m x\|_\infty. \tag{4.5.5}$$

Theorem 4.5.6. (cf. [396, Theorem 4.3.14] and [168]) *Let A be a closed subset of a Banach space E. Then, $A(\Delta^m)$ is also closed in $E(\Delta^m)$.*

Proof. Since $A \subset E$, then $A(\Delta^m) \subset E(\Delta^m)$. Now, we show that $\overline{A(\Delta^m)} = \overline{A}(\Delta^m)$. Let $x \in \overline{A(\Delta^m)}$. Then, there exists a sequence (x^n) in $A(\Delta^m)$ such that

$$\lim_{n\to\infty} \|x^n - x\|_\Delta = 0.$$

Hence,

$$\lim_{n\to\infty} \|(x_k^n) - (x_k)\|_\Delta = 0$$

in $A(\Delta^m)$ so that

$$\lim_{n\to\infty} \left(\sum_{i=1}^{m} |x_i^n - x_i| + \|\Delta^m(x_k^n) - \Delta^m(x_k)\|_\Delta\right) = 0$$

in A. Thus, $\Delta^m x \in \overline{A}$. This implies that $x \in \overline{A}(\Delta^m)$. Conversely, if $x \in \overline{A}(\Delta^m)$, then $x \in \overline{A(\Delta^m)}$. Since A is closed, $\overline{A(\Delta^m)} = A(\Delta^m)$. Hence, $A(\Delta^m)$ is closed subset of $E(\Delta^m)$. □

Theorem 4.5.7. [168] *If E is a separable space, then $E(\Delta^m)$ is also a separable space.*

Proof. This is similar to the proof of Theorem 4.5.6. So, we omit details. □

Theorem 4.5.8. [157] *Let E be a Banach space and let $A \subset E$. Then, the following statements hold:*

(i) *If A is nowhere dense in E, then $A(\Delta^m)$ is nowhere dense in $E(\Delta^m)$.*

(ii) *If A is dense in E, then $A(\Delta^m)$ is dense in $E(\Delta^m)$.*

Proof. Since the proof of Part (ii) is trivial, we omit it and only give the proof of Part (i).

Suppose that $\overline{A}^{\circ} = \emptyset$ but $\overline{A(\Delta^m)}^{\circ} \neq \emptyset$. Then, $\overline{A}$ contains no neighborhood and $B(a) \subset \overline{A(\Delta^m)}$, where $B(a)$ is an open ball of center a and radius r. Hence, $a \in B(a) \subset \overline{A(\Delta^m)} = \left(\overline{A}\right)(\Delta^m)$. This implies that $\Delta^m(a) \in \overline{A}$. So, $B(\Delta^m(a)) \cap A \neq \emptyset$. On the other hand, $B(\Delta^m(a)) \cap A \subset \overline{A}$. This is contrary to the fact $\overline{A}^{\circ} = \emptyset$. Hence, $\overline{\Delta^m(A)}^{\circ} = \emptyset$. □

Corollary 4.5.9. *The following statements hold:*

(i) *The inclusions $c_0(\Delta^m) \subset c(\Delta^m) \subset \ell_\infty(\Delta^m)$ strictly hold.*

(ii) *The inclusions $c_0(\Delta^m) \subset c_0(\Delta^{m+1})$, $c(\Delta^m) \subset c(\Delta^{m+1})$, $\ell_\infty(\Delta^m) \subset \ell_\infty(\Delta^{m+1})$ strictly hold.*

(iii) *$c_0(\Delta^m)$ and $c(\Delta^m)$ are the closed subspaces of $\ell_\infty(\Delta^m)$.*

(iv) *$c_0(\Delta^m)$, $c(\Delta^m)$ and $\ell_p(\Delta^m)$ are separable spaces.*

(v) *$\ell_\infty \cap c(\Delta^m) = \ell_\infty \cap c_0(\Delta^m)$.*

It is well-known that c_0 is a sequence algebra, but $\ell_\infty(\Delta^m)$, $c(\Delta^m)$ and $c_0(\Delta^m)$ need not to be sequence algebra. Indeed, if we consider the sequences $x = (k)$ and $y = (k^{m-1})$, then it is clear that $x, y \in c_0(\Delta^m)$ and $x \cdot y \notin c_0(\Delta^m)$ for $m \in \mathbb{N}_2$.

Now, suppose that $\lambda \subseteq \ell_\infty$, which is a Banach space normed by $\|\cdot\|$. Let us define the operator s by

$$\begin{array}{rcll} s & : & \lambda(\Delta^m) & \longrightarrow \lambda(\Delta^m) \\ & & x = (x_k) & \longmapsto sx = (0, 0, \ldots, x_{m+1}, x_{m+2}, x_{m+3}, \ldots). \end{array}$$

It is trivial that s is a bounded linear operator on $\lambda(\Delta^m)$. Furthermore, the set

$$s[\lambda(\Delta^m)] = s\lambda(\Delta^m) = \{x = (x_k) \in \lambda(\Delta^m) : x_1 = x_2 = \cdots = x_m = 0\}$$

is a subspace of $\lambda(\Delta^m)$ with $\|x\|_\Delta = \|\Delta^m x\|$ in $s\lambda(\Delta^m)$. $s\lambda(\Delta^m)$ and λ are equivalent as topological space since

$$\begin{array}{rcll} \Delta^m & : & s\lambda(\Delta^m) & \longrightarrow X \\ & & x = (x_k) & \longmapsto \Delta^m x = y = \{(\Delta^{m-1}x)_k - (\Delta^{m-1}x)_{k+1}\} \end{array} \tag{4.5.6}$$

is a linear homeomorphism.

Let λ^* and $\{s\lambda(\Delta^m)\}^*$ denote the continuous duals of λ and $s\lambda(\Delta^m)$, respectively. It can be shown that

$$T : \{s\lambda(\Delta^m)\}^* \to \lambda^*, \ f_\Delta \to f_{\Delta^\circ}(\Delta^m)^{-1} = f$$

is a linear isometry. So, $\{s\lambda(\Delta^m)\}^*$ is equivalent to λ^*.

Corollary 4.5.10. *$\ell_\infty(\Delta^m)$ and ℓ_∞, $c(\Delta^m)$ and c, $c_0(\Delta^m)$ and c_0 are equivalent as topological spaces and $\{c(\Delta^m)\}^* = \{c_0(\Delta^m)\}^* \cong l_1$.*

Now, we define

$$\{\Delta^{(m)}x\}_k = \sum_{i=0}^{m}(-1)^i\binom{m}{i}x_{k-i} \text{ for all } k \in \mathbb{N}.$$

It is trivial that $(x_k) \in \lambda(\Delta^{(m)})$ if and only if $\{(\Delta^{(m)}x)_k\} \in \lambda$ for $\lambda \in \{\ell_\infty, c, c_0\}$. In [268], Malkowsky and Parashar defined the sequence spaces

$$\lambda(\Delta^{(m)}) := \left\{x = (x_k) \in \omega : \left\{(\Delta^{(m)}x)_k\right\} \in \lambda\right\}, \quad \lambda \in \{\ell_\infty, c, c_0\};$$

where $m \in \mathbb{N}$, $(\Delta^{(m)}x)_k = \Delta^{(1)} \circ \Delta^{(m-1)}x_k$ and $\Delta^{(m)} = \{\Delta_{nk}^{(m)}\}$ denote the matrix

$$\Delta_{nk}^{(m)} := \begin{cases} (-1)^{n-k}\binom{m}{n-k} & , \quad \max\{0, n-m\} \le k \le n, \\ 0 & , \quad 0 \le k < \max\{0, n-m\} \text{ or } k > n, \end{cases}$$

and they showed that the sequence spaces $\ell_\infty\left(\Delta^{(m)}\right)$, $c\left(\Delta^{(m)}\right)$ and $c_0\left(\Delta^{(m)}\right)$ are BK-spaces endowed with the norm

$$\|x\|_{\Delta^*} = \sup_{k\in\mathbb{N}}\left|(\Delta^{(m)}x)_k\right|. \tag{4.5.7}$$

It can be shown that the norms given by (4.5.5) and (4.5.7) are equivalent.

4.5.1 Dual Spaces of $\ell_\infty(\Delta^m)$, $c(\Delta^m)$ and $c_0(\Delta^m)$

In this subsection, we give the α-, β-, γ- and N-duals of the spaces $\ell_\infty(\Delta^m)$, $c(\Delta^m)$ and $c_0(\Delta^m)$. Also, we show that these spaces are not perfect.

Lemma 4.5.11. *Let (p_n) be a sequence of positive numbers increasing monotonically to infinity. Then, the following statements hold:*

(i) *If $\sup_{n\in\mathbb{N}}\left|p_n\sum_{v=1}^{n}p_v a_v\right| < \infty$, then $\sup_{n\in\mathbb{N}}\left|\sum_{k=n+1}^{\infty}a_k\right| < \infty$.*

(ii) *If $\sum_k p_k a_k$ is convergent, then $p_n\sum_{k=n+1}^{\infty}a_k \to 0$, as $n \to \infty$.*

Lemma 4.5.12. [166] *$x = (x_k) \in \ell_\infty(\Delta^m)$ implies $\sup_{k\in\mathbb{N}} k^{-m}|x_k| < \infty$.*

Theorem 4.5.13. (cf. [268, Theorem 3.13] and [166]) *Let m be a positive integer. Define the sets M^α and $M^{\alpha\alpha}$ by*

$$M^\alpha := \left\{a = (a_k) \in \omega : \sum_k k^m|a_k| < \infty\right\},$$

$$M^{\alpha\alpha} := \left\{a = (a_k) \in \omega : \sup_{k\in\mathbb{N}_1} k^{-m}|a_k| < \infty\right\}.$$

Then, the following statements hold:

(i) $\{c_0(\Delta^m)\}^\alpha = \{c(\Delta^m)\}^\alpha = \{\ell_\infty(\Delta^m)\}^\alpha = M^\alpha$.

(ii) $\{c_0(\Delta^m)\}^{\alpha\alpha} = \{c(\Delta^m)\}^{\alpha\alpha} = \{\ell_\infty(\Delta^m)\}^{\alpha\alpha} = M^{\alpha\alpha}$.

Proof. (i) We give the proof for $\ell_\infty(\Delta^m)$. Let $a = (a_k) \in M^\alpha$. Then,

$$\sum_k |a_k x_k| = \sum_k k^m |a_k| \left(k^{-m} |x_k|\right) < \infty$$

for each $x = (x_k) \in \ell_\infty(\Delta^m)$ by Lemma 4.5.12. Hence, $a \in \{\ell_\infty(\Delta^m)\}^\alpha$.

Let $a = (a_k) \in \{\ell_\infty(\Delta^m)\}^\alpha$. Then, $\sum_k |a_k x_k| < \infty$ for each $x = (x_k) \in \ell_\infty(\Delta^m)$. For the sequence $x = (x_k)$ defined by $x_k = k^m$, we can write

$$\sum_k k^m |a_k| = \sum_k |a_k x_k| < \infty.$$

This implies that $a \in M^\alpha$.

(ii) Let λ denotes any of the classical sequence spaces ℓ_∞, c and c_0. If $a = (a_k) \in M^{\alpha\alpha}$ and $x = (x_k) \in \{\lambda(\Delta^m)\}^\alpha$, then

$$\sum_k |a_k x_k| = \sum_k k^m |a_k| k^{-m} |x_k| \le \sum_k k^m |a_k| \sup_{k\in\mathbb{N}} k^{-m} |x_k| < \infty.$$

Hence, $a \in \{\lambda(\Delta^m)\}^{\alpha\alpha}$.

Now, suppose that $a = (a_k) \in \{\lambda(\Delta^m)\}^{\alpha\alpha} \setminus M^{\alpha\alpha}$. Then, we have

$$\sup_{k\in\mathbb{N}} k^{-m} |a_k| = \infty.$$

Hence, there is a strictly increasing sequence $\{k(i)\}$ of positive integers $k(i)$ such that

$$\frac{1}{[k(i)]^m} \left|a_{k(i)}\right| > i^m.$$

Define the sequence $x = (x_k)$ by

$$x_k := \begin{cases} 1/|a_{k(i)}| & , \quad k = k(i), \\ 0 & , \quad k \neq k(i). \end{cases}$$

Then, we have

$$\sum_k k^m |a_k| = \sum_i [k(i)]^m \left|a_{k(i)}\right|^{-1} \le \sum_i i^{-m} < \infty.$$

Hence, $x \in \{\lambda(\Delta^m)\}^\alpha$ and $\sum_k |a_k x_k| = \infty$. This is contrary to the fact that $a \in \{\lambda(\Delta^m)\}^{\alpha\alpha}$. Hence, $a \in M^{\alpha\alpha}$. □

Lemma 4.5.14. [131] *Let m be a positive integer and $R_k = \sum_{v=k+1}^{\infty} a_v$. Define the sets M^β and M^γ by*

$$\begin{aligned} M^\beta &:= \left\{ a = (a_k) \in \omega : (k^m a_k)_{k\in\mathbb{N}_1} \in cs \text{ and } \sum_k k^{m-1} |R_k| < \infty \right\}, \\ M^\gamma &:= \left\{ a = (a_k) \in \omega : \sup_{n\in\mathbb{N}_1} \left| \sum_{k=1}^n k^m a_k \right| < \infty \text{ and } \sum_k k^{m-1} |R_k| < \infty \right\}. \end{aligned}$$

Then, the following statements hold:

(i) $\{s\ell_\infty(\Delta^m)\}^\beta = M^\beta$.

(ii) $\{s\ell_\infty(\Delta^m)\}^\gamma = M^\gamma$.

Proof. Since Part (ii) can be proved by the same way, we consider only Part (i).

If $x = (x_k) \in s\ell_\infty(\Delta^m)$, then there exists one and only one $y = (y_k) \in \ell_\infty$ such that (4.5.4) holds. Then, we can write

$$\begin{aligned}\sum_{k=1}^{n} a_k x_k &= \sum_{k=1}^{n} a_k \left[\sum_{v=1}^{k-m} (-1)^m \binom{k-v-1}{m-1} y_v\right] \\ &= (-1)^m \sum_{k=1}^{n-m} (k+m-1)^{m-1} R_{k+m-1} \times \\ &\times \left[\frac{1}{(k+m-1)^{m-1}} \sum_{v=1}^{k} \binom{k+m-v-2}{m-2} y_v\right] - n^m R_n n^{-m} x_n.\end{aligned}$$

Since $\sum_k k^{m-1}|R_k| < \infty$, the series $\sum_k (k+m-1)^{m-1} R_{k+m-1} z_k$ is absolutely convergent, where

$$z = (z_k) := \left\{\frac{1}{(k+m-1)^{m-1}} \sum_{v=1}^{k} \binom{k+m-v-2}{m-2} y_v\right\}.$$

Moreover, we have $R_n n^m \to 0$, as $n \to \infty$ by Lemma 4.5.11 and $\sup_{n\in\mathbb{N}_1} n^{-m}|x_n| < \infty$ by Lemma 4.5.12, hence $(a_k x_k) \in cs$ for all $x \in s\ell_\infty(\Delta^m)$. So, $a \in [s\ell_\infty(\Delta^m)]^\beta$.

Let $a \in \{s\ell_\infty(\Delta^m)\}^\beta$. Then, $(a_k x_k) \in cs$ for each $x = (x_k) \in s\ell_\infty(\Delta^m)$. For the sequence $x = (x_k)$ defined by

$$x_k := \begin{cases} 0 & , \quad k \le m, \\ k^m & , \quad k > m, \end{cases}$$

we can write

$$\sum_k k^m a_k = \sum_{k=1}^{n} k^m a_k + \sum_k a_k x_k.$$

Thus, the series $\sum_k k^m a_k$ is convergent. This implies that $R_n n^m = o(1)$ by Part (ii) of Lemma 4.5.11.

Now, let $a = (a_k) \in \{s\ell_\infty(\Delta^m)\}^\beta \setminus M^\beta$. Then, $\sum_k k^{m-1}|R_k|$ is divergent, i.e., $\sum_k k^{m-1}|R_k| = \infty$. We define the sequence $x = (x_k)$ by

$$x_k := \begin{cases} 0 & , \quad k \le m, \\ \sum_{v=1}^{k-1} v^{m-1} \operatorname{sgn} R_v & , \quad k > m, \end{cases}$$

where $a_k > 0$ for all $k \in \mathbb{N}_1$ or $a_k < 0$ for all $k \in \mathbb{N}_1$. Since $|\Delta^m x| = (m-1)!$ for $k > m$, it is trivial that $x = (x_k) \in s\ell_\infty(\Delta^m)$. Then, we derive for $n > m$ that

$$\begin{aligned}\sum_{k=1}^{n} a_k x_k &= -\sum_{k=1}^{m} R_{k-1}(\Delta x)_{k-1} - \\ &\quad -\sum_{k=1}^{n-m} R_{k+m-1}(\Delta x)_{k+m-1} - n^m R_n n^{-m} x_n.\end{aligned}$$

Now, letting $n \to \infty$ we get

$$\begin{aligned}\sum_k a_k x_k &= -\sum_k R_{k+m-1}(\Delta x)_{k+m-1} \\ &= \sum_k (k+m-1)^{m-1} |R_{k+m-1}| = \infty.\end{aligned}$$

This contradicts the fact that $a \in \{s\ell_\infty(\Delta^m)\}^\beta$. Hence, $a \in M^\beta$.

This completes the proof. □

Lemma 4.5.15. [131] $\{s\ell_\infty(\Delta^m)\}^\eta = \{sc(\Delta^m)\}^\eta$ *for* $\eta \in \{\beta, \gamma\}$.

Lemma 4.5.16. [88] *Let* c_0^+ *denotes the set of all positive sequences in* c_0 *and* $v = (v_j) \in c_0^+$. *Define the sets* M_0^β *and* M_0^γ *by*

$$M_0^\beta := \left\{\begin{array}{c} a = (a_k) \in \omega : \sum_k a_k \sum_{j=1}^{k-m} \binom{k-j-1}{m-1} v_j \text{ converges and} \\ \sum_k |R_k| \sum_{j=1}^{k-m+1} \binom{k-j-1}{m-2} v_j < \infty \text{ for all } v \in c_0^+ \end{array}\right\},$$

$$M_0^\gamma := \left\{\begin{array}{c} a = (a_k) \in \omega : \sup \left|\sum_{k=1}^{n} a_k \sum_{j=1}^{k-m} \binom{k-j-1}{m-1} v_j\right| < \infty, \\ \sum_k |R_k| \sum_{j=1}^{k-m+1} \binom{k-j-1}{m-2} v_j < \infty \text{ for all } v \in c_0^+ \end{array}\right\}.$$

Then, the following statements hold:

(i) $\{sc_0(\Delta^m)\}^\beta = M_0^\beta$.

(ii) $\{sc_0(\Delta^m)\}^\gamma = M_0^\gamma$.

Lemma 4.5.17. [131] *Let* $\eta \in \{\beta, \gamma\}$. *Then the following statements hold:*

(i) $\{\ell_\infty(\Delta^m)\}^\eta := \{s\ell_\infty(\Delta^m)\}^\eta$.

(ii) $\{c(\Delta^m)\}^\eta := \{sc(\Delta^m)\}^\eta$.

(iii) $\{c_0(\Delta^m)\}^\eta := \{sc_0(\Delta^m)\}^\eta$.

Proof. We give the proof only for Part (i) with $\eta = \beta$, since the case $\eta = \gamma$ can be proved in the similar way. Since $s\ell_\infty(\Delta^m) \subset \ell_\infty(\Delta^m)$, then $\{\ell_\infty(\Delta^m)\}^\beta \subset \{s\ell_\infty(\Delta^m)\}^\beta$. Let $a = (a_k) \in \{s\ell_\infty(\Delta^m)\}^\beta$. If $x = (x_k) \in \ell_\infty(\Delta^m)$ with

$$x_k := \begin{cases} x_k & , \quad k \le m, \\ x'_k & , \quad k > m, \end{cases} \tag{4.5.8}$$

where $x' = (x'_k) \in s\ell_\infty(\Delta^m)$, then we can write for $n > m$ that

$$\sum_{k=1}^{n} a_k x_k = \sum_{k=1}^{m} a_k x_k + \sum_{k=m+1}^{n} a_k x'_k.$$

Now, letting $n \to \infty$, we obtain the series in the same way as in the proof of Part (i) of Lemma 4.5.14,

$$\sum_{k=1}^{n} a_k x_k = \sum_{k=1}^{m} a_k x_k + (-1)^m \sum_{k} (k+m-1)^{m-1} R_{k+m-1} z_k$$

is convergent. This implies that $a \in \{\ell_\infty(\Delta^m)\}^\beta$. □

Theorem 4.5.18. [131] *Let λ stands for ℓ_∞ or c and $R_k = \sum_{v=k+1}^{\infty} a_v$ for all $k \in \mathbb{N}$. Then, the following statements hold:*

(i) $\{\lambda(\Delta^m)\}^\beta := \{a = (a_k) : (k^m a_k)_{k\in\mathbb{N}_1} \in cs$ and $\sum_k k^{m-1}|R_k| < \infty\}$.

(ii) $\{\lambda(\Delta^m)\}^\gamma := \{a = (a_k) : (k^m a_k)_{k\in\mathbb{N}_1} \in bs$ and $\sum_k k^{m-1}|R_k| < \infty\}$.

Proof. This follows from Lemmas 4.5.14, 4.5.15 and 4.5.17. □

Theorem 4.5.19. [88] *Let c_0^+ denote the set of all positive sequences in c_0 and $v \in c_0^+$. Then, the following statements hold:*

(i) $$\{c_0(\Delta^m)\}^\beta := \left\{ \begin{array}{c} a = (a_k) \in \omega : \sum_k a_k \sum_{j=1}^{k-m} \binom{k-j-1}{m-1} v_j \text{ converges and} \\ \sum_k |R_k| \sum_{j=1}^{k-m+1} \binom{k-j-1}{m-2} v_j < \infty \text{ for all } v \in c_0^+ \end{array} \right\}.$$

(ii) $$\{c_0(\Delta^m)\}^\gamma := \left\{ \begin{array}{c} a = (a_k) \in \omega : \sup \left| \sum_{k=1}^{n} a_k \sum_{j=1}^{k-m} \binom{k-j-1}{m-1} v_j \right| < \infty, \\ \sum_k |R_k| \sum_{j=1}^{k-m+1} \binom{k-j-1}{m-2} v_j < \infty \text{ for all } v \in c_0^+ \end{array} \right\}.$$

Proof. This follows from Lemmas 4.5.16 and 4.5.17. □

Corollary 4.5.20. *The following statements hold:*

(i) *The spaces $\ell_\infty(\Delta^m), c(\Delta^m)$ and $c_0(\Delta^m)$ are not perfect.*

(ii) *The spaces $\ell_\infty(\Delta^m), c(\Delta^m)$ and $c_0(\Delta^m)$ are not normal.*

(iii) *The spaces $\ell_\infty(\Delta^m), c(\Delta^m)$ and $c_0(\Delta^m)$ are not monotone.*

Theorem 4.5.21. [157] *Let $m \in \mathbb{N}_1$. Define the sets M_1 and M_2 by*

$$\begin{aligned} M_1 &= \left\{ a = (a_n) \in \omega : \lim_{n\to\infty} n^m a_n = 0 \right\}, \\ M_2 &= \left\{ a = (a_n) \in \omega : \sup_{n\in\mathbb{N}_1} \left| \sum_{k=0}^{n} \binom{n+m-k-1}{m-1} a_n \right| < \infty \right\}. \end{aligned}$$

Then, $\{\ell_\infty(\Delta^m)\}^N = \{c(\Delta^m)\}^N = M_1$ and $\{c_0(\Delta^m)\}^N = M_2$.

Proof. The proof of the part $\{\ell_\infty(\Delta^m)\}^N = \{c(\Delta^m)\}^N = M_1$ is easy. So, we only show that $\{c_0(\Delta^m)\}^N = M_2$. Let $a = (a_n) \in M_2$ and $x = (x_n) \in c_0(\Delta^m)$. Then, by (4.5.1), (4.5.2) and Lemma 4.5.12, we obtain that

$$\lim_{n\to\infty} a_n x_n = \lim_{n\to\infty} \left[\sum_{k=0}^{n} \binom{n+m-k-1}{m-1}\right] a_n \left[\sum_{k=0}^{n} \binom{n+m-k-1}{m-1}\right]^{-1} x_n = 0.$$

Hence, $a \in \{c_0(\Delta^m)\}^N$.

Now, let $a \in \{c_0(\Delta^m)\}^N$. Then $a_n x_n \to 0$, as $n \to \infty$, for all $x = (x_n) \in c_0(\Delta^m)$. On the other hand, for each $x = (x_n) \in c_0(\Delta^m)$ there exists one and only one $y = (y_k) \in c_0$ such that

$$x_n = \sum_{k=1}^{n} \binom{n+m-k-1}{m-1} y_k = \sum_{k=0}^{n} \binom{n+m-k-1}{m-1} y_k = 0, \quad y_0 = 0$$

by (4.5.6). Hence,

$$\lim_{n\to\infty} a_n x_n = \lim_{n\to\infty} \sum_{k=1}^{n} \binom{n+m-k-1}{m-1} a_n y_k = 0 \text{ for all } y = (y_k) \in c_0.$$

If we define $A = (a_{nk})$ by

$$a_{nk} := \begin{cases} \binom{n+m-k-1}{m-1} a_n & , \quad 1 \le k \le n, \\ 0 & , \quad k > n, \end{cases}$$

for all $n \in \mathbb{N}_1$, then we obtain for all $y = (y_k) \in c_0$ that

$$\lim_{n\to\infty} \sum_k a_{nk} y_k = \lim_{n\to\infty} \sum_{k=1}^{n} \binom{n+m-k-1}{m-1} a_n y_k = 0.$$

Hence, $A \in (c_0 : c_0)$ and so

$$\sup_{n\in\mathbb{N}_1} \sum_{k=1}^{n} |a_{nk}| = \sup_{n\in\mathbb{N}_1} \sum_{k=1}^{n} \binom{n+m-k-1}{m-1} |a_n| < \infty.$$

This completes the proof. □

Now, we give some new kind of duals of sequence spaces.

Definition 4.5.22. [157] *Let λ be any sequence space and $r \ge 1$. Define the sets $\lambda^{r\alpha}$, $\lambda^{r\beta}$, $\lambda^{r\gamma}$ and λ^{rN} as follows:*

$$\begin{aligned}
\lambda^{r\alpha} &= \{a = (a_k) \in \omega : \{(a_k x_k)^r\} \in \ell_1 \text{ for all } x = (x_k) \in \lambda\}, \\
\lambda^{r\beta} &= \{a = (a_k) \in \omega : \{(a_k x_k)^r\} \in cs \text{ for all } x = (x_k) \in \lambda\}, \\
\lambda^{r\gamma} &= \{a = (a_k) \in \omega : \{(a_k x_k)^r\} \in bs \text{ for all } x = (x_k) \in \lambda\}, \\
\lambda^{rN} &= \left\{a = (a_k) \in \omega : \lim_{k\to\infty} (a_k x_k)^r = \lim_{k\to\infty} a_k x_k = 0 \text{ for all } x = (x_k) \in \lambda\right\} = \lambda^N.
\end{aligned}$$

Then $\lambda^{r\alpha}$, $\lambda^{r\beta}$, $\lambda^{r\gamma}$ and λ^{rN} are called $r\alpha$-, $r\beta$-, $r\gamma$- and rN-duals of the space λ, respectively. It can be showed that $\lambda^{r\alpha} \subset \lambda^{r\beta} \subset \lambda^{r\gamma}$ and if $\lambda \subset \mu$ then $\mu^{r\eta} \subset \lambda^{r\eta}$ for $\eta \in \{\alpha, \beta, \gamma\}$. If we take $r = 1$, then we obtain the usual α-, β- and γ-duals of the space λ. If $\lambda = (\lambda^{r\alpha})^{r\alpha}$, then λ is said to be $r\alpha$-perfect.

Theorem 4.5.23. [90] *Let $1 < r < \infty$. Define the sets $U_1^{(r)}$ and $U_2^{(r)}$ by*

$$\begin{aligned} U_1^{(r)} &= \left\{a = (a_k) \in \omega : \sum_k k^{rm}|a_k|^r < \infty\right\}, \\ U_2^{(r)} &= \left\{a = (a_k) \in \omega : \sup_{k\in\mathbb{N}_1} k^{-rm}|a_k|^r < \infty\right\}. \end{aligned}$$

Then, the following statements hold:

(i) $\{\ell_\infty(\Delta^m)\}^{r\alpha} = \{c(\Delta^m)\}^{r\alpha} = \{c_0(\Delta^m)\}^{r\alpha} = U_1^{(r)}$.

(ii) $\{U_1^{(r)}\}^{r\alpha} = U_2^{(r)}$.

Proof. (i) Let $a = (a_k) \in U_1^{(r)}$ and $x = (x_k) \in \ell_\infty(\Delta^m)$, then

$$\begin{aligned} \sum_k |a_k x_k|^r &= \sum_k k^{rm}|a_k|^r k^{-rm}|x_k|^r \\ &\leq \sup_{k\in\mathbb{N}_1} k^{-rm}|x_k|^r \sum_k k^{rm}|a_k|^r < \infty. \end{aligned}$$

Hence, $a \in \{\ell_\infty(\Delta^m)\}^{r\alpha}$.

Conversely suppose that $a = (a_k) \in \{c_0(\Delta^m)\}^{r\alpha} \setminus U_1^{(r)}$. Then, there exists a strictly increasing sequence (n_i) of positive integers such that

$$\sum_{k=n_i+1}^{n_{i+1}} k^{rm}|a_k|^r > i^r.$$

Define $x = (x_k) \in c_0(\Delta^m)$ by

$$x_k := \begin{cases} 0 & , \quad 1 \leq k \leq n_1, \\ k^m(\text{sgn} a_k)/i & , \quad n_i < k \leq n_{i+1}. \end{cases}$$

Then, we have

$$\begin{aligned} \sum_k |a_k x_k|^r &= \sum_{k=n_1+1}^{n_2} |a_k x_k|^r + \cdots + \sum_{k=n_i+1}^{n_{i+1}} |a_k x_k|^r \\ &= \sum_{k=n_1+1}^{n_2} k^{rm}|a_k|^r + \cdots + \frac{1}{i^r}\sum_{k=n_i+1}^{n_{i+1}} k^{rm}|a_k|^r \\ &> 1 + 1 + \cdots = \infty. \end{aligned}$$

This contradicts that $a \in \{c_0(\Delta^m)\}^{r\alpha}$. Hence, $a \in U_1^{(r)}$ as desired.

(ii) Let $a = (a_k) \in U_2^{(r)}$ and $x = (x_k) \in U_1^{(r)}$. Then,

$$\begin{aligned}\sum_k |a_k x_k|^r &= \sum_k k^{rm}|a_k|^r k^{-rm}|x_k|^r \\ &\leq \sup_{k\in\mathbb{N}_1}(k^{-rm}|a_k|^r)\sum_k k^{rm}|x_k|^r < \infty,\end{aligned}$$

which leads us to the fact that $a \in \{U_1^{(r)}\}^{r\alpha}$.

Now, suppose that $a = (a_k) \in \{U_1^{(r)}\}^{r\alpha} \setminus U_2^{(r)}$. Then, we have

$$\sup_{k\in\mathbb{N}_1} k^{-rm}|a_k|^r = \infty.$$

Hence, there is a strictly increasing sequence $\{k(i)\}$ of positive integers $k(i)$ such that

$$[k(i)]^{-rm}|a_{k(i)}|^r > i^m.$$

Consider the sequence $x = (x_k)$ defined by

$$x_k := \begin{cases} |a_{k(i)}|^{-r} & , \quad k = k(i), \\ 0 & , \quad k \neq k(i). \end{cases}$$

Then, we have for $m \in \mathbb{N}_2$ that

$$\sum_k k^{rm}|x_k|^r = \sum_i [k(i)]^{rm}|a_{k(i)}|^{-r} \leq \sum_i i^{-m} < \infty.$$

Hence, $x \in U_1^{(r)}$ and $\sum_k |a_k x_k|^r = \infty$. This leads the contradiction that $a \in \{U_1^{(r)}\}^{r\alpha}$. Hence, $a \in U_2^{(r)}$. □

Corollary 4.5.24. *Let $\lambda \in \{\ell_\infty, c, c_0\}$ and $r \geq 1$. Then, $\lambda^{r\eta} = \ell_r$ for $\eta \in \{\alpha, \beta, \gamma\}$.*

Corollary 4.5.25. *The sequence spaces $\ell_\infty(\Delta^m)$, $c(\Delta^m)$ and $c_0(\Delta^m)$ are not $r\alpha$-perfect for $m \in \mathbb{N}_1$.*

4.5.2 Matrix Transformations

In this subsection, we characterize some matrix classes related to the spaces of mth order difference sequences.

Theorem 4.5.26. *Let λ denotes one of the sequence spaces ℓ_∞ and c. $A = (a_{nk}) \in (\lambda(\Delta^m) : \ell_\infty)$ if and only if*

(i) *$(a_{nj})_{n\in\mathbb{N}_1} \in \ell_\infty$ for all $j \in \{1, 2, \ldots, m\}$ and $\{(Ak^m)_n\} \in \ell_\infty$.*

(ii) *$R_m = (k^{m-1}r_{nk}) \in (\lambda : \ell_\infty)$, where $(Ak^m)_n = \sum_k k^m a_{nk}$ and $r_{nk} = \sum_{v=k+1}^{\infty} a_{nv}$.*

Proof. Let $A = (a_{nk}) \in (\lambda(\Delta^m) : \ell_\infty)$. Then, the series $\sum_k a_{nk}x_k$ is convergent for each $n \in \mathbb{N}_1$ and $\{(Ax)_n\} \in \ell_\infty$ for all $x \in \lambda(\Delta^m)$.

If we take $x = e^{(j)}$ with $1 \leq j \leq m$ and $x = (x_k) = (k^m)$, then we obtain the necessity of (i). If $R_m = (k^{m-1}r_{nk}) \notin (\lambda : \ell_\infty)$, then there exist subsequences (n_i) and (k_i) of positive integers such that

$$\lim_{i\to\infty} \sum_{k=1}^{k_i} k^{m-1} |r_{n_i,k}| = \infty. \tag{4.5.9}$$

From Theorem 4.5.18, we have

$$\sum_k k^{m-1} |r_{nk}| < \infty \quad \text{for each} \quad n \in \mathbb{N}_1. \tag{4.5.10}$$

By (4.5.10), there exists $M > 0$ such that

$$k^{m-1} |r_{nk}| < M \quad \text{for all} \quad k, n \in \mathbb{N}_1. \tag{4.5.11}$$

By (4.5.9), choose $n = n_1$ and $k = s_1$ such that

$$\sum_{k=1}^{s_1} k^{m-1} |r_{n_1,k}| > 1. \tag{4.5.12}$$

Having fixed n_1 by (4.5.10), choose $k_1 > s_1$ such that

$$\sum_{k=k_1+1}^{\infty} k^{m-1} |r_{n_1,k}| < \varepsilon. \tag{4.5.13}$$

Consider the sequence $x = (x_k)$ defined for all $k \in \mathbb{N}_1$ by

$$x_k := \begin{cases} \sum_{v=1}^{k-1} v^{m-1} \text{sgn} r_{nv} & , \quad 1 < n \leq k_1 \text{ and } k_{i-1} < n \leq k_i, \\ 0 & , \quad k = 1, \end{cases} \tag{4.5.14}$$

where $a_{nk} > 0$ or $a_{nk} < 0$ for all $n, k \in \mathbb{N}_1$ and $i \in \mathbb{N}_2$. Then, we see that $x \in \lambda(\Delta^m)$.

On the other hand, we observe from Lemmas 4.5.11 and 4.5.12 that

$$\sum_k a_{nk}x_k = -\sum_k r_{nk}\Delta x_k. \tag{4.5.15}$$

Hence,

$$|(Ax)_{n_1}| \geq \sum_{k=1}^{k_1} k^{m-1} |r_{n_1,k}| - \sum_{k=k_1+1}^{\infty} k^{m-1} |r_{n_1,k}| > 1 - \varepsilon$$

using (4.5.12), (4.5.13), (4.5.14) and (4.5.15).

From (4.5.11), we have for all $n \in \mathbb{N}_1$ that

$$\sum_{k=1}^{k_i} k^{m-1} |r_{n_i,k}| < \sum_{k=1}^{k_i} M = k_i M = C_{k_i}. \tag{4.5.16}$$

By (4.5.9), choose $n = n_2 > n_1$ and $s_2 > k_1$ such that

$$\sum_{k=k_1+1}^{s_2} k^{m-1} |r_{n_2,k}| > 2 + C_{k_1} \tag{4.5.17}$$

Having fixed n_2 by (4.5.10), choose $k_2 > s_2$ such that

$$\sum_{k=k_2+1}^{\infty} k^{m-1} |r_{n_2,k}| < \varepsilon. \tag{4.5.18}$$

Then, we have

$$\begin{aligned} |A_{n_2}(x)| &\geq \sum_{k=k_1+1}^{k_2} k^{m-1} |r_{n_2,k}| - \sum_{k=1}^{k_1} k^{m-1} |r_{n_2,k}| - \\ &\quad - \sum_{k=k_2+1}^{\infty} k^{m-1} |r_{n_2,k}| > 2 - \varepsilon \end{aligned}$$

using (4.5.14), (4.5.15), (4.5.16), (4.5.17) and (4.5.18).

Continuing this process, by (4.5.9), we can choose $n_i > n_{i-1}$ and $s_i > k_{i-1}$ such that

$$\sum_{k=k_{i+1}+1}^{s_i} k^{m-1} |r_{n_i,k}| > i + C_{k_{i-1}}.$$

Having fixed n_i by (4.5.10), choose $k_i > s_i$ such that

$$\sum_{k=k_i+1}^{\infty} k^{m-1} |r_{n_i,k}| < \varepsilon.$$

We can show that $|(Ax)_{n_i}| > i - \varepsilon$. Since ε is arbitrary, $|(Ax)_{n_i}| \to \infty$, as $i \to \infty$. Hence $\{(Ax)_n\} \notin \ell_\infty$. This contradicts the fact that $A \in (\lambda(\Delta^m) : \ell_\infty)$. Hence, $R_m = (k^{m-1} r_{nk}) \in (\lambda : \ell_\infty)$.

Now, suppose that (i) and (ii) hold. We define the sequence $x = (x_k) \in \lambda(\Delta^m)$ by (4.5.8). Then, we write for $m < t$ in the same way as in the proof of Lemma 4.5.14 that

$$\begin{aligned} A_n(t, m, x) &= \sum_{k=1}^{t} a_{nk} x_k \\ &= \sum_{k=1}^{m} a_{nk} x_k + (-1)^m \sum_{k=1}^{t-m} (k+m-1)^{m-1} r_{n,k+m-1} z_k - t^m r_{nt} t^{-m} x'_t, \end{aligned}$$

where $z = (z_k) = \left\{ \sum_{v=1}^{k} \binom{k+m-v-2}{m-2} y_v / (k+m-1)^{m-1} \right\}$ and $y = (y_v) \in \lambda$.

If we consider Lemmas 4.5.11 and 4.5.12, then we have

$$\begin{aligned} \lim_{t \to \infty} A_n(t, m, x) &= (Ax)_n \\ &= \sum_{k=1}^{m} a_{nk} x_k + (-1)^m \sum_{k} (k+m-1)^{m-1} r_{n,k+m-1} z_k \end{aligned}$$

for the sequence $x = (x_k) \in \lambda(\Delta^m)$. This implies that $\{(Ax)_n\} \in \ell_\infty$ for every $x \in \lambda(\Delta^m)$, as desired. □

Theorem 4.5.27. [166] *Let* $\lambda \in \{\ell_\infty, c\}$ *and* $\mu \in \{\ell_\infty, c, c_0\}$. *Then,* $A = (a_{nk}) \in (\lambda : \mu(\Delta^m))$ *if and only if*

(i) $\sum_k |a_{nk}| < \infty$ *for each* $n \in \mathbb{N}_1$.

(ii) $B = (b_{nk}) = (\Delta^{m-1} a_{nk} - \Delta^{m-1} a_{n+1,k}) \in (\lambda : \mu)$.

Proof. Since the proof of the sufficiency is trivial, we only give the proof of necessity part.

Suppose that $A = (a_{nk})$ maps λ into $\mu(\Delta^m)$. Then, the series $\sum_k a_{nk} x_k$ is convergent for each $n \in \mathbb{N}_1$ and for all $x = (x_k) \in \lambda$, and $\{(Ax)_n\} \in \mu(\Delta^m)$. Since $\lambda^\beta = \ell_1$, then we have (i). Furthermore, since $\{(Ax)_n\} \in \mu(\Delta^m)$ we have

$$(\Delta^m Ax)_n = \sum_k \Delta^m a_{nk} x_k = \sum_k b_{nk} x_k = (Bx)_n$$

for all $x \in \lambda$. Therefore, $\{(Bx)_n\} \in \mu$. Hence, $B \in (\lambda : \mu)$.

This completes the proof. □

4.5.3 Δ^m-Statistical Convergence

The concepts of statistical convergence and strongly Cesàro convergence of real sequences have been studied by various mathematicians. The notion of statistical convergence was independently introduced by Fast [170] and Schoenberg [360]. Over the years and under different names, statistical convergence has been discussed in the theory of Fourier analysis, ergodic theory and number theory. It was further investigated from the sequence space point of view and linked with summability theory by Connor [109], Fridy [174], Šalát [341], Tripathy [383] and many others. In recent years, generalizations of statistical convergence have appeared in the study of strong integral summability and the structure of ideals of bounded continuous functions on locally compact spaces. Statistical convergence and its generalizations are also connected with subsets of the Stone-Čech compactification of the natural numbers. Moreover, statistical convergence is closely related to the concept of convergence in probability.

The definitions of statistical convergence and strong p-Cesàro convergence of a sequence of real numbers are introduced in the literature independently of one another and are followed different lines of development since their first appearance. It turns out, however, that the two definitions can be simply related to one another in general and are equivalent for bounded sequences. The idea of statistical convergence depends on the density of subsets of the set $\mathbb{N}$ of natural numbers. The natural density $\delta(A)$ of a subset A of $\mathbb{N}$ is defined by

$$\delta(A) := \lim_{n\to\infty} \frac{1}{n} \sum_{k=1}^{n} \chi(A)(k) \quad \text{provided the limit exists.}$$

It is clear that any finite subset of $\mathbb{N}$ has zero natural density and $\delta(A^c) = 1 - \delta(A)$.

A sequence (x_k) of complex numbers is said to be statistically convergent to l if for every $\varepsilon > 0$, $\delta(\{k \in \mathbb{N} : |x_k - l| \geq \varepsilon\}) = 0$. The set of all statistical convergent sequences is denoted by S. In this case, we write $S - \lim x_k = l$.

If $x = (x_k)$ is a sequence that satisfies some property p for all k except a set of natural density zero, then we say that (x_k) satisfies p for "*almost all k*" and we abbreviate this by "*a.a.k*".

In this subsection, we give the definition of Δ^m-statistical convergence, which reduces to the statistical convergence in the case $m = 0$ and inclusion theorems between the sets of all Δ^m-statistical convergent sequences and other some sequence spaces.

Definition 4.5.28. [168] *A sequence $x = (x_k)$ is said to be Δ^m-statistically convergent if there is a complex number l such that*

$$\lim_{n\to\infty} \frac{|\{k \leq n : |(\Delta^m x)_k - l| \geq \varepsilon\}|}{n} = 0$$

for every $\varepsilon > 0$, in which case we say that x is Δ^m-statistically convergent to l. In this case we write $x_k \to l(S(\Delta^m))$, where the vertical bars indicate the number of elements in the enclosed set. The set of Δ^m-statistically convergent sequences will be denoted by $S(\Delta^m)$. In the case $l = 0$, we shall write $S_0(\Delta^m)$.

Definition 4.5.29. [168] *A sequence $x = (x_k)$ is said to be strongly Δ^m-Cesàro summable if there is a complex number l such that*

$$\lim_{n\to\infty} \frac{\sum_{k=1}^{n} |(\Delta^m x)_k - l|^p}{n} = 0 \ \textit{for some} \ l,$$

where p is a fixed positive number. The number l is unique when it exists. By $\omega_p(\Delta^m)$, we denote the set of all strongly Δ^m-Cesàro convergent sequences. In this case, we write $x_k \to l(\omega_p(\Delta^m))$.

Theorem 4.5.30. *The sequence space $\omega_p(\Delta^m)$ is a BK-space with the norm*

$$\|x\|_{\Delta 1} := \sum_{i=1}^{m} |x_i| + \sup_{n\in\mathbb{N}_1} \left[\frac{1}{n}\sum_{k=1}^{n} |(\Delta^m x)_k|^p\right]^{1/p}, \quad 1 < p < \infty$$

and a complete p-normed space with the p-norm

$$\|x\|_{\Delta 2} := \sum_{i=1}^{m} |x_i|^p + \sup_{n\in\mathbb{N}_1} \frac{1}{n}\sum_{k=1}^{n} |(\Delta^m x)_k|^p, \quad 0 < p \leq 1.$$

Proof. This follows from Theorem 4.5.4. □

We now define the norm in the space $\omega_p(\Delta^m)$, which are different than $\|\cdot\|_{\Delta_1}$ and $\|\cdot\|_{\Delta_2}$ given in Theorem 4.5.30. For $x = (x_k) \in \omega_p(\Delta^m)$ define

$$\begin{aligned}
\|x\|_{\Delta 3} &:= \sum_{i=1}^{m} |x_i| + \sup_{r} \left[2^{-r} \sum_{r} |(\Delta^m x)_k|^p\right]^{1/p}, \ (p \geq 1),\\
\|x\|_{\Delta 4} &:= \sum_{i=1}^{m} |x_i|^p + \sup_{r} 2^{-r} \left[\sum_{k=1}^{n} |(\Delta^m x)_k|^p\right], \ (0 < p \leq 1),
\end{aligned}$$

where $s \geq 0$ and $\sum_r$ denote the sum over $k \in [2^s, 2^{s+1})$. It is trivial that the norms $\|\cdot\|_{\Delta 1}$ and $\|\cdot\|_{\Delta 2}$ are equivalent to the norms $\|\cdot\|_{\Delta 3}$ and $\|\cdot\|_{\Delta 4}$, respectively. The inclusion $\omega_p(\Delta^{m-1}) \subset \omega_p(\Delta^m)$ is strict, since the sequence $x = (k^m)$ belongs to $\omega_p(\Delta^m)$ but does not belong to $\omega_p(\Delta^{m-1})$ for $0 < p < \infty$.

Corollary 4.5.31. *The following statements hold:*

(i) *Let $p, q \in \mathbb{R}$ with $0 \leq p < q < \infty$. Then, $\omega_q(\Delta^m) \subset \omega_p(\Delta^m)$.*

(ii) $\omega_p \subset \omega_p(\Delta^m)$.

(iii) *ω_p and $\omega_p(\Delta^m)$ are the identical topological spaces.*

Theorem 4.5.32. [168] *Let $0 < p < \infty$. If $x_k \to l(\omega_p(\Delta^m))$, then $x_k \to l(S(\Delta^m))$. If $x \in \Delta^m(\ell_\infty)$ and $x_k \to l(S(\Delta^m))$, then $x_k \to l(\omega_p(\Delta^m))$.*

Proof. Let $x_k \to l(\omega_p(\Delta^m))$ and $\varepsilon > 0$. Then, we have

$$\sum_{k=1}^{n} |(\Delta^m x)_k - l|^p \geq |\{k \leq n : |(\Delta^m x)_k - l| \geq \varepsilon\}| \, \varepsilon^p.$$

Hence, $x_k \to l(S(\Delta^m))$.

Suppose that $x = (x_k) \in \ell_\infty(\Delta^m)$ and $x_k \to l(S(\Delta^m))$. Let $\varepsilon > 0$ be given and choose N_ε such that

$$\frac{\left|\left\{k \leq n : |(\Delta^m x)_k - l| \geq \left(\frac{\varepsilon}{2}\right)^{1/p}\right\}\right|}{n} \leq \frac{\varepsilon}{2T^p}$$

for all $n > N_\varepsilon$ and set $L_n = \left\{k \leq n : |(\Delta^m x)_k - l| \geq \left(\frac{\varepsilon}{2}\right)^{1/p}\right\}$, where $T = \|\Delta^m x\|_\infty + |l|$. Now, we have for all $n > N_\varepsilon$ that

$$\begin{aligned}
\frac{1}{n}\sum_{k=1}^{n} |(\Delta^m x)_k - l|^p &= \frac{1}{n}\left(\sum_{k \in L_n} |\Delta^m x_k - l|^p + \sum_{k \notin L_n} |\Delta^m x_k - l|^p\right) \\
&< \frac{1}{n}\left(\frac{n\varepsilon}{2T^p}\right) T^p + \frac{1}{n} n \frac{\varepsilon}{2} = \varepsilon.
\end{aligned}$$

Hence, $x_k \to l(\omega_p(\Delta^m))$. □

Corollary 4.5.33. *The following statements hold:*

(i) $S \cap \ell_\infty \subset S(\Delta^m) \cap \ell_\infty(\Delta^m)$.

(ii) $S(\Delta^m) \cap \ell_\infty(\Delta^m) = \omega_p(\Delta^m)$.

Theorem 4.5.34. [160] *Let $0 < h = \inf_{k \in \mathbb{N}_1} p_k \leq p_k \leq \sup_{k \in \mathbb{N}_1} p_k = H < \infty$, then $\omega(p, \Delta^m) \subset S(\Delta^m)$, where*

$$\omega(p, \Delta^m) := \left\{x = (x_k) : \lim_{n \to \infty} \frac{1}{n} \sum_{k=1}^{n} |(\Delta^m x)_k - l|^{p_k} = 0 \ \textit{ for some } l \in \mathbb{C}\right\}.$$

Proof. Let $x = (x_k) \in \omega(p, \Delta^m)$ and $\varepsilon > 0$ be given. Let $\sum_1$ denote the sum over $k \le n$ such that $|(\Delta^m x)_k - l| \ge \varepsilon$ and $\sum_2$ denote the sum over $k \le n$ such that $|(\Delta^m x)_k - l| < \varepsilon$. Then,

$$\begin{aligned}
\frac{1}{n}\sum_{k=1}^{n}|(\Delta^m x)_k - l|^{p_k} &= \frac{1}{n}\left[\sum_1 |(\Delta^m x)_k - l|^{p_k} + \sum_2 |(\Delta^m x)_k - l|^{p_k}\right] \\
&\ge \frac{1}{n}\sum_1 |(\Delta^m x)_k - l|^{p_k} \\
&\ge \frac{1}{n}\sum_1 \varepsilon^{p_k} \\
&\ge \frac{1}{n}\sum_1 \min\left\{\varepsilon^h, \varepsilon^H\right\} \\
&\ge \frac{1}{n}|\{k \le n : |(\Delta^m x)_k - l| \ge \varepsilon\}| \min\left\{\varepsilon^h, \varepsilon^H\right\}.
\end{aligned}$$

Hence, $x \in S(\Delta^m)$. □

Theorem 4.5.35. [160] $\ell_\infty(\Delta^m) \cap S(\Delta^m) \subset \omega(p, \Delta^m)$.

Proof. Suppose that $x = (x_k)$ is Δ^m-bounded. Let $\varepsilon > 0$, $\sum_1$ and $\sum_2$ be as in the proof of Theorem 4.5.34. Since x is Δ^m-bounded, there exists a positive integer K such that $|(\Delta^m x)_k| < K$ for all $k \in \mathbb{N}_1$. Then,

$$\begin{aligned}
\frac{1}{n}\sum_{k=1}^{n}|(\Delta^m x)_k - l|^{p_k} &= \frac{1}{n}\left[\sum_1 |(\Delta^m x)_k - l|^{p_k} + \sum_2 |(\Delta^m x)_k - l|^{p_k}\right] \\
&\le \max\left\{K^h, K^H\right\}\frac{|\{k \le n : |(\Delta^m x)_k - l| \ge \varepsilon\}|}{n} + \max\left\{\varepsilon^h, \varepsilon^H\right\}.
\end{aligned}$$

Hence, $x \in \omega(p, \Delta^m)$. □

Definition 4.5.36. *The sequence $x = (x_k) \in \omega$ is said to be Δ^m-statistically Cauchy if there exists a number $N = N(\varepsilon)$ such that*

$$\lim_{n\to\infty}\frac{|\{k \le n : |(\Delta^m x)_k - (\Delta^m x)_N| \ge \varepsilon\}|}{n} = 0$$

for every $\varepsilon > 0$.

Now, we can give the following theorem, without proof:

Theorem 4.5.37. [168] *If x is a Δ^m-statistically convergent sequence, then x is a Δ^m-statistically Cauchy sequence.*

Theorem 4.5.38. [168] *If $x = (x_k)$ is a sequence for which there is a Δ^m-statistically convergent sequence $y = (y_k)$ such that $(\Delta^m x)_k = (\Delta^m y)_k$ for almost all $k \in \mathbb{N}$, then x is Δ^m-statistically convergent.*

Proof. Assume that $(\Delta^m x)_k = (\Delta^m y)_k$ for almost all $k \in \mathbb{N}$ and $y_k \to l(S(\Delta^m))$, and $\varepsilon > 0$. Then, for each $n \in \mathbb{N}_1$,

$$\{k \leq n : |(\Delta^m x)_k - l| \geq \varepsilon\} \subseteq \{k \leq n : (\Delta^m x)_k \neq (\Delta^m y)_k\} \cup \{k \leq n : |(\Delta^m x)_k - l| \geq \varepsilon\},$$

since $y_k \to l(S(\Delta^m))$ as $k \to \infty$, the latter set contains a fixed number of integers, say $g = g(\varepsilon)$. Therefore,

$$\lim_{n\to\infty} \frac{|\{k \leq n : |(\Delta^m x)_k - l| \geq \varepsilon\}|}{n} \leq \lim_{n\to\infty} \frac{|\{k \leq n : (\Delta^m x)_k \neq (\Delta^m y)_k\}|}{n} + \lim_{n\to\infty} \frac{g}{n} = 0$$

because $(\Delta^m x)_k = (\Delta^m y)_k$ for almost all $k \in \mathbb{N}$. Hence, $x_k \to l(S(\Delta^m))$ as $k \to \infty$. □

It follows from Lemma 1.2 of [341] and Theorem 4.5.1 that the set of all Δ^m-bounded statistically convergent sequences is a linear subspace of the linear normed space of all Δ^m-bounded sequences.

Now, let $m_0(\Delta^m)$ and $m(\Delta^m)$ be the set of all Δ^m-bounded statistically convergent and the set of all Δ^m-bounded convergent sequences of real numbers, respectively.

Theorem 4.5.39. [168] *The set $m_0(\Delta^m)$ is a closed linear subspace of the linear normed space $m(\Delta^m)$.*

Proof. This follows from Theorem 2.1 of Šalát [341] and Theorem 4.5.6. □

Theorem 4.5.40. [168] *The set $m_0(\Delta^m)$ is nowhere dense in $m(\Delta^m)$.*

Proof. It is well-known that every closed linear subspace of an arbitrary linear normed space E different from E is a nowhere dense set in E. Hence, by taking into account Theorem 4.5.39 it suffices to prove that $m_0(\Delta^m) \neq m(\Delta^m)$. But this is immediate that $x = (x_k) = \{(-1)^k\} \in m(\Delta^m) \setminus m_0(\Delta^m)$, since $\Delta^m x = \{(-1)^k 2^m\}$. □

Corollary 4.5.41. *m_0 and $m_0(\Delta^m)$ are identical topological spaces.*

Theorem 4.5.42. [157] *The following statements hold:*

(i) *The set $S(\Delta^m)$ is dense in the space ω.*

(ii) *The set $S(\Delta^m)$ is a set of the first Baire category in the space ω.*

(iii) *The set $\omega \setminus S(\Delta^m)$ is a set of the second Baire category in the space ω.*

Proof. This follows from Theorem 3.1 of Šalát [341] with Theorem 4.5.8. We recall that the complement A^c of a first category subset A of a complete metric space X is of the second category. □

4.5.4 Paranormed Difference Sequence Spaces

Let $p = (p_k)$ be any sequence of strictly positive real numbers. Now, we give the α-, β- and γ-duals of the sequence spaces $\ell_\infty(p,\Delta^m)$, $c(p,\Delta^m)$ and $c_0(p,\Delta^m)$.

Theorem 4.5.43. [163] *Let $p = (p_k) \in \ell_\infty$. Then, $c_0(p,\Delta^m)$ is a complete paranormed space, paranormed by*

$$g(x) = \sum_{i=1}^{m} |x_i| + \sup_{k\in\mathbb{N}_1} |\Delta^m x|^{p_k/K}, \text{ where } K = \max\left\{1, H = \sup_{k\in\mathbb{N}_1} p_k\right\}. \quad (4.5.19)$$

If $\inf_{k\in\mathbb{N}_1} p_k > 0$, then $c(p,\Delta^m)$ and $\ell_\infty(p,\Delta^m)$ are the complete paranormed spaces paranormed by g given by (4.5.19).

Proof. This follows from Theorem 4.5.4. □

Theorem 4.5.44. [165] *Define the sets $D_1^\alpha(p)$ and $D_1^{\alpha\alpha}(p)$ by*

$$D_1^\alpha(p) := \bigcap_{N=2}^{\infty}\left\{a = (a_k) : \sum_k |a_k| \sum_{j=1}^{k-m} \binom{k-j-1}{m-1} N^{1/p_j} < \infty\right\},$$

$$D_1^{\alpha\alpha}(p) := \bigcup_{N=2}^{\infty}\left\{a = (a_k) : \sup_{k\geq m+1} |a_k| \left[\sum_{j=1}^{k-m} \binom{k-j-1}{m-1} N^{1/p_j}\right]^{-1} < \infty\right\}.$$

Then, for every sequence $p = (p_k)$ of strictly positive real numbers, the following statements hold:

(i) $\{\ell_\infty(p,\Delta^m)\}^\alpha = D_1^\alpha(p)$.

(ii) $\{\ell_\infty(p,\Delta^m)\}^{\alpha\alpha} = D_1^{\alpha\alpha}(p)$.

Proof. (i) Let $a = (a_k) \in D_1^\alpha(p)$ and $x = (x_k) \in \ell_\infty(p,\Delta^m)$. Then, we choose $N > \max\{1, \sup_{n\in\mathbb{N}_1} |(\Delta^m a)_n|^{p_n}\}$. Since

$$\sum_{j=1}^{k-m} \binom{k-j-1}{m-1} N^{1/p_j} > \sum_{j=1}^{m} \binom{k-j-1}{m-j} N^{1/p_j} \geq 1$$

for arbitrary $N > 1$, $(k = 2m, 2m+1, \ldots)$ and $|(\Delta^{m-j}x)_j| \leq M$ $(1 \leq j \leq m)$ for some constant M, $a = (a_k) \in D_1^\alpha(p)$ implies that

$$\sum_k |a_k| \sum_{j=1}^{m} \binom{k-j-1}{m-j} \left|\Delta^{m-j} x_j\right| < \infty.$$

Then,

$$\begin{aligned}
\sum_k |a_k x_k| &= \sum_k |a_k| \left| \sum_{j=1}^{k-m} (-1)^m \binom{k-j-1}{m-1} \Delta^m x_j + \sum_{j=1}^{m} (-1)^{m-j} \binom{k-j-1}{m-j} \Delta^{m-j} x_j \right| \\
&\leq \sum_k |a_k| \sum_{j=1}^{k-m} \binom{k-j-1}{m-1} N^{1/p_j} + \sum_k |a_k| \sum_{j=1}^{m} \binom{k-j-1}{m-j} \left|\Delta^{m-j} x_j\right| < \infty.
\end{aligned}$$

Conversely, suppose that $a = (a_k) \notin D_1^\alpha(p)$. Then, we have

$$\sum_k |a_k| \sum_{j=1}^{k-m} \binom{k-j-1}{m-1} N^{1/p_j} = \infty$$

for some integer $N > 1$. We define the sequence $x = (x_k)$ by

$$x_k = \sum_{j=1}^{k-m} \binom{k-j-1}{m-1} N^{1/p_j} \text{ for all } k \in \mathbb{N}_{m+1}.$$

Then, it is easy to see that $x = (x_k) \in \ell_\infty(p)(\Delta^m)$ and $\sum_k |a_k x_k| = \infty$. Hence, $a \notin \{\ell_\infty(p, \Delta^m)\}^\alpha$. This completes the proof of Part (i).

(ii) Let $a = (a_k) \in D_1^{\alpha\alpha}(p)$ and $x = (x_k) \in \{\ell_\infty(p, \Delta^m)\}^\alpha = D_1^\alpha(p)$. Then, for some $N > 1$, we have

$$\sum_{k=m+1}^{\infty} |a_k x_k| = \sum_{k=m+1}^{\infty} |a_k| \left[\sum_{j=1}^{k-m} \binom{k-j-1}{m-1} N^{1/p_j}\right]^{-1} |x_k| \sum_{j=1}^{k-m} \binom{k-j-1}{m-1} N^{1/p_j}$$

$$\leq \sup_{k \geq m+1} |a_k| \left[\sum_{j=1}^{k-m} \binom{k-j-1}{m-1} N^{1/p_j}\right]^{-1} \sum_{k=m+1}^{\infty} |x_k| \sum_{j=1}^{k-m} \binom{k-j-1}{m-1} N^{1/p_j} < \infty.$$

Conversely, let $a = (a_k) \notin D_1^{\alpha\alpha}(p)$. Then, for all integers $N > 1$, we have

$$\sup_{k \geq m+1} |a_k| \left[\sum_{j=1}^{k-m} \binom{k-j-1}{m-1} N^{1/p_j}\right]^{-1} = \infty.$$

We recall that $\sum_{j=1}^{k-m} \binom{k-j-1}{m-1} y_j = 0$ whenever $k < m+1$ for arbitrary y_j. Hence, there is a strictly increasing sequence $\{k(s)\}$ of integers $k(s) \geq m+1$ such that

$$|a_{k(s)}| \left[\sum_{j=1}^{k(s)-m} \binom{k(s)-j-1}{m-1} s^{1/p_j}\right]^{-1} > s^{m+1} \text{ for all } s \in \mathbb{N}_{m+1}.$$

We define the sequence $x = (x_k)$ by

$$x_k := \begin{cases} |a_{k(s)}|^{-1} & , \quad k = k(s), \\ 0 & , \quad k \neq k(s), \end{cases} \quad k \in \mathbb{N}_{m+1}.$$

Then, for all integers $N > m+1$, we have

$$\begin{aligned}
\sum_k |x_k| \sum_{j=1}^{k-m} \binom{k-j-1}{m-1} N^{1/p_j} &\leq \sum_{s=m+1}^{\infty} \left|a_{k(s)}\right|^{-1} \sum_{j=1}^{k(s)-m} \binom{k(s)-j-1}{m-j} N^{1/p_j} \\
&\leq \sum_{s=m+1}^{N-1} \left|a_{k(s)}\right|^{-1} \sum_{j=1}^{k(s)-m} \binom{k(s)-j-1}{m-1} N^{1/p_j} + \\
&+ \sum_{s=N}^{\infty} \left|a_{k(s)}\right|^{-1} \sum_{j=1}^{k(s)-m} \binom{k(s)-j-1}{m-j} s^{1/p_j}.
\end{aligned}$$

Hence, $x \in \{\ell_\infty(p,\Delta^m)\}^\alpha$ and $\sum_k |a_k x_k| = \sum_{N=1}^{\infty} 1 = \infty$. This leads us to the fact that $a \notin \{\ell_\infty(p,\Delta^m)\}^{\alpha\alpha}$.

This completes the proof. □

Theorem 4.5.45. [87] *Define the sets $M_0^\alpha(p)$ and $M_0^{\alpha\alpha}(p)$ by*

$$M_0^\alpha(p) := \bigcup_{N=2}^{\infty} \left\{ a=(a_k) : \sum_k |a_k| \sum_{j=1}^{k-m} \binom{k-j-1}{m-1} N^{-1/p_j} < \infty \right\},$$

$$M_0^{\alpha\alpha}(p) := \bigcap_{N=2}^{\infty} \left\{ a=(a_k) : \sup_{k\geq m+1} |a_k| \left[\sum_{j=1}^{k-m} \binom{k-j-1}{m-1} N^{-1/p_j} \right]^{-1} < \infty \right\}.$$

Then, for every sequence $p=(p_k)$ of strictly positive real numbers, the following statements hold:

(i) $\{c_0(p,\Delta^m)\}^\alpha = M_0^\alpha(p)$.

(ii) $\{c_0(p,\Delta^m)\}^{\alpha\alpha} = M_0^{\alpha\alpha}(p)$.

Proof. We only give the proof of Part (i).

Let $a=(a_k) \in M_0^\alpha(p)$ and $x=(x_k) \in c_0(p,\Delta^m)$. Then, there is an integer k_0 such that $\sup_{k>k_0} |(\Delta^m x)_k|^{p_k} \leq N^{-1}$, where N is the number in $M_0^\alpha(p)$. We put $M = \max_{1\leq k\leq k_0} |(\Delta^m x)_k|^{p_k}$, $n = \min\{p_k : 1 \leq k \leq k_0\}$, $L=(M+1)N$ and define the sequence $y=(y_k)$ by $y_k = x_k L^{-1/n}$ for all $k \in \mathbb{N}_1$. Then, it is easy to see that $\sup_{k\in\mathbb{N}_1} |(\Delta^m y)_k|^{p_k} \leq N^{-1}$.

Since $\sum_{j=1}^{k-m} \binom{k-j-1}{m-1} N^{-1/p_j} > \sum_{j=1}^{m} \binom{k-j-1}{m-j} N^{-1/p_j}$ for arbitrary $N>1$ for all $k \in \mathbb{N}_{2m}$, $a \in M_0^\alpha(p)$ implies

$$\sum_k |a_k| \sum_{j=1}^{m} \binom{k-j-1}{m-j} |(\Delta^{m-j} y)_j| < \infty.$$

Then,

$$\begin{aligned}
\sum_k |a_k x_k| &= L^{1/n} \sum_k |a_k y_k| \\
&\leq L^{1/n} \sum_k |a_k| \sum_{j=1}^{k-m} \binom{k-j-1}{m-1} N^{-1/p_j} + \\
&+ L^{1/n} \sum_k |a_k| \sum_{j=1}^{m} \binom{k-j-1}{m-j} |(\Delta^{m-j} y)_j| < \infty.
\end{aligned}$$

So, we have $a \in \{c_0(p,\Delta^m)\}^\alpha$. Therefore, $M_0^\alpha(p) \subset \{c_0(p,\Delta^m)\}^\alpha$.

Conversely, let $a=(a_k) \notin M_0^\alpha(p)$. Then, we can determine a strictly increasing sequence $\{k(s)\}$ of positive integers such that $k(1)=1$ and

$$M(s) = \sum_{k=k(s)}^{k(s+1)-1} |a_k| \sum_{j=1}^{k-m} \binom{k-j-1}{m-1} (s+1)^{-1/p_j} > 1 \ \text{ for all } \ s \in \mathbb{N}_1.$$

We define the sequence $x = (x_k)$ by

$$x_k = \sum_{l=1}^{s-1} \sum_{j=k(l)}^{k(l+1)-1} \binom{k-j-1}{m-1} (l+1)^{-1/p_j} + \sum_{j=k(s)}^{k-m} \binom{k-j-1}{m-1} (s+1)^{-1/p_j},$$

where $k(s) \leq k \leq k(s+1)-1$ for all $s \in \mathbb{N}_1$. Then, it is easy to see that $|(\Delta^m x)_k|^{p_k} = 1/(s+1)$ if $k(s) \leq k \leq k(s+1)-1$ for all $s \in \mathbb{N}_1$, hence $x \in c_0(p, \Delta^m)$ and $\sum_k |a_k x_k| \geq \sum_{s=1}^{\infty} 1/(s+1) = \infty$, i.e., $a \notin \{c_0(p, \Delta^m)\}^\alpha$. □

Theorem 4.5.46. [87] *For every strictly positive sequence $p = (p_k)$, we have*

$$\{c(p, \Delta^m)\}^\alpha = M^\alpha(p) = M_0^\alpha(p) \cap \left\{ a = (a_k) : \sum_k |a_k| \sum_{j=1}^{k-m} \binom{k-j-1}{m-1} < \infty \right\}.$$

Proof. Let $a = (a_k) \in M^\alpha(p)$ and $x = (x_k) \in c(p, \Delta^m)$. Then, there is a complex number l such that $|(\Delta^m x)_k - l|^{p_k} \to 0$, as $k \to \infty$. We define $y = (y_k)$ by $y_k = x_k + l\,(-1)^{m+1} \sum_{j=1}^{k-m} \binom{k-j-1}{m-1}$ for all $k \in \mathbb{N}_1$. Then, $y \in c(p, \Delta^m)$ and

$$\begin{aligned} \sum_k |a_k x_k| &\leq \sum_k |a_k| \sum_{j=1}^{k-m} \binom{k-j-1}{m-1} |(\Delta^m y)_j| + \\ &+ \sum_k |a_k| \sum_{j=1}^{m} \binom{k-j-1}{m-j} |(\Delta^{m-j} y)_j| + |l| \sum_k |a_k| \sum_{j=1}^{k-m} \binom{k-j-1}{m-1} \\ &< \infty \end{aligned}$$

by Part (i) of Theorem 4.5.45 and $a \in \{c(p, \Delta^m)\}^\alpha$.

Now, let $a = (a_k) \in \{c(p, \Delta^m)\}^\alpha \subset \{c_0(p, \Delta^m)\}^\alpha = M_0^\alpha(p)$ by Part (i) of Theorem 4.5.45. Since the sequence $x = (x_k)$ defined by $x_k = (-1)^m \sum_{j=1}^{k-m} \binom{k-j-1}{m-1}$ for all $k \in \mathbb{N}_1$ is in $c(p, \Delta^m)$, we have

$$\sum_k |a_k| \sum_{j=1}^{k-m} \binom{k-j-1}{m-1} < \infty.$$

That is to say that $a \in M^\alpha(p)$, which completes the proof. □

Theorem 4.5.47. [87] *Define the sets $M_\infty^\beta(p)$ and $M_\infty^\gamma(p)$, as follows:*

$$M_\infty^\beta(p) := \bigcap_{N>1} \left\{ \begin{array}{r} (a_k) \in \omega : \sum_k a_k \sum_{j=1}^{k-m} \binom{k-j-1}{m-1} N^{1/p_j} \text{ converges and} \\ \sum_k |R_k| \sum_{j=1}^{k-m+1} \binom{k-j-1}{m-2} N^{1/p_j} < \infty \end{array} \right\},$$

$$M_\infty^\gamma(p) := \bigcap_{N>1} \left\{ \begin{array}{r} (a_k) \in \omega : \sup_{n \in \mathbb{N}_1} \left| \sum_{k=1}^{n} a_k \sum_{j=1}^{k-m} \binom{k-j-1}{m-1} N^{1/p_j} \right| < \infty \text{ and} \\ \sum_k |R_k| \sum_{j=1}^{k-m+1} \binom{k-j-1}{m-2} N^{1/p_j} < \infty \end{array} \right\},$$

where $R_k = \sum_{j=k+1}^{\infty} a_j$ for all $k \in \mathbb{N}_1$. Then, for every sequence $p = (p_k)$ of strictly positive real numbers, the following statements hold:

(i) $\{s\ell_\infty(p,\Delta^m)\}^\beta = M_\infty^\beta(p)$.

(ii) $\{s\ell_\infty(p,\Delta^m)\}^\gamma = M_\infty^\gamma(p)$.

Proof. (i) Let $x=(x_k)\in s\ell_\infty(p,\Delta^m)$, that is to say that (4.5.4) holds. Then, there is an integer N such that $N>\max\{1,\sup_{k\in\mathbb{N}_1}|(\Delta^m x)_k|^{p_k}\}$. Let $a=(a_k)\in M_\infty^\beta(p)$. Therefore, one can see that

$$\begin{aligned}\sum_{k=1}^n a_k x_k &= \sum_{k=1}^n a_k\left[\sum_{j=1}^{k-m}(-1)^m\binom{k-j-1}{m-1}y_j\right]\\ &= (-1)^m\sum_{k=1}^{n-m}R_{k+m-1}\sum_{j=1}^{k}\binom{k+m-j-2}{m-2}y_j - R_n\sum_{j=1}^{n-m}(-1)^m\binom{n-j-1}{m-1}y_j.\end{aligned}$$

Since $\sum_k|R_{k+m-1}|\sum_{j=1}^k\binom{k+m-j-2}{m-2}N^{1/p_j}<\infty$, the series $\sum_k R_{k+m-1}\sum_{j=1}^k\binom{k+m-j-2}{m-2}y_j$ is absolutely convergent. Moreover, by Part (ii) of Lemma 4.5.11, the convergence of $\sum_k a_k\sum_{j=1}^{k-m}\binom{k-j-1}{m-1}N^{1/p_j}$ implies $R_n\sum_{j=1}^{n-m}\binom{n-j-1}{m-1}N^{1/p_j}\to 0$, as $n\to\infty$. Hence, $\sum_k a_k x_k$ is convergent for all $x=(x_k)\in s\ell_\infty(p,\Delta^m)$, so $a\in\{s\ell_\infty(p,\Delta^m)\}^\beta$.

Conversely, let $a=(a_k)\in\{s\ell_\infty(p,\Delta^m)\}^\beta$. Then, $\sum_k a_k x_k$ is convergent for all $x=(x_k)\in s\ell_\infty(p,\Delta^m)$. If we take the sequence $x=(x_k)$ defined by

$$x_k := \begin{cases} 0 & , \quad k\le m,\\ \sum_{j=1}^{k-m}\binom{k-j-1}{m-1}N^{1/p_j} & , \quad k>m,\end{cases}$$

then,

$$\sum_k a_k\sum_{j=1}^{k-m}\binom{k-j-1}{m-1}N^{1/p_j} = \sum_k a_k x_k \quad \text{is convergent.}$$

Thus, the series $\sum_k a_k\sum_{j=1}^{k-m}\binom{k-j-1}{m-1}N^{1/p_j}$ is convergent. This implies by Part (ii) of Lemma 4.5.11 that $R_n\sum_{j=1}^{n-m}\binom{n-j-1}{m-1}N^{1/p_j}\to 0$, as $n\to\infty$.

Now, let $a=(a_k)\in\{s\ell_\infty(p,\Delta^m)\}^\beta\setminus M_\infty^\beta(p)$. Then, $\sum_k|R_k|\sum_{j=1}^{k-m+1}\binom{k-j-1}{m-2}N^{1/p_j}$ is divergent, i.e., $\sum_k|R_k|\sum_{j=1}^{k-m+1}\binom{k-j-1}{m-2}N^{1/p_j}=\infty$. Now, define the sequence $x=(x_k)$ by

$$x_k := \begin{cases} 0 & , \quad k\le m,\\ \sum_{i=1}^{k-1}\operatorname{sgn}R_i\sum_{j=1}^{i-m+1}\binom{i-j-1}{m-2}N^{1/p_j} & , \quad k>m,\end{cases}$$

where $a_k>0$ or $a_k<0$ for all $k\in\mathbb{N}_1$. It is trivial that $x=(x_k)\in s\ell_\infty(p,\Delta^m)$. Then, we can write for $n>m$ that

$$\sum_{k=1}^n a_k x_k = -\sum_{k=1}^m R_{k-1}\Delta x_{k-1} - \sum_{k=1}^{n-m} b_{k+m-1}\Delta x_{k+m-1} - R_n x_n.$$

Since $(R_n x_n) \in c_0$, now letting $n \to \infty$ we get

$$\begin{aligned}\sum_k a_k x_k &= -\sum_k R_{k+m-1}\Delta x_{k+m-1} \\ &= \sum_k |R_{k+m-1}| \sum_{j=1}^{k} \binom{k+m-j-2}{m-2} N^{1/p_j} = \infty.\end{aligned}$$

This contradicts the fact $a \in \{s\ell_\infty(p,\Delta^m)\}^\beta$. Hence, $a \in M_\infty^\beta(p)$.

(ii) This part can be proved by the same way used in the proof of Part (i) above, with Part (i) of Lemma 4.5.11. □

Now, we give the following lemma without proof:

Lemma 4.5.48. [87] *$\{s\ell_\infty(p,\Delta^m)\}^\eta = \{sc(p,\Delta^m)\}^\eta$ for $\eta = \beta$ or γ.*

Theorem 4.5.49. [89] *Let c_0^+ denotes the set of all positive null sequences. Define the sets $M_3^\beta(p)$ and $M_4^\gamma(p)$ by*

$$\begin{aligned} M_3^\beta(p) &:= \left\{ \begin{array}{c} a=(a_k)\in\omega : \sum_k a_k \sum_{j=1}^{k-m}\binom{k-j-1}{m-1} N^{1/p_j} v_j \text{ converges and} \\ \sum_k |R_k| \sum_{j=1}^{k-m+1}\binom{k-j-1}{m-2} N^{1/p_j} v_j < \infty \text{ for all } v\in c_0^+ \end{array}\right\}, \\ M_4^\gamma(p) &:= \left\{ \begin{array}{c} a=(a_k)\in\omega : \sup_{n\in\mathbb{N}_1} \left|\sum_{k=1}^{n} a_k \sum_{j=1}^{k-m}\binom{k-j-1}{m-1} N^{1/p_j}\right| < \infty \text{ and} \\ \sum_k |R_k| \sum_{j=1}^{k-m+1}\binom{k-j-1}{m-2} N^{1/p_j} v_j < \infty \text{ for all } v\in c_0^+ \end{array}\right\}. \end{aligned}$$

Then, the following statements hold:

(a) $\{sc_0(p,\Delta^m)\}^\beta = M_3^\beta(p)$.

(b) $\{sc_0(p,\Delta^m)\}^\gamma = M_4^\gamma(p)$.

Proof. Parts (a) and (b) can be proved by the same way used in the proof of Theorem 4.5.47 with Parts (i) and (ii) of Lemma 4.5.11. □

Now, we give the following lemma, without proof:

Lemma 4.5.50. [89] *Let $\eta \in \{\beta,\gamma\}$. Then, the following statements hold:*

(i) $\{\ell_\infty(p,\Delta^m)\}^\eta = \{s\ell_\infty(p,\Delta^m)\}^\eta$.

(ii) $\{c(p,\Delta^m)\}^\eta = \{sc(p,\Delta^m)\}^\eta$.

(iii) $\{c_0(p,\Delta^m)\}^\eta = \{sc_0(p,\Delta^m)\}^\eta$.

4.5.5 The Space of p-Summable Difference Sequences of Order m

In this subsection, following Altay [20], we introduce the difference sequence space $\ell_p(\Delta^{(m)})$ of order m and base on some properties, namely BK-space, monotonicity, AD and AB property and Schauder basis of the space. Since the matrix domain ν_Δ is called as *the difference sequence space* whenever ν is a normed sequence space in the existing literature, we follow the same way and denote it by $\ell_p(\Delta^{(m)})$.

We introduce the sequence space $\ell_p(\Delta^{(m)})$ as the set of all sequences whose $\Delta^{(m)}$-transforms are in the space ℓ_p, i.e.,

$$\ell_p(\Delta^{(m)}) := \left\{ x = (x_k) \in \omega : \sum_k \left| \sum_{j=0}^{m} (-1)^j \binom{m}{j} x_{k-j} \right|^p < \infty \right\}, \quad (1 < p < \infty).$$

With the notation of (4.1.1), we can redefine the space $\ell_p(\Delta^{(m)})$ as

$$\ell_p(\Delta^{(m)}) = (\ell_p)_{\Delta^{(m)}} .$$

It is obvious that the space $\ell_p(\Delta^{(m)})$ is reduced to the space bv_p for $m = 1$ [66].

Define the sequence $y = (y_k)$ as the $\Delta^{(m)}$-transform of a sequence $x = (x_k)$, i.e.,

$$y_k := \left(\Delta^{(m)} x\right)_k = \sum_{j=0}^{m} (-1)^j \binom{m}{j} x_{k-j} \quad \text{for all } k \in \mathbb{N}. \tag{4.5.20}$$

Now, we give the following theorem without proof, which can be proved by the same way used in the proof of Theorem 4.2.1:

Theorem 4.5.51. [20, Theorem 2.1] *The set $\ell_p(\Delta^{(m)})$ is a linear space with the coordinatewise addition and scalar multiplication, which is BK-space endowed with the norm $\|x\|_{\ell_p(\Delta^{(m)})} = \| \Delta^{(m)} x \|_p$.*

It is trivial that $\ell_p(\Delta^{(m)})$ is a sequence space of non-absolute type.

Consider the transformation T defined with the notation of (4.5.20), from $\ell_p(\Delta^{(m)})$ to ℓ_p by $x \mapsto y = Tx$. It is trivial that T is a norm preserving linear bijection. Thus, we have

Theorem 4.5.52. [20, Theorem 2.2] *The space $\ell_p(\Delta^{(m)})$ is norm isomorphic to the space ℓ_p, i.e., $\ell_p(\Delta^{(m)}) \cong \ell_p$.*

We want to derive some inclusion relations concerning with the space $\ell_p(\Delta^{(m)})$.

Theorem 4.5.53. [20, Theorem 2.3] *The inclusion $\ell_p \subset \ell_p(\Delta^{(m)})$ strictly holds.*

Proof. Since $\Delta^{(m)} \in (\ell_1 : \ell_1) \cap (\ell_\infty : \ell_\infty)$, the inclusion $\ell_p \subset \ell_p(\Delta^{(m)})$ holds. Additionally, since $e \in \ell_p(\Delta^{(m)}) \setminus \ell_p$, the inclusion $\ell_p \subset \ell_p(\Delta^{(m)})$ is strict. □

It is well-known that $\ell_p \subset \ell_s$ whenever $1 < p < s$, which leads us to the consequence that $\ell_p(\Delta^{(m)}) \subset \ell_s(\Delta^{(m)})$. Additionally, one can easily show that the inclusion $\ell_p(\Delta^{(m-1)}) \subset \ell_p(\Delta^{(m)})$ also holds.

Before giving the theorem on the monotonicity, perfectness and normality of the space $\ell_p(\Delta^{(m)})$, we present a lemma concerning with the dual spaces of a sequence space.

Lemma 4.5.54. [396, pp. 106, 108] *Let λ be an FK-space, which contains ϕ. Then, the following statements hold:*

(i) $\lambda^\alpha \subseteq \lambda^\beta \subseteq \lambda^\gamma$.

(ii) *If* λ *has AD, then,* $\lambda^\beta = \lambda^\gamma$.

(iii) λ *has AD iff* $\lambda^f = \lambda^*$.

Theorem 4.5.55. [20, Theorem 2.6] *The space* $\ell_p(\Delta^{(m)})$ *is not monotone and so it is neither normal nor perfect.*

Proof. To prove the first part of the theorem, we must find a sequence $y = (y_k) \in m_0$ such that $xy = (x_k y_k) \notin \ell_p(\Delta^{(m)})$ for a sequence $x = (x_k) \in \ell_p(\Delta^{(m)})$.

Let us define the sequence $y = (y_k) \in m_0$ by $y_k = [1 + (-1)^k]/2$ for all $k \in \mathbb{N}$. Then, the sequence $z = (z_k)$ obtained from the co-ordinatewise product of y by the sequence e that $z = ey = y$ is not in the space $\ell_p(\Delta^{(m)})$, which completes the proof of the first part of the theorem.

The proof of the second part of the theorem follows from Lemma 4.3.18. □

Theorem 4.5.56. [20, Theorem 2.7] *Although the sequence space* $\ell_p(\Delta^{(m)})$ *has not AD property for* $p = 1$*, it has AD property for* $1 < p < \infty$.

Proof. Note that if $f \in \{\ell_p(\Delta^{(m)})\}^*$, then $f(x) = g(\Delta^{(m)}x)$ for some $g \in \ell_p^*$. Since ℓ_p has AK property and $\ell_p^* \cong \ell_q$,

$$f(x) = \sum_k t_k \left(\Delta^{(m)}x\right)_k$$

for some $t = (t_k) \in \ell_q$. For any $f \in \{\ell_p(\Delta^{(m)})\}^*$ and $e^{(k)} \in \phi$, we have

$$f(e^{(k)}) = \sum_i t_i \left(\Delta^{(m)}e^{(k)}\right)_i = \left(\Delta^m t\right)_k \quad \text{for all} \quad k \in \mathbb{N}.$$

Hence, from Hahn-Banach theorem, $\phi \subset \ell_p(\Delta^{(m)})$ is dense in $\ell_p(\Delta^{(m)})$ if and only if $\Delta^m t = \theta$ for $t \in \ell_q$ implies $t = \theta$.

For $p = 1$, consider the sequence e which is in $\ell_1^\beta = \ell_\infty$. Since $\Delta^m e = \theta$, $\ell_1(\Delta^{(m)})$ has not *AD* property.

Since the null space of the operator $\Delta^{(m)}$ on ℓ_q is $\{\theta\}$, $\ell_p(\Delta^{(m)})$ has *AD* property. □

Now, as an easy consequence of Theorem 4.5.56 and Part (iii) of Lemma 4.5.54, we have

Corollary 4.5.57. [20, Corollary 2.8] *Let* $1 < p < \infty$*. Then, the following statements hold:*

$$\begin{aligned}\{\ell_p(\Delta^{(m)})\}^f &:= \{\ell_p(\Delta^{(m)})\}^* \\ &:= \left\{a = (a_k) \in \omega : \sum_k \left|\sum_{j=k}^{\infty} \binom{m+j-k-1}{j-k} a_j\right|^q < \infty\right\}.\end{aligned}$$

Theorem 4.5.58. [20, Theorem 2.9] *The following statements hold:*

(a) $\ell_p(\Delta^{(m)})$ *has not AB property for* $1 < p < \infty$.

(b) $\ell_1(\Delta^{(m)})$ *has not AB property for* $m \in \mathbb{N}_2$.

Proof. First, we need to show that ℓ_∞ and $\ell_p(\Delta^{(m)})$ overlap.

Let us consider the sequences $u = (u_k)$ and $x = (x_k)$ defined by $u_k = \sum_{i=0}^{k} 1/(i+1)$ and $x_k = (-1)^k$ for all $k \in \mathbb{N}$, respectively. Then, since $\Delta u = \{1/(k+1)\} \in \ell_p$ which gives that u is in $bv_p \subset \ell_p(\Delta^{(m)})$ but not in ℓ_∞. Nevertheless, x is in ℓ_∞ but not in $\ell_p(\Delta^{(m)})$.

(a) For $x \in \ell_p(\Delta^{(m)})$, we can write

$$\left(\left\|x^{[n]}\right\|_{\ell_p(\Delta^{(m)})}\right)^p = \sum_{k=0}^{n}\left|\sum_{j=0}^{m}(-1)^j\binom{m}{j}x_{k-j}\right|^p + \sum_{i=1}^{m-1}\left|\sum_{j=0}^{m-i}(-1)^{j+i}\binom{m}{j+i}x_{n-j}\right|^p + |x_n|^p.$$

Therefore, one can see that $\sup_{n\in\mathbb{N}}\|x^{[n]}\|_{\ell_p(\Delta^{(m)})} < \infty$ iff $x \in \ell_p(\Delta^{(m)}) \cap \ell_\infty$. This implies that $\ell_p(\Delta^{(m)})$ has not AB property for $1 < p < \infty$, since ℓ_∞ and $\ell_p(\Delta^{(m)})$ overlap.

The proof of Part (b) can be established by the same way. □

We should note that the sequence space $\ell_1(\Delta^{(m)})$ has the AB property for $m = 1$, since $\ell_1(\Delta^{(1)}) = bv_1$.

We will give a sequence of the points of the space $\ell_p(\Delta^{(m)})$ which forms a Schauder basis for the space $\ell_p(\Delta^{(m)})$. Because of the isomorphism T between the sequence spaces $\ell_p(\Delta^{(m)})$ and ℓ_p is onto, the inverse image of the basis of space ℓ_p is the basis of the our new space $\ell_p(\Delta^{(m)})$ by Theorem 4.1.2. Therefore, we have

Theorem 4.5.59. [20, Theorem 2.10] *Define the sequence* $b^{(k)}(m) = \{b_n^{(k)}(m)\}_{n\in\mathbb{N}}$ *of the elements of the space* $\ell_p(\Delta^{(m)})$ *for every fixed* $k \in \mathbb{N}$ *by*

$$b_n^{(k)}(m) := \begin{cases} 0 & , \quad n < k, \\ \binom{m+n-k-1}{n-k} & , \quad n \geq k. \end{cases}$$

Then, the sequence $\{b^{(k)}(m)\}_{k\in\mathbb{N}}$ *is a basis for the space* $\ell_p(\Delta^{(m)})$ *and any* $x \in \ell_p(\Delta^{(m)})$ *has a unique representation of the form* $x = \sum_k \lambda_k(m)b^{(k)}(m)$, *where* $\lambda_k(m) = (\Delta^{(m)}x)_k$ *for all* $k \in \mathbb{N}$.

Now, we can state and prove the theorems determining the α-, β- and γ-duals of the space $\ell_p(\Delta^{(m)})$. Because of the case $p = 1$ may be proved in the similar way, we consider only the case $1 < p < \infty$ in the proof of Theorems 4.5.60–4.5.62, respectively.

Theorem 4.5.60. [20, Theorem 3.4] *Define the sets* d_1 *and* d_q *by*

$$d_1 := \left\{a = (a_k) \in \omega : \sup_{n,k\in\mathbb{N}}\left|\sum_{j=k}^{n}\binom{m+j-k-1}{j-k}a_j\right| < \infty\right\},$$

$$d_q := \left\{a = (a_k) \in \omega : \sup_{n\in\mathbb{N}}\sum_{k=0}^{n}\left|\sum_{j=k}^{n}\binom{m+j-k-1}{j-k}a_j\right|^q < \infty\right\}.$$

Then, the following statements hold: $\{\ell_1(\Delta^{(m)})\}^\gamma = d_1$ *and* $\{\ell_p(\Delta^{(m)})\}^\gamma = d_q$*; where* $1 < p < \infty$.

Proof. Consider the equality

$$\begin{aligned}\sum_{k=0}^{n} a_k x_k &= \sum_{k=0}^{n} a_k \sum_{j=0}^{k} \binom{m+k-j-1}{k-j} y_j \\ &= \sum_{k=0}^{n}\sum_{j=k}^{n} \binom{m+j-k-1}{j-k} a_j y_k = (Cy)_n,\end{aligned} \tag{4.5.21}$$

where $C = (c_{nk})$ defined by

$$c_{nk} := \begin{cases} \sum_{j=k}^{n} \binom{m+j-k-1}{j-k} a_j & , \quad 0 \le k \le n, \\ 0 & , \quad k > n, \end{cases}$$

for all $n, k \in \mathbb{N}$. Thus, we deduce from (4.5.21) that $ax = (a_k x_k) \in bs$ whenever $x = (x_k) \in \ell_p(\Delta^{(m)})$ if and only if $Cy \in \ell_\infty$ whenever $y = (y_k) \in \ell_p$. Thus, $(a_k) \in d_q$ by (4.3.14) which gives us $\{\ell_p(\Delta^{(m)})\}^\gamma = d_q$. □

Theorem 4.5.61. [20, Theorem 3.5] *Define the set* d *by*

$$d := \left\{ a = (a_k) \in \omega : \sum_{j=k}^{\infty} \binom{m+j-k-1}{j-k} a_j = \beta_k \text{ exists for each } k \in \mathbb{N} \right\}.$$

Then, $\{\ell_1(\Delta^{(m)})\}^\beta = d_1 \cap d$ *and* $\{\ell_p(\Delta^{(m)})\}^\beta = d_q$*, where* $1 < p < \infty$.

Proof. For $p = 1$, this may be obtained by the similar way in the proof of Theorem 4.5.60 with Part (ii) of Lemma 4.3.13 instead of Part (ii) of Lemma 4.3.14. So, we omit details.

It is trivial from Theorem 4.5.56 with Part (ii) of Lemma 4.5.54 that

$$\{\ell_p(\Delta^{(m)})\}^\beta = \{\ell_p(\Delta^{(m)})\}^\gamma, \quad (1 < p < \infty).$$

Additionally, if $a \in \{\ell_p(\Delta^{(m)})\}^\beta$, using Banach-Steinhaus theorem, then we have $\sum_k a_k x_k = \sum_k \beta_k y_k$. □

Theorem 4.5.62. [20, Theorem 3.6] *Define the sets* c_1 *and* c_q*, as follows:*

$$\begin{aligned} c_1 &:= \left\{ a = (a_k) \in \omega : \sup_{k\in\mathbb{N}} \sum_n \binom{m+n-1}{n} |a_{n+k}| < \infty \right\}, \\ c_q &:= \left\{ a = (a_k) \in \omega : \sup_{N\in\mathcal{F}} \sum_k \left| \sum_{n\in N^*} \binom{m+n-k-1}{n-k} a_n \right|^q < \infty \right\}, \end{aligned}$$

where $N^* = N \cap \mathbb{N}_k$*. Then, the following statements hold:* $\{\ell_1(\Delta^{(m)})\}^\alpha = c_1$ *and* $\{\ell_p(\Delta^{(m)})\}^\alpha = c_q$*; where* $1 < p < \infty$.

Proof. Let us define the matrix D whose columns are the product of the columns of the matrix S^m, the inverse of the matrix $\Delta^{(m)}$, by the sequence $a = (a_n)$, that is to say that $D^k = (S^m)^k a$, where $(S^m)^k$ and D^k denote the kth columns of the matrices S^m and D, respectively. Bearing in mind the relation (4.5.20), we easily obtain that

$$a_n x_n = \sum_{k=0}^{n} \binom{m+n-k-1}{n-k} a_n y_k = (Dy)_n \text{ for all } n \in \mathbb{N}. \tag{4.5.22}$$

We therefore observe by (4.5.22) that $ax = (a_n x_n) \in \ell_1$ whenever $x \in \ell_p(\Delta^{(m)})$ if and only if $Dy \in \ell_1$ whenever $y \in \ell_p$. Then, we derive by Lemma 4.3.12 that

$$\sup_{N \in \mathcal{F}} \sum_k \left| \sum_{n \in N^*} \binom{m+n-k-1}{n-k} a_n \right|^q < \infty$$

which yields the consequence that $\{\ell_p(\Delta^{(m)})\}^{\alpha} = c_q$. □

4.5.6 Certain Matrix Mappings on the Sequence Space $\ell_p(\Delta^{(m)})$

In this subsection, we desire to characterize the matrix mappings from the sequence space $\ell_p(\Delta^{(m)})$ to some of the known sequence spaces and to itself. We directly prove the theorem characterizing the class $(\ell_p(\Delta^{(m)}) : \mu)$, where μ denotes the one of the sequence spaces ℓ_∞, c, c_0 and ℓ_1, and derive the other characterizations from them by using the suitable relation between the concerning matrix classes.

Theorem 4.5.63. [20, Theorem 4.1] *Let $\mu \in \{\ell_\infty, c, c_0, \ell_1\}$. Then, $A = (a_{nk}) \in (\ell_p(\Delta^{(m)}) : \mu)$ if and only if*

$$A_n \in \{\ell_p(\Delta^{(m)})\}^{\beta} \text{ for all } n \in \mathbb{N}, \tag{4.5.23}$$
$$B \in (\ell_p : \mu), \tag{4.5.24}$$

where $B = (b_{nk})$ with $b_{nk} = \sum_{i=k}^{\infty} \binom{m+i-k-1}{i-k} a_{ni}$ for all n, $k \in \mathbb{N}$ and A_n denotes the nth row of the infinite matrix A.

Proof. Let $A \in (\ell_p(\Delta^{(m)}) : \mu)$ and $x \in \ell_p(\Delta^{(m)})$. Then, $A_n \in \{\ell_p(\Delta^{(m)})\}^{\beta}$ for all $n \in \mathbb{N}$. Writing $y = \Delta^{(m)} x$, we conclude from the fact $A_n \in \{\ell_p(\Delta^{(m)})\}^{\beta}$ that

$$(Ax)_n = (By)_n \text{ for all } n \in \mathbb{N}. \tag{4.5.25}$$

Therefore, $B_n \in \ell_q$ for all $n \in \mathbb{N}$ and $By \in \mu$ for all $y \in \ell_p$. Hence, $B \in (\ell_p : \mu)$.

Conversely, we assume that (4.5.23) and (4.5.24) hold. Then, (4.5.25) again holds. Therefore, $Ax \in \mu$ for all $x \in \ell_p(\Delta^{(m)})$ as desired, which completes the proof. □

We have in the case $\mu = \ell_\infty$, c and c_0 by Theorem 4.5.63:

Corollary 4.5.64. [20, Corollary 4.2] $A = (a_{nk}) \in (\ell_p(\Delta^{(m)}) : \ell_\infty)$ *if and only if*

(i) *For* $1 < p < \infty$,

$$\sup_{t\in\mathbb{N}} \sum_{k}^{t} \left| \sum_{i=k}^{t} \binom{m+i-k-1}{i-k} a_{ni} \right|^q < \infty \text{ for all } n \in \mathbb{N},$$
$$\sup_{n\in\mathbb{N}} \sum_{k} |b_{nk}|^q < \infty,$$

where $B = (b_{nk})$ *is defined as in Theorem 4.5.63.*

(ii) *For* $p = 1$,

$$\lim_{t\to\infty} \sum_{i=k}^{t} \binom{m+i-k-1}{i-k} a_{ni} = b_{nk} \text{ for all } n, k \in \mathbb{N}, \tag{4.5.26}$$
$$\sup_{t,k\in\mathbb{N}} \left| \sum_{i=k}^{t} \binom{m+i-k-1}{i-k} a_{ni} \right| < \infty \text{ for all } n \in \mathbb{N}, \tag{4.5.27}$$
$$\sup_{n,k\in\mathbb{N}} |b_{nk}| < \infty.$$

Corollary 4.5.65. [20, Corollary 4.3] $A = (a_{nk}) \in (\ell_p(\Delta^{(m)}) : c)$ *if and only if* $A \in (\ell_p(\Delta^{(m)}) : \ell_\infty)$ *and*

$$\lim_{n\to\infty} \sum_{i=k}^{\infty} \binom{m+i-k-1}{i-k} a_{ni} = \alpha_k \text{ for all } k \in \mathbb{N}. \tag{4.5.28}$$

Corollary 4.5.66. [20, Corollary 4.4] $A = (a_{nk}) \in (\ell_p(\Delta^{(m)}) : c_0)$ *if and only if* $A \in (\ell_p(\Delta^{(m)}) : \ell_\infty)$ *and (4.5.28) holds with* $\alpha_k = 0$ *for all* $k \in \mathbb{N}$.

Corollary 4.5.67. [20, Corollary 4.5] $A = (a_{nk}) \in (\ell_1(\Delta^{(m)}) : \ell_1)$ *if and only if (4.5.26), (4.5.27) hold and*

$$\sup_{k\in\mathbb{N}} \sum_{n} \left| \sum_{j=k}^{\infty} \binom{m+j-k-1}{j-k} a_{nj} \right| < \infty.$$

It is trivial that Lemma 4.3.24 has several consequences, some of them give the necessary and sufficient conditions of matrix mappings between the spaces of higher order difference sequences. Indeed, combining Lemma 4.3.24 with Corollaries 4.5.64–4.5.67 and choosing B as one of the special matrices $\Delta^{(l)}$ and E^r, one can easily obtain the following results:

Corollary 4.5.68. [20, Corollary 4.7] *Let* $m, l \in \mathbb{N}_1$ *and* A *also be a complex infinite matrix. Define the matrix* $C := \Delta^{(l)}A$. *Then, the necessary and sufficient conditions in order for* A *belongs to any of the classes* $(\ell_p(\Delta^{(m)}) : \ell_\infty(\Delta^l))$, $(\ell_p(\Delta^{(m)}) : c(\Delta^l))$, $(\ell_p(\Delta^{(m)}) : c_0(\Delta^l))$ *and* $(\ell_1(\Delta^{(m)}) : \ell_1(\Delta^l))$ *are obtained from the respective ones in Corollaries 4.5.64–4.5.67 by replacing the entries of the matrix* A *by those of the matrix* C.

Corollary 4.5.69. [20, Corollary 4.8] *Let $m \in \mathbb{N}_1$ and A also be a complex infinite matrix. Define the matrix $C := E^r A$. Then, the necessary and sufficient conditions in order for A belongs to any of the classes $(\ell_p(\Delta^{(m)}) : e_\infty^r)$ and $(\ell_p(\Delta^{(m)}) : e_1^r)$ are obtained from the respective ones in Corollaries 4.5.64 and 4.5.67 by replacing the entries of the matrix A by those of the matrix C.*

4.5.7 v-Invariant Sequence Spaces

Çolak [124] introduced the concept of v-invariance of a sequence space λ and studied the v-invariance of the classical sequence spaces ℓ_∞, c, c_0 and ℓ_p. Following Çolak [124], Asma and Et [41] characterized the v-invariance of the difference spaces $\ell_\infty(\Delta)$, $c(\Delta)$ and $c_0(\Delta)$.

In 1989, Çolak [123] generalized the Kızmaz's spaces to the sequence spaces

$$\lambda(\Delta_v) := \{x = (x_k) \in \omega : \Delta_v x \in \lambda\}$$

and compute α-, β-, γ-duals of those sequence spaces for $\lambda \in \{\ell_\infty, c, c_0\}$, where $v = (v_k) \in U$ and $(\Delta_v x)_k = v_k x_k - v_{k+1} x_{k+1}$ for all $k \in \mathbb{N}$. Recently, Et and Esi [167] generalized the sequence spaces $\lambda(\Delta_v)$ and $\lambda(\Delta^m)$ to the following sequence spaces:

$$\lambda(\Delta_v^m) := \{x = (x_k) : \Delta_v^m x \in \lambda\},$$

where $(\Delta_v^m x)_k = (\Delta_v^{m-1} x)_k - (\Delta_v^{m-1} x)_{k+1} = \sum_{i=0}^m (-1)^i \binom{m}{i} v_{k+i} x_{k+i}$ for all $k \in \mathbb{N}$. If λ is a BK-space normed by $\|\cdot\|$, then $\lambda(\Delta_v^m)$ is also a BK-space normed by

$$\|x\| = \sum_{i=1}^m |v_i x_i| + \|\Delta_v^m x\|.$$

We note that $\lambda(\Delta_v^m)$ and $\lambda(\Delta^m)$ overlap but neither contains the other. For example, we chose $x = (k^m)$ and $v = (k)$, then $x \in \ell_\infty(\Delta^m) \setminus \ell_\infty(\Delta_v^m)$, conversely, if we choose $x = (k^{m+1})$ and $v = (k^{-1})$, then $x \in \ell_\infty(\Delta_v^m) \setminus \ell_\infty(\Delta^m)$. Bektaş et al. [90] determined the α-, β-, γ- and N-duals of the spaces $\ell_\infty(\Delta_v^m)$, $c(\Delta_v^m)$ and $c_0(\Delta_v^m)$, and characterized some classes of matrix maps.

Definition 4.5.70. [159] *Let λ be any sequence space and $v \in U$. Then, we say that the sequence space $\lambda(\Delta^m)$ is v-invariant if $\lambda(\Delta_v^m) = \lambda(\Delta^m)$.*

Theorem 4.5.71. [41] *Let λ denotes one of the sequence spaces ℓ_∞ or c_0. Then, $\lambda(\Delta_v) = \lambda(\Delta)$ if and only if the following statements hold:*

(i) $\sup_{k \in \mathbb{N}_1} |k v_k - (k+1) v_{k+1}| < \infty$.

(ii) $\sup_{k \in \mathbb{N}_1} \left| \frac{k}{v_k} - \frac{k+1}{v_{k+1}} \right| < \infty$.

Theorem 4.5.72. [41] *$c(\Delta_v) = c(\Delta)$ if and only if the following statements hold:*

(i) $\{k v_k - (k+1) v_{k+1}\} \in c$.

(ii) $\left\{\frac{k}{v_k} - \frac{(k+1)}{v_{k+1}}\right\} \in c$.

Theorem 4.5.73. [167] *Let $u = (u_k), v = (v_k) \in U$. Then, the following statements hold:*

(i) $\ell_\infty(\Delta_v^m) \subset \ell_\infty(\Delta_u^m)$ *whenever* $\sup_{k\in\mathbb{N}_1} k^m \left|v_k^{-1} u_k\right| < \infty$.

(ii) *If* $k^m \left|v_k^{-1} u_k\right| \to l$, *as* $k \to \infty$, *then* $c(\Delta_v^m) \subset c(\Delta_u^m)$.

(iii) *If* $k^m \left|v_k^{-1} u_k\right| \to 0$, *as* $k \to \infty$, *then* $c_0(\Delta_v^m) \subset c_0(\Delta_u^m)$.

Proof. The following inequality gives the proof:

$$\begin{aligned}
|\Delta_u^m x_k| &\leq \sum_{i=0}^{m-1} \left|(-1)^i \binom{m-1}{i} \Delta (u_{k+i} x_{k+i})\right| \\
&\leq \sum_{i=0}^{m-1} \binom{m-1}{i} \left[(k+i)^m \left|v_{k+i}^{-1} u_{k+i}\right| (k+i)^{-m} |v_{k+i} x_{k+i}|\right] + \\
&+ \sum_{i=0}^{m-1} \binom{m-1}{i} \left[(k+i+1)^m \left|v_{k+i+1}^{-1} u_{k+i+1}\right| (k+i+1)^{-m} |v_{k+i+1} x_{k+i+1}|\right]
\end{aligned}$$

□

If we take $u = v = e$ in Theorem 4.5.73, then we have the following results:

Corollary 4.5.74. *The following statements hold:*

(i) *If* $\sup_{k\in\mathbb{N}_1} k^m \left|v_k^{-1}\right| < \infty$, *then* $\ell_\infty(\Delta_v^m) \subset \ell_\infty(\Delta^m)$.

(ii) *If* $k^m \left|v_k^{-1}\right| \to l$, *as* $k \to \infty$, *then* $c(\Delta_v^m) \subset c(\Delta^m)$.

(iii) *If* $k^m \left|v_k^{-1}\right| \to 0$, *as* $k \to \infty$, *then* $c_0(\Delta_v^m) \subset c_0(\Delta^m)$.

Corollary 4.5.75. *The following statements hold:*

(i) *If* $\sup_{k\in\mathbb{N}_1} k^m |v_k| < \infty$, *then* $\ell_\infty(\Delta^m) \subset \ell_\infty(\Delta_v^m)$.

(ii) *If* $k^m |v_k| \to l$, *as* $k \to \infty$, *then* $c(\Delta^m) \subset c(\Delta_v^m)$.

(iii) *If* $k^r |v_k| \to 0$, *as* $k \to \infty$, *then* $c_0(\Delta^m) \subset c_0(\Delta_v^m)$.

Now, we put

(i) $\nu_\infty := \left\{v = (v_k) \in \omega : \sup_{k\in\mathbb{N}_1} k^m |v_k| < \infty\right\}$.

(ii) $\nu_c := \left\{v = (v_k) \in \omega : \lim_{k\to\infty} k^m |v_k| = l \text{ for some } l\right\}$.

(iii) $\nu_0 := \left\{ v = (v_k) \in \omega : \lim\limits_{k\to\infty} k^m |v_k| = 0 \right\}$.

(iv) $\nu_\infty^{-1} := \left\{ v = (v_k) \in \omega : \sup\limits_{k\in\mathbb{N}_1} k^m \left|v_k^{-1}\right| < \infty \right\}$.

(v) $\nu_c^{-1} := \left\{ v = (v_k) \in \omega : \lim\limits_{k\to\infty} k^m \left|v_k^{-1}\right| = l \text{ for some } l \right\}$.

(vi) $\nu_0^{-1} := \left\{ v = (v_k) \in \omega : \lim\limits_{k\to\infty} k^m \left|v_k^{-1}\right| = 0 \right\}$.

It can be shown that ν_∞, ν_c and ν_0 are BK-spaces endowed with the norm $\|v\| = \sup\limits_{k\in\mathbb{N}_1} k^m |v_k|$.

Theorem 4.5.76. [167] *Let λ denotes any of the spaces ν_∞, ν_c or ν_0. Then, $\lambda \cap \lambda^{-1} = \phi$.*

Proof. We give the proof only for $\lambda = \nu_\infty$. Let $x \in \nu_\infty \cap \nu_\infty^{-1}$ and $v = (v_k) \in U$. Then, there are positive constants K_1 and K_2 such that $k^m |v_k| \leq K_1$ and $k^m \left|v_k^{-1}\right| \leq K_2$ for all $k \in \mathbb{N}_1$. This implies that $k^{2m} \leq K_1 K_2$ for all $k \in \mathbb{N}_1$, a contradiction, since $m \geq 1$. □

4.5.8 Paranormed Difference Sequence Spaces Generated by Moduli and Orlicz Functions

The notion of a modulus function was introduced in 1953 by Nakano [299]. A modulus f is a function from $[0, \infty)$ to itself such that

(i) $f(x) = 0$ if and only if $x = 0$.

(ii) $f(x + y) \leq f(x) + f(y)$ for $x, y \geq 0$.

(iii) f is increasing.

(iv) f is continuous from the right at 0.

It follows that f must be continuous everywhere on $[0, \infty)$. A modulus may be bounded or unbounded. Ruckle [340] and Maddox [257] used a modulus f to construct some new sequence spaces.

Recall ([211], [230], [311]) that an Orlicz function is a function $M : [0, \infty) \to [0, \infty)$, which is continuous, non-decreasing and convex with $M(0) = 0$, $M(x) > 0$ for $x > 0$ and $M(x) \to \infty$ as $x \to \infty$. If the convexity of Orlicz function M is replaced by $M(x + y) \leq M(x) + M(y)$, then this function is called modulus function. It is well-known that if M is a convex function and $M(0) = 0$, then $M(\alpha x) \leq \alpha M(x)$ for all $\alpha \in (0, 1)$.

The study of Orlicz sequence spaces was initiated with a certain specific purpose in the Banach space theory. Indeed, Lindberg [244] interested in Orlicz spaces in connection with finding Banach spaces with symmetric Schauder bases having complementary subspaces isomorphic to c_0 or ℓ_p, where $1 < p < \infty$. Subsequently, Lindenstrauss and Tzafriri [245] investigated the Orlicz sequence spaces in more detail, and they proved that every Orlicz sequence space ℓ_M contains a subspace isomorphic to ℓ_p $(1 < p < \infty)$, where

$$\ell_M := \left\{x = (x_k) \in \omega : \sum_{k=1}^{\infty} M\left(\frac{|x_k|}{\rho}\right) < \infty \text{ for some } \rho > 0\right\}.$$

The space ℓ_M is a Banach space endowed with the norm

$$\|x\| = \inf\left\{\rho > 0 : \sum_{k=1}^{\infty} M\left(\frac{|x_k|}{\rho}\right) \leq 1\right\}.$$

For $M(t) = t^p$, the space ℓ_M coincides with the classical sequence space ℓ_p, where $1 < p < \infty$.

Many mathematicians [19, 34, 37, 82, 81, 91, 160, 159, 182, 297, 322, 385, 215] introduced the new sequence spaces, using the generalized difference operator Δ^m and an Orlicz function or a modulus function.

Recently, the following results introduced and studied by Et et al. [162]:

Definition 4.5.77. [162] *Let $p = (p_n)$ and $\bar{q} = (\bar{q}_n)$ denote the sequences of positive real numbers, and the sequence $\bar{Q} = (\bar{Q}_n)$ is such that $\bar{q}_1 > 0$ and $\bar{Q}_n = \bar{q}_1 + \bar{q}_2 + \cdots + \bar{q}_n$ for all $n \in \mathbb{N}$. Let $M = (M_k)$ be a sequence of Orlicz functions, X be a locally convex Hausdorff topological linear space whose topology is determined by a set Q of continuous seminorm g and $u \in U$. The symbol $\omega(X)$ denotes the space of all sequences defined over X and for short, we write (Z) instead of $(M, \Delta^m, \bar{q}, p, u, g)$. We define the following sequence spaces:*

$$\omega_1(Z) := \left\{(x_k) \in \omega(X) : \lim_{n\to\infty} \sum_{k=1}^{n} \frac{\bar{q}_k \left[M_k\left(g\left(\frac{u_k(\Delta^m x)_k - l}{\rho}\right)\right)\right]^{p_k}}{\bar{Q}_n} = 0 \right.$$

$$\left. \text{for some } \rho > 0,\ l \in X\right\}$$

$$\omega_0(Z) := \left\{(x_k) \in \omega(X) : \lim_{n\to\infty} \sum_{k=1}^{n} \frac{\bar{q}_k}{\bar{Q}_n}\left[M_k\left(g\left(\frac{u_k(\Delta^m x)_k}{\rho}\right)\right)\right]^{p_k} = 0 \text{ for some } \rho > 0\right\}$$

$$\omega_\infty(Z) := \left\{(x_k) \in \omega(X) : \sup_{n\in\mathbb{N}_1} \sum_{k=1}^{n} \frac{\bar{q}_k}{\bar{Q}_n}\left[M_k\left(g\left(\frac{u_k(\Delta^m x)_k}{\rho}\right)\right)\right]^{p_k} < \infty \text{ for some } \rho > 0\right\}$$

Let $Y \in \{0, 1, \infty\}$. When $\bar{q}_n = 1$, $(\bar{Q}_n = n)$ and $M_n = M$ for all $n \in \mathbb{N}$, we denote the above spaces by $\omega_Y(M, \Delta^m, p, u, g)$. If $x \in \omega_1(M, \Delta^m, \bar{q}, p, u, g)$, we say that x is strongly $\Delta^m_{u_g}$-Cesàro summable with respect to the Orlicz function M. Now, we give the following theorem without proof, since it is a routine verification:

Theorem 4.5.78. [162] *Let* $(p_k) \in \ell_\infty$ *and* $(Z) = (M, \Delta^m, \bar{q}, p, u, g)$. *Then,* $\omega_1(Z)$, $\omega_0(Z)$ *and* $\omega_\infty(Z)$ *are the linear spaces.*

Theorem 4.5.79. [162] *For any sequence* $M = (M_k)$ *of Orlicz functions and a bounded sequence* $p = (p_k)$ *of strictly positive real numbers,* $\omega_0(M, \Delta^m, \bar{q}, p, u, g)$ *is a paranormed space (not necessarily total paranormed) with*

$$g_\Delta(x) = \sum_{i=1}^m g(x_i) + \inf_{\rho>0\ n\geq 1}\left\{\rho^{p_n/K} : \sup_{k\in\mathbb{N}_1}\left[M_k\left(g\left(\frac{u_k(\Delta^m x)_k}{\rho}\right)\right)\right] \leq 1, \rho > 0,\ n \in \mathbb{N}\right\},$$

where $K = \max\left\{1, H = \sup_{k\in\mathbb{N}_1} p_k\right\}$.

Proof. Clearly, $g_\Delta(x) = g_\Delta(-x)$. Since $M_k(0) = 0$ for all $k \in \mathbb{N}$, we get $\inf\{\rho^{p_n/K}\} = 0$ for $x = \theta$. Now, let $x, y \in \omega_0(M, \Delta^m, \bar{q}, p, u, g)$ and let us choose $\rho_1, \rho_2 > 0$ such that

$$\sup_{k\in\mathbb{N}_1}\left[M_k\left(g\left(\frac{u_k(\Delta^m x)_k}{\rho_1}\right)\right)\right] \leq 1 \text{ and } \sup_{k\in\mathbb{N}_1}\left[M_k\left(g\left(\frac{u_k(\Delta^m x)_k}{\rho_2}\right)\right)\right] \leq 1.$$

Let $\rho = \rho_1 + \rho_2$. Then, we get the triangle inequality from the following inequality:

$$\begin{aligned}
&\sup_{k\in\mathbb{N}_1}\left[M_k\left(g\left(\frac{(\Delta^m(x+y))_k}{\rho}\right)\right)\right] \leq \\
&\leq \left(\frac{\rho_1}{\rho_1+\rho_2}\right)\sup_{k\in\mathbb{N}_1}\left[M_k\left(g\left(\frac{(\Delta^m x)_k}{\rho_1+\rho_2}\right)\right)\right] + \left(\frac{\rho_2}{\rho_1+\rho_2}\right)\sup_{k\in\mathbb{N}_1}\left[M_k\left(g\left(\frac{(\Delta^m y)_k}{\rho_1+\rho_2}\right)\right)\right] \\
&\leq 1.
\end{aligned}$$

Finally, let α be a given non-zero scalar, then the continuity of the scalar multiplication follows from the following equality

$$\begin{aligned}
g_\Delta(\alpha x) &= \sum_{i=1}^m g(\alpha x_i) + \inf\left\{\rho^{p_n/K} : \sup_{k\in\mathbb{N}_1}\left[M_k\left(g\left(\frac{u_k(\Delta^m(\alpha x))_k)}{\rho}\right)\right)\right] \leq 1\right\} \\
&= |\alpha|\sum_{i=1}^m g(x_i) + \inf\left\{(|\alpha|s)^{p_n/K} : \sup_{k\in\mathbb{N}_1}\left[M_k\left(g\left(\frac{u_k(\Delta^m x)_k}{s}\right)\right)\right] \leq 1\right\},
\end{aligned}$$

where $s = \rho/|\alpha|$.

This completes the proof. □

The proof of the following results are easy and so we omit details:

Theorem 4.5.80. [162] *Let* (M_k) *and* (T_k) *be sequences of Orlicz functions. For any two sequences* $p = (p_k)$ *and* $t = (t_k)$ *of positive real numbers and for any two seminorms* g_1 *and* g_2, *we have the following:*

(i) *If* g_1 *is stronger than* g_2, *then* $\omega_Y(M, \Delta^m, \bar{q}, p, u, g_1) \subset \omega_Y(M, \Delta^m, \bar{q}, p, u, g_2)$.

(ii) $\omega_Y(M, \Delta^m, \bar{q}, p, u, g_1) \cap \omega_Y(M, \Delta^m, \bar{q}, p, u, g_2) \subset \omega_Y(M, \Delta^m, \bar{q}, p, u, g_1 + g_2)$.

(iii) $\omega_Y(M, \Delta^m, \bar{q}, p, u, g) \cap \omega_Y(T, \Delta^m, \bar{q}, p, u, g) \subset \omega_Y(M + T, \Delta^m, \bar{q}, p, u, g)$.

(iv) $\omega_Y(M, \Delta^m, \bar{q}, p, u, g_1) \cap \omega_Y(M, \Delta^m, \bar{q}, t, u, g_2) \neq \emptyset$.

(v) *The inclusions* $\omega_Y(M, \Delta^{m-1}, \bar{q}, g) \subset \omega_Y(M, \Delta^m, \bar{q}, g)$ *are strict. In general,* $\omega_Y(M, \Delta^i, \bar{q}, g) \subset \omega_Y(M, \Delta^m, \bar{q}, g)$ *for* $i \in \{1, 2, \ldots, m-1\}$ *and the inclusion is strict.*

(vi) *If* M_k *and* T_k *are equivalent for each* $k \in \mathbb{N}$, *then* $\omega_Y(M, \Delta^m, \bar{q}, p, u, g) = \omega_Y(T, \Delta^m, \bar{q}, p, u, g)$.

(vii) *If* $|u_k| \leq 1$, *then* $\omega_Y(M, \Delta^m, \bar{q}, p, g) \subset \omega_Y(M, \Delta^m, \bar{q}, p, u, g)$ *for* $Y \in \{0, \infty\}$.

(viii) *Let* $0 < p_k \leq t_k$ *for all* $k \in \mathbb{N}$ *and* $(t_k/p_k) \in \ell_\infty$. *Then* $\omega_Y(M, \Delta^m, \bar{q}, t, u, g) \subseteq \omega_Y(M, \Delta^m, \bar{q}, p, u, g)$ *for* $Y \in \{0, 1, \infty\}$.

(ix) *The sequence spaces* $\omega_0(M, \bar{q}, p, u, g)$ *and* $\omega_\infty(M, \bar{q}, p, u, g)$ *are monotone.*

(x) *The sequence space* $\omega_1(M, \bar{q}, p, u, g)$ *is neither monotone nor solid.*

(xi) $\omega_0(M, \Delta^m, \bar{q}, p, u, g) \subset \omega_1(M, \Delta^m, \bar{q}, p, u, g) \subset \omega_\infty(M, \Delta^m, \bar{q}, p, u, g)$ *and the inclusions are strict.*

In the following definition, we introduce $\Delta^m_{u_g}$*-statistically convergent* sequences and give some inclusion relations between $\Delta^m_{u_g}$-statistically convergent sequences and $\omega_1(M, \Delta^m, u, g)$-summable sequences with respect to an Orlicz function. We also show that the spaces $S^g_u(\Delta^m) \cap \ell_\infty(u_g, \Delta^m)$ may be represented as the intersection of the spaces $\omega_1(M, \Delta^m, u, g)$ and $\ell_\infty(u_g, \Delta^m)$.

Definition 4.5.81. [162] *A sequence* $x = (x_k)$ *in* X *is said to be* $\Delta^m u_g$*-statistically convergent to* l *if for every* $g \in Q$ *and* $\varepsilon > 0$, $\delta\{k \in \mathbb{N} : g(u_k(\Delta^m x)_k - l) \geq \varepsilon\} = 0$. *In this case, we write* $x_k \to l(S^g_u(\Delta^m))$, *where* $S^g_u(\Delta^m)$ *is the set of all* $\Delta^m u_g$*-statistically convergent sequences.*

If $X = \mathbb{C}$, $g(x) = |x|$, *we write* $S_u(\Delta^m)$ *instead of* $S^g_u(\Delta^m)$ *and if* $u = e$ *we write* $S(\Delta^m)$ *instead of* $S_u(\Delta^m)$. *In the special case* $l = 0$, *we write* $S^g_{0u}(\Delta^m)$.

Theorem 4.5.82. [162] *Let* $0 < p < \infty$. *Then, we have*

(i) *If* $x_k \to l(\omega^g_u(\Delta^m))$, *then* $x_k \to l(S^g_u(\Delta^m))$.

(ii) *If* $x \in \ell_\infty(u_g, \Delta^m)$ *and* $x_k \to l(S^g_u(\Delta^m))$, *then* $x_k \to l(\omega^g_u(\Delta^m))$, *where*

$$\ell_\infty(u_g, \Delta^m) := \left\{ x = (x_k) \in \omega(X) : \sup_{k \in \mathbb{N}_1} g(u_k(\Delta^m x)_k) < \infty \right\},$$

$$\omega^g_u(\Delta^m) := \left\{ x = (x_k) \in \omega(X) : \lim_{n \to \infty} \sum_{k=1}^{n} \frac{[g(u_k(\Delta^m x)_k - l)]^p}{n} = 0 \text{ for some } l \right\}.$$

Proof. Since the proof is similar to that of Theorem 4.5.32, we omit details. □

Theorem 4.5.83. [162] *Let M be an Orlicz function. Then, $\omega_1(M, \Delta^m, p, u, g) \subset S_u^g(\Delta^m)$.*

Proof. This is similar to that of Theorem 4.5.34. □

Theorem 4.5.84. [162] $\ell_\infty(u_g, \Delta^m) \cap S_u^g(\Delta^m) = \ell_\infty(u_g, \Delta^m) \cap \omega_1(M, \Delta^m, u, g)$.

Proof. By Theorem 4.5.83, we need only to show that $\ell_\infty(u_g, \Delta^m) \cap S_u^g(\Delta^m) \subset \ell_\infty(u_g, \Delta^m) \cap \omega_1(M, \Delta^m, u, g)$. Let $t_k = u_k(\Delta^m x)_k - l \to \theta(S^g)$. Let Σ_1 and Σ_2 be as in the proof of Theorem 4.5.34. Since $x = (x_k) \in \ell_\infty(u_g, \Delta^m)$, there exists $T > 0$ such that $[M(g(t_k/\rho))] \leq T$ for all t_k. Then, for a given $\varepsilon > 0$ and for each $n \in \mathbb{N}_1$, we have

$$\begin{aligned}\frac{1}{n}\sum_{k=1}^{n}\left[M\left(g\left(\frac{t_k}{\rho}\right)\right)\right] &\leq \frac{1}{n}\sum_{1} M\left(g\left(\frac{t_k}{\rho}\right)\right) + \frac{1}{n}\sum_{2} M\left(g\left(\frac{t_k}{\rho}\right)\right)\\ &\leq \frac{T}{n}|\{k \leq n : g(t_k) \geq \varepsilon\rho\}| + M\left(\frac{\varepsilon}{\rho}\right).\end{aligned}$$

Hence, $x \in \omega_1(M, \Delta^m, u, g)$. □

Theorem 4.5.85. *The spaces $\omega_Y(M, \Delta^m, \bar{q}, p, u, g)$, $S_u^g(\Delta^m)$ and $S_{0u}^g(\Delta^m)$ are neither solid neither symmetric nor sequence algebra for $m > 0$ with $Y \in \{0, 1, \infty\}$.*

Proof. To show that the spaces are not solid in general, consider the following example:

Let $p_k = u_k = \bar{q}_k = 1$ for all $k \in \mathbb{N}$, $M(x) = x$, $g(x) = |x|$ and $X = \mathbb{C}$. Then, $x = (k^m) \in \omega_1(M, \Delta^m, \bar{q}, p, u, g), \omega_\infty(M, \Delta^m, \bar{q}, p, u, g)$ and $S_u^g(\Delta^m)$. Let $\alpha_k = (-1)^k$ for all $k \in \mathbb{N}$, then $\alpha x \notin \omega_1(M, \Delta^m, \bar{q}, p, u, g), \omega_\infty(M, \Delta^m, \bar{q}, p, u, g)$ and $S_u^g(\Delta^m)$. Hence, $\omega_1(M, \Delta^m, \bar{q}, p, u, g), \omega_\infty(M, \Delta^m, \bar{q}, p, u, g)$ and $S_u^g(\Delta^m)$ are not solid for $m > 0$. To show that $\omega_0(M, \Delta^m, \bar{q}, p, u, g)$ and $S_{0u}^g(\Delta^m)$ are not solid, it is sufficient to consider the sequence $(x_k) = (k^{m-1})$ and $\alpha_k = (-1)^k$ for all $k \in \mathbb{N}$. □

Corollary 4.5.86. *The sequence spaces $\omega_Y(M, \Delta^m, \bar{q}, p, u, g)$, $S_u^g(\Delta^m)$ and $S_{0u}^g(\Delta^m)$ are not perfect for $m > 0$.*

4.6 THE DOMAIN OF THE MATRIX A^r AND CONCERNING DUALITY RELATION

Let λ denotes any of the classical sequence spaces ℓ_∞, c, c_0 and ℓ_p or any of the Maddox's paranormed sequence spaces $\ell_\infty(p)$, $c(p)$, $c_0(p)$ and $\ell(p)$. Then, the domain λ_{A^r} is studied by Aydın and Başar [42, 43, 44, 45, 46, 47].

Let us define the sequence $y = (y_k)$ by the A^r-transform of a sequence $x = (x_k) \in \omega$, i.e.,

$$y_k := (A^r x)_k = \frac{1}{k+1}\sum_{j=0}^{k}(1 + r^j)x_j \tag{4.6.1}$$

for all $k \in \mathbb{N}$. Then, since $\lambda_{A^r} \cong \lambda$ it is trivial that the two-sided implication "$x \in \lambda_{A^r}$ *if and only if* $y \in \lambda$" holds. In this situation, the duality relation between the entries of the matrices $A = (a_{nk})$ and $B = (b_{nk})$ is

$$a_{nk} := \sum_{j=k}^{\infty} \frac{1+r^k}{j+1} b_{nj} \quad \text{or} \quad b_{nk} := (k+1)\left(\frac{a_{nk}}{1+r^k} - \frac{a_{n,k+1}}{1+r^{k+1}}\right) \tag{4.6.2}$$

for all $k, n \in \mathbb{N}$. The pair of such type matrices was used by Aydın and Başar [42, 43, 44, 45, 46, 47] and Rhoades [330].

4.6.1 The Sequence Spaces a_p^r, a_0^r, a_c^r and a_∞^r of Non-absolute Type

We introduce the sequence spaces a_p^r, a_0^r, a_c^r and a_∞^r, as the set of all sequences whose A^r-transforms are in the spaces ℓ_p, c_0, c and ℓ_∞, respectively, i.e.,

$$\begin{aligned}
a_p^r &:= \left\{x = (x_k) \in \omega : \sum_n \left|\sum_{k=0}^{n} \frac{1+r^k}{n+1} x_k\right|^p < \infty\right\}, \quad (0 < p < \infty), \\
a_0^r &:= \left\{x = (x_k) \in \omega : \lim_{n\to\infty} \sum_{k=0}^{n} \frac{1+r^k}{n+1} x_k = 0\right\}, \\
a_c^r &:= \left\{x = (x_k) \in \omega : \lim_{n\to\infty} \sum_{k=0}^{n} \frac{1+r^k}{n+1} x_k \text{ exists}\right\}, \\
a_\infty^r &:= \left\{x = (x_k) \in \omega : \sup_{n\in\mathbb{N}} \left|\sum_{k=0}^{n} \frac{1+r^k}{n+1} x_k\right| < \infty\right\}.
\end{aligned}$$

It is known by Başar [66] that the method A^r is regular for $0 < r < 1$ and is stronger than the Cesàro method C_1. We assume unless stated otherwise that $0 < r < 1$. With the notation of (4.1.1), we can redefine the sequence spaces a_p^r, a_0^r, a_c^r and a_∞^r, as follows:

$$a_p^r := (\ell_p)_{A^r}, \quad a_0^r := (c_0)_{A^r}, \quad a_c^r := (c)_{A^r} \quad \text{and} \quad a_\infty^r := (\ell_\infty)_{A^r}.$$

It is trivial that $a_p^r \subset a_0^r \subset a_c^r \subset a_\infty^r$.

Since it can be proved in the similar way used in the proof of Theorem 4.3.1, we give the following theorem without proof:

Theorem 4.6.1. (cf. [45, Theorem 2.2] and [42, Theorem 2.1]) *The sets a_p^r, a_0^r, a_c^r and a_∞^r are linear spaces with the co-ordinatewise addition and scalar multiplication, which are BK-spaces with the norms $\|x\|_{a_p^r} = \|A^r x\|_p$ and $\|x\|_{a_0^r} = \|x\|_{a_c^r} = \|x\|_{a_\infty^r} = \|A^r x\|_\infty$.*

Because the absolute property does not hold on the spaces a_p^r, a_0^r, a_c^r and a_∞^r, they are the sequence spaces of non-absolute type.

Since the transformation T defined, with the notation of (4.6.1), from a_p^r, a_0^r, a_c^r or a_∞^r to ℓ_p, c_0, c or ℓ_∞ by $x \mapsto y = Tx = A^r x$ is a linear bijection, $a_p^r \cong \ell_p$, $a_0^r \cong c_0$, $a_c^r \cong c$ and $a_\infty^r \cong \ell_\infty$.

4.6.2 The Inclusion Relations

Now, we may give the theorems on the inclusion relations concerning with the spaces a_p^r, a_0^r, a_c^r and a_∞^r.

Theorem 4.6.2. [42, Theorem 2.3] *Although the inclusions $c_0 \subset a_0^r$ and $c \subset a_c^r$ strictly hold, neither of the spaces a_0^r and ℓ_∞ includes the other one.*

Proof. To prove the validity of the inclusion $c_0 \subset a_0^r$, let us take any $y \in c_0$. Then, bearing in mind the regularity of the method A^r we immediately observe that $A^r y \in c_0$, which means that $y \in a_0^r$. Hence, the inclusion $c_0 \subset a_0^r$ holds. Furthermore, let us consider the gap sequence $u = \{u_k(r)\}$ defined by

$$u_k(r) := \begin{cases} \frac{\sqrt[3]{k}}{1+r^k} & , \quad k = m^3, \ m \in \mathbb{N}, \\ 0 & , \quad k \neq m^3, \ m \in \mathbb{N}, \end{cases} \tag{4.6.3}$$

for all $k \in \mathbb{N}$. Then, since

$$\begin{aligned} (A^r u)_n &= \frac{1}{n+1} \sum_{k=0}^{n} (1+r^k) u_k(r) \\ &= \frac{1}{n+1} \sum_{k=0}^{m} k = \frac{m(m+1)}{2(n+1)}; \quad (m^3 \leq n < (m+1)^3, \ m \in \mathbb{N}), \end{aligned}$$

$A^r u \in c_0$ which implies that u is in a_0^r but not in c_0. This shows that the inclusion $c_0 \subset a_0^r$ is strict. By the similar discussion, one can see that the inclusion $c \subset a_c^r$ also strictly holds.

To establish the second part of the theorem, let us consider the sequences $u = \{u_k(r)\}$ defined by (4.6.3) and $x = e$. Then, since

$$(A^r x)_n = 1 + \frac{1 - r^{n+1}}{(1-r)(n+1)},$$

one can easily see that x is in ℓ_∞ but not in a_0^r and u is in a_0^r but not in ℓ_∞. Hence, the sequence spaces a_0^r and ℓ_∞ overlap but neither contains the other.

This completes the proof. □

We continue by giving some inclusion relations concerning with the space a_p^r. Firstly, we give the following lemma which requires in proving Theorem 4.6.4:

Lemma 4.6.3. [200, p. 239] *Let (a_n) be a sequence of non-negative terms and $A_n = a_0 + a_1 + \cdots + a_n$ for all $n \in \mathbb{N}$. Then, the inequality*

$$\sum_n \left(\frac{A_n}{n+1}\right)^p < \left(\frac{p}{p-1}\right) \sum_n a_n^p$$

holds for $p > 1$.

Theorem 4.6.4. [45, Theorem 3.2] *Let $1 < p < \infty$. Then, the inclusion $\ell_p \subset a_p^r$ strictly holds.*

Proof. To prove the validity of the inclusion $\ell_p \subset a_p^r$ for $1 < p < \infty$, it suffices to show the existence of a number $K > 0$ such that $\|x\|_{a_p^r} \leq K\|x\|_p$ for every $x \in \ell_p$.

Let $x = (x_j) \in \ell_p$ and $1 < p < \infty$. Then, combining the fact

$$\left|\sum_{j=0}^{k} \frac{1+r^j}{1+k} x_j\right|^p \leq \left(2\sum_{j=0}^{k} \frac{|x_j|}{1+k}\right)^p$$

by Lemma 4.6.3 that

$$\sum_k \left|\sum_{j=0}^{k} \frac{1+r^j}{1+k} x_j\right|^p < \left(\frac{2p}{p-1}\right)^p \sum_k |x_k|^p,$$

which yields us that $\|x\|_{a_p^r} \leq 2p\|x\|_p/(p-1)$ for $1 < p < \infty$. Hence, the inclusion $\ell_p \subset a_p^r$ holds. Let us consider the sequence defined by $x_k := (-1)^k/(1+r^k)$ for all $k \in \mathbb{N}$. Then, since

$$\begin{aligned} \sum_n \left|\sum_{k=0}^{n} \frac{1+r^k}{1+n} x_k\right|^p &= \sum_n \left|\sum_{k=0}^{n} \frac{(-1)^k}{1+n}\right|^p \\ &= \sum_n \left|\frac{1+(-1)^n}{2(1+n)}\right|^p < \infty, \end{aligned}$$

x is in a_p^r but not in ℓ_p. Because there is at least one sequence in $a_p^r \backslash \ell_p$, the inclusion $\ell_p \subset a_p^r$ is strict.

By the similar discussions, it may easily be proved that the inclusion $\ell_1 \subset a_1^r$ also strictly holds and so we omit details.

This completes the proof. □

Theorem 4.6.5. [45, Theorem 3.3] *If $1 < p < s$, then $a_p^r \subset a_s^r$.*

Proof. Suppose that $1 < p < s$ and $x \in a_p^r$. With the notation of (4.6.1), the isomorphism $a_p^r \cong \ell_p$ implies that $y \in \ell_p$. Then, the well-known inclusion $\ell_p \subset \ell_s$ yields the fact that $y \in \ell_s$. This means that $x \in a_s^r$ and hence the inclusion $a_p^r \subset a_s^r$ holds, as asserted. □

Theorem 4.6.6. [45, Theorem 3.4] *Although the space a_∞^r strictly includes both the space ℓ_∞ and the space a_p^r, the sequence space a_p^r does not include the space ℓ_∞, where $1 < p < \infty$.*

Proof. Let $x = (x_k) \in \ell_\infty$. Then, we have

$$\begin{aligned} \|x\|_{a_\infty^r} &= \sup_{n\in\mathbb{N}} \left|\sum_{k=0}^{n} \frac{1+r^k}{1+n} x_k\right| \\ &\leq \|x\|_\infty \left(\sup_{n\in\mathbb{N}} \sum_{k=0}^{n} \frac{1+r^k}{1+n}\right) \\ &= \|x\|_\infty \left\{\sup_{n\in\mathbb{N}} \left[1 + \frac{1-r^{n+1}}{(1+n)(1-r)}\right]\right\} \leq 2\|x\|_\infty, \end{aligned}$$

which means that $x \in a_\infty^r$, i.e., the inclusion $\ell_\infty \subset a_\infty^r$ holds.

Additionally, it is easily seen that the sequence $x = (x_k)$ defined by

$$x_k := (-1)^k \frac{2k+1}{1+r^k} \quad \text{for all} \quad k \in \mathbb{N}$$

is in a_∞^r but not in ℓ_∞. So, the sequence space a_∞^r strictly includes the sequence space ℓ_∞, as expected.

Let $x \in a_p^r$ and $1 < p < \infty$. Then, it is obvious that $y = A^r x \in \ell_p \subset \ell_\infty$, which leads us to the fact that $x \in a_\infty^r$. Hence, the inclusion $a_p^r \subset a_\infty^r$ holds. Besides, since the sequence $x = e$ belongs to the set $a_\infty^r \setminus a_p^r$, the inclusion $a_p^r \subset a_\infty^r$ is also strict.

It is easy to see that $x = e$ is in $\ell_\infty \setminus a_p^r$, which means that the sequence space a_p^r does not include the sequence space ℓ_∞.

This completes the proof. □

Because of the isomorphism T, defined on the spaces ℓ_p, c_0 or c by $Tx = A^r x$ is onto, the inverse images of the bases of those spaces ℓ_p, c_0 and c are the bases of the spaces a_p^r, a_0^r and a_c^r, by Theorem 4.1.2. Therefore, we have

Theorem 4.6.7. [45, Theorem 4.1] *Let $\alpha_k(r) = (A^r x)_k$ for all $k \in \mathbb{N}$. Define the sequence $b^{(k)}(r) = \{b_n^{(k)}(r)\}_{n\in\mathbb{N}}$ of the elements of the space a_0^r by*

$$b_n^{(k)}(r) := \begin{cases} (-1)^{n-k}\frac{1+k}{1+r^n} & , \quad k \le n \le k+1, \\ 0 & , \quad n < k \text{ or } n > k+1, \end{cases} \tag{4.6.4}$$

for every fixed $k \in \mathbb{N}$. Then, the following statements hold:

(a) *The sequence $\{b^{(k)}(r)\}_{k\in\mathbb{N}}$ is a basis for the spaces a_p^r and a_0^r, and any $x \in a_0^r$ or $x \in a_p^r$ has a unique representation of the form*

$$x = \sum_k \alpha_k(r) b^{(k)}(r).$$

(b) *The set $\left\{b, b^{(k)}(r)\right\}$ is a basis for the space a_c^r and any $x \in a_c^r$ has a unique representation of the form*

$$x := lb + \sum_k [\alpha_k(r) - l] b^{(k)}(r), \text{ where}$$
$$b := \{1/(1+r^k)\} \text{ and } l := \lim_{k\to\infty} (A^r x)_k.$$

4.6.3 The α-, β- and γ-Duals of the Spaces a_p^r, a_0^r, a_c^r and a_∞^r

In this subsection, we state and prove the theorems determining the α-, β- and γ-duals of the sequence spaces a_p^r, a_0^r, a_c^r and a_∞^r of non-absolute type.

Theorem 4.6.8. (cf. [45, Theorem 5.4] and [42, Theorem 4.3]) *Define the sets* d_1^r, d_2^r *and* d_3^r *as follows:*

$$\begin{aligned} d_1^r &:= \left\{a=(a_k)\in\omega : \sup_{k\in\mathbb{N}}\left|\frac{1+k}{1+r^k}a_k\right|<\infty\right\}, \\ d_2^r &:= \left\{a=(a_k)\in\omega : \sum_k\left|\frac{1+k}{1+r^k}a_k\right|^q<\infty\right\}, \\ d_3^r &:= \left\{a=(a_k)\in\omega : \sup_{K\in\mathcal{F}}\sum_n\left|\sum_{k\in K}(-1)^{n-k}\frac{1+k}{1+r^n}a_n\right|<\infty\right\}. \end{aligned}$$

Then, the following statements hold:

(i) *In the case* $0<p\leq 1$, $\left(a_p^r\right)^\alpha = d_1^r$.

(ii) *In the case* $1<p<\infty$, $\left(a_p^r\right)^\alpha = d_2^r$.

(iii) *The* α*-dual of the spaces* a_0^r *and* a_c^r *is the set* d_3^r.

Proof. (iii) Let $a=(a_n)\in\omega$ and define the matrix $B=(b_{nk}^r)$ via the sequence $a=(a_n)$ by

$$b_{nk}^r := \begin{cases} (-1)^{n-k}\frac{1+k}{1+r^n}a_n &, \quad n-1\leq k\leq n, \\ 0 &, \quad 0\leq k<n-1 \text{ or } k>n, \end{cases}$$

for all $k,n\in\mathbb{N}$. Bearing in mind the relation (4.6.1), we immediately derive that

$$a_nx_n = \sum_{k=n-1}^{n}(-1)^{n-k}\frac{1+k}{1+r^n}a_ny_k = (By)_n \quad \text{for all} \quad n\in\mathbb{N}. \tag{4.6.5}$$

We therefore observe by (4.6.5) that $ax=(a_nx_n)\in\ell_1$ whenever $x=(x_n)\in a_0^r$ or a_c^r if and only if $By=\{(By)_n\}\in\ell_1$ whenever $y=(y_k)\in c_0$ or c. Then, we derive by Lemma 4.2.5 that

$$\sup_{K\in\mathcal{F}}\sum_n\left|\sum_{k\in K}(-1)^{n-k}\frac{1+k}{1+r^n}a_n\right|<\infty,$$

which yields the consequence that $(a_0^r)^\alpha=(a_c^r)^\alpha=d_3^r$.

Since Parts (i) and (ii) can be proved by the similar way used in proving Part (iii), we omit their detailed proof. □

Theorem 4.6.9. (cf. [45, Theorem 5.5] and [42, Theorem 4.4]) *Define the sets* d_4^r, d_5^r, d_6^r *and* d_7^r *by*

$$\begin{aligned} d_4^r &:= \left\{a = (a_k) \in \omega : \sum_k \left| \Delta\left(\frac{a_k}{1+r^k}\right)(k+1)\right|^q < \infty\right\}, \\ d_5^r &:= \left\{a = (a_k) \in \omega : \sum_k \left| \Delta\left(\frac{a_k}{1+r^k}\right)(k+1)\right| < \infty\right\}, \\ d_6^r &:= \left\{a = (a_k) \in \omega : \left(\frac{a_k}{1+r^k}\right) \in cs\right\}, \\ d_7^r &:= \left\{a = (a_k) \in \omega : \left(\frac{k+1}{1+r^k}a_k\right) \in \ell_\infty\right\}, \text{ where} \\ \Delta\left(\frac{a_k}{1+r^k}\right) &:= \frac{a_k}{1+r^k} - \frac{a_{k+1}}{1+r^{k+1}} \text{ for all } k \in \mathbb{N}. \end{aligned}$$

Then, the following statements hold:

(i) *In the case* $0 < p \leq 1$, $\left(a_p^r\right)^\beta = d_1^r$.

(ii) *In the case* $1 < p < \infty$, $\left(a_p^r\right)^\beta = d_1^r \cap d_4^r$.

(iii) $(a_0^r)^\beta = d_5^r \cap d_7^r$ *and* $(a_c^r)^\beta = d_5^r \cap d_6^r$.

Proof. To avoid undue repetition in the statements, we do not consider Parts (i) and (ii).

(iii) Because the proof may also be obtained for the space a_0^r in the similar way, we give the proof only for the space a_c^r.

Let $x = (x_k) \in a_c^r$ and $a = (a_k) \in \omega$. Now, consider the equality

$$\begin{aligned} \sum_{k=0}^n a_k x_k &= \sum_{k=0}^n \left[\sum_{j=k-1}^k (-1)^{k-j} \frac{j+1}{1+r^k} y_j\right] a_k \\ &= \sum_{k=0}^{n-1} \Delta\left(\frac{a_k}{1+r^k}\right)(k+1)y_k + \frac{n+1}{1+r^n} a_n y_n = (Ty)_n \end{aligned} \tag{4.6.6}$$

for all $n \in \mathbb{N}$, where $T = (t_{nk}^r)$ is defined by

$$t_{nk}^r := \begin{cases} \Delta\left(\frac{a_k}{1+r^k}\right)(k+1) & , \quad 0 \leq k \leq n-1, \\ \frac{n+1}{1+r^n} a_n & , \quad k = n, \\ 0 & , \quad k > n, \end{cases}$$

for all $k, n \in \mathbb{N}$. Thus, we deduce from Theorem 3.3.3 with (4.6.6) that $ax = (a_k x_k) \in cs$ whenever $x = (x_k) \in a_c^r$ if and only if $Ty \in c$ whenever $y = (y_k) \in c$. It is obvious that the columns of the matrix T are in the space c. Therefore, we derive by (3.3.1) and (3.3.4) that

$$\sum_k \left|\Delta\left(\frac{a_k}{1+r^k}\right)(k+1)\right| < \infty, \tag{4.6.7}$$

$$\left(\frac{k+1}{1+r^k}a_k\right) \in \ell_\infty, \tag{4.6.8}$$

$$\left(\frac{a_k}{1+r^k}\right) \in cs, \tag{4.6.9}$$

respectively. But the condition (4.6.8) is redundant, since it may be obtained by combining the conditions (4.6.7) and (4.6.9). This shows that $(a_c^r)^\beta = d_5^r \cap d_6^r$. □

Theorem 4.6.10. (cf. [45, Theorem 5.5] and [42, Theorem 4.5]) *The following statements hold:*

(i) *In the case* $0 < p \leq 1$, $\left(a_p^r\right)^\gamma = d_1^r$.

(ii) *In the case* $1 < p < \infty$, $\left(a_p^r\right)^\gamma = d_1^r \cap d_4^r$.

(iii) $(a_0^r)^\gamma = (a_c^r)^\gamma = d_5^r \cap d_6^r$.

Proof. This is obtained in the similar way used in the proof of Theorem 4.6.9 and so we omit details. □

By combining Lemma 4.3.18, Theorems 4.6.8 and 4.6.9, we give the following corollary on the monotonicity of the space a_p^r:

Corollary 4.6.11. [45, Corollary 5.8] *The space* a_p^r *is not monotone and so it is neither normal nor perfect.*

4.6.4 Some Matrix Mappings on the Spaces a_p^r and a_c^r

In this subsection, we desire to characterize some matrix mappings on the spaces a_p^r and a_c^r. We directly prove the theorems, characterizing the classes $(a_p^r : \ell_\infty)$, $(a_p^r : c)$, $(a_c^r : \ell_p)$ and $(a_c^r : c)$, where $1 < p < \infty$.

We write for brevity that

$$\widetilde{a}_{nk} := \Delta\left(\frac{a_{nk}}{1+r^k}\right)(k+1) = \left(\frac{a_{nk}}{1+r^k} - \frac{a_{n,k+1}}{1+r^{k+1}}\right)(k+1) \text{ and } a(n,k) := \sum_{j=0}^{n} a_{jk},$$

for all $k, n \in \mathbb{N}$. We will also use the similar notations with other letters.

Theorem 4.6.12. [45, Theorem 6.1] *Let* $1 < p < \infty$. *Then,* $A \in (a_p^r : \ell_\infty)$ *if and only if*

$$\sup_{k\in\mathbb{N}} \left|\frac{1+k}{1+r^k}a_{nk}\right| < \infty \text{ for each } n \in \mathbb{N}, \tag{4.6.10}$$

$$\sup_{n\in\mathbb{N}} \sum_k |\widetilde{a}_{nk}|^q < \infty. \tag{4.6.11}$$

Proof. Suppose that the conditions (4.6.10) and (4.6.11) hold, and take any $x = (x_k) \in a_p^r$. Then, the sequence $(a_{nk})_{k\in\mathbb{N}} \in d_1^r \cap d_4^r$ for all $n \in \mathbb{N}$ and this implies the existence of the A-transform of x.

Let us now consider the following equality derived by using the relation (4.6.1) from the mth partial sum of the series $\sum_k a_{nk}x_k$:

$$\sum_{k=0}^{m} a_{nk}x_k = \sum_{k=0}^{m-1} \tilde{a}_{nk}y_k + \frac{m+1}{1+r^m}a_{nm}y_m \quad \text{for all} \quad m,n \in \mathbb{N}. \tag{4.6.12}$$

Therefore, we obtain from (4.6.12) with (4.6.10), as $m \to \infty$ that

$$\sum_k a_{nk}x_k = \sum_k \tilde{a}_{nk}y_k \quad \text{for all} \quad n \in \mathbb{N}. \tag{4.6.13}$$

Now, by passing to supremum over $n \in \mathbb{N}$ in (4.6.13), we derive by applying the Hölder's inequality with (4.6.11) that

$$\sup_{n\in\mathbb{N}} |(Ax)_n| \leq \|y\|_p \left\{ \sup_{n\in\mathbb{N}} \left(\sum_k |\tilde{a}_{nk}|^q \right)^{1/q} \right\} < \infty,$$

which shows the sufficiency of the conditions (4.6.10) and (4.6.11).

Conversely, suppose that $A \in (a_p^r : \ell_\infty)$. Then, since $(a_{nk})_{k\in\mathbb{N}} \in d_1^r \cap d_4^r$ for all $n \in \mathbb{N}$ by the hypothesis, the necessity of (4.6.10) is trivial and (4.6.13) holds. Consider the continuous linear functionals f_n defined on a_p^r by the sequences $a_n = (a_{nk})_{k\in\mathbb{N}}$ as

$$f_n(x) = \sum_k a_{nk}x_k \quad \text{for all} \quad n \in \mathbb{N}.$$

Since a_p^r and ℓ_p are norm isomorphic, it should follow with (4.6.13) that $\|f_n\| = \|\tilde{a}_n\|_q$. This just says that the functionals defined by the rows of A on a_p^r are pointwise bounded. Hence, by the Banach-Steinhaus theorem, they are uniformly bounded, which yields that there exists a constant $M > 0$ such that $\|f_n\| \leq M$ for all $n \in \mathbb{N}$. It therefore follows, using the complete identification just referred to, that

$$\left(\sum_k |\tilde{a}_{nk}|^q \right)^{1/q} = \|f_n\| \leq M$$

holds for all $n \in \mathbb{N}$, which shows the necessity of the condition (4.6.11).

This completes the proof. □

Theorem 4.6.13. [45, Theorem 6.2] *Let $1 < p < \infty$. Then, $A = (a_{nk}) \in (a_p^r : c)$ if and only if (4.6.10), (4.6.11) hold, and*

$$\lim_{n\to\infty} \tilde{a}_{nk} = \alpha_k \text{ for each } k \in \mathbb{N}. \tag{4.6.14}$$

Proof. Let the conditions (4.6.10), (4.6.11), (4.6.14) hold and take any $x \in a_p^r$. Then, Ax exists and we have by (4.6.14) that

$$\lim_{n\to\infty} |\tilde{a}_{nk}|^q = |\alpha_k|^q$$

for each $k \in \mathbb{N}$, which leads us with (4.6.11) that the inequality

$$\sum_{j=0}^{k} |\alpha_j|^q \leq \sup_{n\in\mathbb{N}} \sum_j |\widetilde{a}_{nj}|^q = M < \infty$$

holds for every $k \in \mathbb{N}$. This gives that $(\alpha_k) \in \ell_q$. Since $x = (x_k) \in a_p^r$ by the hypothesis, the isomorphism $a_p^r \cong \ell_p$ yields that $y = (y_k) \in \ell_p$. Therefore, we derive by applying the Hölder's inequality that $(\alpha_k y_k) \in \ell_1$ for each $y = (y_k) \in \ell_p$. For any given $\varepsilon > 0$, choose a fixed $k_0 \in \mathbb{N}$ such that

$$\left(\sum_{k=k_0+1}^{\infty} |y_k|^p\right)^{1/p} < \frac{\varepsilon}{4M^{1/q}} .$$

Then, there is some $n_0 \in \mathbb{N}$ by (4.6.14) such that

$$\left|\sum_{k=0}^{k_0} (\widetilde{a}_{nk} - \alpha_k)\, y_k\right| < \frac{\varepsilon}{2}$$

for every $n \geq n_0$. Therefore, we have

$$\begin{aligned}
\left|\sum_k a_{nk}x_k - \sum_k \alpha_k y_k\right| &= \left|\sum_k (\widetilde{a}_{nk} - \alpha_k)\, y_k\right| \\
&\leq \left|\sum_{k=0}^{k_0} (\widetilde{a}_{nk} - \alpha_k)\, y_k\right| + \left|\sum_{k=k_0+1}^{\infty} (\widetilde{a}_{nk} - \alpha_k)\, y_k\right| \\
&< \frac{\varepsilon}{2} + \left[\sum_{k=k_0+1}^{\infty} \left(|\widetilde{a}_{nk}| + |\alpha_k|\right)^q\right]^{1/q} \left(\sum_{k=k_0+1}^{\infty} |y_k|^p\right)^{1/p} \\
&< \frac{\varepsilon}{2} + 2M^{1/q}\frac{\varepsilon}{4M^{1/q}} = \varepsilon
\end{aligned}$$

for all sufficiently large n. Hence, $Ax \in c$ which is what we wished to prove.

Conversely, suppose that $A \in (a_p^r : c)$. Then, since $c \subset \ell_\infty$, the necessity of the conditions (4.6.10) and (4.6.11) is immediately obtained from Theorem 4.6.12. To prove the necessity of (4.6.14), consider the sequence $b^{(k)}(r)$ defined by (4.6.4), which is in the space a_p^r for each $k \in \mathbb{N}$. Since Ax exists and is in c for each $x \in a_p^r$ by the hypothesis, it is clear that $Ab^{(k)}(r) = (\widetilde{a}_{nk})_{n\in\mathbb{N}} \in c$ for every $k \in \mathbb{N}$, which gives the necessity of (4.6.14).

This step concludes the proof. □

If the sequence space c is replaced by the space c_0, then Theorem 4.6.13 is reduced to the following:

Corollary 4.6.14. [45, Corollary 6.3] *Let $1 < p < \infty$. Then, $A \in (a_p^r : c_0)$ if and only if (4.6.10), (4.6.11) hold and (4.6.14) also holds with $\alpha_k = 0$ for all $k \in \mathbb{N}$.*

We shall continue by giving a lemma due to Wilansky [396, p. 128], which is needed in the proof of following theorems:

Lemma 4.6.15. *$A = (a_{nk}) \in (c : \ell_p)$ if and only if*

$$\sup_{F\in\mathcal{F}} \sum_n \left|\sum_{k\in F} a_{nk}\right|^p < \infty\,, \quad (1 < p < \infty). \tag{4.6.15}$$

Theorem 4.6.16. [42, Theorem 5.3] *$A = (a_{nk}) \in (a_c^r : \ell_p)$ if and only if*

(i) *For $1 < p < \infty$,*

$$\sup_{F\in\mathcal{F}} \sum_n \left|\sum_{k\in F} \widetilde{a}_{nk}\right|^p < \infty, \tag{4.6.16}$$

$$\sum_k |\widetilde{a}_{nk}| < \infty \text{ for each } n \in \mathbb{N}, \tag{4.6.17}$$

$$\left(\frac{a_{nk}}{1+r^k}\right)_{k\in\mathbb{N}} \in cs \text{ for each } n \in \mathbb{N}. \tag{4.6.18}$$

(ii) *For $p = \infty$, (4.6.18) holds and*

$$\sup_{n\in\mathbb{N}} \sum_k |\widetilde{a}_{nk}| < \infty. \tag{4.6.19}$$

Proof. (i) Suppose the conditions (4.6.16)–(4.6.18) hold and take any $x = (x_k) \in a_c^r$. Then, $(a_{nk})_{k\in\mathbb{N}} \in (a_c^r)^\beta$ for all $n \in \mathbb{N}$ and this implies that Ax exists. Let us define the matrix $B = (b_{nk})$ with $b_{nk} = \widetilde{a}_{nk}$ for all $n, k \in \mathbb{N}$. Then, since (4.6.15) is satisfied for the matrix B, we have $B \in (c : \ell_p)$. Let us now reconsider the equality (4.6.12) obtained from the mth partial sum of the series $\sum_k a_{nk}x_k$. Following the same way that used in the proof of Theorem 4.6.9, one can derive by combining the conditions (4.6.17) and (4.6.18) that

$$\left\{\frac{(1+m)a_{nm}}{1+r^m}\right\}_{m\in\mathbb{N}} \in c_0$$

for each $n \in \mathbb{N}$. Thus, bearing in mind this fact if we pass to limit in (4.6.12) as $m \to \infty$, then the second term on the right-hand side tends to zero and we again obtain the equality (4.6.13), which yields by taking p-norm that $\|Ax\|_p = \|By\|_p < \infty$. This means that $A \in (a_c^r : \ell_p)$.

Conversely, suppose that $A \in (a_c^r : \ell_p)$. Then, since a_c^r and ℓ_p are BK-spaces we have from Lemma 3.1.1 that there exists some real constant $K > 0$ such that

$$\|Ax\|_p \le K\|x\|_{a_c^r} \tag{4.6.20}$$

for all $x \in a_c^r$. Since the inequality (4.6.20) is also satisfied for the sequence $x = (x_k) = \sum_{k\in F} b^{(k)}(r)$ belonging to the space a_c^r, where $b^{(k)}(r) = \{b_n^{(k)}(r)\}$ is defined by (4.6.4), we thus have for any $F \in \mathcal{F}$ that

$$\|Ax\|_p = \left(\sum_n \left|\sum_{k\in F} \widetilde{a}_{nk}\right|^p\right)^{1/p} \le K\|x\|_{a_c^r},$$

which shows the necessity of (4.6.16).

Since A is applicable to the space a_c^r by the hypothesis, the necessity of (4.6.17) and (4.6.18) is trivial. This completes the proof of Part (i) of the theorem.

Since Part (ii) may also be proved in the similar way that of Part (i), we leave the detailed proof to the reader. □

Theorem 4.6.17. [42, Theorem 5.4] $A = (a_{nk}) \in (a_c^r : c)$ *if and only if (4.6.14), (4.6.18) and (4.6.19) hold, and*

$$\lim_{n\to\infty} \sum_k \widetilde{a}_{nk} = \alpha. \tag{4.6.21}$$

Proof. Suppose that $A = (a_{nk})$ satisfies the conditions (4.6.14), (4.6.18), (4.6.19) and (4.6.21). Let us take any $x = (x_k)$ in a_c^r. Then, Ax exists and it is trivial that the sequence $y = (y_k)$ connected with the sequence $x = (x_k)$ by the relation (4.6.1) is in c such that $y_k \to l$, as $k \to \infty$. At this stage, we observe from (4.6.14) and (4.6.19) that

$$\sum_{j=0}^{k} |\alpha_j| \le \sup_{n\in\mathbb{N}} \sum_j |\widetilde{a}_{nj}| < \infty$$

holds for every $k \in \mathbb{N}$. This leads us to the consequence that $(\alpha_k) \in \ell_1$. Considering (4.6.13), let us write

$$\sum_k a_{nk}x_k = \sum_k \widetilde{a}_{nk}(y_k - l) + l\sum_k \widetilde{a}_{nk}. \tag{4.6.22}$$

In this situation, by letting $n \to \infty$ in (4.6.22) we see that the first term on the right-hand side tends to $\sum_k \alpha_k(y_k - l)$ by (4.6.19) and (4.6.14), and the second term tends to $l\alpha$ by (4.6.21). Thus, we have $\lim_{n\to\infty}(Ax)_n = \sum_k \alpha_k(y_k - l) + l\alpha$, which shows that $A \in (a_c^r : c)$.

Conversely, suppose that $A \in (a_c^r : c)$. Then, since the inclusion $c \subset \ell_\infty$ holds, the necessity of (4.6.18) and (4.6.19) is immediately obtained from Theorem 4.6.16. To prove the necessity of (4.6.14), consider the sequence $x^{(k)} = \{x_n^{(k)}(r)\}_{n\in\mathbb{N}} \in a_c^r$ defined by

$$x_n^{(k)}(r) := \begin{cases} (-1)^{n-k}\frac{1+k}{1+r^k} & , \quad k \le n \le k+1, \\ 0 & , \quad 0 \le n \le k-1 \text{ or } n > k+1, \end{cases}$$

for each $k \in \mathbb{N}$. Since Ax exists and is in c for every $x \in a_c^r$, one can easily see that $Ax^{(k)} = (\widetilde{a}_{nk})_{n\in\mathbb{N}} \in c$ for each $k \in \mathbb{N}$, which shows the necessity of (4.6.14).

Similarly by putting $x = e$ in (4.6.13), we also obtain that $Ax = \left(\sum_k \widetilde{a}_{nk}\right)_{n\in\mathbb{N}}$, which belongs to the space c and this shows the necessity of (4.6.21).

This step concludes the proof. □

Let us define the concept of ϱ-multiplicativity of a summability matrix. When there is some notion of limit or sum in the sequence spaces λ and μ, we will say that the method $A \in (\lambda : \mu)$ is *multiplicative* ϱ if every $x \in \lambda$ is A-summable to ϱ times of $\lim x$, for any fixed real number ϱ and denote the class of all ϱ-multiplicative matrices by $(\lambda : \mu)_\varrho$. It is trivial that the class $(a_c^r : c)_\varrho$ of ϱ-multiplicative matrices is reduced to the classes $(a_c^r : c_0)$ and $(a_c^r : c)_{reg}$ in the cases $\varrho = 0$ and $\varrho = 1$, respectively; where $(a_c^r : c)_{reg}$ denotes the class of all matrix mappings A from a_c^r to c such that $A - \lim\ x = \lim\ x$ for all $x \in a_c^r$.

Now, we may give the following corollary to Theorem 4.6.17 without proof:

Corollary 4.6.18. [42, Corollary 5.5] $A = (a_{nk}) \in (a_c^r : c)_\varrho$ *if and only if (4.6.18), (4.6.19) hold, (4.6.14) and (4.6.21) also hold with* $\alpha_k = 0$ *for all* $k \in \mathbb{N}$ *and* $\alpha = \varrho$, *respectively.*

4.7 RIESZ SEQUENCE SPACES AND CONCERNING DUALITY RELATION

Let λ denotes any of the classical sequence spaces ℓ_∞, c, c_0 and ℓ_p or any of the Maddox's paranormed sequence spaces $\ell_\infty(p)$, $c(p)$, $c_0(p)$ and $\ell(p)$. Then, the domain λ_{R^t} is called as *Riesz sequence space* and was studied by Altay and Başar [21, 26].

Let us define the sequence $y = (y_k)$ by the R^t-transform of a sequence $x = (x_k) \in \omega$, i.e.,

$$y_k := (R^t x)_k = \frac{1}{T_k} \sum_{j=0}^{k} t_j x_j \quad \text{for all } k \in \mathbb{N}. \tag{4.7.1}$$

Then, since $\lambda_{R^t} \cong \lambda$ it is trivial that the two-sided implication "$x \in \lambda_{R^t}$ if and only if $y \in \lambda$" holds.

We will assume throughout that $p_k^{-1} + (p_k')^{-1} = 1$ provided $1 < \inf p_k \leq H < \infty$. For the sake of brevity in notation, we shall also write here and after that

$$\tilde{a}_{nk} := \frac{1}{T_n} \sum_{j=0}^{n} t_j a_{jk}, \quad a(n,k) := \sum_{j=0}^{n} a_{jk} \text{ and } \Delta\left(\frac{a_{nk}}{t_k}\right) := \frac{a_{nk}}{t_k} - \frac{a_{n,k+1}}{t_{k+1}}$$

for all $k, n \in \mathbb{N}$.

The main purpose of this section is to introduce the Riesz sequence spaces $r^t(p)$, $r_0^t(p)$, $r_c^t(p)$ and $r_\infty^t(p)$ of non-absolute type, which are the set of all sequences whose R^t-transforms are in the spaces $\ell(p)$, $c_0(p)$, $c(p)$ and $\ell_\infty(p)$, and is to determine their α-, β- and γ-duals. Besides, the basis of the spaces $r^t(p)$, $r_0^t(p)$ and $r_c^t(p)$ is given, and some classes of matrix transformations related to Riesz sequence spaces are characterized.

4.7.1 The Riesz Sequence Spaces $r^t(p)$, $r_0^t(p)$, $r_c^t(p)$ and $r_\infty^t(p)$ of Non–absolute Type

In [105], Choudhary and Mishra have defined the sequence space $\overline{\ell(p)}$, which consists of all sequences whose S-transforms are in the space $\ell(p)$. Başar and Altay [65] have recently examined the space $bs(p)$, which is formerly defined by Başar [60] as the set of all series whose sequences of partial sums are in the space $\ell_\infty(p)$. With the notation of (4.1.1), the spaces $\overline{\ell(p)}$ and $bs(p)$ can be redefined by

$$\overline{\ell(p)} = [\ell(p)]_S \quad \text{and} \quad bs(p) = [\ell_\infty(p)]_S.$$

Following Altay and Başar [21, 26], we define the Riesz sequence spaces $r^t(p)$, $r_0^t(p)$, $r_c^t(p)$ and $r_\infty^t(p)$ by

$$\begin{aligned}
r^t(p) &:= \left\{x=(x_k)\in\omega : \sum_k \left|\frac{1}{T_k}\sum_{i=0}^{k} t_i x_i\right|^{p_k} < \infty\right\}, \quad (0<p_k\le H<\infty),\\
r_0^t(p) &:= \left\{x=(x_k)\in\omega : \lim_{n\to\infty} \left|\frac{1}{T_n}\sum_{k=0}^{n} t_k x_k\right|^{p_n} = 0\right\},\\
r_c^t(p) &:= \left\{x=(x_k)\in\omega : \lim_{n\to\infty} \left|\frac{1}{T_n}\sum_{k=0}^{n} t_k x_k - l\right|^{p_n} = 0 \text{ for some } l\in\mathbb{R}\right\},\\
r_\infty^t(p) &:= \left\{x=(x_k)\in\omega : \sup_{n\in\mathbb{N}} \left|\frac{1}{T_n}\sum_{k=0}^{n} t_k x_k\right|^{p_n} < \infty\right\}.
\end{aligned}$$

Also, if $p_k = p$ for all $k\in\mathbb{N}$, then we write r_p^t instead of $r^t(p)$. In the case $(p_k)=e$, the sequence spaces $r_\infty^t(p)$, $r_c^t(p)$ and $r_0^t(p)$ are, respectively, reduced to the sequence spaces r_∞^t, r_c^t and r_0^t, which are introduced by Malkowsky [263]. It is natural that these spaces may also be defined with the notation of (4.1.1) that

$$\begin{array}{llll}
r^t(p) := \{\ell(p)\}_{R^t}, & r_0^t(p) := \{c_0(p)\}_{R^t}, & r_c^t(p) := \{c(p)\}_{R^t}, & r_\infty^t(p) := \{\ell_\infty(p)\}_{R^t},\\
r_p^t := (\ell_p)_{R^t}, & r_0^t := (c_0)_{R^t}, & r_c^t := (c)_{R^t}, & r_\infty^t := (\ell_\infty)_{R^t}.
\end{array}$$

We should note here in the case $t_n = 1$ for all $n\in\mathbb{N}$ that the sequence spaces r_p^t, r_∞^t and r_c^t, r_0^t are reduced to the Cesàro sequence spaces X_p, X_∞ and $\widetilde{c}$, $\widetilde{c_0}$ of non-absolute type introduced by Ng and Lee [304] and Şengönül and Başar [373].

Now, we begin with the following theorem:

Theorem 4.7.1. (cf. [21, Theorem 2.1] and [26]) *The following statements hold:*

(i) *$r^t(p)$ is a complete linear metric space paranormed by g_3 defined by*

$$g_3(x) := \left(\sum_k \left|\frac{1}{T_k}\sum_{i=0}^{k} t_i x_i\right|^{p_k}\right)^{1/M},$$

where $0<p_k\le H<\infty$.

(ii) *$r_0^t(p)$, $r_c^t(p)$ and $r_\infty^t(p)$ are the complete linear metric spaces paranormed by g_4, defined by*

$$g_4(x) := \sup_{k\in\mathbb{N}} \left|\frac{1}{T_k}\sum_{j=0}^{k} t_j x_j\right|^{p_k/M}.$$

g_4 is a paranorm for the spaces $r_\infty^t(p)$ and $r_c^t(p)$ only in the trivial case $\inf p_k > 0$ *when $r_\infty^t(p) = r_\infty^t$ and $r_c^t(p) = r_c^t$.*

Proof. Since Part (ii) can be proved in the similar way used in the proof of Part (i), we leave the detailed proof of Part (ii) to the reader, and prove only Part (i).

(i) The linearity of $r^t(p)$ with respect to the co-ordinatewise addition and scalar multiplication follows from the inequalities, which are satisfied for $u = (u_j), x = (x_j) \in r^t(p)$ (see [258, p. 30])

$$\left[\sum_k \left|\frac{1}{T_k}\sum_{j=0}^{k} t_j(u_j + x_j)\right|^{p_k}\right]^{1/M} \leq \left(\sum_k \left|\frac{1}{T_k}\sum_{j=0}^{k} t_j u_j\right|^{p_k}\right)^{1/M} + \left(\sum_k \left|\frac{1}{T_k}\sum_{j=0}^{k} t_j x_j\right|^{p_k}\right)^{1/M} \tag{4.7.2}$$

and for any $\alpha \in \mathbb{R}$ (see [251])

$$|\alpha|^{p_k} \leq \max\{1, |\alpha|^M\}. \tag{4.7.3}$$

It is clear that $g_3(\theta) = 0$ and $g_3(x) = g_3(-x)$ for all $x \in r^t(p)$. Again the inequalities (4.7.2) and (4.7.3) yield the subadditivity of g_3 and $g_3(\alpha x) \leq \max\{1, |\alpha|\}g_3(x)$.

Let (x^n) be any sequence of the points of the space $r^t(p)$ such that $g_3(x^n - x) \to 0$ and (α_n) also be any sequence of scalars such that $\alpha_n \to \alpha$, as $n \to \infty$. Then, since the inequality $g_3(x^n) \leq g_3(x) + g_3(x^n - x)$ holds by subadditivity of g_3, $\{g_3(x^n)\}$ is bounded and we thus have

$$\begin{aligned} g_3(\alpha_n x^n - \alpha x) &= \left[\sum_k \left|\frac{1}{T_k}\sum_{j=0}^{k} t_j\left(\alpha_n x_j^{(n)} - \alpha x_j\right)\right|^{p_k}\right]^{1/M} \\ &\leq |\alpha_n - \alpha| g_3(x^n) + |\alpha| g_3(x^n - x), \end{aligned}$$

which tends to zero, as $n \to \infty$. That is to say that the scalar multiplication is continuous. Hence, g_3 is a paranorm on the space $r^t(p)$.

It remains to prove the completeness of the space $r^t(p)$. Let (x^i) be any Cauchy sequence in the space $r^t(p)$, where $x^i = \{x_0^{(i)}, x_1^{(i)}, x_2^{(i)}, \ldots\}$. Then, for a given $\varepsilon > 0$ there exists a positive integer $n_0(\varepsilon)$ such that

$$g_3\left(x^i - x^j\right) < \varepsilon \tag{4.7.4}$$

for all $i, j \geq n_0(\varepsilon)$. Using definition of g_3, we obtain for each fixed $k \in \mathbb{N}$ that

$$\left|\left(R^t x^i\right)_k - \left(R^t x^j\right)_k\right| \leq \left[\sum_k \left|\left(R^t x^i\right)_k - \left(R^t x^j\right)_k\right|^{p_k}\right]^{1/M} < \varepsilon$$

for all $i, j \geq n_0(\varepsilon)$, which leads us to the fact that $\{(R^t x^0)_k, (R^t x^1)_k, (R^t x^2)_k, \ldots\}$ is a Cauchy sequence of real numbers for every fixed $k \in \mathbb{N}$. Since $\mathbb{R}$ is complete, it converges, say $(R^t x^i)_k \to (R^t x)_k$, as $i \to \infty$. Using these infinitely many limits $(R^t x)_0$, $(R^t x)_1$, $(R^t x)_2$, ..., we define the sequence $\{(R^t x)_0, (R^t x)_1, (R^t x)_2, \ldots\}$. From (4.7.4) for each $m \in \mathbb{N}$ and $i, j \geq n_0(\varepsilon)$

$$\sum_{k=0}^{m} \left|\left(R^t x^i\right)_k - \left(R^t x^j\right)_k\right|^{p_k} \leq g_3\left(x^i - x^j\right)^M < \varepsilon^M. \tag{4.7.5}$$

Take any $i \geq n_0(\varepsilon)$. First let $j \to \infty$ in (4.7.5) and then $m \to \infty$ to obtain $g_3(x^i - x) \leq \varepsilon$. Finally, taking $\varepsilon = 1$ in (4.7.5) and letting $i \geq n_0(1)$, we have by Minkowski's inequality for each $m \in \mathbb{N}$ that

$$\left[\sum_{k=0}^{m} \left|(R^t x)_k\right|^{p_k}\right]^{1/M} \leq g_3\left(x^i - x\right) + g_3\left(x^i\right) \leq 1 + g_3\left(x^i\right),$$

which implies that $x \in r^t(p)$. Since $g_3(x - x^i) \leq \varepsilon$ for all $i \geq n_0(\varepsilon)$, it follows that $x^i \to x$, as $i \to \infty$ whence we have shown that $r^t(p)$ is complete. □

Therefore, one can easily check that the absolute property does not hold on the space $r^t(p)$, that is $g_3(x) \neq g_3(|x|)$ for at least one sequence in the space $r^t(p)$, and this says that $r^t(p)$ is a sequence space of non-absolute type; where $|x| = (|x_k|)$.

Now, we can give the following theorem without proof, since it can be proved by the similar way used in the proof of Theorem 4.2.1 with ℓ_p, c_0, c and ℓ_∞, where $1 \leq p < \infty$:

Theorem 4.7.2. *r_p^t, r_0^t, r_c^t and r_∞^t are the linear spaces under the co-ordinatewise addition and scalar multiplication, which are BK-spaces with the norms $\|x\|_{r_p^t} = \|R^t x\|_p$ and $\|x\|_{r_0^t} = \|x\|_{r_c^t} = \|x\|_{r_\infty^t} = \|R^t x\|_\infty$, where $1 < p < \infty$.*

Theorem 4.7.3. (cf. [21, Theorem 2.3] and [26, Theorem 2.2]) *Let $\lambda \in \{\ell(p), c_0(p), c(p), \ell_\infty(p), \ell_p, c_0, c, \ell_\infty\}$. Then, the Riesz sequence space λ_{R^t} of non-absolute type is paranorm or norm isomorphic to the space λ.*

Proof. We prove the theorem for $\lambda = \ell(p)$. To do this, we should show the existence of a paranorm preserving linear bijection between the spaces $r^t(p)$ and $\ell(p)$ for $0 < p_k \leq H < \infty$. With the notation of (4.7.1), define the transformation T from $r^t(p)$ to $\ell(p)$ by $x \mapsto y = Tx = R^t x$. The linearity of T is trivial. Further, it is obvious that $x = \theta$ whenever $Tx = \theta$ and hence, T is injective.

Let $y = (y_k) \in \ell(p)$ and define the sequence $x = (x_k)$ by $x_k = (T_k y_k - T_{k-1} y_{k-1})/t_k$ for all $k \in \mathbb{N}$. Then,

$$\begin{aligned} g_3(x) = \left(\sum_k \left|\frac{1}{T_k}\sum_{i=0}^{k} t_i x_i\right|^{p_k}\right)^{1/M} &= \left(\sum_k \left|\sum_{j=0}^{k} \delta_{kj} y_j\right|^{p_k}\right)^{1/M} \\ &= \left(\sum_k |y_k|^{p_k}\right)^{1/M} = g_1(y) < \infty. \end{aligned}$$

Thus, we obtain that $x \in r^t(p)$. Consequently, T is surjective and is paranorm preserving. Hence, T is a paranorm preserving linear bijection and this says us that the spaces $r^t(p)$ and $\ell(p)$ are paranorm isomorphic, as desired. □

Let us suppose that $1 < p_k \leq s_k$ for every $k \in \mathbb{N}$. Then, it is well-known that $\ell(p) \subset \ell(s)$, which leads us to the immediate consequence that $r^t(p) \subset r^t(s)$.

We shall quote some lemmas that are needed in proving our theorems concerning the α-, β- and γ-duals of the Riesz sequence spaces.

Lemma 4.7.4. [193, Theorem 5.1.0] *Let $A = (a_{nk})$ be an infinite matrix over the complex field. Then, the following statements hold:*

(a) *Let $1 < p_k \le H < \infty$ for every $k \in \mathbb{N}$. Then, $A \in (\ell(p) : \ell_1)$ if and only if there exists an integer $B > 1$ such that $\sup_{N\in\mathcal{F}} \sum_k \left| \sum_{n\in N} a_{nk}B^{-1} \right|^{p'_k} < \infty$.*

(b) *Let $0 < p_k \le 1$ for every $k \in \mathbb{N}$. Then, $A \in (\ell(p) : \ell_1)$ if and only if $\sup_{N\in\mathcal{F}} \sup_{k\in\mathbb{N}} \left| \sum_{n\in N} a_{nk} \right|^{p_k} < \infty$.*

Lemma 4.7.5. [237, Theorem 1(i)-(ii)] *Let $A = (a_{nk})$ be an infinite matrix over the complex field. Then, the following statements hold:*

(a) *Let $1 < p_k \le H < \infty$ for every $k \in \mathbb{N}$. Then, $A = (a_{nk}) \in (\ell(p) : \ell_\infty)$ if and only if there exists an integer $B > 1$ such that*

$$\sup_{n\in\mathbb{N}} \sum_k |a_{nk}B^{-1}|^{p'_k} < \infty. \tag{4.7.6}$$

(b) *Let $0 < p_k \le 1$ for every $k \in \mathbb{N}$. Then, $A = (a_{nk}) \in (\ell(p) : \ell_\infty)$ if and only if*

$$\sup_{n,k\in\mathbb{N}} |a_{nk}|^{p_k} < \infty. \tag{4.7.7}$$

Lemma 4.7.6. [237, Corollary for Theorem 1] *Let $0 < p_k \le H < \infty$ for every $k \in \mathbb{N}$. Then, $A = (a_{nk}) \in (\ell(p) : c)$ if and only if (4.7.6), (4.7.7) hold, and*

$$\lim_{n\to\infty} a_{nk} = \beta_k \quad \text{for all} \quad k \in \mathbb{N}.$$

Theorem 4.7.7. (cf. [21, Theorem 2.7] and [26, Theorem 2.6]) *Let $1 < p_k \le H < \infty$ for every $k \in \mathbb{N}$. Define the sets $D_1(p)$, $D_2(p)$, $D_3(p)$, $D_4(p)$, $D_5(p)$, $D_6(p)$, $D_7(p)$, $D_8(p)$, $D_9(p)$ and $D_{10}(p)$, as follows:*

$$D_1(p) := \bigcup_{B>1} \left\{ a = (a_k) \in \omega : \sup_{N\in\mathcal{F}} \sum_k \left| \sum_{n\in N} (-1)^{n-k} \frac{a_n}{t_n} T_k B^{-1} \right|^{p'_k} < \infty \right\},$$

$$D_2(p) := \bigcup_{B>1} \left\{ a = (a_k) \in \omega : \sum_k \left| \Delta\left(\frac{a_k}{t_k}\right) T_k B^{-1} \right|^{p'_k} < \infty \text{ and } \left\{ \left(\frac{a_k}{t_k} T_k B^{-1} \right)^{p'_k} \right\} \in \ell_\infty \right\},$$

$$D_3(p) := \left\{ a = (a_k) \in \omega : \sup_{N\in\mathcal{F}} \sup_{k\in\mathbb{N}} \left| \sum_{n\in N} (-1)^{n-k} \frac{a_n}{t_n} T_k B^{-1} \right|^{p_k} < \infty \right\},$$

$$D_4(p) := \left\{ a = (a_k) \in \omega : \sup_{k\in\mathbb{N}} \left| \Delta\left(\frac{a_k}{t_k}\right) T_k \right|^{p_k} < \infty \text{ and } \sup_{k\in\mathbb{N}} \left| \frac{a_k}{t_k} T_k \right|^{p_k} < \infty \right\},$$

$$D_5(p) := \bigcap_{B>1} \left\{ a = (a_k) \in \omega : \sup_{N\in\mathcal{F}} \sum_n \left| \sum_{k\in N} (-1)^{n-k} \frac{a_n}{t_n} T_k B^{1/p_k} \right| < \infty \right\},$$

$$D_6(p) := \bigcap_{B>1} \left\{ a = (a_k) \in \omega : \sum_k \left| \Delta\left(\frac{a_k}{t_k}\right) T_k \right| B^{1/p_k} < \infty \text{ and } \left(\frac{a_k}{t_k} T_k B^{1/p_k}\right) \in c_0 \right\},$$

$$D_7(p) := \bigcap_{B>1} \left\{ (a_k) \in \omega : \sum_k \left| \Delta\left(\frac{a_k}{t_k}\right) T_k \right| B^{1/p_k} < \infty \text{ and } \left\{ \Delta\left(\frac{a_k}{t_k}\right) T_k B^{1/p_k} \right\} \in \ell_\infty \right\},$$

$$D_8(p) := \bigcup_{B>1} \left\{ a = (a_k) \in \omega : \sup_{K\in\mathcal{F}} \sum_n \left| \sum_{k\in K} (-1)^{n-k} \frac{a_n}{t_n} T_k B^{-1/p_k} \right| < \infty \right\},$$

$$D_9(p) := \left\{ a = (a_k) \in \omega : \sum_n \left| \sum_k (-1)^{n-k} \frac{a_n}{t_n} T_k \right| < \infty \right\},$$

$$D_{10}(p) := \bigcup_{B>1} \left\{ a = (a_k) \in \omega : \sum_k \left| \Delta\left(\frac{a_k}{t_k}\right) T_k \right| B^{-1/p_k} < \infty \right\}.$$

Then, the following statements hold:

(i) *If* $1 < p_k \leq H < \infty$ *for every* $k \in \mathbb{N}$, $\{r^t(p)\}^\alpha := D_1(p)$ *and* $\{r^t(p)\}^\beta = \{r^t(p)\}^\gamma := D_2(p)$

(ii) *If* $0 < p_k \leq 1$ *for every* $k \in \mathbb{N}$, $\{r^t(p)\}^\alpha := D_3(p)$ *and* $\{r^t(p)\}^\beta := \{r^t(p)\}^\gamma := D_4(p)$.

(iii) $\{r^t_\infty(p)\}^\alpha := D_5(p)$, $\{r^t_\infty(p)\}^\beta := D_6(p)$, $\{r^t_\infty(p)\}^\gamma := D_7(p)$, $\{r^t_c(p)\}^\alpha := D_8(p) \cap D_9(p)$, $\{r^t_c(p)\}^\beta := D_{10}(p) \cap cs$, $\{r^t_c(p)\}^\gamma := D_{10}(p) \cap bs$, $\{r^t_0(p)\}^\alpha := D_8(p)$ *and* $\{r^t_0(p)\}^\beta := \{r^t_0(p)\}^\gamma := D_{10}(p)$.

Proof. To avoid undue repetition in the statements, we give the proof only for the space $r^t(p)$.

(i) Let us take any $a = (a_k) \in \omega$. We easily derive with (4.7.1) that

$$a_n x_n = \sum_{i=n-1}^{n} (-1)^{n-i} \frac{a_n}{t_n} T_i y_i = (Cy)_n \quad \text{for all} \quad n \in \mathbb{N}, \tag{4.7.8}$$

where $C = (c_{nk})$ is defined by $c_{nk} := \begin{cases} (-1)^{n-k} \frac{a_n}{t_n} T_k & , \quad n-1 \leq k \leq n, \\ 0 & , \quad 0 \leq k < n-1 \text{ or } k > n, \end{cases}$
for all $k, n \in \mathbb{N}$. Thus, we observe by combining (4.7.8) with Part (a) of Lemma 4.7.4 that $ax = (a_n x_n) \in \ell_1$ whenever $x = (x_k) \in r^t(p)$ if and only if $Cy \in \ell_1$ whenever $y = (y_k) \in \ell(p)$. This gives the result that $\{r^t(p)\}^\alpha = D_1(p)$.

Consider the equality

$$\sum_{k=0}^{n} a_k x_k = \sum_{k=0}^{n-1} \Delta\left(\frac{a_k}{t_k}\right) T_k y_k + \frac{a_n}{t_n} T_n y_n = (Dy)_n \tag{4.7.9}$$

for all $n \in \mathbb{N}$, where $D = (d_{nk})$ is defined by

$$d_{nk} := \begin{cases} \Delta\left(\frac{a_k}{t_k}\right) T_k & , \quad 0 \le k \le n-1, \\ \frac{a_n}{t_n} T_n & , \quad k = n, \\ 0 & , \quad k > n, \end{cases}$$

for all $k, n \in \mathbb{N}$. Thus, we deduce from Part (a) of Lemma 4.7.6 with (4.7.9) that $ax = (a_k x_k) \in cs$ whenever $x = (x_k) \in r^t(p)$ if and only if $Dy \in c$ whenever $y = (y_k) \in \ell(p)$. Therefore, we derive from (4.7.6) that

$$\sum_k \left| \Delta\left(\frac{a_k}{t_k}\right) T_k B^{-1} \right|^{p_k'} < \infty \text{ and } \sup_{k \in \mathbb{N}} \left| \frac{a_k}{t_k} T_k B^{-1} \right|^{p_k'} < \infty,$$

which shows that $\{r^t(p)\}^\beta = D_2(p)$.

Similarly, we see from Part (a) of Lemma 4.7.5 with (4.7.9) that $ax = (a_k x_k) \in bs$ whenever $x = (x_k) \in r^t(p)$ if and only if $Dy \in \ell_\infty$ whenever $y = (y_k) \in \ell(p)$. This leads us to the desired consequence that $\{r^t(p)\}^\gamma = D_2(p)$.

(ii) This is easily obtained by proceeding as in the proof of Part (i) by using the second parts of Lemmas 4.7.4–4.7.6 instead of the first parts. So, we omit details.

This completes the proof. □

Now, we give the bases for the paranormed spaces $r^t(p)$, $r_0^t(p)$ and $r_c^t(p)$.

Theorem 4.7.8. (cf. [21, Theorem 2.9] and [26, Theorem 3.1]) *Let $\alpha_k(t) = (R^t x)_k$ and $0 < p_k \le H < \infty$ for all $k \in \mathbb{N}$. Define the sequence $b^{(k)}(t) = \{b_n^{(k)}(t)\}_{n \in \mathbb{N}}$ of the elements of the space $r^t(p)$ for every fixed $k \in \mathbb{N}$ by*

$$b_n^{(k)}(t) := \begin{cases} (-1)^{n-k} \frac{T_k}{t_n} & , \quad k \le n \le k+1, \\ 0 & , \quad 0 \le n < k \text{ or } n > k+1. \end{cases} \tag{4.7.10}$$

Then, the following statements hold:

(i) *The sequence $\{b^{(k)}(t)\}_{k \in \mathbb{N}}$ is the basis for the spaces $r^t(p)$ and $r_0^t(p)$, and any $x \in r^t(p)$ or $x \in r_0^t(p)$ has a unique representation of the form*

$$x := \sum_k \alpha_k(t) b^{(k)}(t). \tag{4.7.11}$$

(ii) *The set $\left\{e, b^{(k)}(t)\right\}$ is a basis for the space $r_c^t(p)$ and any $x \in r_c^t(p)$ has a unique representation of the form*

$$x := le + \sum_k [\alpha_k(t) - l] b^{(k)}(t), \text{ where} \tag{4.7.12}$$

$$l := \lim_{k \to \infty} (R^t x)_k. \tag{4.7.13}$$

Proof. (i) It is clear that $\left\{b^{(k)}(t)\right\} \subset r^t(p)$, since

$$R^t b^{(k)}(t) = e^{(k)} \in \ell(p) \text{ for all } k \in \mathbb{N} \text{ with } 0 < p_k \le H < \infty. \tag{4.7.14}$$

Let $x \in r^t(p)$ be given. For every non-negative integer m, we put

$$x^{[m]} = \sum_{k=0}^{m} \alpha_k(q) b^{(k)}(t). \tag{4.7.15}$$

Then, we obtain by applying R^t to (4.7.15) with (4.7.14) that

$$R^t x^{[m]} = \sum_{k=0}^{m} \alpha_k(q) R^t b^{(k)}(t) = \sum_{k=0}^{m} (R^t x)_k e^{(k)},$$

$$\left\{R^t\left(x - x^{[m]}\right)\right\}_i := \begin{cases} 0 & , \quad 0 \leq i \leq m, \\ (R^t x)_i & , \quad i > m, \end{cases}$$

for all $i, m \in \mathbb{N}$. Given $\varepsilon > 0$, there is an integer m_0 such that

$$\left[\sum_{i=m}^{\infty} |(R^t x)_i|^{p_k}\right]^{1/M} < \frac{\varepsilon}{2}$$

for all $m \geq m_0$. Hence,

$$g\left(x - x^{[m]}\right) = \left[\sum_{i=m}^{\infty} |(R^t x)_i|^{p_k}\right]^{1/M} \leq \left[\sum_{i=m_0}^{\infty} |(R^t x)_i|^{p_k}\right]^{1/M} < \varepsilon$$

for all $m \geq m_0$, which proves that $x \in r^t(p)$ is represented as in (4.7.11).

Let us show the uniqueness of the representation for $x \in r^t(p)$ given by (4.7.11). Suppose, on the contrary, that there exists one another representation $x = \sum_k \mu_k(q) b^{(k)}(t)$. Since the linear transformation T, from $r^t(p)$ to $\ell(p)$, used in the proof of Theorem 4.7.3 is continuous, we have at this stage that

$$(R^t x)_n = \sum_k \mu_k(t) \{R^t b^{(k)}(t)\}_n = \sum_k \mu_k(t) e_n^{(k)} = \mu_n(t) \quad \text{for all} \quad n \in \mathbb{N},$$

which contradicts the fact that $(R^t x)_n = \alpha_n(t)$ for all $n \in \mathbb{N}$. Hence, the representation (4.7.11) of $x \in r^t(p)$ is unique. This completes the proof of Part (i) of Theorem.

Since the proof is similar for the space $r_0^t(p)$, we omit details.

(ii) Since $\{b^{(k)}(t)\} \subset r_0^t(p)$ and $e \in c$, the inclusion $\{e, \ b^{(k)}(t)\} \subset r_c^t(p)$ trivially holds. Let us take $x \in r_c^t(p)$. Then, there uniquely exists an l satisfying (4.7.13). We thus have the fact that $u \in r_0^t(p)$ whenever we set $u = x - le$. Therefore, we deduce by Part (i) of the present theorem that the representation of u is unique. This leads us that the representation of x given by (4.7.12) is unique and this step concludes the proof of Part (ii) of Theorem. □

4.7.2 Matrix Mappings Related to the Riesz Sequence Spaces

In this subsection, first we characterize the matrix transformations from the space $r^t(p)$ to the spaces ℓ_∞ and c, and obtain from them the necessary and sufficient conditions on an infinite matrix belonging to the class transforming the space $r^t(p)$

to the spaces bs, cs, c_0, cs_0, r_∞^t, r_c^t and r_0^t, respectively. Both Theorems 4.7.9 and 4.7.11 give the exact conditions of the general case $0 < p_k \leq H < \infty$ by combining the cases $0 < p_k \leq 1$ and $1 < p_k \leq H < \infty$. We consider only the case $1 < p_k \leq H < \infty$ and leave the case $0 < p_k \leq 1$ to the reader because it can be proved in the similar way.

Theorem 4.7.9. [21, Theorem 3.1] *Let $A = (a_{nk})$ be an infinite matrix over the complex field. Then, the following statements hold:*

(a) *Let $1 < p_k \leq H < \infty$ for every $k \in \mathbb{N}$. Then $A = (a_{nk}) \in (r^t(p) : \ell_\infty)$ if and only if there exists an integer $B > 1$ such that*

$$C(B) = \sup_{n\in\mathbb{N}} \sum_k \left| \Delta\left(\frac{a_{nk}}{t_k}\right) T_k B^{-1} \right|^{p'_k} < \infty, \tag{4.7.16}$$

$$\left\{ \left(\frac{a_{nk}}{t_k} T_k B^{-1} \right)^{p'_k} \right\} \in \ell_\infty \text{ for all } n \in \mathbb{N}. \tag{4.7.17}$$

(b) *Let $0 < p_k \leq 1$ for every $k \in \mathbb{N}$. Then, $A = (a_{nk}) \in (r^t(p) : \ell_\infty)$ if and only if*

$$\sup_{k,n\in\mathbb{N}} \left| \Delta\left(\frac{a_{nk}}{t_k}\right) T_k \right|^{p_k} < \infty. \tag{4.7.18}$$

Proof. (a) Let $A = (a_{nk}) \in (r^t(p) : \ell_\infty)$ and $1 < p_k \leq H < \infty$ for every $k \in \mathbb{N}$. Then, Ax exists for every $x \in r^t(p)$ and this implies that $(a_{nk})_{k\in\mathbb{N}} \in \{r^t(p)\}^\beta$ for each $n \in \mathbb{N}$. Now, the necessity of (4.7.16) and (4.7.17) is immediate.

Conversely, suppose that the conditions (4.7.16) and (4.7.17) hold, and $x = (x_k) \in r^t(p)$. In this situation, since $(a_{nk})_{k\in\mathbb{N}} \in \{r^t(p)\}^\beta$ for every fixed $n \in \mathbb{N}$, the A-transform of x exists. Consider the following equality obtained by the relation (4.7.1) that

$$\sum_{k=0}^{m} a_{nk}x_k = \sum_{k=0}^{m-1} \Delta\left(\frac{a_{nk}}{t_k}\right) T_k y_k + \frac{a_{nm}}{t_m} T_m y_m \tag{4.7.19}$$

for all $m, n \in \mathbb{N}$. Taking into account the assumptions we derive from (4.7.19), as $m \to \infty$ that

$$\sum_k a_{nk}x_k = \sum_k \Delta\left(\frac{a_{nk}}{t_k}\right) T_k y_k \text{ for all } n \in \mathbb{N}. \tag{4.7.20}$$

Now, by combining (4.7.20) and the inequality $|ab| \leq B\{|aB^{-1}|^{p'} + |b|^p\}$ which holds for any $B > 0$ and any complex numbers a, b; where $p > 1$ and $p^{-1} + p'^{-1} = 1$ (see [237]), one can easily see that

$$\begin{aligned} \sup_{n\in\mathbb{N}} \left| \sum_k a_{nk}x_k \right| &\leq \sup_{n\in\mathbb{N}} \sum_k \left| \Delta\left(\frac{a_{nk}}{t_k}\right) T_k \right| |y_k| \\ &\leq B[C(B) + g_1^B(y)] < \infty. \end{aligned}$$

This completes the proof. □

Theorem 4.7.10. [21, Theorem 3.2] *Let $A = (a_{nk})$ be an infinite matrix over the complex field. Then, the following statements hold:*

(a) *Let $1 < p_k \leq H < \infty$ for every $k \in \mathbb{N}$. Then, $A = (a_{nk}) \in (r^t(p) : bs)$ if and only if there exists an integer $B > 1$ such that*

$$\sup_{n\in\mathbb{N}} \sum_k \left| \Delta \left\{ \frac{a(n,k)}{t_k} \right\} T_k B^{-1} \right|^{p'_k} < \infty, \tag{4.7.21}$$

$$\sup_{k\in\mathbb{N}} \left| \frac{a(n,k)}{t_k} T_k B^{-1} \right|^{p'_k} < \infty \text{ for all } n \in \mathbb{N}. \tag{4.7.22}$$

(b) *Let $0 < p_k \leq 1$ for every $k \in \mathbb{N}$.* Then *$A = (a_{nk}) \in (r^t(p) : bs)$ if and only if*

$$\sup_{k,n\in\mathbb{N}} \left| \Delta \left\{ \frac{a(n,k)}{t_k} \right\} T_k \right|^{p_k} < \infty. \tag{4.7.23}$$

Proof. Let us define the matrix $E = (e_{nk})$ by $e_{nk} = a(n,k)$ for all $k, n \in \mathbb{N}$. Consider now the following equality derived from the n, m^{th} partial sums of the series $\sum_j \sum_k a_{jk} x_k$, as $m \to \infty$

$$\sum_{j=0}^{n} \sum_k a_{jk} x_k = \sum_k e_{nk} x_k \quad \text{for all} \quad n \in \mathbb{N}.$$

Therefore, bearing in mind the fact that the spaces bs and ℓ_∞ are norm isomorphic one can immediately see that $Ax \in bs$ whenever $x \in r^t(p)$ if and only if $Ex \in \ell_\infty$ whenever $x \in r^t(p)$. Now, the proof directly follows from Theorem 4.7.9 with E instead of A. □

Theorem 4.7.11. [21, Theorem 3.3] *Let $0 < p_k \leq H < \infty$ for every $k \in \mathbb{N}$. Then, $A = (a_{nk}) \in (r^t(p) : c)$ if and only if (4.7.16), (4.7.17) and (4.7.18) hold, and there is a sequence (α_k) of the scalars such that*

$$\lim_{n\to\infty} \Delta \left(\frac{a_{nk} - \alpha_k}{t_k} \right) T_k = 0 \quad \textit{for all} \quad k \in \mathbb{N}. \tag{4.7.24}$$

Proof. (a) Let $A = (a_{nk}) \in (r^t(p) : c)$ and $1 < p_k \leq H < \infty$ for every $k \in \mathbb{N}$. Since $c \subset \ell_\infty$, the necessity of (4.7.16) and (4.7.17) is trivial by Theorem 4.7.9.

Because of A-transform of every $x \in r^t(p)$ exists and is in c by the hypothesis, $Ax^{(k)}$ is also in c for every fixed $k \in \mathbb{N}$, which shows that $\{\Delta(a_{nk}/t_k)T_k\}_{n\in\mathbb{N}} \in c$, where $x^{(k)} = \{x_n^{(k)}(t)\}$ is defined by

$$x_n^{(k)}(t) := \begin{cases} (-1)^{n-k} \frac{T_k}{t_n} & , \quad k \leq n \leq k+1, \\ 0 & , \quad 0 \leq n < k \text{ or } n > k+1, \end{cases}$$

and is in the space $r^t(p)$ for each $k \in \mathbb{N}$. This proves the necessity of (4.7.24).

Conversely, suppose that (4.7.16), (4.7.17) and (4.7.24) hold, and $x \in r^t(p)$. Then, $(a_{nk})_{k\in\mathbb{N}} \in \{r^t(p)\}^\beta$ for each $n \in \mathbb{N}$, which implies that Ax exists. We observe for every $m, n \in \mathbb{N}$ that

$$\sum_{k=0}^{m} \left|\Delta\left(\frac{a_{nk}}{t_k}\right) T_k B^{-1}\right|^{p'_k} \leq \sup_{n\in\mathbb{N}} \sum_k \left|\Delta\left(\frac{a_{nk}}{t_k}\right) T_k B^{-1}\right|^{p'_k},$$

which gives the fact by letting $m, n \to \infty$ with (4.7.24) and (4.7.16) that

$$\sum_k \left|\Delta\left(\frac{\alpha_k}{t_k}\right) T_k B^{-1}\right|^{p'_k} < \infty. \tag{4.7.25}$$

Also, we have from (4.7.17) by letting $n \to \infty$ that $\{ (\alpha_k T_k B^{-1}/t_k)^{p'_k} \} \in \ell_\infty$, which leads us together with (4.7.25) that $(\alpha_k) \in D_2(p)$. That is to say that the series $\sum_k \alpha_k x_k$ converges for every $x \in r^t(p)$.

Let us now consider the equality obtained from (4.7.20) with $a_{nk} - \alpha_k$ instead of a_{nk}:

$$\sum_k (a_{nk} - \alpha_k) x_k = \sum_k \Delta\left(\frac{a_{nk} - \alpha_k}{t_k}\right) T_k y_k \quad \text{for all} \quad n \in \mathbb{N}. \tag{4.7.26}$$

Therefore, we have at this stage from Lemma 4.7.6 with $\beta_k = 0$ for all $k \in \mathbb{N}$ that the matrix $\{\Delta(\frac{a_{nk}-\alpha_k}{t_k}) T_k\}_{n,k\in\mathbb{N}}$ belongs to the class $(\ell(p) : c_0)$. Thus, one can see by (4.7.26) that

$$\lim_{n\to\infty} \sum_k (a_{nk} - \alpha_k) x_k = 0. \tag{4.7.27}$$

Combining (4.7.27) with the above results, one can see that $Ax \in c$, which is what we wished to prove.

(b) Since this can be obtained by the similar way used in the proof of Part (a), we omit details. □

Now, we can give the following corollary characterizing the class $(r^t(p) : c_0)$:

Corollary 4.7.12. [21, Theorem 3.5] *Let $0 < p_k \leq H < \infty$ for every $k \in \mathbb{N}$. Then, $A = (a_{nk}) \in (r^t(p) : c_0)$ if and only if (4.7.16), (4.7.17) and (4.7.18) hold, and (4.7.24) also holds with $\alpha_k = 0$ for each $k \in \mathbb{N}$.*

Theorem 4.7.13. [21, Theorem 3.4] *Let $0 < p_k \leq H < \infty$ for every $k \in \mathbb{N}$. Then, the following statements hold:*

(a) *$A = (a_{nk}) \in (r^t(p) : cs)$ if and only if (4.7.21), (4.7.22) and (4.7.23) hold, and*

$$\lim_{n\to\infty} \Delta\left\{\frac{a(n,k) - \alpha_k}{t_k}\right\} T_k = 0 \text{ for all } k \in \mathbb{N}. \tag{4.7.28}$$

(b) *$A = (a_{nk}) \in (r^t(p) : cs_0)$ if and only if (4.7.21), (4.7.22) and (4.7.23) hold, and (4.7.28) also holds with $\alpha_k = 0$ for each $k \in \mathbb{N}$.*

Proof. This may be obtained by the same kind of argument that has been used in the proof of Theorem 4.7.10, with Theorem 4.7.11 instead of Theorem 4.7.9. □

Now, we may give our final theorem, which also yields the characterizations of the matrix mappings between the Riesz sequence spaces.

Theorem 4.7.14. [21, Theorem 3.6] *Suppose that the elements of the infinite matrices $C = (c_{nk})$ and $D = (d_{nk})$ are connected with the relation*

$$d_{nk} := \frac{1}{T_n} \sum_{j=0}^{n} t_j c_{jk} \text{ for all } k, n \in \mathbb{N} \tag{4.7.29}$$

and λ, μ be any two given sequence spaces. Then, $C \in (\lambda : \mu_{R^t})$ if and only if $D \in (\lambda : \mu)$.

Proof. Let us take any $s = (s_k) \in \lambda$ and consider (4.7.29) with the following equality

$$\sum_{k=0}^{m} d_{nk} s_k = \frac{1}{T_n} \sum_{j=0}^{n} t_j \left(\sum_{k=0}^{m} c_{jk} s_k \right) \quad \text{for all} \quad m, n \in \mathbb{N},$$

which yields, as $m \to \infty$ that $(Ds)_n = \{R^t(Cs)\}_n$ for all $n \in \mathbb{N}$. Now, we immediately deduce from here that $Cs \in \mu_{R^t}$ whenever $s \in \lambda$ if and only if $Ds \in \mu$ whenever $s \in \lambda$.

This step completes the proof. □

Following Altay and Başar [26], we shall employ the duality relation between a pair of infinite matrices such that one of them applied to the sequences in a Riesz space and the other one applied to the sequences in a space, which is isomorphic to the Riesz space. We also give a basic theorem related to the dual summability methods of the new sort as a consequence of an analysis about the such summability methods. Besides, we characterize the class $(\lambda : \mu)$ of infinite matrices; where $\lambda \in \{r_\infty^t(q),\ r_c^t(q),\ r_0^t(q)\}$ and $\mu \in \{\ell_\infty(q),\ r_\infty^t(q),\ bs(q),\ \ell(q),\ r^t(q),\ \overline{\ell(q)},\ c(q),\ r_c^t(q),\ cs(q),\ c_0(q),\ r_0^t(q),\ cs_0(q)\}$.

Our focus now will be on the dual summability methods of the new sort.

Let us suppose that the infinite matrices $A = (a_{nk})$ and $B = (b_{nk})$ transform the sequences $x = (x_k)$ and $y = (y_k)$, which are connected with the relation (4.7.1) to the sequences $u = (u_n)$ and $v = (v_n)$, respectively, i.e.,

$$u_n := (Ax)_n = \sum_k a_{nk} x_k \quad \text{for each} \quad n \in \mathbb{N}, \tag{4.7.30}$$

$$v_n := (By)_n = \sum_k b_{nk} y_k \quad \text{for each} \quad n \in \mathbb{N}. \tag{4.7.31}$$

It is clear here that the method B is applied to the R^t-transform of the sequence $x = (x_k)$, while the method A is directly applied to the terms of the sequence $x = (x_k)$. So, the methods A and B are essentially different.

Let us assume that the matrix product BR^t exists, which is a much weaker assumption than the conditions on the matrix B belonging to any matrix class, in general. We shall say in this situation that the methods A and B in (4.7.30), (4.7.31) are the *dual of the new sort* if u_n becomes v_n (or v_n becomes u_n) under the application of the formal summation by parts. This leads us to the fact that BR^t exists and is equal to A and $(BR^t)x = B(R^t x)$ formally holds, if one side exists. This statement is equivalent to the relation

$$a_{nk} := \sum_{j=k}^{\infty} \frac{t_k}{T_j} b_{nj} \text{ or } b_{nk} := \Delta\left(\frac{a_{nk}}{t_k}\right) T_k \text{ for all } k, n \in \mathbb{N}. \tag{4.7.32}$$

To avoid the danger of confusion, we use the term *dual summability methods of the new sort* instead of the Lorentz's and the others' term *dual summability methods.*

Now, we may give a short analysis on the dual summability methods of the new sort. One can see that v_n reduced to u_n, as follows:

$$v_n = \sum_k b_{nk} y_k = \sum_k b_{nk} \left(\frac{1}{T_k} \sum_{j=0}^{k} t_j x_j\right) = \sum_j \sum_{k=j}^{\infty} \frac{t_j}{T_k} b_{nk} x_j = u_n.$$

But, the order of summation may not be reversed and thus, the methods A and B are not necessarily equivalent.

The partial sums of the series on the right-hand side of (4.7.30) and (4.7.31) are connected with the relation (4.7.19). Hence, if, for a given $n \in \mathbb{N}$, one of the series on the right-hand side of (4.7.30) and (4.7.31) converges, then the other converges if and only if

$$\lim_{m \to \infty} \frac{a_{nm}}{t_m} T_m y_m := z_n \text{ for each fixed } n \in \mathbb{N}. \tag{4.7.33}$$

If (4.7.33) holds, then we have from (4.7.19) by letting $m \to \infty$ that

$$u_n := v_n + z_n \quad \text{for all } \ n \in \mathbb{N}. \tag{4.7.34}$$

Hence, if (y_n) is summable by one of the methods A and B, then it is summable by the other one if and only if (4.7.33) holds, and

$$\lim_{n \to \infty} z_n = \alpha. \tag{4.7.35}$$

Hence, the limits of (u_n) and (v_n) differ by α. Therefore, the A- and B-limits of any sequence summable by one of them agree if and only if B summability implies that (4.7.35) holds with $\alpha = 0$. A similar result holds when A and B are interchanged. It follows by the validity of (4.7.35) with $\alpha \neq 0$ that the methods A and B are inconsistent, and conversely.

The analysis given above leads us to the following basic theorem related to the matrix transformations on the Riesz sequence spaces:

Theorem 4.7.15. [26, Theorem 4.1] *Let λ and μ be any two normed or paranormed sequence spaces, the entries of the matrices $A = (a_{nk})$ and $B = (b_{nk})$ are connected with the relation (4.7.32). Then, $Ax \in \mu$ whenever $x \in \lambda_{R^t}$ if and only if $By \in \mu$ whenever $y \in \lambda$ and*

$$\lim_{k\to\infty} \frac{a_{nk}}{t_k} T_k = 0 \text{ for each } n \in \mathbb{N}. \tag{4.7.36}$$

Because the two-sided implication does not hold, we remark the reader from now on that the condition (4.7.36) of Theorem 4.7.15 cannot be dropped.

It is of course that Theorem 4.7.15 has several consequences depending on the choice of the sequence spaces λ and μ. By Theorem 4.7.15, the necessary and sufficient conditions for $A \in (\lambda_{R^t} : \mu)$ may be derived by replacing the entries of A by those of the entries of $B = A(R^t)^{-1}$, where the necessary and sufficient conditions on the matrix B are read from the concerning results in the existing literature. By interchanging the spaces λ_{R^t} and λ with the space μ, we have

Theorem 4.7.16. [26, Theorem 4.2] *Suppose that the elements of the infinite matrices $A = (a_{nk})$ and $C = (c_{nk})$ are connected with the relation*

$$c_{nk} := \widetilde{a}_{nk} \text{ for all } k, n \in \mathbb{N} \tag{4.7.37}$$

and λ, μ be any two given sequence spaces. Then, $A \in (\mu : \lambda_{R^t})$ if and only if $C \in (\mu : \lambda)$.

Proof. Let us take any $s = (s_k) \in \mu$ and consider the following equality with (4.7.37) that

$$\sum_{k=0}^{m} c_{nk} s_k = \sum_{j=0}^{n} \frac{t_j}{T_n} \sum_{k=0}^{m} a_{jk} s_k \text{ for all } m, n \in \mathbb{N},$$

which yields as $m \to \infty$ that $(Cs)_n = \{R^t(As)\}_n$ for all $n \in \mathbb{N}$. Now, we immediately deduce from here that $As \in \lambda_{R^t}$ whenever $s \in \mu$ if and only if $Cs \in \lambda$ whenever $s \in \mu$.

This step completes the proof. □

Since Theorems 4.7.15 and 4.7.16 are, respectively, related to the matrix transformations on the Riesz sequence spaces and into the Riesz sequence spaces, the characterizations of the matrix mappings between the Riesz sequence spaces may be derived by combining Theorems 4.7.15 and 4.7.16. Now, we may quote our theorems on the characterization of some matrix classes concerning with the Riesz sequence spaces. Before giving the theorems, let us suppose that (q_n) is a non-decreasing bounded sequence of positive real numbers and consider the following conditions:

$$\lim_{k\to\infty} \frac{a_{nk}}{t_k} T_k M^{1/p_k} = 0 \text{ for all } n, M \in \mathbb{N} \tag{4.7.38}$$

$$\sup_{n\in\mathbb{N}} \left[\sum_k \left|\Delta\left(\frac{a_{nk}}{t_k}\right)\right| T_k M^{1/p_k}\right]^{q_n} < \infty \text{ for all } M \in \mathbb{N} \tag{4.7.39}$$

$$\sup_{K\in\mathcal{F}} \sum_n \left|\sum_{k\in K} \Delta\left(\frac{a_{nk}}{t_k}\right) T_k M^{1/p_k}\right|^{q_n} < \infty \text{ for all } M \in \mathbb{N} \tag{4.7.40}$$

$$\sup_{n\in\mathbb{N}} \sum_k \left|\Delta\left(\frac{a_{nk}}{t_k}\right) T_k\right| M^{1/p_k} < \infty \text{ for all } M \in \mathbb{N} \tag{4.7.41}$$

There exists $(\alpha_k) \in \omega$ such that (4.7.42)

$$\lim_{n\to\infty} \left[\sum_k \left|\Delta\left(\frac{a_{nk}}{t_k}\right) T_k - \alpha_k\right| M^{1/p_k}\right]^{q_n} = 0 \text{ for all } M \in \mathbb{N}$$

$$\text{There exists } M \in \mathbb{N} \text{ such that } \sup_{n\in\mathbb{N}} \left[\sum_k \left|\Delta\left(\frac{a_{nk}}{t_k}\right) T_k\right| M^{-1/p_k}\right]^{q_n} < \infty \tag{4.7.43}$$

$$\sup_{n\in\mathbb{N}} \left|\sum_k \Delta\left(\frac{a_{nk}}{t_k}\right) T_k\right|^{q_n} < \infty \tag{4.7.44}$$

$$\text{There exists } M \in \mathbb{N} \text{ such that } \sup_{K\in\mathcal{F}} \sum_n \left|\sum_{k\in K} \Delta\left(\frac{a_{nk}}{t_k}\right) T_k M^{-1/p_k}\right|^{q_n} < \infty \tag{4.7.45}$$

$$\sum_n \left|\sum_k \Delta\left(\frac{a_{nk}}{t_k}\right) T_k\right|^{q_n} < \infty \tag{4.7.46}$$

$$\text{There exists } \alpha \in \mathbb{C} \text{ such that } \lim_{n\to\infty} \left|\sum_k \Delta\left(\frac{a_{nk}}{t_k}\right) T_k - \alpha\right|^{q_n} = 0 \tag{4.7.47}$$

$$\text{There exist} (\alpha_k) \in \omega \text{ such that } \lim_{n\to\infty} \left|\Delta\left(\frac{a_{nk}}{t_k}\right) T_k - \alpha_k\right|^{q_n} = 0 \text{ for all } k \in \mathbb{N} \tag{4.7.48}$$

There exist $(\alpha_k) \in \omega$ and $M \in \mathbb{N}$ such that (4.7.49)

$$\sup_{n\in\mathbb{N}} L^{1/q_n} \sum_k \left|\Delta\left(\frac{a_{nk}}{t_k}\right) T_k - \alpha_k\right| M^{-1/p_k} < \infty \text{ for all } L$$

Theorem 4.7.17. [26, Theorem 4.3] *Let $A = (a_{nk})$ be an infinite matrix over the complex field. Then, the following statements hold:*

(i) *$A \in (r^t_\infty(p) : \ell_\infty(q))$ if and only if (4.7.38) and (4.7.39) hold.*

(ii) *$A \in (r^t_\infty(p) : r^t_\infty(q))$ if and only if (4.7.38) and (4.7.39) hold with $\widetilde{a}_{nk}$ instead of a_{nk}.*

(iii) *$A \in (r^t_\infty(p) : bs(q))$ if and only if (4.7.38) and (4.7.39) hold with $a(n,k)$ instead of a_{nk}.*

(iv) *$A \in (r^t_\infty(p) : \ell(q))$ if and only if (4.7.38) and (4.7.40) hold.*

(v) *$A \in (r^t_\infty(p) : r^t(q))$ if and only if (4.7.38) and (4.7.40) hold with $\widetilde{a}_{nk}$ instead of a_{nk}.*

(vi) *$A \in (r^t_\infty(p) : \overline{\ell(q)})$ if and only if (4.7.38) and (4.7.40) hold with $a(n,k)$ instead of a_{nk}.*

(vii) *$A \in (r^t_\infty(p) : c(q))$ if and only if (4.7.38), (4.7.41) and (4.7.42) hold.*

(viii) *$A \in (r^t_\infty(p) : r^t_c(q))$ if and only if (4.7.38), (4.7.41) and (4.7.42) hold with $\widetilde{a}_{nk}$ instead of a_{nk}.*

(ix) *$A \in (r^t_\infty(p) : cs(q))$ if and only if (4.7.38), (4.7.41) and (4.7.42) hold with $a(n,k)$ instead of a_{nk}.*

(x) *$A \in (r^t_\infty(p) : c_0(q))$ if and only if (4.7.38) holds and (4.7.42) also holds with $\alpha_k = 0$ for all $k \in \mathbb{N}$.*

(xi) *$A \in (r^t_\infty(p) : r^t_0(q))$ if and only if (4.7.38) holds and (4.7.42) also holds with $\alpha_k = 0$ for all $k \in \mathbb{N}$ and $\widetilde{a}_{nk}$ instead of a_{nk}.*

(xii) *$A \in (r^t_\infty(p) : cs_0(q))$ if and only if (4.7.38) holds and (4.7.42) also holds with $\alpha_k = 0$ for all $k \in \mathbb{N}$, and $a(n,k)$ instead of a_{nk}.*

Theorem 4.7.18. [26, Theorem 4.4] *Let $A = (a_{nk})$ be an infinite matrix over the complex field. Then, the following statements hold:*

(i) *$A \in (r^t_c(p) : \ell_\infty(q))$ if and only if (4.7.38), (4.7.43) and (4.7.44) hold.*

(ii) *$A \in (r^t_c(p) : r^t_\infty(q))$ if and only if (4.7.38), (4.7.43) and (4.7.44) hold with $\widetilde{a}_{nk}$ instead of a_{nk}.*

(iii) *$A \in (r^t_c(p) : bs(q))$ if and only if (4.7.38), (4.7.43) and (4.7.44) hold with $a(n,k)$ instead of a_{nk}.*

(iv) *$A \in (r^t_c(p) : \ell(q))$ if and only if (4.7.38), (4.7.45) and (4.7.46) hold.*

(v) *$A \in (r^t_c(p) : r^t(q))$ if and only if (4.7.38), (4.7.45) and (4.7.46) hold with $\widetilde{a}_{nk}$ instead of a_{nk}.*

(vi) *$A \in (r^t_c(p) : \overline{\ell(q)})$ if and only if (4.7.38), (4.7.45) and (4.7.46) hold with $a(n,k)$ instead of a_{nk}.*

(vii) $A \in (r_c^t(p) : c(q))$ *if and only if (4.7.38), (4.7.47), (4.7.48) and (4.7.49) hold, and (4.7.43) also holds with* $q_n = 1$ *for all* $n \in \mathbb{N}$.

(viii) $A \in (r_c^t(p) : r_c^t(q))$ *if and only if (4.7.38), (4.7.47), (4.7.48) and (4.7.49) hold, and (4.7.43) also holds with* $q_n = 1$ *for all* $n \in \mathbb{N}$ *and* $\widetilde{a}_{nk}$ *instead of* a_{nk}.

(ix) $A \in (r_c^t(p) : cs(q))$ *if and only if (4.7.38), (4.7.47), (4.7.48) and (4.7.49) hold, and (4.7.43) also holds with* $q_n = 1$ *for all* $n \in \mathbb{N}$ *and* $a(n,k)$ *instead of* a_{nk}.

(x) $A \in (r_c^t(p) : c_0(q))$ *if and only if (4.7.38) holds, and (4.7.47), (4.7.48) and (4.7.49) also hold with* $\alpha = 0$, $\alpha_k = 0$ *for all* $k \in \mathbb{N}$, *respectively.*

(xi) $A \in (r_c^t(p) : r_0^t(q))$ *if and only if (4.7.38) holds, and (4.7.47), (4.7.48) and (4.7.49) also hold with* $\alpha = 0$, $\alpha_k = 0$ *for all* $k \in \mathbb{N}$, *respectively, and* $\widetilde{a}_{nk}$ *instead of* a_{nk}.

(xii) $A \in (r_c^t(p) : cs_0(q))$ *if and only if (4.7.38) holds, and (4.7.47), (4.7.48) and (4.7.49) also hold with* $\alpha = 0$, $\alpha_k = 0$ *for all* $k \in \mathbb{N}$, *respectively, and* $a(n,k)$ *instead of* a_{nk}.

Theorem 4.7.19. [26, Theorem 4.5] *Let* $A = (a_{nk})$ *be an infinite matrix over the complex field. Then, the following statements hold:*

(i) $A \in (r_0^t(p) : \ell_\infty(q))$ *if and only if (4.7.38) and (4.7.43) hold.*

(ii) $A \in (r_0^t(p) : r_\infty^t(q))$ *if and only if (4.7.38) and (4.7.43) hold with* $\widetilde{a}_{nk}$ *instead of* a_{nk}.

(iii) $A \in (r_0^t(p) : bs(q))$ *if and only if (4.7.38) and (4.7.43) hold with* $a(n,k)$ *instead of* a_{nk}.

(iv) $A \in (r_0^t(p) : \ell(q))$ *if and only if (4.7.38) and (4.7.45) hold.*

(v) $A \in (r_0^t(p) : r^t(q))$ *if and only if (4.7.38) and (4.7.45) hold with* $\widetilde{a}_{nk}$ *instead of* a_{nk}.

(vi) $A \in (r_0^t(p) : \overline{\ell(q)})$ *if and only if (4.7.38) and (4.7.45) hold with* $a(n,k)$ *instead of* a_{nk}.

(vii) $A \in (r_0^t(p) : c(q))$ *if and only if (4.7.38), (4.7.48) and (4.7.49) hold, and (4.7.43) also holds with* $q_n = 1$ *for all* $n \in \mathbb{N}$.

(viii) $A \in (r_0^t(p) : r_c^t(q))$ *if and only if (4.7.38), (4.7.48) and (4.7.49) hold, and (4.7.43) also holds with* $q_n = 1$ *for all* $n \in \mathbb{N}$ *and* $\widetilde{a}_{nk}$ *instead of* a_{nk}.

(ix) $A \in (r_0^t(p) : cs(q))$ *if and only if (4.7.38), (4.7.48) and (4.7.49) hold, and (4.7.43) also holds with* $q_n = 1$ *for all* $n \in \mathbb{N}$ *and* $a(n,k)$ *instead of* a_{nk}.

(x) $A \in (r_0^t(p) : c_0(q))$ *if and only if (4.7.38) holds, (4.7.48) and (4.7.49) also hold with* $\alpha_k = 0$ *for all* $k \in \mathbb{N}$.

(xi) $A \in (r_0^t(p) : r_0^t(q))$ *if and only if (4.7.38) holds, (4.7.48) and (4.7.49) also hold with* $\alpha_k = 0$ *for all* $k \in \mathbb{N}$ *and* $\widetilde{a}_{nk}$ *instead of* a_{nk}.

(xii) $A \in (r_c^t(p) : cs_0(q))$ *if and only if (4.7.38) holds, (4.7.48) and (4.7.49) also hold with* $\alpha_k = 0$ *for all* $k \in \mathbb{N}$ *and* $a(n,k)$ *instead of* a_{nk}.

4.8 EULER SEQUENCE SPACES AND CONCERNING DUALITY RELATION

Let λ denotes any of the classical sequence spaces ℓ_∞, c, c_0 and ℓ_p. Then, the domain λ_{E^r} is called as *Euler sequence space* and was studied by Altay and Başar [23], Altay et al. [31] and Mursaleen et al. [289].

Let us define the sequence $y = (y_k)$ by the E^r-transform of a sequence $x = (x_k) \in \omega$, i.e.,

$$y_k := (E^r x)_k = \sum_{j=0}^{k} \binom{k}{j} (1-r)^{k-j} r^j x_j \quad \text{for all } k \in \mathbb{N}. \tag{4.8.1}$$

Then, since $\lambda_{E^r} \cong \lambda$ it is trivial that the two-sided implication "$x \in \lambda_{E^r}$ if and only if $y \in \lambda$" holds. In this situation, the duality relation between the entries of the matrices $A = (a_{nk})$ and $B = (b_{nk})$ is

$$a_{nk} := \sum_{j=k}^{\infty} \binom{j}{k} (1-r)^{j-k} r^k b_{nj} \quad \text{or} \quad b_{nk} := \sum_{j=k}^{\infty} \binom{j}{k} (r-1)^{j-k} r^{-j} a_{nj} \tag{4.8.2}$$

for all $k, n \in \mathbb{N}$. The pair of such type matrices was used by Altay and Başar [23], Altay et al. [31] and Mursaleen et al. [289].

4.8.1 Euler Sequence Spaces of Non-absolute Type

We introduce the Euler sequence spaces e_p^r, e_0^r, e_c^r and e_∞^r, as the set of all sequences such that their E^r-transforms are in the spaces ℓ_p, c_0, c and ℓ_∞, i.e.,

$$\begin{aligned}
e_p^r &:= \left\{ x = (x_k) \in \omega : \sum_n \left| \sum_{k=0}^{n} \binom{n}{k} (1-r)^{n-k} r^k x_k \right|^p < \infty \right\}, \quad (1 < p < \infty), \\
e_0^r &:= \left\{ x = (x_k) \in \omega : \lim_{n\to\infty} \sum_{k=0}^{n} \binom{n}{k} (1-r)^{n-k} r^k x_k = 0 \right\}, \\
e_c^r &:= \left\{ x = (x_k) \in \omega : \lim_{n\to\infty} \sum_{k=0}^{n} \binom{n}{k} (1-r)^{n-k} r^k x_k \text{ exists} \right\}, \\
e_\infty^r &:= \left\{ x = (x_k) \in \omega : \sup_{n\in\mathbb{N}} \left| \sum_{k=0}^{n} \binom{n}{k} (1-r)^{n-k} r^k x_k \right| < \infty \right\}.
\end{aligned}$$

With the notation of (4.1.1), we can redefine the sequence spaces e_p^r, e_0^r, e_c^r and e_∞^r, as follows:

$$e_p^r := (\ell_p)_{E^r}, \quad e_0^r := (c_0)_{E^r}, \quad e_c^r := (c)_{E^r} \quad \text{and} \quad e_\infty^r := (\ell_\infty)_{E^r}.$$

Now, we can give the following theorem without proof, since it will be proved by the similar way used in the proof of Theorem 4.2.1 with ℓ_p, c_0, c and ℓ_∞ instead of c_0 and c:

Theorem 4.8.1. (cf. Theorem 2.1 of [31] and [23]) *Each of the sets e_p^r, e_0^r, e_c^r and e_∞^r becomes a linear space with the co-ordinatewise addition and scalar multiplication, which are BK-spaces with the norms $\|x\|_{e_p^r} = \|E^r x\|_p$ and $\|x\|_{e_0^r} = \|x\|_{e_c^r} = \|x\|_{e_\infty^r} = \|E^r x\|_\infty$, where $1 < p \leq \infty$.*

Therefore, one can easily check that the absolute property does not hold on the Euler spaces e_p^r, e_0^r, e_c^r and e_∞^r, i.e. $\|x\|_{\lambda_{E^r}} \neq \||x|\|_{\lambda_{E^r}}$, for at least one sequence in the space λ_{E^r}, and this indicates that λ_{E^r} is a sequence space of non-absolute type; where $|x| = (|x_k|)$, $\lambda \in \{\ell_p, c_0, c, \ell_\infty\}$ and $1 < p < \infty$. It is immediate by the well-known inclusion $\ell_p \subset \ell_s$ that the inclusion $e_p^r \subset e_s^r$ holds whenever $1 < p \leq s$.

One may expect the similar result for the space e_p^r, as observed for the space ℓ_p, and ask the question: Isn't the space e_p^r a Hilbert space with $p \neq 2$? The answer is positive and is given by the following theorem:

Theorem 4.8.2. [31, Theorem 2.3] *Except the case $p = 2$, the space e_p^r is not an inner product space, hence not a Hilbert space for $1 \leq p < \infty$.*

Proof. We prove that the space e_2^r is the only Hilbert space among the e_p^r spaces for $1 < p < \infty$, firstly. Since the space e_2^r is BK-space endowed with the norm $\|x\|_{e_2^r} = \|E^r x\|_2$ by Theorem 4.8.1 and its norm can be obtained from an inner product, i.e., $\|x\|_{e_2^r} = \sqrt{< E^r x, E^r x >}$ holds, the space e_2^r is a Hilbert space.

Let us now consider the sequences $u = \{u_k(r)\}$ and $v = \{v_k(r)\}$ defined by

$$u_k(r) := \frac{r+k-1}{r}\left(1 - \frac{1}{r}\right)^{k-1} \quad \text{and} \quad v_k(r) := \frac{r-k-1}{r}\left(1 - \frac{1}{r}\right)^{k-1}$$

for all $k \in \mathbb{N}$. Then, one can easily see that

$$\|u+v\|_{e_p^r}^2 + \|u-v\|_{e_p^r}^2 = 8 \neq 4(2^{2/p}) = 2\left(\| u \|_{e_p^r}^2 + \| v \|_{e_p^r}^2\right), \quad (p \neq 2),$$

that is, the norm of the space e_p^r does not satisfy the parallelogram equality, which means that the norm cannot be obtained from an inner product. Hence, the space e_p^r with $p \neq 2$ is a Banach space, which is not a Hilbert space.

This completes the proof. □

Now, we can give some inclusion relations concerning with the Euler spaces.

Theorem 4.8.3. [31, Theorem 3.1] *The inclusion $\ell_p \subset e_p^r$ strictly holds for $1 \leq p < \infty$.*

Proof. To prove the validity of the inclusion $\ell_p \subset e_p^r$ for $1 \leq p < \infty$, it suffices to show the existence of a number $K > 0$ such that $\|x\|_{e_p^r} \leq K\|x\|_p$ for every $x \in \ell_p$.

Let us take any $x = (x_k) \in \ell_p$. Then, we obtain, with the notation of (4.8.1), by applying the Hölder's inequality for $1 < p < \infty$ that

$$\begin{aligned} |y_k(r)|^p &= \left| \sum_{j=0}^{k} \binom{k}{j} (1-r)^{k-j} r^j x_j \right|^p \qquad (4.8.3) \\ &\leq \left[\sum_{j=0}^{k} \binom{k}{j} (1-r)^{k-j} r^j |x_j|^p \right] \left[\sum_{j=0}^{k} \binom{k}{j} (1-r)^{k-j} r^j \right]^{p-1} \\ &= \sum_{j=0}^{k} \binom{k}{j} (1-r)^{k-j} r^j |x_j|^p. \end{aligned}$$

Thus, we have by using the fact in (4.8.3) that

$$\begin{aligned} \sum_k |y_k(r)|^p &\leq \sum_k \sum_{j=0}^{k} \binom{k}{j} (1-r)^{k-j} r^j |x_j|^p \\ &= \sum_j |x_j|^p \sum_{k=j}^{\infty} \binom{k}{j} (1-r)^{k-j} r^j = \frac{1}{r} \sum_j |x_j|^p, \end{aligned}$$

which yields us that

$$\|x\|_{e_p^r} \leq \left(\frac{1}{r}\right)^{1/p} \|x\|_p \qquad (4.8.4)$$

for $1 < p < \infty$, as expected. Besides, let us consider the sequence $u = \{u_k(r)\}$ defined by $u_k(r) = (-r)^{-k}$ for all $k \in \mathbb{N}$. Then, since $E^r u = \{(-r)^k\} \in \ell_p$ we immediately observe that u is in e_p^r but not in ℓ_p. Because, there is at least one sequence in $e_p^r \setminus \ell_p$, the inclusion $\ell_p \subset e_p^r$ is strict.

By the similar discussions, it may be easily proved that the inequality (4.8.4) also holds in the case $p = 1$ and so we omit details.

This completes the proof. □

Theorem 4.8.4. [31, Theorem 3.2] *Neither of the spaces e_p^r and ℓ_∞ includes the other one, where $1 < p < \infty$.*

Proof. It is trivial by Theorem 4.8.3 that the sequence spaces e_p^r and ℓ_∞ are not disjoint. Let us now consider the sequences $u = \{u_k(r)\}$ defined as in the proof of Theorem 4.8.3, above, and $x = e$. Then, $u \in e_p^r \setminus \ell_\infty$ and $x \in \ell_\infty \setminus e_p^r$. Hence, the sequence spaces e_p^r and ℓ_∞ overlap but neither contains the other.

This completes the proof. □

Theorem 4.8.5. [31, Theorem 3.3] *The space e_∞^r strictly includes both the space ℓ_∞ and the space e_p^r, where $1 < p < \infty$.*

Proof. Since the inequality

$$\|x\|_{e_\infty^r} = \sup_{k\in\mathbb{N}} \left| \sum_{j=0}^{k} \binom{k}{j} (1-r)^{k-j} r^j x_j \right| \leq \|x\|_\infty$$

holds for any $x = (x_k) \in \ell_\infty$, we easily obtain the inclusion $\ell_\infty \subset e^r_\infty$. Additionally, again the sequence $u = \{u_k(r)\}$, used in the proof of Theorem 4.8.3, above, is in e^r_∞ but not in ℓ_∞. So, the sequence space e^r_∞ strictly includes the sequence space ℓ_∞, as asserted.

Let $x \in e^r_p$ and $1 < p < \infty$. Then, we derive from the fact "*the Euler space* e^r_p *is norm isomorphic to the space* ℓ_p" that $y = E^r x \in \ell_p \subset \ell_\infty$, which yields that $x \in e^r_\infty$. Hence, the inclusion $e^r_p \subset e^r_\infty$ holds. Besides, since the sequence $x = e$ belongs to the set $e^r_\infty \setminus e^r_p$, the inclusion $e^r_p \subset e^r_\infty$ is also strict, which concludes the proof. □

One can observe that the inequality $\|x\|_{e^t_p} \le \left(\frac{r}{t}\right)^{1/p} \|x\|_{e^r_p}$ derived in the similar way to obtaining (4.8.4) holds for $0 < t \le r < 1$, which leads us together with Theorem 4.8.3 to the following:

Theorem 4.8.6. [31, Theorem 3.4] *If* $0 < t \le r < 1$, *then* $e^r_p \subset e^t_p$.

Theorem 4.8.7. *Although the inclusions* $c_0 \subset e^r_0$ *and* $c \subset e^r_c$ *strictly hold, neither of the spaces* e^r_0 *and* ℓ_∞ *includes the other one.*

Proof. Let us take any $y \in c_0$. Then, bearing in mind the regularity of the Euler mean of order r, we immediately observe that $E^r y \in c_0$, which means that $y \in e^r_0$. Hence, the inclusion $c_0 \subset e^r_0$ holds. Furthermore, let us consider the sequence $u = \{u_k(r)\}$ defined by $u_k(r) = (-r)^{-k}$ for all $k \in \mathbb{N}$. Then, since $E^r u = \{(-r)^k\} \in c_0$, u is in e^r_0 but not in c_0. By the similar discussion, one can see that the inclusion $c \subset e^r_c$ also strictly holds.

To establish the second part of the theorem, consider that sequence $u = \{u_k(r)\}$ defined above, and $x = e$. Then, u is in e^r_0 but not in ℓ_∞ and x is in ℓ_∞ but not in e^r_0. Hence, neither of the spaces e^r_0 and ℓ_∞ includes the other one.

This completes the proof. □

Theorem 4.8.8. *If* $0 < t < r < 1$, *then* $e^r_0 \subset e^t_0$ *and* $e^r_c \subset e^t_c$.

Proof. Let us take $x = (x_i) \in e^r_0$. Then, we observe that

$$z_k = \sum_{i=0}^{k} e^t_{ki} x_i = \sum_{i=0}^{k} e^t_{ki} \left(\sum_{j=0}^{i} e^{1/r}_{ij} y_j \right) = \sum_{j=0}^{k} e^{t/r}_{kj} y_j.$$

Since $0 < \frac{t}{r} < 1$, the method $E^{t/r}$ is regular, which implies that $z = (z_k) \in c_0$ whenever $y = (y_k) \in c_0$ and we thus see that $x \in e^t_0$. This means that the inclusion $e^r_0 \subset e^t_0$ holds.

Now, one can show in the similar way that the validity of the inclusion $e^r_c \subset e^t_c$ and so we leave it to the reader. □

Let $\lambda \in \{\ell_p, c_0, c\}$ and $1 < p \le \infty$. Because of the isomorphism T, defined on the Euler spaces λ_{E^r} by $Tx = E^r x$ is onto, the inverse images of the bases of those spaces ℓ_p, c_0 and c are the bases of the spaces e^r_p, e^r_0 and e^r_c by Theorem 4.1.2. Therefore, we have

Theorem 4.8.9. (cf. [31, Theorem 4.7] and [23, Theorem 3.1]) *Let $\alpha_k(r) = (E^r x)_k$ for all $k \in \mathbb{N}$. Define the sequence $b^{(k)}(r) = \{b_n^{(k)}(r)\}_{n\in\mathbb{N}}$ of the elements of the spaces e_p^r or e_0^r by*

$$b_n^{(k)}(r) := \begin{cases} 0 & , \quad n < k, \\ \binom{n}{k}(r-1)^{n-k}r^{-n} & , \quad n \geq k, \end{cases} \tag{4.8.5}$$

for every fixed $k \in \mathbb{N}$. Then, the following statements hold:

(a) *The sequence $\{b^{(k)}(r)\}_{k\in\mathbb{N}}$ is a basis for the spaces e_p^r and e_0^r, and any $x \in e_p^r$ or $x \in e_0^r$ has a unique representation of the form $x = \sum_k \alpha_k(r) b^{(k)}(r)$.*

(b) *The set $\left\{e, b^{(k)}(r)\right\}$ is a basis for the space e_c^r and any $x \in e_c^r$ has a unique representation of the form*

$$x = le + \sum_k [\alpha_k(r) - l] b^{(k)}(r), \quad \text{where } l = \lim_{k\to\infty} (E^r x)_k. \tag{4.8.6}$$

We continue this section by giving the α-, β- and γ-duals of the Euler sequence spaces.

Theorem 4.8.10. (cf. [31, Theorem 4.4] and [23, Theorem 4.1]) *Define the sets h_∞^r, h_r and h_q^r as follows:*

$$\begin{aligned} h_\infty^r &:= \left\{a = (a_k) \in \omega : \sup_{k\in\mathbb{N}} \sum_n \left| \binom{n}{k}(r-1)^{n-k} r^{-n} a_n \right| < \infty \right\}, \\ h_r &:= \left\{a = (a_k) \in \omega : \sup_{K\in\mathcal{F}} \sum_n \left| \sum_{k\in K} \binom{n}{k}(r-1)^{n-k} r^{-n} a_n \right| < \infty \right\}, \\ h_q^r &:= \left\{a = (a_k) \in \omega : \sup_{K\in\mathcal{F}} \sum_k \left| \sum_{n\in K} \binom{n}{k}(r-1)^{n-k} r^{-n} a_n \right|^q < \infty \right\}. \end{aligned}$$

Then, $(e_1^r)^\alpha := h_\infty^r$, $(e_0^r)^\alpha := (e_c^r)^\alpha := h_r$ and $\left(e_p^r\right)^\alpha := h_q^r$; where $1 < p \leq \infty$.

Proof. Since the proof for the spaces e_1^r, e_0^r and e_c^r is obtained by analogy, we consider only the space e_p^r.

Let us define the matrix B^r whose rows are the product of the rows of the matrix $E^{1/r}$ by the sequence $a = (a_n)$. Bearing in mind the relation (4.8.1), we easily obtain that

$$a_n x_n = \sum_{k=0}^{n} \binom{n}{k}(r-1)^{n-k} r^{-n} a_n y_k = (B^r y)_n \quad \text{for all} \quad n \in \mathbb{N}. \tag{4.8.7}$$

We therefore observe by (4.8.7) that $ax = (a_n x_n) \in \ell_1$ whenever $x = (x_k) \in e_p^r$ if and only if $B^r y = \{(B^r y)_n\} \in \ell_1$ whenever $y = (y_k) \in \ell_p$. Then, we derive by Lemma 4.3.12 that

$$\sup_{K\in\mathcal{F}} \sum_k \left| \sum_{n\in K} \binom{n}{k}(r-1)^{n-k} r^{-n} a_n \right|^q < \infty.$$

This yields that $\left(e_p^r\right)^\alpha = h_q^r$. □

Theorem 4.8.11. (cf. [31, Theorem 4.5] and [23, Theorem 4.2]) *Define the sets b_q^r, d_1^r, d_2^r, d_3^r, d_4^r, d_5^r and d_6^r by*

$$b_q^r := \left\{ a = (a_k) \in \omega : \sup_{n\in\mathbb{N}} \sum_{k=0}^{n} \left| \sum_{j=k}^{n} \binom{j}{k} (r-1)^{j-k} r^{-j} a_j \right|^q < \infty \right\}, \quad (1 < q < \infty),$$

$$d_1^r := \left\{ a = (a_k) \in \omega : \sum_{j=k}^{\infty} \binom{j}{k} (r-1)^{j-k} r^{-j} a_j \text{ exists for each } k \in \mathbb{N} \right\},$$

$$d_2^r := \left\{ a = (a_k) \in \omega : \sup_{n,k\in\mathbb{N}} \left| \sum_{j=k}^{n} \binom{j}{k} (r-1)^{j-k} r^{-j} a_j \right| < \infty \right\},$$

$$d_3^r := \left\{ a = (a_k) \in \omega : \sup_{n\in\mathbb{N}} \sum_{k=0}^{n} \left| \sum_{j=k}^{n} \binom{j}{k} (r-1)^{j-k} r^{-j} a_j \right| < \infty \right\},$$

$$d_4^r := \left\{ a = (a_k) \in \omega : \sum_{j=k}^{\infty} \binom{j}{k} (r-1)^{j-k} r^{-j} a_j \text{ exists for each } k \in \mathbb{N} \right\},$$

$$d_5^r := \left\{ a = (a_k) \in \omega : \lim_{n\to\infty} \sum_{k=0}^{n} \sum_{j=k}^{n} \binom{j}{k} (r-1)^{j-k} r^{-j} a_j \text{ exists} \right\},$$

$$d_6^r := \left\{ a = (a_k) \in \omega : \lim_{n\to\infty} \sum_{k} \left| \sum_{j=k}^{n} \binom{j}{k} (r-1)^{j-k} r^{-j} a_j \right| = \sum_{k} \left| \sum_{j=k}^{\infty} \binom{j}{k} (r-1)^{j-k} r^{-j} a_j \right| \right\}.$$

Then, $(e_1^r)^\beta := d_1^r \cap d_2^r$, $\left(e_p^r\right)^\beta := d_1^r \cap b_q^r$, $(e_0^r)^\beta := d_3^r \cap d_4^r$, $(e_c^r)^\beta := d_3^r \cap d_4^r \cap d_5^r$ *and* $(e_\infty^r)^\beta := d_1^r \cap d_6^r$*; where* $1 < p < \infty$.

Proof. We give the proof only for the space e_p^r. Consider the equality

$$\begin{aligned} \sum_{k=0}^{n} a_k x_k &= \sum_{k=0}^{n} \left[\sum_{j=0}^{k} \binom{k}{j} (r-1)^{k-j} r^{-k} y_j \right] a_k \\ &= \sum_{k=0}^{n} \left[\sum_{j=k}^{n} \binom{j}{k} (r-1)^{j-k} r^{-j} a_j \right] y_k = (T^r y)_n, \end{aligned} \tag{4.8.8}$$

where $T^r = (t_{nk}^r)$ is defined by

$$t_{nk}^r := \begin{cases} \sum_{j=k}^{n} \binom{j}{k} (r-1)^{j-k} r^{-j} a_j &, \quad 0 \le k \le n, \\ 0 &, \quad k > n, \end{cases} \tag{4.8.9}$$

for all $k, n \in \mathbb{N}$. Thus, we deduce from Lemma 4.3.13 with (4.8.8) that $ax = (a_k x_k) \in cs$ whenever $x = (x_k) \in e_p^r$ if and only if $T^r y = \{(T^r y)_n\} \in c$ whenever $y = (y_k) \in \ell_p$.

Therefore, we derive from (3.3.3) and (4.3.14) that

$$\sum_{j=k}^{\infty}\binom{j}{k}(r-1)^{j-k}r^{-j}a_j \text{ exists for each } k\in\mathbb{N},$$

$$\sup_{n\in\mathbb{N}}\sum_{k=0}^{n}\left|\sum_{j=k}^{n}\binom{j}{k}(r-1)^{j-k}r^{-j}a_j\right|^q<\infty,$$

which shows that $\left(e_p^r\right)^\beta = d_1^r \cap b_q^r$. □

Theorem 4.8.12. (cf. [31, Theorem 4.6] and [23, Theorem 4.3]) *Define the set b_1^r by*

$$b_1^r := \left\{a=(a_k)\in\omega:\ \sup_{n\in\mathbb{N}}\sum_{k=0}^{n}\left|\sum_{j=k}^{n}\binom{j}{k}(r-1)^{j-k}r^{-j}a_j\right|<\infty\right\}.$$

Then, $(e_1^r)^\gamma := d_2^r$, $\left(e_p^r\right)^\gamma := b_q^r$, $(e_0^r)^\gamma := (e_c^r)^\gamma := d_1^r$ *and* $(e_\infty^r)^\gamma := b_1^r$*; where* $1<p<\infty$.

Proof. It is natural that the present theorem may be proved by the technique used in the proof of Theorems 4.8.10 and 4.8.11. But, we prefer here the following classical way and give the proof only for the space e_p^r.

Let $a=(a_k)\in b_q^r$ and $x=(x_k)\in e_p^r$. Then, we obtain by applying the Hölder's inequality that

$$\begin{aligned}\left|\sum_{k=0}^{n}a_kx_k\right| &= \left|\sum_{k=0}^{n}\left[\sum_{j=0}^{k}\binom{k}{j}(r-1)^{k-j}r^{-k}y_j\right]a_k\right|\\ &= \left|\sum_{k=0}^{n}t_{nk}^r y_k\right| \le \left(\sum_{k=0}^{n}|t_{nk}^r|^q\right)^{1/q}\left(\sum_{k=0}^{n}|y_k|^p\right)^{1/p},\end{aligned}$$

which gives us by taking supremum over $n\in\mathbb{N}$ that

$$\begin{aligned}\sup_{n\in\mathbb{N}}\left|\sum_{k=0}^{n}a_kx_k\right| &\le \sup_{n\in\mathbb{N}}\left[\left(\sum_{k=0}^{n}|t_{nk}^r|^q\right)^{1/q}\left(\sum_{k=0}^{n}|y_k|^p\right)^{1/p}\right]\\ &\le \|y\|_p\left(\sup_{n\in\mathbb{N}}\sum_{k=0}^{n}|t_{nk}^r|^q\right)^{1/q}<\infty.\end{aligned}$$

This means that $a=(a_k)\in\left(e_p^r\right)^\gamma$. Hence, we have

$$b_q^r\subset\left(e_p^r\right)^\gamma. \tag{4.8.10}$$

Conversely, let $a=(a_k)\in\left(e_p^r\right)^\gamma$ and $x=(x_k)\in e_p^r$. Then, one can easily see that $\left(\sum_{k=0}^{n}t_{nk}^r y_k\right)_{n\in\mathbb{N}}\in\ell_\infty$ whenever $(a_kx_k)\in bs$. This shows that the triangle matrix

$T^r = (t^r_{nk})$, defined by (4.8.9), is in the class $(\ell_p : \ell_\infty)$. Hence, the condition (4.3.14) holds with t^r_{nk} instead of a_{nk}, which yields that $a = (a_k) \in b^r_q$. That is to say that

$$\left(e^r_p\right)^\gamma \subset b^r_q. \tag{4.8.11}$$

Therefore, by combining the inclusions (4.8.10) and (4.8.11), we deduce that the γ-dual of the space e^r_p is the set b^r_q.

This step completes the proof. □

Theorem 4.8.13. [23, Theorem 4.4] $(e^r_c)^*$ *and* $(e^r_0)^*$ *are isometrically isomorphic to* ℓ_1.

Proof. We only give the proof for the space e^r_c. Suppose that $f \in (e^r_c)^*$. Since by Part (b) of Theorem 4.8.9, $\{e, b^k(r)\}$ is a basis for the space e^r_c and so any element $x \in e^r_c$ can be expressed as in the form of (4.8.6). By the linearity and the continuity of f, we get from (4.8.6) that

$$f(x) = lf(e) + \sum_k [\alpha_k(r) - l] f(b^{(k)}(r))$$

for all $x \in e^r_c$. Define the sequence $x = \{x_k(r)\} \in e^r_c$ such that $\|x\|_{e^r_c} = 1$ by

$$x_k(r) := \begin{cases} \sum_{j=0}^k \binom{k}{j}(r-1)^{k-j} r^{-k} \text{sgn} f(b^{(j)}(r)) & , \quad 0 \le k \le n, \\ \sum_{j=0}^n \binom{k}{j}(r-1)^{k-j} r^{-k} \text{sgn} f(b^{(j)}(r)) & , \quad k > n, \end{cases}$$

for all $k \in \mathbb{N}$. Therefore, we have

$$|f(x)| = \sum_{k=0}^n |f(b^{(k)}(r))| \le \|f\|.$$

It follows from above inequality that

$$\sum_k |f(b^{(k)}(r))| = \sup_{n\in\mathbb{N}} \sum_{k=0}^n |f(b^{(k)}(r))| \le \|f\|.$$

Write $f(x) = al + \sum_k a_k \alpha_k(r)$; where $a = f(e) - \sum_k f(b^{(k)}(r))$, $a_k = f(b^{(k)}(r))$, the series $\sum_k f(b^{(k)}(r))$ being absolutely convergent. Since $|\lim_{k\to\infty} (E^r x)_k| \le \|x\|_{e^r_c}$, one can observe that

$$|f(x)| \le \|x\|_{e^r_c} \left(|a| + \sum_k |a_k|\right)$$

whence

$$\|f\| \le |a| + \sum_k |a_k|. \tag{4.8.12}$$

Also, for $\|x\|_{e_c^r} = 1$, we have $|f(x)| \leq \|f\|$, so we define for any $n \in \mathbb{N}$ that

$$x_k(r) := \begin{cases} \sum_{j=0}^{k} \binom{k}{j}(r-1)^{k-j}r^{-k}\mathrm{sgn}a_j & , \quad 0 \leq k \leq n, \\ \sum_{j=0}^{n} \binom{k}{j}(r-1)^{k-j}r^{-k}\mathrm{sgn}a_j + \sum_{j=n+1}^{k} \binom{k}{j}(r-1)^{k-j}r^{-k}\mathrm{sgn}a & , \quad k > n. \end{cases}$$

Then $x \in e_c^r$ with $\|x\|_{e_c^r} = 1$, $\lim_{k\to\infty} (E^r x)_k = \mathrm{sgn}a$ and so

$$|f(x)| = \left| |a| + \sum_{k=0}^{n} |a_k| + \sum_{k=n+1}^{\infty} a_k \mathrm{sgn}a \right| \leq \|f\|. \tag{4.8.13}$$

Since $(a_k) \in \ell_1$ we have $\sum_{k=n+1}^{\infty} a_k \to 0$ as $n \to \infty$, and thus, we obtain by letting $n \to \infty$ in (4.8.13) that

$$|a| + \sum_k |a_k| \leq \|f\|. \tag{4.8.14}$$

Combining the results (4.8.12) and (4.8.14) we see that

$$\|f\| = |a| + \sum_k |a_k|$$

which is the norm on ℓ_1.

Now, let $T : (e_c^r)^* \to \ell_1$ be defined by $f \mapsto (a, a_0, a_1, \ldots)$. Then, we have

$$\|T(f)\| = |a| + |a_0| + |a_1| + \cdots = \|f\|.$$

$\|T(f)\|$ being the ℓ_1-norm. Thus, T is norm preserving. T is obviously surjective and linear, and hence, is an isomorphism from $(e_c^r)^*$ to ℓ_1.

This completes the proof. □

4.8.2 Certain Matrix Transformations Related to the Euler Sequence Spaces

In this subsection, we desire to characterize the matrix mappings on the Euler sequence spaces e_p^r and e_c^r. Because the cases $p = 1$ and $p = \infty$ can be proved in the similar way, we omit the proof of those cases and consider only the case $1 < p < \infty$ in the proof of theorems given, below. Additionally, we directly prove the theorems characterizing the classes $(e_c^r : \ell_p)$, $(e_c^r : c)$ and derive the other characterizations from them by means of a given basic lemma, where $1 < p \leq \infty$. Furthermore, we give a Steinhaus type theorem concerning with the disjointness of the classes $(e_\infty^r : c)$ and $(e_c^r : c)_\varrho$.

We write for brevity that

$$\widetilde{a}_{nk} := \sum_{j=k}^{\infty} \binom{j}{k}(r-1)^{j-k}r^{-j}a_{nj} \quad \text{for all} \quad n, k \in \mathbb{N},$$

$$a(n,k) := \sum_{j=0}^{n} a_{jk} \quad \text{and} \quad a(n,k,m) := \frac{1}{m+1}\sum_{j=0}^{m} a_{n+j,k} \quad \text{for all} \quad k, m, n \in \mathbb{N}.$$

We will also use the similar notations with other letters.

Theorem 4.8.14. [289, Theorem 2.2] *Let $A = (a_{nk})$ be an infinite matrix over the complex field. Then, the following statements hold:*

(i) $A = (a_{nk}) \in (e_1^r : \ell_\infty)$ *if and only if*

$$\sup_{k,n\in\mathbb{N}} \left|\tilde{a}_{nk}\right| < \infty. \tag{4.8.15}$$

(ii) *Let* $1 < p < \infty$. *Then,* $A = (a_{nk}) \in (e_p^r : \ell_\infty)$ *if and only if*

$$\sup_{n\in\mathbb{N}} \sum_k \left|\tilde{a}_{nk}\right|^q < \infty, \tag{4.8.16}$$

$$(a_{nk})_{k\in\mathbb{N}} \in b_q^r \text{ for each } n \in \mathbb{N}. \tag{4.8.17}$$

(iii) $A = (a_{nk}) \in (e_\infty^r : \ell_\infty)$ *if and only if*

$$\sup_{n\in\mathbb{N}} \sum_k \left|\tilde{a}_{nk}\right| < \infty, \tag{4.8.18}$$

$$\lim_{m\to\infty} \sum_k \left| \sum_{j=k}^{m} \binom{j}{k} (r-1)^{j-k} r^{-j} a_{nj} \right| = \sum_k \left|\tilde{a}_{nk}\right| \text{ for all } n \in \mathbb{N}. \tag{4.8.19}$$

Proof. (ii) Suppose that the conditions (4.8.16) and (4.8.17) hold, and take any $x = (x_k) \in e_p^r$. Then, the sequence $(a_{nk})_{k\in\mathbb{N}} \in \left(e_p^r\right)^\beta$ for all $n \in \mathbb{N}$ and this implies the existence of the A-transform of x.

Let us now consider the following equality derived by using the relation (4.8.1) from the mth partial sum of the series $\sum_k a_{nk}x_k$:

$$\sum_{k=0}^{m} a_{nk}x_k = \sum_{k=0}^{m} \sum_{j=k}^{m} \binom{j}{k} (r-1)^{j-k} r^{-j} a_{nj} y_k \text{ for all } m, n \in \mathbb{N}. \tag{4.8.20}$$

Therefore, we obtain from (4.8.20) as $m \to \infty$ that

$$\sum_k a_{nk}x_k = \sum_k \tilde{a}_{nk} y_k \text{ for all } n \in \mathbb{N}. \tag{4.8.21}$$

Now, by passing to supremum over $n \in \mathbb{N}$ in (4.8.21) we derive by applying the Hölder's inequality that

$$\begin{aligned} \|Ax\|_\infty &= \sup_{n\in\mathbb{N}} \left| \sum_k a_{nk}x_k \right| \\ &\leq \sup_{n\in\mathbb{N}} \left(\sum_k \left|\tilde{a}_{nk}\right|^q \right)^{1/q} \left(\sum_k |y_k|^p \right)^{1/p} < \infty, \end{aligned}$$

which shows the sufficiency of the conditions (4.8.16) and (4.8.17).

Conversely, suppose that $A \in (e_p^r : \ell_\infty)$. Then, since $(a_{nk})_{k\in\mathbb{N}} \in \left(e_p^r\right)^\beta$ for all $n \in \mathbb{N}$ by the hypothesis, the necessity of (4.8.17) is immediate and $(\widetilde{a}_{nk})_{n,k\in\mathbb{N}}$ exists. Since $(a_{nk})_{k\in\mathbb{N}} \in \left(e_p^r\right)^\beta$, (4.8.21) holds and the sequences $a_n = (a_{nk})_{k\in\mathbb{N}}$ define the continuous linear functionals f_n on e_p^r by $f_n(x) = \sum_k a_{nk}x_k$ for all $n \in \mathbb{N}$. Since e_p^r and ℓ_p are norm isomorphic, it should follow with (4.8.21) that $\| f_n \|=\| \widetilde{a}_n \|_q$. This just says that the functionals defined by the rows of A on e_p^r are pointwise bounded. Hence, by the Banach-Steinhaus theorem, they are uniformly bounded, which yields that there exists a constant $M > 0$ such that $\| f_n \|\leq M$ for all $n \in \mathbb{N}$. It therefore follows, using the complete identification just referred to, that

$$\left(\sum_k |\tilde{a}_{nk}|^q\right)^{1/q} =\| f_n \|\leq M$$

holds for all $n \in \mathbb{N}$, which shows the necessity of the condition (4.8.16).

This completes the proof. □

Before giving the theorem characterizing the class $(e_1^r : \ell_p)$, we record the following lemma due to Stieglitz and Tietz [368]:

Lemma 4.8.15. *Let $1 < p < \infty$. Then, $A \in (\ell_1 : \ell_p)$ if and only if*

$$\sup_{k\in\mathbb{N}} \sum_n |a_{nk}|^p < \infty. \tag{4.8.22}$$

Theorem 4.8.16. [289, Theorem 2.4] *Let $1 < p < \infty$. Then, $A \in (e_1^r : \ell_p)$ if and only if*

$$\sup_{k\in\mathbb{N}} \sum_n |\widetilde{a}_{nk}|^p < \infty\ . \tag{4.8.23}$$

Proof. Suppose that the condition (4.8.23) holds and $x \in e_1^r$. Then, $y \in \ell_1$ and $(a_{nk})_{k\in\mathbb{N}} \in (e_1^r)^\beta$ for all $n \in \mathbb{N}$, i.e., A-transform of x exists. Hence, the series $\sum_k \widetilde{a}_{nk}y_k$ absolutely converges for every fixed $n \in \mathbb{N}$ and each $y = (y_k) \in \ell_1$. Therefore, we obtain from (4.8.21) by applying the Minkowski's inequality that

$$\left[\sum_n |(Ax)_n|^p\right]^{1/p} \leq \sum_k |y_k| \left(\sum_n |\widetilde{a}_{nk}|^p\right)^{1/p} < \infty.$$

This shows that $Ax \in \ell_p$.

Conversely, suppose that $A \in (e_1^r : \ell_p)$ and $1 < p < \infty$. Then, since Ax exists and is in the space ℓ_p for all $x \in e_1^r$, $(a_{nk})_{k\in\mathbb{N}} \in (e_1^r)^\beta$ for all $n \in \mathbb{N}$, which leads us to the relation (4.8.21). Let us define the matrix $B = (b_{nk})$ by $b_{nk} := \widetilde{a}_{nk}$ for all $n, k \in \mathbb{N}$. Therefore, one can easily see that $B \in (\ell_1 : \ell_p)$ and hence, B satisfies the condition (4.8.22), which yields the necessity of (4.8.23).

This step completes the proof. □

Theorem 4.8.17. [289, Theorem 2.5] *Let $A = (a_{nk})$ be an infinite matrix over the complex field. Then, the following statements hold:*

(i) *$A \in (e_1^r : f)$ if and only if (4.8.15) holds, and*

$$f - \lim \widetilde{a}_{nk} = \alpha_k \text{ for each } k \in \mathbb{N}. \tag{4.8.24}$$

(ii) *Let $1 < p < \infty$. Then, $A \in (e_p^r : f)$ if and only if (4.8.16), (4.8.17) and (4.8.24) hold.*

(iii) *$A \in (e_\infty^r : f)$ if and only if (4.8.18), (4.8.19) and (4.8.24) hold, and*

$$\lim_{m\to\infty} \sum_k |\widetilde{a}(n,k,m) - \alpha_k| = 0 \text{ uniformly in } n. \tag{4.8.25}$$

Proof. (ii) Let the conditions (4.8.16), (4.8.17), (4.8.24) hold and $x \in e_p^r$. Then, Ax exists and we have by (4.8.24) that

$$\lim_{m\to\infty} |\widetilde{a}(n,k,m)|^q = |\alpha_k|^q \text{ uniformly in } n,$$

for each $k \in \mathbb{N}$, which leads us with (4.8.16) that the inequality

$$\begin{aligned} \sum_{j=0}^{k} |\alpha_j|^q &= \lim_{m\to\infty} \sum_{j=0}^{k} |\widetilde{a}(n,j,m)|^q, \quad \text{(uniformly in } n) \\ &\leq \sup_{n,m\in\mathbb{N}} \sum_j |\widetilde{a}(n,j,m)|^q = M < \infty \end{aligned}$$

holds for every $k \in \mathbb{N}$. This gives that $(\alpha_k) \in \ell_q$. Since $x \in e_p^r$ by the hypothesis, and $e_p^r \cong \ell_p$, we have $y \in \ell_p$. Therefore, we derive by applying the Hölder's inequality that $(\alpha_k y_k) \in \ell_1$ for each $y \in \ell_p$. For any given $\varepsilon > 0$, choose a fixed $k_0 \in \mathbb{N}$ such that

$$\left(\sum_{k=k_0+1}^{\infty} |y_k|^p \right)^{1/p} < \frac{\varepsilon}{4M^{1/q}}.$$

Then, there is some $m_0 \in \mathbb{N}$ by (4.8.24) such that

$$\left| \sum_{k=0}^{k_0} [\widetilde{a}(n,k,m) - \alpha_k]\, y_k \right| < \frac{\varepsilon}{2}$$

for every $m \geq m_0$, uniformly in n. Therefore, we have

$$\begin{aligned} \left| \sum_{i=0}^{m} \frac{(Ax)_{n+i}}{m+1} - \sum_k \alpha_k y_k \right| &= \left| \sum_k [\widetilde{a}(n,k,m) - \alpha_k]\, y_k \right| \\ &\leq \left| \sum_{k=0}^{k_0} [\widetilde{a}(n,k,m) - \alpha_k]\, y_k \right| + \left| \sum_{k=k_0+1}^{\infty} [\widetilde{a}(n,k,m) - \alpha_k]\, y_k \right| \\ &< \frac{\varepsilon}{2} + \left\{ \sum_{k=k_0+1}^{\infty} [|\widetilde{a}(n,k,m)| + |\alpha_k|]^q \right\}^{1/q} \left(\sum_{k=k_0+1}^{\infty} |y_k|^p \right)^{1/p} \\ &< \frac{\varepsilon}{2} + 2M^{1/q} \frac{\varepsilon}{4M^{1/q}} = \varepsilon \end{aligned}$$

for all sufficiently large m uniformly in n. Hence, $Ax \in f$ which is what we wished to prove.

Conversely, suppose that $A \in (e_p^r : f)$. Then, since $f \subset \ell_\infty$, the necessity of (4.8.16) and (4.8.17) is immediately obtained from Theorem 4.8.14. To prove the necessity of (4.8.24), consider the sequence $b^{(k)}(r) = \{b_n^{(k)}(r)\}_{n\in\mathbb{N}} \in e_p^r$, defined by (4.8.5) for every fixed $k \in \mathbb{N}$. Since Ax exists and is in f for each $x \in e_p^r$, one can easily see that

$$Ab^{(k)}(r) := \left\{\sum_{j=k}^{\infty}\binom{j}{k}(1-r)^{j-k}r^{-j}a_{nj}\right\}_{n\in\mathbb{N}} \in f$$

for all $k \in \mathbb{N}$, which shows the necessity of (4.8.24).

Since Parts (i) and (iii) can also be proved in the similar way that of Part (ii), we omit details.

This concludes the proof. □

If f-limit is replaced by the ordinary limit in Theorem 4.8.17, then, we have

Corollary 4.8.18. [289, Corollary 2.6] *Let $A = (a_{nk})$ be an infinite matrix over the complex field. Then, the following statements hold:*

(i) *$A \in (e_1^r : c)$ if and only if (4.8.15) holds, and*

$$\lim_{n\to\infty} \widetilde{a}_{nk} = \alpha_k \text{ for each } k \in \mathbb{N}. \tag{4.8.26}$$

(ii) *Let $1 < p < \infty$. Then, $A \in (e_p^r : c)$ if and only if (4.8.16), (4.8.17) and (4.8.26) hold.*

(iii) *$A \in (e_\infty^r : c)$ if and only if (4.8.18), (4.8.19) and (4.8.26) hold, and*

$$\lim_{n\to\infty}\sum_k |\widetilde{a}_{nk} - \alpha_k| = 0. \tag{4.8.27}$$

Lemma 4.8.19. *$A \in (c : \ell_p)$ if and only if*

$$\sup_{F\in\mathcal{F}}\sum_n\left|\sum_{k\in F} a_{nk}\right|^p < \infty, \quad (1 < p < \infty). \tag{4.8.28}$$

Theorem 4.8.20. [23, Theorem 5.1] *$A \in (e_c^r : \ell_p)$ if and only if*

(i) *For $1 < p < \infty$,*

$$\sup_{F\in\mathcal{F}}\sum_n\left|\sum_{k\in F}\widetilde{a}_{nk}\right|^p < \infty, \tag{4.8.29}$$

$$\widetilde{a}_{nk} \text{ exists for all } k, n \in \mathbb{N}, \tag{4.8.30}$$

$$\sum_k \widetilde{a}_{nk} \text{ converges for each } n \in \mathbb{N}, \tag{4.8.31}$$

$$\sup_{m\in\mathbb{N}}\sum_{k=0}^{m}\left|\sum_{j=k}^{m}\binom{j}{k}(r-1)^{j-k}r^{-j}a_{nj}\right| < \infty \text{ for each } n \in \mathbb{N}. \tag{4.8.32}$$

(ii) *For $p = \infty$; (4.8.18), (4.8.30) and (4.8.32) hold.*

Proof. Suppose the conditions (4.8.29)–(4.8.32) hold and take any $x \in e_c^r$. Then, $(a_{nk})_{k\in\mathbb{N}} \in (e_c^r)^\beta$ for all $n \in \mathbb{N}$ and this implies that Ax exists. Let us define the matrix $B = (b_{nk})$ with $b_{nk} = \widetilde{a}_{nk}$ for all $n, k \in \mathbb{N}$. Then, since (4.8.28) is satisfied by the matrix B, we have $B \in (c : \ell_p)$. Thus, we have (4.8.20), which shows as $m \to \infty$ that (4.8.21) also holds. Therefore, (4.8.21) yields by taking ℓ_p-norm that $\|Ax\|_p = \|By\|_p < \infty$. This means that $A \in (e_c^r : \ell_p)$.

Conversely, suppose that $A \in (e_c^r : \ell_p)$. Then, since e_c^r and ℓ_p are BK-spaces, we have from Lemma 3.1.1 that there exists some real constant $K > 0$ such that

$$\|Ax\|_p \leq K\|x\|_{e_c^r} \tag{4.8.33}$$

for all $x \in e_c^r$. Since the inequality (4.8.33) is also satisfied for the sequence $x = (x_k) = \sum_{k\in F} b^{(k)}(r)$ belonging to the space e_c^r, where $b^{(k)}(r) = \{b_n^{(k)}(r)\}$ is defined by (4.8.5), we thus have for any $F \in \mathcal{F}$ that

$$\|Ax\|_p = \left(\sum_n \left|\sum_{k\in F} \widetilde{a}_{nk}\right|^p\right)^{1/p} \leq K\|x\|_{e_c^r} = K,$$

which shows the necessity of (4.8.29).

Since A is applicable to the space e_c^r by the hypothesis, the necessity of the conditions (4.8.30)–(4.8.32) is trivial. This completes the proof of Part (i) of the theorem.

Since Part (ii) can also be proved in the similar way used in the proof of Part (i), we leave the detailed proof to the reader. □

Theorem 4.8.21. [23, Theorem 5.2] $A \in (e_c^r : c)$ *if and only if (4.8.18), (4.8.26), (4.8.30), (4.8.32) hold and*

$$\lim_{n\to\infty} \sum_k \widetilde{a}_{nk} = \alpha. \tag{4.8.34}$$

Proof. Suppose that A satisfies the conditions (4.8.18), (4.8.26), (4.8.30), (4.8.32) and (4.8.34). Let us take any $x = (x_k)$ in e_c^r such that $x_k \to l$, as $k \to \infty$. Then, Ax exists and it is trivial that the sequence $y = (y_k)$ connected with the sequence $x = (x_k)$ by the relation (4.8.1) is in c such that $y_k \to l$, as $k \to \infty$. At this stage, we observe from (4.8.26) and (4.8.18) that

$$\sum_{j=0}^{k} |\alpha_j| \leq \sup_{n\in\mathbb{N}} \sum_j |\widetilde{a}_{nj}| < \infty$$

holds for every $k \in \mathbb{N}$. This leads us to the consequence that $(\alpha_k) \in \ell_1$. Considering (4.8.21), let us write

$$\sum_k a_{nk}x_k = \sum_k \widetilde{a}_{nk}(y_k - l) + l\sum_k \widetilde{a}_{nk}. \tag{4.8.35}$$

In this situation, by letting $n \to \infty$ in (4.8.35) we see that the first term on the right-hand side tends to $\sum_k \alpha_k(y_k - l)$ by (4.8.18) and (4.8.26), and the second term tends to $l\alpha$ by (4.8.34). Now, in the light of these facts we obtain from (4.8.35) that

$$\lim_{n\to\infty}(Ax)_n = \sum_k \alpha_k(y_k - l) + l\alpha.$$

This shows that $A \in (e_c^r : c)$.

Conversely, suppose that $A \in (e_c^r : c)$. Then, since the inclusion $c \subset \ell_\infty$ holds, the necessity of (4.8.30), (4.8.32) and (4.8.18) is immediately obtained from Part (ii) of Theorem 4.8.20. To prove the necessity of (4.8.26), consider the sequence $x = x^{(k)} = \{x_n^{(k)}(r)\}_{n\in\mathbb{N}} \in e_c^r$ for each $k \in \mathbb{N}$ defined by

$$x_n^{(k)}(r) := \begin{cases} \binom{n}{k}(r-1)^{n-k}r^{-n} & , \quad n \geq k, \\ 0 & , \quad 0 \leq n \leq k-1. \end{cases}$$

Since Ax exists and is in c for every $x \in e_c^r$, one can easily see that $Ax^{(k)} = (\widetilde{a}_{nk})_{n\in\mathbb{N}} \in c$ for each $k \in \mathbb{N}$, which shows the necessity of (4.8.26).

Similarly by putting $x = e$ in (4.8.21), we also obtain that $Ax = \left(\sum_k \widetilde{a}_{nk}\right)_{n\in\mathbb{N}}$, which belongs to the space c and this shows the necessity of (4.8.34).

This step concludes the proof. □

It is of course that the class $(e_c^r : c)_\varrho$ of ϱ-multiplicative matrices reduces to the classes $(e_c^r : c_0)$ and $(e_c^r : c)_{reg}$ in the cases $\varrho = 0$ and $\varrho = 1$, respectively; where $(e_c^r : c)_{reg}$ denotes the class of all matrix mappings A from e_c^r to c such that $A - \lim\, x = \lim\, x$ for all $x \in e_c^r$. Now, we may give a corollary to Theorem 4.8.21, without proof.

Corollary 4.8.22. [23, Corollary 5.1] *$A \in (e_c^r : c)_\varrho$ if and only if (4.8.30), (4.8.32), (4.8.18) hold, (4.8.26) and (4.8.34) also hold with $\alpha_k = 0$ for each $k \in \mathbb{N}$ and $\alpha = \varrho$, respectively.*

The Steinhaus type theorems were formulated by Maddox [249], as follows: Consider the class $(\lambda : \mu)_1$ of 1-multiplicative matrices and ν be a sequence space such that $\nu \supset \lambda$. Then, the result of the form $(\lambda : \mu)_1 \cap (\nu : \mu) = \emptyset$ is called a theorem of Steinhaus type, where $\emptyset$ denotes the empty set. Now, we may give a Steinhaus type theorem:

Theorem 4.8.23. [23, Theorem 5.3] *There is no matrix belonging to the classes both $(e_c^r : c)_\varrho$ and $(e_\infty^r : c)$.*

Proof. Suppose that the classes $(e_c^r : c)_\varrho$ and $(e_\infty^r : c)$ are not disjoint. Then, there is at least one matrix A satisfying the conditions of both (iii) of Corollaries 4.8.18 and 4.8.22. Combining the condition (4.8.26) of Corollary 4.8.22 with (4.8.19), one can easily see that $\lim_{n\to\infty} \sum_k |\widetilde{a}_{nk}| = 0$, which contradicts the condition (4.8.34) of Corollary 4.8.22.

This completes the proof. □

4.8.3 Some Geometric Properties of the Space e_p^r

In this subsection, we study some geometric properties of the space e_p^r.

A Banach space X is said to have the *Banach-Saks property* if every bounded sequence (x_n) in X admits a subsequence (z_n) such that the sequence $\{t_k(z)\}$ is convergent in the norm in X [150]; where $t_k(z) = (C_1 z)_k$ for all $k \in \mathbb{N}$.

A Banach space X is said to have the *weak Banach-Saks property* whenever given any weakly null sequence $(x_n) \subset X$ there exists a subsequence (z_n) of (x_n) such that the sequence $\{t_k(z)\}$ strongly converges to zero.

A Banach space X has the *weak fixed point property* if every non-expansive self-mapping defined on a weakly compact convex subset of X has a fixed point.

In [179], Garcia-Falset introduces the following coefficient

$$R(X) := \sup\left\{\liminf_{n\to\infty} \|x_n - x\| : (x_n) \subset B(X),\ x_n \xrightarrow{w} 0,\ x \in B(X)\right\};$$

where $B(X)$ denotes the *unit ball* of X.

Remark 4.8.24. *A Banach space X with $R(X) < 2$ has the weak fixed point property* [180].

Let $1 < p < \infty$. A Banach space is said to have the *Banach-Saks type p* or *property (BS_p)*, if every weakly null sequence (x_k) has a subsequence (x_{k_l}) such that $\|\sum_{l=0}^{n} x_{k_l}\| \le C(n+1)^{1/p}$ for some $C > 0$ and for all $n \in \mathbb{N}$ [221].

For a normed linear space E, the Gurarii's modulus of convexity is defined by

$$\beta_E(\varepsilon) := \inf\left\{1 - \inf_{0\le\alpha\le1} \|\alpha x + (1-\alpha)x\| : x, y \in S(E),\ \|x - y\| = \varepsilon\right\};$$

where $S(E)$ denotes the *unit sphere* of E and $0 \le \varepsilon \le 2$ [196, 342].

Now, we may give the following results related to some geometric properties of the space e_p^r, mentioned above:

Theorem 4.8.25. [289, Theorem 3.2] *The space e_p^r has the Banach-Saks type p.*

Proof. Let (ε_n) be a sequence of positive numbers for which $\sum_n \varepsilon_n \le 1/2$ and (x_n) also be a weakly null sequence in $B(e_p^r)$. Set $z_0 = x_0 = 0$ and $z_1 = x_{n_1} = x_1$. Then, there exists $s_1 \in \mathbb{N}$ such that

$$\left\|\sum_{i=s_1+1}^{\infty} z_1(i)e^{(i)}\right\|_{e_p^r} < \varepsilon_1.$$

Since $x_n \xrightarrow{w} 0$ implies $x_n \to 0$ co-ordinatewise, there is an $n_2 \in \mathbb{N}$ such that

$$\left\|\sum_{i=0}^{s_1} x_n(i)e^{(i)}\right\|_{e_p^r} < \varepsilon_1,$$

when $n \ge n_2$. Set $z_2 = x_{n_2}$. Then, there exists a $s_2 > s_1$ such that

$$\left\|\sum_{i=s_2+1}^{\infty} z_2(i)e^{(i)}\right\|_{e_p^r} < \varepsilon_2.$$

Again using the fact $x_n \to 0$ co-ordinatewise, there exists an $n_3 > n_2$ such that

$$\left\| \sum_{i=0}^{s_2} x_n(i)e^{(i)} \right\|_{e_p^r} < \varepsilon_2,$$

when $n \geq n_3$.

Using this process, we can find two increasing sequences (s_i) and (n_i) such that

$$\left\| \sum_{i=0}^{s_j} x_n(i)e^{(i)} \right\|_{e_p^r} < \varepsilon_j$$

for each $n \geq n_{j+1}$ and

$$\left\| \sum_{i=s_j+1}^{\infty} z_j(i)e^{(i)} \right\|_{e_p^r} < \varepsilon_j,$$

where $z_j = x_{n_j}$. Hence,

$$\begin{aligned} \left\| \sum_{j=0}^{n} z_j \right\|_{e_p^r} &= \left\| \sum_{j=0}^{n} \left(\sum_{i=0}^{s_{j-1}} z_j(i)e^{(i)} + \sum_{i=s_{j-1}+1}^{s_j} z_j(i)e^{(i)} + \sum_{i=s_j+1}^{\infty} z_j(i)e^{(i)} \right) \right\|_{e_p^r} \\ &\leq \left\| \sum_{j=0}^{n} \left(\sum_{i=s_{j-1}+1}^{s_j} z_j(i)e^{(i)} \right) \right\|_{e_p^r} + 2\sum_{j=0}^{n} \varepsilon_j \end{aligned}$$

and

$$\begin{aligned} \left[\left\| \sum_{j=0}^{n} \sum_{i=s_{j-1}+1}^{s_j} z_j(i)e^{(i)} \right\|_{e_p^r} \right]^p &= \sum_{j=0}^{n} \sum_{i=s_{j-1}+1}^{s_j} \left| \sum_{k=0}^{i} \binom{i}{k} (1-r)^{i-k} r^k z_j(k) \right|^p \\ &\leq \sum_{j=0}^{n} \sum_{i=0}^{\infty} \left| \sum_{k=0}^{i} \binom{i}{k} (1-r)^{i-k} r^k z_j(k) \right|^p \leq n. \end{aligned}$$

Therefore,

$$\left\| \sum_{j=0}^{n} z_j \right\|_{e_p^r} \leq n^{1/p} + 1 \leq 2n^{1/p},$$

i.e., e_p^r has the Banach-Saks type p. □

Remark 4.8.26. *Note that $R(e_p^r) = R(\ell_p) = 2^{1/p}$, since e_p^r is norm isomorphic to ℓ_p.*

Hence, by Remarks 4.8.24 and 4.8.26, we have the following:

Theorem 4.8.27. *The space e_p^r has the weak fixed point property, where $1 < p < \infty$.*

Theorem 4.8.28. [289, Theorem 3.5] *The modulus of convexity for the space* e_p^r *is* $\beta_{e_p^r}(\varepsilon) \leq 1 - [1-(\varepsilon/2)^p]^{1/p}$, *where* $0 \leq \varepsilon \leq 2$.

Proof. Let $s \in e_p^r$. Then,

$$\|s\|_{e_p^r} = \|E^r s\|_p = \left[\sum_n |(E^r s)_n|^p\right]^{1/p}.$$

Let $0 \leq \varepsilon \leq 2$ and consider the sequences

$$\begin{aligned} u &:= \left(\left(E^{1/r}\left(1-(\varepsilon/2)^p\right)\right)^{1/p}, E^{1/r}(\varepsilon/2), 0, 0, \ldots\right), \\ v &:= \left(\left(E^{1/r}\left(1-(\varepsilon/2)^p\right)\right)^{1/p}, E^{1/r}(-\varepsilon/2), 0, 0, \ldots\right). \end{aligned}$$

Then, $\|E^r u\|_p = \|u\|_{e_p^r} = 1$, $\|E^r v\|_p = \|v\|_{e_p^r} = 1$, i.e., $u, v \in S(e_p^r)$ and $\|E^r u - E^r v\|_p = \|u-v\|_{e_p^r} = \varepsilon$. For $0 \leq \alpha \leq 1$,

$$\begin{aligned} \left[\|\alpha u + (1-\alpha)v\|_{e_p^r}\right]^p &= \left[\|\alpha E^r u + (1-\alpha)E^r v\|_p\right]^p \\ &= 1 - \left(\frac{\varepsilon}{2}\right)^p + |\,2\alpha - 1\,| \left(\frac{\varepsilon}{2}\right)^p. \end{aligned}$$

Hence,

$$\inf_{0 \leq \alpha \leq 1} \left[\|\alpha u + (1-\alpha)v\|_{e_p^r}\right]^p = 1 - \left(\frac{\varepsilon}{2}\right)^p. \tag{4.8.36}$$

Therefore, for $p \geq 1$

$$\beta_{e_p^r}(\varepsilon) \leq 1 - \left[1 - \left(\frac{\varepsilon}{2}\right)^p\right]^{1/p}.$$

This completes the proof. □

From (4.8.36), we have the following:

Corollary 4.8.29. [289, Corollary 3.6] *Let* $A = (a_{nk})$ *be an infinite matrix over the complex field. Then, the following statements hold:*

(i) *For* $\varepsilon = 2$, $\beta_{e_p^r}(\varepsilon) = 1$. *Hence,* e_p^r *is strictly convex.*

(ii) *For* $0 < \varepsilon \leq 2$, $0 < \beta_{e_p^r}(\varepsilon) \leq 1$. *Hence,* e_p^r *is uniformly convex.*

Theorem 4.8.30. [289, Theorem 3.7] *The space* e_1^r *has the Schur property, i.e., every weakly convergent sequence is norm convergent in* e_1^r.

Proof. Since the space ℓ_1 has the Schur property and $\|x\|_{e_1^r} \leq r^{-1}\|x\|_{\ell_1}$ for all $x \in e_1^r$, it easily follows that the space e_1^r has the Schur property too. □

4.9 DOMAIN OF THE GENERALIZED WEIGHTED MEAN AND CONCERNING DUALITY RELATION

Let λ denotes any of the classical sequence spaces ℓ_∞, c, c_0 and ℓ_p or any of the Maddox's paranormed sequence spaces $\ell_\infty(p)$, $c(p)$, $c_0(p)$ and $\ell(p)$. The domain $\lambda_{G(u,v)}$ is studied by Altay and Başar [27, 28].

Let us define the sequence $y = (y_k)$ by the $G(u,v)$-transform of a sequence $x = (x_k) \in \omega$, i.e.,

$$y_k := \{G(u,v)x\}_k = \sum_{j=0}^{k} u_k v_j x_j \quad \text{for all } k \in \mathbb{N}.$$

Then, since $\lambda_{G(u,v)} \cong \lambda$ it is trivial that the two-sided implication "$x \in \lambda_{G(u,v)}$ if and only if $y \in \lambda$" holds. In this situation, the duality relation between the entries of the matrices $A = (a_{nk})$ and $B = (b_{nk})$ is

$$a_{nk} := \sum_{j=k}^{\infty} u_j v_k b_{nj} \quad \text{or} \quad b_{nk} := \frac{1}{u_k}\left(\frac{a_{nk}}{v_k} - \frac{a_{n,k+1}}{v_{k+1}}\right) \tag{4.9.1}$$

for all $k, n \in \mathbb{N}$. The pair of such type matrices was used by Altay and Başar [27, 28].

Following Choudhary and Mishra [105], Başar and Altay [65] and Altay and Başar [26], we define the sequence spaces $\lambda(u,v;p)$ by

$$\lambda(u,v;p) := \left\{ x = (x_k) \in \omega : \left(u_k \sum_{j=0}^{k} v_j x_j \right) \in \lambda(p) \right\}.$$

If $p_k = 1$ for all $k \in \mathbb{N}$, we write $\lambda(u,v)$ instead of $\lambda(u,v;p)$, which are introduced by Malkowsky and Savaş [272]. If λ is any normed or paranormed sequence space, then we call the matrix domain $\lambda_{G(u,v)}$ as the *generalized weighted means sequence space, in short GWM.* It is natural that these spaces can also be defined with the notation of (4.1.1) that $\lambda(u,v;p) = \{\lambda(p)\}_{G(u,v)}$.

This definition includes the following special cases:

(i) If $p = e$ and $\lambda = \ell_p$ or $\lambda = \ell_\infty$, c, c_0, then $\ell(u,v;p) = \ell(p)$ or $\lambda(u,v;p) = Z(u,v;\lambda)$; (cf. [105, 272]).

(ii) If $u = v = e$ and $\lambda = \ell_\infty$ or $\lambda = \ell_p$, then $\lambda(u,v;p) = bs(p)$, (cf. [65]), or $\lambda(u,v;p) = \ell(p)$.

(iii) If $v = e$, $u = (1/(n+1))$, $p = e$ and $\lambda = \ell_p$, ℓ_∞ or $\lambda = c$, c_0, then $\ell(u,v;p) = X_p$ or $\lambda(u,v;p) = \tilde{c}$, $\tilde{c}_0$; (cf. [304, 373]); where $1 < p \leq \infty$.

(iv) If $v = (1 + r^k)$, $u = (1/(n+1))$, $p = e$ and $\lambda = \ell_p$ or $\lambda = \ell_\infty$, c, c_0, then $\ell(u,v;p) = a_p^r$ or $\lambda(u,v;p) = a_\infty^r$, a_c^r, a_0^r; (cf. [45, 42]); where $1 < p < \infty$.

(v) If $v = q$, $u = (1/\sum_{k=0}^{n} q_k)$ and $\lambda = \ell_\infty$, c and c_0, then $\lambda(u,v;p) = (\overline{N},q)_\infty$, $(\overline{N},q)$ and $(\overline{N},q)_0$; (cf. [263]).

(vi) If $v = ((1 + r^k)z_k)$, $u = (1/(n + 1))$, and $\lambda = c$ or c_0, then $\lambda(u, v; p) = a_c^r(z, p)$ or $a_0^r(z, p)$; (cf. [43]).

(vii) If $v = t$ and $u = (1/\sum_{k=0}^{n} t_k)$, then $\ell(u, v; p) = r^t(p)$, (cf. [43]), $\ell_\infty(u, v) = (\overline{N}, t)_\infty$, $c(u, v) = (\overline{N}, t)$ and $c_0(u, v) = (\overline{N}, t)_0$; (cf. [263]).

Define the sequence $y = (y_k)$ by the $G(u, v)$-transform of a sequence $x = (x_k)$, i.e.,

$$y_k := \sum_{j=0}^{k} u_k v_j x_j \quad \text{for all } \ k \in \mathbb{N}. \tag{4.9.2}$$

Although Theorems 4.9.1, 4.9.2 and 4.9.3, below, are related to the sequence spaces $\ell_\infty(u, v; p)$, $c(u, v; p)$, $c_0(u, v; p)$ and $\ell(u, v; p)$, we give the proof only for one of those spaces. Since the proof can also be obtained in the similar way for the other spaces, to avoid undue repetition in the statements, we leave the detail to the reader. Now, we may begin with the following theorem:

Theorem 4.9.1. [27, Theorem 2.1] *The following statements hold:*

(a) *$\ell_\infty(u, v; p)$, $c(u, v; p)$, $c_0(u, v; p)$ and $\ell(u, v; p)$ are complete linear metric spaces paranormed by h_1 and h_2, defined by*

$$h_1(x) := \sup_{k\in\mathbb{N}} \left| \sum_{j=0}^{k} u_k v_j x_j \right|^{p_k/M} \quad \textit{and} \quad h_2(x) := \left(\sum_k \left| \sum_{j=0}^{k} u_k v_j x_j \right|^{p_k} \right)^{1/M}.$$

h_1 is a paranorm for the spaces $\ell_\infty(u, v; p)$ and $c(u, v; p)$ only in the trivial case $\inf p_k > 0$ when $\ell_\infty(u, v; p) = \ell_\infty(u, v)$, $c(u, v; p) = c(u, v)$.

(b) *The sets $\lambda(u, v)$ are Banach spaces endowed with the norm $\|x\|_{\lambda(u,v)} = \| y \|_\lambda$.*

Proof. We prove only Part (a) for the space $c_0(u, v; p)$ and omit Part (b).

The linearity of $c_0(u, v; p)$ with respect to the co-ordinatewise addition and scalar multiplication follows from the following inequalities, which are satisfied for $s = (s_j), x = (x_j) \in c_0(u, v; p)$ (see [258, p. 30])

$$\sup_{k\in\mathbb{N}} \left| \sum_{j=0}^{k} u_k v_j (s_j + x_j) \right|^{p_k/M} \le \sup_{k\in\mathbb{N}} \left| \sum_{j=0}^{k} u_k v_j s_j \right|^{p_k/M} + \sup_{k\in\mathbb{N}} \left| \sum_{j=0}^{k} u_k v_j x_j \right|^{p_k/M} \tag{4.9.3}$$

and for any $\alpha \in \mathbb{R}$ (see [251])

$$|\alpha|^{p_k} \le \max\{1, |\alpha|^M\}. \tag{4.9.4}$$

It is clear that $h_1(\theta) = 0$ and $h_1(x) = h_1(-x)$ for all $x \in c_0(u, v; p)$. Again the inequalities (4.9.3) and (4.9.4) yield the subadditivity of h_1 and $h_1(\alpha x) \le \max\{1, |\alpha|\}h_1(x)$.

Let (x^n) be any sequence of the points in the space $c_0(u, v; p)$ such that $h_1(x^n - x) \to 0$ and (α_n) also be any sequence of scalars such that $\alpha_n \to \alpha$, as $n \to \infty$.

Then, since the inequality $h_1(x^n) \leq h_1(x) + h_1(x^n - x)$ holds by subadditivity of h_1, $\{h_1(x^n)\}$ is bounded and we thus have

$$\begin{aligned} h_1(\alpha_n x^n - \alpha x) &= \sup_{k\in\mathbb{N}} \left| \sum_{j=0}^{k} u_k v_j \left(\alpha_n x_j^{(n)} - \alpha x_j \right) \right|^{p_k/M} \\ &\leq |\alpha_n - \alpha| h_1(x^n) + |\alpha| h_1(x^n - x), \end{aligned}$$

which tends to zero, as $n \to \infty$. That is to say that the scalar multiplication is continuous. Hence, h_1 is a paranorm on the space $c_0(u, v; p)$.

It remains to prove the completeness of the space $c_0(u, v; p)$. Let (x^i) be any Cauchy sequence in the space $c_0(u, v; p)$, where $x^i = \{x_0^{(i)}, x_1^{(i)}, x_2^{(i)}, \ldots\}$. Then, for a given $\varepsilon > 0$ there exists a positive integer $n_0(\varepsilon)$ such that $h_1(x^i - x^j) < \varepsilon$ for all $i, j \geq n_0(\varepsilon)$. Using the definition of h_1, we obtain for each fixed $k \in \mathbb{N}$ and for all $i, j \geq n_0(\varepsilon)$ that

$$\left|\left(G(u,v)x^i\right)_k - \left(G(u,v)x^j\right)_k\right| \leq \sup_{k\in\mathbb{N}} \left|\left(G(u,v)x^i\right)_k - \left(G(u,v)x^j\right)_k\right|^{p_k/M} < \varepsilon, \tag{4.9.5}$$

which leads us to the fact that $\{(G(u,v)x^0)_k, (G(u,v)x^1)_k, (G(u,v)x^2)_k, \ldots\}$ is a Cauchy sequence of real numbers for every fixed $k \in \mathbb{N}$. Since $\mathbb{R}$ is complete, it converges, say $(G(u,v)x^i)_k \to (G(u,v)x)_k$, as $i \to \infty$. Using these infinitely many limits $(G(u,v)x)_0$, $(G(u,v)x)_1$, $(G(u,v)x)_2, \ldots$, we define the sequence $\{(G(u,v)x)_0, (G(u,v)x)_1, (G(u,v)x)_2, \ldots\}$. From (4.9.5) with $j \to \infty$ we have

$$\left|\left(G(u,v)x^i\right)_k - (G(u,v)x)_k\right| \leq \varepsilon, \quad (i \geq n_0(\varepsilon)) \tag{4.9.6}$$

for every fixed $k \in \mathbb{N}$. Since $x^i = \{x_k^{(i)}\} \in c_0(u, v; p)$, $\left|\left(G(u,v)x^i\right)_k\right|^{p_k/M} < \varepsilon$ for all $k \in \mathbb{N}$. Therefore, we obtain by (4.9.6) that

$$\begin{aligned} \left|(G(u,v)x)_k\right|^{p_k/M} &\leq \left|(G(u,v)x)_k - \left(G(u,v)x^i\right)_k\right|^{p_k/M} \\ &\quad + \left|\left(G(u,v)x^i\right)_k\right|^{p_k/M} < 2\varepsilon, \quad (i \geq n_0(\varepsilon)). \end{aligned}$$

This shows that the sequence $G(u,v)x$ belongs to the space $c_0(p)$. Since (x^i) was an arbitrary Cauchy sequence, the space $c_0(u, v; p)$ is complete.

This terminates the proof. □

Therefore, one can easily check that the absolute property does not hold on the spaces $\ell_\infty(u, v; p)$, $c(u, v; p)$, $c_0(u, v; p)$ and $\ell(u, v; p)$, that is $h_i(x) \neq h_i(|x|)$ for at least one sequence in those spaces, and this says that $\ell_\infty(u, v; p)$, $c(u, v; p)$, $c_0(u, v; p)$ and $\ell(u, v; p)$ are the sequence spaces of non-absolute type; where $|x| = (|x_k|)$ and $i = 1, 2$.

Theorem 4.9.2. [27, Theorem 2.2] *The GWM sequence spaces $\ell_\infty(u, v; p)$, $c(u, v; p)$, $c_0(u, v; p)$ and $\ell(u, v; p)$ of non-absolute type are paranorm isomorphic to the spaces $\ell_\infty(p)$, $c(p)$, $c_0(p)$ and $\ell(p)$, respectively; where $0 < p_k \leq H < \infty$.*

Proof. We establish this for the space $\ell_\infty(u,v;p)$. To prove the theorem, we should show the existence of a paranorm preserving linear bijection between the spaces $\ell_\infty(u,v;p)$ and $\ell_\infty(p)$ for $1 \le p_k \le H < \infty$. With the notation of (4.9.2), define the transformation T from $\ell_\infty(u,v;p)$ to $\ell_\infty(p)$ by $x \mapsto y = Tx$. The linearity of T is trivial. Further, it is obvious that $x = \theta$ whenever $Tx = \theta$ and hence T is injective.

Let $y = (y_k) \in \ell_\infty(p)$ and define the sequence $x = (x_k)$ by $x_k = \left(\frac{y_k}{u_k} - \frac{y_{k-1}}{u_{k-1}}\right)/v_k$ for all $k \in \mathbb{N}$. Then,

$$\begin{aligned} h_1(x) = \sup_{k\in\mathbb{N}} \left| u_k \sum_{j=0}^{k} v_j x_j \right|^{p_k/M} &= \sup_{k\in\mathbb{N}} \left| \sum_{j=0}^{k} \delta_{kj} y_j \right|^{p_k/M} \\ &= \sup_{k\in\mathbb{N}} |y_k|^{p_k/M} = g_1(y) < \infty. \end{aligned}$$

Thus, we observe that $x \in \ell_\infty(u,v;p)$ and consequently T is surjective, and is paranorm preserving. Hence, T is a paranorm preserving linear bijection and this says us that the spaces $\ell_\infty(u,v;p)$ and $\ell_\infty(p)$ are norm isomorphic, as desired. □

Theorem 4.9.3. (cf. Theorem 2.3 of [27] and [28]) *Let $u \in \ell_1$. The sequence spaces $\ell_p(u,v)$ and $c_0(u,v)$ have AD property.*

Proof. Note that if $f \in [\ell_p(u,v)]^*$, then $f(x) = g(Ax)$ for some $g \in \ell_p^*$. Since ℓ_p has AK property and $\ell_p^* \cong \ell_q$, $f(x) = \sum_k a_k\{G(u,v)x\}_k$ for some $a = (a_k) \in \ell_q$. For any $f \in [\ell_p(u,v)]^*$ and $e^{(k)} \in \phi$, we have

$$f(e^{(k)}) = \sum_i a_i \{G(u,v)e^{(k)}\}_i = \{G'(u,v)a\}_k,$$

for all $k \in \mathbb{N}$; where $G'(u,v)$ denotes the transpose of the matrix $G(u,v)$. Hence, from Hahn-Banach theorem, $\phi \subset \ell_p(u,v)$ is dense in $\ell_p(u,v)$ if and only if $G'(u,v)a = \theta$ for $a \in \ell_q$ implies $a = \theta$.

Since the null space of the operator $G'(u,v)$ on ω is $\{\theta\}$, $\ell_p(u,v)$ has *AD* property.

Because, it can be established by analogy to the proof of the sequence space $\ell_p(u,v)$ that the sequence space $c_0(u,v)$ has *AD* property, we omit details. □

Now, as an easy consequence of Theorem 4.9.3 and Part (iii) of Lemma 4.5.54, we have

Corollary 4.9.4. (cf. [27, Corollary 2.5] and [28]) *The following statements hold:*

(i) $c_0^f(u,v) := c_0^*(u,v) := \left\{(a_k) \in \omega : \left(\frac{a_k}{u_k v_k} - \frac{a_{k+1}}{u_k v_{k+1}}\right) \in \ell_1\right\}$.

(ii) $\ell_p^f(u,v) := \ell_p^*(u,v) := \left\{(a_k) \in \omega : \left(\frac{a_k}{u_k v_k} - \frac{a_{k+1}}{u_k v_{k+1}}\right) \in \ell_p^*\right\}$.

Now, we may give the theorem determining the α- and β-duals of the sequence space $\lambda(u,v;p)$. In proving the theorem, we apply the technique used by Başar and Altay [66].

Theorem 4.9.5. (cf. Theorem 2.6 of [27] and [28]) *Let λ denotes any of the Maddox's paranormed sequence spaces $\ell_\infty(p)$, $c(p)$, $c_0(p)$ and $\ell(p)$, $u, v \in \mathcal{U}$ and $z = (z_k)$ be any sequence in the space ω. Define the sequences $t^1 = (t_k^1)$, $t^2 = (t_k^2)$ and the matrix $B = (b_{nk})$ by*

$$t_k^1 := \frac{1}{u_k}\left(\frac{z_k}{v_k} - \frac{z_{k+1}}{v_{k+1}}\right), \quad t_k^2 := \frac{z_k}{u_k v_k},$$
$$b_{nk} := \begin{cases} t_k^1 & , \quad k < n, \\ t_k^2 & , \quad k = n, \\ 0 & , \quad k > n, \end{cases} \tag{4.9.7}$$

for all $k, n \in \mathbb{N}$. Then, the following statements hold:

$$\begin{aligned} \lambda(u, v; p)^\beta &:= \{z = (z_k) \in \omega : B \in (\lambda(p) : c)\}, \\ \lambda(u, v; p)^\gamma &:= \{z = (z_k) \in \omega : B \in (\lambda(p) : \ell_\infty)\}. \end{aligned}$$

Proof. Consider the equality

$$\sum_{k=0}^{n} z_k x_k = \sum_{k=0}^{n-1} t_k^1 y_k + t_n^2 y_n = (By)_n \quad \text{for all} \quad n \in \mathbb{N}, \tag{4.9.8}$$

where $B = (b_{nk})$ defined by (4.9.7).

Thus, one can easily deduce from (4.9.8) that $zx = (z_k x_k) \in cs$ or bs whenever $x = (x_k) \in \lambda(u, v; p)$ if and only if $By = \{(By)_n\} \in c$ or ℓ_∞ whenever $y = (y_k) \in \lambda(p)$. This means that $z = (z_k) \in \lambda^\beta(u, v; p)$ or $z = (z_k) \in [\lambda(u, v; p)]^\gamma$ if and only if $B \in (\lambda(p) : c)$ or $B \in (\lambda(p) : \ell_\infty)$, which is what we wished to prove.

This completes the proof. □

As a direct consequence of Theorem 4.9.5, we have

Corollary 4.9.6. (cf. [27, Corollary 2.7] and [28]) *Let $u, v \in \mathcal{U}$ and $s = (s_k)$ with $s_k = M^{1/p_k}$ for all $k \in \mathbb{N}$ and $M \in \mathbb{N}$. Put $p_k' = p_k/(p_k - 1)$ for $1 < p_k < \infty$ and for all $k \in \mathbb{N}$. Then, the following statements hold:*

(i) $\ell_\infty(u, v; p)^\beta := \bigcap_{M \geq 1} \{z = (z_k) \in \omega : t^1 s \in \ell_1, t^2 s \in \ell_\infty, t^3 s \in c_0\}$.

(ii) $c(u, v; p)^\beta := \bigcup_{M \geq 1} \{z = (z_k) \in \omega : t^1 s^{-1} \in \ell_1, t^2 s^{-1}, t^3 s^{-1} \in \ell_\infty, t^4 \in cs\}$.

(iii) $c_0(u, v; p)^\beta := \bigcup_{M \geq 1} \{z = (z_k) \in \omega : t^1 s^{-1} \in \ell_1, t^2 s^{-1}, t^3 s^{-1} \in \ell_\infty\}$.

(iv) *For* $p_k \leq 1$, $\ell(u, v; p)^\beta := \bigcap_{M > 1} \{z = (z_k) \in \omega : t^1 M, t^3 M, t^2 \in \ell_\infty(p)\}$.

(v) *For* $p_k > 1$, $\ell(u, v; p)^\beta := \bigcup_{M > 1} \{z = (z_k) \in \omega : t^1 M^{-1} \in \ell(p'), t^2 M^{-1}, t^3 M^{-1} \in \ell_\infty(p)\}$.

(vi) $\ell_\infty(u, v; p)^\gamma := \bigcap_{M \geq 1} \{z = (z_k) \in \omega : t^1 s \in \ell_1, t^2 s \in \ell_\infty\}$.

(vii) $c(u, v; p)^\gamma := \bigcup_{M \geq 1} \{z = (z_k) \in \omega : t^4 \in bs, t^1 s^{-1} \in \ell_1, t^2 s^{-1} \in \ell_\infty\}$; *where* $t^3 = (t_k^3)$ *and* $t^4 = (t_k^4)$ *are defined by* $t_k^3 = z_k v_k^{-1} u_{k-1}^{-1}$ *and* $t_k^4 = z_k v_k^{-1}(u_k^{-1} - u_{k-1}^{-1})$.

(viii) $c_0(u,v;p)^\gamma := \bigcup_{M\geq 1}\{z=(z_k)\in\omega : t^1 s^{-1}\in\ell_1, t^2 s^{-1}\in\ell_\infty\}$.

(ix) *For* $p_k \leq 1$, $\ell(u,v;p)^\gamma := \bigcup_{M>1}\{z=(z_k)\in\omega : t^1 M^{-1}, t^2 M^{-1}\in\ell_\infty(p)\}$.

(x) *For* $p_k > 1$, $\ell(u,v;p)^\gamma := \bigcup_{M>1}\{z=(z_k)\in\omega : t^1 M^{-1}\in\ell(p'), t^2 M^{-1} \in\ell_\infty(p')\}$; *where* $t^3=(t^3_k)$ *is defined by* $t^3_k = z_k v_k^{-1} u_{k-1}$.

Now, we may give a sequence of the points of the spaces $c_0(u,v;p)$, $\ell(u,v;p)$ and $c(u,v;p)$, which forms a Schauder basis for the spaces. Because of the isomorphism T between the sequence spaces $c_0(u,v;p)$ and $c_0(p)$, $\ell(u,v;p)$ and $\ell(p)$, $c(u,v;p)$ and $c(p)$ is onto, by Theorem 4.1.2, the inverse image of the bases of the spaces $c_0(p)$, $\ell(p)$ and $c(p)$ are the bases of the new spaces $c_0(u,v;p)$, $\ell(u,v;p)$ and $c(u,v;p)$, respectively. Therefore, we have

Theorem 4.9.7. (cf. Theorem 2.8 of [27] and [28]) *Let* $\alpha_k = \{G(u,v)x\}_k$ *and* $0 < p_k \leq H < \infty$ *for all* $k\in\mathbb{N}$. *Define the sequence* $b^{(k)} = \{b_n^{(k)}\}_{n\in\mathbb{N}}$ *for every fixed* $k\in\mathbb{N}$ *by*

$$b_n^{(k)} := \begin{cases} \frac{(-1)^{n-k}}{u_k v_n} &, \quad k\leq n\leq k+1, \\ 0 &, \quad 0\leq n<k \text{ or } n>k+1. \end{cases}$$

Then, the following statements hold:

(a) *The sequence* $\{b^{(k)}\}_{k\in\mathbb{N}}$ *is a basis for the spaces* $c_0(u,v;p)$ *and* $\ell(u,v;p)$. *Any* x *in* $c_0(u,v;p)$ *or* $\ell(u,v;p)$ *has a unique representation of the form* $x := \sum_k \alpha_k b^{(k)}$.

(b) *The set* $\left\{t, b^{(k)}\right\}$ *is a basis for the space* $c(u,v;p)$ *and any* $x\in c(u,v;p)$ *has a unique representation of the form* $x := lt + \sum_k[\alpha_k - lt_k]b^{(k)}$, *where* $t=(t_k)$ *with* $t_k := v_k^{-1}\Delta u_k^{-1}$ *for all* $k\in\mathbb{N}$ *and* $l := \lim_{k\to\infty}\{G(u,v)x\}_k$.

4.9.1 Some Matrix Transformations Related to the Sequence Spaces $\lambda(u,v;p)$

In this subsection, subsequent to defining the pair of summability methods such that one of them applied to the sequences in the space $\lambda(u,v;p)$ and the other one applied to the sequences in the space $\lambda(p)$, we give a basic theorem as a consequence of an analysis related to the pair of this type summability methods. Besides, we characterize the classes $(\lambda(u,v;p):\ell_\infty)$ and $(\lambda(u,v;p):c)$ of infinite matrices and derive the characterization of some other classes from them by using a suitable relation. We write throughout for brevity that

$$\widetilde{a}_{nk} := \frac{1}{u_k}\left(\frac{a_{nk}}{v_k} - \frac{a_{n,k+1}}{v_{k+1}}\right) \text{ and } b_{nk} := \sum_{j=0}^{n} u_n v_j a_{jk} \text{ for all } k,n\in\mathbb{N}. \tag{4.9.9}$$

We will also employ the similar notation with other letters and use the convention that any term with negative subscript is equal to zero.

Let us suppose that the infinite matrices $E = (e_{nk})$ and $F = (f_{nk})$ map the sequences $x = (x_k)$ and $y = (y_k)$, which are connected with the relation (4.9.2) to the sequences $r = (r_n)$ and $s = (s_n)$, respectively, i.e.,

$$r_n := (Ex)_n = \sum_k e_{nk}x_k \text{ for all } n \in \mathbb{N}, \tag{4.9.10}$$

$$s_n := (Fy)_n = \sum_k f_{nk}y_k \text{ for all } n \in \mathbb{N}. \tag{4.9.11}$$

One can immediately deduce here that the method F is applied to the $G(u, v)$-transform of the sequence $x = (x_k)$, while the method E is directly applied to the terms of the sequence $x = (x_k)$. So, the methods E and F are essentially different.

Let us assume that the matrix product $FG(u, v)$ exists, which is a much weaker assumption than the conditions on the matrix F belonging to any matrix class, in general. We shall say in this situation that the methods E and F in (4.9.10) and (4.9.11) are the *pair of summability methods* (PSM), if r_n becomes s_n (or s_n becomes r_n) under the application of the formal summation by parts. This leads us to the fact that $FG(u, v)$ exists and is equal to E and $[FG(u, v)]x = F[G(u, v)x]$ formally holds, if one side exists. This statement is equivalent to the relation

$$e_{nk} := \sum_{j=k}^{\infty} u_j v_k f_{nj} \text{ or } f_{nk} := \frac{1}{u_k}\left(\frac{e_{nk}}{v_k} - \frac{e_{n,k+1}}{v_{k+1}}\right) \text{ for all } k, n \in \mathbb{N}. \tag{4.9.12}$$

Now, we may give a short analysis on the PSM. One can see that s_n is reduced to r_n, as follows:

$$s_n = \sum_k f_{nk}y_k = \sum_k f_{nk}\left(\sum_{j=0}^{k} u_k v_j x_j\right) = \sum_j \left(\sum_{k=j}^{\infty} u_k v_j f_{nk}\right) x_j = r_n.$$

But the order of summation may not be reversed and thus the methods E and F are not necessarily equivalent.

The partial sums of the series on the right-hand side of (4.9.10) and (4.9.11) are connected with the relation

$$\sum_{k=0}^{m} e_{nk}x_k = \sum_{k=0}^{m-1}\left(\frac{e_{nk}}{v_k} - \frac{e_{n,k+1}}{v_{k+1}}\right)\frac{y_k}{u_k} + \frac{e_{nm}}{u_m v_m}y_m \text{ for all } m, n \in \mathbb{N}. \tag{4.9.13}$$

Hence if, for a given $n \in \mathbb{N}$, one of the series on the right-hand side of (4.9.10) and (4.9.11) converges, then the other converges if and only if

$$\lim_{m\to\infty} \frac{e_{nm}}{u_m v_m} y_m = z_n \tag{4.9.14}$$

for every fixed $n \in \mathbb{N}$. If (4.9.14) holds, then we have from (4.9.13) by letting $m \to \infty$ that $r_n = s_n + z_n$ for all $n \in \mathbb{N}$. Hence, if (y_n) is summable by one of the methods E and F, then it is summable by the other one if and only if (4.9.14) holds, and

$$\lim_{n\to\infty} z_n = \alpha. \tag{4.9.15}$$

Hence, the limits of (r_n) and (s_n) differ by α. Therefore, the E- and F-limits of any sequence summable by one of them agree if and only if F summability implies that (4.9.15) holds with $\alpha = 0$. The similar result holds when E and F are interchanged. It follows by the validity of (4.9.15) with $\alpha \neq 0$ that the methods E and F are inconsistent and conversely.

The analysis given above leads us to the following basic theorem related to the matrix mappings on the sequence space $\lambda(u, v; p)$:

Theorem 4.9.8. [27, Theorem 3.1] *Let the matrices $E = (e_{nk})$ and $F = (f_{nk})$ are connected with the relation (4.9.12) and μ be any given sequence space. Then, $E \in (\lambda(u, v; p) : \mu)$ if and only if $F \in (\lambda(p) : \mu)$, and*

$$F^{(n)} \in (\lambda(p) : c) \tag{4.9.16}$$

for every fixed $n \in \mathbb{N}$, where $F^{(n)} = (f_{mk}^{(n)})$ is defined by

$$f_{mk}^{(n)} := \begin{cases} \frac{1}{u_k}\left(\frac{e_{nk}}{v_k} - \frac{e_{n,k+1}}{v_{k+1}}\right) & , \quad k < m, \\ \frac{e_{nm}}{u_m v_m} & , \quad k = m, \\ 0 & , \quad k > m, \end{cases} \quad \text{for all } k, m \in \mathbb{N}.$$

Proof. Suppose that E and F be PSM, that is to say that (4.9.12) holds, μ be any given sequence space and take into account that the spaces $\lambda(u, v; p)$ and $\lambda(p)$ are norm isomorphic.

Let $E \in (\lambda(u, v; p) : \mu)$ and take any $y \in \lambda(p)$. Then, $FG(u, v)$ exists and $E_n = (e_{nk})_{k\in\mathbb{N}} \in \lambda^{\beta}(u, v; p)$, which yields that (4.9.16) is necessary and $F_n = (f_{nk})_{k\in\mathbb{N}} \in \lambda^{\beta}(p)$ for each $n \in \mathbb{N}$. Hence, Fy exists for each $y \in \lambda(p)$ and thus by letting $m \to \infty$ in the equality (4.9.13), we have by (4.9.12) that $Ex = Fy$, which leads us to the consequence $F \in (\lambda(p) : \mu)$.

Conversely, let $F \in (\lambda(p) : \mu)$ and (4.9.16) holds, and take any $x = (x_k) \in \lambda(u, v; p)$. Then, we have $F_n = (f_{nk})_{k\in\mathbb{N}} \in \lambda^{\beta}(p)$, which gives together with (4.9.16) that $E_n = (e_{nk})_{k\in\mathbb{N}} \in \lambda^{\beta}(u, v; p)$ for each $n \in \mathbb{N}$. Hence, Ex exists. Therefore, we obtain from the equality

$$\sum_{k=0}^{m} f_{nk} y_k = \sum_{k=0}^{m}\sum_{j=k}^{m} u_j v_k f_{nj} x_k \quad \text{for all} \quad m, n \in \mathbb{N},$$

by letting $m \to \infty$ that $Fy = Ex$ and this shows that $E \in (\lambda(u, v; p) : \mu)$.

This completes the proof. □

By changing the roles of the spaces $\lambda(u, v; p)$ and $\lambda(p)$ with μ, we have

Theorem 4.9.9. [27, Theorem 3.2] *Suppose that the elements of the infinite matrices $A = (a_{nk})$ and $B = (b_{nk})$ are connected with the relation (4.9.9) and μ be any given sequence space. Then, $A \in (\mu : \lambda(u, v; p))$ if and only if $B \in (\mu : \lambda(p))$.*

Proof. Let $z = (z_k) \in \mu$ and consider the following equality with (4.9.9):

$$\sum_{k=0}^{m} b_{nk} z_k = \sum_{j=0}^{n}\sum_{k=0}^{m} u_n v_j a_{jk} z_k \quad \text{for all} \quad m, n \in \mathbb{N},$$

which yields by letting $m \to \infty$ that $(Bz)_n = \{G(u,v)(Az)\}_n$. Therefore, one can immediately observe from here that $Az \in \lambda(u,v;p)$ whenever $z \in \mu$ if and only if $Bz \in \lambda(p)$ whenever $z \in \mu$.

This step concludes the proof. □

It is clear that Theorems 4.9.8 and 4.9.9 have several consequences depending on the choice of the sequence spaces λ, μ and the sequences $u = (u_n)$, $v = (v_k)$ and $p = (p_k)$. Therefore, by Theorems 4.9.8 and 4.9.9, the necessary and sufficient conditions for $(\lambda(u,v;p) : \mu)$ and $(\mu : \lambda(u,v;p))$ may be derived by replacing the entries of E and A by those of the entries of $F = EG^{-1}(u,v)$ and $B = G(u,v)A$, respectively; where the necessary and sufficient conditions on the matrices F and B are read from the concerning results in the existing literature.

Now, we can quote our theorems on the characterization of some matrix classes concerning with the sequence space $\lambda(u,v;p)$. The necessary and sufficient conditions characterizing the matrix mappings between the sequence spaces $\ell_\infty(p)$, $c(p)$ and $c_0(p)$ of Maddox are determined by Grosse-Erdmann [193]. Let N and K be the finite subsets of $\mathbb{N}$, and $L, M \in \mathbb{N}_1$. Define the sets K_1 and K_2 by $K_1 = \{k \in \mathbb{N} : p_k \leq 1\}$ and $K_2 = \{k \in \mathbb{N} : p_k > 1\}$. Prior to giving the theorems, let us suppose that (q_n) is a non-decreasing bounded sequence of positive real numbers and consider the following conditions:

$$\sup_N \sup_{k\in K_1} \left|\sum_{n\in N} a_{nk}\right|^{p_k} < \infty \tag{4.9.17}$$

$$\text{There exists } M \text{ such that } \sup_N \sum_{k\in K_2} \left|\sum_{n\in N} a_{nk}M^{-1}\right|^{p'_k} < \infty \tag{4.9.18}$$

$$\text{There exists } M \text{ such that } \sup_K \sum_n \left|\sum_{k\in K} a_{nk}M^{-1/p_k}\right|^{q_n} < \infty, \quad (q_n \geq 1) \tag{4.9.19}$$

$$\sum_n \left|\sum_k a_{nk}\right|^{q_n} < \infty, \quad (q_n \geq 1) \tag{4.9.20}$$

$$\sup_K \sum_n \left|\sum_{k\in K} a_{nk}M^{1/p_k}\right|^{q_n} < \infty \text{ for all } M, \quad (q_n \geq 1) \tag{4.9.21}$$

$$\lim_{n\to\infty} |a_{nk}|^{q_n} = 0 \text{ for all } k \in \mathbb{N} \tag{4.9.22}$$

$$\sup_{n\in\mathbb{N}} \sup_{k\in K_1} \left|a_{nk}L^{1/q_n}\right|^{p_k} < \infty \text{ for all } L \tag{4.9.23}$$

$$\text{There exists } M \text{ such that } \sup_{n\in\mathbb{N}} \sum_{k\in K_2} \left|a_{nk}L^{1/q_n}M^{-1}\right|^{p'_k} < \infty \text{ for all } L \tag{4.9.24}$$

$$\text{There exists } M \text{ such that } \sup_{n\in\mathbb{N}} L^{1/q_n} \sum_k |a_{nk}|\, M^{-1/p_k} < \infty \text{ for all } L \tag{4.9.25}$$

$$\lim_{n\to\infty}\left|\sum_k a_{nk}\right|^{q_n}=0 \tag{4.9.26}$$

$$\lim_{n\to\infty}\left(\sum_k |a_{nk}|\,M^{1/p_k}\right)^{q_n}=0 \text{ for all } M \tag{4.9.27}$$

$$\sup_{n\in\mathbb{N}}\sup_{k\in K_1}|a_{nk}|^{p_k}<\infty \tag{4.9.28}$$

$$\text{There exists } M \text{ such that } \sup_{n\in\mathbb{N}}\sum_{k\in K_2}\left|a_{nk}M^{-1}\right|^{p'_k}<\infty \tag{4.9.29}$$

$$\text{There exists } (\alpha_k)\in\omega \text{ such that } \lim_{n\to\infty}|a_{nk}-\alpha_k|^{q_n}=0 \text{ for all } k\in\mathbb{N} \tag{4.9.30}$$

$$\text{There exists } (\alpha_k)\in\omega \text{ such that } \sup_{n\in\mathbb{N}}\sup_{k\in K_1}\left(|a_{nk}-\alpha_k|\,L^{1/q_n}\right)^{p_k}<\infty \tag{4.9.31}$$

for all L and for all $k\in\mathbb{N}$

There exist M and $(\alpha_k)\in\omega$ such that (4.9.32)

$$\sup_{n\in\mathbb{N}}\sum_{k\in K_2}\left(|a_{nk}-\alpha_k|\,L^{1/q_n}M^{-1}\right)^{p'_k}<\infty \text{ for all } L$$

There exist M and $(\alpha_k)\in\omega$ such that (4.9.33)

$$\sup_{n\in\mathbb{N}}\sum_k |a_{nk}-\alpha_k|\,L^{1/q_n}M^{-1/p_k}<\infty \text{ for all } L$$

$$\text{There exists } M \text{ such that } \sup_{n\in\mathbb{N}}\sum_k |a_{nk}|\,M^{-1/p_k}<\infty \tag{4.9.34}$$

$$\text{There exists } \alpha\in\mathbb{C} \text{ such that } \lim_{n\to\infty}\left|\sum_k a_{nk}-\alpha\right|^{q_n}=0 \tag{4.9.35}$$

$$\text{There exists } (\alpha_k)\in\omega \text{ such that } \lim_{n\to\infty}\left(\sum_k |a_{nk}-\alpha_k|\,M^{1/p_k}\right)^{q_n}=0 \tag{4.9.36}$$

for all M

$$\sup_{n\in\mathbb{N}}\sum_k |a_{nk}|\,M^{1/p_k}<\infty \text{ for all } M \tag{4.9.37}$$

$$\text{There exists } L \text{ such that } \sup_{n\in\mathbb{N}}\sup_{k\in K_1}\left|a_{nk}L^{-1/q_n}\right|^{p_k}<\infty \tag{4.9.38}$$

$$\text{There exists } L \text{ such that } \sup_{n\in\mathbb{N}}\sum_{k\in K_2}\left|a_{nk}L^{-1/q_n}\right|^{p'_k}<\infty \tag{4.9.39}$$

$$\text{There exists } M \text{ such that } \sup_K \sum_n \left| \sum_{k\in K} a_{nk} M^{-1/p_k} \right|^{q_n} < \infty, \quad (q_n \geq 1) \tag{4.9.40}$$

$$\sum_n \left| \sum_k a_{nk} \right|^{q_n} < \infty, \quad (q_n \geq 1) \tag{4.9.41}$$

$$\sup_K \sum_n \left| \sum_{k\in K} a_{nk} M^{1/p_k} \right|^{q_n} < \infty \text{ for all } M, \quad (q_n \geq 1) \tag{4.9.42}$$

$$\text{There exists } L \text{ such that } \sup_{n\in\mathbb{N}} \sup_{k\in K_2} \left| a_{nk} L^{-1/q_n} \right|^{p'_k} < \infty \tag{4.9.43}$$

$$\text{There exists } M \text{ such that } \sup_{n\in\mathbb{N}} \left(\sup_{k\in\mathbb{N}_1} |a_{nk}| M^{-1/p_k} \right)^{q_n} < \infty \tag{4.9.44}$$

$$\sup_{n\in\mathbb{N}} \left| \sum_k a_{nk} \right|^{q_n} < \infty \tag{4.9.45}$$

$$\sup_{n\in\mathbb{N}} \left(\sum_k |a_{nk}| M^{1/p_k} \right)^{q_n} < \infty \text{ for all } M \tag{4.9.46}$$

Theorem 4.9.10. [28, Theorem 3.3] *Let $A = (a_{nk})$ be an infinite matrix over the complex field. Then, the following statements hold:*

(i) $A \in (\ell(u, v; p) : \ell_1)$ *if and only if (4.9.17) and (4.9.18) hold with $\widetilde{a}_{nk}$ instead of a_{nk} and (4.9.16) also holds.*

(ii) $A \in (\ell(u, v; p) : c_0(q))$ *if and only if (4.9.22)–(4.9.24) hold with $\widetilde{a}_{nk}$ instead of a_{nk} and (4.9.16) also holds.*

(iii) $A \in (\ell(u, v; p) : c(q))$ *if and only if (4.9.28)–(4.9.32) hold with $\widetilde{a}_{nk}$ instead of a_{nk} and (4.9.16) also holds.*

(iv) $A \in (\ell(u, v; p) : \ell_\infty(q))$ *if and only if (4.9.38) and (4.9.39) hold with $\widetilde{a}_{nk}$ instead of a_{nk} and (4.9.16) also holds.*

Theorem 4.9.11. [28, Theorem 3.4] *Let $A = (a_{nk})$ be an infinite matrix over the complex field. Then, the following statements hold:*

(i) $A \in (\ell(p) : \ell_1(u, v))$ *if and only if (4.9.17) and (4.9.18) hold with b_{nk} instead of a_{nk}.*

(ii) $A \in (\ell(p) : c_0(u, v; q))$ *if and only if (4.9.22)–(4.9.24) hold with b_{nk} instead of a_{nk}.*

(iii) $A \in (\ell(p) : c(u, v; q))$ *if and only if (4.9.28)–(4.9.32) hold with b_{nk} instead of a_{nk}.*

(iv) $A \in (\ell(p) : \ell_\infty(u, v; q))$ *if and only if (4.9.38) and (4.9.39) hold with b_{nk} instead of a_{nk}.*

(v) $A \in (c_0(p) : \ell(u, v; q))$ *if and only if (4.9.40) holds with b_{nk} instead of a_{nk}.*

(vi) $A \in (c(p) : \ell(u, v; q))$ *if and only if (4.9.40) and (4.9.41) hold with* b_{nk} *instead of* a_{nk}.

(vii) $A \in (\ell_\infty(p) : \ell(u, v; q))$ *if and only if (4.9.42) holds with* b_{nk} *instead of* a_{nk}.

Theorem 4.9.12. [27, Theorem 3.3] *Let* $A = (a_{nk})$ *be an infinite matrix over the complex field. Then, the following statements hold:*

(i) $A \in (c_0(u, v; p) : \ell(q))$ *if and only if (4.9.19) holds with* $\widetilde{a}_{nk}$ *instead of* a_{nk} *and (4.9.16) also holds with* $\lambda = c_0$.

(ii) $A \in (c_0(u, v; p) : c(q))$ *if and only if (4.9.30), (4.9.33) and (4.9.34) hold with* $\widetilde{a}_{nk}$ *instead of* a_{nk} *and (4.9.16) also holds with* $\lambda = c_0$.

(iii) $A \in (c_0(u, v; p) : \ell_\infty(q))$ *if and only if (4.9.44) holds with* $\widetilde{a}_{nk}$ *instead of* a_{nk} *and (4.9.16) also holds with* $\lambda = c_0$.

Theorem 4.9.13. [27, Theorem 3.4] *Let* $A = (a_{nk})$ *be an infinite matrix over the complex field. Then, the following statements hold:*

(i) $A \in (c(u, v; p) : \ell(q))$ *if and only if (4.9.19) and (4.9.20) hold with* $\widetilde{a}_{nk}$ *instead of* a_{nk} *and (4.9.16) also holds with* $\lambda = c$.

(ii) $A \in (c(u, v; p) : c(q))$ *if and only if (4.9.30), (4.9.33)–(4.9.35) hold with* $\widetilde{a}_{nk}$ *instead of* a_{nk} *and (4.9.16) also holds with* $\lambda = c$.

(iii) $A \in (c(u, v; p) : \ell_\infty(q))$ *if and only if (4.9.44) and (4.9.45) hold with* $\widetilde{a}_{nk}$ *instead of* a_{nk} *and (4.9.16) also holds with* $\lambda = c$.

Theorem 4.9.14. [27, Theorem 3.5] *Let* $A = (a_{nk})$ *be an infinite matrix over the complex field. Then, the following statements hold:*

(i) $A \in (\ell_\infty(u, v; p) : \ell(q))$ *if and only if (4.9.22) holds with* $\widetilde{a}_{nk}$ *instead of* a_{nk} *and (4.9.16) also holds with* $\lambda = \ell_\infty$.

(ii) $A \in (\ell_\infty(u, v; p) : c_0(q))$ *if and only if (4.9.27) holds with* $\widetilde{a}_{nk}$ *instead of* a_{nk} *and (4.9.16) also holds with* $\lambda = \ell_\infty$.

(iii) $A \in (\ell_\infty(u, v; p) : c(q))$ *if and only if (4.9.36) and (4.9.37) hold with* $\widetilde{a}_{nk}$ *instead of* a_{nk} *and (4.9.16) also holds with* $\lambda = \ell_\infty$.

(iv) $A \in (\ell_\infty(u, v; p) : \ell_\infty(q))$ *if and only if (4.9.46) holds with* $\widetilde{a}_{nk}$ *instead of* a_{nk} *and (4.9.16) also holds with* $\lambda = \ell_\infty$.

Theorem 4.9.15. [27, Theorem 3.6] *Let* $A = (a_{nk})$ *be an infinite matrix over the complex field. Then, the following statements hold:*

(i) $A \in (c_0(p) : c(u, v; q))$ *if and only if (4.9.30), (4.9.33) and (4.9.34) hold with* b_{nk} *instead of* a_{nk}.

(ii) $A \in (c_0(p) : \ell_\infty(u, v; q))$ *if and only if (4.9.44) holds with* b_{nk} *instead of* a_{nk}.

Theorem 4.9.16. [27, Theorem 3.7] *Let $A = (a_{nk})$ be an infinite matrix over the complex field. Then, the following statements hold:*

(i) $A \in (c(p) : c(u, v; q))$ *if and only if (4.9.30), (4.9.33)–(4.9.35) hold with* b_{nk} *instead of* a_{nk}.

(ii) $A \in (c(p) : \ell_\infty(u, v; q))$ *if and only if (4.9.44) and (4.9.45) hold with* b_{nk} *instead of* a_{nk}.

Theorem 4.9.17. [27, Theorem 3.8] *Let $A = (a_{nk})$ be an infinite matrix over the complex field. Then, the following statements hold:*

(i) $A \in (\ell_\infty(p) : c_0(u, v; q))$ *if and only if (4.9.27) holds with* b_{nk} *instead of* a_{nk}.

(ii) $A \in (\ell_\infty(p) : c(u, v; q))$ *if and only if (4.9.36) and (4.9.37) hold with* b_{nk} *instead of* a_{nk}.

(iii) $A \in (\ell_\infty(p) : \ell_\infty(u, v; q))$ *if and only if (4.9.46) holds with* b_{nk} *instead of* a_{nk}.

Theorem 4.9.18. [27, Theorem 3.9] *Let $A = (a_{nk})$ be an infinite matrix over the complex field. Then, the following statements hold:*

(i) $A \in (\ell(p) : c_0(u, v; q))$ *if and only if (4.9.22)–(4.9.24) hold with* b_{nk} *instead of* a_{nk}.

(ii) $A \in (\ell(p) : c(u, v; q))$ *if and only if (4.9.28)–(4.9.32) hold with* b_{nk} *instead of* a_{nk}.

(iii) $A \in (\ell(p) : \ell_\infty(u, v; q))$ *if and only if (4.9.38) and (4.9.43) hold with* b_{nk} *instead of* a_{nk}.

4.10 DOMAINS OF TRIANGLES IN THE SPACES OF STRONGLY C_1–SUMMABLE ...

Following Başar et al. [68], we deal with the matrix domains $w_0^p(U) = (w_0^p)_U$, $w^p(U) = (w^p)_U$ and $w_\infty^p(U) = (w_\infty^p)_U$ $(1 \leq p < \infty)$ of arbitrary triangles U in the spaces w_0^p, w^p and w_∞^p, determine their β-duals, and characterize matrix transformations on them into the spaces c_0, c and ℓ_∞.

Maddox [252] introduced and studied the following sets of sequences that are strongly summable and bounded with index p $(1 \leq p < \infty)$ by the Cesàro method of order 1

$$
\begin{aligned}
w_0^p &:= \left\{ x = (x_k) \in \omega : \lim_{n\to\infty} \frac{1}{n} \sum_{k=1}^{n} |x_k|^p = 0 \right\}, \\
w^p &:= \left\{ x = (x_k) \in \omega : \lim_{n\to\infty} \frac{1}{n} \sum_{k=1}^{n} |x_k - \xi|^p = 0 \text{ for some } \xi \in \mathbb{C} \right\}, \\
w_\infty^p &:= \left\{ x = (x_k) \in \omega : \sup_{n\in\mathbb{N}_1} \frac{1}{n} \sum_{k=1}^{n} |x_k|^p < \infty \right\}.
\end{aligned}
$$

Throughout we use the convention that every term with a subscript less than one is equal to zero. We write $\sum_v = \sum_{k=2^\nu}^{2^{\nu+1}-1}$ and $\max_v = \max_{2^\nu \le k \le 2^{\nu+1}-1}$ for $\nu \in \mathbb{N}$.

The following result is known.

Proposition 4.10.1. [269, Proposition 3.44, p. 207] *Let $1 \le p < \infty$. Then, the sets w_0^p, w^p and w_∞^p are BK spaces with the (equivalent) norms*

$$\|x\|_{w_\infty^p} = \sup_{\nu \in \mathbb{N}_1} \left(\frac{1}{2^\nu} \sum_v |x_k|^p \right)^{1/p} \quad \text{and} \quad \|x\|_{w_\infty^p}^{\dagger} = \sup_{n \in \mathbb{N}_1} \left(\frac{1}{n} \sum_{k=1}^{n} |x_k|^p \right)^{1/p};$$

w_0^p is a closed subspace of w^p and w^p is a closed subspace of w_∞^p; w_0^p has AK; every sequence $x = (x_k)_{k=1}^\infty \in w^p$ has a unique representation $x = \xi e + \sum_{k=1}^\infty (x_k - \xi) e^{(k)}$, where $\xi \in \mathbb{C}$ is the strong C_1–limit *of the sequence λ, i.e.,*

$$\lim_{n\to\infty} \frac{1}{n} \sum_{k=1}^{n} |x_k - \xi|^p = 0. \tag{4.10.1}$$

Corollary 4.10.2. [68, Corollary 1.2] *The following statements hold:*

(a) *The sets $w_0^p(U)$, $w^p(U)$ and $w_\infty^p(U)$ are BK spaces with*

$$\|x\|_{w_\infty^p(U)} = \sup_\nu \left(\frac{1}{2^\nu} \Sigma_\nu |(Ux)_k|^p \right)^{1/p};$$

$w_0^p(U)$ is a closed subspace of $w^p(U)$, and $w^p(U)$ is a closed subspace of $w_\infty^p(U)$.

(b) *We put $c^{(n)} = \{c_k^{(n)}\}_{k=1}^\infty := U^{-1} e^{(n)} = V e^{(n)}$ for $n = 1, 2, \ldots$, that is*

$$c_k^{(n)} := \begin{cases} 0 & , \quad 1 \le k \le n-1, \\ v_{kn} & , \quad k \ge n. \end{cases}$$

Every sequence $z = (z_n)_{n=1}^\infty \in w_0^p(U)$ has a unique representation

$$z := \sum_{n=1}^{\infty} (Uz)_n \, c^{(n)}. \tag{4.10.2}$$

(c) *We put the sequence $c^{(0)} = \{c_k^{(0)}\}_{k=1}^\infty$ defined by $c_k^{(0)} = \sum_{j=1}^{k} v_{kj}$ for all $k \in \mathbb{N}_1$. Every sequence $z = (z_n)_{n=1}^\infty \in w^p(U)$ has a unique representation*

$$w := \xi c^{(0)} + \sum_{n=1}^{\infty} [(Uz)_n - \xi] c^{(n)}, \tag{4.10.3}$$

where $\xi \in \mathbb{C}$ is the strong limit of z in $w^p(U)$, *i.e.,*

$$\lim_{n\to\infty} \left[\frac{1}{n} \sum_{k=1}^{n} |(Uz)_k - \xi|^p \right] = 0. \tag{4.10.4}$$

Proof. (a) This part is an immediate consequence of Proposition 4.10.1 and [395, Theorems 4.3.12 and 4.3.14, pp. 63 and 64].

(b) For every fixed n, we have

$$\left(Ve^{(n)}\right)_k = \sum_{j=1}^{k} v_{kj}e^{(n)} = \begin{cases} 0 & , \quad 1 \leq k \leq n-1, \\ v_{kn} & , \quad k \geq n, \end{cases}$$

hence $c^{(n)} = Ve^{(n)}$. Since w_0^p has AK by Proposition 4.10.1, we have $e^{(n)} \in w_0^p$ for all $n \in \mathbb{N}_1$, and it follows from $Uc^{(n)} = U(Ve^{(n)}) = (UV)e^{(n)} = e^{(n)} \in w_0^p$ that $c^{(n)} \in w_0^p(U)$ for all $n \in \mathbb{N}_1$. Now, let $z = (z_n)_{n=1}^{\infty} \in w_0^p(U)$ be given, that is $x = Uz \in w_0^p$. We obtain for $x^{[m]} = \sum_{k=1}^{m} x_k e^{(k)}$

$$\begin{aligned}
\lim_{m\to\infty} \left\| x - x^{[m]} \right\|_{w_\infty^p} &= \lim_{m\to\infty} \left\| x - \sum_{[n=1}^{m} x_n e^{(n)} \right\|_{w_\infty^p} \\
&= \lim_{m\to\infty} \left\| x - \sum_{n=1}^{m} (Uz)_n \, e^{(n)} \right\|_{w_\infty^p} = 0.
\end{aligned}$$

We put $z^{<m>} = \sum_{n=1}^{m} (Uz)_n \, c^{(n)}$ for all m. Then, we have

$$z^{<m>} \in w_0^p(U), \quad Uz^{<m>} = \sum_{n=1}^{m} (Uz)_n \, Uc^{(n)} = \sum_{n=1}^{m} x_n e^{(n)} = x^{[m]}$$

and so by Part (a)

$$\begin{aligned}
\lim_{m\to\infty} \|z - z^{<m>}\|_{w_\infty^p(U)} &= \lim_{m\to\infty} \|U(z - z^{<m>})\|_{w_\infty^p} \\
&= \lim_{m\to\infty} \|Uz - Uz^{<m>}\|_{w_\infty^p} = \lim_{m\to\infty} \left\| x - x^{[m]} \right\|_{w_\infty^p} = 0,
\end{aligned}$$

that is the representation in (4.10.2) holds, which is obviously unique.

(c) We have $(Ve)_k = \sum_{j=1}^{k} v_{kj} e_j = \sum_{j=1}^{k} v_{kj} = c_k$ for all $k \in \mathbb{N}_1$, hence $c = Ve$ and $Uc = U(Ve) = (UV)e = e \in w^p$ implies $c \in w^p(U)$. Let $z = (z_n)_{n=1}^{\infty} \in w^p(U)$ be given. Then, $x = Uz \in w^p$ and by Proposition 4.10.1 there exists a unique complex number ξ that satisfies (4.10.1). We write $x^{(0)} = x - \xi e$ and put $z^{(0)} = z - \xi c$. Then, we have $x^{(0)} \in w_0^p$, and it follows from $Uz^{(0)} = Uz - \xi Uc = x - \xi e = x^{(0)} \in w_0^p$ that $z^{(0)} \in w_0^p(U)$. So, $z^{(0)}$ has a unique representation $z^{(0)} = \sum_{n=1}^{\infty} (Uz)_n^{(0)} \, c^{(n)} = \sum_{n=1}^{\infty} (U_n - \xi e) c^{(n)}$ by Part (b), and so

$$z = \xi c + z^{(0)} = \xi c + \sum_{n=1}^{\infty} (U_n - \xi) \, c^{(n)}.$$

This establishes the unique representation in (4.10.3). □

Example 4.10.3. [68, Example 1.3] *Let $\mathcal{U}$ be the set of all sequences $u = (u_k)_{k=1}^{\infty}$ with $u_k \neq 0$ for all $k \in \mathbb{N}_1$. If $u, v \in \mathcal{U}$, then we write $u/v = (u_k/v_k)_{k=1}^{\infty}$. Let $u, v \in \mathcal{U}$*

be given and $U = (u_{nk})_{n,k=1}^{\infty}$ *be the factorable matrix with* $u_{nk} = u_n v_k$ *for* $1 \le k \le n$ $(n = 1, 2, \dots)$. *Then, the inverse* $V = (v_{nk})_{n,k=1}^{\infty}$ *of* U *is obviously given by*

$$v_{nk} := \begin{cases} \dfrac{1}{u_n v_n} & , \quad k = n, \\ -\dfrac{1}{u_{n-1} v_n} & , \quad k = n-1, \\ 0 & , \quad \text{otherwise,} \end{cases} \qquad (n = 1, 2, \dots),$$

and so we have

$$c^{(n)} = \frac{1}{u_n} \left[\frac{1}{v_n} e^{(n)} - \frac{1}{v_{n+1}} e^{(n+1)} \right] \text{ for } n = 1, 2, \dots$$

and

$$c^{(0)} = \sum_{j=1}^{k} v_{kj} = \left\{ \frac{1}{v_k} \{\Delta(1/u)\}_k \right\}_{k=1}^{\infty}.$$

So, it follows from (4.10.2) and (4.10.3) in Corollary 4.10.2 that every sequence $z = (z_n)_{n=1}^{\infty} \in w_0^p(U)$ *has a unique representation*

$$z := \sum_{n=1}^{\infty} \left(\sum_{j=1}^{k} v_j z_j \right) \left[\frac{1}{v_n} e^{(n)} - \frac{1}{v_{n+1}} e^{(n+1)} \right],$$

and that every sequence $z = (z_n)_{n=1}^{\infty} \in w^p(U)$ *has a unique representation*

$$z := \xi \left\{ \frac{1}{v_k} \{\Delta(1/u)\}_k \right\}_{k=1}^{\infty} + \sum_{n=1}^{\infty} \left(\sum_{j=1}^{k} v_j z_k - \frac{\xi}{u_n} \right) \left[\frac{1}{v_n} e^{(n)} - \frac{1}{v_{n+1}} e^{(n+1)} \right],$$

with ξ *from (4.10.4).*

4.10.1 Matrix Transformations on w_0^p, w^p and w_∞^p

We will show in Subsection 4.10.2 that the determination of the β-duals of the sets $w_0^p(U)$, $w^p(U)$ and $w_\infty^p(U)$ can be reduced to that of the β-duals of the sets w_0^p, w^p and w_∞^p, and the characterizations of the classes $(w_0^p : c_0)$, $(w^p : c)$ and $(w_\infty^p : c_0)$. The β-duals of the sets w_0^p, w^p and w_∞^p are known. In this section, we characterize the classes $(\lambda : \mu)$, where λ is any of the sets w_0^p, w^p and w_∞^p, and μ is any of the sets c_0, c and ℓ_∞.

Throughout, let $1 \le p < \infty$ and q be the conjugate number of p, that is $q = \infty$ for $p = 1$ and $q = p/(p-1)$ for $1 < p < \infty$. If $p = 1$, then we omit the index p, that is, we write $w_0 = w_0^1$ etc., for short.

We put $\mathcal{M}_p := \{a \in \omega : \|a\|_{\mathcal{M}_p} < \infty\}$, where

$$\|a\|_{\mathcal{M}_p} := \begin{cases} \sum_{\nu=0}^{\infty} 2^\nu \max_\nu |a_k| & , \quad p = 1, \\ \sum_{\nu=0}^{\infty} 2^{\nu/p} \left(\sum_\nu |a_k|^q\right)^{1/q} & , \quad 1 < p < \infty. \end{cases}$$

Given $a \in \omega$, we write

$$\|a\|_X^* = \sup_{x \in S_\lambda} \left| \sum_{k=1}^{\infty} a_k x_k \right|$$

provided the expression on the right-hand side is defined and finite, which is the case whenever λ is a BK space and $a \in \lambda^\beta$ ([395, Theorem 7.2.9, p. 107]).

The next results are known. They give the β-duals of w_0^p, w^p and w_∞^p, the continuous duals of w_0^p and w^p, and the characterization of the class $(\lambda : \ell_\infty)$ for arbitrary BK spaces λ.

Proposition 4.10.4. ([252] and [269, Proposition 3.47, p. 208]) *We have*

(a) $(w_0^p)^\beta = (w^p)^\beta = (w_\infty^p)^\beta = \mathcal{M}_p$;

(b) $(w_0^p, \|\cdot\|)^* \equiv \mathcal{M}_p$, *that is* $(w_0^p)^*$ *and* $\mathcal{M}_p$ *are norm isomorphic;*

(c) $f \in (w^p)^*$ *if and only if there exist* $a_0 \in \mathbb{C}$ *and a sequence* $a = (a_k)_{k=1}^\infty \in \mathcal{M}_p$ *such that*

$$f(x) := \xi a_0 + \sum_{k=1}^{\infty} a_k x_k \text{ for all } x \in w^p \text{ with } \xi \textit{ from (4.10.1);}$$

moreover

$$\|f\| := |a_0| + \|a\|_{\mathcal{M}_p} \text{ for all } f \in (w^p)^*;$$

(d) $\|a\|_{w_\infty^p}^* = \|a\|_{\mathcal{M}_p}$ *for all* $a \in (w_\infty^p)^\beta$.

Proposition 4.10.5. [207, Theorem 1.8] *Let* λ *be a BK space. Then, we have* $A \in (\lambda : \ell_\infty)$ *if and only if* $A_n \in X^\beta$, $(n \in \mathbb{N}_1)$ *and* $\|A\|_{(\lambda:\ell_\infty)} = \sup_{n \in \mathbb{N}_1} \|A_n\|_\lambda^* < \infty$.

We also need the next lemma.

Lemma 4.10.6. [68, Lemma 2.3] *Let* $B = (b_{nk})_{n,k=1}^\infty$ *be an infinite matrix. If* $\|B_n\|_{\mathcal{M}_p} < \infty$ *for all* $n \in \mathbb{N}_1$ *and* $\lim_{n\to\infty} \|B_n\|_{\mathcal{M}_p} = 0$, *then* $\|B_n\|_{\mathcal{M}_p}$ *converges uniformly in* $n \in \mathbb{N}_1$.

Proof. Let $\varepsilon > 0$ be given. Since $\lim_{n\to\infty} \|B_n\|_{\mathcal{M}_p} = 0$, there exists $N \in \mathbb{N}_1$ such that $\|B_n\|_{\mathcal{M}_p} < \varepsilon$ for all $n > N$. For $\rho \in \mathbb{N}_1$ and $\mu \in \mathbb{N}_1 \cup \{\infty\}$, we write

$$\|B_n\|_{\mathcal{M}_p}^{<\rho,\mu>} := \begin{cases} \sum_{\nu=\rho}^{\mu} 2^\nu \max_\nu |b_{nk}| & , \quad p = 1, \\ \sum_{\nu=\rho}^{\mu} 2^{\nu/p} \left(\sum_\nu |b_{nk}|^q\right)^{1/q} & , \quad 1 < p < \infty. \end{cases}$$

Since $\|B_n\|_{\mathcal{M}_p} < \infty$ for all $n \in \mathbb{N}_1$, for each n with $1 \le n \le N$, there exists $\nu(n) \in \mathbb{N}_0$ such that $\|B_n\|_{\mathcal{M}_p}^{<\nu(n),\infty>} < \varepsilon$. We choose $\rho = \max_{1 \le n \le N} \nu(n)$. Then, we have

$$\|B_n\|_{\mathcal{M}_p}^{<\nu,\infty>} \le \|B_n\|_{\mathcal{M}_p}^{<\rho,\infty>} < \varepsilon \quad \text{for all} \quad \nu \ge \rho \quad \text{and for all} \quad n \in \mathbb{N}_1.$$

□

Now, we characterize the classes $(\lambda : \mu)$ for $\lambda \in \{w_0^p, w^p, w_\infty^p\}$ and $\mu \in \{\ell_\infty, c, c_0\}$.

Theorem 4.10.7. [68, Theorem 2.4] *The necessary and sufficient conditions for $A \in (\lambda : \mu)$ when $\lambda \in \{w_0^p, w^p, w_\infty^p\}$ and $\mu \in \{\ell_\infty, c, c_0\}$ can be read from the following table:*

To \ From	w_∞^p	w_0^p	w^p
ℓ_∞	**1.**	**1.**	**1.**
c_0	**2.**	**3.**	**4.**
c	**5.**	**6.**	**7.**

where

1. *(1.1)* $\sup_{n\in\mathbb{N}_1} \|A_n\|_{\mathcal{M}_p} < \infty$

2. *(2.1)* $\lim_{n\to\infty} \|A_n\|_{\mathcal{M}_p} = 0$

3. *(1.1) and (3.1), where*
(3.1) $\lim_{n\to\infty} a_{nk} = 0$ for all $k \in \mathbb{N}_1$

4. *(1.1), (3.1) and (4.1), where*
(4.1) $\lim_{n\to\infty} \sum_{k=1}^{\infty} a_{nk} = 0$

5. *(5.1), (5.2) and (5.3), where*
(5.1) $\alpha_k = \lim_{n\to\infty} a_{nk}$ exists for all $k \in \mathbb{N}_1$
(5.2) $(\alpha_k)_{k=1}^{\infty}, A_n \in \mathcal{M}_p$ for all $n \in \mathbb{N}_1$
(5.3) $\lim_{n\to\infty} \|A_n - (\alpha_k)_{k=1}^{\infty}\|_{\mathcal{M}_p} = 0$

6. *(1.1) and (5.1)*

7. *(1.1), (5.1) and (7.1), where*
(7.1) $\alpha = \lim_{n\to\infty} \sum_{k=1}^{\infty} a_{nk}$ exists.

We remark that the conditions for $A \in (w_\infty^p : c_0)$ and $A \in (w_\infty^p : c)$ can be replaced by the conditions

2′. *(2.1′) and (3.1), where*
(2.1′) $\|A_n\|_{\mathcal{M}_p}$ converges uniformly in $n \in \mathbb{N}_1$

5′. *(2.1′) and (5.1).*

Proof. **1.** Condition (1.1) for $A \in (w_0^p : \ell_\infty)$ follows from Propositions 4.10.5 and 4.10.4 (b) and (d). Then, (1.1) for $A \in (w^p : \ell_\infty)$ and $A \in (w_\infty^p : \ell_\infty)$ follows from the fact that $(w_\infty^p : \ell_\infty) \supset (w^p : \ell_\infty) \supset (w_0^p : \ell_\infty)$ by Proposition 4.10.1.

3. and **6.** Since w_0^p is a BK space with AK by Proposition 4.10.1, and c_0 and c are closed subspaces of ℓ_∞, the conditions follow from the characterization of $(w_0^p : \ell_\infty)$ and [396, 8.3.6, p. 123].

4. and **7.** The conditions follow from those in **3.** and **6.** and [396, 8.3.7, p. 123].

5. and **5′.** First we show that the conditions in **5.** imply those in **5′.**. We assume that the conditions in **5.** are satisfied, and define the matrix $B = (b_{nk})_{n,k=1}^{\infty}$ by

$b_{nk} = a_{nk} - \alpha_k$ for all $n, k \in \mathbb{N}_1$. It follows from (5.2) and (5.3) that $B_n \in \mathcal{M}_p$ for all $n \in \mathbb{N}_1$ and $\lim_{n\to\infty} \|B_n\|_{\mathcal{M}_p} = 0$. Hence, $\|B_n\|_{\mathcal{M}_p}$ is uniformly convergent in $n \in \mathbb{N}_1$ by Lemma 4.10.6, and so $\|A_n\|_{\mathcal{M}_p} = \|B_n + (\alpha_k)_{k=1}^\infty\|_{\mathcal{M}_p}$ is uniformly convergent in n. This shows that the conditions in **5.** imply those in **5′.**.

Now, we show the sufficiency of the conditions in **5′.** for $A \in (w_\infty^p : c)$. We assume that (2.1′) and (5.1) are satisfied. It follows from (2.1′) and (5.1) that there exists $\rho \in \mathbb{N}$ such that $\|A_n\|_{\mathcal{M}_p}^{<\rho+1,\infty>} < 1$ for all $n \in \mathbb{N}_1$, and $(a_{nk})_{n=1}^\infty \in c \subset \ell_\infty$ for every $k \in \mathbb{N}_1$, hence for every $k \in \mathbb{N}_1$, there exists a constant $M_k > 0$ such that $|a_{nk}| \le M_k$ for all $n \in \mathbb{N}_1$. We put $M = 1 + \|(M_k)_{k=1}^\infty\|_{\mathcal{M}_p}^{<0,\rho>}$. Then, we have $\|A_n\|_{\mathcal{M}_p} = \|A_n\|_{\mathcal{M}_p}^{<0,\rho>} + \|A_n\|_{\mathcal{M}_p}^{<\rho+1,\infty>} \le \|(M_k)_{k=1}^\infty\|_{\mathcal{M}_p}^{<0,\rho>} + 1 = M$ for all $n \in \mathbb{N}_1$, hence, (1.1) holds.

Now we show that (1.1) and (5.1) together imply $(\alpha_k)_{k=1}^\infty \in \mathcal{M}_p$. Let $\mu \in \mathbb{N}_0$ be given. Then, we have

$$\begin{aligned}\|(\alpha_k)_{k=1}^\infty\|_{\mathcal{M}_p}^{<0,\mu>} &\le \|A_n - (\alpha_k)_{k=1}^\infty\|_{\mathcal{M}_p}^{<0,\mu>} + \|A_n\|_{\mathcal{M}_p}^{<0,\mu>} \\ &\le \|A_n - (\alpha_k)_{k=1}^\infty\|_{\mathcal{M}_p}^{<0,\mu>} + \sup_{n\in\mathbb{N}_1} \|A_n\|_{\mathcal{M}_p} \text{ for all } n \in \mathbb{N}_1.\end{aligned}$$

Since the first term on the right-hand side of the inequality converges to zero for $n \to \infty$, we obtain from (5.1) and (1.1)

$$\|(\alpha_k)_{k=1}^\infty\|_{\mathcal{M}_p}^{<0,\mu>} \le \sup_{n\in\mathbb{N}_1} \|A_n\|_{\mathcal{M}_p} < \infty.$$

Since $\mu \in \mathbb{N}_0$ was arbitrary, it follows that $\|(\alpha_k)_{k=1}^\infty\|_{\mathcal{M}_p} < \infty$, hence $(\alpha_k)_{k=1}^\infty \in \mathcal{M}_p$, and so $(\alpha_k)_{k=1}^\infty \in (w_\infty^p)^\beta$ by Proposition 4.10.4 (a).

Furthermore, (1.1) and (2.1′) imply that $(Ax)_n$ is absolutely and uniformly convergent in n for each $x \in w_\infty^p$, since $\sum_{k=1}^\infty |a_{nk}x_k| \le \sup_{n\in\mathbb{N}_1} \|A_n\|_{\mathcal{M}_p} \|x\|_{w_\infty^p}$. This implies

$$\lim_{n\to\infty} (Ax)_n = \sum_{k=1}^\infty \left(\lim_{n\to\infty} a_{nk}\right) x_k = \sum_{k=1}^\infty \alpha_k x_k \text{ for each } x \in w_\infty^p,$$

that is, $Ax \in c$ for all $x \in w_\infty^p$, hence $A \in (\ell_\infty : c)$. Thus, we have proved the sufficiency of conditions (2.1′) and (5.1).

Now, we show the necessity of the conditions in **5.** and **5′.**. We assume $A \in (w_\infty^p : c)$. Since $e^{(k)} \in w_\infty^p$ for every $k \in \mathbb{N}_1$, it follows that $Ae^{(k)} = (a_{nk})_{n=1}^\infty \in c$, hence (5.1) holds. Also, $w^p \subset w_\infty^p$ implies $(w_\infty^p : c) \subset (w^p : c)$, hence (1.1) holds by **1.**. Obviously (1.1) implies $A_n \in \mathcal{M}_p$ for all $n \in \mathbb{N}_1$, and as in the sufficiency part of the proof, (1.1) and (5.1) imply $(\alpha_k)_{k=1}^\infty \in \mathcal{M}_p$, so the conditions in (5.2) hold. Now, $A \in (w_\infty^p : c)$ and $(\alpha_k)_{k=1}^\infty \in \mathcal{M}_p = (w_\infty^p)^\beta$ trivially imply $B \in (w_\infty^p : c)$, where the matrix B is defined as above. We show that this implies

$$\lim_{n\to\infty} \|B_n\|_{\mathcal{M}_p} = 0, \tag{I}$$

that is (5.3). Then, it will follow from Lemma 4.10.6 that $\|B_n\|_{\mathcal{M}_p}$ converges uniformly in n, whence $\|A_n\|_{\mathcal{M}_p} = \|B_n + (\alpha_k)_{k=1}^\infty\|_{\mathcal{M}_p}$ converges uniformly in n, that is (2.1′). Thus, all the conditions in **5.** and **5′.** hold.

To show that (I) is necessary, we assume that it is not satisfied and construct a sequence $x \in w_{\infty}^{p}$ with $Bx \notin c$, which is a contradiction to $B \in (w_{\infty}^{p} : c)$. If $\|B_n\|_{\mathcal{M}_p} \not\to 0$ $(n \to \infty)$, then there exists a real $c > 0$ such that $\limsup_{n\to\infty} \|B_n\|_{\mathcal{M}_p} = c$, hence $\lim_{j\to\infty} \|B_{n_j}\|_{\mathcal{M}_p} = c$ for some subsequence (n_j). We omit the indices j, that is, we assume without loss of generality

$$\lim_{n\to\infty} \|B_n\|_{\mathcal{M}_p} = c. \tag{II}$$

It follows from (5.1) that

$$\lim_{n\to\infty} b_{nk} = 0 \text{ for every } k \in \mathbb{N}_1. \tag{III}$$

By (II) and (III), there exists an integer $n(1)$ such that

$$\Big| \|B_{n(1)}\|_{\mathcal{M}_p} - c \Big| < \frac{c}{10} \quad \text{and} \quad |b_{n(1),1}| < \frac{c}{10}.$$

Since $\|B_{n(1)}\|_{\mathcal{M}_p} < \infty$, we can choose an integer $\nu(2) > 0$ such that

$$\|B_{n(1)}\|_{\mathcal{M}_p}^{<\nu(2),\infty>} < \frac{c}{10},$$

and it follows that

$$\Big| \|B_{n(1)}\|_{\mathcal{M}_p}^{<0,\nu(2)>} - c \Big| \le \Big| \|B_{n(1)}\|_{\mathcal{M}_p} - c \Big| + \|B_{n(1)}\|_{\mathcal{M}_p}^{<\nu(2)+1,\infty>} + |b_{n(1),1}| < \frac{3c}{10}.$$

Now, we choose an integer $n(2) > n(1)$ such that

$$\|B_{n(2)}\|_{\mathcal{M}_p}^{<0,\nu(2)>} < \frac{c}{10} \quad \text{and} \quad \Big| \|B_{n(2)}\|_{\mathcal{M}_p} - c \Big| < \frac{c}{10},$$

and an integer $\nu(3) > \nu(2)$ such that $\|B_{n(2)}\|_{\mathcal{M}_p}^{<\nu(3)+1,\infty>} < c/10$. Again, it follows that

$$\Big| \|B_{n(2)}\|_{\mathcal{M}_p}^{<\nu(2)+1,\nu(3)>} - c \Big| < \frac{3c}{10}.$$

Continuing in this way, we can determine sequences $\{n(r)\}_{r=1}^{\infty}$ and $\{\nu(r)\}_{r=1}^{\infty}$ of integers $n(1) < n(2) < \cdots$ and $0 = \nu(1) < \nu(2) < \cdots$ such that for all $r = 1, 2, \ldots$

$$\|B_{n(r)}\|_{\mathcal{M}_p}^{<0,\nu(r)>} < \frac{c}{10}, \quad \|B_{n(r)}\|_{\mathcal{M}_p}^{<\nu(r+1)+1,\infty>} < \frac{c}{10}$$

and

$$\Big| \|B_{n(r)}\|_{\mathcal{M}_p}^{<\nu(r)+1,\nu(r+1)>} - c \Big| < \frac{3c}{10}.$$

If $p = 1$, we define the sequence $x = (x_k)$ by

$$x_k := \begin{cases} 0 & , \ k = 1, \\ (-1)^r 2^{\nu} \operatorname{sgn}(b_{n(r),k(\nu)}) & , \ k = k(\nu), \\ & \quad \text{where } k(\nu) \in [2^{\nu}, 2^{\nu+1} - 1] \text{ is the smallest} \\ & \quad \text{integer with } |b_{n(r),k(\nu)}| = \max_{\nu} |b_{n(r),k}|, \\ 0 & , \ k \neq k(\nu), \\ & (\nu(r) + 1 \le \nu \le \nu(r+1);\ r = 1, 2, \ldots). \end{cases}$$

Then, we obviously have $x \in w_\infty$ and $\|x\|_{w_\infty} \leq 1$, and

$$\begin{aligned}
\left|(Bx)_{n(r)} - (-1)^r c\right| &\leq \sum_{\nu=0}^{\nu(r)} \Sigma_\nu |b_{n(r),k}|\,|x_k| + \sum_{\nu=\nu(r+1)+1}^{\infty} \Sigma_\nu |b_{n(r),k}|\,|x_k| + \\
&+ \left| \sum_{\nu=\nu(r)+1}^{\nu(r+1)} \Sigma_\nu b_{n(r),k} x_k - c \right| \\
&\leq \sum_{\nu=0}^{\nu(r)} 2^\nu \max{}_\nu |b_{n(r),k}| + \sum_{\nu=\nu(r+1)+1}^{\infty} 2^\nu \max{}_\nu |b_{n(r),k}| + \\
&+ \left| (-1)^r \left(\sum_{\nu=\nu(r)+1}^{\nu(r+1)} 2^\nu \max{}_\nu |b_{n(r),k}| - c \right) \right| \\
&= \|B_{n(r)}\|_{\mathcal{M}_1}^{<0,\nu(r)>} + \|B_{n(r)}\|_{\mathcal{M}_1}^{<\nu(r+1)+1,\infty>} + \\
&+ \left| \|B_{n(r)}\|_{\mathcal{M}_1}^{<\nu(r)+1,\nu(r+1)>} - c \right| \\
&< \frac{c}{10} + \frac{c}{10} + \frac{3c}{10} = \frac{c}{2} \text{ for all } r \in \mathbb{N}_1.
\end{aligned}$$

Consequently, $\{(Bx)_n\}_{n=1}^\infty$ is not a Cauchy sequence, hence not convergent.

If $1 < p < \infty$, we define the sequence $x = (x_k)$ by

$$x_k := \begin{cases} 0 & , \quad k = 1, \\ 2^{\nu/p}(-1)^r \operatorname{sgn}(b_{n(r),k})|b_{n(r),k}|^{q-1} \left(\Sigma_\nu |b_{n(r),k}|^q\right)^{-1/p} & , \\ \qquad (2^\nu \leq k \leq 2^{\nu+1} - 1) \\ \qquad \nu(r) + 1 \leq \nu \leq \nu(r+1);\; r = 1, 2, \ldots. \end{cases}$$

Let $\nu \in \mathbb{N}_0$ be given. Then, there exists r such that $\nu(r) + 1 \leq \nu \leq \nu(r+1)$ and

$$\frac{1}{2^\nu} \Sigma_\nu |x_k|^p = \frac{1}{2^\nu} \Sigma_\nu 2^\nu |b_{n(r),k}|^{pq-p} \left(\Sigma_\nu |b_{n(r),k}|^q\right)^{-1} = 1,$$

that is $x \in w_\infty^p$ and $\|x\|_{w_\infty^p} \leq 1$. We also have by Hölder's inequality

$$\begin{aligned}
\left|B_{n(r)}x - (-1)^r c\right| &\leq \sum_{\nu=0}^{\nu(r)} 2^{\nu/p} \left(\Sigma_\nu |b_{n(r),k}|^q\right)^{1/q} + \sum_{\nu=\nu(r+1)+1}^{\infty} 2^{\nu/p} \left(\Sigma_\nu |b_{n(r),k}|^q\right)^{1/q} + \\
&+ \left| (-1)^r \left(\sum_{\nu=\nu(r)+1}^{\nu(r+1)} 2^{\nu/p} \left(\Sigma_\nu |b_{n(r),k}|^q\right)^{1/q} - c \right) \right| \\
&= \|B_{n(r)}\|_{\mathcal{M}_p}^{<0,\nu(r)>} + \|B_{n(r)}\|_{\mathcal{M}_p}^{<\nu(r+1)+1,\infty>} + \\
&+ \left| \|B_{n(r)}\|_{\mathcal{M}_p}^{<\nu(r)+1,\nu(r+1)>} - c \right| \\
&< \frac{c}{10} + \frac{c}{10} + \frac{3c}{10} = \frac{c}{2} \text{ for all } r \in \mathbb{N}_1.
\end{aligned}$$

Consequently, $(B_n x)_{n=1}^{\infty}$ is not a Cauchy sequence, hence not convergent. So, we have $x \in w_{\infty}^p$, but $Bx \notin c$ in both cases, which is a contradiction to $B \in (w_{\infty}^p : c)$. This completes the proof of **5.** and **5′.**.
2. and **2′.** are proved in the same way as **5.** and **5′.** with $\alpha_k = 0$ for all $k \in \mathbb{N}_1$. □

4.10.2 The β-Duals of $w_0^p(U)$, $w^p(U)$ and $w_{\infty}^p(U)$

Now, we determine the β-duals of $w_0^p(U)$, $w^p(U)$ and $w_{\infty}^p(U)$.

Let $U = S$, the summation matrix defined by (3.4.8). Then, $cs = (c)_S$ for the set of all convergent series, and $R = V'$ for the transpose of the inverse V of the triangle S.

The following results are helpful.

Lemma 4.10.8. [68, Lemma 3.1] *We have*

(a) $a \in \{w_0^p(U)\}^{\beta}$ *if and only if* $a \in (\mathcal{M}_p)_R$ *and* $W \in (w_0^p : c_0)$, *where* $W = (w_{mk})_{m,k=1}^{\infty}$ *is the triangle with* $w_{mk} := \sum_{j=m}^{\infty} v_{jk} a_j$ *for* $1 \le k \le m$ $(m = 1, 2, \ldots)$;

(b) $a \in \{w^p(U)\}^{\beta}$ *if and only if* $a \in (\mathcal{M}_p)_R$ *and* $W \in (w^p : c)$;

(c) $a \in \{w_{\infty}^p(U)\}^{\beta}$ *if and only if* $a \in (\mathcal{M}_p)_R$ *and* $W \in (w_{\infty}^p : c_0)$.

Proof. Let $a = (a_n)_{n=1}^{\infty} \in \omega$. We define the triangles $B = (b_{nk})_{n,k=1}^{\infty}$ and $C = (c_{nk})_{n,k=1}^{\infty}$ by $b_{nk} := a_n v_{nk}$ and $c_{nk} := \sum_{j=k}^{n} a_j v_{jk}$ for $1 \le k \le n$ $(n = 1, 2, \ldots)$, hence $C = SB$. Let λ be any of the sets w_0^p, w^p or w_{∞}^p. Since $x \in \lambda$ if and only if $z = Vx \in \lambda_U$, and since

$$a_n z_n = a_n (Vx)_n = a_n \sum_{k=1}^{n} v_{nk} x_k = \sum_{k=1}^{n} a_n v_{nk} x_k = (Bx)_n$$

for all $n \in \mathbb{N}_1$, we observe that $a \in (\lambda_U)^{\beta}$ if and only if $B \in (\lambda : cs)$, and this is the case by [269, Theorem 3.8, p. 180] if and only if $C \in (\lambda : c)$.

(a) First we assume $a \in \{w_0^p(U)\}^{\beta}$. Then, $C \in (w_0^p : c)$ which is the case by **6.** in Theorem 4.10.7 if and only if

$$(Ra)_k := \lim_{n\to\infty} c_{nk} = \sum_{j=k}^{\infty} a_j v_{jk} \text{ exists for every } k \in \mathbb{N}_1, \tag{4.10.5}$$

$$\|C\|_{\mathcal{M}_p} := \sup_{n\in\mathbb{N}_1} \|C_n\|_{\mathcal{M}_p} < \infty. \tag{4.10.6}$$

We show that (4.10.5) and (4.10.6) imply

$$Ra := \{(Ra)_k\}_{k=1}^{\infty} \in \mathcal{M}_p = (w_0^p)^{\beta}. \tag{4.10.7}$$

Let $\mu \in \mathbb{N}_0$ be given. It follows from (4.10.6) that there exists a constant $M > 0$ such that $\|(\sum_{j=k}^{n} a_j v_{jk})_{k=1}^{\infty}\|_{\mathcal{M}_p}^{<0,\mu>} \le M$. Letting $n \to \infty$ and using (4.10.5), we obtain $\|Ra\|_{\mathcal{M}_p}^{<0,\mu>} \le M$, and (4.10.7) follows, since μ was arbitrary.

We note that the matrix W is defined in view of (4.10.5). Furthermore, we also have for all $z \in \omega$ and for all $m \in \mathbb{N}_1$

$$\begin{aligned}\sum_{k=1}^{m}(Ra)_k(Uz)_k - \{W(Uz)\}_m &= \sum_{k=1}^{m}\left(\sum_{j=k}^{\infty} a_j v_{jk}\right)(Uz)_k - \sum_{k=1}^{m}\left(\sum_{j=m}^{\infty} a_j v_{jk}\right)(Uz)_k \\ &= \sum_{k=1}^{m}\left(\sum_{j=k}^{m-1} a_j v_{jk}\right)(Uz)_k = \sum_{k=1}^{m-1}\left(\sum_{j=k}^{m-1} a_j v_{jk}\right)(Uz)_k \\ &= \sum_{j=1}^{m-1} a_j \sum_{k=1}^{j} v_{jk}(Uz)_k = \sum_{j=1}^{m-1} a_j\{V(Uz)\}_j = \sum_{j=1}^{m-1} a_j z_j,\end{aligned}$$

that is,

$$\sum_{k=1}^{m-1} a_k z_k = \sum_{k=1}^{m}(R_k a)(Uz)_k - \{W(Uz)\}_m \text{ for all } m \in \mathbb{N}_1. \tag{4.10.8}$$

It follows from (4.10.8), $a \in \{w_0^p(U)\}^\beta$ and (4.10.7) that $\{(W(Uz))_m\}_{m=1}^\infty \in c$ for all $z \in \lambda_U$, which is equivalent to $W \in (w_0^p : c)$. Now, (4.10.5) implies

$$\lim_{m\to\infty} w_{mk} = \lim_{m\to\infty} \sum_{j=m}^{\infty} a_j v_{jk} = 0 \text{ for all } k \in \mathbb{N}_1 \tag{4.10.9}$$

and $W \in (w_0^p : c)$ and (4.10.9) imply $W \in (w_0^p : c_0)$ by **3.** and **6.** in Theorem 4.10.7.

Conversely, if $a \in (\mathcal{M}_p)_R$ and $W \in (w_0^p : c_0)$, then $Ra \in \mathcal{M}_p = (w_0^p)^\beta$ by Proposition 4.10.4 (a), and then $a \in \{w_0^p(U)\}^\beta$ follows from (4.10.8).

(b) and (c) Let $\lambda \in \{w^p, w_\infty^p\}$. First, we assume $a \in (\lambda_U)^\beta$. Since $w_0^p(U) \subset \lambda_U$ by Corollary 4.10.2 (a), we have $a \in \{w_0^p(U)\}^\beta$, and $Ra \in \mathcal{M}_p = \lambda^\beta$ follows from Part (a). Now, (4.10.8) implies $W \in (\lambda : c)$. Again, $Ra \in \mathcal{M}_p$ implies (4.10.9), and $W \in (w_\infty^p : c)$ and (4.10.9) imply $W \in (w_\infty^p : c_0)$ by **2.** and **5.** in Theorem 4.10.7.

The proof of the converse part is analogous to that of the converse part of (a). □

Theorem 4.10.9. [68, Theorem 3.2] *For every $m \in \mathbb{N}_1$, let $\nu(m)$ be the uniquely defined number with $2^{\nu(m)} \le m \le 2^{\nu(m)+1} - 1$. We have*

$$\|W_m\|_{\mathcal{M}_p} := \begin{cases} \sum_{\nu=0}^{\nu(m)-1} 2^\nu \max_v \left|\sum_{j=m}^\infty a_j v_{jk}\right| \\ \quad +2^{\nu(m)} \max_{2^{\nu(m)}\le k\le m} \left|\sum_{j=m}^\infty a_j v_{jk}\right| < \infty & , \quad p=1, \\ \\ \sum_{\nu=0}^{\nu(m)-1} 2^{\nu/p} \left(\sum_\nu \left|\sum_{j=m}^\infty a_j v_{jk}\right|^q\right)^{1/q} \\ \quad +2^{\nu(m)/p} \left(\sum_{k=2^{\nu(m)}}^m \left|\sum_{j=m}^\infty a_j v_{jk}\right|^q\right)^{1/q} < \infty & , \quad 1<p<\infty, \end{cases} \tag{4.10.10}$$

and

(a) $a \in \{w_0^p(U)\}^\beta$ *if and only if*

$$\|Ra\|_{\mathcal{M}_p} := \begin{cases} \sum_{\nu=0}^\infty 2^\nu \max_\nu \left|\sum_{j=k}^\infty a_j v_{jk}\right| < \infty & , \quad p = 1, \\ \sum_{\nu=0}^\infty 2^{\nu/p} \left(\sum_\nu \left|\sum_{j=k}^\infty a_j v_{jk}\right|^q\right)^{1/q} < \infty & , \quad 1 < p < \infty, \end{cases} \tag{4.10.11}$$

and

$$\sup_{m \in \mathbb{N}_1} \|W_m\|_{\mathcal{M}_p} < \infty; \tag{4.10.12}$$

(b) $a \in \{w^p(U)\}^\beta$ *if and only if (4.10.11) and (4.10.12) hold and*

$$\eta := \lim_{m \to \infty} \sum_{k=1}^m \sum_{j=m}^\infty a_j v_{jk} \text{ exists;} \tag{4.10.13}$$

(c) $a \in \{w_\infty^p(U)\}^\beta$ *if and only if (4.10.11) holds and*

$$\lim_{m \to \infty} \|W_m\|_{\mathcal{M}_p} = 0. \tag{4.10.14}$$

(d) *Let* $\lambda \in \{w_0^p, w_\infty^p\}$. *If* $a \in (\lambda_U)^\beta$, *then*

$$\sum_{k=1}^\infty a_k z_k = \sum_{k=1}^\infty (Ra)_k (Uz)_k \text{ for all } z \in \lambda_U; \text{ also } \|a\|_{\lambda_U}^* = \|Ra\|_{\mathcal{M}_p}. \tag{4.10.15}$$

If $a \in \{w^p(U)\}^\beta$, *then*

$$\sum_{k=1}^\infty a_k z_k = \sum_{k=1}^\infty (Ra)_k (Uz)_k - \xi\eta \text{ for all } z \in w^p(U),$$

where ξ *and* η *are from (4.10.1) and (4.10.14);* (4.10.16)

also

$$\|a\|_{w^p(U)}^* = |\eta| + \|Ra\|_{\mathcal{M}_p} \text{ for all } a \in \{w^p(U)\}^\beta. \tag{4.10.17}$$

Proof. We apply Lemma 4.10.8 and Theorem 4.10.7.

Condition (4.10.11) is $Ra \in \mathcal{M}_p = (w_0^p)^\beta = (w^p)^\beta = (w_\infty^p)^\beta$ by Proposition 4.10.4 (a).

Condition (4.10.12) comes from $W \in (w_0^p : c_0)$ and $W \in (w^p : c)$ and is (1.1) in Theorem 4.10.7 **3.** and **7.**; the conditions $\lim_{m\to\infty} w_{mk} = 0$ and $\lim_{m\to\infty} w_{mk} = \beta_k$, which are (3.1) and (5.1) in **3.** and **7.**, are redundant. Condition (4.10.13) for $W \in (w^p : c)$ comes from (7.1) in Theorem 4.10.7 **7.**.

Condition (4.10.14) comes from $W \in (w_\infty^p : c_0)$ and is (2.1) in Theorem 4.10.7 **2.**.

Thus, we have shown Parts (a), (b) and (c).

(d) The first condition in (4.10.15) follows from (4.10.8) and the fact that $W \in (\lambda : c_0)$; the second condition follows from Proposition 4.10.4 (b) and (d).

Now, let $a \in \{w^p(U)\}^\beta$ and $z \in w^p(U)$. Then, $x = Tz \in w^p$ and ξ from (4.10.1) exists, hence there exists $x^{(0)} \in w_0^p$ such that $x = x^{(0)} + \xi e$. We put $z^{(0)} = Vx^{(0)}$. Then, it follows that $z^{(0)} \in w_0^p(U)$ and $z = Vx = V(x^{(0)} + \xi e) = z^{(0)} + \xi Ve$, and we obtain as in (4.10.8) for all $m \in \mathbb{N}_1$

$$\begin{aligned}\sum_{k=1}^{m-1} a_k z_k &= \sum_{k=1}^{m} (Ra)_k (Uz)_k - \{W(U(z^{(0)} + \xi Ve)\}_m \\ &= \sum_{k=1}^{m} (Ra)_k (Uz)_k - \{W(Uz^{(0)})\}_m - \xi (We)_m.\end{aligned}$$

The first term on the right-hand side of the last equation converges since $Ra \in \mathcal{M}_p$. The second term tends to 0, since $a \in \{w^p(U)\}^\beta \subset \{w_0^p(U)\}^\beta$ implies $W \in (w_0^p : c_0)$. Furthermore, since $W \in (w^p : c)$ implies that $\eta = \lim_{m\to\infty} W_m e$ exists by (7.1) in Theorem 4.10.7 **7.**, the identity in (4.10.16) follows. Finally, (4.10.17) follows from Proposition 4.10.4 (c). □

We apply Theorem 4.10.9 to the matrix T of Example 4.10.3.

Example 4.10.10. [68, Example 3.3] *Let U be the matrix of Example 4.10.3. Then, it is easy to see that*

(a) $a \in \{w_0^p(U)\}^\beta$ *if and only if*

$$\|Ra\|_{\mathcal{M}_p} := \begin{cases} \sum_{\nu=0}^{\infty} 2^\nu \max_\nu \left| \dfrac{a_k}{u_k v_k} - \dfrac{a_{k+1}}{u_k v_{k+1}} \right| < \infty & , \quad p = 1, \\ \sum_{\nu=0}^{\infty} 2^{\nu/p} \left(\sum_\nu \left| \dfrac{a_k}{u_k v_k} - \dfrac{a_{k+1}}{u_k v_{k+1}} \right|^q \right)^{1/q} < \infty & , \quad 1 < p < \infty \end{cases} \tag{4.10.18}$$

and

$$\sup_{m \in \mathbb{N}_1} \left(2^{\nu(m)/p} \left| \frac{a_{m+1}}{u_m v_{m+1}} \right| \right) < \infty; \tag{4.10.19}$$

(b) $a \in \{w^p(U)\}^\beta$ *if and only if (4.10.18), (4.10.19) and*

$$\eta := \lim_{m\to\infty} \left(\frac{a_m}{u_{m-1} v_m} + \frac{a_m}{u_m v_m} - \frac{a_{m+1}}{u_m v_{m+1}} \right) \text{ exists;}$$

(c) $a \in \{w_\infty^p(U)\}^\beta$ *if and only if (4.10.18) and*

$$\lim_{m\to\infty} \left(2^{\nu(m)/p} \frac{a_{m+1}}{u_m v_{m+1}} \right) = 0.$$

4.10.3 Matrix Transformations on the Spaces $w_0^p(U)$, $w^p(U)$ and $w_\infty^p(U)$

Now, we characterize the classes $(\lambda : \mu)$, where λ is any of the spaces $w_0^p(U)$, $w^p(U)$ and $w_\infty^p(U)$, and μ is any of the spaces c_0, c and ℓ_∞.

The following results are useful.

Lemma 4.10.11. [68, Lemma 4.1] *The following statements hold:*

(a) *Let* $\lambda \in \{w_0^p, w_\infty^p\}$ *and* μ *be an arbitrary subset of* ω. *Then, we have* $A \in (\lambda_U : \mu)$ *if and only if* $\hat{A} \in (\lambda : \mu)$ *and* $W^{(n)} \in (\lambda : c_0)$ *for all* $n = 1, 2, \ldots,$ *where the matrix* $\hat{A} = (\hat{a}_{nk})_{n,k=1}^\infty$ *and the triangles* $W^{(n)} = \{w_{mk}^{(n)}\}_{m,k=1}^\infty$ *are defined by*

$$\hat{a}_{nk} := \sum_{j=k}^{\infty} a_{nj} v_{jk} \text{ for all } n, k \in \mathbb{N}_1 \quad \text{and} \quad w_{mk}^{(n)} := \sum_{j=m}^{\infty} a_{nj} v_{jk} \text{ for } 1 \leq k \leq m.$$

Moreover, we also have if $A \in (\lambda_U : \mu)$, *then*

$$Az = \hat{A}(Uz) \text{ for all } z \in \lambda_U. \tag{4.10.20}$$

(b) *Let* μ *be an arbitrary linear subspace of* ω. *Then, we have* $A \in (w^p(U) : \mu)$ *if and only if*

$$\hat{A} \in (w_0^p : \mu), \tag{4.10.21}$$

$$W^{(n)} \in (w^p : c) \text{ for all } n \in \mathbb{N}_1 \tag{4.10.22}$$

and

$$\hat{A}e - \{\rho^{(n)}\}_{n=1}^\infty \in \mu, \textit{ where, } \rho^{(n)} = \lim_{m\to\infty} \sum_{k=1}^{m} w_{mk}^{(n)} \text{ for all } n \in \mathbb{N}_1. \tag{4.10.23}$$

Moreover, if $A \in (w^p(U) : \mu)$, *then we also have*

$$Az = \hat{A}(Uz) - \xi\{\rho^{(n)}\}_{n=1}^\infty \textit{ for all } z \in w^p(U), \textit{ where, } \xi \text{ is from } (4.10.4). \tag{4.10.24}$$

Proof. (a) First, we assume $A \in (\lambda_U : \mu)$. Then, it follows that $A_n \in (\lambda_U)^\beta$ for all $n \in \mathbb{N}_1$, hence $\hat{A}_n \in \lambda^\beta$ and $W^{(n)} \in (\lambda : c_0)$ for all $n \in \mathbb{N}_1$ by Lemma 4.10.8 (a) and (c). Let $x \in \lambda$ be given, hence $z = Vx \in \lambda_U$. Since $A_n \in (\lambda_U)^\beta$ implies

$$(Az)_n = \{\hat{A}(Uz)\}_n = (\hat{A}x)_n$$

for all $n \in \mathbb{N}_1$ by (4.10.15) in Theorem 4.10.9 (d), that is $\hat{A}x = Az$, $Az \in \mu$ for all $z \in \lambda_U$ implies $Ax \in \mu$, that is (4.10.20) holds, and we have $\hat{A} \in (\lambda : \mu)$ since $x \in \lambda$ was arbitrary.

Conversely, we assume $\hat{A} \in (\lambda : \mu)$ and $W^{(n)} \in (\lambda : c_0)$ for all $n \in \mathbb{N}_1$. Then, we have $\hat{A}_n \in \lambda^\beta$, and this and $W^{(n)} \in (\lambda : c_0)$ together implies $A_n \in \lambda_U^\beta$ by Lemma 4.10.8 (a) and (c). Now, let $z \in \lambda_U$ be given, hence $x = Uz \in \lambda$. Again, we have $(Az)_n = (\hat{A}x)_n$ for all $n \in \mathbb{N}_1$ by (4.10.15) in Theorem 4.10.9 (d), hence $Az = \hat{A}x \in \mu$, and $\hat{A}x \in \mu$ for all $x \in \lambda$ implies $Az \in \mu$. Thus, we have $\hat{A} \in (\lambda : \mu)$, since $z \in \lambda_U$ was arbitrary.

(b) First, we assume that $A \in (w^p(U) : \mu)$. Then, it follows that $A \in (w_0^p(U) : \mu)$ and so $\hat{A} \in (w_0^p : \mu)$ by Part (a). Also, $A_n \in \{w^p(U)\}^\beta$ implies $W^{(n)} \in (w^p : c)$ by Lemma 4.10.8 (b), and also (4.10.16) by Theorem 4.10.9 (d). Since $z = Ve \in w^p(U)$, hence $Az \in \mu$, and $\xi = 1$, we obtain (4.10.23) from (4.10.16).

Conversely, we assume that the conditions in (4.10.21), (4.10.22) and (4.10.23) are satisfied. First, $\hat{A}_n \in (w_0^p)^\beta = \mathcal{M}_p$ and $W^n \in (w^p : c)$ together imply $A_n \in \{w^p(U)\}^\beta$ by Lemma 4.10.8 (b), and again (4.10.16) follows by Theorem 4.10.9 (d). Let $z \in w^p(U)$ be given. Then, we have $x = Uz \in w^p$. We put $x^{(0)} = x - \xi e$, where ξ is from

$$\lim_{n\to\infty} \frac{1}{n} \sum_{k=1}^{n} |x_k - \xi|^p = \lim_{n\to\infty} \frac{1}{n} \sum_{k=1}^{n} |(Uz)_k - \xi|^p = 0,$$

that is from (4.10.4). Then, we have $x^{(0)} \in w_0^p$ and it follows from (4.10.16) that

$$Az = \hat{A}(Uz) - \xi\{\rho^{(n)}\}_{n=1}^\infty = \hat{A}(x^{(0)}) + \xi[\hat{A}e - \{\rho^{(n)}\}_{n=1}^\infty] \in \mu,$$

since $\hat{A} \in (w_0^p : \mu)$, $\hat{A}e - \{\rho^{(n)}\}_{n=1}^\infty \in \mu$ and μ is a linear space. □

Theorem 4.10.12. [68, Theorem 4.2] *The necessary and sufficient conditions for the entries of $A \in (\lambda_U : \mu)$ when $\lambda \in \{w_0^p, w^p, w_\infty^p\}$ and $\mu \in \{\ell_\infty, c_0, c\}$ can be read from the following table: where*

TABLE 4.4 The characterization of the class $(\lambda_U : \mu)$

From / To	$w_\infty^p(U)$	$w_0^p(U)$	$w^p(U)$
ℓ_∞	**1.**	**2.**	**3.**
c_0	**4.**	**5.**	**6.**
c	**7.**	**8.**	**9.**

1. *(1.1)* $\sup_{n\in\mathbb{N}_1} \|\hat{A}_n\|_{\mathcal{M}_p} < \infty$
(1.2) $\lim_{m\to\infty} \|W_m^{(n)}\|_{\mathcal{M}_p} = 0$ for all $n \in \mathbb{N}_1$
with $\|\hat{A}\|_{\mathcal{M}_p}$ and $\|W_m^{(n)}\|_{\mathcal{M}_p}$ defined as in *(4.10.11)* and *(4.10.10)* with a_j replaced by a_{nj}

2. *(1.1) and (2.1), where*
(2.1) $\sup_m \|W_m^{(n)}\|_{\mathcal{M}_p} < \infty$ for all $n \in \mathbb{N}_1$

3. *(1.1), (2.1) (3.1) and (3.2), where*
(3.1) $\rho^{(n)} = \lim_{m\to\infty} \sum_{k=1}^m \sum_{j=m}^\infty a_{nj}v_{jk}$ exists for all $n \in \mathbb{N}_1$
(3.2) $\sup_{n\in\mathbb{N}_1} \left|\sum_{k=1}^\infty \sum_{j=k}^\infty a_{nj}v_{jk} - \rho^{(n)}\right| < \infty$

4. *(4.1) and (1.2), where*
(4.1) $\lim_{n\to\infty} \|\hat{A}_n\|_{\mathcal{M}_p} = 0$

5. *(1.1), (5.1) and (2.1), where*
(5.1) $\lim_{n\to\infty} \hat{a}_{nk} = 0$ for all $k \in \mathbb{N}_1$

6. *(1.1), (5.1), (2.1), (3.1) and (6.1) where*
(6.1) $\lim_{n\to\infty} \left(\sum_{k=1}^{\infty} \sum_{j=k}^{\infty} a_{nj} v_{jk} - \rho^{(n)}\right) = 0$

7. *(1.2), (7.1), (7.2) and (7.3) where*
(7.1) $\hat{\alpha}_k = \lim_{n\to\infty} \hat{a}_{nk}$ exists for all $k \in \mathbb{N}_1$
(7.2) $(\hat{\alpha}_k)_{k=1}^{\infty}$, $\hat{A}_n \in \mathcal{M}_p$ for all $n \in \mathbb{N}_1$
(7.3) $\lim_{n\to\infty} \|\hat{A}_n - (\hat{\alpha}_k)_{k=1}^{\infty}\|_{\mathcal{M}_p} = 0$

8. *(1.1), (7.1) and (2.1)*

9. *(1.1), (7.1), (2.1), (3.1) and (9.2), where*
(9.2) $\beta = \lim_{n\to\infty} \left(\sum_{k=1}^{\infty} \sum_{j=k}^{\infty} a_{nj} v_{jk} - \rho^{(n)}\right)$ exists.

Proof. We apply Lemma 4.10.11 and Theorems 4.10.9 and 4.10.7.

If $\lambda \in \{w_{\infty}^p, w_0^p\}$, then we apply Lemma 4.10.11 (a), i.e., we get the conditions for $\hat{A} \in (\lambda : \mu)$ and $W^{(n)} \in (\lambda : c_0)$. Applying Theorem 4.10.7 **1.**, **2.**, **3.**, **5.** and **6.**, we obtain (1.1) in **1.** and **2.**, (4.1) in **4.**, (1.1) and (5.1) in **5.**, (7.1), (7.2) and (7.3) in **7.**, and (1.1) and (7.1) in **8.** for $\hat{A} \in (\lambda : \mu)$. We obtain the conditions (1.2) in **1.**, **4.** and **7.** for $W^{(n)} \in (w_{\infty}^p : c_0)$ and (2.1) in **2.**, **5.** and **8.** from Theorem 4.10.7 **2.** and **3.**, taking into account that the condition $\lim_{m\to\infty} w_{mk}^{(n)} = 0$ for all $k \in \mathbb{N}_1$ is redundant as in the proof of Theorem 4.10.9.

If $\lambda = w^p$, we apply Lemma 4.10.11 (b), that is we get the conditions for $\hat{A} \in (w_0^p : \mu)$, $W^{(n)} \in (w^p : c)$ for all n and $\hat{A}e - \{\rho^{(n)}\}_{n=1}^{\infty} \in \mu$. Applying Theorem 4.10.7 **1.**, **3.** and **6.** for $\hat{A} \in (w_0^p : \mu)$, we obtain (1.1) in **3.**, (1.1) and (5.1) in **6.** and (1.1) and (7.1) in **9.**. We obtain the conditions (2.1) and (3.1) in **3.**, **6.** and **9.** for $W^{(n)} \in (w^p : c)$; again the condition $\lim_{m\to\infty} w_{mk}^{(n)}$ exists for all $k \in \mathbb{N}_1$ is redundant. Finally, $\hat{A}e - \{\rho^{(n)}\}_{n=1}^{\infty} \in \mu$ yields (3.2) in **3.**, (6.1) in **6.** and (9.2) in **9.**. □

4.10.4 Conclusion

Let λ denotes the anyone of the spaces w_0^p, w^p or w_{∞}^p. We have introduced the sequence space $\lambda(U)$, which is the domain of a triangle matrix $U = (u_{nk})$ in the sequence space λ. We have essentially concerned with two subjects: Determination of the β-dual of the space $\lambda(U)$ and the characterization of the certain matrix transformations defined on the sequence space $\lambda(U)$.

Although the domain of summability matrices in the classical spaces ℓ_{∞}, c, c_0 and ℓ_p of sequences was studied by several authors (see Altay [20], Altay and Başar [21, 23, 31], Aydın and Başar [42, 44], Başar and Altay [66], Çolak and Et [131], Çolak et al. [133], Et [155], Et and Çolak [166], Et and Başarır [165], Malkowsky et al. [267], Malkowsky and Parashar [268], Malkowsky and Savaş [272], Mursaleen [283], Mursaleen et al. [289], Ng and Lee [304], Polat and Başar [318], Şengönül and Başar [373], Wang [393]), the matrix domain of the spaces w_0^p , w^p and w_{∞}^p has not been examined. This section is filled up this gap in the existing literature. It is obvious that the α- and γ-dulas of the new spaces $w_0^p(U)$, $w^p(U)$ and $w_\infty^p(U)$ are still open. Besides, one can try to characterize the classes of infinite matrices from the spaces $w_0^p(U)$, $w^p(U)$ or $w_\infty^p(U)$ to a sequence space μ, which is different than that of Subsection 4.10.3. We should note from now on that a new paper can be based on the extension of the new spaces $w_0^p(U)$, $w^p(U)$ and $w_\infty^p(U)$ to the paranormed case.

4.11 CHARACTERIZATIONS OF SOME OTHER CLASSES OF MATRIX TRANSFORMATIONS

It is trivial that Lemma 4.3.24 has several consequences. Indeed, combining Lemma 4.3.24 with Theorems 4.3.20–4.3.23, Theorems 4.6.12, 4.6.13, 4.6.16, 4.6.17 and Corollaries 4.6.14 and 4.6.18, Theorems 4.7.9–4.7.13, Theorems 4.8.14–4.8.17, 4.8.20, 4.8.21 and Corollaries 4.8.18 and 4.8.22 by choosing B as one of the special matrices C_1, E^r, R^t, Δ, $\Delta^{(1)}$, A^r or S, one can easily obtain the following results:

Corollary 4.11.1. [66, Corollary 5.1] *Let $A = (a_{nk})$ be an infinite matrix over the complex field. Then, the following statements hold:*

(i) $E = (e_{nk}) \in (bv_p : bv_\infty)$ *if and only if (4.3.17) and (4.3.18) hold with d_{nk} instead of a_{nk}, where $d_{nk} := e_{nk} - e_{n-1,k}$ for all $k, n \in \mathbb{N}$.*

(ii) $E = (e_{nk}) \in (bv_p : e_\infty^r)$ *if and only if (4.3.17) and (4.3.18) hold with d_{nk} instead of a_{nk}, where $d_{nk} := \sum_{j=0}^{n} \binom{n}{j}(1-r)^{n-j}r^j e_{jk}$ for all $k, n \in \mathbb{N}$.*

(iii) $E = (e_{nk}) \in (bv_p : X_\infty)$ *if and only if (4.3.17) and (4.3.18) hold with d_{nk} instead of a_{nk}, where $d_{nk} := e(n,k)/(n+1)$ for all $k, n \in \mathbb{N}$.*

(iv) $E = (e_{nk}) \in (bv_p : r_\infty^t)$ *if and only if (4.3.17) and (4.3.18) hold with d_{nk} instead of a_{nk}, where $d_{nk} := \sum_{j=0}^{n} t_j e_{jk}/T_n$ for all $k, n \in \mathbb{N}$.*

(v) $E = (e_{nk}) \in (bv_p : bs)$ *if and only if (4.3.17) and (4.3.18) hold with d_{nk} instead of a_{nk}, where $d_{nk} := e(n,k)$ for all $k, n \in \mathbb{N}$.*

Corollary 4.11.2. [66, Corollary 5.2] *Let $A = (a_{nk})$ be an infinite matrix over the complex field. Then, the following statements hold:*

(i) $E = (e_{nk}) \in (bv_\infty : bv_p)$ *if and only if (4.3.20)–(4.3.22) hold with d_{nk} instead of a_{nk}, where $d_{nk} := e_{nk} - e_{n-1,k}$ for all $k, n \in \mathbb{N}$.*

(ii) $E = (e_{nk}) \in (bv_\infty : e_p^r)$ *if and only if (4.3.20)–(4.3.22) hold with d_{nk} instead of a_{nk}, where $d_{nk} := \sum_{j=0}^{n} \binom{n}{j}(1-r)^{n-j}r^j e_{jk}$ for all $k, n \in \mathbb{N}$.*

(iii) $E = (e_{nk}) \in (bv_\infty : X_p)$ *if and only if (4.3.20)–(4.3.22) hold with d_{nk} instead of a_{nk}, where $d_{nk} := e(n,k)/(n+1)$ for all $k, n \in \mathbb{N}$.*

(iv) $E = (e_{nk}) \in (bv_\infty : r_p^t)$ *if and only if (4.3.20)–(4.3.22) hold with d_{nk} instead of a_{nk}, where $d_{nk} := \sum_{j=0}^{n} t_j e_{jk}/T_n$ for all $k, n \in \mathbb{N}$.*

Corollary 4.11.3. [66, Corollary 5.3] *Let $A = (a_{nk})$ be an infinite matrix over the complex field. Then, the following statements hold:*

(i) *$E = (e_{nk}) \in (bv_p : bv_1)$ if and only if (4.3.17) and (4.3.23) hold with d_{nk} instead of a_{nk}, where $d_{nk} := e_{nk} - e_{n-1,k}$ for all $k, n \in \mathbb{N}$.*

(ii) *$E = (e_{nk}) \in (bv_p : e_1^r)$ if and only if (4.3.17) and (4.3.23) hold with d_{nk} instead of a_{nk}, where $d_{nk} := \sum_{j=0}^{n} \binom{n}{j}(1-r)^{n-j} r^j e_{jk}$ for all $k, n \in \mathbb{N}$.*

(iii) *$E = (e_{nk}) \in (bv_p : X_1)$ if and only if (4.3.17) and (4.3.23) hold with d_{nk} instead of a_{nk}, where $d_{nk} := e(n,k)/(n+1)$ for all $k, n \in \mathbb{N}$.*

(iv) *$E = (e_{nk}) \in (bv_p : r_1^t)$ if and only if (4.3.17) and (4.3.23) hold with d_{nk} instead of a_{nk}, where $d_{nk} := \sum_{j=0}^{n} t_j e_{jk}/T_n$ for all $k, n \in \mathbb{N}$.*

Corollary 4.11.4. [45, Corollary 6.5] *Let $A = (a_{nk})$ be an infinite matrix over the complex field and d_{nk} be defined by $d_{nk} := \sum_{j=0}^{n} \binom{n}{j}(1-r)^{n-j} r^j e_{jk}$ for all $k, n \in \mathbb{N}$. Then, the following statements hold:*

(i) *$E = (e_{nk}) \in (a_p^r : e_\infty^r)$ if and only if (4.6.10) and (4.6.11) hold with d_{nk} instead of a_{nk}.*

(ii) *$E = (e_{nk}) \in (a_p^r : e_c^r)$ if and only if (4.6.10), (4.6.11) and (4.6.14) hold with d_{nk} instead of a_{nk}.*

(iii) *$E = (e_{nk}) \in (a_p^r : e_0^r)$ if and only if (4.6.10) and (4.6.11) hold, and (4.6.14) also holds with $\alpha_k = 0$ for all $k \in \mathbb{N}$ and d_{nk} instead of a_{nk} for all $k, n \in \mathbb{N}$.*

(iv) *$E = (e_{nk}) \in (a_c^r : e_p^r)$ if and only if (4.6.16)–(4.6.19) hold with d_{nk} instead of a_{nk}.*

(v) *$E = (e_{nk}) \in (a_c^r : e_c^r)$ if and only if (4.6.14), (4.6.18), (4.6.19) and (4.6.21) hold with d_{nk} instead of a_{nk}.*

(vi) *$E = (e_{nk}) \in (a_c^r : e_c^r)_\varrho$ if and only if (4.6.18), (4.6.19) hold, (4.6.14) and (4.6.21) also hold with $\alpha_k = 0$ for each $k \in \mathbb{N}$ and $\alpha = \varrho$, respectively, and d_{nk} instead of a_{nk}.*

Corollary 4.11.5. [45, Corollary 6.6] *Let $A = (a_{nk})$ be an infinite matrix over the complex field and d_{nk} be defined by $d_{nk} := \sum_{j=0}^{n} t_j e_{jk}/T_n$ for all $k, n \in \mathbb{N}$. Then, the following statements hold:*

(i) *$E = (e_{nk}) \in (a_p^r : r_\infty^t)$ if and only if (4.6.10) and (4.6.11) hold with d_{nk} instead of a_{nk}.*

(ii) *$E = (e_{nk}) \in (a_p^r : r_c^t)$ if and only if (4.6.10), (4.6.11) and (4.6.14) hold with d_{nk} instead of a_{nk}.*

(iii) *$E = (e_{nk}) \in (a_p^r : r_0^t)$ if and only if (4.6.10) and (4.6.11) hold, and (4.6.14) also holds with $\alpha_k = 0$ for all $k \in \mathbb{N}$ and d_{nk} instead of a_{nk} for all $k, n \in \mathbb{N}$.*

(iv) $E = (e_{nk}) \in (a_c^r : r_p^t)$ *if and only if (4.6.16)–(4.6.19) hold with* d_{nk} *instead of* a_{nk}.

(v) $E = (e_{nk}) \in (a_c^r : r_c^t)$ *if and only if (4.6.14), (4.6.18), (4.6.19) and (4.6.21) hold with* d_{nk} *instead of* a_{nk}.

(vi) $E = (e_{nk}) \in (a_c^r : r_c^t)_\varrho$ *if and only if (4.6.18), (4.6.19) hold, (4.6.14) and (4.6.21) also hold with* $\alpha_k = 0$ *for each* $k \in \mathbb{N}$ *and* $\alpha = \varrho$, *respectively, and* d_{nk} *instead of* a_{nk}.

Corollary 4.11.6. [45, Corollary 6.7] *Let* $A = (a_{nk})$ *be an infinite matrix over the complex field and* d_{nk} *be defined by* $d_{nk} := e_{nk} - e_{n+1,k}$ *or* $d_{nk} := e_{nk} - e_{n-1,k}$ *for all* $k, n \in \mathbb{N}$. *Then, the following statements hold:*

(i) $E = (e_{nk}) \in (a_p^r : r_\infty(\Delta))$ *if and only if (4.6.10) and (4.6.11) hold with* d_{nk} *instead of* a_{nk}.

(ii) $E = (e_{nk}) \in (a_p^r : c(\Delta))$ *if and only if (4.6.10), (4.6.11) and (4.6.14) hold with* d_{nk} *instead of* a_{nk}.

(iii) $E = (e_{nk}) \in (a_p^r : c_0(\Delta))$ *if and only if (4.6.10) and (4.6.11) hold, and (4.6.14) also holds with* $\alpha_k = 0$ *for all* $k \in \mathbb{N}$ *and* d_{nk} *instead of* a_{nk} *for all* $k, n \in \mathbb{N}$.

(iv) $E = (e_{nk}) \in (a_c^r : bv_p)$ *if and only if (4.6.16)–(4.6.19) hold with* d_{nk} *instead of* a_{nk}.

(v) $E = (e_{nk}) \in (a_c^r : c(\Delta))$ *if and only if (4.6.14), (4.6.18), (4.6.19) and (4.6.21) hold with* d_{nk} *instead of* a_{nk}.

(vi) $E = (e_{nk}) \in (a_c^r : c(\Delta))_\varrho$ *if and only if (4.6.18), (4.6.19) hold, (4.6.14) and (4.6.21) also hold with* $\alpha_k = 0$ *for each* $k \in \mathbb{N}$ *and* $\alpha = \varrho$, *respectively, and* d_{nk} *instead of* a_{nk}.

Corollary 4.11.7. [45, Corollary 6.8] *Let* $A = (a_{nk})$ *be an infinite matrix over the complex field and* d_{nk} *be defined by* $d_{nk} := \sum_{j=0}^{n}(1 + r^k)e_{jk}/(j + 1)$ *for all* $k, n \in \mathbb{N}$. *Then, the following statements hold:*

(i) $E = (e_{nk}) \in (a_p^r : a_\infty^r)$ *if and only if (4.6.10) and (4.6.11) hold with* d_{nk} *instead of* a_{nk}.

(ii) $E = (e_{nk}) \in (a_p^r : a_c^r)$ *if and only if (4.6.10), (4.6.11) and (4.6.14) hold with* d_{nk} *instead of* a_{nk}.

(iii) $E = (e_{nk}) \in (a_p^r : a_0^r)$ *if and only if (4.6.10) and (4.6.11) hold, and (4.6.14) also holds with* $\alpha_k = 0$ *for all* $k \in \mathbb{N}$ *and* d_{nk} *instead of* a_{nk} *for all* $k, n \in \mathbb{N}$.

(iv) $E = (e_{nk}) \in (a_c^r : a_p^r)$ *if and only if (4.6.16)–(4.6.19) hold with* d_{nk} *instead of* a_{nk}.

(v) $E = (e_{nk}) \in (a_c^r : a_c^r)$ *if and only if (4.6.14), (4.6.18), (4.6.19) and (4.6.21) hold with* d_{nk} *instead of* a_{nk}.

(vi) $E = (e_{nk}) \in (a_c^r : a_c^r)_\varrho$ *if and only if (4.6.18), (4.6.19) hold, (4.6.14) and (4.6.21) also hold with* $\alpha_k = 0$ *for each* $k \in \mathbb{N}$ *and* $\alpha = \varrho$, *respectively, and* d_{nk} *instead of* a_{nk}.

Corollary 4.11.8. [45, Corollary 6.9] *Let* $A = (a_{nk})$ *be an infinite matrix over the complex field and* d_{nk} *be defined by* $d_{nk} := e(n,k)$ *for all* $k, n \in \mathbb{N}$. *Then, the following statements hold:*

(i) $E = (e_{nk}) \in (a_p^r : bs)$ *if and only if (4.6.10) and (4.6.11) hold with* d_{nk} *instead of* a_{nk}.

(ii) $E = (e_{nk}) \in (a_p^r : cs)$ *if and only if (4.6.10), (4.6.11) and (4.6.14) hold with* d_{nk} *instead of* a_{nk}.

(iii) $E = (e_{nk}) \in (a_p^r : cs_0)$ *if and only if (4.6.10) and (4.6.11) hold, and (4.6.14) also holds with* $\alpha_k = 0$ *for all* $k \in \mathbb{N}$ *and* d_{nk} *instead of* a_{nk} *for all* $k, n \in \mathbb{N}$.

(iv) $E = (e_{nk}) \in (a_c^r : \overline{\ell_p})$ *if and only if (4.6.16)–(4.6.19) hold with* d_{nk} *instead of* a_{nk}, *where* $\overline{\ell_p}$ *denotes the normed sequence space obtained in the case* $p_k = p$ *for all* $k \in \mathbb{N}$ *from the paranormed sequence space* $\overline{\ell(p)}$ *introduced by Choudhary and Mishra* [105].

(v) $E = (e_{nk}) \in (a_c^r : cs)$ *if and only if (4.6.14), (4.6.18), (4.6.19) and (4.6.21) hold with* d_{nk} *instead of* a_{nk}.

(vi) $E = (e_{nk}) \in (a_c^r : cs)_\varrho$ *if and only if (4.6.18), (4.6.19) hold, (4.6.14) and (4.6.21) also hold with* $\alpha_k = 0$ *for each* $k \in \mathbb{N}$ *and* $\alpha = \varrho$, *respectively, and* d_{nk} *instead of* a_{nk}.

Corollary 4.11.9. *Let* $A = (a_{nk})$ *be an infinite matrix over the complex field and* d_{nk} *be defined as in Corollary 4.11.4. Then, the following statements hold:*

(i) $E = (e_{nk}) \in (r^t(p) : e_\infty^r)$ *if and only if (4.7.16)–(4.7.18) hold with* d_{nk} *instead of* a_{nk}.

(ii) $E = (e_{nk}) \in (r^t(p) : e_c^r)$ *if and only if (4.7.16)–(4.7.18) and (4.7.25) hold with* d_{nk} *instead of* a_{nk}.

(iii) $E = (e_{nk}) \in (r^t(p) : e_0^r)$ *if and only if (4.7.16)–(4.7.18) hold and (4.7.25) also holds with* $\alpha_k = 0$ *for each* $k \in \mathbb{N}$ *and* d_{nk} *instead of* a_{nk}.

Corollary 4.11.10. *Let* $A = (a_{nk})$ *be an infinite matrix over the complex field and* d_{nk} *be defined by* $d_{nk} := e_{nk} - e_{n+1,k}$ *for all* $k, n \in \mathbb{N}$. *Then, the following statements hold:*

(i) $E = (e_{nk}) \in (r^t(p) : \ell_\infty(\Delta))$ *if and only if (4.7.16)–(4.7.18) hold with* d_{nk} *instead of* a_{nk}.

(ii) $E = (e_{nk}) \in (r^t(p) : c(\Delta))$ *if and only if (4.7.16)–(4.7.18) and (4.7.25) hold with* d_{nk} *instead of* a_{nk}.

(iii) $E = (e_{nk}) \in (r^t(p) : c_0(\Delta))$ *if and only if (4.7.16)–(4.7.18) hold and (4.7.25) also holds with* $\alpha_k = 0$ *for each* $k \in \mathbb{N}$ *and* d_{nk} *instead of* a_{nk}.

Corollary 4.11.11. *Let* $A = (a_{nk})$ *be an infinite matrix over the complex field and* d_{nk} *be defined as in Corollary 4.11.7. Then, the following statements hold:*

(i) $E = (e_{nk}) \in (r^t(p) : a_\infty^r)$ *if and only if (4.7.16)–(4.7.18) hold with* d_{nk} *instead of* a_{nk}.

(ii) $E = (e_{nk}) \in (r^t(p) : a_c^r)$ *if and only if (4.7.16)–(4.7.18) and (4.7.25) hold with* d_{nk} *instead of* a_{nk}.

(iii) $E = (e_{nk}) \in (r^t(p) : a_0^r)$ *if and only if (4.7.16)–(4.7.18) hold and (4.7.25) also holds with* $\alpha_k = 0$ *for each* $k \in \mathbb{N}$ *and* d_{nk} *instead of* a_{nk}.

Corollary 4.11.12. [289, Corollary 2.8] *Let* $A = (a_{nk})$ *be an infinite matrix over the complex field and* d_{nk} *be defined as in Corollary 4.11.4. Then, the following statements hold:*

(i) $E = (e_{nk}) \in (e_1^r : e_\infty^r)$ *if and only if (4.8.15) holds with* d_{nk} *instead of* a_{nk}.

(ii) $E = (e_{nk}) \in (e_p^r : e_\infty^r)$ *if and only if (4.8.16) and (4.8.17) hold with* d_{nk} *instead of* a_{nk}.

(iii) $E = (e_{nk}) \in (e_\infty^r : e_\infty^r)$ *if and only if (4.8.18) holds with* d_{nk} *instead of* a_{nk}.

(iv) $E = (e_{nk}) \in (e_1^r : e_p^r)$ *if and only if (4.8.23) holds with* d_{nk} *instead of* a_{nk}.

(v) $E = (e_{nk}) \in (e_p^r : e_c^r)$ *if and only if (4.8.15) and (4.8.26) hold with* d_{nk} *instead of* a_{nk}.

(vi) $E = (e_{nk}) \in (e_p^r : e_c^r)$ *if and only if (4.8.16), (4.8.17) and (4.8.26) hold with* d_{nk} *instead of* a_{nk}.

(vii) $E = (e_{nk}) \in (e_\infty^r : e_c^r)_\varrho$ *if and only if (4.8.16), (4.8.17) and (4.8.27) hold with* d_{nk} *instead of* a_{nk}.

(viii) $E = (e_{nk}) \in (e_c^r : e_p^r)$ *if and only if (4.8.29)–(4.8.32) hold with* d_{nk} *instead of* a_{nk}.

(ix) $E = (e_{nk}) \in (e_c^r : e_\infty^r)$ *if and only if (4.8.30), (4.8.32) and (4.8.18) hold with* d_{nk} *instead of* a_{nk}.

(x) $E = (e_{nk}) \in (e_c^r : e_c^r)$ *if and only if (4.8.30), (4.8.32), (4.8.18), (4.8.26) and (4.8.34) hold with* d_{nk} *instead of* a_{nk}.

(xi) $E = (e_{nk}) \in (e_c^r : e_c^r)_\varrho$ *if and only if (4.8.30), (4.8.32), (4.8.18) hold, (4.8.26) and (4.8.34) also hold with* $\alpha_k = 0$ *for each* $k \in \mathbb{N}$ *and* $\alpha = \varrho$*, respectively, and* d_{nk} *instead of* a_{nk}.

Corollary 4.11.13. [289, Corollary 2.9] *Let $A = (a_{nk})$ be an infinite matrix over the complex field and d_{nk} be defined as in Corollary 4.11.5. Then, the following statements hold:*

(i) $E = (e_{nk}) \in (e_1^r : r_\infty^t)$ *if and only if (4.8.15) holds with d_{nk} instead of a_{nk}.*

(ii) $E = (e_{nk}) \in (e_p^r : r_\infty^t)$ *if and only if (4.8.16) and (4.8.17) hold with d_{nk} instead of a_{nk}.*

(iii) $E = (e_{nk}) \in (e_\infty^r : r_\infty^t)$ *if and only if (4.8.18) holds with d_{nk} instead of a_{nk}.*

(iv) $E = (e_{nk}) \in (e_1^r : r_p^t)$ *if and only if (4.8.23) holds with d_{nk} instead of a_{nk}.*

(v) $E = (e_{nk}) \in (e_p^r : r_c^t)$ *if and only if (4.8.15) and (4.8.26) hold with d_{nk} instead of a_{nk}.*

(vi) $E = (e_{nk}) \in (e_p^r : r_c^t)$ *if and only if (4.8.16), (4.8.17) and (4.8.26) hold with d_{nk} instead of a_{nk}.*

(vii) $E = (e_{nk}) \in (e_\infty^r : r_c^t)$ *if and only if (4.8.16), (4.8.17) and (4.8.27) hold with d_{nk} instead of a_{nk}.*

(viii) $E = (e_{nk}) \in (e_c^r : r_p^t)$ *if and only if (4.8.29)–(4.8.32) hold with d_{nk} instead of a_{nk}.*

(ix) $E = (e_{nk}) \in (e_c^r : r_\infty^t)$ *if and only if (4.8.30), (4.8.32) and (4.8.18) hold with d_{nk} instead of a_{nk}.*

(x) $E = (e_{nk}) \in (e_c^r : r_c^t)$ *if and only if (4.8.30), (4.8.32), (4.8.18), (4.8.26) and (4.8.34) hold with d_{nk} instead of a_{nk}.*

(xi) $E = (e_{nk}) \in (e_c^r : r_c^t)_\varrho$ *if and only if (4.8.30), (4.8.32), (4.8.18) hold, (4.8.26) and (4.8.34) also hold with $\alpha_k = 0$ for each $k \in \mathbb{N}$ and $\alpha = \varrho$, respectively, and d_{nk} instead of a_{nk}.*

Corollary 4.11.14. [289, Corollary 2.10] *Let $A = (a_{nk})$ be an infinite matrix over the complex field and d_{nk} be defined as in Corollary 4.11.6. Then, the following statements hold:*

(i) $E = (e_{nk}) \in (e_1^r : \ell_\infty(\Delta))$ *if and only if (4.8.15) holds with d_{nk} instead of a_{nk}.*

(ii) $E = (e_{nk}) \in (e_p^r : \ell_\infty(\Delta))$ *if and only if (4.8.16) and (4.8.17) hold with d_{nk} instead of a_{nk}.*

(iii) $E = (e_{nk}) \in (e_\infty^r : \ell_\infty(\Delta))$ *if and only if (4.8.18) holds with d_{nk} instead of a_{nk}.*

(iv) $E = (e_{nk}) \in (e_1^r : bv_p)$ *if and only if (4.8.23) holds with d_{nk} instead of a_{nk}.*

(v) $E = (e_{nk}) \in (e_1^r : c(\Delta))$ *if and only if (4.8.15) and (4.8.26) hold with d_{nk} instead of a_{nk}.*

(vi) $E = (e_{nk}) \in (e_p^r : c(\Delta))$ *if and only if (4.8.16), (4.8.17) and (4.8.26) hold with* d_{nk} *instead of* a_{nk}.

(vii) $E = (e_{nk}) \in (e_\infty^r : c(\Delta))$ *if and only if (4.8.16), (4.8.17) and (4.8.27) hold with* d_{nk} *instead of* a_{nk}.

(viii) $E = (e_{nk}) \in (e_c^r : bv_p)$ *if and only if (4.8.29)–(4.8.32) hold with* d_{nk} *instead of* a_{nk}.

(ix) $E = (e_{nk}) \in (e_c^r : \ell_\infty(\Delta))$ *if and only if (4.8.30), (4.8.32) and (4.8.18) hold with* d_{nk} *instead of* a_{nk}.

(x) $E = (e_{nk}) \in (e_c^r : c(\Delta))$ *if and only if (4.8.30), (4.8.32), (4.8.18), (4.8.26) and (4.8.34) hold with* d_{nk} *instead of* a_{nk}.

(xi) $E = (e_{nk}) \in (e_c^r : c(\Delta))_\varrho$ *if and only if (4.8.30), (4.8.32), (4.8.18) hold, (4.8.26) and (4.8.34) also hold with* $\alpha_k = 0$ *for each* $k \in \mathbb{N}$ *and* $\alpha = \varrho$, *respectively, and* d_{nk} *instead of* a_{nk}.

Corollary 4.11.15. *Let* $A = (a_{nk})$ *be an infinite matrix over the complex field and* d_{nk} *be defined as in Corollary 4.11.7. Then, the following statements hold:*

(i) $E = (e_{nk}) \in (e_1^r : a_\infty^r)$ *if and only if (4.8.15) holds with* d_{nk} *instead of* a_{nk}.

(ii) $E = (e_{nk}) \in (e_p^r : a_\infty^r)$ *if and only if (4.8.16) and (4.8.17) hold with* d_{nk} *instead of* a_{nk}.

(iii) $E = (e_{nk}) \in (e_\infty^r : a_\infty^r)$ *if and only if (4.8.18) holds with* d_{nk} *instead of* a_{nk}.

(iv) $E = (e_{nk}) \in (e_1^r : a_p^r)$ *if and only if (4.8.23) holds with* d_{nk} *instead of* a_{nk}.

(v) $E = (e_{nk}) \in (e_p^r : a_c^r)$ *if and only if (4.8.15) and (4.8.26) hold with* d_{nk} *instead of* a_{nk}.

(vi) $E = (e_{nk}) \in (e_p^r : a_c^r)$ *if and only if (4.8.16), (4.8.17) and (4.8.26) hold with* d_{nk} *instead of* a_{nk}.

(vii) $E = (e_{nk}) \in (e_\infty^r : a_c^r)$ *if and only if (4.8.16), (4.8.17) and (4.8.27) hold with* d_{nk} *instead of* a_{nk}.

(viii) $E = (e_{nk}) \in (e_c^r : a_p^r)$ *if and only if (4.8.29)–(4.8.32) hold with* d_{nk} *instead of* a_{nk}.

(ix) $E = (e_{nk}) \in (e_c^r : a_\infty^r)$ *if and only if (4.8.30), (4.8.32) and (4.8.18) hold with* d_{nk} *instead of* a_{nk}.

(x) $E = (e_{nk}) \in (e_c^r : a_c^r)$ *if and only if (4.8.30), (4.8.32), (4.8.18), (4.8.26) and (4.8.34) hold with* d_{nk} *instead of* a_{nk}.

(xi) $E = (e_{nk}) \in (e_c^r : a_c^r)_\varrho$ *if and only if (4.8.30), (4.8.32), (4.8.18) hold, (4.8.26) and (4.8.34) also hold with* $\alpha_k = 0$ *for each* $k \in \mathbb{N}$ *and* $\alpha = \varrho$, *respectively, and* d_{nk} *instead of* a_{nk}.

Corollary 4.11.16. [289, Corollary 2.11] *Let $A = (a_{nk})$ be an infinite matrix over the complex field and d_{nk} be defined as in Corollary 4.11.8. Then, the following statements hold:*

(i) $E = (e_{nk}) \in (e_1^r : bs)$ *if and only if (4.8.15) holds with d_{nk} instead of a_{nk}.*

(ii) $E = (e_{nk}) \in (e_p^r : bs)$ *if and only if (4.8.16) and (4.8.17) hold with d_{nk} instead of a_{nk}.*

(iii) $E = (e_{nk}) \in (e_\infty^r : bs)$ *if and only if (4.8.18) holds with d_{nk} instead of a_{nk}.*

(iv) $E = (e_{nk}) \in (e_1^r : \overline{\ell_p})$ *if and only if (4.8.23) holds with d_{nk} instead of a_{nk}.*

(v) $E = (e_{nk}) \in (e_p^r : fs)$ *if and only if (4.8.15) and (4.8.24) hold with d_{nk} instead of a_{nk}.*

(vi) $E = (e_{nk}) \in (e_p^r : fs)$ *if and only if (4.8.16), (4.8.17) and (4.8.24) hold with d_{nk} instead of a_{nk}.*

(vii) $E = (e_{nk}) \in (e_\infty^r : fs)$ *if and only if (4.8.18), (4.8.19), (4.8.24) and (4.8.25) hold with d_{nk} instead of a_{nk}.*

(viii) $E = (e_{nk}) \in (e_1^r : cs)$ *if and only if (4.8.15) and (4.8.26) hold with d_{nk} instead of a_{nk}.*

(ix) $E = (e_{nk}) \in (e_p^r : cs)$ *if and only if (4.8.16), (4.8.17) and (4.8.26) hold with d_{nk} instead of a_{nk}.*

(x) $E = (e_{nk}) \in (e_\infty^r : cs)$ *if and only if (4.8.16), (4.8.17) and (4.8.27) hold with d_{nk} instead of a_{nk}.*

(xi) $E = (e_{nk}) \in (e_c^r : \overline{\ell_p})$ *if and only if (4.8.29)–(4.8.32) hold with d_{nk} instead of a_{nk}.*

(xii) $E = (e_{nk}) \in (e_c^r : bs)$ *if and only if (4.8.30), (4.8.32) and (4.8.18) hold with d_{nk} instead of a_{nk}.*

(xiii) $E = (e_{nk}) \in (e_c^r : cs)$ *if and only if (4.8.30), (4.8.32), (4.8.18), (4.8.26) and (4.8.34) hold with d_{nk} instead of a_{nk}.*

(xiv) $E = (e_{nk}) \in (e_c^r : cs)_\varrho$ *if and only if (4.8.30), (4.8.32), (4.8.18) hold, (4.8.26) and (4.8.34) also hold with $\alpha_k = 0$ for each $k \in \mathbb{N}$ and $\alpha = \varrho$, respectively, and d_{nk} instead of a_{nk}.*

4.12 CONCLUSION

Let λ be a sequence space and T also be a triangle matrix. Then, since $\lambda_T \cong \lambda$; $x \in \lambda_T$ whenever $y = Tx \in \lambda$, and conversely. Under certain formal hypothesis, this leads us to the similar consequence

$$A := BT \quad \text{or equivalently } B := AT^{-1} \tag{4.12.1}$$

between the matrices A and B. This is the most general form of the duality relation between the summability matrices. The relation (4.12.1) can be expressed in the coordinate form, as follows:

$$a_{nk} := \sum_{j=k}^{\infty} b_{nj} t_{jk} \quad \text{or equivalently } b_{nk} := \sum_{j=k}^{\infty} a_{nj} z_{jk}, \tag{4.12.2}$$

where $T^{-1} = (z_{nk})_{n,k \in \mathbb{N}}$. It is trivial that the relation (4.12.1) is reduced to the relations (3.4.12), (4.2.2), (4.3.2), (4.6.2), (4.7.32), (4.8.2) and (4.9.1) if T is replaced by $S, C, \Delta^{(1)}, A^r, R^t, E^r$ or $G(u,v)$, respectively. The relation (4.12.2) is very useful for obtaining the characterization of the matrix transformations on the domain λ_T of the matrix T whenever the characterization of the matrix transformations on the sequence space λ is known.

Recently, Altay and Başar have used the relation (4.12.2), in their studies, for obtaining the α-, β- and γ-duals of the new sequence space λ_T and the basis of λ_T if λ has.

Quite recently, Altay and Başar [29] have benefited from the relation (4.12.2) for deriving some topological properties, such as monotonicity, solidity, AB-, AD- and KB-properties of the new sequence space λ_T.

Let $\widetilde{r} = (r_n)_{n=0}^{\infty}$, $\widetilde{s} = (s_n)_{n=0}^{\infty}$ and $\widetilde{u} = (u_n)_{n=0}^{\infty}$ be given convergent sequences having certain properties. Define the sequential triple band matrix $B(\widetilde{r}, \widetilde{s}, \widetilde{u}) = \{b_{nk}(r,s,u)\}$ by

$$b_{nk}(r,s,u) := \begin{cases} r_n &, \quad k = n, \\ s_n &, \quad k = n-1, \\ u_n &, \quad k = n-2, \\ 0 &, \quad 0 \leq k \leq n-3 \text{ or } k > n \end{cases}$$

for all $k, n \in \mathbb{N}$. It is clear that the sequential triple band matrix $B(\widetilde{r}, \widetilde{s}, \widetilde{u})$ is most general form of the difference matrix $\Delta^{(1)}$, generalized difference matrix $B(r,s)$, sequential generalized difference matrix $B(\widetilde{r}, \widetilde{s})$ and the triple band matrix $B(r,s,t)$. So, to work on the domain $\lambda_{B(\widetilde{r},\widetilde{s},\widetilde{u})}$ of sequential triple band matrix, $B(\widetilde{r}, \widetilde{s}, \widetilde{u})$ is meaningful if λ is chosen any of the normed sequence spaces $\ell_\infty, f, f_0, c, c_0, \ell_p, \omega_0^p, \omega^p$ and ω_∞^p or any of the paranormed spaces $\ell_\infty(p), c(p), c_0(p), \ell(p), \omega_0(p), \omega(p)$ and $\omega_\infty(p)$.

Let $\lambda(F)$ and T, respectively, denote any of the classical sets $\ell_\infty(F)$, $c(F)$, $c_0(F)$ and $\ell_p(F)$ consisting of the bounded, convergent, null and absolutely p-summable sequences of fuzzy numbers and any of the triangle matrices $S = (s_{nk})$, $C = (c_{nk})$, $\Delta = (\delta_{nk})$, $A^r = (a_{nk}^r)$ and $R^t = (r_{nk}^t)$ of fuzzy numbers. Quite recently, Talo and Başar [379] have determined the α-, β- and γ-duals of the classical set $\lambda(F)$ and have given the necessary and sufficient conditions on an infinite matrix of fuzzy numbers transforming one of the classical sets to the another one. Although the fuzzy analogous of some results related to the space bv_p have been studied by Talo and Başar [378], one can derive the similar results about Cesàro, Riesz, Euler, etc., spaces. Indeed following Talo and Başar [378, 381, 379], since $\lambda(F) \cong \lambda(F)_T$, the reader may achieve to derive the fuzzy analogues of the duality relation (4.12.2) between the infinite matrices of fuzzy numbers such that one of them is applied to the points of $\lambda(F)$, while the other one is applied to the points of $\lambda(F)_T$. This will be a new development leading us to the significant topological results concerned with the isomorphic sets of fuzzy sequences.

CHAPTER 5

Spectrum of Some Particular Matrices

In this chapter, the spectrum and the fine spectrum of the Cesàro operator C_1, the difference operator $\Delta^{(1)}$, the generalized difference operator $B(r, s)$ and the operator generated by the triple band matrix $B(r, s, t)$ acting on the sequence spaces c_0, c, ℓ_p and bv_p with respect to Goldberg's classification are determined, where $1 \leq p < \infty$.

Keywords: Spectrum and resolvent of an operator, fine spectrum of an operator with respect to Goldberg's classification, subdivisions of spectrum, point spectrum, continuous spectrum and residual spectrum, approximate point spectrum, defect spectrum and compression spectrum.

5.1 PRELIMINARIES, BACKGROUND AND NOTATIONS

Let X and Y be Banach spaces and $T : X \to Y$ also be a bounded linear operator. By $R(T)$, we denote *the range of* T, i.e., $R(T) = \{y \in Y : y = Tx,\ x \in X\}$. The *kernel* or *null space* $Ker(T)$ of T consists of all $x \in D(T)$ such that $Tx = \theta$. By $B(X)$, we also denote the *set of all bounded linear operators* on X into itself. If X is any Banach space and $T \in B(X)$, then the *adjoint* T^* of T is a bounded linear operator on the dual X^* of X defined by $(T^*f)(x) = f(Tx)$ for all $f \in X^*$ and $x \in X$.

Let $X \neq \{\theta\}$ be a non-trivial complex normed space and $T : \mathcal{D}(T) \to X$ a linear operator defined on subspace $\mathcal{D}(T) \subseteq X$. We do not assume that $D(T)$ is dense in X, or that T has closed graph $\{(x, Tx) : x \in D(T)\} \subseteq X \times X$. We mean by the expression "T is *invertible*" that there exists a bounded linear operator $S : R(T) \to X$ for which $ST = I$ on $D(T)$ and $\overline{R(T)} = X$; such that $S = T^{-1}$ is necessarily uniquely determined, and linear; the boundedness of S means that T must be *bounded below*, in the sense that there is $k > 0$ for which $\|Tx\| \geq k\|x\|$ for all $x \in D(T)$. Associated with each complex number α is the perturbed operator $T_\alpha = T - \alpha I$, defined on the same domain $D(T)$ as T. The *spectrum* $\sigma(T, X)$ consists of those $\alpha \in \mathbb{C}$ for which T_α is not invertible, and the *resolvent* is the mapping from the complement $\sigma(T, X)$ of the spectrum into the algebra of bounded linear operators on X defined by $\alpha \mapsto T_\alpha^{-1}$.

DOI: 10.1201/9781003294153-5

5.2 SUBDIVISIONS OF THE SPECTRUM

In this section, we give the definitions of the parts point spectrum, continuous spectrum, residual spectrum, approximate point spectrum, defect spectrum and compression spectrum. There are many different ways to subdivide the spectrum of a bounded linear operator. Some of them are motivated by applications to physics, in particular, quantum mechanics.

5.2.1 The Point Spectrum, Continuous Spectrum and Residual Spectrum

The name resolvent is appropriate, since T_α^{-1} helps to solve the equation $T_\alpha x = y$. Thus, $x = T_\alpha^{-1} y$ provided T_α^{-1} exists. More important, the investigation of properties of T_α^{-1} will be basic for an understanding of the operator T itself. Naturally, many properties of T_α and T_α^{-1} depend on α, and spectral theory is concerned with these properties. For instance, we shall be interested in the set of all α's in the complex plane such that T_α^{-1} exists. Boundedness of T_α^{-1} is another property that will be essential. We shall also ask for what α's the domain of T_α^{-1} is dense in X, to name just a few aspects. A *regular value* α of T is a complex number such that T_α^{-1} exists and is bounded, and whose domain is dense in X. For our investigation of T, T_α and T_α^{-1}, we need some basic concepts in the spectral theory, which are given as follows (see [231, pp. 370–371]):

The *resolvent set* $\rho(T, X)$ of T is the set of all regular values α of T. Furthermore, the spectrum $\sigma(T, X)$ is partitioned into three disjoint sets as follows:

The *point (discrete) spectrum* $\sigma_p(T, X)$ is the set such that T_α^{-1} does not exist. An $\alpha \in \sigma_p(T, X)$ is called an *eigenvalue* of T.

The *continuous spectrum* $\sigma_c(T, X)$ is the set such that T_α^{-1} exists and is unbounded and the domain of T_α^{-1} is dense in X.

The *residual spectrum* $\sigma_r(T, X)$ is the set such that T_α^{-1} exists (and may be bounded or not) but the domain of T_α^{-1} is not dense in X.

Therefore, these three parts of spectrum form a disjoint subdivisions such that

$$\sigma(T, X) = \sigma_p(T, X) \cup \sigma_c(T, X) \cup \sigma_r(T, X). \tag{5.2.1}$$

To avoid trivial misunderstandings, let us say that some of the sets defined above may be empty. This is an existence problem, which we shall have to discuss. Indeed, it is well-known that $\sigma_c(T, X) = \sigma_r(T, X) = \emptyset$ and the spectrum $\sigma(T, X)$ consists of only the set $\sigma_p(T, X)$ in the finite-dimensional case.

5.2.2 The Approximate Point Spectrum, Defect Spectrum and Compression Spectrum

In this subsection, following Appell et al. [38], we give the definitions of three more subdivisions of the spectrum called as the *approximate point spectrum*, *defect spectrum* and *compression spectrum*.

Given a bounded linear operator T in a Banach space X, we call a sequence (x_k) in X as a *Weyl sequence* for T if $\|x_k\| = 1$ and $\|Tx_k\| \to 0$, as $k \to \infty$.

In what follows, we call the set

$$\sigma_{ap}(T, X) = \{\lambda \in \mathbb{C} : \text{there exists a Weyl sequence for } \lambda I - T\} \tag{5.2.2}$$

the *approximate point spectrum of* T. Moreover, the subspectrum

$$\sigma_{\delta}(T, X) = \{\lambda \in \mathbb{C} : \lambda I - T \text{ is not surjective}\} \tag{5.2.3}$$

is called *defect spectrum* of T.

The two subspectra given by (5.2.2) and (5.2.3) form a (not necessarily disjoint) subdivisions $\sigma(T, X) = \sigma_{ap}(T, X) \cup \sigma_{\delta}(T, X)$ of the spectrum. There is another subspectrum $\sigma_{co}(T, X) = \{\lambda \in \mathbb{C} : \overline{R(\lambda I - T)} \neq X\}$, which is often called *compression spectrum* in the literature. The compression spectrum gives rise to another (not necessarily disjoint) decomposition $\sigma(T, X) = \sigma_{ap}(T, X) \cup \sigma_{co}(T, X)$ of the spectrum. Clearly, $\sigma_p(T, X) \subseteq \sigma_{ap}(T, X)$ and $\sigma_{co}(T, X) \subseteq \sigma_{\delta}(T, X)$. Moreover, comparing these subspectrum with those in (5.2.1), we note that

$$\begin{aligned} \sigma_r(T, X) &= \sigma_{co}(T, X) \backslash \sigma_p(T, X), \\ \sigma_c(T, X) &= \sigma(T, X) \backslash [\sigma_p(T, X) \cup \sigma_{co}(T, X)]. \end{aligned}$$

Sometimes it is useful to relate the spectrum of a bounded linear operator to that of its adjoint. Building on classical existence and uniqueness results for linear operator equations in Banach spaces and their adjoints.

Proposition 5.2.1. [38, Proposition 1.3, p. 28] *Spectrum and subspectrum of an operator $T \in B(X)$ and its adjoint $T^* \in B(X^*)$ are related by the following relations:*

(a) $\sigma(T^*, X^*) = \sigma(T, X)$.

(b) $\sigma_c(T^*, X^*) \subseteq \sigma_{ap}(T, X)$.

(c) $\sigma_{ap}(T^*, X^*) = \sigma_{\delta}(T, X)$.

(d) $\sigma_{\delta}(T^*, X^*) = \sigma_{ap}(T, X)$.

(e) $\sigma_p(T^*, X^*) = \sigma_{co}(T, X)$.

(f) $\sigma_{co}(T^*, X^*) \supseteq \sigma_p(T, X)$.

(g) $\sigma(T, X) = \sigma_{ap}(T, X) \cup \sigma_p(T^*, X^*) = \sigma_p(T, X) \cup \sigma_{ap}(T^*, X^*)$.

The relations (c)–(f) show that the approximate point spectrum is in a certain sense dual to the defect spectrum, and the point spectrum dual to the compression spectrum. The equality (g) implies, in particular, that $\sigma(T, X) = \sigma_{ap}(T, X)$ if X is a Hilbert space and T is normal. Roughly speaking, this shows that normal (in particular, self-adjoint) operators on Hilbert spaces are most similar to matrices in finite-dimensional spaces (see [38]).

5.2.3 Goldberg's Classification of Spectrum

From Goldberg [187, pp. 58-71], if X is a Banach space and $T \in B(X)$, then there are three possibilities for $R(T)$ and T^{-1}:

(I) $R(T) = X$.

(II) $R(T) \neq \overline{R(T)} = X$.

(III) $\overline{R(T)} \neq X$ and

(1) T^{-1} exists and is continuous.

(2) T^{-1} exists but is discontinuous.

(3) T^{-1} does not exist.

Applying Golberg's classification to T_α, we have three possibilities for T_α and T_α^{-1}:

(I) T_α is surjective.

(II) $R(T_\alpha) \neq \overline{R(T_\alpha)} = X$.

(III) $\overline{R(T_\alpha)} \neq X$ and

(1) T_α is injective and T_α^{-1} is continuous.

(2) T_α is injective and T_α^{-1} is discontinuous.

(3) T_α is not injective.

If these possibilities are combined in all possible ways, nine different states are created. These are labeled by I_1, I_2, I_3, II_1, II_2, II_3, III_1, III_2 and III_3. If α is a complex number such that $T_\alpha \in I_1$ or $T_\alpha \in II_1$, then α is in the resolvent set $\rho(T, X)$ of T. The further classification gives rise to the *fine spectrum* of T. If an operator is in state II_2 for example, then $R(T) \neq \overline{R(T)} = X$ and T^{-1} exists but is discontinuous and we write $\alpha \in II_2\sigma(T, X)$ (see [187]).

If λ is a complex number such that $T_\lambda = \lambda I - T \in I_1$ or $T_\lambda = \lambda I - T \in II_1$, then $\lambda \in \rho(T, X)$. All scalar values of λ not in $\rho(T, X)$ comprise the spectrum of T. The further classification of $\sigma(T, X)$ gives rise to the fine spectrum of T. That is, $\sigma(T, X)$ can be divided into the subsets $I_2\sigma(T, X) = \emptyset$, $I_3\sigma(T, X)$, $II_2\sigma(T, X)$, $II_3\sigma(T, X)$, $III_1\sigma(T, X)$, $III_2\sigma(T, X)$, $III_3\sigma(T, X)$. For example, if $T_\lambda = \lambda I - T$ is in a given state, III_2 (say), then we write $\lambda \in III_2\sigma(T, X)$.

By the definitions given above, we can illustrate the subdivisions (5.2.1) in 5.2: Observe that the case in the first row and second column cannot occur in a Banach space X, by the closed graph theorem. If we are not in the third column, i.e., if λ is not an eigenvalue of T, we may always consider the resolvent operator T_λ^{-1} (on a possibly "thin" domain of definition) as "algebraic" inverse of $\lambda I - T$.

Now, we may give the following lemmas, which are needed in the proof of theorems in this chapter:

TABLE 5.1 State diagram for $B(X)$ and $B(X^*)$ for a non-reflective Banach space X.

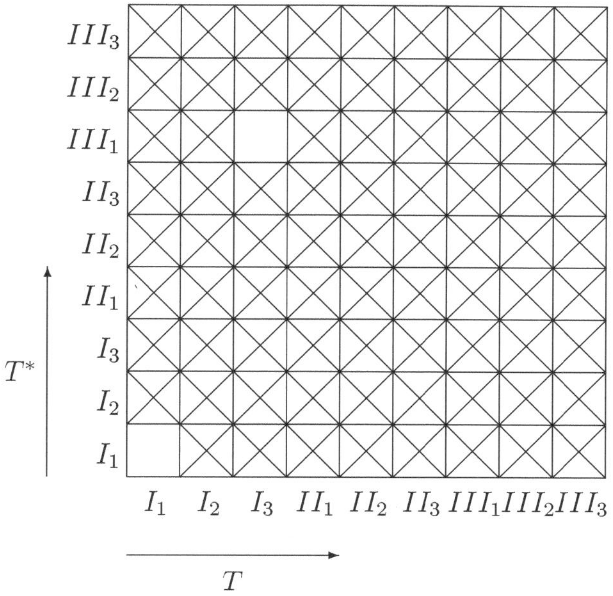

TABLE 5.2 Subdivisions of spectrum of a linear operator.

		1	2	3
		T_λ^{-1} exists and is bounded	T_λ^{-1} exists and is unbounded	T_λ^{-1} does not exist
I	$R(\lambda I - T) = X$	$\lambda \in \rho(T, X)$	–	$\lambda \in \sigma_p(T, X)$ $\lambda \in \sigma_{ap}(T, X)$
II	$\overline{R(\lambda I - T)} = X$	$\lambda \in \rho(T, X)$	$\lambda \in \sigma_c(T, X)$ $\lambda \in \sigma_{ap}(T, X)$ $\lambda \in \sigma_\delta(T, X)$	$\lambda \in \sigma_p(T, X)$ $\lambda \in \sigma_{ap}(T, X)$ $\lambda \in \sigma_\delta(T, X)$
III	$\overline{R(\lambda I - T)} \neq X$	$\lambda \in \sigma_r(T, X)$ $\lambda \in \sigma_\delta(T, X)$ $\lambda \in \sigma_{co}(T, X)$	$\lambda \in \sigma_r(T, X)$ $\lambda \in \sigma_{ap}(T, X)$ $\lambda \in \sigma_\delta(T, X)$ $\lambda \in \sigma_{co}(T, X)$	$\lambda \in \sigma_p(T, X)$ $\lambda \in \sigma_{ap}(T, X)$ $\lambda \in \sigma_\delta(T, X)$ $\lambda \in \sigma_{co}(T, X)$

Lemma 5.2.2. [187, p. 59] *T has a dense range if and only if T^* is one-to-one, where T^* denotes the adjoint operator of the operator T.*

Lemma 5.2.3. [187, p. 60] *The adjoint operator T^* is onto if and only if T has a bounded inverse.*

We should remark the reader from now on that the index p has different meanings in the notation of the spaces ℓ_p, ℓ_p^*, bv_p, bv_p^* and in the point spectrum $\sigma_p(\Delta^{(1)}, \ell_p)$, $\sigma_p(\Delta^{(1)^*}, \ell_p^*)$, $\sigma_p(\Delta^{(1)}, bv_p)$, $\sigma_p(\Delta^{(1)^*}, bv_p^*)$, $\sigma_p(C_1, bv_p)$, $\sigma_p(C_1^*, bv_p^*)$ which occur in the next sections.

We give a short survey concerning with the spectrum and the fine spectrum of the linear operators defined by some particular summability matrices over certain sequence spaces. Wenger [394] examined the fine spectrum of the integer power of the Cesàro operator in c and Rhoades [327] generalized this result to the weighted mean methods. Gonzàlez [189] studied the fine spectrum of the Cesàro operator on the sequence space ℓ_p, where $1 < p < \infty$. Reade [325], Akhmedov and Başar [13] and Okutoyi [308] investigated the spectrum of the Cesàro operator on the sequence spaces c_0 and bv, respectively. Yıldırım [407] examined the fine spectrum of the Rhally operators on the sequence spaces c_0 and c. Furthermore, Coşkun [112] studied the spectrum and fine spectrum for p-Cesàro operator acting on the space c_0. More recently, de Malafosse [259] and Altay and Başar [22] have respectively studied the spectrum and the fine spectrum of the difference operator on the sequence spaces s_r and c_0, c; where s_r denotes the Banach space of all sequences $x = (x_k)$ normed by

$$\|x\|_{s_r} := \sup_{k\in\mathbb{N}} \frac{|x_k|}{r^k}, \quad (r > 0).$$

Altay and Karakuş [32] have determined the fine spectrum of the Zweier matrix, which is a band matrix as an operator over the sequence spaces ℓ_p and bv. In 2010, Srivastava and Kumar [366] have determined the spectra and the fine spectra of generalized difference operator Δ_ν on ℓ_1, where Δ_ν is defined by $(\Delta_\nu)_{nn} = \nu_n$ and $(\Delta_\nu)_{n+1,n} = -\nu_n$ for all $n \in \mathbb{N}$, under certain conditions on the sequence $\nu = (\nu_k)$. At this stage, Table 5.3 may be useful:

5.3 THE FINE SPECTRUM OF THE CESÀRO OPERATOR IN THE SPACES c_0 AND c

In this section, following Akhmedov and Başar [13], we examine the fine spectrum of the Cesàro operator in the sequence spaces c_0 and c. Furthermore, we also state and prove a Mercerian theorem.

The Cesàro operator is represented by the matrix $C_1 = (c_{nk})$ given by (1.1.5). It is known that

(i) $C_1 : c_0 \to c_0$ is a linear operator.

(ii) $\| C_1 \|_{(c_0:c_0)} = 1$.

TABLE 5.3 Spectrum and fine spectrum of some triangle matrices in certain sequence spaces.

$\sigma(A,\lambda)$	$\sigma_p(A,\lambda)$	$\sigma_c(A,\lambda)$	$\sigma_r(A,\lambda)$	Refer to:
$\sigma(C_1^p, c)$	-	-	-	[394]
$\sigma(W, c)$	-	-	-	[327]
$\sigma(C_1, c_0)$	-	-	-	[325]
$\sigma(C_1, c_0)$	$\sigma_p(C_1, c_0)$	$\sigma_c(C_1, c_0)$	$\sigma_r(C_1, c_0)$	[13]
$\sigma(C_1, bv)$	-	-	-	[308]
$\sigma(R, c_0)$	$\sigma_p(R, c_0)$	$\sigma_c(R, c_0)$	$\sigma_r(R, c_0)$	[407]
$\sigma(R, c)$	$\sigma_p(R, c)$	, $\sigma_c(R, c)$	$\sigma_r(R, c)$	[407]
$\sigma(C_1^p, c_0)$	-	-	-	[112]
$\sigma(\Delta, s_r)$	-	-	-	[259]
$\sigma(\Delta, c_0)$	-	-	-	[259]
$\sigma(\Delta, c)$	-	-	-	[259]
$\sigma(\Delta^{(1)}, c)$	$\sigma_p(\Delta^{(1)}, c)$	$\sigma_c(\Delta^{(1)}, c)$	$\sigma_r(\Delta^{(1)}, c)$	[22]
$\sigma(\Delta^{(1)}, c_0)$	$\sigma_p(\Delta^{(1)}, c_0)$	$\sigma_c(\Delta^{(1)}, c_0)$	$\sigma_r(\Delta^{(1)}, c_0)$	[22]
$\sigma(B(r,s), \ell_p)$	$\sigma_p(B(r,s), \ell_p)$	$\sigma_c(B(r,s), \ell_p)$	$\sigma_r(B(r,s), \ell_p)$	[93]
$\sigma(B(r,s), bv_p)$	$\sigma_p(B(r,s), bv_p)$	$\sigma_c(B(r,s), bv_p)$	$\sigma_r(B(r,s), bv_p)$	[93]
$\sigma(B(r,s,t), \ell_p)$	$\sigma_p(B(r,s,t), \ell_p)$	$\sigma_c(B(r,s,t), \ell_p)$	$\sigma_r(B(r,s,t), \ell_p)$	[178]
$\sigma(B(r,s,t), bv_p)$	$\sigma_p(B(r,s,t), bv_p)$	$\sigma_c(B(r,s,t), bv_p)$	$\sigma_r(B(r,s,t), bv_p)$	[178]

(iii) $C_1^* : \ell_1 \to \ell_1$ and is represented by the matrix

$$C_1^* = \begin{bmatrix} 1 & \frac{1}{2} & \frac{1}{3} & \cdots & \frac{1}{n+1} & \cdots \\ 0 & \frac{1}{2} & \frac{1}{3} & \cdots & \frac{1}{n+1} & \cdots \\ 0 & 0 & \frac{1}{3} & \cdots & \frac{1}{n+1} & \cdots \\ \vdots & \vdots & \vdots & \ddots & \vdots & \vdots \\ 0 & 0 & 0 & \cdots & \frac{1}{n+1} & \cdots \\ \vdots & \vdots & \vdots & \cdots & \vdots & \ddots \end{bmatrix},$$

where C_1^* denotes the dual operator of C_1.

We shall begin with quoting some lemmas that the first of them is *Weierstrass criterion* and is needed in the proof of Theorem 5.3.4, and the others summarize the consequences on the spectrum of the Cesàro operator in the sequence space c_0, in the existing literature.

Lemma 5.3.1. [226, Weierstrass criterion; pp. 398–399] *A series $\sum_n a_n$ of complex terms, for which*

$$\frac{a_{n+1}}{a_n} = 1 - \frac{b}{n} - \frac{A_n}{n^d}$$

with A_n bounded, is absolutely convergent if and only if $Re(b) > 1$; where b is a complex number, $d > 1$ and $Re(b)$ denotes the real part of $b \in \mathbb{C}$. For $Re(b) \leq 0$ the *series is invariably divergent. If* $0 < Re(b) \leq 1$, *both the series* $\sum_n |a_n - a_{n+1}|$ *and* $\sum_n (-1)^n a_n$ *are convergent.*

Lemma 5.3.2. [325, Theorem 3] $\sigma(C_1, c_0) := \{\alpha : |\alpha - \frac{1}{2}| \leq \frac{1}{2}\}$.

Lemma 5.3.3. [327] $\sigma_p(C_1, c_0) := \emptyset$.

Theorem 5.3.4. [13, Theorem 2.4] $\sigma_p(C_1^*, c_0^*) := \left\{\alpha \in \mathbb{C} : \left|\alpha - \frac{1}{2}\right| < \frac{1}{2}\right\} \cup \{1\}$.

Proof. Suppose that $C_1^* f = \alpha f$ for $f = (f_0, f_1, f_2, \ldots) \neq \theta$ in $c_0^* \cong \ell_1$. Then, by solving the system of linear equations

$$\left.\begin{array}{rcl} f_0 + \frac{1}{2}f_1 + \frac{1}{3}f_2 + \frac{1}{4}f_3 + \cdots & = & \alpha f_0 \\ \frac{1}{2}f_1 + \frac{1}{3}f_2 + \frac{1}{4}f_3 + \cdots & = & \alpha f_1 \\ \frac{1}{3}f_2 + \frac{1}{4}f_3 + \cdots & = & \alpha f_2 \\ & \vdots & \end{array}\right\}$$

we obtain that

$$f_n = \left[\prod_{k=1}^{n}\left(1 - \frac{1}{k\alpha}\right)\right] f_0 \quad \text{for all } \ n \in \mathbb{N}_1.$$

(i) If $m+1$ is the first integer for which $f_{m+1} = 0$, then $\alpha = 1/(m+1)$. So, $\alpha = 1/(m+1)$ is an eigenvalue with the corresponding eigenvector

$$f = \left[\sum_{k=0}^{m}(-1)^{k-1}\binom{m}{k}e^{(k)}\right] f_m \in \ell_1.$$

(ii) If $f_n \neq 0$ for all $n \in \mathbb{N}$, then

$$\frac{f_{n+1}}{f_n} = 1 - \frac{1/\alpha}{n} - \frac{A_n}{n^2},$$

where $A_n = -n/\alpha(n+1)$. Since (A_n) is a bonded sequence, it follows from the Weierstrass criterion that $f = (f_0, f_1, f_2, \ldots)$ is in ℓ_1 if and only if $Re(1/\alpha) > 1$.

(iii) If $\alpha = 1$, then it is easy to show that the corresponding eigenvector is $(f_0, 0, 0, \ldots) \in \ell_1$.

Thus, one can easily see by combining the discussions in Parts (i)-(iii), above, that

$$\begin{array}{rcl} \sigma_p(C_1^*, c_0^*) & = & \left\{\alpha \in \mathbb{C} : Re\left(\frac{1}{\alpha}\right) > 1\right\} \cup \{1\} \\ & \equiv & \left\{\alpha \in \mathbb{C} : \left|\alpha - \frac{1}{2}\right| < \frac{1}{2}\right\} \cup \{1\}, \end{array}$$

which is what we wished to prove. □

Theorem 5.3.5. [13, Theorem 2.5] $\sigma_r(C_1, c_0) := \sigma_p(C_1^*, \ell_1)$.

Proof. If $\alpha \neq 1/(m+1)$ for each $m \in \mathbb{N}$, then $C_1 - \alpha I$ is a triangle matrix. Hence, $(C_1 - \alpha I)^{-1}$ exists. In the case $\alpha = 1/(m+1)$, $(C_1 - \alpha I)x = \theta$ yields that

$$x_1 = x_2 = \cdots = x_{m-1} = 0 \text{ and } x_{m+k} = \binom{m+k}{m} x_m$$

for all $k \in \mathbb{N}$. Thus, we have

$$x = (x_m) \in c_0 \iff x_m = 0,$$

which means that $x = \theta$. Hence, $C_1 - \alpha I$ is one-to-one operator on the space c_0. Therefore, $C_1 - \alpha I$ has an inverse. Now, it is deduced from the arguments, above, that the operator $C_1 - \alpha I$ has an inverse for $\alpha \in \sigma_p(C_1^*, \ell_1)$. Since $C_1^* - \alpha I$ is not one-to-one operator by Theorem 5.3.4, Lemma 5.2.2 gives the fact that the range $C_1 - \alpha I$ is not dense in c_0. □

Theorem 5.3.6. [13, Theorem 2.6] $\sigma_c(C_1, c_0) := \left\{\alpha \in \mathbb{C} : \left|\alpha - \frac{1}{2}\right| = \frac{1}{2},\ \alpha \neq 1\right\}$.

Proof. Let $\alpha \in \mathbb{C} \setminus \{1\}$ such that $\left|\alpha - \frac{1}{2}\right| = \frac{1}{2}$. Since $\alpha \neq 1/(m+1)$ for each $m \in \mathbb{N}$, $C_1 - \alpha I$ is triangle and hence it has an inverse.

Consider the adjoint operator $C_1^* - \alpha I$. Then, we obtain by the linear system of homogenous equations in the matrix form $(C_1^* - \alpha I)f = \theta$ that

$$f_n := \left\{\prod_{k=0}^{n-1}\left[1 - \frac{1}{(k+1)\alpha}\right]\right\} f_0 \quad \text{for} \quad n \in \mathbb{N}_1.$$

Since $Re(1/\alpha) = 1$, we have

$$f = (f_0, f_1, f_2, \ldots) \in \ell_1 \iff f = \theta.$$

Hence, $C_1^* - \alpha I$ is one-to-one. From Lemma 5.2.2, the range of $C_1 - \alpha I$ is dense in c_0.

This step completes the proof. □

Combining Theorems 5.3.4–5.3.6, we have the following main result:

Theorem 5.3.7. [13, Theorem 2.7] *For the Cesàro operator $C_1 : c_0 \to c_0$, the following statements hold:*

(i) $\sigma(C_1, c_0) := \left\{\alpha \in \mathbb{C} : \left|\alpha - \frac{1}{2}\right| \leq \frac{1}{2}\right\}$.

(ii) $\sigma_p(C_1, c_0) := \emptyset$.

(iii) $\sigma_c(C_1, c_0) := \left\{\alpha \in \mathbb{C} : \left|\alpha - \frac{1}{2}\right| = \frac{1}{2},\ \alpha \neq 1\right\}$.

(iv) $\sigma_r(C_1, c_0) := \left\{\alpha \in \mathbb{C} : \left|\alpha - \frac{1}{2}\right| < \frac{1}{2}\right\} \cup \{1\}$.

Now, we give the corresponding theorems to parts of Theorem 5.3.7 on the fine spectrum of the Cesàro operator in the sequence space c, which can be proved in the similar way to that of the space c_0, above. We should record here that the adjoint operator C_1^* on the dual space $c^* = \mathbb{C} \oplus \ell_1$ of the sequence space c is represented by the matrix of the form

$$\begin{bmatrix} 1 & 0 \\ 0 & C_1' \end{bmatrix},$$

where C_1' denotes the transpose of the matrix C_1. So, we omit details.

Theorem 5.3.8. [13, Theorem 2.8] *For the Cesàro operator* $C_1 : c \to c$, *the following statements hold:*

(i) $\sigma(C_1, c) := \sigma(C_1, c_0)$.

(ii) $\sigma_p(C_1, c) := \{1\}$.

(iii) $\sigma_c(C_1, c) := \{\alpha \in \mathbb{C} : |\alpha - \frac{1}{2}| = \frac{1}{2}\}$.

(iv) $\sigma_r(C_1, c) := \{\alpha \in \mathbb{C} : |\alpha - \frac{1}{2}| < \frac{1}{2}\}$.

Since $\sigma(A, \ell_\infty) = \sigma(A, c)$ by Cartlidge [101] whenever a matrix operator A is bounded on c, as an easy consequence of Theorem 5.3.8, we have

Theorem 5.3.9. [13, Theorem 2.9] *For the Cesàro operator* $C_1 : \ell_\infty \to \ell_\infty$, *we have:* $\sigma(C_1, \ell_\infty) := \sigma(C_1, c)$.

Remark 5.3.10. [13, Remark] *We should remark the reader that Parts (a) and (b) of Theorem 5.6.7 are established in [325] and [328]. In contrast to [325, Theorem 2], we show that* $\sigma_p(C_1^*, c_0^* \cong \ell_1)$ *does not consist of only the points of* $\alpha \in \mathbb{C}$ *such that* $|\alpha - \frac{1}{2}| < \frac{1}{2}$, *and the point* $\alpha = 1$ *must also be added to the set.*

After stating the *Mercerian theorem*, we give a Mercerian theorem. Let A be an infinite matrix and the set c_A denotes the convergence field of that matrix A. A theorem which proves that $c_A = c$ is called a Mercerian theorem, after Mercer, who proved a significant theorem of this type [94, p. 91].

Now, we may present our final theorem of this section.

Theorem 5.3.11. [13, Theorem 2.10] *Suppose that* α *is in the set*

$$\{\alpha \in \mathbb{C} : |\alpha + 1| > |\alpha - 1|\} = \{\alpha \in \mathbb{C} : Re(\alpha) > 0\}.$$

Then, the convergence field of $A = \alpha I + (1 - \alpha)C_1$ *is the space* c.

Proof. Let us write $A = (\beta I - C_1)/(\beta - 1)$ with $\beta = \alpha/(\alpha - 1)$. Then, the inverse of the operator $A = (\beta I - C_1)/(\beta - 1)$ exists and continuous whenever $\beta \in \rho(C_1, c)$. This yields the desired fact that $|\alpha + 1| > |\alpha - 1|$, which completes the proof. □

5.4 THE FINE SPECTRA OF THE DIFFERENCE OPERATOR $\Delta^{(1)}$ ON THE SPACE ℓ_p

In this section, following Akhmedov and Başar [14], we examine the fine spectrum with respect to the Goldberg's classification of the difference operator $\Delta^{(1)}$ on ℓ_p, where $1 < p < \infty$. We shall begin with stating and proving the basic result concerning the spectrum of difference operator $\Delta^{(1)}$ over ℓ_p. The difference operator $\Delta^{(1)}$ is represented by the matrix

$$\Delta^{(1)} = \begin{bmatrix} 1 & 0 & 0 & 0 & \cdots \\ -1 & 1 & 0 & 0 & \cdots \\ 0 & -1 & 1 & 0 & \cdots \\ 0 & 0 & -1 & 1 & \cdots \\ \vdots & \vdots & \vdots & \vdots & \ddots \end{bmatrix}.$$

One can trivially see that

$$\left\|\Delta^{(1)}x\right\|_p \leq 2\|x\|_p$$

for any $x \in \ell_p$ which leads us to the boundedness of the linear operator $\Delta^{(1)}$ on ℓ_p.

Theorem 5.4.1. [14, Theorem 2.1] *Define the set C by $C := \{\alpha \in \mathbb{C} : |\alpha - 1| \leq 1\}$. Then, $\sigma(\Delta^{(1)}, \ell_p) = C$.*

Proof. Firstly, we show that the set C includes $\sigma(\Delta^{(1)}, \ell_p)$. Let $y = (y_k) \in \ell_p$ and solve the equation $(\Delta^{(1)} - \alpha I)x = y$. One can calculate that

$$x_k := \sum_{j=0}^{k} \frac{y_j}{(1-\alpha)^{k+1-j}} \quad \text{for all} \quad k \in \mathbb{N}.$$

Therefore, we obtain that

$$|x_k|^p \leq \frac{(k+1)^{p-1}}{|1-\alpha|^p} \sum_{j=0}^{k} \frac{|y_j|^p}{|1-\alpha|^{p(k-j)}}$$

for all $k \in \mathbb{N}$. Thus, we have

$$\begin{aligned} \sum_k |x_k|^p &\leq \frac{1}{|1-\alpha|^p} \sum_k \sum_{j=k}^{\infty} \frac{(j+1)^{p-1}}{|1-\alpha|^{jp}} |y_k|^p \\ &\leq \frac{1}{|1-\alpha|^p} \left[\sum_{j=0}^{k} \frac{(j+1)^{p-1}}{|1-\alpha|^{jp}} \right] \left(|y_0|^p + |y_1|^p + |y_2|^p + \cdots \right). \end{aligned}$$

Now, we can investigate the convergence of the series

$$\sum_k a_k = \sum_k \frac{(k+1)^{p-1}}{|1-\alpha|^{kp}}. \tag{5.4.1}$$

One can see by d'Alembert criterion that

$$\lim_{k\to\infty}\frac{a_{k+1}}{a_k}=\frac{1}{|1-\alpha|^p},$$

which implies that the series in (5.4.1) as well as $\sum_k |x_k|^p$ converges if $|\alpha-1|>1$. So, the equation $(\Delta^{(1)}-\alpha I)x=y$ has an unique solution in the space ℓ_p if $|1-\alpha|>1$. Hence, $\alpha\notin C$, i.e., $\alpha\notin\sigma(\Delta^{(1)},\ell_p)$. That is to say that the inclusion

$$\sigma(\Delta^{(1)},\ell_p)\subset C \tag{5.4.2}$$

holds.

Conversely, suppose that $\alpha\notin\sigma(\Delta^{(1)},\ell_p)$. Then, $(\Delta^{(1)}-\alpha I)^{-1}\in B(\ell_p)$. Since the $(\Delta^{(1)}-\alpha I)^{-1}$-transform of the unit sequence $e^{(0)}$ is in ℓ_p, we have $|1-\alpha|>1$. This leads us to the inclusion

$$C\subset\sigma(\Delta^{(1)},\ell_p). \tag{5.4.3}$$

Therefore, by combining the inclusions (5.4.2) and (5.4.3) we obtain the desired result.

This completes the proof. □

Prior to giving our theorem related to the norm of the operator $\Delta^{(1)}$ acting on the sequence space ℓ_p, we wish to define the spectral radius of a bounded linear operator, which is needed in the proof.

The *spectral radius* $r_\sigma(T)$ of an operator $T\in B(X)$ on a complex Banach space X is the radius of the smallest closed disk centered at the origin of the complex α-plane, i.e.,

$$r_\sigma(T)=\sup_{\alpha\in\sigma(T,X)}|\alpha|$$

and containing $\sigma(T,X)$ (see [231, p. 378]). It is obvious that the inequality

$$r_\sigma(T)\le\|T\|$$

holds for the spectral radius of a bounded linear operator T on a complex Banach space.

Theorem 5.4.2. [14, Theorem 2.2] $\Delta^{(1)}\in B(\ell_p)$ *endowed with the norm* $\|\Delta^{(1)}\|_{(\ell_p:\ell_p)}=2$.

Proof. The linearity of the operator $\Delta^{(1)}:\ell_p\to\ell_p$ is trivial and so, is omitted. Let us take any $x=(x_k)\in\ell_p$. Since

$$\left(\left\|\Delta^{(1)}x\right\|_p\right)^p\le 2^p\|x\|_p^p$$

then we have

$$\left\|\Delta^{(1)}x\right\|_{(\ell_p:\ell_p)}\le 2. \tag{5.4.4}$$

Besides, because of $r_\sigma(\Delta^{(1)}) = 2$ by Theorem 5.3.4 we observe that

$$\left\|\Delta^{(1)}x\right\|_{(\ell_p:\ell_p)} \geq 2. \tag{5.4.5}$$

Therefore, by combining the inequalities (5.4.4) and (5.4.5) we obtain that

$$\left\|\Delta^{(1)}x\right\|_{(\ell_p:\ell_p)} = 2,$$

as desired, which concludes the proof. □

Theorem 5.4.3. [14, Theorem 2.3] $\sigma_p(\Delta^{(1)}, \ell_p) := \emptyset$.

Proof. This is immediate by combining the well-known inclusion $\ell_p \subset c_0$ by the fact given Theorem 2.2 of Altay and Başar [22]. □

Theorem 5.4.4. [14, Theorem 2.4] *The following statements hold:*

$$\sigma_p(\Delta^{(1)^*}, \ell_p^*) := \begin{cases} \{\alpha \in \mathbb{C} : |\alpha - 1| < 1\} & , \quad 1 < p < \infty, \\ \{\alpha \in \mathbb{C} : |\alpha - 1| \leq 1\} & , \quad p = 1. \end{cases}$$

Proof. Suppose $\Delta^{(1)^*} f = \alpha f$ for $f \neq \theta$ in ℓ_p^*; where $\ell_p^* \cong \ell_q$ if $1 < p < \infty$ and $\ell_1^* \cong \ell_\infty$. Solving the system of linear equations

$$\left.\begin{array}{rcl} f_0 - f_1 & = & \alpha f_0 \\ f_1 - f_2 & = & \alpha f_1 \\ f_2 - f_3 & = & \alpha f_2 \\ & \vdots & \\ f_k - f_{k+1} & = & \alpha f_k \\ & \vdots & \end{array}\right\} \tag{5.4.6}$$

we obtain that $f_k = (1 - \alpha)^k f_0$ for all $k \in \mathbb{N}$. This shows that $f = (f_k) \in \ell_p^*$ if and only if $|\alpha - 1| < 1$ and $f = (f_k) \in \ell_1^*$ if and only if $|\alpha - 1| \leq 1$. This means that $f \in \ell_p^*$ if $f_0 \neq 0$ and $|\alpha - 1| < 1$ or $|\alpha - 1| \leq 1$, as asserted.

This completes the proof. □

Theorem 5.4.5. [14, Theorem 2.7] *The following statements hold:*

$$\sigma_r(\Delta^{(1)}, \ell_p) := \begin{cases} \{\alpha \in \mathbb{C} : |\alpha - 1| < 1\} & , \quad 1 < p < \infty, \\ \{\alpha \in \mathbb{C} : |\alpha - 1| \leq 1\} & , \quad p = 1. \end{cases}$$

Proof. For $|\alpha - 1| < 1$, the operator $\Delta^{(1)} - \alpha I$ is one-to-one in the space ℓ_p, whenever $1 < p < \infty$. Hence, it has an inverse. But $\Delta^{(1)^*} - \alpha I$ is not one-to-one by Theorem 5.4.4. Now, Lemma 5.2.2 yields the fact that the range of the operator $\Delta^{(1)} - \alpha I$ is not dense in ℓ_p and this step concludes the proof of the first part of the theorem.

Analogously, one can obtain the proof of the case $p = 1$. □

Theorem 5.4.6. [14, Theorem 2.8] $1 \in III_1\sigma(\Delta^{(1)}, \ell_p)$.

Proof. By Theorem 5.4.4 and Lemma 5.2.2, we have $\Delta^{(1)} - I \in III$. Additionally, 1 is not in $\sigma_p(\Delta^{(1)}, \ell_p)$ by Theorem 5.4.3. Hence, $\Delta^{(1)} - I$ has an inverse. Thus, $\Delta^{(1)} - I \in 1 \cup 2$.

For establishing the fact $\Delta^{(1)} - I \in 1$, it is enough to show, by Lemma 5.2.3, that $\Delta^{(1)^*} - I$ is onto. For a given $y = (y_k) \in \ell_q$, we have to find $x = (x_k) \in \ell_q$ such that $(\Delta^{(1)^*} - I)x = y$. A direct calculation yields that $x_n = -y_{n-1}$ for all $n \in \mathbb{N}$. This means that $\Delta^* - I$ is onto, as desired. □

Theorem 5.4.7. [14, Theorem 2.9] $1 \in III_1\sigma(\Delta^{(1)}, \ell_p)$.

Proof. Since $\alpha \neq 1$, the operator $\Delta^{(1)} - \alpha I$ is triangle. Hence, it has an inverse. Take $e^{(0)} \in \ell_p$. Then, $(\Delta^{(1)} - \alpha I)^{-1}$-transform of $e^{(0)}$ is $y = \big((1-\alpha)^{-1}, (1-\alpha)^{-2}, (1-\alpha)^{-3}, \dots\big)$, which is not in ℓ_p. This shows that $(\Delta^{(1)} - \alpha I)^{-1}$ is discontinuous. Therefore, $\Delta^{(1)} - \alpha I \in 2$. It is clear that $\alpha \in III$ by Theorem 5.4.5.

This completes the proof. □

Theorem 5.4.8. [14, Theorem 2.10] *The following statements hold:*

$$\sigma_c(\Delta^{(1)}, \ell_p) := \begin{cases} \{\alpha \in \mathbb{C} : |\alpha - 1| = 1\} & , \quad 1 < p < \infty, \\ \emptyset & , \quad p = 1. \end{cases}$$

Proof. Because the parts $\sigma_p(T, X)$, $\sigma_r(T, X)$ and $\sigma_c(T, X)$ of the spectrum $\sigma(T, X)$ of $T \in B(X)$ are disjoint and their union is $\sigma(T, X)$, we always have the relation

$$\sigma_c(T, X) = \sigma(T, X) \setminus \{\sigma_p(T, X) \cup \sigma_r(T, X)\}.$$

Now, the proof of the theorem immediately follows from Theorem 5.4.1, Theorem 5.4.4 and Theorem 5.4.5. □

Theorem 5.4.9. [14, Theorem 2.11] *If* $\alpha \in \sigma_c(\Delta^{(1)}, \ell_p)$, *then* $\alpha \in II_2$.

Proof. Let us take $\alpha \in \sigma_c(\Delta^{(1)}, \ell_p)$. In order to prove that $\alpha \in II_2$, we must show that $\Delta^{(1)} - \alpha I$ is not onto, $\overline{R(\Delta^{(1)} - \alpha I)} = \ell_p$ and $(\Delta^{(1)} - \alpha I)^{-1}$ is discontinuous.

In the case $\alpha \neq 0$, taking $y = e^{(0)}$ leads us to the sequence $x = (x_k)$ with $x_k = 1/(1-\alpha)^{k+1}$ for all $k \in \mathbb{N}$, which is not in the space, ℓ_p. Also in the case $\alpha = 0$, the sequence $y = \{(k+1)^{-1}\}$ yields $x = (x_k)$ with $x_k = \sum_{j=0}^{k} 1/(j+1)$ for all $k \in \mathbb{N}$, which is not in the space ℓ_p. This shows that $(\Delta^{(1)} - \alpha I)^{-1}$ is discontinuous and $\Delta^{(1)} - \alpha I$ is not onto. Therefore, $\Delta^{(1)} - \alpha I$ is in 2 but not in I. Besides, by Theorem 5.4.8, we have that $\overline{R(\Delta^{(1)} - \alpha I)} = \ell_p$.

This step completes the proof. □

5.5 THE FINE SPECTRA OF THE DIFFERENCE OPERATOR $\Delta^{(1)}$ ON THE SPACE bv_p

In this section, following Akhmedov and Başar [15], we examine the fine spectrum with respect to the Goldberg's classification of the difference operator $\Delta^{(1)}$ on the sequence space bv_p and find the norm of the difference operator $\Delta^{(1)}$, where $1 < p < \infty$.

Theorem 5.5.1. [15, Theorem 3.1] $\sigma(\Delta^{(1)}, bv_p) := \{\alpha \in \mathbb{C} : |\alpha - 1| \leq 1\}$.

Proof. Solving the equation

$$\left(\Delta^{(1)} - \alpha I\right)x = y \tag{5.5.1}$$

for $y \in bv_1$, we can formally derive that

$$\left(\Delta^{(1)} - \alpha I\right)^{-1} = \begin{bmatrix} 1/(1-\alpha) & 0 & 0 & \cdots \\ 1/(1-\alpha)^2 & 1/(1-\alpha) & 0 & \cdots \\ 1/(1-\alpha)^3 & 1/(1-\alpha)^2 & 1/(1-\alpha) & \cdots \\ \vdots & \vdots & \vdots & \ddots \end{bmatrix}.$$

Therefore, one can obtain by (5.5.1) that $x = \left(\Delta^{(1)} - \alpha I\right)^{-1} y$. Thus, we observe that

$$\left\|\left(\Delta^{(1)} - \alpha I\right)^{-1}\right\|_{(bv_1:bv_1)} = \sum_k \frac{1}{|(1-\alpha)^{k+1}|} < \infty$$

if and only if $|\alpha - 1| > 1$. At this situation, it is easy to show that $x_k - x_{k-1} = \sum_{j=0}^{k}(y_j - y_{j-1})/(1-\alpha)^{k-j+1}$ for all $k \in \mathbb{N}$ and $\left(\Delta^{(1)} - \alpha I\right)^{-1}$ is bounded if and only if $|\alpha - 1| > 1$. It follows from here that $\sigma(\Delta^{(1)}, bv_1) := \{\alpha \in \mathbb{C} : |\alpha - 1| \leq 1\}$.

Let $y = (y_k) \in bv_p$ and $p > 1$. Solve the equation $(\Delta^{(1)} - \alpha I)x = y$. Then, one can calculate that $x_k = \sum_{j=0}^{k} y_j/(1-\alpha)^{k+1-j}$ for all $k \in \mathbb{N}$. Therefore, we obtain for all $k \in \mathbb{N}$ that

$$|x_k - x_{k-1}|^p \leq (k+1)^{p-1} \sum_{j=0}^{k} \frac{|y_j - y_{j-1}|^p}{|1-\alpha|^{p(k+1-j)}}.$$

Thus, the d'Alambert criterion yields the fact that the series $\sum_k (k+1)^{p-1}/|1-\alpha|^{kp}$ converges if $|1-\alpha| > 1$. Hence, the equation $(\Delta^{(1)} - \alpha I)x = y$ has an unique solution $x \in bv_p$ if $|1-\alpha| > 1$ and $\left(\Delta^{(1)} - \alpha I\right)^{-1}$ is a bounded linear operator for such α's, which is what we wished to prove. □

Now, we can give the theorem related to the norm of the operator $\Delta^{(1)}$ acting on the sequence space bv_p.

Theorem 5.5.2. [15, Theorem 3.2] $\Delta^{(1)} \in B(bv_p)$ *endowed with the norm* $\|\Delta^{(1)}\|_{(bv_p:bv_p)} = 2$.

Proof. The linearity of the operator $\Delta^{(1)} : bv_p \to bv_p$ is trivial. Let us take any $x = (x_k) \in bv_p$. Then, since

$$\begin{aligned} \left\|\Delta^{(1)} x\right\|_{bv_p} &= \left(\sum_k |x_k - x_{k-1} - (x_{k-1} - x_{k-2})|^p\right)^{1/p} \\ &\leq \left(\sum_k |x_k - x_{k-1}|^p\right)^{1/p} + \left(\sum_{k=1}^{\infty} |x_{k-1} - x_{k-2}|^p\right)^{1/p} \\ &= 2\|x\|_{bv_p} \end{aligned}$$

we derive from here that

$$\left\|\Delta^{(1)}\right\|_{(bv_p:bv_p)} \leq 2. \tag{5.5.2}$$

Further, because of $r_\sigma(\Delta^{(1)}) = 2$ by Theorem 5.5.1 we have with (5.4.4) that

$$2 \leq \left\|\Delta^{(1)}\right\|_{(bv_p:bv_p)}. \tag{5.5.3}$$

Therefore, the desired result is obtained by combining the inequalities (5.5.2) and (5.5.3).

This step concludes the proof. □

Theorem 5.5.3. [15, Theorem 3.3] $\sigma_p(\Delta^{(1)}, bv_p) = \emptyset$.

Proof. Suppose that $\Delta^{(1)}x = \alpha x$ for $x \neq \theta = (0,0,0,\ldots)$ in bv_p. Then, by solving the system of the linear equations

$$\left.\begin{array}{rcl} x_0 & = & \alpha x_0 \\ x_1 - x_0 & = & \alpha x_1 \\ x_2 - x_1 & = & \alpha x_2 \\ & \vdots & \\ x_k - x_{k-1} & = & \alpha x_k \\ & \vdots & \end{array}\right\}$$

we find that if x_{n_0} is the first non-zero entry of the sequence $x = (x_n)$, then $\alpha = 1$ and

$$x_{n_0+1} - x_{n_0} = x_{n_0+1},$$

which implies that $x_{n_0} = 0$. This contradicts the fact that $x_{n_0} \neq 0$, which means that $\sigma_p(\Delta^{(1)}, bv_p) = \emptyset$.

This completes the proof. □

Theorem 5.5.4. [15, Theorem 3.4] $\sigma_p(\Delta^{(1)^*}, bv_p^*) := \{\alpha \in \mathbb{C} : |\alpha - 1| < 1\}$.

Proof. Suppose $\Delta^{(1)^*}f = \alpha f$ for $f \neq \theta$ in bv_p^*, which is isometrically isomorphic to D_q by Part (ii) of Theorem 4.3.8. Then, by solving the system of the linear equations

$$\left.\begin{array}{rcl} f_0 - f_1 & = & \alpha f_0 \\ f_1 - f_2 & = & \alpha f_1 \\ f_2 - f_3 & = & \alpha f_2 \\ & \vdots & \\ f_k - f_{k+1} & = & \alpha f_k \\ & \vdots & \end{array}\right\} \tag{5.5.4}$$

we obtain that $f_k = (1-\alpha)^k f_0$ for all $k \in \mathbb{N}$. This shows that

$$\begin{aligned}\left|\sum_{j=k}^{\infty} f_j\right|^q &= |f_0|^q \left|\sum_{j=k}^{\infty}(1-\alpha)^j\right|^q \\ &= |f_0|^q\,|1-\alpha|^{kq}\left|\sum_j (1-\alpha)^j\right|^q \leq \left|\frac{f_0(1-\alpha)^k}{\alpha}\right|^q\end{aligned}$$

if $|\alpha - 1| < 1$. Therefore, one can see that

$$\sum_k \left|\sum_{j=k}^{\infty} f_j\right|^q \leq \left|\frac{f_0}{\alpha}\right|^q \sum_k |\alpha-1|^{kq} < \infty$$

if $|\alpha - 1| < 1$. This means that $f \in bv_p^*$ if $f_0 \neq 0$ and $|\alpha - 1| < 1$, as asserted.

Since the case $p = 1$ may be proved by analogy to the case $p > 1$, we omit the detailed proof. □

Theorem 5.5.5. [15, Theorem 3.7] $\sigma_r(\Delta^{(1)}, bv_p) := \{\alpha \in \mathbb{C} : |\alpha - 1| < 1\}$.

Proof. For $|\alpha - 1| < 1$, the operator $\Delta^{(1)} - \alpha I$ is one-to-one, hence has an inverse. However, $\Delta^{(1)^*} - \alpha I$ is not one-to-one by Theorem 5.5.4. Now, Lemma 5.2.2 yields the fact that $\overline{R(\Delta^{(1)} - \alpha I)} \neq bv_p$.

This step concludes the proof. □

Theorem 5.5.6. [15, Theorem 3.8] $1 \in III_1\sigma(\Delta^{(1)}, bv_p)$.

Proof. By Theorem 5.5.5, $\Delta^{(1)} - I \in III$. Additionally, $\alpha = 1$ is not in $\sigma_p(\Delta^{(1)}, bv_p)$ by Theorem 5.5.3. Hence, $\Delta^{(1)} - I$ has an inverse. Therefore, $\Delta^{(1)} - I \in 1 \cup 2$.

For establishing the fact $\Delta^{(1)} - I \in 1$, it is enough to show by Lemma 5.2.3 that $\Delta^{(1)^*} - I$ is onto. For a given $y = (y_k) \in bv_p^*$ we must find that $x = (x_k) \in bv_p^*$ such that $(\Delta^{(1)^*} - I)x = y$. A direct calculation yields that $x_n = -y_{n-1}$ for all $n \in \mathbb{N}$. This means that $\Delta^* - I$ is onto, as desired. □

Theorem 5.5.7. [15, Theorem 3.9] *If $\alpha \neq 1$ and $\alpha \in \sigma_r(\Delta^{(1)}, bv_p)$, then $\alpha \in III_2\sigma(\Delta^{(1)}, bv_p)$.*

Proof. Since $\alpha \neq 1$, the operator $\Delta^{(1)} - \alpha I$ is a triangle, hence has an inverse. Take $e^{(0)} \in bv_p$. Then, $(\Delta^{(1)} - \alpha I)^{-1}$-transform of the sequence $e^{(0)}$ is the sequence $y = \{(1-\alpha)^{-1}, (1-\alpha)^{-2}, (1-\alpha)^{-3}, \ldots\}$, which is not in bv_p. This shows that $(\Delta^{(1)} - \alpha I)^{-1}$ is discontinuous. Therefore, $\Delta^{(1)} - \alpha I \in 2$.

Furthermore, $\Delta^{(1)^*} - \alpha I$ is not one-to-one by Theorem 5.5.4 and $\Delta^{(1)} - \alpha I$ does not have a dense range by Lemma 5.2.2. Hence, $\Delta^{(1)} - \alpha I \in III$.

This completes the proof of Theorem. □

Theorem 5.5.8. [15, Theorem 3.10] $\sigma_c(\Delta^{(1)}, bv_p) := \{\alpha \in \mathbb{C} : |\alpha - 1| = 1\}$.

Proof. Since $\alpha \neq 1$, $\Delta^{(1)} - \alpha I$ is triangle and has an inverse. Therefore, $\Delta^{(1)^*} - \alpha I$ is one-to-one from Lemma 5.2.2, which is what we wished to prove. □

Theorem 5.5.9. [15, Theorem 3.11] *If* $\alpha \in \sigma_c(\Delta^{(1)}, bv_p)$, *then* $\alpha \in II_2$.

Proof. Let us take $\alpha \in \sigma_c(\Delta^{(1)}, bv_p)$. In order to prove that $\alpha \in II_2$ we must show that $\Delta^{(1)} - \alpha I$ is not onto, $\overline{R(\Delta^{(1)} - \alpha I)} = bv_p$ and $(\Delta^{(1)} - \alpha I)^{-1}$ is discontinuous.

In the case $\alpha \neq 0$, since the proof is similar to the proof of Theorem 5.5.7, we omit details. Also, in the case $\alpha = 0$, $(\Delta^{(1)} - \alpha I)^{-1}$-transform of the sequence e is the sequence $y = (k+1)$, which is not in the space bv_p. That is to say that $\Delta^{(1)} - \alpha I$ is in 2 but not in I. Besides, we have by Theorem 5.5.8 that $\overline{R(\Delta^{(1)} - \alpha I)} = bv_p$.

This step completes the proof. □

5.6 THE FINE SPECTRA OF THE CESÀRO OPERATOR C_1 ON THE SPACE bv_p

In this section, following Akhmedov and Başar [16], we study the fine spectra of the Cesàro operator C_1 over the sequence space bv_p, where $1 < p < \infty$. We shall begin with giving the basic result concerning endowed with the norm of Cesàro operator C_1 on the space bv_p.

Theorem 5.6.1. [16, Theorem 3.1] $C_1 \in B(bv_p)$ *endowed with the norm* $\| C_1 \|_{(bv_p:bv_p)} = 1$.

Proof. Since the linearity and the boundedness of the operator $C_1 : bv_p \to bv_p$ are obvious, we omit details. Let us take any $x = (x_k) \in bv_p$. Then, since one can observe that

$$\left| \frac{x_0 + x_1 + \cdots + x_k}{k+1} - \frac{x_0 + x_1 + \cdots + x_{k-1}}{k} \right|^p = \left| \frac{\sum_{j=0}^{k-1}(x_k - x_j)}{k(k+1)} \right|^p$$

and the inequalities

$$\begin{aligned}
|x_k - x_0| &\leq |x_k - x_{k-1}| + |x_{k-1} - x_{k-2}| + \cdots + |x_2 - x_1| + |x_1 - x_0| \\
|x_k - x_1| &\leq |x_k - x_{k-1}| + |x_{k-1} - x_{k-2}| + \cdots + |x_2 - x_1| \\
&\vdots \\
|x_k - x_{k-2}| &\leq |x_k - x_{k-1}| + |x_{k-1} - x_{k-2}|
\end{aligned}$$

hold, we see by using the following known inequalities

$$\begin{aligned}
\left(\sum_{n=1}^{k} |a_n| \right)^p &\leq k^{p-1} \sum_{n=1}^{k} |a_n|^p \, , \ (p \geq 1), \\
\frac{n^p}{(k+1)^p} &\leq \frac{k}{k+1}; \ \ (1 \leq n \leq k, \ p \geq 1)
\end{aligned}$$

that

$$\begin{aligned}
\left|\frac{x_0+x_1+\cdots+x_k}{k+1}-\frac{x_0+x_1+\cdots+x_{k-1}}{k}\right|^p &\leq \frac{\left(\sum_{j=1}^k j|x_j-x_{j-1}|\right)^p}{k^p(k+1)^p} \\
&\leq \frac{k^{p-1}\left(\sum_{j=1}^k j^p|x_j-x_{j-1}|^p\right)}{k^p(k+1)^p} \\
&= \frac{\sum_{j=1}^k j^p|x_j-x_{j-1}|^p}{k(k+1)^p} \\
&\leq \frac{\sum_{j=1}^k j|x_j-x_{j-1}|^p}{k(k+1)}.
\end{aligned}$$

Now, we obtain by applying Proposition I of Knopp [226, p. 143] that

$$\begin{aligned}
(\|C_1x\|_{bv_p})^p &\leq |x_0|^p+\sum_{k=1}^{\infty}\frac{\sum_{j=1}^k j|x_j-x_{j-1}|^p}{k(k+1)} \\
&= \sum_k |x_k-x_{k-1}|^p = \|x\|_{bv_p}^p.
\end{aligned}$$

So, $\|C_1x\|_{bv_p} \leq \|x\|_{bv_p}$ which leads us to the consequence

$$\|C_1\|_{(bv_p:bv_p)} \leq 1. \tag{5.6.1}$$

Now, let us consider the element $b^{(0)} = e$. It is clear that $C_1b^{(0)} = b^{(0)}$ and $\|b^{(0)}\|_{bv_p} = 1$. Hence,

$$\|C_1\|_{(bv_p:bv_p)} \geq \left\|C_1b^{(0)}\right\|_{bv_p} = \left\|b^{(0)}\right\|_{bv_p} = 1,$$

which yields the fact that

$$\|C_1\|_{(bv_p:bv_p)} \geq 1. \tag{5.6.2}$$

The desired result obtained by combining the inequalities (5.6.1) and (5.6.2). □

Theorem 5.6.2. [16, Theorem 3.2] *$C_1^* \in B(D_q)$ endowed with the norm $\|C_1^*\|_{(D_q:D_q)} = 1$, where $1 < q \leq \infty$.*

Proof. Let $1 < p < \infty$. Since $D_\infty \cong bv_1^*$ and $D_q \cong bv_p^*$ by Parts (i) and (ii) of Theorem 4.3.8 and $\|C_1^*\|_{(D_q:D_q)} = \|C_1\|_{(bv_p:bv_p)}$, the proof is immediate by Theorem 5.6.1. □

Theorem 5.6.3. [16, Theorem 3.3] $\sigma(C_1, bv_p) := \{\alpha \in \mathbb{C} : |\alpha - \frac{1}{2}| \leq \frac{1}{2}\}$.

Proof. To prove the theorem, it is enough to show that $(C_1 - \alpha I)^{-1}$ is bounded for all α's such that $|\alpha - 1/2| > 1/2$. Suppose $y = (y_k) \in bv_p$. Solving the equation $(C_1 - \alpha I)x = y$ for x in terms of y, we derive that

$$\begin{aligned}
x_0 &= \tfrac{1}{1-\alpha} y_0 \\
x_1 &= \tfrac{-1}{(1-\alpha)(1-2\alpha)} y_0 + \tfrac{2}{1-2\alpha} y_1 \\
x_2 &= \tfrac{2\alpha}{\prod_{k=1}^{3} 1-k\alpha} y_0 - \tfrac{2}{\prod_{k=2}^{3} 1-k\alpha} y_1 + \tfrac{3}{1-3\alpha} y_2 \\
&\vdots \\
x_n &= \tfrac{1}{n+1} \textstyle\sum_{k=0}^{n-1} (-1)^{n-k} \left(\prod_{j=k+1}^{n+1} \tfrac{j}{1-j\alpha} \right) \alpha^{n-k-1} y_k + \tfrac{n+1}{1-(n+1)\alpha} y_n \\
&\vdots
\end{aligned}$$

Therefore, we have

$$(C_1 - \alpha I)^{-1} = \begin{bmatrix}
\frac{1}{1-\alpha} & 0 & \cdots & 0 & \cdots \\
\frac{-1}{(1-\alpha)(1-2\alpha)} & \frac{2}{1-2\alpha} & \cdots & 0 & \cdots \\
\frac{2\alpha}{(1-\alpha)(1-2\alpha)(1-3\alpha)} & \frac{-2}{(1-2\alpha)(1-3\alpha)} & \cdots & 0 & \cdots \\
\vdots & \vdots & \cdots & \vdots & \cdots \\
\frac{(-1)^{n-1}}{n} \prod_{k=1}^{n} \frac{k}{1-k\alpha} \alpha^{n-2} & \frac{(-1)^{n}}{n} \prod_{k=2}^{n} \frac{k}{1-k\alpha} \alpha^{n-3} & \cdots & \frac{n}{1-n\alpha} & \cdots \\
\vdots & \vdots & \cdots & \vdots & \ddots
\end{bmatrix}.$$

Thus, it is seen by [368] that $\|C_1 - \alpha I\|_{(bv_1:bv_1)} < \infty$ if $|\alpha - 1/2| > 1/2$, which is equivalent to the fact that $Re(1/\alpha) < 1$.

Furthermore, if $p > 1$, then

$$|x_n - x_{n-1}|^p = |x_n - x_{n-1}|^{p-1} \cdot |x_n - x_{n-1}| . \tag{5.6.3}$$

We can show that

$$\lim_{n \to \infty} |x_n - x_{n-1}| = 0, \tag{5.6.4}$$

if $Re(1/\alpha) < 1$. Indeed,

$$\begin{aligned}
x_n - x_{n-1} &= \frac{(-1)^n}{n+1} \prod_{k=1}^{n+1} \frac{k}{1-k\alpha} \alpha^{n-1} y_0 + \\
&+ \left[\frac{(-1)^{n+1}}{n+1} \prod_{k=2}^{n+1} \frac{k}{1-k\alpha} \alpha^{n-2} y_1 - \frac{(-1)^{n-1}}{n} \prod_{k=1}^{n} \frac{k}{1-k\alpha} \alpha^{n-2} y_0 \right] + \\
&+ \cdots + \left[\frac{n+1}{1-(n+1)\alpha} y_n - \frac{n}{1-n\alpha} y_{n-1} \right].
\end{aligned}$$

If we use Lemma 7 in [325, p. 266] with the last relation, then we have (5.6.4). Thus, (5.6.3) and (5.6.4) yield that

$$\| (C_1 - \alpha I)^{-1} \|_{(bv_p:bv_p)} < \infty,$$

if $Re(1/\alpha) < 1$. This completes the proof. □

Theorem 5.6.4. [16, Theorem 3.4] $\sigma_p(C_1, bv_p) = \{1\}$.

Proof. Suppose that $C_1 x = \alpha x$ for $x \neq \theta = (0,0,0,\ldots)$ in bv_p. Consider the system of the linear equations

$$\left.\begin{array}{rcl} x_0 & = & \alpha x_0 \\ \frac{1}{2}x_0 + \frac{1}{2}x_1 & = & \alpha x_1 \\ \frac{1}{3}x_0 + \frac{1}{3}x_1 + \frac{1}{3}x_2 & = & \alpha x_2 \\ & \vdots & \\ \frac{1}{k+1}x_0 + \frac{1}{k+1}x_1 + \frac{1}{k+1}x_2 + \cdots + \frac{1}{k+1}x_k & = & \alpha x_k \\ & \vdots & \end{array}\right\}. \tag{5.6.5}$$

If x_0 is the non-zero entry of the sequence $x = (x_n)$, then $\alpha = 1$ and we obtain from (5.6.5) that $x_k = x_0$ for any $k \in \mathbb{N}_1$. Hence, $x = (x_k) \in bv_p$ such that $x \neq \theta$ for $p \geq 1$.

If x_{n_0} is the first non-zero entry of the sequence $x = (x_n)$, then we find that $x_{n_0}/(n_0+1) = \alpha x_{n_0}$, which yields the fact $\alpha = 1/(n_0+1)$. Therefore, we also get by (5.6.5) that

$$x_{n_0+k} = \frac{(n_0+1)(n_0+2)\cdots(n_0+k)}{k!} x_{n_0}$$

for any $k \in \mathbb{N}_1$. Furthermore,

$$\begin{aligned} |x_{n_0+k} - x_{n_0+k-1}|^p &= \frac{n_0^p (n_0+1)^p \cdots (n_0+k-1)^p}{(k!)^p} |x_{n_0}|^p \\ &= \frac{1}{[(n_0-1)!]^p}(k+1)^p(k+2)^p \cdots (n_0+k-1)^p |x_{n_0}|^p, \end{aligned}$$

which shows that $x \notin bv_p$.

This completes the proof. □

Prior to giving Theorem 5.6.6 we shall quote the following lemma, which is needed in proving:

Lemma 5.6.5. [226, p. 115] *All harmonic series $\sum_{n=1}^{\infty} n^{-\alpha}$ for $\alpha \leq 1$ are divergent, and for $\alpha > 1$, convergent.*

Theorem 5.6.6. [16, Theorem 3.6] $\sigma_p(C_1^*, bv_p^*) := \{\alpha \in \mathbb{C} : |\alpha - \frac{1}{2}| < \frac{1}{2}\} \cup \{1\}$.

Proof. Suppose $C_1^* f = \alpha f$ for $f \neq \theta$ in bv_p^*. Then, by solving the system of the linear equations

$$\left.\begin{array}{rcl} f_0 + \frac{1}{2}f_1 + \frac{1}{3}f_2 + \cdots & = & \alpha f_0 \\ \frac{1}{2}f_1 + \frac{1}{3}f_2 + \cdots & = & \alpha f_1 \\ \frac{1}{3}f_2 + \cdots & = & \alpha f_2 \\ & \vdots & \\ \frac{1}{k+1}f_k + \cdots & = & \alpha f_k \\ & \vdots & \end{array}\right\}$$

we obtain that $f_k = \prod_{n=1}^{k}(1-\frac{1}{n\alpha}) f_0$ for all $k \in \mathbb{N}_1$, if $\alpha \neq 0$. Since $f = (f_0, 0, 0, \ldots) \neq \theta$ in bv_p^* for $\alpha = 1$, it is clear that $1 \in \sigma_p(C_1^*, bv_p^*)$. Define the sequence $z = (z_k)$ by $z_k = \prod_{n=1}^{k}(1-\frac{1}{n\alpha})$ for all $k \in \mathbb{N}_1$. Okutoyi [308, Lemma 1.4] has proved that

$$z_k = A \cdot k^{-1/\alpha} + O\left(k^{-Re(1/\alpha)-1}\right),$$

where A is a constant and the series $\sum_{k=1}^{\infty} z_k$ is bounded if $Re(1/\alpha) > 1$, diverges if $Re(1/\alpha) \leq 1$. Consider the sequence $s = (s_k)$ defined by $s_k = \sum_{j=k}^{\infty} j^{-1/\alpha}$ for all $k \in \mathbb{N}$. It is known that $|s_1| < \infty$ if and only if $Re(1/\alpha) > 1$. Denote $Re(1/\alpha) = \beta$ and let $\beta > 1$. Therefore, using the fact given by Lemma 5.6.5 for the convergence of the series $s_1 = \sum_{j=1}^{\infty} j^{-\beta}$ we obtain that

$$s_k \leq \left[\frac{1}{2^{(m-1)\beta}} + \cdots + \frac{1}{(2^m-1)^\beta}\right] + \left[\frac{1}{2^{m\beta}} + \cdots + \frac{1}{(2^{m+1}-1)^\beta}\right] + \cdots, \tag{5.6.6}$$

where $2^{m-1} \leq k \leq 2^m - 1$. Now, replacing any separate term by the first term in each parenthesis in (5.6.6), we derive that

$$s_j \leq \frac{2^{\beta-1}}{2^{\beta-1}-1} \cdot \frac{1}{2^{(m-1)(\beta-1)}}. \tag{5.6.7}$$

It is clear by (5.6.7) that

$$\left|\sum_{j=k}^{\infty} \frac{1}{j^{1/\alpha}}\right| \leq \frac{2^{\beta-1}}{2^{(m-1)(\beta-1)}\left(2^{\beta-1}-1\right)} \quad \text{for all } k \in \mathbb{N}, \tag{5.6.8}$$

if $Re(1/\alpha) > 1$. Similarly, one can show that

$$\left|\sum_{j=k}^{\infty} j^{-Re(1/\alpha)-1}\right| \leq \frac{B}{2^{(m-1)(\beta-1)}} \quad \text{for all } k \in \mathbb{N}, \tag{5.6.9}$$

where B is a positive constant. It follows from (5.6.8) and (5.6.9) that

$$\sup_{k,n\in\mathbb{N}} \left|\sum_{j=k}^{n} f_j\right| < \infty. \tag{5.6.10}$$

If $f_0 \neq 0$ and $Re(1/\alpha) > 1$, then $f \in bv_1^*$ and $f \neq \theta$ whenever $f_0 \neq 0$.

It is clear that

$$\left\{\alpha \in \mathbb{C} : Re\left(\frac{1}{\alpha}\right) > 1\right\} = \left\{\alpha \in \mathbb{C} : \left|\alpha - \frac{1}{2}\right| < \frac{1}{2}\right\}.$$

Now, using (5.6.8) and (5.6.9) we obtain that

$$\left|\sum_{j=k}^{n} f_j\right|^q \leq \frac{M}{2^{(m-1)(\beta-1)q}}, \tag{5.6.11}$$

where M is a positive constant. Consequently, we derive from (5.6.11) that

$$\sum_k \left| \sum_{j=k}^{\infty} f_j \right|^q < \infty.$$

This means that $(f_k) \in bv_p^*$ and (5.6.10) together with (5.6.11) complete the proof. □

Theorem 5.6.7. [16, Theorem 3.8] $\sigma_c(C_1, bv_p) := \{\alpha \in \mathbb{C} : |\alpha - \frac{1}{2}| = \frac{1}{2} \ , \ \alpha \neq 1\}$.

Proof. It is not hard to show that

$$\left\{\alpha \in \mathbb{C} : \left|\alpha - \frac{1}{2}\right| = \frac{1}{2}\right\} = \left\{\alpha \in \mathbb{C} : Re\left(\frac{1}{\alpha}\right) = 1\right\} \cup \{0\}.$$

Suppose that $\alpha \neq 1$. Then, it follows from Theorem 5.6.6 that $\alpha \notin \sigma_p(C_1^*, bv_p^*)$. Hence, $Ker(C_1^* - \alpha I^*) = \{\theta\}$ for such α's, which shows that

$$\overline{R(C_1 - \alpha I)} = bv_p. \tag{5.6.12}$$

Now, suppose that $\alpha = 0$ and consider the equation $C_1 x = \theta$. Then, it is easy to see that $x = \theta$, i.e., $Ker(C_1) = \{\theta\}$ and C_1 has an inverse. One can also see that $Ker(C_1^*) = \{\theta\}$ and we thus have

$$\overline{R(C_1)} = bv_p. \tag{5.6.13}$$

Therefore, we obtain by combining (5.6.12), (5.6.13) and Lemma 5.2.2 that

$$\sigma_c(C_1, bv_p) = \{0\} \cup \left\{\alpha \in \mathbb{C} : Re\left(\frac{1}{\alpha}\right) = 1 \ , \ \alpha \neq 1\right\},$$

which completes the proof. □

Theorem 5.6.8. [16, Theorem 3.9] $\sigma_r(C_1, bv_p) := \{\alpha \in \mathbb{C} : |\alpha - \frac{1}{2}| < \frac{1}{2}\}$.

Proof. This immediately follows from Theorems 5.6.3, 5.6.4 and 5.6.7 by taking into account the definition of the concept of the spectrum of a bounded linear operator acting in a Banach space. □

5.7 THE SPECTRUM OF THE OPERATOR $B(r, s)$ ON THE SPACES c_0 AND c

In this section, following Altay and Başar [25], we examine the fine spectrum with respect to the Goldberg's classification of the operator $B(r, s)$ defined by a double band matrix over the sequence spaces c_0 and c. Finally, we also give a Mercerian theorem. The main results of the present section are more general than the fine spectrum of the difference operator Δ on the spaces c_0 and c derived by Altay and Başar [22].

5.7.1 The Generalized Difference Operator $B(r,s)$

Our main focus in this subsection is the band matrix $B(r,s)$ given by

$$B(r,s)=\begin{bmatrix} r & 0 & 0 & 0 & \dots \\ s & r & 0 & 0 & \dots \\ 0 & s & r & 0 & \dots \\ 0 & 0 & s & r & \dots \\ \vdots & \vdots & \vdots & \vdots & \ddots \end{bmatrix}, \quad (s \neq 0).$$

We begin with determining when a matrix A induces a bounded operator from the space c into itself:

Lemma 5.7.1. (cf. [396, Theorem 1.3.6, p. 6]) *The matrix $A=(a_{nk})$ gives raise to a bounded linear operator $T \in B(c)$ from c to itself if and only if*

1. *The rows of A in ℓ_1 and their ℓ_1 norms are bounded.*

2. *The columns of A in c.*

3. *The sequence of row sums of A in c.*

The operator norm of T is the supremum of the ℓ_1 norms of the rows.

Corollary 5.7.2. *$B(r,s): c \to c$ is a bounded linear operator and $\| B(r,s) \|_{(c:c)} := |r|+|s|$.*

Lemma 5.7.3. (cf. [396, Example 8.4.5A, p. 129]) *The matrix $A=(a_{nk})$ gives raise to a bounded linear operator $T \in B(c_0)$ from c_0 to itself if and only if*

1. *The rows of A in ℓ_1 and their ℓ_1 norms are bounded.*

2. *The columns of A in c_0.*

The operator norm of T is the supremum of the ℓ_1 norms of the rows.

Corollary 5.7.4. [25, Corollary 1.4] *$B(r,s): c_0 \to c_0$ is a bounded linear operator such that $\| B(r,s) \|_{(c_0:c_0)} := \| B(r,s) \|_{(c:c)}$.*

Theorem 5.7.5. [25, Theorem 2.1] $\sigma(B(r,s),c_0) := \{\alpha \in \mathbb{C} : |\alpha - r| \leq |s|\}$.

Proof. Firstly, we prove that $(B(r,s)-\alpha I)^{-1}$ exists and is in $(c_0 : c_0)$ for $\alpha \notin \sigma(B(r,s),c_0)$ and next the operator $B(r,s)-\alpha I$ is not invertible for $\alpha \in \sigma(B(r,s),c_0)$.

Let $\alpha \notin \sigma(B(r,s),c_0)$. Since $B(r,s)-\alpha I$ is triangle, $(B(r,s)-\alpha I)^{-1}$ exists and solving $(B(r,s)-\alpha I)x = y$ for x in terms of y gives the matrix $(B(r,s)-\alpha I)^{-1}$. The nth row turns out to be

$$\frac{(-s)^{n-k}}{(r-\alpha)^{n-k+1}}$$

in the kth place for $k \leq n$ and zero otherwise. Thus, we observe that

$$\| (B(r,s) - \alpha I)^{-1} \|_{(c_0:c_0)} = \sup_{n\in\mathbb{N}} \sum_{k=0}^{n} \left| \frac{(-s)^{n-k}}{(r-\alpha)^{n-k+1}} \right| = \left| \frac{1}{r-\alpha} \right| \sup_{n\in\mathbb{N}} \sum_{k=0}^{n} \left| \frac{s}{r-\alpha} \right|^{k} < \infty, \tag{5.7.1}$$

i.e., $(B(r,s) - \alpha I)^{-1} \in (c_0 : c_0)$.

Let $\alpha \in \sigma(B(r,s), c_0)$ and $\alpha \neq r$. Since $B(r,s) - \alpha I$ is triangle, $(B(r,s) - \alpha I)^{-1}$ exists but one can see by (5.7.1) that

$$\left\| (B(r,s) - \alpha I)^{-1} \right\|_{(c_0:c_0)} = \infty,$$

whenever $\alpha \in \sigma(B(r,s), c_0)$, i.e., $(B(r,s) - \alpha I)^{-1}$ is not in $B(c_0)$.

If $\alpha = r$, then the operator $B(r,s) - \alpha I = B(0,s)$ is represented by the matrix

$$B(0,s) = \begin{bmatrix} 0 & 0 & 0 & 0 & \cdots \\ s & 0 & 0 & 0 & \cdots \\ 0 & s & 0 & 0 & \cdots \\ 0 & 0 & s & 0 & \cdots \\ \vdots & \vdots & \vdots & \vdots & \ddots \end{bmatrix}.$$

Since $\overline{R(B(0,s))} \neq c_0$, $B(0,s)$ is not invertible.

This completes the proof. □

Theorem 5.7.6. [25, Theorem 2.2] $\sigma_p(B(r,s), c_0) := \emptyset$.

Proof. Suppose $B(r,s)x = \alpha x$ for $x \neq \theta = (0,0,0,\ldots)$ in c_0. Then, by solving the system of linear equations

$$\left.\begin{array}{rcl} rx_0 &=& \alpha x_0 \\ sx_0 + rx_1 &=& \alpha x_1 \\ sx_1 + rx_2 &=& \alpha x_2 \\ &\vdots& \\ sx_k + rx_{k+1} &=& \alpha x_{k+1} \\ &\vdots& \end{array}\right\}$$

we find that if x_{n_0} is the first non-zero entry of the sequence $x = (x_n)$, then $\alpha = r$ and $x_{n_0+k} = 0$ for all $k \in \mathbb{N}$. This contradicts the fact that $x_{n_0} \neq 0$, which completes the proof. □

If $T : c_0 \to c_0$ is a bounded linear operator with the matrix A, then it is known that the adjoint operator $T^* : c_0^* \to c_0^*$ is defined by the transpose A' of the matrix A. It should be noted that the dual space c_0^* of c_0 is isometrically isomorphic to the Banach space ℓ_1 of absolutely summable sequences normed by $\|x\| = \sum_k |x_k|$.

Theorem 5.7.7. [25, Theorem 2.3] $\sigma_p(B(r,s)^*, c_0^*) := \{\alpha \in \mathbb{C} : |\alpha - r| < |s|\}$.

Proof. Suppose $B(r,s)^*x = \alpha x$ for $x \neq \theta$ in $c_0^* \cong \ell_1$. Then, by solving the system of linear equations

$$\left.\begin{array}{rcl} rx_0 + sx_1 & = & \alpha x_0 \\ rx_1 + sx_2 & = & \alpha x_1 \\ rx_2 + sx_3 & = & \alpha x_2 \\ & \vdots & \\ rx_k + sx_{k+1} & = & \alpha x_k \\ & \vdots & \end{array}\right\}$$

we observe that

$$x_n := \left(\frac{\alpha - r}{s}\right)^n x_0 \text{ for all } n \in \mathbb{N}_1.$$

This shows that $x \in \ell_1$ if and only if $|\alpha - r| < |s|$, as asserted. □

Theorem 5.7.8. [25, Theorem 2.5] $\sigma_r(B(r,s), c_0) := \{\alpha \in \mathbb{C} : |\alpha - r| < |s|\}$.

Proof. We show that the operator $B(r,s) - \alpha I$ has an inverse and for α satisfying $|\alpha - r| < |s|$, $\overline{R(B(r,s) - \alpha I)} \neq c_0$. For $\alpha \neq r$, the operator $B(r,s) - \alpha I$ is triangle, hence has an inverse. For $\alpha = r$, the operator $B(r,s) - \alpha I$ is one-to-one, hence has an inverse. But $B(r,s)^* - \alpha I$ is not one-to-one by Theorem 5.7.7. Now, Lemma 5.2.2 yields the fact that $\overline{R(B(r,s) - \alpha I)} \neq c_0$.

This step concludes the proof. □

Theorem 5.7.9. [25, Theorem 2.6] *If $\alpha = r$, then $\alpha \in III_1\sigma(B(r,s), c_0)$.*

Proof. Since the operator $B(r,s) - \alpha I$ is reduced to the operator $B(0,s)$ for $\alpha = r$, $B(0,s) \in III_1$ or $\in III_2$ by Theorem 5.7.8. To verify the fact $B(0,s)$ has a bounded inverse, it is enough to show that $B(0,s)$ is bounded below. Indeed one can easily see for all $x \in c_0$ that

$$\|B(0,s)x\| \geq \frac{|s|}{2}\|x\|,$$

which means that $B(0,s)$ is bounded below.

This completes the proof. □

Theorem 5.7.10. [25, Theorem 2.7] *If $\alpha \neq r$ and $\alpha \in \sigma_r(B(r,s), c_0)$, then $\alpha \in III_2\sigma(B(r,s), c_0)$.*

Proof. By Theorem 5.7.8, $B(r,s) - \alpha I \in III_1$ or $\in III_2$. Hence, by (5.7.1), the inverse of the operator $B(r,s) - \alpha I$ is discontinuous. Therefore, $B(r,s) - \alpha I$ has an unbounded inverse. □

Theorem 5.7.11. [25, Theorem 2.8] $\sigma_c(B(r,s), c_0) := \{\alpha \in \mathbb{C} : |\alpha - r| = |s|\}$.

Proof. For this, we prove that the operator $B(r,s) - \alpha I$ has an inverse and

$$\overline{R(B(r,s) - \alpha I)} = c_0, \tag{5.7.2}$$

if $\alpha \in \sigma_c(B(r,s), c_0)$. Since $\alpha \neq r$, $B(r,s) - \alpha I$ is triangle and has an inverse. Therefore, $B(r,s)^* - \alpha I$ is one-to-one by Theorem 5.7.7 and (5.7.2) holds from Lemma 5.2.2. This is what we wished to prove. □

Theorem 5.7.12. [25, Theorem 2.9] *If* $\alpha \in \sigma_c(B(r,s), c_0)$*, then* $\alpha \in II_2\sigma(B(r,s), c_0)$.

Proof. By (5.7.1), the inverse of the operator $B(r,s) - \alpha I$ is discontinuous. Therefore, $B(r,s) - \alpha I$ has an unbounded inverse.

By Theorem 5.7.7, $B(r,s)^* - \alpha I$ is one-to-one. By Lemma 5.2.2, $B(r,s) - \alpha I$ has a dense range.

To verify the operator $B(r,s) - \alpha I$ is not surjective, it is sufficient to show that there is no sequence $x = (x_n)$ in c_0 such that $(B(r,s) - \alpha I)x = y$ for some $y \in c_0$. Let us consider the sequence $y = e^{(0)} \in c_0$. For this sequence, we obtain $x_n = \{s/(\alpha - r)\}^n/(r - \alpha)$ for all $n \in \mathbb{N}$. This yields that $x \notin c_0$, i.e., $B(r,s) - \alpha I$ is not onto.

This completes the proof. □

Theorem 5.7.13. [25, Theorem 2.10] $\sigma(B(r,s), c) := \{\alpha \in \mathbb{C} : |\alpha - r| \leq |s|\}$.

Proof. This is obtained by the similar way that used in the proof of Theorem 5.7.5. □

Theorem 5.7.14. [25, Theorem 2.11] $\sigma_p(B(r,s), c) := \emptyset$.

Proof. This can be obtained by proceeding as in proving Theorem 5.7.6, above. So, we omit details. □

If $T : c \to c$ is a bounded matrix operator with the matrix A, then $T^* : c^* \to c^*$ acting on $\mathbb{C} \oplus \ell_1$ has a matrix representation of the form

$$\begin{bmatrix} \chi & 0 \\ b & A' \end{bmatrix};$$

where χ is the limit of the sequence of row sums of A minus the sum of the limit of the columns of A, and b is the column vector whose k^{th} entry is the limit of the kth column of A for each $k \in \mathbb{N}$. For $B(r,s) : c \to c$, the matrix $B(r,s)^* \in B(\ell_1)$ is of the form

$$B(r,s)^* := \begin{bmatrix} r+s & 0 \\ 0 & B(r,s)' \end{bmatrix}.$$

Theorem 5.7.15. [25, Theorem 2.12]
$\sigma_p(B(r,s)^*, c^*) := \{\alpha \in \mathbb{C} : |\alpha - r| < |s|\} \cup \{r+s\}$.

Proof. Suppose $B(r,s)^*x = \alpha x$ for $x \neq \theta$ in ℓ_1. Then, by solving the system of linear equations

$$\left.\begin{array}{rcl} (r+s)x_0 &=& \alpha x_0 \\ rx_1 + sx_2 &=& \alpha x_1 \\ rx_2 + sx_3 &=& \alpha x_2 \\ &\vdots& \\ rx_k + sx_{k+1} &=& \alpha x_k \\ &\vdots& \end{array}\right\}$$

we obtain that

$$x_n := \left(\frac{\alpha - r}{s}\right)^{n-1} x_1 \text{ for all } n \in \mathbb{N}_2. \tag{5.7.3}$$

If $x_0 \neq 0$, then $\alpha = r + s$. So, $\alpha = r + s$ is an eigenvalue with the corresponding eigenvector $x = (x_0, 0, 0, \ldots)$. If $\alpha \neq r + s$, then $x_0 = 0$ and one can see by (5.7.3) that $x \in \ell_1$ if and only if $|\alpha - r| < |s|$. □

Theorem 5.7.16. [25, Theorem 2.13] $\sigma_r(B(r,s), c) := \sigma_p(B(r,s)^*, c^*)$.

Proof. The proof is obtained by the analogy with the proof of Theorem 5.7.8. □

Theorem 5.7.17. [25, Theorem 2.14] $\sigma_c(B(r,s), c) := \{\alpha \in \mathbb{C} : |\alpha - r| = |s|\} \backslash \{r+s\}$.

Proof. This is similar to the proof of Theorem 5.7.11 with $\alpha \neq r + s$ such that $|\alpha - r| = |s|$. □

Since the fine spectrum of the operator $B(r,s)$ on the space c can be derived by analogously to the space c_0, we omit details and give it without proof. Therefore, the following statements hold:

$$\begin{cases} B(r,s) - \alpha I \in I_1 & , \quad \alpha \notin \sigma(B(r,s), c), \\ \alpha \in III_1\sigma(B(r,s), c) & , \quad \alpha = r, \\ \alpha \in II_2\sigma(B(r,s), c) & , \quad \alpha \in \sigma_c(B(r,s), c), \\ \alpha \in III_2\sigma(B(r,s), c) & , \quad \alpha \in \sigma_r(B(r,s), c) \setminus \{r\}. \end{cases}$$

Theorem 5.7.18. [25, Theorem 2.15] $\sigma(B(r,s), \ell_\infty) := \{\alpha \in \mathbb{C} : |\alpha - r| \leq |s|\}$.

Proof. It is known by Cartlidge [101] that if a matrix operator A is bounded on the space c, then $\sigma(A, c) = \sigma(A, \ell_\infty)$. Now, the proof is immediate from Theorem 5.7.13 with $A := B(r,s)$. □

Theorem 5.7.19. [25, Theorem 2.16] *Suppose that $\alpha \in \mathbb{C}$ satisfies the inequality $|\alpha(1-r) + r| > |s(1-\alpha)|$. Then, the convergence field of $A = \alpha I + (1-\alpha)B(r,s)$ is c.*

Proof. If $\alpha = 1$, there is nothing to prove. Let us suppose that $\alpha \neq 1$. Then, one can observe by Theorem 5.7.13 and the choice of α that $B(r,s) - [\alpha/(\alpha-1)]I$ has an inverse in $B(c)$. That is to say that

$$A^{-1} = \frac{1}{1-\alpha}\left(B(r,s) - \frac{\alpha}{\alpha-1}I\right)^{-1} \in B(c).$$

Since A is a triangle and is in $B(c)$, A^{-1} is also conservative, which implies that $c_A = c$; [396, p. 12]. □

Now, we can give our final results of the present section concerning the subdivisions of the approximate point spectrum, defect spectrum and compression spectrum of the matrix operator $B(r,s)$ on the spaces c_0 and c:

Theorem 5.7.20. [71, Theorem 3.1] *The following results hold:*

(a) $\sigma_{ap}(B(r,s), c_0) := \{\lambda \in \mathbb{C} : |\lambda - r| \leq |s|\} \setminus \{r\}$.

(b) $\sigma_{\delta}(B(r,s), c_0) := \{\lambda \in \mathbb{C} : |\lambda - r| \leq |s|\}$.

(c) $\sigma_{co}(B(r,s), c_0) := \{\lambda \in \mathbb{C} : |\lambda - r| < |s|\}$.

Proof. (a) Since, $\sigma_{ap}(B(r,s), c_0) = \sigma(B(r,s), c_0) \setminus III_1\sigma(B(r,s), c_0)$,

$$\sigma_{ap}(B(r,s), c_0) := \{\lambda \in \mathbb{C} : |\lambda - r| \leq |s|\} \setminus \{r\}$$

is obtained by Theorems 2.1 and 2.6 of Altay and Başar [25].

(b) Since $\sigma_{\delta}(B(r,s), c_0) = \sigma(B(r,s), c_0) \setminus I_3\sigma(B(r,s), c_0)$ from Table 5.2 and

$$I_3\sigma(B(r,s), c_0) = II_3\sigma(B(r,s), c_0) = III_3\sigma(B(r,s), c_0) = \emptyset$$

is obtained by Theorem 2.2 of Altay and Başar [25], we therefore derive that $\sigma_{\delta}(B(r,s), c_0) = \sigma(B(r,s), c_0)$.

(c) Since the equality

$$\sigma_{co}(B(r,s), c_0) := III_1\sigma(B(r,s), c_0) \cup III_2\sigma(B(r,s), c_0) \cup III_3\sigma(B(r,s), c_0)$$

holds from Table 5.2, it can be easily seen by Theorems 2.2 and 2.3 of Altay and Başar [25] that $\sigma_{\delta}(B(r,s), c_0)$ is the set of $\lambda \in \mathbb{C}$ such that $|\lambda - r| < |s|$. □

The next corollary is an immediate consequence of Proposition 5.2.1:

Corollary 5.7.21. [71, Corollary 3.2] *The following results hold:*

(a) $\sigma_{ap}(B(r,s)^*, \ell_1) := \{\lambda \in \mathbb{C} : |\lambda - r| \leq |s|\}$.

(b) $\sigma_{\delta}(B(r,s)^*, \ell_1) := \{\lambda \in \mathbb{C} : |\lambda - r| \leq |s|\} \setminus \{r\}$.

(c) [25, Theorem 2.3] $\sigma_p(B(r,s)^*, \ell_1) := \{\lambda \in \mathbb{C} : |\lambda - r| < |s|\}$.

Theorem 5.7.22. [71, Theorem 3.3] *The following results hold:*

(a) $\sigma_{ap}(B(r,s),c) := \{\lambda \in \mathbb{C} : |\lambda - r| \leq |s|\} \setminus \{r\}$.

(b) $\sigma_{\delta}(B(r,s),c) := \{\lambda \in \mathbb{C} : |\lambda - r| \leq |s|\}$.

(c) $\sigma_{co}(B(r,s),c) := \{\lambda \in \mathbb{C} : |\lambda - r| < |s|\} \cup \{r+s\}$.

Proof. (a) $\sigma_{ap}(B(r,s),c) = \sigma(B(r,s),c) \backslash III_1\sigma(B(r,s),c)$ is obtained from Table 5.2. Now, the validity of the present part of the theorem follows from Theorem 2.10 of Altay and Başar [25].

(b) $\sigma_{\delta}(B(r,s),c) := \sigma(B(r,s),c) \backslash I_3\sigma(B(r,s),c)$ is obtained from Table 5.2. Moreover, since

$$\sigma_p(B(r,s),c) := I_3\sigma(B(r,s),c) \cup II_3\sigma(B(r,s),c) \cup III_3\sigma(B(r,s),c) = \emptyset$$

then $I_3\sigma(B(r,s),c) := \emptyset$ by Theorem 2.11 of Altay and Başar [25]. Hence, $\sigma_{\delta}(B(r,s),c) := \sigma(B(r,s),c)$.

(c) From Table 5.2

$$\sigma_{co}(B(r,s),c) := III_1\sigma(B(r,s),c) \cup III_2\sigma(B(r,s),c) \cup III_3\sigma(B(r,s),c),$$

$$III_1\sigma(B(r,s),c) \cup III_2\sigma(B(r,s),c) := \sigma_r(B(r,s),c)$$

and $III_3\sigma(B(r,s),c) := \emptyset$ from Theorems 2.11 and 2.13 of Altay and Başar [25], we have

$$\sigma_{co}(B(r,s),c) := \{\lambda \in \mathbb{C} : |\lambda - r| < |s|\} \cup \{r+s\}.$$

□

As a consequence of Proposition 5.2.1, we have

Corollary 5.7.23. [71, Corollary 3.4] *The following results hold:*

(a) $\sigma_{ap}(B(r,s)^*, \ell_1) := \{\lambda \in \mathbb{C} : |\lambda - r| \leq |s|\}$.

(b) $\sigma_{\delta}(B(r,s)^*, \ell_1) := \{\lambda \in \mathbb{C} : |\lambda - r| \leq |s|\} \backslash \{r\}$.

(c) [25, Theorem 2.12] $\sigma_p(B(r,s)^*, \ell_1) := \{\lambda \in \mathbb{C} : |\lambda - r| < |s|\} \cup \{r+s\}$.

Theorem 5.7.24. [71, Theorem 3.5] *The following results hold:*

(a) $\sigma_{ap}(B(r,s), \ell_p) := \{\lambda \in \mathbb{C} : |\lambda - r| \leq |s|\} \setminus \{r\}$.

(b) $\sigma_{\delta}(B(r,s), \ell_p) := \{\lambda \in \mathbb{C} : |\lambda - r| \leq |s|\}$.

(c) $\sigma_{co}(B(r,s), \ell_p) := \{\lambda \in \mathbb{C} : |\lambda - r| < |s|\}$.

Proof. (a) Since, $\sigma_{ap}\left(B(r,s),\ell_p\right) = \sigma\left(B(r,s),\ell_p\right) \backslash III_1\sigma\left(B(r,s),\ell_p\right)$, one can derive from Theorems 2.3 and 2.9 of Bilgiç and Furkan [93] that

$$\sigma_{ap}\left(B(r,s),\ell_p\right) := \{\lambda \in \mathbb{C} : |\lambda - r| \leq |s|\} \setminus \{r\}.$$

(b) Since $\sigma_\delta\left(B(r,s),\ell_p\right) := \sigma\left(B(r,s),\ell_p\right) \backslash I_3\sigma\left(B(r,s),\ell_p\right)$ from Table 5.2 and

$$I_3\sigma\left(B(r,s),\ell_p\right) = II_3\sigma\left(B(r,s),\ell_p\right) = III_3\sigma\left(B(r,s),\ell_p\right) = \emptyset$$

is obtained by Theorem 2.4 of Bilgiç and Furkan [93], one can therefore see that $\sigma_\delta\left(B(r,s),\ell_p\right) := \sigma\left(B(r,s),\ell_p\right)$.

(c) Since the equality

$$\sigma_{co}\left(B(r,s),\ell_p\right) := III_1\sigma\left(B(r,s),\ell_p\right) \cup III_2\sigma\left(B(r,s),\ell_p\right) \cup III_3\sigma\left(B(r,s),\ell_p\right)$$

holds from Table 5.2, it can be easily seen by Theorems 2.3 and 2.4 of Bilgiç and Furkan [93] that $\sigma_{co}(B(r,s),\ell_p)$ consists of $\lambda \in \mathbb{C}$ such that $|\lambda - r| < |s|$. □

As a consequence of Proposition 5.2.1, we also have the following:

Corollary 5.7.25. [71, Corollary 3.6] *Let $p^{-1} + q^{-1} = 1$. Then, the following results hold:*

(a) $\sigma_{ap}\left(B(r,s)^*,\ell_q\right) := \{\lambda \in \mathbb{C} : |\lambda - r| \leq |s|\}$.

(b) $\sigma_\delta\left(B(r,s)^*,\ell_q\right) := \{\lambda \in \mathbb{C} : |\lambda - r| \leq |s|\} \setminus \{r\}$.

(c) [93, Theorem 2.5] $\sigma_p\left(B(r,s)^*,\ell_q\right) := \{\lambda \in \mathbb{C} : |\lambda - r| < |s|\}$.

Since the subdivisions of the spectrum of the operator $B(r,s)$ on the sequence space bv_p can be derived by analogy to that space ℓ_p, we omit details and give the related results without proof.

Theorem 5.7.26. [71, Theorem 3.7] *The following results hold:*

(a) $\sigma_{ap}\left(B(r,s),bv_p\right) := \{\lambda \in \mathbb{C} : |\lambda - r| \leq |s|\} \setminus \{r\}$.

(b) $\sigma_\delta\left(B(r,s),bv_p\right) := \{\lambda \in \mathbb{C} : |\lambda - r| \leq |s|\}$.

(c) $\sigma_{co}\left(B(r,s),bv_p\right) := \{\lambda \in \mathbb{C} : |\lambda - r| < |s|\}$.

The following corollary is also a consequence of Proposition 5.2.1:

Corollary 5.7.27. [71, Corollary 3.8] *The following results hold:*

(a) $\sigma_{ap}\left(B(r,s)^*,(bv_p)^*\right) := \{\lambda \in \mathbb{C} : |\lambda - r| \leq |s|\}$.

(b) $\sigma_\delta\left(B(r,s)^*,(bv_p)^*\right) := \{\lambda \in \mathbb{C} : |\lambda - r| \leq |s|\} \setminus \{r\}$.

(c) [93, Theorem 3.4.(ii)] $\sigma_p\left(B(r,s)^*,(bv_p)^*\right) := \{\lambda \in \mathbb{C} : |\lambda - r| < |s|\}$.

5.8 THE FINE SPECTRA OF THE OPERATOR $B(r,s,t)$ ON THE SPACES ℓ_p AND bv_p

In this section, following Furkan et al. [178], we determine the fine spectrum with respect to the Goldberg's classification of the operator $B(r,s,t)$ defined by a triple band matrix over the sequence spaces ℓ_p and bv_p with $1<p<\infty$. These results are more general than the spectrum of the second-order difference operator Δ^2 and include some other special cases such as the spectrum of generalized difference operator $B(r,s)$ over ℓ_p and bv_p of Bilgiç and Furkan [93], the spectrum of difference operator $\Delta^{(1)}$ over ℓ_p of Akhmedov and Başar [14], the spectrum of the same operator over bv_p of Akhmedov and Başar [15], the right shift and Zweier matrices.

Our main focus in this section is on the triple band matrix $B(r,s,t)$, where

$$B(r,s,t):=\begin{bmatrix} r & 0 & 0 & 0 & 0 & \dots \\ s & r & 0 & 0 & 0 & \dots \\ t & s & r & 0 & 0 & \dots \\ 0 & t & s & r & 0 & \dots \\ 0 & 0 & t & s & r & \dots \\ \vdots & \vdots & \vdots & \vdots & \vdots & \ddots \end{bmatrix}, \quad (s,t\neq 0).$$

5.8.1 The Fine Spectrum of the Operator $B(r,s,t)$ on the Sequence Space ℓ_p, $(1<p<\infty)$.

In this subsection, we determine the fine spectrum with respect to the Goldberg's classification of the operator $B(r,s,t)$ over the sequence space ℓ_p, where $1<p<\infty$. We begin with a theorem concerning bounded linearity of the operator $B(r,s,t)$ acting on the sequence space ℓ_p:

Theorem 5.8.1. *$B(r,s,t):\ell_p\to\ell_p$ is a bounded linear operator satisfying the inequalities*

$$(|r|^p+|s|^p+|t|^p)^{1/p}\leq\|B(r,s,t)\|_p\leq|r|+|s|+|t|. \tag{5.8.1}$$

Proof. The linearity of $B(r,s,t)$ is trivial and so, is omitted. Let us take $e^{(0)}=(1,0,0,\dots)$ in ℓ_p. Then, $B(r,s,t)e^{(0)}=(r,s,t,0,0,\dots)$ and we observe that

$$\frac{\|B(r,s,t)e^{(0)}\|_p}{\|e^{(0)}\|_p}=(|r|^p+|s|^p+|t|^p)^{1/p}$$

which gives the fact that

$$(|r|^p+|s|^p+|t|^p)^{1/p}\leq\|B(r,s,t)\|_p \tag{5.8.2}$$

for any $p > 1$. Now, take any $x = (x_k) \in \ell_p$. Then, using Minkowski's inequality, we have

$$\begin{aligned}
\|B(r,s,t)x\|_p &= \left(\sum_k |sx_{k-1} + rx_k + tx_{k+1}|^p\right)^{1/p} \\
&\leq \left(\sum_k |sx_{k-1}|^p\right)^{1/p} + \left(\sum_k |rx_k|^p\right)^{1/p} + \left(\sum_k |tx_{k+1}|^p\right)^{1/p} \\
&= \left(|s|^p \sum_k |x_{k-1}|^p\right)^{1/p} + \left(|r|^p \sum_k |x_k|^p\right)^{1/p} + \left(|t|^p \sum_k |x_{k+1}|^p\right)^{1/p} \\
&= (|s| + |r| + |t|)\ \|x\|_p
\end{aligned}$$

which gives

$$\|B(r,s,t)\|_p \leq |s| + |r| + |t|. \tag{5.8.3}$$

Therefore, by combining the inequalities (5.8.2) and (5.8.3), one can easily see the required result (5.8.1), which completes the proof. □

Before giving the main theorem of this subsection, we should note the following remark. In this section, here and in what follows, if z is a complex number, then by $\sqrt{z}$ we always mean the square root of z with a non-negative real part. If $Re(\sqrt{z}) = 0$, then $\sqrt{z}$ represents the square root of z with $Im(\sqrt{z}) \geq 0$. Same results are obtained if $\sqrt{z}$ represents the other square root.

We begin with the following lemma, which is needed in proving the next theorem:

Lemma 5.8.2. [107, p. 254, Theorem 34.18] *Let $1 < p < \infty$ and $A \in (\ell_\infty : \ell_\infty) \cap (\ell_1 : \ell_1)$. Then, $A \in (\ell_p : \ell_p)$.*

Theorem 5.8.3. *Let s be a complex number such that $\sqrt{s^2} = -s$ and define the set S by*

$$S := \left\{\alpha \in \mathbb{C} : |2(r-\alpha)| \leq \left|-s + \sqrt{s^2 - 4t(r-\alpha)}\right|\right\}.$$

Then, $\sigma(B(r,s,t), \ell_p) = S$.

Proof. First, we prove that $(B(r,s,t) - \alpha I)^{-1}$ exists and is in $B(\ell_p)$ for $\alpha \notin S$ and then, the operator $B(r,s,t) - \alpha I$ is not invertible for $\alpha \in S$.

Let $\alpha \notin S$. It is easy to see that we must have $\alpha \neq r$ and so $B(r,s,t) - \alpha I$ is triangle, hence has an inverse. We can calculate that

$$(B(r,s,t) - \alpha I)^{-1} := \begin{bmatrix} a_1 & 0 & 0 & 0 & \cdots \\ a_2 & a_1 & 0 & 0 & \cdots \\ a_3 & a_2 & a_1 & 0 & \cdots \\ a_4 & a_3 & a_2 & a_1 & \cdots \\ \vdots & \vdots & \vdots & \vdots & \ddots \end{bmatrix},$$

where

$$\begin{aligned} a_1 &= \tfrac{1}{r-\alpha} \\ a_2 &= -\tfrac{s}{(r-\alpha)^2} \\ a_3 &= \tfrac{s^2-(r-\alpha)t}{(r-\alpha)^3} \\ &\vdots \end{aligned}$$

In fact this sequence is obtained recursively by letting

$$a_1 := \frac{1}{r-\alpha}, \quad a_2 := \frac{-s}{(r-\alpha)^2} \quad \text{and} \quad a_n := \frac{-(sa_{n-1}+ta_{n-2})}{(r-\alpha)} \quad \text{for all } \ n \in \mathbb{N}_3.$$

We are led to solve the equation $(r-\alpha)u^2 + su + t = 0$, whose roots are

$$u_1 := \frac{-s+\sqrt{s^2-4t(r-\alpha)}}{2(r-\alpha)} \quad \text{and} \quad u_2 := \frac{-s-\sqrt{s^2-4t(r-\alpha)}}{2(r-\alpha)}.$$

Elementary calculations on recurrent sequences give

$$a_n := \frac{1}{\sqrt{s^2-4t(r-\alpha)}}\left(u_1^n - u_2^n\right) \quad \text{for all } \ n \in \mathbb{N}_1. \tag{5.8.4}$$

If $s^2 = 4t(r-\alpha)$, then

$$a_n := \left(-\frac{2n}{s}\right)\left[\frac{-s}{2(r-\alpha)}\right]^n \tag{5.8.5}$$

and simple calculations give that $(a_n) \in \ell_p$ if and only if $|-s/2(r-\alpha)| < 1$. Therefore, $\alpha \notin S$ implies $(a_n) \notin \ell_p$. So, we may assume that $s^2 \neq 4t(r-\alpha)$. Since $\alpha \notin S$, we have $|u_1| < 1$. Now, we show that $|u_2| < 1$. Since $|u_1| < 1$, we have

$$\left|1+\sqrt{1-\frac{4t(r-\alpha)}{s^2}}\right| < \left|\frac{2(r-\alpha)}{-s}\right|.$$

Since $|1-\sqrt{z}| \leq |1+\sqrt{z}|$ for any $z \in \mathbb{C}$ we must have

$$\left|1-\sqrt{1-\frac{4t(r-\alpha)}{s^2}}\right| < \left|\frac{2(r-\alpha)}{-s}\right|.$$

It follows that $|u_2| < 1$. This proves that $\alpha \notin S \Rightarrow a_n \to 0$. Now,

$$\begin{aligned} \|(B(r,s,t)-\alpha I)^{-1}\|_{(\ell_1:\ell_1)} &= \sup_{n\in\mathbb{N}} \sum_{k=n}^{\infty} |a_k| = \sum_{k=1}^{\infty} |a_k| \\ &\leq \frac{1}{|\sqrt{s^2-4t(r-\alpha)}|}\left(\sum_{k=1}^{\infty} |u_1|^n + \sum_{k=1}^{\infty} |u_2|^n\right) < \infty \end{aligned}$$

since $|u_1| < 1$ and $|u_2| < 1$. This shows that $(B(r,s,t) - \alpha I)^{-1} \in (\ell_1 : \ell_1)$. Similarly, we can show that $(B(r,s,t) - \alpha I)^{-1} \in (\ell_\infty : \ell_\infty)$, since

$$\|(B(r,s,t) - \alpha I)^{-1}\|_{(\ell_\infty:\ell_\infty)} = \sup_{n\in\mathbb{N}} \sum_{k=1}^{n} |a_k| = \sum_{k=1}^{\infty} |a_k|.$$

By Lemma 5.8.2, we have $(B(r,s,t) - \alpha I)^{-1} \in (\ell_p : \ell_p)$. This shows that $\sigma(B(r,s,t), \ell_p) \subseteq S$.

Now, let $\alpha \in S$. If $\alpha = r$, then $B(r,s,t) - \alpha I$ is represented by the matrix

$$B(0,s,t) - \alpha I = \begin{bmatrix} 0 & 0 & 0 & 0 & \dots \\ s & 0 & 0 & 0 & \dots \\ t & s & 0 & 0 & \dots \\ 0 & t & s & 0 & \dots \\ \vdots & \vdots & \vdots & \vdots & \ddots \end{bmatrix}.$$

Since $B(r,s,t) - rI = B(0,s,t)$ does not have a dense range, it is not invertible.

If $s^2 = 4t(r-\alpha)$ for all $n \in \mathbb{N}_1$, by the identity (5.8.5), where $|-s/[2(r-\alpha)]| \geq 1$, we deduce that $(B(r,s,t) - \alpha I)^{-1} \notin B(\ell_p)$.

So, we may assume that $\alpha \neq r$ and $s^2 \neq 4t(r-\alpha)$.

Now, since $\alpha \neq r$, then $B(r,s,t) - \alpha I$ is triangle but since $s^2 \neq 4t(r-\alpha)$ we must have $|u_1| > |u_2|$ and $|u_1| > 1$ from which it follows that $a_n \nrightarrow 0$ and so $\sum_{n=1}^{\infty} |a_n|^p$ diverges. Now, $y = (1,0,0,\dots) \in \ell_p$ gives $x = (a_1, a_2, a_3, \dots)$, which is not in ℓ_p. Therefore, $(B(r,s,t) - \alpha I)^{-1}$ is not in $B(\ell_p)$. This shows that $S \subseteq \sigma(B(r,s,t), \ell_p)$.

This completes the proof. □

Remark 5.8.4. *If $\sqrt{s^2} = s$, then we obtain the same sequence and*

$$\sigma(B(r,s,t), \ell_p) := \left\{ \alpha \in \mathbb{C} : |2(r-\alpha)| \leq \left| s + \sqrt{s^2 - 4t(r-\alpha)} \right| \right\}.$$

Theorem 5.8.5. $\sigma_p(B(r,s,t), \ell_p) := \emptyset$.

Proof. Suppose $B(r,s,t)x = \alpha x$ for $x \neq \theta = (0,0,0,\dots)$ in ℓ_p. Then, by solving the system of linear equations

$$\left. \begin{array}{rcl} rx_0 & = & \alpha x_0 \\ sx_0 + rx_1 & = & \alpha x_1 \\ tx_0 + sx_1 + rx_2 & = & \alpha x_2 \\ tx_1 + sx_2 + rx_3 & = & \alpha x_3 \\ tx_2 + sx_3 + rx_4 & = & \alpha x_4 \\ & \vdots & \end{array} \right\}$$

we find that $\alpha = r$ from the equation

$$tx_{n_0-2} + sx_{n_0-1} + rx_{n_0} = \alpha x_{n_0}$$

if x_{n_0} is the first non-zero term of the sequence $x = (x_k)$ and we also obtain $x_{n_0} = 0$ from the next equation, which contradicts the fact that $x_{n_0} \neq 0$.

This completes the proof. □

If $T : \ell_p \to \ell_p$ is a bounded linear operator with the matrix A, then it is known that the adjoint operator $T^* : \ell_p^* \to \ell_p^*$ is defined by the transpose of the matrix A. The dual space ℓ_p^* of ℓ_p is isomorphic to ℓ_q with $p^{-1} + q^{-1} = 1$.

Theorem 5.8.6. *Let* $S_1 := \left\{\alpha \in \mathbb{C} : |2(r-\alpha)| < \left|-s + \sqrt{s^2 - 4t(r-\alpha)}\right|\right\}$. *Then,* $\sigma_p(B(r,s,t)^*, \ell_p^*) := S_1$.

Proof. We first show that $S_1 \subseteq \sigma_p(B(r,s,t)^*, \ell_p^*)$. Suppose $B(r,s,t)^* x = \alpha x$ for $\theta \neq x \in \ell_p^* \cong \ell_q$ with $p^{-1} + q^{-1} = 1$. Then, we solve the system of linear equations

$$\left.\begin{array}{rcl} rx_0 + sx_1 + tx_2 & = & \alpha x_0 \\ rx_1 + sx_2 + tx_3 & = & \alpha x_1 \\ rx_2 + sx_3 + tx_4 & = & \alpha x_2 \\ & \vdots & \end{array}\right\}$$

It is clear that if $\alpha = r$, then we may choose $x_0 \neq 0$ and so $x = (x_0, 0, 0, \dots)$ is an eigenvector corresponding to $\alpha = r$. Assume $\alpha \neq r$. The formula for x_n, in terms of x_0 and x_1, can be obtained by solving the recurrence relation

$$tx_n + sx_{n-1} + (r-\alpha)x_{n-2} = 0.$$

Since a_n's satisfy the recurrence relation $(r-\alpha)a_n + sa_{n-1} + ta_{n-2} = 0$, we notice that

$$x_n := \frac{a_{n-1}(r-\alpha)^k}{t^{k-1}} c_1 + \frac{a_n(r-\alpha)^k}{t^{k-1}} c_2$$

satisfies the relation for x_n for any constants c_1, c_2 and $k \in \mathbb{N}_1$. In order to derive a correct formula for x_n in terms of x_0, x_1 and a_n's, we need to let $c_1 = -x_0, c_2 = x_1$ and $k = n$, obtaining

$$x_n = -\frac{a_{n-1}(r-\alpha)^n}{t^{n-1}} x_0 + \frac{a_n(r-\alpha)^n}{t^{n-1}} x_1 \quad \text{for all} \quad n \in \mathbb{N}_2. \tag{5.8.6}$$

Now, we choose

$$x_0 = 1 \quad \text{and} \quad x_1 = \frac{2(r-\alpha)}{-s + \sqrt{s^2 - 4t(r-\alpha)}}.$$

Then, one can verify that $x_n = x_1^n$ for all $n \in \mathbb{N}_2$. Now $x = (x_k) \in \ell_p^*$ since $|x_1| < 1$. This shows that

$$S_1 \subseteq \sigma_p(B(r,s,t)^*, \ell_p^*).$$

Now, assume α is a number such that

$$\left|\frac{2(r-\alpha)}{-s + \sqrt{s^2 - 4t(r-\alpha)}}\right| \geq 1,$$

i.e., we have $|u_1| \leq 1$. We must show that $\alpha \notin \sigma_p(B(r,s,t)^*, \ell_q)$. Using (5.8.6) we obtain

$$\frac{x_{n+1}}{x_n} = \left(\frac{r-\alpha}{t}\right)\left(\frac{a_n}{a_{n-1}}\right)\left(\frac{-x_0 + \frac{a_{n+1}}{a_n}x_1}{-x_0 + \frac{a_n}{a_{n-1}}x_1}\right).$$

Notice that $(r-\alpha)/t = 1/(u_1u_2)$. Now, we examine the following three cases:

Case (i): $|u_2| < |u_1| \leq 1$.

In this situation, we have $s^2 \neq 4t(r-\alpha)$ and

$$\lim_{n\to\infty} \frac{a_n}{a_{n-1}} = \lim_{n\to\infty} \frac{a_{n+1}}{a_n} = \lim_{n\to\infty} \frac{u_1^{n+1} - u_2^{n+1}}{u_1^n - u_2^n} = \lim_{n\to\infty} \frac{u_1^{n+1}\left[1 - \left(\frac{u_2}{u_1}\right)^{n+1}\right]}{u_1^n\left[1 - \left(\frac{u_2}{u_1}\right)^n\right]} = u_1.$$

Now, if $-x_0 + u_1x_1 = 0$, then we get $x_n = x_0/u_1^n$, which is not in ℓ_q since $|u_1| \leq 1$. Otherwise

$$\lim_{n\to\infty}\left|\frac{x_{n+1}}{x_n}\right| = \frac{1}{|u_1||u_2|}|u_1| = \frac{1}{|u_2|} > 1.$$

Case (ii): $|u_2| = |u_1| < 1$.

In this situation, $s^2 = 4t(r-\alpha)$ and using the formula

$$a_n := \left(-\frac{2n}{s}\right)\left[\frac{-s}{2(r-\alpha)}\right]^n$$

we obtain

$$\lim_{n\to\infty} \frac{a_{n+1}}{a_n} = \frac{-s}{2(r-\alpha)} = u_2 = u_1$$

and so

$$\lim_{n\to\infty}\left|\frac{x_{n+1}}{x_n}\right| = \frac{1}{|u_1||u_2|}|u_1| = \frac{1}{|u_2|} > 1.$$

Case (iii): $|u_2| = |u_1| = 1$.

In this situation, $s^2 = 4t(r-\alpha)$ and so we have $|-s/2t| = 1$. Assume $\alpha \in \sigma_p(B(r,s,t)^*, \ell_q)$. This implies that $x \in \ell_q$ and $x \neq \theta$. From (5.8.6) we have

$$x_n := \left(\frac{-s}{2t}\right)^{n-1}\left[(1-n)\left(\frac{-s}{2t}\right)x_0 + nx_1\right].$$

Since $\lim_{n\to\infty}|x_n| = 0$, we must have either $x_0 = x_1 = 0$ or $x_1 = (-s)/(2t)x_0$. Since $(-s)/(2t) = 1/u_1$, as in Case (i), we get a contradiction. So, we must have $x_0 = x_1 = 0$. But, this implies that $x = \theta$, a contradiction: Hence, we must have $\alpha \notin \sigma_p(B(r,s,t)^*, \ell_q)$.

In Cases (i) and (ii) by d'Alembert test $x \notin \ell_q$. In Case (iii) $\alpha \in \sigma_p(B(r,s,t)^*, \ell_q)$ leads a contradiction. This is what we wished to prove. □

Theorem 5.8.7. *Let S_1 be defined as in Theorem 5.8.6 and*

$$S_2 := \left\{ \alpha \in \mathbb{C} : \left| \frac{2(r-\alpha)}{-s+\sqrt{s^2-4t(r-\alpha)}} \right| = 1 \right\}.$$

Then, the following statements hold:

(i) $\sigma_r(B(r,s,t),\ell_p) := S_1$.

(ii) $\sigma_c(B(r,s,t),\ell_p) := S_2$.

Proof. (i) Since $\sigma_p(B(r,s,t)^*,\ell_p^*) = S_1$, $B(r,s,t)^* - \alpha I$ is not one-to-one for all $\alpha \in S_1$. Therefore, by Lemma 5.2.2, $B(r,s,t) - \alpha I$ does not have a dense range for all $\alpha \in S_1$.

(ii) Since $\sigma(B(r,s,t),\ell_p)$ is the disjoint union of the parts $\sigma_p(B(r,s,t),\ell_p)$, $\sigma_r(B(r,s,t),\ell_p)$ and $\sigma_c(B(r,s,t),\ell_p)$, we must have $\sigma_c(B(r,s,t),\ell_p) = S_2$. □

Theorem 5.8.8. *If $|t| < |s|$, then $r \in III_1\sigma(B(r,s,t),\ell_p)$. If $|t| \geq |s|$, then $r \in III_2\sigma(B(r,s,t),\ell_p)$.*

Proof. If $\alpha = r$ then, by Part (i) of Theorem 5.8.7, $B(r,s,t) - \alpha I$ is in state III_1 or III_2. A left inverse ${}^{(-1)}B(0,s,t) = (b_{nk})$ of $B(0,s,t)$ is

$${}^{(-1)}B(0,s,t) := \begin{bmatrix} 0 & \frac{1}{s} & 0 & 0 & \cdots \\ 0 & \frac{-t}{s^2} & \frac{1}{s} & 0 & \cdots \\ 0 & \frac{(-t)^2}{s^3} & \frac{-t}{s^2} & \frac{1}{s} & \cdots \\ 0 & \frac{(-t)^3}{s^4} & \frac{(-t)^2}{s^3} & \frac{-t}{s^2} & \cdots \\ \vdots & \vdots & \vdots & \vdots & \ddots \end{bmatrix}$$

which can be formulated, as follows;

$$b_{nk} := \begin{cases} \frac{(-t)^{n+1-k}}{s^{n+2-k}} & , \quad 1 \leq k \leq n+1, \\ 0 & , \quad k = 0 \text{ or } k \geq n+2. \end{cases}$$

Then, ${}^{(-1)}B(0,s,t) \in (\ell_1 : \ell_1) \cap (\ell_\infty : \ell_\infty)$ for $|t| < |s|$. By Lemma 5.8.2, $B(0,s,t)$ has a continuous inverse for $|t| < |s|$. For the case $|t| \geq |s|$, we take $e^{(1)} = (0,1,0,\ldots) \in \ell_p$. Now,

$$\left\{ {}^{(-1)}B(0,s,t)e^{(1)} \right\}_n := \frac{(-t)^n}{s^{n+1}} \quad \text{for all} \quad n \in \mathbb{N}$$

is not in ℓ_p. This shows that $B(0,s,t)$ does not have a continuous inverse for $|t| \geq |s|$. Therefore, $r \in III_1\sigma(B(r,s,t),\ell_p)$ for $|t| < |s|$ and $r \in III_2\sigma(B(r,s,t),\ell_p)$ for $|t| \geq |s|$.

This completes the proof. □

Theorem 5.8.9. *If $\alpha \neq r$ and $\alpha \in \sigma_r(B(r,s,t),\ell_p)$, then $\alpha \in III_2\sigma(B(r,s,t),\ell_p)$.*

Proof. Since $\alpha \neq r$, the operator $B(r,s,t) - \alpha I$ is triangle, hence it has an inverse. Now, we show that the inverse of the operator $B(r,s,t) - \alpha I$ is discontinuous. Since $\alpha \in \sigma_r(B(r,s,t), \ell_p)$, we have $|u_1| > 1$. If $s^2 = 4t(r-\alpha)$, then using (5.8.5), we have $\lim_{n\to\infty} |a_n| = \infty$, since $|s| > |2(r-\alpha)|$. If $s^2 \neq 4t(r-\alpha)$, then using (5.8.4) we obtain that $\lim_{n\to\infty} |a_n| = \infty$, since $|u_1| > 1$ and $|u_1| > |u_2|$. So, $B(r,s,t) - \alpha I$ has an unbounded inverse. Therefore, $B(r,s,t) - \alpha I \in 2$.

By Theorem 5.8.6, $B(r,s,t)^* - \alpha I$ is not one-to-one. By Lemma 5.2.2, $B(r,s,t) - \alpha I$ does not have a dense range. Therefore, $B(r,s,t) - \alpha I \in III$.

This completes the proof. □

Theorem 5.8.10. *If $\alpha \in \sigma_c(B(r,s,t), \ell_p)$, then $\alpha \in II_2$.*

Proof. Let $\alpha \in \sigma_c(B(r,s,t), \ell_p)$, i.e., we have $|u_1| = 1$. In order to show that $\alpha \in II_2$, we must establish that $B(r,s,t) - \alpha I$ is not onto. $y = (1,0,0,\dots) \in \ell_p$ gives $x = (a_1, a_2, a_3, \dots)$. If $s^2 = 4t(r-\alpha)$, then

$$|a_n| := \frac{2n}{|s|}$$

which shows that x is not in ℓ_p. Otherwise, we get $|u_2| < |u_1| = 1$, which implies that $\lim_{n\to\infty} a_n \neq 0$ and so $x \notin \ell_p$.

This completes the proof. □

In this subsection, we conclude by giving the results on the subdivisions of the approximate point spectrum, defect spectrum and compression spectrum of the matrix operator $B(r,s,t)$ on the space ℓ_p. Here and after, we suppose that S is defined as in Theorem 5.8.3.

Theorem 5.8.11. [72, Theorem 3.5] *The following results hold:*

(a) $\sigma_{ap}[B(r,s,t), \ell_1] := \widetilde{S}$.

(b) $\sigma_{\delta}[B(r,s,t), \ell_1] := S$.

(c) $\sigma_{co}[B(r,s,t), \ell_1] := S$.

Proof. (a) Since

$$\sigma_{ap}[B(r,s,t), \ell_1] := \sigma[B(r,s,t), \ell_1] \backslash III_1\sigma[B(r,s,t), \ell_1],$$

$\sigma_{ap}[B(r,s,t), \ell_1] := \widetilde{S}$ is obtained by Theorems 2.1 and 2.4 of Bilgiç and Furkan [93].

(b) Since $\sigma_{\delta}[B(r,s,t), \ell_1] = \sigma[B(r,s,t), \ell_1] \backslash I_3\sigma[B(r,s,t), \ell_1]$ from Table 5.2 and

$$I_3\sigma[B(r,s,t), \ell_1] = II_3\sigma[B(r,s,t), \ell_1] = III_3\sigma[B(r,s,t), \ell_1] = \emptyset$$

by Theorem 2.2 of Bilgiç and Furkan [93], we obtain from Theorem 2.1 of Bilgiç and Furkan [93] that $\sigma_{\delta}[B(r,s,t), \ell_1] = \sigma[B(r,s,t), \ell_1]$.

(c) Since the equality

$$\sigma_{co}[B(r,s,t), \ell_1] = III_1\sigma[B(r,s,t), \ell_1] \cup III_2\sigma[B(r,s,t), \ell_1] \cup III_3\sigma[B(r,s,t), \ell_1]$$

holds from Table 5.2, Theorems 2.2–2.4 of Bilgiç and Furkan [93] show that the compression spectrum of $B(r,s,t)$ over the sequence space ℓ_1 is the set S.

The following corollary is an easy consequence of Proposition 5.2.1:

Corollary 5.8.12. [72, Corollary 3.6] *The following results hold:*

(a) $\sigma_{ap}[B(r,s,t)^*, \ell_\infty] := S$.

(b) $\sigma_\delta[B(r,s,t)^*, \ell_\infty] := \widetilde{S}$.

(c) [93, Theorem 2.5] $\sigma_p[B(r,s,t)^*, \ell_\infty] := S$.

Theorem 5.8.13. [72, Theorem 3.7] *The following results hold:*

(a) $\sigma_{ap}[B(r,s,t), \ell_p] := \widetilde{S}$.

(b) $\sigma_\delta[B(r,s,t), \ell_p] := S$.

(c) $\sigma_{co}[B(r,s,t), \ell_p] := Int\ S$.

Proof. (a) Since

$$\sigma_{ap}[B(r,s,t), \ell_p] := \sigma[B(r,s,t), \ell_p] \backslash III_1\sigma[B(r,s,t), \ell_p],$$

$\sigma_{ap}[B(r,s,t), \ell_p] := \widetilde{S}$ is obtained by Theorems 2.2 and 2.7 of Furkan et al. [178].

(b) Since $\sigma_\delta[B(r,s,t), \ell_p] := \sigma[B(r,s,t), \ell_p] \backslash I_3\sigma[B(r,s,t), \ell_p]$ from Table 5.2 and

$$I_3\sigma[B(r,s,t), \ell_p] = II_3\sigma[B(r,s,t), \ell_p] = III_3\sigma[B(r,s,t), \ell_p] = \emptyset$$

is observed by Theorem 2.3 of Furkan et al. [178] whose Theorem 2.2 gives that $\sigma_\delta[B(r,s,t), \ell_p] := \sigma[B(r,s,t), \ell_p]$.

(c) Since the equality

$$\sigma_{co}[B(r,s,t), \ell_p] := III_1\sigma[B(r,s,t), \ell_p] \cup III_2\sigma[B(r,s,t), \ell_p] \cup III_3\sigma[B(r,s,t), \ell_p]$$

holds from Table 5.2, the desired result can easily be seen by Theorems 2.2–2.7 of Furkan et al. [178]. □

The following corollary is a consequence of Proposition 5.2.1:

Corollary 5.8.14. [72, Corollary 3.8] *Let $p^{-1} + q^{-1} = 1$, then we have*

(a) $\sigma_{ap}[B(r,s,t)^*, \ell_q] := S$.

(b) $\sigma_\delta[B(r,s,t)^*, \ell_q] := \widetilde{S}$.

(c) [178, Theorem 2.4] $\sigma_p[B(r,s,t)^*, \ell_q] := Int\ S$.

5.8.2 The Spectrum of the Operator $B(r,s,t)$ on the Sequence Space bv_p, $(1 < p < \infty)$

In this subsection, we determine the fine spectrum with respect to the Goldberg's classification of the operator $B(r,s,t)$ over the sequence space bv_p, where $1 < p < \infty$. First, we give a theorem concerning bounded linearity of the operator $B(r,s,t)$ acting on the sequence space bv_p:

Theorem 5.8.15. *$B(r,s,t) \in B(bv_p)$.*

Proof. The linearity of $B(r,s,t)$ is trivial. Let us take any $x = (x_k) \in bv_p$. Then, using Minkowski's inequality, we have

$$\|B(r,s,t)x\|_{bv_p} := \left[\sum_k |r(x_k - x_{k-1}) + s(x_{k-1} - x_{k-2}) + t(x_{k-2} - x_{k-3})|^p\right]^{1/p}$$

$$\leq \left(|r|^p \sum_k |x_k - x_{k-1}|^p\right)^{1/p} + \left(|s|^p \sum_k |x_{k-1} - x_{k-2}|^p\right)^{1/p} +$$
$$+ \left(|t|^p \sum_k |x_{k-2} - x_{k-3}|^p\right)^{1/p}$$
$$= (|s| + |r| + |t|)\ \|x\|_{bv_p}$$

which leads us to the desired result $\|B(r,s,t)\|_{(bv_p,bv_p)} \leq |s| + |r| + |t|$. □

Theorem 5.8.16. *$\sigma(B(r,s,t), bv_p) := S$, where S is defined as in Theorem 5.8.3.*

Proof. First, we prove that $(B(r,s,t) - \alpha I)^{-1}$ exists and is in $B(bv_p)$ for $\alpha \notin S$ and next the operator $B(r,s,t) - \alpha I$ is not invertible for $\alpha \in S$.

Let $\alpha \notin S$. Since $s \neq 0$, we have $\alpha \neq r$ and so $B(r,s,t) - \alpha I$ is triangle, hence $(B(r,s,t) - \alpha I)^{-1}$ exists. Let $y = (y_k) \in bv_p$. This implies that $(y_k - y_{k-1}) \in \ell_p$. Solving the equation $(B(r,s,t) - \alpha I)x = y$, we find the matrix in the proof of Theorem 5.8.3. We therefore have

$$x_k - x_{k-1} = \sum_{j=0}^{k} a_{k-j+1}(y_j - y_{j-1}) \quad \text{for all} \quad k \in \mathbb{N},$$

i.e., $(x_k - x_{k-1}) = (B(r,s,t) - \alpha I)^{-1}(y_k - y_{k-1})$. Since $(B(r,s,t) - \alpha I)^{-1} \in (\ell_p : \ell_p)$ by Theorem 5.8.3, $(x_k - x_{k-1}) \in \ell_p$. This implies that $x = (x_k) \in bv_p$. This shows that $\sigma(B(r,s,t), bv_p) \subseteq S$.

Now, let $\alpha \in S$. If $\alpha \neq r$, then $B(r,s,t) - \alpha I$ is triangle, hence $(B(r,s,t) - \alpha I)^{-1}$ exists but $y = (1,1,1,\dots) \in bv_p$ gives $x = (x_k)$ with $x_k = \sum_{j=0}^{k} a_{j+1}$ for all $k \in \mathbb{N}$. Now, $\sum_k |x_k - x_{k-1}|^p = \sum_k |a_k|^p$ is divergent since it was proved in Theorem 5.8.3 that $a_n \nrightarrow 0$ as $n \to \infty$. When $s^2 = 4t(r - \alpha)$, we may use the same argument. Hence, $(B(r,s,t) - \alpha I)^{-1}$ is not in $B(bv_p)$.

If $\alpha = r$, then similar arguments in the proof of Theorem 5.8.3 show that the operator $B(r,s,t) - \alpha I = B(0,s,t)$ is not invertible. This shows that $S \subseteq \sigma(B(r,s,t), bv_p)$.

This completes the proof. □

In the light of Part (ii) of Theorem 4.3.8, we have

Theorem 5.8.17. *$\sigma_p(B(r,s,t)^*, bv_p^*) := S_1$, where S_1 is the set as in Theorem 5.8.6.*

Proof. Suppose $B(r,s,t)^*x = \alpha x$ for $x \neq \theta$ in $bv_p^* \cong D_q$ with $p^{-1}+q^{-1}=1$. Consider the system of linear equations

$$\left.\begin{array}{rcl} rx_0 + sx_1 + tx_2 & = & \alpha x_0 \\ rx_1 + sx_2 + tx_3 & = & \alpha x_1 \\ rx_2 + sx_3 + tx_4 & = & \alpha x_2 \\ & \vdots & \end{array}\right\}$$

It is clear that if $\alpha = r$, then we may choose $x_0 \neq 0$ and so $x = (x_0, 0, 0, \ldots)$ is an eigenvector corresponding to $\alpha = r$. Assume $\alpha \neq r$, then we obtain (5.8.6). Let $\alpha \in S_1$ so that

$$\left|\frac{2(r-\alpha)}{-s+\sqrt{s^2-4t(r-\alpha)}}\right| < 1.$$

Then, we may choose $x_0 = 1$, $x_1 = 2(r-\alpha)/[-s+\sqrt{s^2-4t(r-\alpha)}]$. Then, one can verify that $x_2 = x_1^2, x_3 = x_1^3, \ldots, x_n = x_1^n, \ldots$ for all $n \geq 2$. Now, since $|x_1| < 1$, we have

$$\sum_{j=k}^{\infty} x_j = \sum_{j=k}^{\infty} x_1^j = \frac{x_1^k}{1-x_1}.$$

Hence,

$$\sum_k \left|\sum_{j=k}^{\infty} x_j\right|^q = \frac{1}{|1-x_1|^q}\sum_k |x_1|^{kq} < \infty,$$

since $|x_1| < 1$. Therefore, $x \in D_q$. This shows that $S_1 \subseteq \sigma_p(B(r,s,t)^*, bv_p^*)$.

Now, assume that α is a number such that

$$\left|\frac{2(r-\alpha)}{-s+\sqrt{s^2-4t(r-\alpha)}}\right| \geq 1,$$

i.e., we have $|u_1| \leq 1$. We must show that $\alpha \notin \sigma_p(B(r,s,t)^*, bv_p^*)$. Assume $\alpha \in \sigma_p(B(r,s,t)^*, bv_p^*)$. Then, $x \in D_q$ with $x \neq \theta$ and so $\lim_{n\to\infty} x_n = 0$. We examine the following two cases:

Case (i): $s^2 = 4t(r-\alpha)$.

In this state, we have $|(-s)/(2t)| = 1$. As in Case (iii) of Theorem 5.8.6, we obtain that $x = \theta$, which is a contradiction.

Case (ii): $s^2 \neq 4t(r-\alpha)$.

In this state, we have $u_1 \neq u_2$. Using (5.8.6), we obtain that

$$x_n = \left(\frac{1}{u_1u_2}\right)^{n-1} \frac{(u_1^n - u_2^n)x_1 - (u_1^{n-1} - u_2^{n-1})x_0}{u_1 - u_2} \quad \text{for all } n \in \mathbb{N}_1.$$

Then, we must have either $x_0 = x_1 = 0$ or

$$\lim_{n\to\infty} \frac{u_1^n - u_2^n}{u_1^{n-1} - u_2^{n-1}} = \frac{x_0}{x_1},$$

since $\lim_{n\to\infty} x_n = 0$. As in Case (iii) of Theorem 5.8.6, the above limit is equal to u_1. But then, we get $u_1 x_1 = x_0$. Then, we get $x_n = x_0/u_1^n$ which is not in D_q, since $|u_1| \leq 1$. So, we must have $x_0 = x_1 = 0$.

In Cases (i) and (ii) above, $\alpha \in \sigma_p(B(r,s,t)^*, D_q)$ leads a contradiction. This is what we wished to prove. □

Since the spectrum and fine spectrum of the matrix $B(r,s,t)$ as an operator on the sequence space bv_p are similar to that of the space ℓ_p in Subsection 5.8.1, to avoid from the repetition of the similar statements, we give the results by the following theorem without proof:

Theorem 5.8.18. *Let S_1 and S_2 be the sets defined in Theorems 5.8.6 and 5.8.7. Then, the following statements hold:*

(i) $\sigma_p(B(r,s,t), bv_p) := \emptyset$.

(ii) $\sigma_r(B(r,s,t), bv_p) := S_1$.

(iii) $\sigma_c(B(r,s,t), bv_p) := S_2$.

(iv) *If $|t| < |s|$, then $r \in III_1\sigma(B(r,s,t), bv_p)$. If $|t| \geq |s|$, then $r \in III_2\sigma(B(r,s,t), bv_p)$.*

(v) *If $\alpha \neq r$ and $\alpha \in \sigma_r(B(r,s,t), bv_p)$, then $\alpha \in III_2\sigma(B(r,s,t), bv_p)$.*

(vi) *If $\alpha \in \sigma_c(B(r,s,t), bv_p)$, then $\alpha \in II_2$.*

Finally, we give the results on the subdivisions of the approximate point spectrum, defect spectrum and compression spectrum of the matrix operator $B(r,s,t)$ on the space bv_p. Since the subdivisions of the spectrum of the operator $B(r,s,t)$ on the space bv_p can be derived by analogy to the space ℓ_p, we omit details and give the concerning results without proof.

Theorem 5.8.19. [72, Theorem 3.9] *The following results hold:*

(a) $\sigma_{ap}[B(r,s,t), bv] := \widetilde{S}$.

(b) $\sigma_\delta[B(r,s,t), bv] := S$.

(c) $\sigma_{co}[B(r,s,t), bv] := S$.

As a consequence of Proposition 5.2.1, we also have the following:

Corollary 5.8.20. [72, Corollary 3.10] *The following results hold:*

(a) $\sigma_{ap}[B(r,s,t)^*, bv^*] := S$.

(b) $\sigma_\delta[B(r,s,t)^*, bv^*] := \widetilde{S}$.

(c) [93, Theorem 2.8.(i)] $\sigma_p[B(r,s,t)^*, bv^*] := S$.

Theorem 5.8.21. [72, Theorem 3.11] *The following results hold:*

(a) $\sigma_{ap}[B(r,s,t), bv_p] := \widetilde{S}$.

(b) $\sigma_\delta[B(r,s,t), bv_p] := S$.

(c) $\sigma_{co}[B(r,s,t), bv_p] := Int\ S$.

The following corollary is derived from Proposition 5.2.1:

Corollary 5.8.22. [72, Corollary 3.12] *Let $p^{-1} + q^{-1} = 1$. Then, we have*

(a) $\sigma_{ap}\left(B\left(r,s,t\right)^*, (bv_p)^*\right) := S$.

(b) $\sigma_\delta[B(r,s,t)^*, (bv_p)^*] := \widetilde{S}$.

(c) [178, Theorem 3.3] $\sigma_p[B(r,s,t)^*, (bv_p)^*] := Int\ S$.

5.9 CONCLUSION

It is immediate that our results given in Subsection 5.8.2 concerning the triple band matrix $B(r,s,t)$ are reduced to the spectrum of $B(r,s)$ studied by Bilgiç and Furkan [93], and to the spectrum of $\Delta^{(1)}$ determined by Akhmedov and Başar [14] and [15] over the sequence spaces ℓ_p and bv_p, respectively, since $B(r,s,0) = B(r,s)$ and $B(1,-1,0) = \Delta^{(1)}$, respectively. Additionally, the present work includes several special cases such as the right shift and Zweier matrices. By this reason, the study is more general and more comprehensive than the works in the existing literature.

Furthermore, although by applying the spectral mapping theorem for polynomials (cf. [231, p. 381]) to Theorems 2.1 and 3.1 of Akhmedov and Başar [14], [15] over the sequence spaces ℓ_p and bv_p, the spectrum of Δ^2, which is the composition of the operator $\Delta^{(1)}$ with itself, over the sequence spaces ℓ_p and bv_p may be derived as the set

$$\{\alpha \in \mathbb{C} : |1 - \sqrt{\alpha}| \leq 1\},$$

one can directly produce the same result from the Subsection 5.8.2, since $\Delta^2 = B(1,-2,1)$. It is interesting that the spectrum of operator Δ^2 over the sequence spaces ℓ_p and bv_p is the region enclosed by the cardioid $r = 4\cos^2(\theta/2)$. Nevertheless, the spectrum of several special summability matrices over the sequence spaces ℓ_p and bv_p is the region enclosed by a circle (cf. [14] and [15]).

CHAPTER 6

Core of a Sequence

In this chapter, the Knopp core, σ-core, $\mathcal{I}$-core and $F_{\mathcal{B}}$-core of a sequence are studied. Also, a short survey for the results related to the core of a sequence is given.

Keywords: Core of a sequence, K-core, σ-core, $\mathcal{I}$-core and $F_{\mathcal{B}}$-core, directed set of a sequence, Knopp's core theorem.

6.1 KNOPP CORE

The concept of core of a sequence was introduced firstly by Knopp in 1929–1930 [111, p. 137]. So, this first type of core was called the Knopp core or $\mathcal{K}$-core. The following notes can be found in [111, p. 137]. Now, we may look at the definition given by Knopp:

Definition 6.1.1. [111, 6.1] *Let $x = (x_k) \in \omega$ and R_k be the least closed convex region in the complex plane which contains the points $x_k, x_{k+1}, x_{k+2}, \ldots$. Then,*

$$\mathcal{K} - core(x) := \bigcap_{k=0}^{\infty} R_k. \tag{6.1.1}$$

Note that the set $\mathcal{K} - core(x)$ is necessarily convex and closed, and it is clear that $R_1 \supseteq R_2 \supseteq R_3 \supseteq \cdots$. Also, there exists a connection between the core of a sequence and its limit points.

Theorem 6.1.2. [111, 6.1] *Let D be the set of all limit points of $x = (x_k) \in \omega$. Then, $D \subset \mathcal{K} - core(x)$.*

Proof. Let α be a limit point of $x = (x_k) \in \omega$. Then, there exists an increasing sequence of integers (k_i) such that $\lim_{i\to\infty} x_{k_i} = \alpha$. Now, choose any fixed natural number p and determine q so that $k_q > p$. Then, $x_{k_q}, x_{k_{q+1}}, x_{k_{q+2}}, \ldots$ are in R_p. Since R_p is closed, it contains the limit points of the sequence in itself. Hence, α is in R_p. Since p is arbitrary, this proves the result. □

It can be derived from the definition that $\mathcal{K} - core(x)$ contains just one point if and only if $x = (x_k)$ is convergent. If $\mathcal{K} - core(x)$ is empty, i.e., it does not contain finite point, then the sequence $x = (x_k)$ is called *definitely divergent* and is denoted by $x_k \sim \infty$.

DOI: 10.1201/9781003294153-6

Example 6.1.3. [111, 6.1] *Define the sequence* $x = (x_k)$ *by*

$$x_k := \begin{cases} k & , \quad k \text{ even}, \\ ki & , \quad k \text{ odd}. \end{cases}$$

Then, each set R_k *consists of the first quadrant of the complex plane remaining after removal of a triangular section with right angle at the origin. So, it is easy to see that the* $\mathcal{K} - core(x)$ *is empty. Hence,* $x_k \sim \infty$.

Example 6.1.4. [111, 6.1] *Define* $x = (x_k)$ *by* $x_k := ki^k$ *for all* $k \in \mathbb{N}$. *Then, each* R_k *contains all points of the complex plane. So,* $\mathcal{K} - core(x) = \mathbb{C}$. *Therefore, it is not true that* $x_k \sim \infty$.

Example 6.1.5. [111, 6.1] *Define* $x = (x_k)$ *by* $x_k := (-1)^k$ *for all* $k \in \mathbb{N}$. *Then, since* $\mathcal{K} - core(x) = [-1, 1]$, *again it is not true that* $x_k \sim \infty$.

An alternative definition of the $\mathcal{K} - core$ is given by Knopp as follows. Firstly, we have to give the following definition:

Definition 6.1.6. [111, 6.1, I] *Every line* L *divides the plane into two half-planes. If a set of the points* S *lies entirely in such a half-plane (some or all of the points may lie on the line* L*), we say that* L *is a barrier line for* S.

Now, we can give the alternative definition of $\mathcal{K} - core$.

Definition 6.1.7. [111, 6.1, I] *The following statements are satisfied:*

(a) *If the set* S *containing the entries of* $x = (x_k)$ *has no barrier lines, then* $\mathcal{K} - core(x)$ *is the whole plane.*

(b) *If the set* S *containing the entries of* $x = (x_k)$ *has barrier lines, then* $\mathcal{K} - core(x)$ *is the intersection of all half-planes containing the set* D *of the limit points of* $x = (x_k)$.

The following theorem is due to Allen [18], which shows the equivalence of the two definitions for $\mathcal{K} - core$:

Theorem 6.1.8. [111, 6.1, I] *Definitions 6.1.1 and 6.1.6 of the* $\mathcal{K} - core$ *are equivalent.*

Proof. Let $x = (x_k) \in \omega$, E be the $\mathcal{K} - core(x)$ with respect to Definition 6.1.1, D be the set of all limit points of $x = (x_k)$ and F be the intersection of all half-planes which contains D. It is known by the definition that $D \subseteq E$ and it is also obvious that $D \subseteq F$. For the proof, we should show that $E = F$. Firstly, suppose that $\alpha \notin E$. Then, $\alpha \notin R_k$ for some fixed value of k. We can draw a barrier line L separating α from R_k. Since R_k is closed, $D \subseteq R_k$ and so, L separates α from D. Hence, $\alpha \notin F$. This means that $F \subseteq E$.

Now, draw a half-plane P containing D and call the barrier line L. All except a finite number of the elements of (x_k) lie on the same side of L as D. Otherwise, there would be at least one limit point on the side of L remote from D. Consequently, there is an n such that $x_n, x_{n+1}, \ldots$ are in P. Hence, $R_n \subset P$ and so, $E \subset P$. Thus, $E \subseteq F$.

By combining these two inclusions, the proof of the theorem is completed. □

In [361], it is shown that for every bounded sequence $x = (x_k)$,

$$\mathcal{K} - core(x) := \bigcap_{z \in \mathbb{C}} B_x(z),$$

where $B_x(z) := \{w \in \mathbb{C} : |w - z| \leq \limsup_{k \in \mathbb{N}} |x_k - z|\}$.

Now, we may give a useful result. Firstly, let us call the set of limit points of a sequence as its *derived set.*

Remark 6.1.9. [111, 6.1, II] *If the two sequences have the same derived set, then they have the same* $\mathcal{K} - core$. *But, the converse of this is not true, in general.*

The first part of Remark 6.1.9 is clear from Definition 6.1.6. For the converse, let us consider the sequences

$$(x_k) := (0, 1, 0, 1, 0, 1, \ldots) \quad \text{and} \quad (y_k) := (0, 1/2, 1, 0, 1/2, 1, 0, 1/2, 1, \ldots).$$

It is obvious that $\mathcal{K} - core(x) = \mathcal{K} - core(y) = [0, 1]$. But, the derived sets D_x and D_y of the sequences $x = (x_k)$ and $y = (y_k)$ are $D_x = \{0, 1\}$ and $D_y = \{0, 1/2, 1\}$.

Here, the question "*If two sequences have the same core, what can be said about their derived sets?*" arises. The answer of this question may be stated in two ways with respect to Definition 6.1.1 or Definition 6.1.6 of the concept of the $\mathcal{K} - core$. So, we can give the following theorems:

Theorem 6.1.10. [111, 6.1, II] *Let* $x = (x_k), y = (y_k) \in \omega$. *Then,* $\mathcal{K} - core(x) = \mathcal{K} - core(y)$ *if and only if every half-plane which contains the derived set* D_x *of* x *also contains the derived set* D_y *of* y *and vice versa.*

Proof. Firstly, suppose that $\mathcal{K} - core(x) = \mathcal{K} - core(y)$. Let P be a half-plane containing D_x. Then, by Definition 6.1.7, $\mathcal{K} - core(x) = \mathcal{K} - core(y) \subseteq P$. Since $D_y \subseteq \mathcal{K} - core(y)$, it follows that $D_y \subseteq P$.

Conversely, suppose that every half-plane which contains D_x also contains D_y and vice versa. If an α is in $\mathcal{K} - core(x)$, then by Definition 6.1.7, α is in every half-plane which contains D_x. Therefore, α is in every half-plane which contains D_y. Hence, α is in $\mathcal{K} - core(y)$. So, $\mathcal{K} - core(x) \subseteq \mathcal{K} - core(y)$.

The converse inclusion can be shown by the similar way. Therefore, the proof is completed. □

Theorem 6.1.11. [111, 6.1, II] *Let* $x, y \in \omega$. *Then,* $\mathcal{K} - core(x) = \mathcal{K} - core(y)$ *if and only if every convex region which contains* D_x *also contains* D_y *and vice versa.*

Proof. Firstly, suppose that $\mathcal{K} - core(x) = \mathcal{K} - core(y)$ for $x, y \in \omega$. Let P be a convex region containing D_x. If P does not contain D_y, then there exists at least one point α of D_y outside P, and there exists a barrier line separating α from D_x. This means that there exists a half-plane which contains D_x but not D_y. This contradicts the fact that $\mathcal{K} - core(x) = \mathcal{K} - core(y)$ by Theorem 6.1.10.

Conversely, suppose that every convex region which contains D_x also contains D_y and vice versa. Then, every half-plane which contains D_x also contains D_y and conversely. Hence, $\mathcal{K} - core(x) = \mathcal{K} - core(y)$. □

Next, theorem determines a condition for the equality $\mathcal{K} - core$ of two sequences.

Theorem 6.1.12. [111, 6.3, II] *Let $x = (x_k)$, $y = (y_k) \in \omega$ such that $|x_k - y_k| \to 0$, as $k \to \infty$. Then, $\mathcal{K} - core(x) = \mathcal{K} - core(y)$.*

Proof. Take a point z such that $z \in \mathcal{K} - core(y) \setminus \mathcal{K} - core(x)$. Then, there is a suffix n such that $z \notin R_n$. Let z' be the unique point of R_n nearest to z. Take $\alpha = z + (z' - z)/3$, $\beta = z + 2(z' - z)/3$ and $|z' - z| = 3d$. Then,

$$|\alpha - z| = |\beta - \alpha| = |z' - \beta| = d.$$

Choose $q > n$ so that $|x_k - y_k| < d$ when $k \in \mathbb{N}_q$. Let L_α and L_β be the lines perpendicular to the line zz' at α and β, respectively. The points of R_n and hence, the points $x_q, x_{q+1}, x_{q+2}, \ldots$ are all separated from z by the line L_β. Also, it follows that the points $y_q, y_{q+1}, y_{q+2}, \ldots$ are all separated from z by the line L_α. Hence, $z \notin \mathcal{K} - core(y)$, a contradiction. This proves that $\mathcal{K} - core(y) \subseteq \mathcal{K} - core(x)$.

Similarly, one can show the reverse inclusion.

This completes the proof. □

Now, we may give Knopp's Core Theorem as follows:

Knopp's Core Theorem. [111, 6.1, I] *Let $A = (a_{nk})$ be a regular matrix with positive real entries. Then, $\mathcal{K} - core(Ax) \subseteq \mathcal{K} - core(x)$ for every sequence $x = (x_k)$ of real numbers.*

Proof. Let $A = (a_{nk})$ be a regular matrix with positive real entries and $x = (x_k)$ be a sequence of real numbers. By the definition of Knopp core, we have $\mathcal{K} - core(x) = [\liminf x, \limsup x]$ and $\mathcal{K} - core(Ax) = [\liminf Ax, \limsup Ax]$. To prove the theorem, it is enough to show that $\limsup Ax \leq \limsup x$. If $\limsup x = \infty$, there is nothing to prove. Now, let $\limsup x = b < \infty$. Then, for $\varepsilon > 0$ there exists an $m \in \mathbb{N}$ such that $x_k \leq b + \varepsilon$ whenever $k > m$. Therefore, since $A = (a_{nk})$ is positive, we can write

$$\begin{aligned}
(Ax)_n &= \sum_{k=0}^{m} a_{nk}x_k + \sum_{k=m+1}^{\infty} a_{nk}x_k \\
&\leq \sum_{k=0}^{m} a_{nk}x_k + (b+\varepsilon) \sum_{k=m+1}^{\infty} a_{nk} \\
&\leq \sum_{k=0}^{m} a_{nk}x_k + (b+\varepsilon) \sum_{k} a_{nk}.
\end{aligned}$$

Then, the conditions of regularity of $A = (a_{nk})$ imply as $n \to \infty$ that $a_{nk} \to 0$ for each fixed $k \in \mathbb{N}$ and $\sum_k a_{nk} \to 1$. Therefore, one can see for arbitrary ε that $\limsup Ax \leq b + \varepsilon$.

This completes the proof. □

In 1960, Rhoades [326] gave a slight generalization of Knopp's Core Theorem. To give his result, we need some concepts and symbols. An infinite matrix $A = (a_{nk})$ is called *multiplicative* if $A = (a_{nk})$ is conservative and $\lim_{n\to\infty} a_{nk} = \alpha_k = 0$ for each $k \in \mathbb{N}$. For brevity, we shall write $L(x) := \limsup_{k\to\infty} x_k$ and $l(x) := \liminf_{k\to\infty} x_k$, where $x = (x_k) \in \omega$.

Theorem 6.1.13. [326, Theorem 4] *Let $A = (a_{nk})$ be a matrix for which $\chi(A)$ is defined. Then, the condition*

$$\text{There exists a positive integer } q \text{ such that } a_{nk} \geq 0 \text{ for all } k \in \mathbb{N}_q \tag{6.1.2}$$

is sufficient for

$$\liminf_{n\to\infty} (Ax)_n \geq \sum_k a_k x_k + \chi(A) \cdot l(x), \tag{6.1.3}$$

$$\limsup_{n\to\infty} (Ax)_n \leq \sum_k a_k x_k + \chi(A) \cdot L(x) \tag{6.1.4}$$

whenever the series $\sum_k a_k x_k$ is convergent.
(Note that (6.1.3) is true without (6.1.2) provided that $\chi(A) > 0$ when $l(x) = -\infty$ and (6.1.4) is also true without (6.1.2) provided that $\chi(A) > 0$ when $L(x) = \infty$.)

Proof. Assume $l(x) > -\infty$ and fix an $\varepsilon > 0$. Then, there exists a positive integer N such that $x_k \geq l(x) - \varepsilon$ and $a_{nk} \geq 0$ for all $k \geq N$. If $r \geq N$, then

$$(Ax)_n := [l(x) - \varepsilon] \sum_k a_{nk} + \sum_{k=0}^{r} a_{nk}[x_k - l(x) + \varepsilon] + \sum_{k=r+1}^{\infty} a_{nk}[x_k - l(x) + \varepsilon].$$

The third series is non-negative, and

$$\lim_{n\to\infty} \sum_k a_{nk} = a,$$
$$\lim_{n\to\infty} \sum_{k=0}^{r} a_{nk}[x_k - l(x) + \varepsilon] = \sum_{k=0}^{r} a_k[x_k - l(x) + \varepsilon].$$

Therefore,

$$\lim_{n\to\infty} (Ax)_n \geq [l(x) - \varepsilon] \left(a - \sum_{k=0}^{r} a_k\right) + \sum_{k=0}^{r} a_k x_k.$$

Since $r \geq N$ is arbitrary,

$$\liminf_{n\to\infty} (Ax)_n \geq [l(x) - \varepsilon]\chi(A) + \sum_k a_k x_k$$

and (6.1.3) follows.

(6.1.4) is obtained from (6.1.3) by considering $-x = (-x_k)$ instead of $x = (x_k)$. □

To state the following theorem, we need the next definition: We shall say that a matrix has property q if there exists a positive integer q such that $a_k = 0$ for $k \in \mathbb{N}_q$.

Theorem 6.1.14. [326, Theorem 5] *If $A = (a_{nk})$ is a coregular triangular matrix with property q, then (6.1.2) is the necessary and sufficient condition for (6.1.3) and (6.1.4).*

The sufficiency is included in Theorem 6.1.13. The necessity follows from [201, Theorem II'] by taking $k_1 > q$.

There is a class of matrices for which (6.1.3) and (6.1.4) hold for bounded sequences, but which do not necessarily satisfy (6.1.2), and is considered in next two theorems.

Theorem 6.1.15. [326, Theorem 6] *Let $A = (a_{nk})$ be a matrix for which $\chi(A)$ is defined. Then,*

$$\lim_{n\to\infty} \sum_k |a_{nk}| = a \tag{6.1.5}$$

is a sufficient condition for (6.1.3) and (6.1.4) to hold for all bounded sequences $x = (x_k)$ for which $(a_k x_k) \in cs$ (in particular, for all bounded sequences $x = (x_k)$ if $A = (a_{nk})$ is conservative).

Proof. If we write for all $k, n \in \mathbb{N}$

$$b_{nk} := \frac{|a_{nk}| + a_{nk}}{2} \quad \text{and} \quad c_{nk} := \frac{|a_{nk}| - a_{nk}}{2},$$

then $a_{nk} := b_{nk} - c_{nk}$. By hypothesis, it is clear that

$$\lim_{n\to\infty} \sum_k b_{nk} = a \quad \text{and} \quad \lim_{n\to\infty} \sum_k c_{nk} = 0.$$

Since $x = (x_k)$ is bounded, there exists a number $K > 0$ such that $|x_k| < K$ for all $k \in \mathbb{N}$. For any fixed $\varepsilon > 0$, there exist integers $M, N > q$ such that $l(x) - \varepsilon \leq x_k \leq L(x) + \varepsilon$ and for all $k > N$ and $n > M$,

$$\sum_k c_{nk} < \frac{\varepsilon}{K + H + \varepsilon},$$

where $H = \max\{|l(x)|, |L(x)|\}$. Let $r > \max\{M, N\}$ and write

$$\begin{aligned}(Ax)_n &= [l(x) - \varepsilon]\sum_k a_{nk} + \sum_{k=0}^{r} a_{nk}[x_k - l(x) + \varepsilon] + \\ &+ \sum_{k=r+1}^{\infty} b_{nk}[x_k - l(x) + \varepsilon] - \sum_{k=r+1}^{\infty} c_{nk}[x_k - l(x) + \varepsilon].\end{aligned}$$

The third sum on the right-hand side of above equality is non-negative, and

$$\sum_{k=r+1}^{\infty} c_{nk}[x_k - l(x) + \varepsilon] < [K + |l(x)| + \varepsilon] \sum_{k=r+1}^{\infty} c_{nk} < \varepsilon,$$

$$\lim_{n\to\infty} \sum_{k=0}^{r} a_{nk}[x_k - l(x) + \varepsilon] = \sum_{k=0}^{r} a_k x_k - [l(x) - \varepsilon] \sum_{k=0}^{r} a_k.$$

Therefore,

$$\liminf_{n\to\infty} (Ax)_n \geq [l(x) - \varepsilon]\left(a - \sum_{k=0}^{r} a_k\right) + \sum_{k=0}^{r} a_k x_k - \varepsilon$$

for each $r > \max\{M, N\}$. Letting $r \to \infty$ and then, since ε was arbitrary, we obtain (6.1.3).

Since the proof of (6.1.4) is similar to that of (6.1.3), we omit details. □

Theorem 6.1.16. [326, Theorem 7] *Let $A = (a_{nk})$ be coregular triangular multiplicative matrix. Then,*

$$\limsup_{n\to\infty} (Ax)_n \leq \chi(A) \cdot L(x) \quad \text{and} \quad \liminf_{n\to\infty} (Ax)_n \geq \chi(A) \cdot l(x)$$

for each $x \in \ell_\infty$ if and only if (6.1.5) holds.

Proof. The sufficiency follows from Theorem 6.1.15.

Conversely, suppose that there exists a number $\lambda > 0$ such that, repeatedly,

$$\sum_{k=0}^{n} |a_{nk}| > |a|(1 + 3\lambda).$$

By the conservativeness of the matrix $A = (a_{nk})$, this can be written as

$$\liminf_{n\to\infty} \sum_{k=0}^{n} |a_{nk}| \geq \left| \lim_{n\to\infty} \sum_{k=0}^{n} a_{nk} \right| = |a|.$$

Now, choose n_1 so that

$$\sum_{k=0}^{n} |a_{n_1,k}| > |a|(1 + 3\lambda).$$

Since $A = (a_{nk})$ is multiplicative, choose a positive integer n_2 so that

$$\sum_{k=0}^{n_1} |a_{n_2,k}| < |a|\lambda, \quad \sum_{k=0}^{n_2} |a_{n_2,k}| > |a|(1+3\lambda),$$

and generally n_p so that

$$\sum_{k=0}^{n_{p-1}} |a_{n_p,k}| < |a|\lambda, \quad \sum_{k=0}^{n_p} |a_{n_p,k}| > |a|(1+3\lambda).$$

Define the sequence $x = (x_k)$ by $x_k := (-1)^p \text{sgn} a_{n_p,k}$ if $n_{p-1} < k \leq n_p$, where $n_{-1} = 0$. Then,

$$\begin{aligned} (-1)^p (Ax)_{n_p} &:= (-1)^p \sum_{k=0}^{n_{p-1}} a_{n_p,k} x_k - \sum_{k=0}^{n_{p-1}} |a_{n_p,k}| + \sum_{k=0}^{n_p} |a_{n_p,k}| \\ &> -2\lambda|a| + |a|(1+3\lambda) = |a|(1+\lambda). \end{aligned}$$

Hence, $(Ax)_{n_p} > |a|(1+\lambda) \geq a(1+\lambda)$ for even p and $(Ax)_{n_p} = -|a|(1+\lambda) \leq -a(1+\lambda)$ for odd p. Therefore, $\limsup_{n\to\infty} (Ax)_n > a$ and $\liminf_{n\to\infty} (Ax)_n < -a$. Since the sequence $x = (x_k)$ is defined in terms of the sign of $a_{n_p,k}$, there are two possibilities:

(i) $x = (x_k)$ is a sequence containing an infinite number of 1's and -1's.

(ii) All but a finite number of the x_k's are of the same sign.

If (i) is true, then $l = -1$ and $L = 1$, and (6.1.3) and (6.1.4) do not hold.

If (ii) is true, then $x = (x_k) \in c$ and the above discussion shows that $A = (a_{nk})$ is not conservative, and hence is not coregular, since Ax is not convergent.

This completes the proof of necessity part. □

Following example shows that the condition "multiplicativeness" in the necessity part cannot be replaced by "$A = (a_{nk})$ has property q."

Example 6.1.17. [326] *Define* $A = (a_{nk})$ *by*

$$a_{nk} := \begin{cases} -1 & , \quad k = 1, n \in \mathbb{N}, \\ 1 & , \quad k = n > 1, \\ 0 & , \quad \text{otherwise.} \end{cases}$$

Then, $a = 0$, $a_1 = -1$, $a_k = 0$ *for* $k > 1$ *and* $\chi(A) = 1$. *Therefore,* $A = (a_{nk})$ *is coregular. Since for any* $x = (x_k) \in \ell_\infty$, $(Ax)_n = -x_1 + x_n$; $\limsup_{n\to\infty} (Ax)_n = -x_1 + L$ *and* $\liminf_{n\to\infty} (Ax)_n = -x_1 + l$. *But,* $\sum_{k=0}^{n} |a_{nk}| = 2$ *for all* $n \in \mathbb{N}$.

In 1965, P. Schaefer [358] extended the above results due to Rhoades to the complex sequences, as follows. Firstly, he proved the following lemma:

Lemma 6.1.18. [358, Lemma 2.5] *Let $A = (a_{nk})$ be a coregular matrix and define the matrix $B = (b_{nk})$ by*

$$b_{nk} := \frac{a_{nk} - a_k}{\chi(A)} \quad \text{for all} \quad k, n \in \mathbb{N}.$$

Then, B is regular and $(Ax)_n = \chi(A) \cdot (Bx)_n + \sum_k a_k x_k$ for all sequences $x = (x_k)$ for which $(a_k x_k) \in cs$ and $\{(Ax)_n\}$ exists.

Proof. Since $b_{nk} \to 0$, as $n \to \infty$, for each $k \in \mathbb{N}$,

$$\sum_k |b_{nk}| \leq \frac{1}{\chi(A)} \left(\sum_k |a_{nk}| + \sum_k |a_k| \right) \text{ for every } n \in \mathbb{N}$$

and $\sum_k b_{nk} \to 1$, as $n \to \infty$. Clearly, $(Ax)_n = \chi(A) \cdot (Bx)_n + \sum_k a_k x_k$. □

Theorem 6.1.19. [358, Theorem 2.1] *Let $A = (a_{nk})$ be a complex coregular matrix and $x = (x_k) \in \ell_\infty$. Then,*

$$\mathcal{K} - core(Ax) \subseteq \mathcal{K} - core(u) \tag{6.1.6}$$

if and only if

$$\lim_{n \to \infty} \sum_k |a_{nk} - a_k| = |\chi(A)|,$$

where $u = (u_k)$ defined by $u_k = z \cdot \chi(A) + \sum_k a_k x_k$ for $z \in \mathbb{C}$ and for all $k \in \mathbb{N}$.

Proof. Agnew proved that (see [111, Theorem 6.4 II]) if B is a regular matrix and $x \in \ell_\infty$, then the core of Bx is contained in the core of x if and only if $\sum_k |b_{nk}| \to 1$, as $n \to \infty$. Thus, the proof of the theorem follows from Agnew's result by choosing the matrix B as in Lemma 6.1.18. □

In the case of regularity of $A = (a_{nk})$, we have the following:

Corollary 6.1.20. [358, Corollary 2.2] *Let $A = (a_{nk})$ be a real coregular matrix with $\chi(A) > 0$ and let $x = (x_k)$ be a bounded real sequence. Then, the inequalities in (6.1.3) and (6.1.4) hold if and only if*

$$\lim_{n \to \infty} \sum_k |a_{nk} - a_k| := \chi(A).$$

Theorem 6.1.21. [358, Theorem 2.3] *Let $A = (a_{nk})$ be a complex coregular matrix and let $(x_k) \in \omega$ such that Ax exists and $(a_k x_k) \in cs$. Then, the inclusion in (6.1.6) holds if and only if there exists a number K such that*

$$\frac{a_{nk} - a_k}{\chi(A)} := Re\left[\frac{a_{nk} - a_k}{\chi(A)}\right] \geq 0, \tag{6.1.7}$$

for all n and for all $k \geq K$, where $Re[z]$ denotes the real part of $z \in \mathbb{C}$.

Proof. Cooke [111, p. 145] remarks that the condition that there exists a K such that $b_{nk} = Re[b_{nk}] \geq 0$ for all $n \in \mathbb{N}$ and $k \geq K$ is a sufficient condition that the core of $\{(Bx)_n\}$ be contained in the core of $x = (x_k)$ when B is regular and $x = (x_k)$ is arbitrary. The proof follows from this result by choosing $B = (b_{nk})$ as in Lemma 6.1.18. □

Corollary 6.1.22. [358, Corollary 2.4] *Let $A = (a_{nk})$ be a real coregular matrix with $\chi(A) > 0$ and $x = (x_k)$ be a real sequence such that Ax exists and $(a_k x_k) \in cs$. If there exists a number K such that $a_{nk} \geq a_k$ for all $n \in \mathbb{N}$ and all $k \geq K$ then, the inequalities (6.1.3) and (6.1.4) hold.*

Theorem 6.1.23. [358, Theorem 2.6] *Let $A = (a_{nk})$ be a triangular coregular matrix. Then, (6.1.6) holds for every sequence $x = (x_k)$ such that $(a_k x_k) \in cs$ if and only if (6.1.7) holds for all $n \geq k \geq K$.*

Proof. Define the triangular matrix $B = (b_{nk})$ by

$$b_{nk} := \begin{cases} (a_{nk} - a_k)/\chi(A) & , \quad n \geq k, \\ 0 & , \quad \text{otherwise}, \end{cases}$$

for all $k, n \in \mathbb{N}$. Then, as in the proof of Lemma 6.1.18, B is regular and

$$(Ax)_n := \chi(A) \cdot (Bx)_n + \sum_{k=0}^{n} a_k x_k.$$

By a result of Agnew [111, Theorem 6.4 I], the core of $\{(Bx)_n\}$ is contained in the core of $x = (x_k)$ if and only if there exists a K such that $b_{nk} = Re[b_{nk}] \geq 0$ for all $n \in \mathbb{N}$ and for all $k \geq K$, where $Re[b_{nk}]$ denotes the real part of $b_{nk} \in \mathbb{C}$. Hence, if

$$(Wx)_n := \chi(A) \cdot (Bx)_n + \sum_k a_k x_k,$$

the core of Wx is contained in the image of the core of $x = (x_k)$ under the transformation

$$u_n := \chi(A) \cdot (Bx)_n + \sum_k a_k x_k.$$

On the other hand, since

$$\lim_{n\to\infty} |(Wx)_n - (Ax)_n| = \lim_{n\to\infty} \left| \sum_{k \geq n+1} a_k x_k \right| = 0,$$

it follows from Theorem 6.1.12 that the core of Ax is the same as the core of Wx. □

Corollary 6.1.24. [358, Corollary 2.7] *Let $A = (a_{nk})$ be a real coregular triangular matrix with $\chi(A) > 0$ and $x = (x_k)$ be a bounded real sequence. Then, the inequalities in (6.1.3) and (6.1.4) hold for the sequence $x = (x_k)$ such that $(a_k x_k) \in cs$ if and only if there exists a K such that $a_{nk} \geq a_k$ for all $n \geq k \geq K$.*

In 1981, Rath and Tripathy [324] proved some inequalities for some classes of regular matrices related to the lim sup and lim inf. Therefore, they generalized the result due to Goffman and Petersen [186]. Firstly, we introduce some notations. We will write $I = [0, 1]$ and for each $\alpha \in I$, by $Q(\alpha)$, we denote the class of regular matrices $A = (a_{nk})$ such that $a_{nk} \geq 0$ for all $k, n \in \mathbb{N}$ and in addition, for each $n \in \mathbb{N}$ there is $k \in \mathbb{N}$ such that $a_{nk} = \alpha$ and also for each $k \in \mathbb{N}$ there is $n \in \mathbb{N}$ such that $a_{nk} = \alpha$.

Theorem 6.1.25. [324, Theorem 1] *For $x = (x_k) \in \ell_\infty$, $\alpha \in [0, 1/2]$ and $\lambda \in I$, there is $A \in Q(\alpha)$ which limits $x = (x_k)$ to $\lambda a + (1-\lambda)b$; where $a = \alpha L(x) + (1-\alpha)l(x)$ and $b = \alpha l(x) + (1-\alpha)L(x)$.*

Proof. We have $\lambda a + (1-\lambda)b = cl(x) + (1-c)L(x)$, where $c = \lambda + \alpha - 2\lambda\alpha \in I$. One can find a null sequence (ε_k) so that $l(x) - \varepsilon_k \leq x_k \leq L(x) + \varepsilon_k$ for every $k \in \mathbb{N}$. So, we write $x_k = \delta_k[l(x) - \varepsilon_k] + \lambda_k[L(x) + \varepsilon_k]$, where $\delta_k \geq 0$, $\lambda_k \geq 0$ and $\delta_k + \lambda_k = 1$ for each $k \in \mathbb{N}$. Putting $\alpha_k = c - \alpha\delta_k$ and $\beta_k = 1 - c - \alpha\lambda_k$, one can see that $1 - \alpha \geq \alpha_k \geq 0$ and $\beta_k = 1 - \alpha - \alpha_k \geq 0$.

Let (p_k) and (q_k) be increasing sequences of integers such that $p_k \neq k$, $p_k \neq q_k$, $q_k \neq k$ for each $k \in \mathbb{N}$ and $\lim_{k\to\infty} x_{p_k} = l(x)$, $\lim_{k\to\infty} x_{q_k} = L(x)$. Define $A \in Q(\alpha)$ by $a_{kk} = \alpha$, $a_{kp_k} = \alpha_k$, $a_{kq_k} = \beta_k$, $a_{kn} = 0$, otherwise, $k \in \mathbb{N}$. Then,

$$\begin{aligned}
\sum_n a_{kn}x_n &= \alpha\delta_k[l(x) - \varepsilon_k] + \alpha\lambda_k[L(x) + \varepsilon_k] + \alpha_k l(x) + \\
&\quad + \beta_k L(x) + \alpha_k[x_{p_k} - l(x)] + \beta_k[x_{q_k} - L(x)] \\
&= cl(x) + (1-c)L(x) - \varepsilon_k\alpha\delta_k + \varepsilon_k\alpha\lambda_k + \\
&\quad + \alpha_k[x_{p_k} - l(x)] + \beta_k[x_{q_k} - L(x)].
\end{aligned}$$

Now, we derive from here that $(Ax)_k \to cl(x) + (1-c)L(x)$, as $k \to \infty$.

This completes the proof. □

Theorem 6.1.26. [324, Theorem 2] *If $\alpha \in I$ and $x = (x_k) \in \ell_\infty$ which is limitable by a matrix $A \in Q(\alpha)$, then $a \leq A-\lim x_k \leq b$; where a, b are defined as in Theorem 6.1.25, above.*

Proof. If $\alpha = 0$, $a = l(x)$, $b = L(x)$, then the conclusion is well-known. Now, suppose that $\alpha > 0$. Let (p_k) and (q_k) be increasing sequences of integers such that $x_{p_k} \to l(x)$ and $x_{q_k} \to L(x)$, as $k \to \infty$. Then, we can find integers r_k and m_k for each $k \in \mathbb{N}$ such that $a_{r_k,p_k} = \alpha$, $a_{m_k,p_k} = \alpha$. It follows by the regularity of $A = (a_{nk})$ that $r_k \to \infty$, $m_k \to \infty$ and $\sum_{n\neq p_k} a_{r_k,n} \to 1-\alpha$, as $k \to \infty$. For any $\varepsilon > 0$, we can find the positive integers n_1, k_1 so that

$$\sum_{k\neq p_k} a_{r_k,k} < 1 - \alpha + \varepsilon$$

for $n > n_1$, $k > k_1$ and $x_k < L(x) + \varepsilon$. Then, for $k > k_1$,

$$\begin{aligned}
\sum_k a_{r_k,k}x_k &= \alpha x_{p_k} + \sum_{k\neq p_k, k\leq n_1} a_{r_k,k}x_k + \sum_{k\neq p_k, k>n_1} a_{r_k,k}x_k \\
&< \alpha x_{p_k} + \sum_{k\neq p_k, k\leq n_1} a_{r_k,k}x_k + [L(x) + \varepsilon](1 - \alpha + \varepsilon).
\end{aligned}$$

Letting $k \to \infty$ and then $\varepsilon \to 0$, one can see that

$$A - \lim x_k \leq \alpha l(x) + (1-\alpha)L(x).$$

Proceeding with (m_k) and (q_k) in place of (r_k) and (p_k), one can similarly also show that

$$A - \lim x_k \geq \alpha L(x) + (1-\alpha)l(x),$$

as was desired. □

In 1979, Maddox [256] achieved giving the relation by an inequality between the cores of the sequences $Ax = \{(Ax)_n\}$ and $x = (x_k)$, as follows:

Theorem 6.1.27. [256, Theorem 1] $\limsup_{n\to\infty} (Ax)_n \leq \limsup_{n\to\infty} x_n$ *for every* $x \in \ell_\infty$ *if and only if* $A = (a_{nk})$ *is regular and* $\sum_k |a_{nk}| \to 1$, *as* $n \to \infty$.

Proof. For the necessity, let $x \in c$. Then, since $\liminf_{n\to\infty} x_n = \limsup_{n\to\infty} x_n = \lim_{n\to\infty} x_n$ and $\limsup_{n\to\infty} \{A(-x)\}_n \leq \limsup_{n\to\infty} - x_n$, we get that $\lim_{n\to\infty} x_n \leq \liminf_{n\to\infty} (Ax)_n \leq \limsup_{n\to\infty} (Ax)_n \leq \limsup_{n\to\infty} x_n = \lim_{n\to\infty} x_n$. So, $Ax \in c$ with $\lim_{n\to\infty} (Ax)_n = \lim_{n\to\infty} x_n$. This means that $A = (a_{nk})$ is regular. By the conditions of regularity, there exists a $y \in \ell_\infty$ such that $\|y\| \leq 1$ and $\limsup_{n\to\infty} (Ay)_n = \limsup_{n\to\infty} \sum_k |a_{nk}|$, (see [6]). Hence,

$$1 \leq \liminf_{n\to\infty} \sum_k |a_{nk}| \leq \limsup_{n\to\infty} \sum_k |a_{nk}| = \limsup_{n\to\infty} (Ay)_n \leq \limsup_{n\to\infty} y_n \leq \|y\| \leq 1$$

which implies that $\sum_k |a_{nk}| \to 1$, as $n \to \infty$.

For the sufficiency, take $x = (x_k) \in \ell_\infty$. Then, we can write

$$(Ax)_n \leq \|x\| \sum_{k<m} |a_{nk}| + \left(\sup_{k\geq m} x_k\right) \sum_k |a_{nk}| + \|x\| \sum_k (|a_{nk}| - a_{nk}).$$

Applying the operator lim sup, we obtain $\limsup_{n\to\infty} (Ax)_n \leq \limsup_{k\to\infty} x_k$ which completes the proof. □

After Theorem 6.1.27, Maddox proved the next interesting result which states the impossibility of the inequality $\limsup_{n\to\infty} (Ax)_n \leq \liminf_{n\to\infty} (Bx)_n$ for two infinite matrices $A = (a_{nk})$ and $B = (b_{nk})$. Firstly, we should state that a matrix $A = (a_{nk})$ is called *almost positive* if $\sum_k a_{nk}^- \to 0$ as $n \to \infty$, where $a_{nk}^- = \max\{-a_{nk}, 0\}$.

Theorem 6.1.28. [256, Theorem 3] *Let* B *be any regular and almost positive matrix. Then, there is no matrix* A *such that*

$$\limsup_{n\to\infty} (Ax)_n \leq \liminf_{n\to\infty} (Bx)_n \quad \text{for } x \in \ell_\infty. \tag{6.1.8}$$

Proof. Suppose that there exists a matrix $A = (a_{nk})$ such that (6.1.8) holds. Then, $\limsup_{n\to\infty} (Bx)_n \leq \limsup_{n\to\infty} x_n$ by Theorem 6.1.27 and so we have

$$\limsup_{n\to\infty} (Ax)_n \leq \liminf_{n\to\infty} (Bx)_n \leq \limsup_{n\to\infty} (Bx)_n \leq \limsup_{n\to\infty} x_n,$$

whence $A = (a_{nk})$ is regular. By the Steinhaus theorem, given by Theorem 3.3.14, there exists $z \in \ell_\infty$ such that $\liminf_{n\to\infty} (Az)_n < \limsup_{n\to\infty} (Az)_n$. Since, $\limsup_{n\to\infty} (Ax)_n \leq \limsup_{n\to\infty} (Bx)_n$, we get $\liminf_{n\to\infty} (Bz)_n \leq \liminf_{n\to\infty} (Az)_n$. So, $\liminf_{n\to\infty} (Bz)_n < \limsup_{n\to\infty} (Az)_n \leq \liminf_{n\to\infty} (Bz)_n$, a contradiction.

This completes the proof of the theorem. □

Theorem 6.1.28 reduces in the case B is either C_1 or the unit matrix to the following result:

Corollary 6.1.29. [256] *For any* $x \in \ell_\infty$,

(a) *There is no matrix A such that* $\limsup_{n\to\infty} (Ax)_n \leq \liminf_{n\to\infty} (C_1x)_n$.

(b) *There is no matrix A such that* $\limsup_{n\to\infty} (Ax)_n \leq \liminf_{n\to\infty} x_n$.

Now, we can give the following theorem due to Maddox [256] which is more general than Theorem 6.1.27:

Theorem 6.1.30. [256, Theorem 4] *For all $x \in \ell_\infty$, the following statements are equivalent:*

$$\limsup_{n\to\infty} |(Ax)_n| \leq \liminf_{n\to\infty} |x_n|. \tag{6.1.9}$$

$$A = (a_{nk}) \in (\ell_\infty : c_0). \tag{6.1.10}$$

$$\lim_{n\to\infty} \sum_k |a_{nk}| = 0 \textit{ and } A^k \in c_0 \text{ for each fixed } n \in \mathbb{N}. \tag{6.1.11}$$

Proof. The equivalence of (6.1.10) and (6.1.11) is well-known, (see [396, 1.7.19, p. 17]). We shall prove that (6.1.9) is equivalent to (6.1.11).

If (6.1.11) holds, then $\limsup_{n\to\infty} |(Ax)_n| = 0$ for all $x \in \ell_\infty$ which implies (6.1.9).

Conversely, suppose that (6.1.9) holds. Then, Ax is bounded on the Banach space ℓ_∞ whence $\sup_{n\in\mathbb{N}} \sum_k |a_{nk}| < \infty$ by the Banach-Steinhaus theorem. Also, choosing $x = e^{(k)}$, we must have $a_{nk} \to 0$, as $n \to \infty$. Now, suppose if possible, that $\limsup_{n\to\infty} \sum_k |a_{nk}| = d > 0$. Choose $m(1) > 1$ such that $|a_{m(1),1}| < d/10$ and

$$\left| \sum_k |a_{m(1),k}| - d \right| < \frac{d}{10}.$$

Define $k(1) = 1$ and choose $k(2) > 2 + k(1)$ such that

$$\sum_{k=k(2)}^{\infty} |a_{m(1),k}| < \frac{d}{10}.$$

Let choose $m(2) > m(1)$ such that

$$\sum_{k=0}^{k(2)} |a_{m(2),k}| < \frac{d}{10}; \quad \left|\sum_{k} |a_{m(2),k}| - d\right| < \frac{d}{10}.$$

Now, choose $k(3) > 2 + k(2)$ such that

$$\sum_{k=k(3)}^{\infty} |a_{m(2),k}| < \frac{d}{10}.$$

Continuing this process, we can define a sequence $x = (x_k)$ by

$$x_k := \begin{cases} \operatorname{sgn} a_{m(r),k} & , \quad k(r) < k < k(r+1); \ r \geq 1, \\ 0 & , \quad k = k(r+1); \ r \geq 0. \end{cases}$$

Then, $\|x\| \leq 1$ and $\liminf_{k\to\infty} |x_k| = 0$, so (6.1.9) implies $\limsup_{n\to\infty} |(Ax)_n| = 0$. But, for $m = m(r)$ with $r > 1$, we have

$$\left|\sum_{k} a_{mk} x_k\right| > \sum_{k=k(r)+1}^{k(r+1)-1} |a_{mk}| - \frac{d}{5}.$$

Also, we have

$$\left|\sum_{k=k(r)+1}^{k(r+1)-1} |a_{mk}| - d\right| < \frac{3d}{10}$$

and so,

$$\left|\sum_{k} a_{mk} x_k\right| > d - \frac{3d}{10} - \frac{d}{5} = \frac{d}{2}.$$

Since the last inequality holds for infinitely many m, it follows that $\limsup_{n\to\infty} |(Ax)_n| \geq d/2$. But, this contradicts the fact that $\limsup_{n\to\infty} |(Ax)_n| = 0$.

This completes the proof. □

In 1987, Das [141] proved some theorems which generalize the above results due to Rhoades and Schaefer. Firstly, we give some required concepts.

Let $\mathcal{A}$ be a sequence of infinite matrices $A^i = \{a_{nk}(i)\}$. For a given sequence $x = (x_k)$, we write $(Ax)_n^i = \sum_k a_{nk}(i) x_k$ if it exits for each $i, n \in \mathbb{N}$. We also write $\mathcal{A}x$ for $\{(Ax)_n^i\}_{i,n=0}^{\infty}$. A sequence $x = (x_k)$ is said to be $\mathcal{A}$-summable to the value s if $\lim_{n\to\infty} (Ax)_n^i = s$ uniformly in i. The method $\mathcal{A}$ is called conservative if $x_k \to s$, as $k \to \infty$ implies $(Ax)_n^i \to s'$, as $n \to \infty$, and is regular if $s = s'$. The characterization of sequences of conservative matrices is due to Stieglitz [367]. In what follows, we shall write $\sup_{n,i\in\mathbb{N}} \sum_k |a_{nk}(i)| = \|\mathcal{A}\| < \infty$ to mean that there exists a constant K such that $\sum_k |a_{nk}(i)| \leq K$ for all $n, i \in \mathbb{N}$ and the series $\sum_k |a_{nk}(i)|$ converges uniformly in i for each $n \in \mathbb{N}$.

To prove the theorems, Das gave the following lemmas:

Lemma 6.1.31. [141, Lemma 1] *Let $\mathcal{A}$ be conservative and $\beta \geq 0$. Then,*

$$\limsup_{n\to\infty} \sup_{i\in\mathbb{N}} \sum_k |a_{nk}(i) - a_k| \leq \beta \tag{6.1.12}$$

if and only if

$$\limsup_{n\to\infty} \sup_{i\in\mathbb{N}} \sum_k [a_{nk}(i) - a_k]^+ \leq \frac{\beta + \chi(\mathcal{A})}{2}, \tag{6.1.13}$$

$$\limsup_{n\to\infty} \sup_{i\in\mathbb{N}} \sum_k [a_{nk}(i) - a_k]^- \leq \frac{\beta - \chi(\mathcal{A})}{2}, \tag{6.1.14}$$

and the condition (6.1.12) implies that $\beta \geq |\chi(\mathcal{A})|$.

Proof. By the conservativeness of $\mathcal{A}$, we obtain

$$\limsup_{n\to\infty} \sup_{i\in\mathbb{N}} \sum_k [a_{nk}(i) - a_k] = \chi(\mathcal{A}).$$

Hence, the equivalence of (6.1.12) to the conditions (6.1.13) and (6.1.14) follows from the relations $\lambda = \lambda^+ + \lambda^-$ and $\lambda = \lambda^+ - \lambda^-$. It follows from (6.1.13) and (6.1.14) that $\beta + \chi(\mathcal{A}) \geq 0$, $\beta - \chi(\mathcal{A}) \geq 0$ and this yields $\beta \geq |\chi(\mathcal{A})|$. □

Lemma 6.1.32. [141, Lemma 2] *Let $\|\mathcal{A}\| < \infty$ and $\sup_{i\in\mathbb{N}}|a_{nk}(i)| \to 0$, as $n \to \infty$. Then, there exists a $y = (y_k) \in \ell_\infty$ such that $\|y\| \leq 1$ and*

$$\limsup_{n\to\infty} \sup_{i\in\mathbb{N}} \sum_k a_{nk}(i) y_k = \limsup_{n\to\infty} \sup_{i\in\mathbb{N}} \sum_k |a_{nk}(i)|.$$

Proof. Suppose that

$$\limsup_{n\to\infty} \sup_{i\in\mathbb{N}} \sum_k |a_{nk}(i)| := \lambda \quad \text{and} \quad N(\varepsilon) := \left\{ n \in \mathbb{N} : \sup_{i\in\mathbb{N}} \sum_k |a_{nk}(i)| > 1 - \lambda \right\}$$

for any $\varepsilon > 0$. Then, there exist the increasing sequences of integers $(n_r) \subset N(1/r)$ and (k_r) independent of i such that

$$\sum_{k\leq k_{r-1}} |a_{n_r,k}(i)| < \frac{1}{r} \quad \text{and} \quad \sum_{k>k_r} |a_{n_r,k}(i)| < \frac{1}{r} \quad \text{for all } i. \tag{6.1.15}$$

Now, define $y = (y_k) \in \ell_\infty$ for $k_{r-1} < k < k_r$ by

$$y_k := \begin{cases} 1 & , \quad a_{n_r,k}(i) \geq 0, \\ -1 & , \quad a_{n_r,k}(i) < 0. \end{cases}$$

Then, for all i and $n_r \in N(1/r)$,

$$\begin{aligned}
\sum_k a_{n_r,k}(i)y_k &:= \left(\sum_{k\le k_{r-1}} + \sum_{k_{r-1}<k\le k_r} + \sum_{k>k_r}\right) a_{n_r,k}(i)y_k \\
&\ge \sum_{k_{r-1}<k\le k_r} a_{n_r,k}(i)y_k - \sum_{k\le k_{r-1}} |a_{n_r,k}|\|y\| - \sum_{k>k_r} |a_{n_r,k}|\|y\| \\
&\ge \sum_{k_{r-1}<k\le k_r} a_{n_r,k}(i)y_k - \frac{2}{r} \\
&= \sum_{k_{r-1}<k\le k_r} |a_{n_r,k}(i)| - \frac{2}{r} \\
&= \left(\sum_k - \sum_{k\le k_{r-1}} - \sum_{k>k_r}\right) |a_{n_r,k}| - \frac{2}{r} \\
&\ge \sum_k |a_{n_r,k}| - \frac{2}{r}.
\end{aligned}$$

Hence,

$$\limsup_{n\to\infty} \sup_{i\in\mathbb{N}} \sum_k a_{nk}(i)y_k \ge \lambda.$$

Also, since for such y,

$$\limsup_{n\to\infty} \sup_{i\in\mathbb{N}} \sum_k a_{nk}(i)y_k \le \lambda.$$

This step completes the proof. □

Remark 6.1.33. [141] *The proof of Lemma 6.1.32 does not hold if the condition* $\|\mathcal{A}\| < \infty$ *is replaced by the condition*

$$\sup_{r,i\in\mathbb{N}} \sum_k |a_{n_r,k}(i)| < \infty;$$

for, in that case we can produce a matrix $\mathcal{A}$ *such that (6.1.15) holds but there does not exist any sequence* (k_r) *independent of* i *such that*

$$\sum_{k>k_r} |a_{n_r,k}| < \frac{1}{r}$$

and this is invalid for the proof in the step (6.1.15).

The following example explains this situation:

Example 6.1.34. [141] *Define* x *and* p *by*

$$x := 1 - \frac{1}{i+1}; \quad (i \in \mathbb{N}_1), \quad p := \frac{1}{n+1}; \quad (n \in \mathbb{N}_1).$$

Then, $1/2 \le x < 1$, $0 < p \le 1/2$. *Let* $a_{nk}(i) := p^{x^k} - p^{x^{k-1}}$ *for all* x, p *and* $k \in \mathbb{N}$. *Thus,*

$$\sum_k |a_{nk}(i)| = \sum_k a_{nk}(i) = \sum_k (p^{x^k} - p^{x^{k-1}}) = \lim_{k\to\infty} (p^{x^k} - p) = 1 - p.$$

Hence,

$$\sup_{i,n\in\mathbb{N}} \sum_k a_{nk}(i) = \sup(p-1) = 1 < \infty.$$

We now show that $\lim_{n\to\infty} \sup_{i\in\mathbb{N}} |a_{nk}(i)| = 0$. *For* $1/2 \le x < 1$, $0 < p \le 1$ *write*

$$f(k) := a_{nk}(i) = p^{x^k} - p^{x^{k-1}}.$$

Then, for $k \ge 2$

$$0 \le p^{x^k} - p^{x^{k-1}} \le p^{(1/2)^k} - p^{(1)^{k-1}} \to 0, \quad as \quad p \to 0^+.$$

For k=1,

$$0 \le p^x - p^1 \le p^{1/2} - p \to 0, \quad as \quad p \to 0^+.$$

Thus,

$$\sup_{1/2 \le x < 1} (p^{x^k} - p^{x^{k-1}}) \to 0, \quad as \quad p \to 0^+.$$

Therefore, since $x := 1 - 1/(i+1)$, $p := 1/(n+1)$, *we see that* $\lim_{n\to\infty} \sup_{i\in\mathbb{N}} |a_{nk}(i)| = 0$. *Now, let* (n_r) *be any sequence of natural numbers which increases with* r. *We have*

$$\begin{aligned} \sum_{k=m+1}^{\infty} |a_{nk}(i)| &= \sum_{k=m+1}^{\infty} (p^{x^k} - p) \\ &= \lim_{k\to\infty} (1 - p^m), \quad \left(p = \frac{1}{n_r + 1}\right) \end{aligned}$$

thus

$$\begin{aligned} \sum_{k=m+1}^{\infty} |a_{nk}(i)| < \frac{1}{r} \quad &\Rightarrow \quad 1 - p^{x^m} < \frac{1}{r} \\ &\Rightarrow \quad p^{x^m} > \frac{r-1}{r} \\ &\Rightarrow \quad \left(\frac{1}{n_r+1}\right)^{x^m} > \frac{r-1}{r} \\ &\Rightarrow \quad (n_r+1)^{x^m} > \frac{r-1}{r} \\ &\Rightarrow \quad x^m \log(n_r+1) < \log\frac{r}{r-1} \\ &\Rightarrow \quad x^m < \frac{\log\frac{r}{r-1}}{\log(n_r+1)}, \end{aligned} \tag{6.1.16}$$

where $1 < r/(r-1) < 2$. *The right-hand side of (6.1.16) is smaller than* 1 *if and only if*

$$\begin{aligned} \frac{r}{r-1} < n_r + 1 &\Rightarrow n_r > \frac{r}{r-1} - 1 \\ &\Rightarrow n_r > \frac{1}{r-1}, \ (r \in \mathbb{N}_2) \\ &\Rightarrow n_r \geq 2 \end{aligned}$$

which is true for every chosen increasing sequence (n_r). *Now, (6.1.16) is equivalent to*

$$\begin{aligned} &\Big(\frac{1}{x}\Big)^m > \frac{\log(n_r+1)}{\log \frac{r}{r-1}} \Rightarrow \Big(\frac{1+i}{i}\Big)^m > \frac{\log(n_r+1)}{\log \frac{r}{r-1}} \\ &m\Big(\frac{1+i}{i}\Big) > \log \frac{\log(n_r+1)}{\log \frac{r}{r-1}} \\ &m > \frac{1}{\log \frac{i+1}{i}} \log \frac{\log(n_r+1)}{\log \frac{r}{r-1}}, \ r \in \mathbb{N}_2. \end{aligned}$$

Thus, the least value of $m-1$ *is the integer part of*

$$\frac{1}{\log \frac{i+1}{i}} \log \frac{\log(n_r+1)}{\log \frac{r}{r-1}}, \ r \in \mathbb{N}_2,$$

which tends to ∞ *as* $i \to \infty$. *Hence, the value of* m *must depend on* i.

Theorem 6.1.35. [141, Theorem 1] *We have:*

(a) *Let* $\|\mathcal{A}\| < \infty$ *and suppose that the inequality*

$$K_\theta(x) = \limsup_{n\to\infty} \sup_{i\in\mathbb{N}} \sum_k [a_{nk}(i) - \theta_k] x_k \leq uL(x) - vl(x) \tag{6.1.17}$$

holds for some sequence $(\theta_k) \in cs$ *and for some non-negative constants* u *and* v. *Then, the following statements hold:*

(i) $\mathcal{A}$ *is conservative.*

(ii) $a_k = \theta_k$.

(iii) $\chi(\mathcal{A}) = u - v$.

(iv)
$$\begin{aligned} \chi(\mathcal{A}) &\leq \liminf_{n\to\infty} \sup_{i\in\mathbb{N}} \sum_k |a_{nk}(i) - a_k| \\ &\leq \limsup_{n\to\infty} \sup_{i\in\mathbb{N}} \sum_k |a_{nk}(i) - a_k| \leq u + v. \end{aligned} \tag{6.1.18}$$

(b) *Let $\mathcal{A}$ be conservative. If*

$$\limsup_{n\to\infty} \sup_{i\in\mathbb{N}} \sum_k |a_{nk}(i) - a_k| \le \beta \tag{6.1.19}$$

for some constant $\beta \ge 0$, then $\beta \ge |\chi(\mathcal{A})|$ and

$$K_\theta(x) \le \frac{\beta + \chi(\mathcal{A})}{2} L(x) - \frac{\beta - \chi(\mathcal{A})}{2} l(x) \text{ for all } x \in \ell_\infty. \tag{6.1.20}$$

(c) *Let $\mathcal{A}$ be conservative and $\|\mathcal{A}\| < \infty$. In order that for some constant $\beta \ge |\chi(\mathcal{A})|$, the inequality (6.1.20) holds it is necessary and sufficient that (6.1.19) holds.*

We should note here that the hypothesis $\beta \ge |\chi(\mathcal{A})|$ is not needed in the sufficiency part of (c).

Proof. (a) Since $K_\theta(x)$ and $h(x) = uL(x) - vl(x)$ are finite valued sublinear functionals on ℓ_∞, we have

$$-h(-x) \le -K_\theta(-x) \le K_\theta(x) \le h(x), \ x \in \ell_\infty. \tag{6.1.21}$$

If $x \in c$, it follows from (6.1.21) that $K_\theta(x) = -K_\theta(-x) = (u - v) \lim x$ which is equivalent to, while $\sum_k \theta_k x_k$ converges,

$$\lim_{n\to\infty} \sum_k a_{nk}(i) x_k = \sum_k \theta_k x_k + (u - v) \lim_{k\to\infty} x_k, \tag{6.1.22}$$

uniformly in i. This proves that $\mathcal{A}$ is conservative. Therefore, we have (see [368])

$$\lim_{n\to\infty} \sum_k a_{nk}(i) x_k = \sum_k a_k x_k + \chi(\mathcal{A}) \lim_{k\to\infty} x_k, \tag{6.1.23}$$

uniformly in i. By considering the special sequences $e^{(k)}$ and e in (6.1.22) and (6.1.23), we get the results (ii) and (iii) of Part (a).

By applying Lemma 6.1.32 to the matrix sequence $(a_{nk}(i) - a_k)$ we obtain $y = (y_k) \in \ell_\infty$ such that $\|y\| \le 1$ and

$$\begin{aligned} \limsup_{n\to\infty} \sup_{i\in\mathbb{N}} \sum_k |a_{nk}(i) - a_k| &= \limsup_{n\to\infty} \sup_{i\in\mathbb{N}} \sum_k [a_{nk}(i) - a_k] y_k \\ &\le l(y), \text{ (by the hypothesis (6.1.17))} \\ &\le u + v. \end{aligned} \tag{6.1.24}$$

On the other hand,

$$\begin{aligned} \liminf_{n\to\infty} \inf_{i\in\mathbb{N}} \sum_k |a_{nk}(i) - a_k| &\ge \liminf_{n\to\infty} \inf_{i\in\mathbb{N}} \sum_k [a_{nk}(i) - a_k] \\ &= \chi(\mathcal{A}). \end{aligned} \tag{6.1.25}$$

Now, (iv) follows by combining (6.1.24) and (6.1.25).

(b) Let $\mathcal{A}$ be conservative. Hence, $\sum_k |a_k| = P < \infty$, (see [368]). Thus,

$$\limsup_{n\to\infty} \sup_{i\in\mathbb{N}} \sum_k |a_{nk}(i) - a_k| \le \|\mathcal{A}\| + P < \infty.$$

So, we can always choose a $\beta \ge 0$ satisfying condition (6.1.19). By Lemma 6.1.31, $\beta \ge |\chi(\mathcal{A})|$.

Let $\varepsilon > 0$. Then, there exists a positive integer k_0 such that

$$l(x) - \varepsilon < x_k < L(x) + \varepsilon, \tag{6.1.26}$$

for $k > k_0$. Now,

$$\begin{aligned} K_\theta(x) &\le \limsup_{n\to\infty} \sup_{i\in\mathbb{N}} \sum_{k\le k_0} [a_{nk}(i) - a_k]x_k + \\ &+ \limsup_{n\to\infty} \sup_{i\in\mathbb{N}} \sum_{k>k_0} [a_{nk}(i) - a_k]^+ x_k + \\ &+ \limsup_{n\to\infty} \sup_{i\in\mathbb{N}} \Big\{ -\sum_{k>k_0} [a_{nk}(i) - a_k]^- x_k \Big\} \\ &= I_1(x) + I_2(x) + I_3(x). \end{aligned} \tag{6.1.27}$$

By the conservativeness conditions of $\mathcal{A}$, $I_1(x) = 0$ for all $x \in \ell_\infty$. By Lemma 6.1.31 and (6.1.26), $I_2(x)$ and $I_3(x)$ do not exceed

$$\frac{\beta + \chi(\mathcal{A})}{2}[L(x) + \varepsilon] \text{ and } \frac{-\beta - \chi(\mathcal{A})}{2}[l(x) - \varepsilon], \tag{6.1.28}$$

respectively. Since ε is arbitrary, the result (6.1.20) follows from (6.1.27) and (6.1.28).

Part (c) follows by combining Parts (a) and (b), and taking

$$u := \frac{\beta + \chi(\mathcal{A})}{2}, \quad v := \frac{\beta - \chi(\mathcal{A})}{2},$$

both of which are non-negative by Part (d) of Lemma 6.1.31. □

Theorem 6.1.36. [141, Theorem 2] *Let $\mathcal{A}$ be conservative and $\|\mathcal{A}\| < \infty$. Then, the following statements hold:*

(a) *For all $x \in \ell_\infty$*

$$K_\theta(x) \le \frac{|\chi(\mathcal{A})| + \chi(\mathcal{A})}{2} L(x) - \frac{|\chi(\mathcal{A})| - \chi(\mathcal{A})}{2} l(x) \tag{6.1.29}$$

if and only if

$$\limsup_{n\to\infty} \sup_{i\in\mathbb{N}} \sum_k |a_{nk}(i) - a_k| = |\chi(\mathcal{A})|. \tag{6.1.30}$$

(b) *Let $\chi(\mathcal{A}) \ge 0$. Then,*

$$K_\theta(x) \le \chi(\mathcal{A}) L(x) \tag{6.1.31}$$

for all $x \in \ell_\infty$ if and only if

$$\limsup_{n\to\infty} \sup_{i\in\mathbb{N}} \sum_k |a_{nk}(i) - a_k| = \chi(\mathcal{A}). \tag{6.1.32}$$

(c) *Let $\mathcal{A}$ be conull. Then, for some $\beta \geq 0$ and all $x \in \ell_\infty$*

$$K_\theta(x) \leq \frac{\beta}{2}[L(x) - l(x)] \tag{6.1.33}$$

if and only if (6.1.19) holds. In particular

$$\limsup_{n\to\infty} \sup_{i\in\mathbb{N}} \sum_k a_{nk}(i)x_k = \sum_k a_k x_k$$

if and only if

$$\limsup_{n\to\infty} \sup_{i\in\mathbb{N}} \sum_k |a_{nk}(i) - a_k| = 0.$$

Remark 6.1.37. [141] *In the case $\mathcal{A} = A = (a_{nk})$ and $\chi(\mathcal{A}) > 0$, Part (b) of Theorem 6.1.36 is reduced to a result of Schaefer* [358]. *He has deduced it by applying Knopp's core theorem to the matrix $\delta(A)^{-1}(a_{nk} - a_k)$. Note that similar deduction is not possible in Part (c) of Theorem 6.1.35 and Part (a) of Theorem 6.1.36, since $\chi(\mathcal{A})$ can be zero or even negative. For example, take $(Ax)_n = x_{n-1} - 2x_n$. Then, clearly $A = (a_{nk})$ is conservative and $\chi(\mathcal{A}) = -1$. We may take $\beta = 3$ and apply Part (c) of Theorem 6.1.35 to obtain*

$$\limsup_{n\to\infty} Ax \leq L(x) - 2l(x),$$

which is also evident otherwise. Note that Part (a) of Theorem 6.1.36 is not applicable. In the case $(Ax)_n = 3x_{n-1} - x_n$, then $A = (a_{nk})$ is conservative and $\chi(\mathcal{A}) = 2$, and we may take $\beta = 4$. In this case even if $\chi(\mathcal{A}) > 0$, Part (b) of Theorem 6.1.36 is not applicable, since the condition (6.1.32) is not satisfied. But, we make use Part (c) of Theorem 6.1.35 to obtain

$$\limsup_{n\to\infty} Ax \leq 3L(x) - l(x),$$

which is also evident otherwise. In the case $(Ax)_n = x_{n-1} - x_n$, A is conull. We may take $\beta = 2$ so that by Part (c) of Theorem 6.1.35,

$$\limsup_{n\to\infty} Ax \leq L(x) - l(x),$$

which is also evident otherwise.

We choose the above examples for the cases $\chi(\mathcal{A}) < 0$, $\chi(\mathcal{A}) > 0$ and $\chi(\mathcal{A}) = 0$, where Part (c) of Theorem 6.1.35 is applicable while Parts (a) and (b) of Theorem 6.1.36 are not.

Proof. The proof of Part (a) of Theorem 6.1.36 follows from Part (c) of Theorem 6.1.35 by taking $\beta = |\chi(\mathcal{A})|$. When $\chi(\mathcal{A}) \geq 0$, (6.1.29) is reduced to (6.1.31) and the inequality (6.1.30) together with the first inequality in (6.1.18) are reduced to (6.1.32). Thus, Part (b) follows. Part (c) follows immediately from Part (c) of Theorem 6.1.35 by taking $\chi(\mathcal{A}) = 0$. □

In 1979, Davydov [142] gave a slight generalization of Theorem 6.1.27. Davydov proved the following theorems:

Theorem 6.1.38. [142, Theorem 1] *If a lower triangular matrix $A = (a_{nk})$ satisfies the condition $\Delta^r a_{nk} \geq 0$; $(r \in \mathbb{N},\ k = 0, 1, 2, \ldots, n)$ for all $n > n_0$ and $a_{n\mu} = 0$ for $\mu > n$, then $\mathcal{K} - core(Ax) \subseteq \mathcal{K} - core(C^r x)$ for each sequence $x = (x_k)$; where j is a positive number, $r \in \mathbb{N}_1$ and*

$$\Delta^r a_{nk} := \sum_{j=0}^{r} (-1)^j C_r^j a_{n,k+j}, \text{ where}$$

$$C_r^j := \frac{1}{E_r^j} \sum_{k=0}^{r} E_{r-k}^{j-1} x_k \text{ with } E_r^j := \frac{1}{r!}(j+1)(j+2)\cdots(j+r).$$

Theorem 6.1.39. [142, Theorem 2] *Let $A = (a_{nk})$ be a regular and positive matrix satisfying the following two conditions for all $n > n_0$ and $r \in \mathbb{N}$:*

$$\Delta^r a_{nk} \geq 0 \text{ for all } k \in \mathbb{N} \text{ and } \lim_{k\to\infty} k^r a_{nk} = 0.$$

Then, the inclusion $\mathcal{K} - core(Ax) \subseteq \mathcal{K} - core(C^r x)$ holds for each sequence $x = (x_k)$ for which the series $\sum_k a_{nk} x_k$ is absolutely convergent, in particular, for each $x = (x_k) \in \ell_\infty$.

From Theorems 6.1.38 and 6.1.39, Davydov derived the following results:

Corollary 6.1.40. [142, Corollary 1] *If a lower triangular positive regular matrix $A = (a_{nk})$ satisfies the condition $a_{nk} \geq 0$; $(k = 0, 1, 2, \ldots, n-1)$, $a_{nn} \geq 0$ for all $n > n_0$, then $\mathcal{K} - core(Ax) \subseteq \mathcal{K} - core(C_1 x)$ for each sequence $x = (x_k)$.*

Corollary 6.1.41. [142, Corollary 2] *Let the sequence (q_n) of real numbers satisfies the following conditions for $r \in \mathbb{N}$:*

$$q_n > 0 \text{ for all } n \in \mathbb{N} \text{ and } \lim_{n\to\infty} \tfrac{q_n}{Q_n} = 0,$$
$$\textstyle\sum_{j=0}^{r} (-1)^j C_r^j q_{n-k-j} \geq 0;\ (n \in \mathbb{N},\ k \in \{0, 1, 2, \ldots, n\}),$$

where $q_\mu = 0$ for $\mu < 0$. Then, $\mathcal{K} - core(Bx) \subseteq \mathcal{K} - core(C^r x)$; where

$$Bx = \{(Bx)_n\} \text{ with } (Bx)_n = \frac{\sum_{k=0}^{n} q_{n-k} x_k}{\sum_{k=0}^{n} q_k} \text{ for all } n \in \mathbb{N}.$$

In 1988, Choudhary [104] generalized the above results given by Davydov. Firstly, note that the existence of the inverse $A^{-1} = (a_{nk}^{-1})$ of a triangle matrix $A = (a_{nk})$ is guaranteed.

For proving the main result, Choudhary [104] used the following lemma:

Lemma 6.1.42. [104, Lemma 2] *Let $n \in \mathbb{N}$ be fixed. In order that whenever Bx is bounded, $\{(Ax)_n\}$ should be defined for that particular n it is necessary and sufficient that*

$$\lim_{J\to\infty} \sum_{k=0}^{J} \left| \sum_{j=J+1}^{\infty} a_{nj} b_{jk}^{-1} \right| = 0, \tag{6.1.34}$$

$$C = AB^{-1} \text{ exists,} \tag{6.1.35}$$

$$\sum_k |c_{nk}| < \infty, \text{ where } c_{nk} = \sum_{j=k}^{\infty} a_{nj} b_{jk}^{-1} \text{ for all } k, n \in \mathbb{N}. \tag{6.1.36}$$

For the proof of Lemma 6.1.42, the reader may refer to Choudhary [104].

Theorem 6.1.43. [104, Theorem 1] *Let $B = (b_{jk})$ be a triangle matrix with the inverse $B^{-1} = (b_{jk}^{-1})$ and $A = (a_{nk})$ be any infinite matrix. In order that whenever Bx is bounded Ax should be exist and be bounded, and satisfy* $\limsup Ax \leq \limsup Bx$ *it is necessary and sufficient that (6.1.34) and (6.1.35) hold and*

$$C \text{ is regular,} \tag{6.1.37}$$

$$\lim_{n\to\infty} \sum_k |c_{nk}| = 1. \tag{6.1.38}$$

Proof. If assume merely that $(Ax)_n$ exists for every $n \in \mathbb{N}$ whenever Bx is bounded, then it follows from Lemma 6.1.42 that the conditions (6.1.34) and (6.1.35) are satisfied, and also (6.1.36) of Lemma 6.1.42 holds. Further, for every bounded Bx we have $(Ax)_n := \sum_k c_{nk} y_k$, where $y_k := (Bx)_k$.

Conversely, if the conditions of the theorem are satisfied, then (6.1.37) implies, in particular, that (6.1.36) holds for every $n \in \mathbb{N}$. Thus, the conditions of Lemma 6.1.42 are satisfied for every $n \in \mathbb{N}$ so that

$$(Ax)_n := \sum_k c_{nk} y_k \tag{6.1.39}$$

holds. Thus, whether we are considering necessity or sufficiency, we can assume (6.1.39), and the result now follows from Knopp's core theorem. □

Remark 6.1.44. [104, Remark] *When $A = (a_{nk})$ is row finite, the expression inside the modulus in (6.1.36) is 0 for sufficiently large J and for all $k \in \mathbb{N}$. Thus, (6.1.36) is necessarily satisfied, so that this is reduced to the following result:*

Theorem 6.1.45. [104, Theorem 2] *Let B be triangle and $A = (a_{nk})$ be a row finite matrix. Then,* $\limsup Ax \leq \limsup Bx$ *for all $x \in \ell_\infty$ if and only if AB^{-1} is regular and almost positive.*

In what follows, we state the results related to the core of sequences with complex entries. The first result is a generalization of Knopp's core theorem for the complex sequences [111, p. 140].

Theorem 6.1.46. [111, 6.1, II] *If $A = (a_{nk})$ is a non-negative regular matrix, then $\mathcal{K} - core(Az) \subseteq \mathcal{K} - core(z)$ for all $z = (z_k) \in \omega$.*

Proof. If $\mathcal{K} - core(z)$ is the whole plane, then there is nothing to prove.

If (z_k) has a barrier line of type $x = a$, i.e., if $\liminf_{k\to\infty} x_k = a$, then Knopp's core theorem shows that $\liminf_{n\to\infty} (Ax)_n > a$ and so, $x = a$ is a barrier line for Az. It follows that if any line L, inclined at an angle θ with the imaginary axis (in the positive sense), is a barrier line for (z_k) then, $(e^{-i\theta}x_k)$ has a barrier line of type $x = a$ which is also a barrier line for $e^{-i\theta}Az$; and thus L is a barrier line for Az.

Hence, every half-plane containing D also contains D'; where D and D' are the sets of the limit points of (z_k) and $\{(Az)_n\}$, respectively.

This completes the proof of the theorem. □

Theorem 6.1.46 for sequences of complex numbers was extended to lower semi-matrices by Agnew [5]. Later, his result was generalized to the general infinite matrices by Robinson [111, p. 149], as follows. Prior to giving them, we need the following two lemmas:

Lemma 6.1.47. [111, 6.4, II, Lemma 1] *If every circular region in the complex plane which includes a given closed convex set S also includes another given convex set G, then $G \subset S$.*

Proof. Suppose that there is a point g such that $g \in G$ but $g \notin S$. Then, by the properties of convex sets, there is a straight line l which separates g from S; i.e., g is in one of the half-planes determined by l, and S is in the other half-plane, and l has no common points with S. We will prove that there is a circle $\overline{C}$ touching l and containing S. Then, $\overline{C}$ cannot contain g, and its existence therefore establishes Lemma 6.1.47.

Since S is closed, it has a positive distance δ from l. Let l_1 be the straight line parallel to l at a distance $\delta/2$, and on the same side of l as S. Since S is bounded, it may be enclosed in a square $ABCD$ having part of l_1 for one of its sides, say AD. Then, AB being perpendicular to AD, choose a point E in AB produced, such that $AE = d(4a^2 - \delta^2)/4\delta$, where a is the side of the square. The circle $\overline{C}$ with center E and touching l is the required circle.

If r is the radius of $\overline{C}$, then $r = d+\delta/2$ and the points A and B are clearly inside $\overline{C}$. Also, $DE = \sqrt{a^2+d^2}$. But, $d\delta > a^2 - \delta^2/4$ so that $r^2 = (d+\delta/2)^2 > a^2 + d^2$, i.e., $r^2 > DE^2$. Thus, D and hence C are inside $\overline{C}$. Hence, $\overline{C}$ touches l and contains S which completes the proof of Lemma 6.1.47. □

Lemma 6.1.48. [111, 6.4, II, Lemma 2] *Let $A = (a_{nk})$ be a regular matrix and the sequence of sequences $S = \{x_k^{(n)}\}$ be bounded, i.e., there exists an $M > 0$ such that $|x_k^{(n)}| < M$ for all $k, n \in \mathbb{N}$. Also, let Y be an arbitrary limit point of the sequence $(y_n) = \left\{\sum_k a_{nk}x_k^{(n)}\right\}_{n\in\mathbb{N}}$. Then, there is a sequence $(\widetilde{x}_k)$ whose elements are taken from sequences of S such that Y is a limit point of $(\overline{y}_n) = \{\sum_k a_{nk}\widetilde{x}_k\}_{n\in\mathbb{N}}$.*

Proof. By the hypothesis, there is an increasing sequence of positive integers $(n_r)_{r=1}^{\infty}$ such that $y_{n_r} \to Y$, as $r \to \infty$. In order to define $(\widetilde{x}_k)$, we first fix k_1 so that

$$\sum_{k\geq k_1+1} |a_{n_1,k}| < \frac{1}{2M}.$$

Then, define $(\widetilde{x}_k)$ for $0 \leq k \leq k_1$ by $\widetilde{x}_k = x_k^{(n_1)}$. We next choose r_2 so that

$$|a_{n_{r_2},k}| < \frac{1}{4Mk_1} \quad \text{for} \quad 0 \leq k \leq k_1$$

and determine $k_2 > k_1$ so that

$$\sum_{k \geq k_2+1} |a_{n_{r_2},k}| < \frac{1}{4M}.$$

We then define $(\widetilde{x}_k)$ for $k_1 < k \leq k_2$ by $\widetilde{x}_k = x_k^{(n_{r_2})}$.

We now choose $r_3 > r_2$ so that

$$|a_{n_{r_3},k}| < \frac{1}{8Mk_2} \quad \text{for} \quad 0 \leq k \leq k_2$$

and determine $k_3 > k_2$ so that

$$\sum_{k \geq k_3+1} |a_{n_{r_3},k}| < \frac{1}{8M}.$$

We then define $(\widetilde{x}_k)$ for $k_2 < k \leq k_3$ by $\widetilde{x}_k = x_k^{(n_{r_3})}$, and so on.

Then, $(\widetilde{x}_k)$ is the composition of the elements taken from sequences of S, and we have

$$\overline{y}_{n_{r_p}} = \sum_k a_{n_{r_p},k}\widetilde{x}_k = \left(\sum_{k=0}^{k_{p-1}} + \sum_{k=k_{p-1}+1}^{k_p} + \sum_{k=k_p+1}^{\infty} \right) a_{n_{r_p},k}\widetilde{x}_k.$$

By the definition of $\widetilde{x}_k$, we have

$$\widetilde{x}_k = x_k^{(n_{r_p})} \quad \text{for} \quad k_{p-1} + 1 \leq k \leq k_p,$$

where $k_0 = 0, r_1 = 1$; hence

$$\begin{aligned} |\overline{y}_{n_{r_p}} - y_{n_{r_p}}| &= \left| \sum_k a_{n_{r_p},k} \left(\widetilde{x}_k - x_k^{(n_{r_p})} \right) \right| \\ &= \left| \left(\sum_{k=0}^{k_{p-1}} + \sum_{k=k_p+1}^{\infty} \right) a_{n_{r_p},k}\widetilde{x}_k \right|, \end{aligned}$$

so that

$$|\overline{y}_{n_{r_p}} - y_{n_{r_p}}| \leq \frac{2Mk_{p-1}}{2^p k_{p-1} M} + \frac{2M}{2^p M} = \frac{1}{2^{p-2}} \to 0, \quad \text{as} \quad p \to \infty.$$

Hence, $\lim_{p\to\infty} \overline{y}_{n_{r_p}} = \lim_{p\to\infty} y_{n_{r_p}} = Y$.

This completes the proof of the lemma. □

Theorem 6.1.49. [111, 6.4, II] *Let $A = (a_{nk})$ be a regular matrix with complex entries. Then, $\mathcal{K} - core(Az) \subseteq \mathcal{K} - core(z)$ for all $z = (z_k) \in \ell_\infty$ if and only if*

$$\lim_{n\to\infty} \sum_k |a_{nk}| = 1. \tag{6.1.40}$$

Proof. Sufficiency. Let the core of (z_k) be contained in a circle of radius R with the center origin; for any given $\varepsilon > 0$, $|z_k| \leq R+\varepsilon$ for every $k > k(\varepsilon)$. For a given $\delta > 0$, we can choose $n(\delta, \varepsilon)$ so that

$$|a_{nk}| \leq \frac{\delta}{k(\varepsilon)\max_{k\leq k(\varepsilon)}|z_k|} \tag{6.1.41}$$

for every $k \leq k(\varepsilon)$ and $n > n(\delta, \varepsilon)$. Now, if (6.1.40) is satisfied, then there is a number n_0 such that

$$\sum_k |a_{nk}| \leq 1+\delta \tag{6.1.42}$$

for every $n > n_0$. Hence, for every $n > \max\{n_0, n(\delta, \varepsilon)\}$, we have by (6.1.40) and (6.1.42) that

$$\begin{aligned} |(Az)_n| &\leq \left|\sum_{k=0}^{k(\varepsilon)} a_{nk}z_k\right| + \left|\sum_{k=k(\varepsilon)+1}^{\infty} a_{nk}z_k\right| \\ &\leq \delta + (1+\delta)(R+\varepsilon) = R+\varepsilon+\delta(1+R+\varepsilon). \end{aligned}$$

But, for a fixed ε, $\delta(1+R+\varepsilon) \to 0$ with δ, and therefore any limit point of Az is contained in the circle $|z| = R$, since ε is arbitrary. So, the same is true for the core of Az.

Now, consider a circle of radius R with arbitrary center u_0, which includes the core of $z = (z_k)$. The circle of radius R with the center origin will then include the core of $(\widetilde{z}_k) = (z_k - u_0)$. Hence, the core of $\widetilde{Az}$, the A-transform of $(\widetilde{z}_k)$, is also included in the circle $|z| = R$. But,

$$(Az)_n := (\widetilde{Az})_n + u_0 \sum_k a_{nk},$$

whence, since $A = (a_{nk})$ is regular, $(Az)_n \to 0$, as $n \to \infty$. This also implies that the derived sets of Az and of $\{(\widetilde{Az})_n + u_0\}$ are identical. Since the core of $\widetilde{Az}$ is included in $|z| = R$, it follows that the core of $\{(\widetilde{Az})_n + u_0\}$ is included in the circle with radius R and center u_0. The same is also true for the core of Az.

So, we have shown that every circle including the core of $z = (z_k)$ in the complex plane also includes the core of $\{(Az)_n\}$. Hence, by Lemma 6.1.47, $\mathcal{K} - core(Az) \subseteq \mathcal{K} - core(z)$.

Necessity. Define the sequence $\{z_k^{(n)}\}$ by

$$z_k^{(n)} := \begin{cases} e^{-i\arg a_{nk}} & , \quad a_{nk} \neq 0, \\ 1 & , \quad a_{nk} = 0, \end{cases}$$

for all $k, n \in \mathbb{N}$. Then, $y_n = \sum_k a_{nk} z_k^{(n)} = \sum_k |a_{nk}|$ for all $n \in \mathbb{N}$. If (6.1.40) is false, then there exists a limit point Y of (y_n) for which $Y > 1$. On the other hand, by Lemma 6.1.48, there is a sequence $(\widetilde{z}_k)$, composed of suitably selected elements of $\{z_k^{(n)}\}$, whose A-transform is $(\widetilde{y}_n)$ and has Y as a limit point. But, $|z_k^{(n)}| = 1$ for all $k, n \in \mathbb{N}$; hence all members of $\{z_k^{(n)}\}$ belong to the unit circle, with center at the origin and so does the core of $\{z_k^{(n)}\}$.

This is a contradiction, since the A-transform $(\widetilde{y}_n)$ of a bounded sequence whose core is contained in the unit circle cannot have a limit point outside this circle by the hypothesis. □

The next theorem is due to Robinson and gives the connection between the Knopp's core and the absolute equivalence of matrices [111, p. 152].

Theorem 6.1.50. [111, 6.5, I] *Let $A = (a_{nk})$ be a regular matrix. Then, $\mathcal{K} - core(Ax) \subseteq \mathcal{K} - core(x)$ for any $x = (x_k) \in \ell_\infty$ if and only if $A = (a_{nk})$ is absolutely equivalent to a non-negative matrix $B = (b_{nk})$ for all bounded sequences.*

Proof. Sufficiency. Let $(y_n) = \{(Ax)_n\}$ and $(y'_n) = \{(Bx)_n\}$. Then,

$$\lim_{n\to\infty} |y_n - y'_n| \le M \lim_{n\to\infty} \sum_k |a_{nk} - b_{nk}| = 0.$$

Hence, by Theorem 6.1.12, the core of (y_n) is the same as the core of (y'_n). But, by Knopp's core theorem, the core of (y'_n) is contained in the core of $x = (x_k)$ since B is a non-negative regular matrix.

Necessity. Let $Re[\widetilde{z}] = Re[z] - Re[z]_p$ and

$$Re[z]_p := \begin{cases} Re[z] & , \quad Re[z] \ge 0, \\ 0 & , \quad Re[z] < 0. \end{cases}$$

If $b_{nk} = Re[a_{nk}]$, which is a regular matrix, then A and B are absolutely equivalent for all bounded sequences. If this is not so, then

$$\limsup_{n\to\infty} \sum_k |Re[a_{nk}]| > 0. \tag{6.1.43}$$

Define the sequence $\{x_k^{(n)}\}$ by

$$x_k^{(n)} := \begin{cases} 1 & , \quad Re[a_{nk}] \ge 0, \\ -1 & , \quad Re[a_{nk}] < 0. \end{cases}$$

Then, $\sigma_n = \sum_k a_{nk} x_k^{(n)}$ such that $\mathbb{R}(\sigma_n) = \sum_k |Re[a_{nk}]|$ and (6.1.43) implies that (σ_n) has a limit point with non-zero imaginary part. It follows by Lemma 6.1.48 that a sequence $(\widetilde{x}_k)$ whose entries are ± 1 can be determined so that its A-transform has a non-real limit point, although the core of $(\widetilde{x}_k)$ clearly consists of real points only. This is a contrary to the hypothesis that the core of the A-transform and therefore the set of all its limit points is contained in the core of the original sequence.

We next show that (b_{nk}) and $(c_{nk}) = \{Re[a_{nk}]_p\}$ are absolutely equivalent for all bounded sequences. If this is not so, then

$$\limsup_{n\to\infty}\sum_k |Im[a_{nk}]| > 0. \tag{6.1.44}$$

Define the sequence $\{x_k^{(n)}\}$ by

$$x_k^{(n)} := \begin{cases} 1 \;, & Im[a_{nk}] < 0, \\ 0 \;, & Im[a_{nk}] = 0, \end{cases}$$

for all $k, n \in \mathbb{N}$. Then,

$$u_n = \sum_k b_{nk} x_k^{(n)} = -\sum_k |Im[a_{nk}]|$$

and (6.1.44) implies that (u_n) has a negative limit point. It again follows by Lemma 6.1.48 that $(\widetilde{x}_k)$ can be determined so that whose elements are 0 or 1, and its B-transform, hence its A-transform, since A and B are absolutely equivalent for all bounded sequences, has a negative limit point, although the core of $(\widetilde{x}_k)$ clearly contains non-negative points only. Hence, B and C, and therefore C and A are absolutely equivalent for all bounded sequences. Also, C is a non-negative regular matrix.

This step concludes the proof of the theorem. □

Next theorems are related to some applications of core and are due to Agnew [3]:

Theorem 6.1.51. [111, 6.2, I] *Let (z_n) be a sequence of the points of the core of (x_n). Then, there is a real regular matrix $A = (a_{nk})$ satisfying the conditions*

$$a_{nk} = 0 \text{ for all } k \in \mathbb{N} \text{ and almost all } n \in \mathbb{N} \tag{6.1.45}$$

$$\sum_k a_{nk} = 1 \text{ for each } n \in \mathbb{N} \tag{6.1.46}$$

$$a_{nk} \geq 0 \text{ for all } k, n \in \mathbb{N}, \tag{6.1.47}$$

which carries (x_n) into a sequence (y_n) such that $|z_n - y_n| < \frac{1}{n}$ for all $n \in \mathbb{N}$.

Proof. Consider the least closed convex region $R_{n,p}$ containing the points x_n, $x_{n+1}, \ldots, x_{n+p}$. Let $R_{n,\infty}$ denote the union of the sets $R_{n,1}, R_{n,2}, R_{n,3}, \ldots$; then $R_{n,p} \subseteq R_{n,p+1} \subseteq R_{n,\infty}$ for every $p \in \mathbb{N}$. It can be seen that $R_{n,\infty} \subseteq R_n$ (R_n being defined as in the Knoop Core) and that $R_n \subseteq R^0_{n,\infty}$ consisting of the points and limit points of $R_{n,\infty}$, where $R^0_{n,\infty}$ is the closure of $R_{n,\infty}$.

Also, since R_n is closed by definition,

$$R^0_{n,\infty} = R_n. \tag{6.1.48}$$

Since z_n is a point of R, $z_n \in R_n$ for every $n \in \mathbb{N}$. Hence, by (6.1.48) we may choose a point y_n and a suffix p_n such that $|z_n - y_n| < 1/(n+1)$ and $y_n \in R_{n,p_n}$ for every $n \in \mathbb{N}$. Also, since R_{n,p_n} is a convex closed polygon whose vertices are some or all of the points $x_n, x_{n+1}, \ldots, x_{n,p_n}$, we may choose three of its vertices, say $x_{n,p_n}, x_{n,r_n}, x_{n,s_n}$, so that the resulting triangle contains y_n as an inner or as a boundary point. If $x_n, x_{n+1}, \ldots, x_{n,p_n}$ are collinear or coincident, the polygon is a straight line or a point, and the vertices of the triangle are collinear or coincident.

Now, we can choose α_n and β_n such that $0 \le \alpha_n \le 1$, $0 \le \beta_n \le 1$, and

$$y_n := (1 - \beta_n)x_{n,p_n} + \alpha_n \beta_n x_{n,r_n} + \beta_n(1 - \alpha_n)x_{n,s_n}.$$

The transformation

$$x'_n := (1 - \beta_n)x_{n,p_n} + \alpha_n \beta_n x_{n,r_n} + \beta_n(1 - \alpha_n)x_{n,s_n}$$

has all the required properties and is therefore, the proof is completed. □

Theorem 6.1.51 leads us to the following theorem:

Theorem 6.1.52. [111, 6.2, II] *Let G be a closed set contained in the core of a sequence (x_n). Then, there is a real regular transformation satisfying the conditions (6.1.45)–(6.1.47) which carries (x_n) into a sequence including the set G of its limit points.*

Since the core of any sequence is closed, the set G may be the core itself. By taking the set G consisting of a single point s, we have the following result:

Theorem 6.1.53. [111, 6.2, III] *Let s be a constant contained in the core of a sequence (x_n). Then, there is a real regular transformation satisfying the conditions (6.1.45)–(6.1.47) which sums (x_n) to s.*

Next, we give two more results due to Agnew [4] related to the Steinhaus theorem which may be restated as follows:

Theorem 6.1.54. [111, 6.6] *For every regular matrix A, there exists a $x \in \ell_\infty$ such that $\mathcal{K}$-core (Ax) contains more than one point.*

Now, we can state the results (see [111, p. 157]) related to Theorem 6.1.54 and indicate that it is impossible to specialize a regular matrix A such that $\mathcal{K}$-core(Ax) must be a subset of $\mathcal{K}$-core(x).

Theorem 6.1.55. [111, 6.6, I] *Let $A = (a_{nk})$ be a lower semi-regular matrix and E is a closed non-empty set in the complex plane. Then, there is a sequence $x = (x_k)$ of points of E such that $E = D_x \subset D_{Ax}$; where D_x and D_{Ax} are the sets of limit points of $x = (x_k)$ and $Ax = \{(Ax)_n\}$, respectively.*

Proof. Let $(\alpha_k)_{k \in \mathbb{N}_1}$ be a sequence whose elements form a subset of E everywhere dense in E. Let the elements of the sequence

$$\alpha_1;\ \alpha_1, \alpha_2;\ \alpha_1, \alpha_2, \alpha_3;\ \ldots;\ \alpha_1, \alpha_2, \ldots, \alpha_k;\ \ldots$$

be denoted in order by $\beta_1, \beta_2, \beta_3, \ldots$. Thus, the set of the limit points of the sequence $(\beta_1, \beta_2, \beta_3, \ldots)$ consists of the set $\{\alpha_1, \alpha_2, \alpha_3, \ldots\}$ and its limit points. So, this set is identical to the set E, since $\{\alpha_1, \alpha_2, \alpha_3, \ldots\}$ is everywhere dense in E.

Now, we can choose n_1 so that

$$\left|\beta_1 - \sum_{k=0}^{n_1} a_{n_1,k}\beta_1\right| < \frac{1}{2}$$

and is then, we put $x_k = \beta_1$ for $1 \leq k \leq n_1$. In general, when $n_1 < n_2 < \cdots < n_{p-1}$ are defined and x_k is defined for $1 \leq k \leq n_{p-1}$, we choose $n_p > n_{p-1}$ so that

$$\left|\beta_p - \sum_{k=0}^{n_{p-1}} a_{n_p,k}x_k - \sum_{k=n_{p-1}+1}^{n_p} a_{n_p,k}\beta_p\right| < \frac{1}{2^p}$$

and put $x_k = \beta_p$ for $n_{p-1} < k \leq n_p$. Thus, we observe for $x = (x_k)$ defined by induction that $|\beta_p - (Ax)_{n_p}| < 2^{-p}$ and that $x = (x_k)$ and $Ax = \{(Ax)_n\}$ have the required properties to establish the theorem. □

Note that if the set E is convex as well as non-empty and closed, the $\mathcal{K}$-core of the sequence $x = (x_k)$ constructed in the proof is the set E, and so, we have $\mathcal{K}$-core$(Ax) \supset D_A \supset D_x = E = \mathcal{K}$-core$(x)$.

If, moreover, $A = (a_{nk})$ is a lower semi-regular matrix which satisfies the condition

$$a_{nk} = Re[a_{nk}] \geq 0 \text{ for every } n \text{ and for } k \geq K, \tag{6.1.49}$$

then we have $\mathcal{K}$-core$(Ax) \subset \mathcal{K}$-core(x) in addition to $\mathcal{K}$-core$(Ax) \supset \mathcal{K}$-core(x). Hence, we have:

Theorem 6.1.56. [111, 6.6, II] *If $A = (a_{nk})$ is a lower semi-regular matrix which satisfies the condition (6.1.49), then corresponding to each closed convex non-empty set E in the complex plane there is a sequence $x = (x_k)$ such that $\mathcal{K}$-core$(Ax) = \mathcal{K}$-core$(x) = E$.*

A different generalization of this result was given by Allen [18]:

Theorem 6.1.57. [18] *Let $A = (a_{nk})$ be a regular matrix. Then, $\mathcal{K}$-core$(Ax) = \mathcal{K}$-core(x) for every $x \in \ell_\infty$ if and only if (6.1.40) holds and*

$$1 \text{ is a limit point of } (u_n) \text{ with } u_n = \sum_i a_{n,p_i} \text{ for every } (p_i)_{i\in\mathbb{N}_1}. \tag{6.1.50}$$

Proof. Necessity. The necessity of (6.1.40) follows from Theorem 6.1.49. Prior to showing that (6.1.50) is necessary, we observe that if $q_1, q_2, \ldots, q_N$ are the column suffixes with fixed N, a finite number, and if we omit the entries in these columns from the sums of the rows of $A = (a_{nk})$, we obtain a sequence $v = (v_n)$ with $v_n = \sum_k a_{nk}^{(N)}$, which tends to 1 as $n \to \infty$ for every regular matrix.

Now, let $(p_i)_{i=1}^{\infty}$ be any given sequence of suffixes and let $x = (x_k)$ be the sequence in which $x_{p_i} = 1$ for every i, and all the other entries are zero. Then, the core of $x = (x_k)$ is the interval $(0, 1)$ if $x = (x_k)$ has an infinite number of zero entries, and is the point 1 if the number of zero entries in $x = (x_k)$ is finite. Since the core of (u_n)

is the core of $x = (x_k)$, it follows that (6.1.50) is necessary, where $u_n = \sum_i a_{n,p_i}$ for all $n \in \mathbb{N}$.

Sufficiency. Suppose that $x = (x_k) \in \ell_\infty$ and a be its limit point. Let $x_{p_i} \to a$, as $i \to \infty$ and (q_J) be the sequence obtained by omitting the numbers p_i from the sequence $(1, 2, 3, \ldots)$. Then,

$$(Ax)_n := \sum_i a_{n,p_i} x_{p_i} + \sum_J a_{n,q_J} x_{q_J}.$$

We see from (6.1.50) that there is a sequence of suffixes (m_r) such that

$$\lim_{r\to\infty} \sum_i a_{m_r,p_i} = \lim_{r\to\infty} u_{m_r} = 1$$

and we have

$$(Ax)_{m_r} = \sum_i a_{m_r,p_i} x_{p_i} + \sum_J a_{m_r,q_J} x_{q_J}.$$

Given any $\varepsilon > 0$, we put $x_{p_i} = a + \varepsilon_{p_i}$ and determine N so that

$$|\varepsilon_{p_i}| \leq \frac{\varepsilon}{8} \quad \text{for} \quad i \geq N.$$

Then, we have

$$\begin{aligned}(Ax)_{m_r} - a = \sum_{i=1}^{N} a_{m_r,p_i} x_{p_i} \quad &+ \quad a\left(\sum_{i=N+1}^{\infty} a_{m_r,p_i} - 1\right) + \\ &+ \sum_{i=N+1}^{\infty} a_{m_r,p_i} \varepsilon_{p_i} + \sum_J a_{m_r,q_J} x_{q_J}.\end{aligned}$$

We will prove that $(Ax)_{m_r} \to a$, as $r \to \infty$.

Since $|x_k| \leq M$ for every $k \in \mathbb{N}$, we have

$$\begin{aligned}|(Ax)_{m_r} - a| \leq M \sum_{i=1}^{N} |a_{m_r,p_i}| \quad &+ \quad |a| \left|\sum_{i=N+1}^{\infty} a_{m_r,p_i} - 1\right| + \\ &+ \frac{\varepsilon}{8} \sum_{i=N+1}^{\infty} |a_{m_r,p_i}| + M \sum_J |a_{m_r,q_J}|.\end{aligned} \tag{6.1.51}$$

Columns of the regular matrix $A = (a_{nk})$ are the null sequences and, having fixed N, we can determine N_1 so that for $r \geq N_1$,

$$\sum_{i=1}^{N} |a_{m_r,p_i}| \leq \begin{cases} \min\left\{\frac{\varepsilon}{4M}, \frac{\varepsilon}{8|a|}\right\} & , \quad a \neq 0, \\ \frac{\varepsilon}{4M} & , \quad a = 0. \end{cases}$$

We also have

$$u_{m_r} = \sum_{i=1}^{N} a_{m_r,p_i} + \sum_{i=N+1}^{\infty} a_{m_r,p_i}$$

and, if $a \neq 0$, we determine $N_2 \geq N_1$ so that

$$|u_{m_r} - 1| \leq \frac{\varepsilon}{8|a|}$$

for $r \geq N_2$. Then,

$$\begin{aligned}\left|\sum_{i=N+1}^{\infty} a_{m_r,p_i} - 1\right| &\leq |u_{m_r} - 1| + \sum_{i=1}^{N} |a_{m_r,p_i}| \\ &\leq \frac{\varepsilon}{8|a|} + \frac{\varepsilon}{8|a|} = \frac{\varepsilon}{4|a|}\end{aligned}$$

for $r \geq N_2$. It follows from (6.1.40) that there is a positive integer N_3, greater than N_2 if $a \neq 0$, and greater than N_1 if $a = 0$, such that

$$\sum_{i=N+1}^{\infty} |a_{m_r,p_i}| \leq 2$$

for $r \geq N_3$. If $a \neq 0$, we obtain from (6.1.51),

$$|(Ax)_{m_1} - a| \leq \frac{\varepsilon}{4} + \frac{\varepsilon}{4} + \frac{\varepsilon}{4} + M \sum_J |a_{m_r,q_J}| \tag{6.1.52}$$

for $r \geq N_3$. If $a = 0$ we have,

$$|(Ax)_{m_1}| \leq \frac{\varepsilon}{4} + \frac{\varepsilon}{4} + M \sum_J |a_{m_r,q_J}|. \tag{6.1.53}$$

Now, let

$$s_n := \sum_k |a_{nk}|, \; z_n := \sum_i |a_{n,p_i}| \text{ and } v_n := \sum_i |a_{n,q_i}|$$

so that $s_n := z_n + v_n$. It follows from (6.1.40) that there is an N_1 such that $s_{m_r} \leq 1+\varepsilon$ for $r \geq N_1$ and therefore $z_{m_r} \leq 1+\varepsilon$ for $r \geq N_1$. Since $u_{m_r} \to 1$, as $r \to \infty$, we can determine $N_2 \geq N_1$ so that $1-\varepsilon \leq |u_{m_r}|$ for $r \geq N_2$, and thus $1-\varepsilon \leq z_{m_r} \leq 1+\varepsilon$ for $r \geq N_2$. Therefore, $z_{m_r} \to 1$, as $r \to \infty$, and since $s_{m_r} = z_{m_r} + v_{m_r}$ and $s_{m_r} \to 1$, it follows that $v_{m_r} \to 0$, as $r \to \infty$. Hence, one can determine $N_4 \geq N_2$ so that for $r \geq N_4$,

$$v_{m_r} = \sum_i |a_{m_r,q_i}| \leq \begin{cases} \frac{\varepsilon}{4M} & , \quad a \neq 0, \\ \frac{\varepsilon}{2M} & , \quad a = 0. \end{cases}$$

From (6.1.52) and (6.1.53), we have $|(Ax)_{m_r} - a| \leq \varepsilon$ for $r \geq N_4$. Therefore, a is in the derived set of Ax and so, the derived set of Ax contains the derived set of x. It follows that $\mathcal{K}$-core(Ax) contains the derived set of x. Since x is bounded, its core is the convex cover of its derived set. So, $\mathcal{K}$-core$(Ax) \supseteq \mathcal{K}$-core$(x)$. Since by (6.1.40) it is known that $\mathcal{K}$-core$(Ax) \subseteq \mathcal{K}$-core$(x)$, the proof of theorem is completed. □

Here, let us quote some matrices which satisfy the conditions of the above theorem:

(a) Any diagonal matrix $diag(d_0, d_1, d_2, \ldots, d_n, \ldots)$ such that $d_n \to 1$, as $n \to \infty$.

(b) Matrices in which there exists one non-zero element only each row and column and whose non-zero elements are all 1.

(c) The matrix $A = (a_{nk})$ defined by

$$a_{nk} := \begin{cases} 1 & , \quad k = n, \\ 0 & , \quad k < n, \\ (n+1)^{n-k} & , \quad k > n, \end{cases}$$

for all $k, n \in \mathbb{N}$.

We conclude this section by giving the core theorems concerning definitely divergent sequences, (see [111, p. 154]). Firstly, we state and prove the following lemma which is necessary:

Lemma 6.1.58. [111, 6.5, II, Lemma] *Let the sequence of sequences $S = \{x_k^{(n)}\}_{k,n\in\mathbb{N}}$ satisfies $|x_k^{(n)}| \leq v_k$ for every $n \in \mathbb{N}$ with $v_k \to \infty$, as $k \to \infty$; and $A = (a_{nk})$ be a matrix such that $a_{nk} \to 0$, as $n \to \infty$ for each $k \in \mathbb{N}$ and $\sum_k |a_{nk}|v_k$ exists for each $n \in \mathbb{N}$. Also, let Y be an arbitrary finite limit point of $(y_n) := \left\{\sum_k a_{nk}x_k^{(n)}\right\}$, if there is one. Then, there is a sequence $(\widetilde{x}_k)$ whose elements are taken from S such that Y is a limit point of $(\widetilde{y}_n) := \{\sum_k a_{nk}\widetilde{x}_k\}$.*

Proof. Firstly note that $\sum_k a_{nk}\widetilde{x}_k$ converges for every $n \in \mathbb{N}$, since the elements of $(\widetilde{x}_k)$ are taken from $x_k^{(n)}$, $(n \in \mathbb{N})$ and $|x_k^{(n)}| \leq v_k$, and also $\sum_k |a_{nk}|v_k$ exists for each $n \in \mathbb{N}$.

Now, by the hypothesis, there is an increasing sequence of positive integers (p_n) such that $y_{p_n} \to Y$, as $n \to \infty$. Let M_N be the upper bound of (v_k) for $0 \leq k \leq N$. In order to define $(\widetilde{x}_k)$, we first fix k_1 so that

$$\sum_{k=k_1+1}^{\infty} |a_{p_1,k}|v_k < \frac{1}{2}$$

and then define $(\widetilde{x}_k)$ for $0 \leq k \leq k_1$ by $\widetilde{x}_k = x_k^{p_1}$. We next choose $r_2 > 1$ so that

$$|a_{p_{r_2},k}| < \frac{1}{4k_1 M_{k_1}} \quad \text{for} \quad 0 \leq k \leq k_1,$$

which is possible since $a_{nk} \to 0$, as $n \to \infty$, for each $k \in \mathbb{N}$. Now, determine $k_2 > k_1$ so that

$$\sum_{k=k_2+1}^{\infty} |a_{p_{r_2},k}|v_k < \frac{1}{4}$$

and define $(\widetilde{x}_k)$ for $k_1 \leq k \leq k_2$ by $\widetilde{x}_k = x_k^{p_{r_2}}$. We next choose $r_3 > r_2$ so that

$$|a_{p_{r_3},k}| < \frac{1}{8k_2 M_{k_2}} \quad \text{for} \quad 1 \leq k \leq k_2$$

and then determine $k_3 > k_2$ so that

$$\sum_{k=k_3+1}^{\infty} |a_{p_{r_3},k}| v_k < \frac{1}{8}$$

and define $(\widetilde{x}_k)$ for $k_2 \leq k \leq k_3$ by $\widetilde{x}_k = x_k^{p_{r_3}}$, and so on.

All the elements $(\widetilde{x}_k)$ satisfy the condition that there is an $n \in \mathbb{N}$ for each of them such that $(\widetilde{x}_k) = x_k^{(n)}$. Also, we have

$$y_{p_{r_m}} = \left(\sum_{k=1}^{k_{m-1}} + \sum_{k=k_{m-1}+1}^{k_m} + \sum_{k=k_m+1}^{\infty} \right) a_{p_{r_m},k} \widetilde{x}_k.$$

However, by definition of $(\widetilde{x}_k)$, we have $(\widetilde{x}_k) = \{x_k^{p_{r_m}}\}$ for $k_{m-1} + 1 \leq k \leq k_m$, where $k_0 = 0$, $p_1 = 1$. Hence,

$$\begin{aligned} |\widetilde{y}_{p_{r_m}} - y_{p_{r_m}}| &= \left| \sum_k a_{p_{r_m},k} (\widetilde{x}_k - x_k^{p_{r_m}}) \right| \\ &= \left| \left(\sum_{k=0}^{k_{m-1}} + \sum_{k=k_m+1}^{\infty} \right) a_{p_{r_m},k} (\widetilde{x}_k - x_k^{p_{r_m}}) \right| \\ &\leq k_{m-1} \frac{2M_{k_{m-1}}}{2^m k_{m-1} M_{k_{m-1}}} + \frac{2}{2^m} \\ &= \frac{1}{2^{m-2}} \to 0, \text{ as } m \to \infty. \end{aligned}$$

Hence,

$$\lim_{m\to\infty} \widetilde{y}_{p_{r_m}} = \lim_{m\to\infty} y_{p_{r_m}} = Y,$$

which completes the proof of the lemma. □

Theorem 6.1.59. [111, 6.5, II] *Let $x = (x_k) \in \omega$ such that $|x_k| \leq v_k$ for all $k \in \mathbb{N}$ with $v_k \to \infty$, as $k \to \infty$ and $A = (a_{nk})$ be a real regular matrix. Then, $\mathcal{K}$-core$(Ax) \subseteq$ $\mathcal{K}$-core(x) if and only if $A = (a_{nk})$ is absolutely equivalent to a non-negative matrix B for all sequences $x = (x_k)$; where*

$$\sum_k |a_{nk}| v_k, \ \sum_k b_{nk} v_k \text{ and } \sum_k |a_{nk} - b_{nk}| v_k \tag{6.1.54}$$

exist for each $n \in \mathbb{N}$.

Proof. Sufficiency. We have

$$|(Ax)_n - (Bx)_n| \leq \sum_k |a_{nk} - b_{nk}| v_k \to 0, \text{ as } n \to \infty.$$

Now, the rest of the proof of sufficiency follows by the similar way used in the proof of Theorem 6.1.50.

Necessity. With the notation used in the proof of necessity of Theorem 6.1.50, $a_{nk} \equiv b_{nk}$, since $A = (a_{nk})$ assumed real. Also, A and C are absolutely equivalent for all $x = (x_k)$ such that $|x_k| \leq v_k$, (v_k) satisfying the conditions of the present theorem. If, this is not so, then

$$\limsup_{n\to\infty} \sum_k |Re[\widetilde{a}_{nk}]| \, v_k > 0. \tag{6.1.55}$$

Define the sequence $\{x_k^{(n)}\}$ by $x_k^{(n)} := \begin{cases} v_k & , \quad Re[\widetilde{a}_{nk}] < 0, \\ 0 & , \quad Re[\widetilde{a}_{nk}] = 0. \end{cases}$ Then,

$$y_n = \sum_k b_{nk} x_k^{(n)} = -\sum_k |Re[\widetilde{a}_{nk}]| \, v_k. \tag{6.1.56}$$

(a) Suppose that the sequence $(\alpha_n) = \{\sum_k |Re[\widetilde{a}_{nk}]| v_k\}$ has a finite positive limit point. Then, by (6.1.56), (y_n) has a finite negative limit point. By Lemma 6.1.58, we can determine $(\widetilde{x}_k)$ so that $\widetilde{x}_k = v_k$ or $= 0$, and so that its B-transform, hence the A-transform, has a negative limit point, although the core of $(\widetilde{x}_k)$ clearly contains non-negative limit points only.

Hence, A and C are absolutely equivalent for all $x = (x_k)$ such that $|x_k| \leq v_k$ for all $k \in \mathbb{N}$ with (v_k) satisfying the conditions of the theorem, and C is a non-negative regular matrix.

(b) Suppose that (α_n) has ∞ and (if possible) 0 as its only limit points. If 0 is the only limit point, then the proof is trivial. If not, by hypothesis, there is an increasing sequence of positive integers (n_r) such that $\alpha_{n_r} \to \infty$, as $r \to \infty$. We now take $x_k^{(n)} = 0$ for all $k \in \mathbb{N}$ if n does not belong to the sequence (n_r) and

$$x_k^{n_r} := \begin{cases} v_k/\alpha_{n_r} & , \quad Re[\widetilde{a}_{nk}] < 0, \\ 0 & , \quad Re[\widetilde{a}_{nk}] = 0. \end{cases}$$

Then, since

$$y_{n_r} = \sum_k a_{n_r,k} x_k^{n_r} = -\sum_k \frac{|Re[\widetilde{a}_{nk}]| \, v_k}{\alpha_{n_r}} = -1,$$

the sequence $(y_n) = \left\{\sum_k a_{nk} x_k^{(n)}\right\}$ has -1 as a limit point while all the terms of $\{x_k^{(n)}\}$ are non-negative, a contradiction by Lemma 6.1.58.

This step completes the proof of the theorem. □

Note that the condition of Theorem 6.1.59 is not necessary when $A = (a_{nk})$ is complex. For, if A and B are not absolutely equivalent for all $x = (x_k)$ such that $|x_k| \leq v_k$ for all $k \in \mathbb{N}$, (v_k) satisfying the conditions of the theorem, then

$$\limsup_{n\to\infty} \sum_k |Re[a_{nk}]|\, v_k > 0. \tag{6.1.57}$$

Define the sequence $\{x_k^{(n)}\}$ by $x_k^{(n)} := \begin{cases} v_k & , \quad Re[a_{nk}] \geq 0, \\ -v_k & , \quad Re[a_{nk}] < 0. \end{cases}$ Then, $Re[y_n] = \sum_k |Re[a_{nk}]|\, v_k$ and (6.1.57) implies that (y_n) has a limit point with non-zero imaginary part. But, $Re(y_n)$ may tend to ∞, in which case $\mathcal{K}$-core of (y_n) is empty, so no contradiction would be established in this case an applying Lemma 6.1.58 with $\widetilde{x}_k = \pm v_k$.

The next interesting result is given by Cooke [111, p. 147]:

Theorem 6.1.60. [111, 6.3, III, Lemma 3] *Let $A = (a_{nk})$ be a lower semi-regular matrix which satisfies*

$$Re[a_{nk}] = 0 \text{ for every } n \in \mathbb{N} \text{ and for } k \geq K_2 \tag{6.1.58}$$

but not the condition (6.1.49). Then, there exists a sequence $x = (x_k)$ of non-negative real numbers tending to ∞ for which $\mathcal{K}$-core(Ax) is the entire real axis.

Proof. Choose K_2 so that (6.1.58) holds, and let $x_k = 0$ for $0 \leq k < K_2$. Suppose that suffixes $K_2 = n_0 < v_1 < n_1 < \cdots < v_{p-1} < n_{p-1}$ are defined and also $x = (x_k)$ is defined for $0 \leq k \leq n_{p-1}$. Now, choose $v_p > n_{p-1}$ so that

$$\left|\sum_{k=0}^{n_{p-1}} a_{v_p,k} x_k\right| < \frac{1}{2p}, \quad \left|1 - \sum_{k=n_{p-1}+1}^{v_p} a_{v_p,k}\right| < \frac{1}{2p^2},$$

and put $x_k = p$ for $n_{p-1} < k \leq v_p$. Then,

$$(Ax)_{v_p} = \sum_{k=0}^{n_{p-1}} a_{v_p,k} x_k + p \sum_{k=n_{p-1}+1}^{v_p} a_{v_p,k}.$$

So,

$$|(Ax)_{v_p} - p| \leq \left|\sum_{k=0}^{n_{p-1}} a_{v_p,k} x_k\right| + p \left|1 - \sum_{k=n_{p-1}+1}^{v_p} a_{v_p,k}\right| < \frac{1}{p}.$$

Now, choose k_p and n_p so that $v_p < k_p \leq n_p$ and $a_{n_p,k_p} < 0$. Let $x_k = p$ for $v_p < k \leq n_p$, $k \neq k_p$, and $x_{k_p} = p + s$, where s is chosen so that $s \geq 0$ and

$$(Ax)_{n_p} := \sum_{k=0}^{v_p} a_{n_p,k} x_k + p \sum_{k=v_p+1}^{n_p} a_{n_p,k} + a_{n_p,k_p} s < -p.$$

This procedure create suffixes $K_2 = n_0 < v_1 < n_1 < v_2 < n_2 < \cdots$, and a sequence $x = (x_k)$ of non-negative real numbers for which

$$x_k \geq p \ (n_{p-1} < k \leq n_p), \quad (Ax)_{v_p} > p - \frac{1}{p}, \quad (Ax)_{n_p} < -p \text{ for all } p \in \mathbb{N}_1.$$

It is easy to see that $x_k \to \infty$, as $k \to \infty$ and $\mathcal{K}$-core$(Ax) = \mathbb{R}$. □

6.2 σ-CORE

Let σ be a one to one mapping on $\mathbb{N}$ into itself and $T : \ell_\infty \to \ell_\infty$ be a linear operator defined by $Tx = (x_{\sigma(k)})$. An element $\phi \in \ell'_\infty$, the conjugate space of ℓ_∞, is called an *invariant mean* [323] or a σ-mean if and only if

(i) $\phi(x) \geq 0$ for the sequence $x = (x_k)$ of non-negative reals.

(ii) $\phi(e) = 1$.

(iii) $\phi(Tx) = \phi(x)$ for all $x \in \ell_\infty$.

Let M be the set of all σ-means on ℓ_∞.

A sublinear functional P on ℓ_∞ is said to generate σ-means if $\phi \in \ell'_\infty$ and $\phi \leq P$ implies ϕ is a σ-mean, to dominate σ-means if $\phi \leq P$ for all $\phi \in M$, where $\phi \leq P$ means that $\phi(x) \leq P(x)$ for all $x \in \ell_\infty$. If P is a sublinear functional on ℓ_∞, then we write $\{\ell_\infty, P\}$ to denote the set of all linear functionals ϕ on ℓ_∞ such that $\phi \leq P$.

It is shown in [279] that the sublinear functional

$$V(x) := \sup_{n\in\mathbb{N}} \limsup_{p\to\infty} t_{pn}(x)$$

both generates and dominates σ-means, where

$$t_{pn}(x) := \frac{x_n + x_{\sigma(n)} + \cdots + x_{\sigma^p(n)}}{p+1}, \quad t_{-1,n}(x) = 0.$$

In [296], it is shown that

$$V(x) := \limsup_{p\to\infty} \sup_{n\in\mathbb{N}} t_{pn}(x) = q_\sigma(x) \quad \text{for all} \quad x = (x_n) \in \ell_\infty.$$

A bounded sequence x is called σ-convergent [323] to s if $q_\sigma(x) = q_\sigma(-x) = s$. In this case, we write $\sigma - \lim x = s$. By V_σ, we denote the set of all σ-convergent sequences. We assume throughout that $\sigma^p(n) \neq n$ for all $n \in \mathbb{N}$ and $p \geq 1$, where $\sigma^p(n)$ is the pth iterate of σ at $n \in \mathbb{N}$. Thus, a σ-mean extends the limit functional onto c in the sense that $\phi(x) = \lim x$ for all $x \in c$ [282]. In the case $\sigma(n) = n+1$, σ-limit is reduced to the famous Banach limits and V_σ corresponds to the space f of all almost convergent sequences introduced by Lorentz [246].

Definition 6.2.1. [279] *σ-core of a real valued sequence $x = (x_k)$ is the closed interval $[-q_\sigma(-x), q_\sigma(x)]$.*

In the case $\sigma(n) = n+1$, σ-core is said to be Banach Core, (cf. [141, 144, 309]). Following theorem related to the $\mathcal{K}$-core and σ-core was proved by Mishra et al. [279]:

Theorem 6.2.2. [279, Theorem 2] *$q_\sigma(Ax) \leq L(x)$ for all $x \in \ell_\infty$ if and only if $A \in (c : V_\sigma)_{reg}$ and*

$$\lim_{p\to\infty} \sum_k a(p,n,k) = 1 \text{ uniformly in } n, \text{ where} \tag{6.2.1}$$

$$a(p,n,k) := \frac{1}{p+1} \sum_{i=0}^{p} a_{\sigma^i(n),k} \text{ for all } k,n,p \in \mathbb{N}.$$

Proof. Sufficiency. Since $x = (x_k) \in \ell_\infty$, it is known that for any given $\varepsilon > 0$, there exists a positive integer k_0 such that $x_k \leq L(x) + \varepsilon$ whenever $k \geq k_0$. Now, let us write

$$\sum_k a(p,n,k)x_k = \sum_{k<k_0} a(p,n,k)x_k + \sum_{k\geq k_0} a(p,n,k)^+ x_k - \sum_{k\geq k_0} a(p,n,k)^- x_k. \tag{6.2.2}$$

Then, by the hypothesis, the first and the last sums on the right-hand side of (6.2.2) tend to zero, as $p \to \infty$, uniformly in n. Therefore, we obtain by applying the operator $\limsup_{p\to\infty} \sup_{n\in\mathbb{N}}$ to the equality (6.2.2) that

$$q_\sigma(Ax) \leq [L(x) + \varepsilon] \limsup_{p\to\infty} \sup_{n\in\mathbb{N}} \sum_k a(p,n,k).$$

Since (6.2.1) holds, we have $q_\sigma(Ax) \leq L(x)$.

Necessity. We observe by inserting $-x = (-x_k)$ in place of $x = (x_k)$ in the inequality $q_\sigma(Ax) \leq L(x)$ that

$$l(x) \leq -q_\sigma(-Ax) \leq q_\sigma(Ax) \leq L(x).$$

If $x \in c$, then $l(x) = L(x) = \lim x$ and so $-q_\sigma(-Ax) = q_\sigma(Ax)$ which means that $\sigma - \lim Ax = \lim x$. Hence, $A \in (c : V_\sigma)_{reg}$.

It is clear that

$$\liminf_{p\to\infty} \sum_k |a(p,n,k)| \geq \liminf_{p\to\infty} \sum_k a(p,n,k) = 1, \tag{6.2.3}$$

uniformly in n. On the other hand, if we choose the sequence of matrices $\mathcal{A} = \{a_{pk}(n)\}$ defined by $a_{pk}(n) = a(p,n,k)$ for all $k, n, p \in \mathbb{N}$, then clearly $\mathcal{A}$ satisfies the conditions of Lemma 6.1.32. So, there exists a $y = (y_k) \in \ell_\infty$ with $\|y\| \leq 1$ and

$$\limsup_{p\to\infty} \sup_{n\in\mathbb{N}} \sum_k |a(p,n,k)| = \limsup_{p\to\infty} \sup_{n\in\mathbb{N}} \sum_k a(p,n,k)y_k.$$

Thus, we derive by inserting this y in the hypothesis that

$$\limsup_{p\to\infty} \sup_{n\in\mathbb{N}} \sum_k |a(p,n,k)| = q_\sigma(Ay) \leq L(y) \leq \|y\| \leq 1.$$

Combining this result by the fact in (6.2.3), we obtain the required condition. □

Prior to stating the following theorem which is more general (see [135]) and gives a connection between the σ- and μ-cores, we present two lemmas, where μ is a mapping from $\mathbb{N}$ into itself, as σ.

Lemma 6.2.3. [144, Lemma 1] *Let P and Q be sublinear functionals on a linear space X. Then, $\{X, P\} \subset \{X, Q\}$ if and only if $P(x) \leq Q(x)$ for all $x \in X$.*

Lemma 6.2.4. [135, Theorem 5] *$A \in (V_\mu : V_\sigma)_{reg}$ if and only if $A \in (c : V_\sigma)_{reg}$* (cf. [359]) *and*

$$\lim_{p\to\infty} \sum_k \frac{1}{p+1} \left| \sum_{i=0}^{p} \left[a_{\sigma^i(n),k} - a_{\sigma^i(n),\mu(k)}\right] \right| = 0, \text{ uniformly in } n. \tag{6.2.4}$$

Proof. Sufficiency. Define $z = (z_k)$ by $z_k = x_{\mu(k)} - x_k$ for all $k \in \mathbb{N}$. Then, clearly $z \in Z_\mu$ and we can write (cf. [323])

$$(Az)_n = \sum_k (a_{nk} - a_{n,\mu(k)})x_{\mu(k)} + \sum_{k\notin\mu(\mathbb{N})} a_{nk}x_k.$$

If we choose a sequence $x = (x_k) \in \ell_\infty$ such that $x_k = 0$ for $k \notin \mu(\mathbb{N})$, then

$$(Az)_n = \sum_k (a_{nk} - a_{n,\mu(k)})x_{\mu(k)}. \tag{6.2.5}$$

Thus, (6.2.4) implies that $Az \in Z_\sigma$ for all $z \in Z_\mu$.

Now, for $y \in V_\mu$ with $\mu - \lim y = s$, there exists an element $z \in Z_\mu$ such that $y := z + se$, (see [323]). Hence, $Ay := Az + sAe$. Since, $Ae = \left(\sum_k a_{nk}\right)_{n\in\mathbb{N}}$ and $A \in (c : V_\sigma)_{reg}$, $\sigma - \lim Ay = s$ and so $A \in (V_\mu : V_\sigma)_{reg}$.

Necessity. Since $c \subset V_\mu$, we conclude from the assumption $A \in (V_\mu : V_\sigma)_{reg}$ that $A \in (c : V_\sigma)_{reg}$. Define a matrix $B = (b_{nk})$ by $b_{nk} := a_{nk} - a_{n,\mu(k)}$ for all $k, n \in \mathbb{N}$. Then, (6.2.5) implies that $B \in (\ell_\infty : V_\sigma)$. Therefore, the condition (6.2.4) follows from the conditions which characterized the class $(\ell_\infty : V_\sigma)$, (cf. [359]). □

Theorem 6.2.5. [135, Theorem 9] *$q_\sigma(Ax) \leq q_\mu(x)$ for all $x \in \ell_\infty$ if and only if $A \in (V_\mu : V_\sigma)_{reg}$ and (6.2.1) holds.*

Proof. Necessity. By the similar way used in the proof of necessity of Theorem 6.2.2, one can see that $A \in (V_\mu : V_\sigma)_{reg}$. To show the necessity of (6.2.1), again as in the proof of necessity of Theorem 6.2.2, choose the sequence of matrices $\mathcal{A} = \{a_{pk}(n)\}$ with $a_{pk}(n) = a(p, n, k)$. Then, Lemma 6.1.32 and hypothesis give (6.2.1).

Sufficiency. Since $A \in (V_\mu : V_\sigma)_{reg}$ implies that $A \in (c : V_\sigma)_{reg}$, from Theorem 6.2.2,

$$u(x) = \inf_{z\in Z_\mu} q_\sigma(Ax + z) \leq \inf_{z\in Z_\mu} L(x + z) = v(x). \tag{6.2.6}$$

On the other hand, since $q_\sigma(Az) = -q_\sigma(-Az) = 0$ for $z \in Z_\mu$, we have

$$u(x) \geq \inf_{z\in Z_\mu} \{q_\sigma(Ax) + [-q_\sigma(-Az)]\} = q_\sigma(Ax). \tag{6.2.7}$$

Now, combining (6.2.6) and (6.2.7), we obtain $q_\sigma(Ax) \leq v(x)$ for all $x \in \ell_\infty$. Since $v(x) = q_\mu(x)$ by Lemma 6.2.3, the proof is completed. □

6.3 $\mathcal{I}$-CORE

Let $X \neq \emptyset$. A class $S \subset 2^X$ of subsets of X is said to be an *ideal* in X if S satisfies the following conditions:

(i) $\emptyset \in S$.

(ii) $Y \cup Z \in S$ whenever $Y, Z \in S$.

(iii) $Y \in S$ and $Z \subseteq Y$ implies $Z \in S$.

An ideal is called *non-trivial* if $X \notin S$. A non-trivial ideal is called *admissible* if $\{x\} \in S$ for each $x \in X$ [229].

Let $\mathcal{I}$ be a non-trivial ideal in $\mathbb{N}$. A sequence $x = (x_k)$ is said to be $\mathcal{I}$-convergent to a number l if for every $\varepsilon > 0$, $\{k \in \mathbb{N} : |x_k - l| > \varepsilon\} \in \mathcal{I}$ [229]. In this case, we write $\mathcal{I}\text{-}\lim x = l$. It is clear that a $\mathcal{I}$-convergent sequence need not be bounded. Let $F_{\mathcal{I}}(b)$ be the set of all $\mathcal{I}$-convergent and bounded sequences.

Note that in the cases $\mathcal{I}_\delta = \{E \subseteq \mathbb{N} : \delta(E) = 0\}$, $\mathcal{I}_{\delta_A} = \{E \subseteq \mathbb{N} : \delta_A(E) = 0\}$ and $\mathcal{I}_{\delta_{\mathcal{B}}} = \{E \subseteq \mathbb{N} : \delta_{\mathcal{B}}(E) = 0\}$, the $\mathcal{I}$-convergence is reduced to the statistical convergence [170], A-statistical convergence [99, 109] and $\mathcal{B}$-statistical convergence [295], respectively.

An admissible ideal $\mathcal{I}$ in $\mathbb{N}$ is said to be satisfy the additive property if for every countable system $\{Y_1, Y_2, \ldots\}$ of mutually disjoint sets in $\mathcal{I}$, there exist sets $Z_j \subseteq \mathbb{N}$ such that the symmetric differences $Y_j \Delta Z_j$ are finite for all $j \in \mathbb{N}$ and $\cup_j Z_j \in \mathcal{I}$ [229].

Demirci [146] has introduced the concepts $\mathcal{I}$-limit superior and inferior. For a real number sequence $x = (x_k)$, let B_x and A_x denote the sets $\{b \in \mathbb{R} : \{k : x_k > b\} \notin \mathcal{I}\}$ and $\{a \in \mathbb{R} : \{k : x_k < a\} \notin \mathcal{I}\}$, respectively, and also $\mathcal{I}$ be admissible. Then,

$$\mathcal{I}\text{-}\limsup x := \begin{cases} \sup B_x & , \quad B_x \neq \emptyset, \\ -\infty & , \quad B_x = \emptyset \end{cases}$$

and

$$\mathcal{I}\text{-}\liminf x := \begin{cases} \inf A_x & , \quad A_x \neq \emptyset, \\ \infty & , \quad A_x = \emptyset. \end{cases}$$

It is shown by Demirci [146] that $\mathcal{I}\text{-}\limsup x = \beta$ if and only if for every $\varepsilon > 0$, $\{k : x_k < \beta - \varepsilon\} \notin \mathcal{I}$ and $\{k : x_k > \beta + \varepsilon\} \in \mathcal{I}$. Also, $\mathcal{I}\text{-}\liminf x = \alpha$ if and only if for every $\varepsilon > 0$, $\{k : x_k < \alpha + \varepsilon\} \notin \mathcal{I}$ and $\{k : x_k < \alpha - \varepsilon\} \in \mathcal{I}$. Recall that a sequence $x = (x_k)$ is said to be *$\mathcal{I}$-bounded* if there exists a $N > 0$ such that $\{k : |x_k| > N\} \in \mathcal{I}$. It is proved in [146] that a $\mathcal{I}$-bounded sequence $x = (x_k)$ is $\mathcal{I}$-convergent if and only if $\mathcal{I}\text{-}\limsup x = \mathcal{I}\text{-}\liminf x$.

Definition 6.3.1. [146] *$\mathcal{I}$-core of a real bounded sequence is the closed interval* $[\mathcal{I}\text{-}\liminf x, \mathcal{I}\text{-}\limsup x]$.

In the cases $\mathcal{I}_\delta = \{E \subseteq \mathbb{N} : \delta(E) = 0\}$ and $\mathcal{I}_{\delta_A} = \{E \subseteq \mathbb{N} : \delta_A(E) = 0\}$, the $\mathcal{I}$-core is reduced to the statistical core [176] and A-statistical core [177], respectively. For some inequalities related to the $\mathcal{I}$-core, the reader may refer to [145] and [147].

To give some inequalities involving the $\mathcal{I}$-core, we begin with the following lemma:

Lemma 6.3.2. [118, Lemma 2.1] *$A = (a_{nk}) \in (c : F_{\mathcal{I}}(b))$ if and only if*

$$\sup_{n\in\mathbb{N}} \sum_k |a_{nk}| < \infty, \tag{6.3.1}$$

$$\mathcal{I} - \lim_{n\to\infty} a_{nk} = \alpha_k \text{ for each } k \in \mathbb{N}, \tag{6.3.2}$$

$$\mathcal{I} - \lim_{n\to\infty} \sum_k a_{nk} = \alpha. \tag{6.3.3}$$

Proof. Suppose that $A = (a_{nk}) \in (c : F_{\mathcal{I}}(b))$. Then, (6.3.1) follows from the fact that $(c : F_{\mathcal{I}}(b)) \subset (\ell_\infty : \ell_\infty)$. For the necessity of the conditions (6.3.2) and (6.3.3), it is enough to consider the sequences $e^{(k)}$ and e, respectively.

Conversely, suppose that the conditions (6.3.1)–(6.3.3) hold. Let $x = (x_k) \in c$ with $x_k \to l$, as $k \to \infty$. Then, for any given $\varepsilon > 0$, there exists a $k_0 \in \mathbb{N}$ such that $|x_k - l| \leq \varepsilon$ whenever $k \geq k_0$. Now, we have the equality (3.3.5). By a straightforward calculation, one can see that

$$\mathcal{I} - \lim_{n\to\infty} \sum_k a_{nk}(x_k - l) = \sum_k \alpha_k(x_k - l),$$

which leads the result that

$$\mathcal{I} - \lim_{n\to\infty} (Ax)_n = l\alpha + \sum_k \alpha_k(x_k - l).$$

This completes the proof. □

A matrix A belonging to the class $(c : F_{\mathcal{I}}(b))$ is said to be $\mathcal{I}$-conservative. In the case $A = (a_{nk})$ is $\mathcal{I}$-conservative, the number

$$K_{\mathcal{I}} := K_{\mathcal{I}}(A) = \alpha - \sum_k \alpha_k$$

is said to be $\mathcal{I}$-characteristic of $A = (a_{nk})$.

To prove our main results, we need two lemmas which can be proved by the same technique used in Lemma 2.3 and Lemma 2.4 of [137], respectively.

Lemma 6.3.3. [118, Lemma 2.2] *Let $A = (a_{nk})$ be $\mathcal{I}$-conservative and $\lambda > 0$. Then,*

$$\mathcal{I} - \limsup_{n\to\infty} \sum_k |a_{nk} - \alpha_k| \leq \lambda$$

if and only if

$$\mathcal{I} - \limsup_{n\to\infty} \sum_k (a_{nk} - \alpha_k)^+ \leq \frac{\lambda + K_{\mathcal{I}}}{2},$$

$$\mathcal{I} - \limsup_{n\to\infty} \sum_k (a_{nk} - \alpha_k)^- \leq \frac{\lambda - K_{\mathcal{I}}}{2}.$$

Lemma 6.3.4. [118, Lemma 2.3] *Let $\|A\| < \infty$ and $\mathcal{I} - \lim_{n\to\infty} |a_{nk}| = 0$. Then, there exists a $y = (y_k) \in \ell_\infty$ such that $\|y\| \leq 1$ and*

$$\mathcal{I} - \limsup_{n\to\infty} \sum_k a_{nk}y_k := \mathcal{I} - \limsup_{n\to\infty} \sum_k |a_{nk}|.$$

Following the results on $\mathcal{I}$-conservative matrices given by Çakan and Çoşkun [118], we have:

Theorem 6.3.5. [118, Theorem 2.4] *Let $A = (a_{nk})$ be $\mathcal{I}$-conservative. Then, for some constant $\lambda \geq |K_{\mathcal{I}}|$ and for all $x = (x_k) \in \ell_\infty$,*

$$\mathcal{I} - \limsup_{n\to\infty} \sum_k (a_{nk} - \alpha_k)x_k \leq \frac{\lambda + K_{\mathcal{I}}}{2} L(x) - \frac{\lambda - K_{\mathcal{I}}}{2} l(x) \tag{6.3.4}$$

if and only if

$$\mathcal{I} - \limsup_{n\to\infty} \sum_k |a_{nk} - \alpha_k| \leq \lambda. \tag{6.3.5}$$

Proof. Let (6.3.4) holds. Define $B = (b_{nk})$ by $b_{nk} = a_{nk} - \alpha_k$ for all $k, n \in \mathbb{N}$. Then, since $A = (a_{nk})$ is $\mathcal{I}$ conservative, the matrix B satisfies the hypothesis of Lemma 6.3.4. Hence, we have from (6.3.4) for a $y = (y_k) \in \ell_\infty$ with $\|y\| \leq 1$ that

$$\begin{aligned}
\mathcal{I} - \limsup_{n\to\infty} \sum_k |b_{nk}| &= \mathcal{I} - \limsup_{n\to\infty} \sum_k b_{nk}y_k \\
&\leq \frac{\lambda + K_{\mathcal{I}}}{2} L(y) - \frac{\lambda - K_{\mathcal{I}}}{2} l(y) \\
&\leq \left(\frac{\lambda + K_{\mathcal{I}}}{2} + \frac{\lambda - K_{\mathcal{I}}}{2}\right)\|y\| = \lambda,
\end{aligned}$$

which is equivalent to (6.3.5).

Conversely, let (6.3.5) holds and $x = (x_k) \in \ell_\infty$. Then, for any $\varepsilon > 0$ there exits a $k_0 \in \mathbb{N}$ such that $l(x) - \varepsilon < x_k < L(x) + \varepsilon$ whenever $k > k_0$. Now, we can write

$$\sum_k (a_{nk} - \alpha_k)x_k = \sum_{k \leq k_0} (a_{nk} - \alpha_k)x_k + \sum_{k > k_0} (a_{nk} - \alpha_k)^+ x_k - \sum_{k > k_0} (a_{nk} - \alpha_k)^- x_k.$$

Since $A = (a_{nk})$ is $\mathcal{I}$-conservative, we obtain by Lemma 6.3.3

$$\begin{aligned}
\mathcal{I} - \limsup_{n\to\infty} \sum_k (a_{nk} - \alpha_k)x_k &\leq [L(x) + \varepsilon]\left(\frac{\lambda + K_{\mathcal{I}}}{2}\right) - [l(x) - \varepsilon]\left(\frac{\lambda - K_{\mathcal{I}}}{2}\right) \\
&= \frac{\lambda + K_{\mathcal{I}}}{2} L(x) - \frac{\lambda - K_{\mathcal{I}}}{2} l(x) + \lambda\varepsilon,
\end{aligned}$$

which yields (6.3.4), since ε is arbitrary. □

When $K_{\mathcal{I}} > 0$ and $\lambda = K_{\mathcal{I}}$, Theorem 6.3.5 is reduced to the following theorem:

Theorem 6.3.6. [118, Theorem 2.5] *Let $A = (a_{nk})$ be $\mathcal{I}$-conservative. Then, for all $x = (x_k) \in \ell_\infty$,*

$$\mathcal{I} - \limsup_{n\to\infty} \sum_k (a_{nk} - \alpha_k)x_k \leq K_{\mathcal{I}} L(x)$$

if and only if

$$\mathcal{I} - \lim_{n\to\infty} \sum_k |a_{nk} - \alpha_k| \leq K_{\mathcal{I}}. \tag{6.3.6}$$

In the cases $\mathcal{I} = \mathcal{I}_{\delta_{\mathcal{B}}}$ and $\mathcal{I} = \mathcal{I}_{\delta_A}$, we respectively derive the following results from Theorem 6.3.5:

Theorem 6.3.7. [118, Theorem 2.6] *The following statements hold:*

(a) *Let $A = (a_{nk}) \in (c : st_{\mathcal{B}} \cap \ell_\infty)$. Then, for some constant $\lambda \geq |K_{\mathcal{B}}|$ and for all $x = (x_k) \in \ell_\infty$,*

$$st_{\mathcal{B}} - \limsup_{n\to\infty} \sum_k (a_{nk} - \alpha_k)x_k \leq \frac{\lambda + K_{\mathcal{B}}}{2} L(x) - \frac{\lambda - K_{\mathcal{B}}}{2} l(x)$$

if and only if

$$st_{\mathcal{B}} - \limsup_{n\to\infty} \sum_k |a_{nk} - \alpha_k| \leq \lambda. \tag{6.3.7}$$

(b) *Let $A = (a_{nk}) \in (c : st_A \cap \ell_\infty)$. Then, for some constant $\lambda \geq |K_A|$ and for all $x = (x_k) \in \ell_\infty$,*

$$st_A - \limsup_{n\to\infty} \sum_k (a_{nk} - \alpha_k)x_k \leq \frac{\lambda + K_A}{2} L(x) - \frac{\lambda - K_A}{2} l(x)$$

if and only if

$$st_A - \limsup_{n\to\infty} \sum_k |a_{nk} - \alpha_k| \leq \lambda. \tag{6.3.8}$$

Also; if $\mathcal{I} = \mathcal{I}_\delta$, Theorem 6.3.5 is reduced to Theorem 2.5 of Çoşkun and Çakan [137].

Theorem 6.3.8. [118, Theorem 2.7] *Let $A = (a_{nk})$ and λ be as in Theorem 6.3.5. Then, for all $x = (x_k) \in \ell_\infty$*

$$\mathcal{I} - \limsup_{n\to\infty} \sum_k (a_{nk} - \alpha_k)x_k \leq \frac{\lambda + K_{\mathcal{I}}}{2} V(x) + \frac{\lambda - K_{\mathcal{I}}}{2} V(-x) \tag{6.3.9}$$

if and only if (6.3.5) holds and

$$\mathcal{I} - \lim_{n\to\infty} \sum_k |a_{nk} - a_{n,\sigma(k)} - \alpha_k + \alpha_{\sigma(k)}| = 0. \tag{6.3.10}$$

Proof. Suppose that (6.3.9) holds. Then, since $V(x) \leq L(x)$ and $V(-x) \leq -l(x)$ for all $x \in \ell_\infty$, (6.3.5) follows from Theorem 6.3.5.

Define the matrix $C = (c_{nk})$ by $c_{nk} := b_{nk} - b_{n,\sigma(k)}$ for all $k, n \in \mathbb{N}$; where b_{nk} is defined as in Theorem 6.3.5. Then, the hypothesis of Lemma 6.3.4 are satisfied. Now, choose the sequence $y = (y_k)$ such that $y_k = 0$ for $k \notin \sigma(\mathbb{N})$. Then, $(y_k - y_{\sigma(k)}) \in Z$ and also, by the same argument used in Theorem 23 of [323], one can easily see that

$$\sum_k b_{nk}[y_k - y_{\sigma(k)}] = \sum_k c_{nk} y_{\sigma(k)}. \tag{6.3.11}$$

Hence, (6.3.9) implies that

$$\begin{aligned}\mathcal{I}-\limsup_{n\to\infty}\sum_k |c_{nk}| &= \mathcal{I}-\limsup_{n\to\infty}\sum_k c_{nk}y_{\sigma(k)}\\ &= \mathcal{I}-\limsup_{n\to\infty}\sum_k b_{nk}[y_k-y_{\sigma(k)}]\\ &\le \frac{\lambda+K_{\mathcal{I}}}{2}V[y_k-y_{\sigma(k)}]+\frac{\lambda-K_{\mathcal{I}}}{2}V[y_{\sigma(k)}-y_k]=0,\end{aligned}$$

which is (6.3.10).

Conversely, suppose that (6.3.9) and (6.3.10) hold. Then, for any $x=(x_k)\in\ell_\infty$, we have (6.3.11). Hence, since $\{x_k-x_{\sigma(k)}\}\in Z$, (6.3.10) implies that $B\in(Z:F_{\mathcal{I}}(b))$ with $\mathcal{I}$-$\lim Bz=0$, $(z\in Z)$. We also see from the assumption that (6.3.4) holds. Thus, by taking infimum over $z\in Z$ in (6.3.4), we observe that

$$\begin{aligned}\inf_{z\in Z}\left\{\mathcal{I}-\limsup_{n\to\infty}\sum_k b_{nk}(x_k+z_k)\right\} &\le \frac{\lambda+K_{\mathcal{I}}}{2}L(x+z)-\frac{\lambda-K_{\mathcal{I}}}{2}l(x+z)\\ &= \frac{\lambda+K_{\mathcal{I}}}{2}W(x)+\frac{\lambda-K_{\mathcal{I}}}{2}W(-x).\end{aligned}$$

On the other hand, since $\mathcal{I}$-$\lim Bz=0$,

$$\begin{aligned}\inf_{z\in Z}\left\{\mathcal{I}-\limsup_{n\to\infty}\sum_k b_{nk}(x_k+z_k)\right\} &\ge \mathcal{I}-\limsup_{n\to\infty}\sum_k b_{nk}x_k+\\ &+\inf_{z\in Z}\left(\mathcal{I}-\limsup_{n\to\infty}\sum_k b_{nk}z_k\right)\\ &= \mathcal{I}-\limsup_{n\to\infty}\sum_k b_{nk}x_k.\end{aligned}$$

Since $W(x)=V(x)$ for all $x\in\ell_\infty$, we conclude that (6.3.9) holds and the proof is completed. □

When $K_{\mathcal{I}}>0$ and $\lambda=K_{\mathcal{I}}$, we have the following result:

Theorem 6.3.9. [118, Theorem 2.8] *Suppose that $A=(a_{nk})$ be $\mathcal{I}$-conservative. Then,*

$$\mathcal{I}-\limsup_{n\to\infty}\sum_k(a_{nk}-\alpha_k)x_k\le K_{\mathcal{I}}\,V(x)$$

for all $x=(x_k)\in\ell_\infty$ if and only if (6.3.6) and (6.3.10) hold.

Following results can be derived from Theorem 6.3.8 for the special cases $\mathcal{I}=\mathcal{I}_{\delta_{\mathcal{B}}}$ and $\mathcal{I}=\mathcal{I}_{\delta_A}$:

Theorem 6.3.10. [118, Theorem 2.9] *The following statements hold:*

(a) *Let $A=(a_{nk})\in(c:st_{\mathcal{B}}\cap\ell_\infty)$. Then, for some constant $\lambda\ge|K_{\mathcal{B}}|$ and for all $x=(x_k)\in\ell_\infty$,*

$$st_{\mathcal{B}}-\limsup_{n\to\infty}\sum_k(a_{nk}-\alpha_k)x_k\le\frac{\lambda+K_{\mathcal{B}}}{2}V(x)+\frac{\lambda-K_{\mathcal{B}}}{2}V(-x)$$

if and only if (6.3.7) holds and

$$st_{\mathcal{B}} - \lim_{n\to\infty} \sum_k \left| a_{nk} - a_{n,\sigma(k)} - \alpha_k + \alpha_{\sigma(k)} \right| = 0.$$

(b) *Let $A = (a_{nk}) \in (c : st_A \cap \ell_\infty)$. Then, for some constant $\lambda \geq |K_A|$ and for all $x = (x_k) \in \ell_\infty$,*

$$st_A - \limsup_{n\to\infty} \sum_k (a_{nk} - \alpha_k)x_k \leq \frac{\lambda + K_A}{2} V(x) + \frac{\lambda - K_A}{2} V(-x)$$

if and only if (6.3.8) holds and

$$st_A - \lim_{n\to\infty} \sum_k \left| a_{nk} - a_{n,\sigma(k)} - \alpha_k + \alpha_{\sigma(k)} \right| = 0.$$

Further; for $\mathcal{I} = \mathcal{I}_\delta$, Theorem 6.3.8 is reduced to Theorem 2.7 of Çoşkun and Çakan [137].

Theorem 6.3.11. [118, Theorem 2.10] *Let $A = (a_{nk})$ and λ be as in Theorem 6.3.5. Then, for all $x = (x_k) \in \ell_\infty$*

$$\mathcal{I} - \limsup_{n\to\infty} \sum_k (a_{nk} - \alpha_k)x_k \leq \frac{\lambda + K_{\mathcal{I}}}{2} \gamma(x) + \frac{\lambda - K_{\mathcal{I}}}{2} \gamma(-x) \tag{6.3.12}$$

if and only if (6.3.5) holds and

$$\mathcal{I} - \lim_{n\to\infty} \sum_{k\in E} |a_{nk} - \alpha_k| = 0 \tag{6.3.13}$$

for every $E \in \mathcal{I}$, where $\gamma(x) = \mathcal{I} - \limsup_{k\to\infty} x_k$.

Proof. If (6.3.12) holds, since $\gamma(x) \leq L(x)$ and $\gamma(-x) \leq -l(x)$, (6.3.5) follows from Theorem 6.3.5. To show the necessity of (6.3.13), for any $E \in \mathcal{I}$, let us define the matrix $D = (d_{nk})$ by $d_{nk} := a_{nk} - \alpha_k$, $k \in E$; otherwise $= 0$ for all $n \in \mathbb{N}$. Then, clearly, D satisfies the conditions of Lemma 6.3.4 and therefore, there exists a $y = (y_k) \in \ell_\infty$ such that $\|y\| \leq 1$ and

$$\mathcal{I} - \limsup_{n\to\infty} \sum_k d_{nk} y_k = \mathcal{I} - \limsup_{n\to\infty} \sum_k |d_{nk}|.$$

Now, for the same E we define the sequence $y = (y_k)$ by

$$y_k := \begin{cases} 1 & , \quad k \in E, \\ 0 & , \quad k \notin E. \end{cases}$$

Then, since $\mathcal{I}\text{-}\lim y = \gamma(y) = \gamma(-y) = 0$, (6.3.12) implies that

$$\mathcal{I} - \limsup_{n\to\infty} \sum_{k\in E} |d_{nk}| \leq \frac{\lambda + K_{\mathcal{I}}}{2} \gamma(y) + \frac{\lambda - K_{\mathcal{I}}}{2} \gamma(-y) = 0,$$

which is (6.3.13).

Conversely, suppose that the conditions of the theorem hold and $x = (x_k) \in \ell_\infty$. Let $E_1 = \{k : x_k > \gamma(x) + \varepsilon\}$ and $E_2 = \{k : x_k < \gamma(x) - \varepsilon\}$. Then, since $E_1, E_2 \in \mathcal{I}$, $E = E_1 \cap E_2 \in \mathcal{I}$. Now, we can write

$$\sum_k (a_{nk} - \alpha_k)x_k := \sum_{k\in E} (a_{nk} - \alpha_k)x_k + \sum_{k\notin E} (a_{nk} - \alpha_k)^+ x_k - \sum_{k\notin E} (a_{nk} - \alpha_k)^- x_k.$$

Thus, by (6.3.13) and Lemma 6.3.3, (6.3.12) is obtained since

$$\mathcal{I} - \limsup_{n\to\infty} \sum_k (a_{nk} - \alpha_k)x_k \le \frac{\lambda + K_{\mathcal{I}}}{2}\gamma(x) + \frac{\lambda - K_{\mathcal{I}}}{2}\gamma(-x) + \lambda\varepsilon$$

and ε is arbitrary. □

When $K_{\mathcal{I}} > 0$ and $\lambda = K_{\mathcal{I}}$, we have the following result:

Theorem 6.3.12. [118, Theorem 2.11] *Let $A = (a_{nk})$ be $\mathcal{I}$-conservative. Then, for all $x = (x_k) \in \ell_\infty$,*

$$\mathcal{I} - \limsup_{n\to\infty} \sum_k (a_{nk} - \alpha_k)x_k \le K_{\mathcal{I}}\, \gamma(x)$$

if and only if (6.3.6) and (6.3.13) hold.

If we choose $\mathcal{I} = \mathcal{I}_{\delta_{\mathcal{B}}}$ and $\mathcal{I} = \mathcal{I}_{\delta_A}$ in Theorem 6.3.11 then, we have

Theorem 6.3.13. [118, Theorem 2.12] *The following statements hold:*

(a) *Let $A = (a_{nk}) \in (c : st_{\mathcal{B}} \cap \ell_\infty)$. Then, for some constant $\lambda \ge |K_{\mathcal{B}}|$ and for all $x = (x_k) \in \ell_\infty$,*

$$st_{\mathcal{B}} - \limsup_{n\to\infty} \sum_k (a_{nk} - \alpha_k)x_k \le \frac{\lambda + K_{\mathcal{B}}}{2}\gamma(x) + \frac{\lambda - K_{\mathcal{B}}}{2}\gamma(-x)$$

if and only if (6.3.7) holds and

$$st_{\mathcal{B}} - \lim_{n\to\infty} \sum_{k\in E} |a_{nk} - \alpha_k| = 0 \quad \text{for every} \quad E \in \mathcal{I}.$$

(b) *Let $A = (a_{nk}) \in (c : st_A \cap \ell_\infty)$. Then, for some constant $\lambda \ge |K_A|$ and for all $x = (x_k) \in \ell_\infty$,*

$$st_A - \limsup_{n\to\infty} \sum_k (a_{nk} - \alpha_k)x_k \le \frac{\lambda + K_A}{2}\gamma(x) + \frac{\lambda - K_A}{2}\gamma(-x)$$

if and only if (6.3.8) holds and

$$st_A - \lim_{n\to\infty} \sum_{k\in E} |a_{nk} - \alpha_k| = 0 \text{ for every } E \in \mathcal{I}.$$

Moreover, Theorem 6.3.11 corresponds to Theorem 2.6 of Çoşkun and Çakan [137] for $\mathcal{I} = \mathcal{I}_\delta$.

6.4 $F_{\mathcal{B}}$-CORE

Let $\mathcal{B}$ be a sequence of infinite matrices $B^i = \{b_{mn}(i)\}$. For a given sequence $x = (x_n)$, we write $(Bx)^i_m = \sum_n b_{mn}(i)x_n$ if it exits for each $m, i \in \mathbb{N}$. We also write $\mathcal{B}x$ for $\{(Bx)^i_m\}^{\infty}_{i,m=0}$. A sequence $x = (x_k) \in \ell_\infty$ is said to be $F_{\mathcal{B}}$-convergent (cf. Stieglitz [367]) to a number s if

$$\lim_{m\to\infty}(Bx)^i_m = \lim_{m\to\infty}\sum_n b_{mn}(i)x_n = s, \text{ uniformly in } i.$$

In this case, we write $F_{\mathcal{B}} - \lim x = s$. By $F_{\mathcal{B}}$ and $F_{0\mathcal{B}}$, we denote the space of all $F_{\mathcal{B}}$-convergent and $F_{\mathcal{B}}$-null sequences, respectively. The space $F_{\mathcal{B}}$ depends on choosing the sequence $\mathcal{B} = (B^i)$ of infinite matrices. For example, if we define $B^i = I$ for all $i \in \mathbb{N}$, then $F_{\mathcal{B}} = c$. If we define $\mathcal{B} = \{b_{mn}(i)\}$ by

$$b_{mn}(i) := \begin{cases} \frac{1}{m+1} & , \quad n = \sigma^j(i),\ 0 \le j \le m, \\ 0 & , \quad \text{otherwise}, \end{cases} \tag{6.4.1}$$

then $F_{\mathcal{B}} = V_\sigma$.

Let $\mathcal{B} = \{b_{mn}(i)\}$ be a sequence of matrices such that

$$t_{mi}(x) = \sum_n b_{mn}(i)x_n$$

exists for all $i, m \in \mathbb{N}$. Now, we can define $F_{\mathcal{B}}$-*core* of a sequence of complex numbers as follows (see [121]):

Definition 6.4.1. [121, Definition 3.1] *Let $C_{mi}(x)$ be a smallest closed convex subset of $\mathbb{C}$ containing $\{t_{mi}(x), t_{m+1,i}(x), \ldots\}$. Then,*

$$F_{\mathcal{B}}\text{-}core(x) := \bigcap_{m=1}^{\infty}\left[\bigcup_{i=1}^{\infty} C_{mi}(x)\right].$$

We should note here that if we define $F_{\mathcal{B}}$-limit point of a sequence $x = (x_k)$ as the $F_{\mathcal{B}}$-limit of a subsequence of $x = (x_k)$, then clearly $F_{\mathcal{B}}$-*core*(x) contains all of $F_{\mathcal{B}}$-limit points of $x = (x_k)$.

Now, we characterize the $F_{\mathcal{B}}$-*core* of a complex bounded sequence:

Theorem 6.4.2. [121, Theorem 3.2] *For any $z \in \mathbb{C}$, define the set $G_x(z)$ by*

$$G_x(z) := \left\{u \in \mathbb{C} : |u - z| \le \limsup_{m\to\infty}\sup_{i\in\mathbb{N}}|t_{mi}(x) - z|\right\}.$$

Then, $F_{\mathcal{B}} - core(x) = \bigcap_{z\in\mathbb{C}} G_x(z)$.

Proof. Let $E := \bigcap_{p\in\mathbb{N}}\left[\bigcap_{n\in\mathbb{N}} G_{mi}(x)\right]$ and $F := \bigcap_{z\in\mathbb{C}} G_x(z)$. We will show that $E = F$ for all $x = (x_k) \in \ell_\infty$.

Let $u \notin E$. Then, $u \notin G_{mi}(x)$ for all $m \in \mathbb{N}$. Hence, for any $z \in \mathbb{C}$, $|u - z| \geq |t_{mi}(x) - z|$ and $u \notin G_x(z)$. Thus, $u \notin F$ and consequently $E \subset F$.

Now, let $u \notin F$. Then, for any $z \in \mathbb{C}$, $|u - z| \geq \limsup_{m\to\infty} \sup_{i\in\mathbb{N}} |t_{mi}(x) - z|$. Hence, since $t_{mi}(x) \in G_{mi}(x)$, $u \notin G_{mi}(x)$ for all $p, n \in \mathbb{N}$. Thus, $u \notin E$ and $F \subset E$.

This completes the proof. □

Remark 6.4.3. [121, Remark 3.3] *We note the following special cases:*

(i) *In the case $\mathcal{B} = (I)$, $F_{\mathcal{B}}$-core is reduced to the Knopp core.*

(ii) *When $\mathcal{B}$ is chosen as in (6.4.1), $F_{\mathcal{B}}$-core is reduced to σ-core.*

Now, we may give the inclusion theorems.

Theorem 6.4.4. [121, Theorem 3.4] *$F_{\mathcal{B}}$-core$(Ax) \subseteq \mathcal{K} -$ core(x) for all $x = (x_k) \in \ell_\infty$ if and only if $A = (a_{nk}) \in (c : F_{\mathcal{B}})_{reg}$* (see [281]) *and*

$$\lim_{m\to\infty} \sup_{i\in\mathbb{N}} \sum_k \left| \sum_n b_{mn}(i) a_{nk} \right| = 1. \tag{6.4.2}$$

Proof. Let $F_{\mathcal{B}}$-core$(Ax) \subseteq \mathcal{K} -$ core(x) and take $x = (x_k) \in c$ with $\lim x = l$. Then, since $\mathcal{K} -$ core$(x) = \{l\}$, $F_{\mathcal{B}}$-core$(Ax) \subseteq \{l\}$. So, $F_{\mathcal{B}} - \lim Ax = l$ which means that $A \in (c : F_{\mathcal{B}})_{reg}$.

Now, if it is possible, assume that (6.4.2) does not hold. Then, since $A = (a_{nk}) \in (c : F_{\mathcal{B}})_{reg}$,

$$\lim_{m\to\infty} \sup_{i\in\mathbb{N}} \sum_k \left| \sum_n b_{mn}(i) a_{nk} \right| > 1. \tag{6.4.3}$$

Using the conditions of Schur's theorem, we can choose two strictly increasing sequences $\{n(j)\}$ and $\{k[n(j)]\}$ of positive integers such that

$$\sum_{k=0}^{k[n(j-1)]} \left| \sum_n b_{m,n(j)}(i) a_{nk} \right| < \frac{1}{j}, \qquad \sum_{k=k[n(j-1)]+1}^{k[n(j)]} \left| \sum_n b_{m,n(j)}(i) a_{nk} \right| > 1 + \frac{2}{j}$$

and

$$\sum_{k>k[n(j-1)]+1} \left| \sum_n b_{m,n(j)}(i) a_{nk} \right| < \frac{1}{j}.$$

If we define a sequence $x = (x_k)$ by

$$x_k = \operatorname{sgn}\left\{ \sum_n b_{m,n(j)}(i) a_{nk} \right\}, \quad k[n(j-1)] \leq k \leq k[n(j)], \ j \in \mathbb{N}$$

then, $\mathcal{K} - core(x) \subseteq B_1(0)$ and

$$\begin{aligned}\left|\sum_k\sum_n b_{m,n(j)}(i)a_{n(j),k}x_k\right| &= \left|\sum_k\sum_n b_{m,n(j)}(i)a_{n(j),k}x_k\right| \\ &\geq \sum_{k=k[n(j-1)]+1}^{k[n(j)]}\left|\sum_n b_{m,n(j)}(i)a_{nk}\right| - \\ &- \sum_{k=0}^{k[n(j-1)]}\left|\sum_n b_{m,n(j)}(i)a_{nk}\right| - \\ &- \sum_{k>k[n(j-1)]+1}\left|\sum_n b_{m,n(j)}(i)a_{nk}\right| \\ &> 1+\frac{2}{j}-\frac{1}{j}-\frac{1}{j}=1.\end{aligned}$$

This means that $F_{\mathcal{B}}$-limit point of $\{(Ax)_{n(j)}\}$ cannot be in $C_{mi}(x)$ and so in $B_1(0)$ which contradicts the assumption. Therefore, the condition (6.4.2) must hold.

Conversely, let $u \in F_{\mathcal{B}}\text{-}core(Ax)$. Then, from Theorem 6.4.2 and by the hypothesis, we have

$$\begin{aligned}|u-z| &\leq \limsup_{m\to\infty}\sup_{i\in\mathbb{N}}\left|\sum_k\sum_n b_{mn}(i)a_{nk}x_k - z\right| \qquad (6.4.4)\\ &= \limsup_{m\to\infty}\sup_{i\in\mathbb{N}}\left|z-\sum_k\sum_n b_{mn}(i)a_{nk}x_k\right| \\ &\leq \limsup_{m\to\infty}\sup_{i\in\mathbb{N}}\left|\sum_k\sum_n b_{mn}(i)a_{nk}(z-x_k)\right| + \\ &+ |z|\limsup_{m\to\infty}\sup_{i\in\mathbb{N}}\left|1-\sum_k\sum_n b_{mn}(i)a_{nk}\right| \\ &= \limsup_{m\to\infty}\sup_{i\in\mathbb{N}}\left|\sum_k\sum_n b_{mn}(i)a_{nk}(z-x_k)\right|.\end{aligned}$$

Now, if $L(x) = \limsup_{k\to\infty}|x_k - z|$, for any given $\varepsilon > 0$, then we can write $|x_k - z| \leq L(x) + \varepsilon$ whenever $k \geq k_0$. Hence,

$$\begin{aligned}&\left|\sum_k\sum_n b_{mn}(i)a_{nk}(z-x_k)\right| = \\ &= \left|\sum_{k<k_0}\sum_n b_{mn}(i)a_{nk}(z-x_k) + \sum_{k\geq k_0}\sum_n b_{mn}(i)a_{nk}(z-x_k)\right| \\ &\leq \|(z-x_k)\|\sum_{k<k_0}\left|\sum_n b_{mn}(i)a_{nk}\right| + [L(x)+\varepsilon]\sum_{k\geq k_0}\left|\sum_n b_{mn}(i)a_{nk}\right|\end{aligned}$$

and by the hypothesis, we have $|u - z| \leq L(x) + \varepsilon$.

Since ε is arbitrary, this step completes the proof. □

Theorem 6.4.5. [121, Theorem 3.5] *$F_{\mathcal{B}}$-core$(Ax) \subseteq \sigma$-core(x) for all $x = (x_k) \in \ell_\infty$ if and only if $A = (a_{nk}) \in (V_\sigma : F_{\mathcal{B}})_{reg}$* (see [136]) *and (6.4.2) holds.*

Proof. Suppose that $F_{\mathcal{B}}$-core$(Ax) \subseteq \sigma$-core(x) for all $x = (x_k) \in \ell_\infty$. Since σ-core$(x) = \{l\}$ whenever $\sigma - \lim x = l$, $F_{\mathcal{B}}$-core$(Ax) \subseteq \{l\}$. So, $F_{\mathcal{B}} - \lim Ax = l$ and therefore $A = (a_{nk}) \in (V_\sigma : F_{\mathcal{B}})_{reg}$.

On the other hand, since σ-core$(x) \subseteq \mathcal{K} - core(x)$ for any sequence $x = (x_k)$, the condition (6.4.2) follows from Theorem 6.4.4.

Conversely; suppose that $A = (a_{nk}) \in (V_\sigma : F_{\mathcal{B}})_{reg}$ and (6.4.2) holds. Take $u \in F_{\mathcal{B}}$-core(Ax). Then, by Lemma 6.3.3, since $A = (a_{nk}) \in (c : F_{\mathcal{B}})_{reg}$, Theorem 6.4.4 implies that $u \in \mathcal{K} - core(x)$. This means that

$$|u - z| \leq \limsup_{k\to\infty} |x_k - z| \tag{6.4.5}$$

for any $z \in \mathbb{C}$. On the other hand, since

$$|x_k - z| \leq |x_k - t_{pn}(x)| + |t_{pn}(x) - z|$$

and $\{x_k - x_{\sigma(k)}\} \in Z$ whenever $x = (x_k) \in \ell_\infty$, we have

$$\limsup_{k\to\infty} |x_k - z| \leq \limsup_{p\to\infty} \sup_{n\in\mathbb{N}} |t_{pn}(x) - z|. \tag{6.4.6}$$

Combining (6.4.5) with (6.4.6), we get that

$$|u - z| \leq \limsup_{p\to\infty} \sup_{n\in\mathbb{N}} |t_{pn}(x) - z|.$$

This means that $u \in \sigma$-core(x) which completes the proof. □

Theorem 6.4.6. [121, Theorem 3.6] *$F_{\mathcal{B}}$-core$(Ax) \subseteq st$-core(x) for all $x = (x_k) \in \ell_\infty$ if and only if $A = (a_{nk}) \in (st \cap \ell_\infty : F_{\mathcal{B}})_{reg}$* (see [136]) *and (6.4.2) holds, where st denotes the space of all statistically convergent sequences.*

Proof. Firstly, we assume that $F_{\mathcal{B}}$-core$(Ax) \subseteq st$-core(x) for all $x = (x_k) \in \ell_\infty$. By taking $x = (x_k) \in st \cap \ell_\infty$, one can see that $A = (a_{nk}) \in (st \cap \ell_\infty : F_{\mathcal{B}})_{reg}$. Also, since st-core$(x) \subseteq \mathcal{K} - core(x)$, (cf. [177]) for any $x = (x_k)$, the necessity of the condition (6.4.2) follows from Theorem 6.4.4.

Conversely; let $A = (a_{nk}) \in (st \cap \ell_\infty : F_{\mathcal{B}})_{reg}$, (6.4.2) holds and take $u \in F_{\mathcal{B}}$-core. Then, we again have the relation (6.4.4).

Now, let $\beta = st - \limsup_{k\to\infty} |z - x_k|$. If we write $E = \{k : |z - x_k| \geq \beta + \varepsilon\}$, then $\delta(E) = 0$ and $|z - x_k| \leq \beta + \varepsilon$ whenever $k \notin E$. Hence, we have

$$\left|\sum_k \sum_n b_{mn}(i)a_{nk}(z - x_k)\right| =$$

$$= \left|\sum_{k\in E} \sum_n b_{mn}(i)a_{nk}(z - x_k) + \sum_{k\notin E} \sum_n b_{mn}(i)a_{nk}(z - x_k)\right|$$

$$\leq \sum_{k\in E} \left|\sum_n b_{mn}(i)a_{nk}\right| |z - x_k| + \sum_{k\notin E} \left|\sum_n b_{mn}(i)a_{nk}\right| |z - x_k|$$

$$\leq \|(z - x_k)\| \sum_{k\in E} \left|\sum_n b_{mn}(i)a_{nk}\right| + (\beta + \varepsilon) \sum_{k\notin E} \left|\sum_n b_{mn}(i)a_{nk}\right|.$$

Applying the operator $\limsup_{m\to\infty} \sup_{i\in\mathbb{N}}$ to the above inequality and using the hypothesis, we obtain that

$$\limsup_{m\to\infty} \sup_{i\in\mathbb{N}} \left|\sum_k \sum_n b_{mn}(i)a_{nk}(z - x_k)\right| \leq \beta + \varepsilon. \tag{6.4.7}$$

Thus, (6.4.4) and (6.4.7) imply that $|u - z| \leq \beta + \varepsilon$. Since ε is arbitrary, this means $u \in$ *st-core*(x) which completes the proof. □

Finally, we should note that

(i) Theorem 6.4.4 gives the inclusion $\mathcal{K}-core(Ax) \subseteq \mathcal{K}-core(x)$ (see Shcherbakov [361]), when $\mathcal{B} = (I)$.

(ii) Theorem 6.4.4 leads the inclusion σ-$core(Ax) \subseteq \mathcal{K} - core(x)$, when $\mathcal{B}$ is chosen as in (6.4.1).

(iii) Theorem 6.4.5 provides the inclusion σ-$core(Ax) \subseteq \sigma$-$core(x)$, when $\mathcal{B}$ is chosen as in (6.4.1).

(iv) Theorem 6.4.6 implies the inclusion $\mathcal{K} - core(Ax) \subseteq st$-$core(x)$, (see Fridy and Orhan [177]), when $\mathcal{B} = (I)$.

(v) Theorem 6.4.6 yields the inclusion σ-$core(Ax) \subseteq st$-$core(x)$, when $\mathcal{B}$ is chosen as in (6.4.1).

For more results on the core of sequences, the reader may refer to [11, 114, 116, 117, 120, 119, 136, 134, 138, 153, 242, 292, 309] and [365].

CHAPTER 7

Double Sequences

In this chapter, the fundamental results on double sequences and related topics are given. In particular, the concept of convergence of double series in the Pringsheim's sense is defined, certain spaces of double sequences are introduced, and their α- and β-duals are determined. Additionally, some classes of four-dimensional matrices are characterized.

Keywords: Double sequence, convergence of a double sequence in the Pringsheim's sense, boundedness of a double sequence, regular convergence of a double sequence, Pringsheim, regular, bounded and absolute convergence of a double series, Silverman-Toeplitz and Steinhaus theorems for double sequences, convergence tests for double series, Cauchy product of double series, α- and β-duals of a space of double sequences, characterization of some classes of four-dimensional matrices.

7.1 PRELIMINARIES, BACKGROUND AND NOTATIONS

By Ω, we denote the set of all complex valued double sequences, i.e.,

$$\Omega := \{x = (x_{mn}) : x_{mn} \in \mathbb{C} \text{ for all } m, n \in \mathbb{N}\},$$

which is a vector space with co-ordinatewise addition and scalar multiplication of double sequences. Any vector subspace of Ω is called as a *double sequence space.* The *space $\mathcal{M}_u$ of all bounded double sequences* is defined by

$$\mathcal{M}_u := \left\{x = (x_{mn}) \in \Omega : \|x\|_\infty = \sup_{m,n\in\mathbb{N}} |x_{mn}| < \infty\right\},$$

which is a Banach space endowed with the norm $\|\cdot\|_\infty$. Consider the sequence $x = (x_{mn}) \in \Omega$. If for every $\varepsilon > 0$ there exists $n_0 = n_0(\varepsilon) \in \mathbb{N}$ and $l \in \mathbb{C}$ such that

$$|x_{mn} - l| < \varepsilon$$

for all $m, n > n_0$, then we call that the double sequence x is convergent in the Pringsheim's sense to the limit l and write $p-\lim_{m,n\to\infty} x_{mn} = l$. By $\mathcal{C}_p$, we denote the *space of all convergent double sequences in the Pringsheim's sense.* It is well-known that there are such sequences in the space $\mathcal{C}_p$ but not in the space $\mathcal{M}_u$. So,

DOI: 10.1201/9781003294153-7

we may mention the space $\mathcal{C}_{bp}$ of the double sequences which are both convergent in the Pringsheim's sense and bounded, i.e., $\mathcal{C}_{bp} := \mathcal{C}_p \cap \mathcal{M}_u$. Móricz [280] proved that $\mathcal{C}_{bp}$ is a Banach space with the norm $\|\cdot\|_\infty$. A sequence in the space $\mathcal{C}_p$ is said to be *regularly convergent* if it is a single convergent sequence with respect to each index and denote the set of all such sequences by $\mathcal{C}_r$.

Let us consider a double sequence $x = (x_{mn})$ and define the sequence $s = (s_{mn})$ via x by

$$s_{mn} := \sum_{i,j=0}^{m,n} x_{ij} \quad \text{for all} \quad m, n \in \mathbb{N}, \tag{7.1.1}$$

which will be used throughout. For the sake of brevity, here and in what follows, we abbreviate the summations $\sum_{i=0}^{\infty}\sum_{j=0}^{\infty}$ and $\sum_{i=0}^{m}\sum_{j=0}^{n}$ by $\sum_{i,j}$ and $\sum_{i,j=0}^{m,n}$, and use the convention that any term with negative subscript is equal to naught. Then, the pair (x, s) and the sequence $s = (s_{mn})$ are called as a *double series* and the *sequence of partial sums* of the double series, respectively. Let λ be the space of double sequences, converging with respect to some linear convergence rule $v - \lim : \lambda \to \mathbb{C}$. The sum of a double series $\sum_{i,j} x_{ij}$ with respect to this rule is defined by $v - \sum_{ij} x_{ij} = v - \lim_{m,n\to\infty} s_{mn}$. Let λ, μ be two spaces of double sequences, converging with respect to the linear convergence rules $v_1 - \lim$ and $v_2 - \lim$, respectively, and $A = (a_{mnkl})$ also be a four-dimensional matrix over the real or complex field. Define the set

$$\lambda_A^{(v_2)} := \left\{ x = (x_{kl}) \in \Omega : Ax = \left(v_2 - \sum_{k,l} a_{mnkl} x_{kl} \right)_{m,n\in\mathbb{N}} \text{ exists and is in } \lambda \right\}. \tag{7.1.2}$$

Then, we say, with the notation of (7.1.2), that A maps the space λ into the space μ if $\lambda \subset \mu_A$ and denote the set of all four-dimensional matrices, mapping the space λ into the space μ, by $(\lambda : \mu)$. It is trivial that for any matrix $A \in (\lambda : \mu)$, $(a_{mnkl})_{k,l\in\mathbb{N}}$ is in the $\beta(v_2)$-dual $\lambda^{\beta(v_2)}$ of the space λ for all $m, n \in \mathbb{N}$. An infinite matrix A is said to be $\mathcal{C}_v$*-conservative* if $\mathcal{C}_v \subset (\mathcal{C}_v)_A$. Also by $(\lambda : \mu; p)$, we denote the class of all four-dimensional matrices (a_{mnkl}) in the class $(\lambda : \mu)$ such that $v_2 - \lim Ax = v_1 - \lim x$ for all $x \in \lambda$. We shall write throughout for simplicity in notation for all $m, n, k, l \in \mathbb{N}$ that

$$\begin{aligned}
\Delta_{10} a_{mn} &:= a_{mn} - a_{m+1,n}, & \Delta_{10}^{kl} a_{mnkl} &:= a_{mnkl} - a_{mn,k+1,l}, \\
\Delta_{01} a_{mn} &:= a_{mn} - a_{m,n+1}, & \Delta_{01}^{kl} a_{mnkl} &:= a_{mnkl} - a_{mnk,l+1}, \\
\Delta_{11} a_{mn} &:= \Delta_{01}(\Delta_{10} a_{mn}), & \Delta_{11}^{kl} a_{mnkl} &:= \Delta_{01}^{kl}(\Delta_{10}^{kl} a_{mnkl}) \\
&:= \Delta_{10}(\Delta_{01} a_{mn}) & &:= \Delta_{10}^{kl}(\Delta_{01}^{kl} a_{mnkl}).
\end{aligned}$$

Let us define the following sets of double sequences:

$$\begin{aligned}
\mathcal{M}_u(t) &:= \left\{x = (x_{mn}) \in \Omega : \sup_{m,n\in\mathbb{N}} |x_{mn}|^{t_{mn}} < \infty\right\}, \\
\mathcal{C}_p(t) &:= \left\{x = (x_{mn}) \in \Omega : p - \lim_{m,n\to\infty} |x_{mn} - \ell|^{t_{mn}} = 0 \text{ for some } \ell \in \mathbb{C}\right\}, \\
\mathcal{C}_{0p}(t) &:= \left\{x = (x_{mn}) \in \Omega : p - \lim_{m,n\to\infty} |x_{mn}|^{t_{mn}} = 0\right\}, \\
\mathcal{L}_u(t) &:= \left\{x = (x_{mn}) \in \Omega : \sum_{m,n} |x_{mn}|^{t_{mn}} < \infty\right\}, \\
\mathcal{C}_{bp}(t) &:= \mathcal{C}_p(t) \cap \mathcal{M}_u(t) \text{ and } \mathcal{C}_{0bp}(t) := \mathcal{C}_{0p}(t) \cap \mathcal{M}_u(t);
\end{aligned}$$

where $t = (t_{mn})$ is the sequence of strictly positive reals t_{mn}. In the case $t_{mn} = 1$ for all $m, n \in \mathbb{N}$; $\mathcal{M}_u(t)$, $\mathcal{C}_p(t)$, $\mathcal{C}_{0p}(t)$, $\mathcal{L}_u(t)$, $\mathcal{C}_{bp}(t)$ and $\mathcal{C}_{0bp}(t)$ reduce to the sets $\mathcal{M}_u$, $\mathcal{C}_p$, $\mathcal{C}_{0p}$, $\mathcal{L}_u$, $\mathcal{C}_{bp}$ and $\mathcal{C}_{0bp}$, respectively. Now, we may survey the information given in some document about the double sequences. Türkmenoğlu [389] showed under which conditions that $\mathcal{C}_{0p}(t)$, $\mathcal{L}_u(t)$ and $\mathcal{C}_{0bp}(t)$ are the paranormed double sequence spaces, determined their duals and gave some inclusion relations between those spaces. The summability of double sequences defining by the product of the complex sequences $x = (x_n)$ and $y = (y_k)$, Jardas and Sarapa [205] proved the Silverman-Toeplitz and Steinhaus type theorems for three-dimensional matrices. Boos et al. [95] defined the concept of $\mathcal{V}$-SM-method by the application domain of a matrix sequence $\mathcal{A} = (\mathcal{A}^{(v)})$ of infinite matrices and gave the consistency theory for such type methods and introduced the notion of $\mathcal{C}_e$-convergence for double sequences. By $\mathcal{C}_e$ and $\mathcal{C}_{be}$, we denote the spaces of all $\mathcal{C}_e$-convergent and of all bounded $\mathcal{C}_e$-convergent double sequences, respectively. By using the gliding hump method, Zeltser [409] has recently characterized the classes of four-dimensional matrix mappings from λ into μ; where $\lambda, \mu \in \{\mathcal{C}_e, \mathcal{C}_{be}\}$. Also employing the same arguments, Zeltser [411] gives the theorems determining the necessary and sufficient conditions for $\mathcal{C}_e$-SM and $\mathcal{C}_{be}$-SM-methods to be conservative and coercive. Zeltser [412] considers the dual pairs $\langle E, E^{\beta(v)}\rangle$ of double sequence spaces E and $E^{\beta(v)}$, where $E^{\beta(v)}$ denotes the β-dual of E with respect to v-convergence of double sequences for $v \in \{p, bp, r\}$ and introduces two oscillating properties for a double sequence space E. Also, Zeltser [413] emphasizes two types of summability methods of double sequences defined by four-dimensional matrices which preserves the regular convergence and the $\mathcal{C}_c$-convergence of double sequences and extends some well-known facts of summability to four-dimensional matrices. The reader may refer to Robison [335] and Hamilton [197] for the characterizations of some four-dimensional matrix transformations between double sequence spaces. By using the definitions of limit inferior, limit superior and the core of a double sequence with the notion of the regularity of four-dimensional matrices, Patterson [314] proves an invariant core theorem. Also, Patterson [315] determines the sufficient conditions on a four-dimensional matrix in order to be stronger than the convergence in the Pringsheim's sense and derives some results concerning with the summability of double sequences. Mursaleen and Edely [293] recently introduce the statistical convergence and Cauchy for double sequences and give the relation between statistical convergent and strongly Cesàro summable double sequences. Next, Mursaleen [284] and Mursaleen and Edely [294] define the almost strong regularity of matrices for double sequences, and apply these matrices to establish a core theorem and introduce the M-core for double sequences and determined those four-dimensional matrices transforming every bounded double sequence $x = (x_{jk})$ into one whose core is a subset of the M-core of x. Quite recently, Gökhan and Çolak [190, 191] prove that $\mathcal{M}_u(t)$ and $\mathcal{C}_p(t)$, $\mathcal{C}_{bp}(t)$ are complete paranormed double sequence spaces and give the α-, β-, γ-duals of the spaces $\mathcal{M}_u(t)$ and $\mathcal{C}_{bp}(t)$.

Let us consider the isomorphism T which plays an essential role for the next section and enables carrying the problem related with a double sequence space to the corresponding single sequence space, defined by Zeltser [410, p. 36] as

$$\begin{array}{rcll} T & : & \Omega & \longrightarrow \ \omega \\ & & x & \longmapsto \ z = (z_i) = (x_{\psi^{-1}(i)}), \end{array} \tag{7.1.3}$$

where $\psi : \mathbb{N} \times \mathbb{N} \to \mathbb{N}$ is a bijection defined by

$$\begin{aligned} \psi[(0,0)] &= 1, \\ \psi[(0,1)] &= 2, \psi[(1,1)] = 3, \psi[(1,0)] = 4, \\ &\vdots \\ \psi[(0,n)] &= (n-1)^2 + 1, \psi[(1,n)] = (n-1)^2 + 2, \ldots, \\ \psi[(n,n)] &= (n-1)^2 + n, \psi[(n,n-1)] = n^2 - n + 2, \ldots, \psi[(n,0)] = n^2, \\ &\vdots \end{aligned}$$

7.2 PRINGSHEIM CONVERGENCE OF DOUBLE SERIES

In this section, following Limaye and Zeltser [243], we give some results on the Pringsheim convergence of double series. Since Pringsheim introduces the notion of convergence of a numerical double series in terms of the convergence of the double sequence of its rectangular partial sums in [320], several authors contribute to this topic during the last century. However, an exhaustive treatment giving analogues of all well-known convergence aspects of single series seems to be unavailable. The purpose of this section is to fill in some of the gaps in such a treatment, and also to point out some errors in previous attempts to obtain results exactly analogous to those of a single series. In first, we give some tests for absolute convergence of a double series including analogues of Cauchy's condensation test, Abel's kth term test, limit comparison test, ratio test, ratio comparison test, and Raabe's test. Next, we give necessary and sufficient conditions on a double sequence (a_{kl}) in order that the Cauchy product double series $\sum_{k,l} a_{kl} * b_{kl}$ would be convergent/boundedly convergent/regularly convergent whenever a double series $\sum_{k,l} b_{kl}$ is convergent/boundedly convergent/regularly convergent. It is also shown that if two double series are boundedly convergent, then the Cauchy product double series is Cesàro summable and its Cesàro sum is equal to the product of the sums of the given double series. Some results derived in this section are compared with those obtained previously and given several examples to which the new results apply. Although it is considered, for simplicity, only double series with real terms, the consequences can be extended to double series with complex terms.

Throughout this section, by $\mathbb{N}_1^2$ and $\mathbb{N}^2$, we denote the set of all pairs of positive integers and of all pairs of non-negative integers, respectively. We shall use the partial order on $\mathbb{N}^2$ given by "$(k_1, l_1) \leq (k_2, l_2)$ if and only if $k_1 \leq k_2$ and $l_1 \leq l_2$". Monotonicity of a double sequence is defined in terms of this partial order. We shall adopt Pringsheim's definition of convergence of a double series $\sum_{k,l} a_{kl}$ of real numbers: $\sum_{k,l} a_{kl}$ is said to be convergent if the double sequence (s_{mn}) of its partial sums is convergent in the sense of Pringsheim, that is, there is $s \in \mathbb{R}$ such that for every $\varepsilon > 0$, there is $(m_0, n_0) \in \mathbb{N}^2$ satisfying $(m, n) \geq (m_0, n_0) \Rightarrow |s_{mn} - s| < \varepsilon$. When every a_{kl} is non-negative, $\sum_{k,l} a_{kl}$ is convergent if and only if (s_{mn}) is bounded, above. For each fixed $k \in \mathbb{N}$, the series $\sum_l a_{kl}$ is called a *row-series*, and for each fixed $l \in \mathbb{N}$, the series $\sum_k a_{kl}$ is called a *column-series* corresponding to the double series $\sum_{k,l} a_{kl}$.

7.2.1 Absolute Convergence

If a double series is absolutely convergent, then evidently the corresponding row-series and the column-series are all absolutely convergent. However, the converse is not true, as can be seen by considering $\sum_{k,l} a_{kl}$, where $a_{kl} := \begin{cases} 1 & , \quad k = l, \\ 0 & , \quad k \neq l. \end{cases}$ The following result gives the necessary and sufficient conditions for the absolute convergence of a double series:

Lemma 7.2.1. [243, Lemma 2.1] *A double sequence $(a_{kl}) \in \mathcal{L}_u$ if and only if the following conditions hold:*

(i) *There are $(k_0, l_0) \in \mathbb{N}^2$ and $\alpha_0 > 0$ such that $\sum_{k,l=k_0,l_0}^{m,n} |a_{kl}| \leq \alpha_0$ for all $(m, n) \geq (k_0, l_0)$.*

(ii) *Each row-series as well as each column-series is absolutely convergent.*

We shall provide a variety of conditions each of which imply condition (i) of Lemma 7.2.1. These yield convergence tests for double series which are analogous to well-known convergence tests for single series. (See, for example, Chapter 9 of [183] and the exercises therein). The following test shows that we can study the convergence of certain double series by considering only some of its terms.

Theorem 7.2.2 (Cauchy's condensation test)**.** [243, Theorem 2.2] *Let (a_{kl}) be a monotonically decreasing double sequence of non-negative numbers. Then, $\sum_{k,l=1}^{\infty} a_{kl}$ converges if and only if $\sum_{k,l} 2^{k+l} a_{2^k,2^l}$ converges.*

Proof. Given $(m, n) \in \mathbb{N}^2$, let $i, j \in \mathbb{N}$ be such that $2^i \leq m < 2^{i+1}$ and $2^j \leq n < 2^{j+1}$. Since $a_{kl} \geq 0$ for all $(k, l) \in \mathbb{N}_1^2$, we have

$$\sum_{k,l=0}^{i-1,j-1} \left(\sum_{u,v=2^k,2^l}^{2^{k+1}-1,2^{l+1}-1} a_{uv} \right) \leq \sum_{k,l=1}^{m,n} a_{kl} \leq \sum_{k,l=0}^{i,j} \left(\sum_{u,v=2^k,2^l}^{2^{k+1}-1,2^{l+1}-1} a_{uv} \right),$$

and since (a_{kl}) is monotonically decreasing, we obtain

$$\frac{1}{4} \sum_{k,l=1}^{i,j} 2^{k+l} a_{2^k,2^l} = \sum_{k,l=0}^{i-1,j-1} 2^{k+l} a_{2^{k+1},2^{l+1}} \leq \sum_{k,l=1}^{m,n} a_{kl} \leq \sum_{k,l=0}^{i,j} 2^{k+l} a_{2^k,2^l}.$$

This shows that if the partial sums of $\sum_{k,l} 2^{k+l} a_{2^k,2^l}$ are bounded, then so are the partial sums of $\sum_{k,l=1}^{\infty} a_{kl}$ and if the partial sums of $\sum_{k,l=1}^{\infty} a_{kl}$ are bounded, then so are the partial sums $\sum_{k,l=1}^{\infty} 2^{k+l} a_{2^k,2^l}$. Further, if $\sum_{k,l=1}^{\infty} a_{kl}$ is convergent, then the row-series $\sum_{l=1}^{\infty} a_{kl}$ and the column-series $\sum_{k=1}^{\infty} a_{kl}$ are convergent, where (a_{k1}) and (a_{1l}) are monotonically decreasing sequences of non-negative numbers. In this case, the series $\sum_k 2^k a_{2^k,1}$ and $\sum_l 2^l a_{1,2^l}$ are convergent by Cauchy's condensation test for single series. Hence, the desired result follows. □

Example 7.2.3. [243, Example 2.3] *Let $p \in \mathbb{R}$ and $a_{kl} := 1/(k+l)^p$ for $(k,l) \in \mathbb{N}_1^2$. By Theorem 7.2.2, $\sum_{k,l=1}^{\infty} a_{kl}$ converges if and only if the double series $\sum_{k,l} b_{kl}$ converges, where $b_{kl} := 2^{k+l}/(2^k + 2^l)^p$. If $p \leq 2$, then*

$$b_{kk} := \frac{2^{2k}}{(2 \cdot 2^k)^p} = \frac{2^{k(2-p)}}{2^p} \geq \frac{1}{2^p}$$

for $k \in \mathbb{N}$ and so, the double series $\sum_{k,l} b_{kl}$ diverges. If $p > 2$, then

$$b_{kl} := \frac{2^{k+l}}{(2^k + 2^l)^p} \leq \frac{2^{k+l}}{2^p (2^{k+l})^{p/2}} = \frac{1}{2^p} \left(2^{(2-p)/2}\right)^{k+l}$$

for $(k,l) \in \mathbb{N}_1^2$ and so, the double series $\sum_{k,l} b_{kl}$ converges. It follows that the double series $\sum_{k,l=1}^{\infty} 1/(k+l)^p$ converges if and only if $p > 2$.

If $\sum_{k,l} a_{kl}$ is convergent, then $a_{kl} \to 0$ as $k, l \to \infty$. The $(k,l)^{th}$ term test is useful for establishing the divergence of a double series. The following variant of this test is analogous to Abel's k^{th} term test for single series:

Theorem 7.2.4 (Abel's $(k,l)^{th}$ term test)**.** [243, Theorem 2.4] *Suppose (a_{kl}) is a monotonically decreasing double sequence of non-negative numbers. If the double series $\sum_{k,l} a_{kl}$ is convergent, then $kla_{kl} \to 0$ as $k, l \to \infty$.*

Proof. Given $(k,l) \in \mathbb{N}^2$, let $i_k, j_l \in \mathbb{N}$ be such that $2^{i_k} \leq k < 2^{i_k+1}$ and $2^{j_l} \leq l < 2^{j_l+1}$, and note that

$$0 \leq kla_{kl} \leq 2^{i_k+1} 2^{j_l+1} a_{2^{i_k},2^{j_l}} = 4\left(2^{i_k} 2^{j_l} a_{2^{i_k},2^{j_l}}\right).$$

By Theorem 7.2.2, $\sum_{i,j} 2^{i+j} a_{2^i,2^j}$ is convergent, and so $2^{i+j} a_{2^i,2^j} \to 0$ as $i, j \to \infty$. Hence, $kla_{kl} \to 0$ as $k, l \to \infty$. □

Example 7.2.5. [243, Examples 2.5] *The following statements hold:*

(i) *Let $p, q \in \mathbb{R}$ satisfy $p > 0$, $q > 0$ and $p^{-1} + q^{-1} \geq 1$, and define $a_{kl} := 1/(k^p + l^q)$ for $(k,l) \in \mathbb{N}_1^2$. For $k \in \mathbb{N}_1$ and $l = [k^{p/q}]$, the integer part of $k^{p/q}$, we have*

$$kla_{kl} = \frac{kl}{k^p + l^q} > \frac{k(k^{p/q} - 1)}{k^p + (k^{p/q})^q} = \frac{1}{2} k^{1-p+(p/q)} (1 - k^{-p/q}),$$

which does not tend to 0 as $k \to \infty$ since $1 - p + (p/q) \geq 0$ and $p/q > 0$. Hence, by Theorem 7.2.4, the double series $\sum_{k,l} 1/(k^p + l^q)$ diverges.

(ii) *The converse of Theorem 7.2.4 does not hold. Define* $a_{kl} = 1/kl(\ln k)(\ln l)$ *for* $(k,l) \in \mathbb{N}_1^2$. *Then,* (a_{kl}) *is a monotonically decreasing double sequence of non-negative numbers and* $kla_{kl} \to 0$ *as* $k,l \to \infty$. *However,*

$$\lim_{m,n\to\infty} \sum_{k,l=1}^{m,n} a_{kl} = \lim_{m,n\to\infty} \left(\sum_{k=1}^{m} \frac{1}{k\ln k}\right)\left(\sum_{l=1}^{n} \frac{1}{l\ln l}\right) = \infty.$$

Theorem 7.2.6 (Limit comparison test). [243, Theorem 2.6] *Let* $(a_{kl}), (b_{kl}) \in \Omega$ *such that* $a_{kl} > 0$, $b_{kl} > 0$ *for all* $(k,l) \in \mathbb{N}^2$, *each row-series as well as each column-series corresponding to both* $\sum_{k,l} a_{kl}$ *and* $\sum_{k,l} b_{kl}$ *is convergent, and* $\lim_{k,l\to\infty} a_{kl}/b_{kl} = r$, *where* $r \in \mathbb{R} \setminus \{0\}$. *Then,* $\sum_{k,l} a_{kl}$ *is convergent if and only if* $\sum_{k,l} b_{kl}$ *is convergent.*

Proof. Let the double series $\sum_{k,l} b_{kl}$ be convergent. Then, there is $\beta > 0$ such that $\sum_{k,l=1}^{m,n} b_{kl} \le \beta$ for all $(m,n) \in \mathbb{N}_1^2$. Since $a_{kl}/b_{kl} \to r$ as $k,l \to \infty$, there is $(k_0,l_0) \in \mathbb{N}_1^2$ such that $a_{kl} \le (r+1)b_{kl}$ for all $(k,l) \ge (k_0,l_0)$. Hence, for all $(m,n) \ge (k_0,l_0)$, we have

$$\sum_{k,l=k_0,l_0}^{m,n} a_{kl} \le (r+1) \sum_{k,l=k_0,l_0}^{m,n} b_{kl} \le (r+1)\beta.$$

By Lemma 7.2.1, it follows that the double series $\sum_{k,l} a_{kl}$ is convergent.

Conversely, let the double series $\sum_{k,l} a_{kl}$ be convergent. Since $b_{kl}/a_{kl} \to 1/r$ as $k,l \to \infty$, the convergence of the double series $\sum_{k,l} b_{kl}$ follows from the first part of the proof by interchanging a_{kl} and b_{kl}. □

We shall now develop several convergence tests involving ratios of 'consecutive' terms of a double series.

Theorem 7.2.7 (Ratio test). [243, Theorem 2.7] *Let* (a_{kl}) *be a double sequence of non-zero numbers such that either* $|a_{k,l+1}|/|a_{kl}| \to a$ *or* $|a_{k+1,l}|/|a_{kl}| \to \tilde{a}$ *as* $k,l \to \infty$, *where* $a, \tilde{a} \in \mathbb{R} \cup \{\infty\}$. *The following statements hold:*

(i) *Suppose each row-series as well as each column-series corresponding to* $\sum_{k,l} a_{kl}$ *is absolutely convergent. If* $a < 1$ *or* $\tilde{a} < 1$, *then* $\sum_{k,l} a_{kl}$ *is absolutely convergent.*

(ii) *If* $a > 1$ *or* $\tilde{a} > 1$, *then* $\sum_{k,l} a_{kl}$ *is divergent.*

Proof. (i) Assume that $a < 1$. Then, there are $\alpha \in (0,1)$ and $(k_0,l_0) \in \mathbb{N}^2$ such that $|a_{k,l+1}| \le \alpha|a_{kl}|$ for all $(k,l) \ge (k_0,l_0) \in \mathbb{N}^2$. Hence,

$$|a_{kl}| \le \alpha|a_{k,l-1}| \le \cdots \le \alpha^{l-l_0}|a_{k,l_0}|$$

for all $(k,l) \ge (k_0, l_0+1)$. Since $\alpha < 1$, we have $\sum_{l=1}^{n} \alpha^l \le 1/(1-\alpha)$ for all $n \in \mathbb{N}_1$. Also, since the series $\sum_k a_{k,l_0}$ is assumed to be absolutely convergent, there is $\beta > 0$ such that $\sum_{k=1}^{m} |a_{k,l_0}| \le \beta$ for all $m \in \mathbb{N}_1$. Hence, we obtain

$$\sum_{k,l=k_0,l_0+1}^{m,n} |a_{kl}| \le \frac{\alpha^{-l_0}\beta}{1-\alpha} \quad \text{for all} \quad (m,n) \ge (k_0, l_0+1).$$

By Lemma 7.2.1, it follows that $\sum_{k,l} a_{kl}$ is absolutely convergent. A similar argument holds if $\widetilde{a} < 1$ instead of $a < 1$.

(ii) Assume that $a \in \mathbb{R}$ with $a > 1$ or $a = \infty$. Then, there are $\alpha \in (1, \infty)$ and $(k_0, l_0) \in \mathbb{N}_1^2$ such that $|a_{k,l+1}|/|a_{kl}| \geq \alpha$ for all $(k, l) \geq (k_0, l_0)$. Hence,

$$|a_{kl}| \geq \alpha |a_{k,l-1}| \geq \cdots \geq \alpha^{l-l_0} |a_{k,l_0}| > 0$$

for all $(k, l) \geq (k_0, l_0 + 1)$. For a fixed $k \in \mathbb{N}$ with $k \geq k_0$, there is $l_1 \in \mathbb{N}_1$ such that $|a_{kl}| \geq \alpha^{l-l_0}|a_{k,l_0}| \geq 1$ for all $l \geq l_1$. Hence, $a_{kl} \not\to 0$ as $k \to \infty$, so that $\sum_{k,l} a_{kl}$ is divergent. The same conclusion holds if $\widetilde{a} \in \mathbb{R}$ with $\widetilde{a} > 1$ or $\widetilde{a} = \infty$. □

Remark 7.2.8. [243, Remarks 2.8] *The following statements hold:*

(i) *The proof of Part (ii) of Theorem 7.2.7 shows that if $a > 1$, then the row-series $\sum_l a_{kl}$ diverges for each fixed large k. Hence, if each row-series converges and the limit a exists, then $a \leq 1$. Similarly, if $\widetilde{a} > 1$, then the column-series $\sum_k a_{kl}$ diverges for each fixed large l. Hence, if each column-series converges and the limit $\widetilde{a}$ exists, then $\widetilde{a} \leq 1$.*

(ii) *Suppose each row-series as well as each column-series corresponding to $\sum_{k,l} a_{kl}$ is absolutely convergent. If $a = \widetilde{a} = 1$, then the double series may converge or may diverge, as Example 7.2.3 shows.*

(iii) *For a double sequence (a_{kl}) of positive numbers, consider the limits $b_k = lim_{l\to\infty} a_{k,l+1}/a_{kl}$ for a fixed $k \in \mathbb{N}_1$ and $c_l = lim_{k\to\infty} a_{k+1,l}/a_{kl}$ for a fixed $l \in \mathbb{N}$, whenever they exist. Biermann* [92, p. 123] *and Vorob'ev* [392, §4 of Chapter 13] *claimed that if b_k exists and is less than 1 for each $k \in \mathbb{N}$, and if c_l exists and is less than 1 for each $l \in \mathbb{N}_1$, then the double series $\sum_{k,l} a_{kl}$ is absolutely convergent. Although the Ratio test for single series shows that each row-series as well as each column-series corresponding to $\sum_{k,l} a_{kl}$ is absolutely convergent, the double series $\sum_{k,l} a_{kl}$ may not be absolutely convergent. For example, define $a_{kl} := 1/(2^k 2^{l/2^k})$ for (k, l) in $\mathbb{N}_1^2$. Then, $b_k < 1$ for each $k \in \mathbb{N}$ and $c_l < 1$ for each $l \in \mathbb{N}_1$. However, since for each fixed $k \in \mathbb{N}_1$,*

$$\sum_{l=1}^{\infty} a_{kl} := \frac{1}{2^k} \sum_{l=1}^{\infty} \left(\frac{1}{2^{1/2^k}} \right)^l = \frac{1}{2^k (2^{1/2^k} - 1)},$$

and since $1/2^k(2^{1/2^k} - 1) \to 1/\ln 2$ as $k \to \infty$, we see that the iterated series $\sum_k \sum_l a_{kl}$ diverges, and so the double series $\sum_{k,l} a_{kl}$ also diverges. Thus, the claims of Biermann and Vorob'ev are incorrect.

(iv) *Suppose $a_{kl} > 0$ for all $(k, l) \in \mathbb{N}_1^2$ and each row-series as well as each column-series corresponding to $\sum_{k,l} a_{kl}$ is convergent. A rather involved version of the Ratio test is given by Baron* [56, in §2] *as follows: If $\lim_{k,l\to\infty} a_{kl}$ exists and if the limit*

$$d := \lim_{k,l\to\infty} \frac{a_{k+1,l} + a_{k,l+1} - a_{k+1,l+1}}{a_{kl}}$$

exists with $d < 1$, then the double series $\sum_{k,l} a_{kl}$ is convergent. Let us compare Baron's version of the Ratio test with Theorem 7.2.7. Suppose both the limits a and $\widetilde{a}$ stated in Theorem 7.2.7 exist and are in $\mathbb{R}$. Then, the limit d exists and

$$d := \lim_{k,l\to\infty}\left(\frac{a_{k+1,l}}{a_{kl}} + \frac{a_{k,l+1}}{a_{kl}} - \frac{a_{k+1,l+1}}{a_{k+1,l}}\frac{a_{k+1,l}}{a_{kl}}\right) = \widetilde{a} + a - a\widetilde{a}.$$

Since, in view of (i) above, we have $a \le 1$ and $\widetilde{a} \le 1$, and since $1 - d = (1-a)(1-\widetilde{a})$, we see that $d < 1$ if and only if $a < 1$ and $\widetilde{a} < 1$. Thus, if one of a and $\widetilde{a}$ is equal to 1 and the other is not, then $d = 1$, and hence, Theorem 7.2.7 is applicable, but Baron's version of the Ratio test is not. For example, if $a_{kl} = 1/k^2 2^l$ for $(k,l) \in \mathbb{N}_1^2$, then $a = 1/2$, $\widetilde{a} = 1$, and $d = 1$. Now, suppose that the limit a exists and it is a real number other than 1. If the limit d exists, then it can be seen that the limit $\widetilde{a}$ exists and is equal to $(d-a)/(1-a)$. Thus, if $a < 1$ and $\widetilde{a}$ does not exist, then the limit d cannot exist, and hence, Theorem 7.2.7 is applicable, but Baron's version of the Ratio test is not. For example, if $a_{kl} := 1/2^{(k^2+kl+l)/k}$, then $a = 1/2$, while the limits $\widetilde{a}$ and d do not exist.

Now we consider an analogue of the Ratio comparison test for single series. (See, for example [100, Theorem 6 in Chapter 5].)

Theorem 7.2.9 (Ratio comparison test). [243, Theorem 2.9] *Let $(a_{kl}), (b_{kl}) \in \Omega$ with $b_{kl} > 0$ for all $(k,l) \in \mathbb{N}^2$. Then, the following statements hold:*

(i) *Suppose each row-series as well as each column-series corresponding to $\sum_{k,l} |a_{kl}|$ is convergent. If $|a_{k,l+1}|b_{kl} \le |a_{kl}|b_{k,l+1}$ and $|a_{k+1,l}|b_{kl} \le |a_{kl}|b_{k+1,l}$ whenever k and l are large, and if $\sum_{k,l} b_{kl}$ is convergent, then so is $\sum_{k,l} |a_{kl}|$.*

(ii) *If $|a_{k,l+1}|b_{kl} \ge |a_{kl}|b_{k,l+1} > 0$ whenever l is large and $k \in \mathbb{N}$, and $|a_{k+1,l}|b_{kl} \ge |a_{kl}|b_{k+1,l} > 0$ whenever k is large and $l \in \mathbb{N}$, and if $\sum_{k,l} b_{kl}$ is divergent, then so is $\sum_{k,l} |a_{kl}|$.*

Proof. (i) Let $k_0, l_0 \in \mathbb{N}$ be such that $|a_{k,l+1}|b_{kl} \le |a_{kl}|b_{k,l+1}$ and $|a_{k+1,l}|b_{kl} \le |a_{kl}|b_{k+1,l}$ for all $(k,l) \ge (k_0,l_0)$. Then,

$$\frac{|a_{kl}|}{b_{kl}} \le \frac{|a_{k,l-1}|}{b_{k,l-1}} \le \cdots \le \frac{|a_{k,l_0}|}{b_{k,l_0}} \le \frac{|a_{k-1,l_0}|}{b_{k-1,l_0}} \le \cdots \le \frac{|a_{k_0,l_0}|}{b_{k_0,l_0}}$$

for $(k,l) \ge (k_0,l_0)$. Let $\sum_{k,l} b_{kl}$ be convergent. Since

$$\beta := \sup\left\{\sum_{k,l=0}^{m,n} b_{kl} : (m,n) \in \mathbb{N}^2\right\} < \infty,$$

we obtain

$$\sum_{k,l=k_0,l_0}^{m,n} |a_{kl}| \le \frac{|a_{k_0,l_0}|}{b_{k_0,l_0}} \sum_{k,l=k_0,l_0}^{m,n} b_{k,l} \le \beta\frac{|a_{k_0,l_0}|}{b_{k_0,l_0}}$$

for all $m \geq k_0$ and $n \geq l_0$. Hence, by Lemma 7.2.1, the double series $\sum_{k,l} |a_{kl}|$ converges.

(ii) Suppose $k_0 \in \mathbb{N}$ is such that $|a_{k+1,l}|b_{kl} \geq |a_{kl}|b_{k+1,l} > 0$ for $k \geq k_0$ and $l \in \mathbb{N}$, and $l_0 \in \mathbb{N}$ is such that $|a_{k,l+1}|b_{kl} \geq |a_{kl}|b_{k,l+1} > 0$ for $l \geq l_0$ and $k \in \mathbb{N}$. Let $\sum_{k,l} b_{kl}$ be divergent. If $\sum_k b_{kl}$ diverges for some $l \in \mathbb{N}$, then by the Ratio comparison test for single series, $\sum_l |a_{kl}|$ also diverges for that l. Similarly, if $\sum_l b_{kl}$ diverges for some $k \in \mathbb{N}$, then $\sum_l |a_{kl}|$ also diverges for that k. In these cases, condition (ii) of Lemma 7.2.1 is not satisfied, and so the double series $\sum_{k,l} |a_{kl}|$ diverges. In the remaining case, the set $\left\{\sum_{k,l=k_0,l_0}^{m,n} b_{kl} : (m,n) \in \mathbb{N}^2\right\}$ is unbounded. Reversing the inequality signs in (i) above, we obtain

$$\sum_{k,l=k_0,l_0}^{m,n} |a_{kl} \geq \frac{|a_{k_0,l_0}|}{b_{k_0,l_0}} \sum_{k,l=k_0,l_0}^{m,n} b_{k,l},$$

which tends to ∞ as $m, n \to \infty$. Hence, by Lemma 7.2.1, $\sum_l |a_{kl}|$ diverges. □

Remark 7.2.10. [243, Remarks 2.10]

(i) *The following example shows that both the inequalities*

$$|a_{k,l+1}|b_{kl} \leq |a_{kl}|b_{k,l+1} \quad \text{and} \quad |a_{k+1,l}|b_{kl} \leq |a_{kl}|b_{k+1,l}$$

are needed in Part (i) of Theorem 7.2.9. Define

$$a_{kl} := \frac{1}{(k+l)^2} \quad \text{and} \quad b_{kl} := \frac{1}{2^k(k+l)^2} \quad \text{for} \quad (k,l) \in \mathbb{N}_1^2.$$

Although each row-series as well as each column-series converges, the double series $\sum_{k,l} a_{kl}$ diverges, as we have seen in Example 7.2.3. However, the double series $\sum_{k,l} b_{kl}$ converges, since $1/[2^k(k+l)^2] \leq 1/(2^k l^2)$ for all $(k,l) \in \mathbb{N}_1^2$. Here the first inequality mentioned above holds but the second does not. To see that both the inequalities

$$|a_{k,l+1}|b_{kl} \geq |a_{kl}|b_{k,l+1} \quad \text{and} \quad |a_{k+1,l}|b_{kl} \geq |a_{kl}|b_{k+1,l}$$

are needed in Part (ii) of Theorem 7.2.9, we just interchange the roles of a_{kl} and b_{kl} in (i) above.

(ii) *The requirement "$|a_{k,l+1}|b_{kl} \leq |a_{kl}|b_{k,l+1}$ and $|a_{k+1,l}|b_{kl} \leq |a_{kl}|b_{k+1,l}$ whenever k and l are large", in Part (i) of Theorem 7.2.9, is less stringent than the requirement*

$$\begin{aligned} \text{“}|a_{k+p,l+1+q}|b_{k+r,l+s} &\leq |a_{k+p,l+q}|b_{k+r,l+1+s}, \\ |a_{k+1+p,l+q}|b_{k+r,l+s} &\leq |a_{k+1+p,l+q}|b_{k+1+r,l+s} \quad \text{for all} \quad k,l,p,q,r,s \in \mathbb{N}\text{”} \end{aligned}$$

imposed for a similar result given by Biermann [92, p. 124].

As a consequence of the Ratio comparison test, we obtain an analogue of *Raabe's test* for single series. It is useful in some cases when $a = \tilde{a} = 1$ in Theorem 7.2.7.

Theorem 7.2.11. [243, Theorem 2.11] *Let $(a_{kl}) \in \Omega$. Then, the following statements hold:*

(i) *Suppose each row-series as well as each column-series corresponding to $\sum_{k,l} |a_{kl}|$ is convergent. If there is $p > 1$ such that*

$$|a_{k,l+1}| \leq \left(1 - \frac{p}{l}\right) |a_{kl}| \quad \textit{and} \quad |a_{k+1,l}| \leq \left(1 - \frac{p}{k}\right) |a_{kl}|$$

whenever k and l are large, then $\sum_{k,l} |a_{kl}|$ is convergent.

(ii) *If $|a_{k,l+1}| \leq (1 - 1/l)|a_{kl}| > 0$ for some $k \in \mathbb{N}$ and all large $l \in \mathbb{N}$, or if $|a_{k+1,l}| \geq (1 - 1/k)|a_{kl}| > 0$ for some $l \in \mathbb{N}$ and all large $k \in \mathbb{N}$, then $\sum_{k,l} |a_{kl}|$ is divergent.*

Proof. (i) Suppose there is $p > 1$ with the stated properties. Using the inequality $1 - px \leq (1 - x)^p$ for $x \in [0, 1]$, we obtain

$$|a_{k,l+1}| \leq \left(1 - \frac{p}{l}\right) |a_{kl}| \leq \left(1 - \frac{1}{l}\right)^p |a_{kl}| \leq \left(\frac{l}{l+1}\right)^p |a_{kl}|$$

whenever k and l are large, and hence

$$\frac{|a_{k,l+1}|}{k^p l^p} \leq \frac{|a_{kl}|}{k^p (l+1)^p}.$$

Similarly, we obtain

$$\frac{|a_{k+1,l}|}{k^p l^p} \leq \frac{|a_{kl}|}{l^p (k+1)^p}$$

whenever k and l are large. By Part (i) of Theorem 7.2.9 with $b_{kl} = 1/(k^p l^p)$ for $(k, l) \in \mathbb{N}_1^2$, we obtain the desired result.

(ii) Suppose that $|a_{k,l+1}| \leq (1 - 1/l)|a_{kl}| > 0$ for some $k \in \mathbb{N}$ and all large $l \in \mathbb{N}$, or $|a_{k+1,l}| \geq (1 - 1/k)|a_{kl}| > 0$ for some $l \in \mathbb{N}$ and all large $k \in \mathbb{N}$. Then, by Raabe's test for single series, $\sum_k |a_{kl}|$ diverges for some $l \in \mathbb{N}$ or $\sum_l |a_{kl}|$ diverges for some $k \in \mathbb{N}$. In any case, condition (ii) of Lemma 7.2.1 is not satisfied, and so the double series $\sum_{k,l} |a_{kl}|$ is divergent. □

Example 7.2.12. [243, Examples 2.12] *Consider $(a_{kl}) \in \Omega$. Then, the following statements hold:*

(i) *Let $a_{11} = 1$, $a_{k+1,1} = (2k - 1)a_{k,1}/(2k + 2)$ for $k \in \mathbb{N}$ and $a_{k,l+1} = l(2l - 1)a_{kl}/(2l + 2)$ for $(k, l) \in \mathbb{N}_1^2$. By Part (i) of Theorem 7.2.11 with $p = 5/4$, we see that the double series $\sum_{k,l} a_{kl}$ is convergent.*

(ii) *Let $a_{11} = 1$, $a_{k+1,1} = ka_{k,1}/(k + 1)$ for $k \in \mathbb{N}_1$ and $a_{k,l+1} = la_{kl}/(l + 1)$ for $(k, l) \in \mathbb{N}_1^2$. By Part (ii) of Theorem 7.2.11, we see that the double series $\sum_{k,l} a_{kl}$ is divergent.*

In both examples given in Parts (i) and (ii), Theorem 7.2.7 is not applicable, since $a = \tilde{a} = 1$.

We deduce the following *limit version of Raabe's test* from Theorem 7.2.11:

Theorem 7.2.13. [243, Theorem 2.13] *Let* (a_{kl}) *be a double sequence of non-zero numbers. Then, the following statements hold:*

(i) *Suppose* $l(1 - |a_{k,l+1}/a_{kl}|) \to \alpha$ *and* $k(1 - |a_{k+1,l}/a_{kl}|) \to \tilde{\alpha}$ *as* $k, l \to \infty$, *where* $\alpha, \tilde{\alpha} \in \mathbb{R} \cup \{\infty\}$. *Suppose each row-series as well as each column-series corresponding to* $\sum_{k,l} |a_{kl}|$ *is convergent. If* $\alpha > 1$ *and* $\tilde{\alpha} > 1$, *then* $\sum_{k,l} |a_{kl}|$ *is convergent.*

(ii) *If for some* $k \in \mathbb{N}$, *the limit* $\lim_{l\to\infty} l(1 - |a_{k,l+1}/a_{kl}|)$ *exists and is less than 1, or if for some* $l \in \mathbb{N}$, *the limit* $\lim\limits_{k\to\infty} k(1 - |a_{k+1,l}/a_{kl}|)$ *exists and is less than 1, then* $\sum_k |a_{kl}|$ *is divergent.*

Remark 7.2.14. [243, Remark 2.14] *Suppose* $a_{kl} > 0$ *for all* $(k, l) \in \mathbb{N}^2$ *and each row-series as well as each column-series corresponding to* $\sum_{k,l} a_{kl}$ *is convergent. A rather involved version of Raabe's test for double series is given by Baron* [56, in §3] *as follows: If* $\lim_{k,l\to\infty} a_{kl}$ *exists and if the limit*

$$r := \lim_{k,l\to\infty} \left[(k + l) \left(1 - \frac{a_{k+1,l} + a_{k,l+1} - a_{k+1,l+1}}{a_{kl}} \right) + \frac{a_{k+1,l+1}}{a_{kl}} \right]$$

exists with $r > 1$, *then the double series* $\sum_{k,l} a_{kl}$ *is convergent. Let us compare Baron's version of Raabe's test with Theorem 7.2.13. Suppose both the limits* α *and* $\tilde{\alpha}$ *stated in Theorem 7.2.13 exist and are in* $\mathbb{R}$. *Then, since* $a_{k,l+1}/a_{kl} \to 1$ *and* $a_{k+1,l}/a_{kl} \to 1$ *as* $k, l \to \infty$, *we see that the limit* r *exists and*

$$\begin{aligned} r \quad &:= \lim_{k,l\to\infty} \left[k \left(1 - \frac{a_{k+1,l}}{a_{kl}} \right) + l \left(1 - \frac{a_{k,l+1}}{a_{kl}} \right) - \frac{a_{k+1,l}}{a_{kl}} l \left(1 - \frac{a_{k+1,l+1}}{a_{k+1,l}} \right) - \right. \\ &\quad \left. - \frac{a_{k,l+1}}{a_{kl}} k \left(1 - \frac{a_{k+1,l+1}}{a_{k,l+1}} \right) + \frac{a_{k+1,l+1}}{a_{k,l+1}} \frac{a_{k,l+1}}{a_{kl}} \right] \\ &= \tilde{\alpha} + \alpha - \alpha - \tilde{\alpha} + 1 = 1. \end{aligned}$$

In all such cases, Theorem 7.2.13 is applicable, but Baron's version of Raabe's test is not. For instance, in Part (i) of Example 7.2.12, we have $\alpha = 2 = \tilde{\alpha}$, *but* $r = 1$.

7.2.2 Cauchy Product

The *Cauchy product of single sequences* (a_k) and (b_k) with $k \in \mathbb{N}$ is defined to be the sequence $(a_k * b_k)$, where $a_k * b_k = \sum_{j=0}^{k} a_j b_{k-j}$ for $k \in \mathbb{N}$, and the *Cauchy product of single series* $\sum_k a_k$ and $\sum_k b_k$ is defined to be the double series $\sum_k a_k * b_k$. Analogously, the *Cauchy product of double sequences* (a_{kl}) and (b_{kl}) with $(k, l) \in \mathbb{N}^2$ is defined to be the sequence $(a_{kl} * b_{kl})$, where

$$a_{kl} * b_{kl} := \sum_{i,j=0}^{k,l} a_{ij} b_{k-i,l-j}$$

for $(k,l) \in \mathbb{N}^2$ and the *Cauchy product of double series* $\sum_{k,l} a_{kl}$ and $\sum_{k,l} b_{kl}$ is defined to be the double series $\sum_{k,l} a_{kl} * b_{kl}$. A classical result of Mertens states that if one of the given single series is absolutely convergent and the other is convergent, then their Cauchy product series is convergent. Another result due to Abel states that if both the given single series and their Cauchy product series are convergent, then the sum of the Cauchy product series is equal to the product of the sums of the given series. It has been known for a long time that the exact analogue of Mertens' result does not hold for double series. To see the examples, the reader may refer to [362, p. 1036] and [102, p. 190]. The following example shows that the exact analogue of Abel's result does not hold for double series:

Example 7.2.15. [243, Example 3.1] *Consider* $(a_{kl}), (b_{kl}) \in \Omega$ *defined by*

$$a_{kl} := \begin{cases} 1 & , \; k=0, \; l \in \mathbb{N} \\ -1 & , \; k=1, \; l \in \mathbb{N} \\ 0 & , \; k \in \mathbb{N}_2, \; l \in \mathbb{N} \end{cases} \quad \text{and} \quad b_{kl} := \begin{cases} 1 & , \; l=0, \; k \in \mathbb{N} \\ -1 & , \; l=1, \; k \in \mathbb{N} \\ 0 & , \; l \in \mathbb{N}_2, \; k \in \mathbb{N} \end{cases}$$

Then, $\sum_{k,l} a_{kl}$ *and* $\sum_{k,l} b_{kl}$ *are convergent and the sum of each is equal to 0. Also, it is easy to see that* $a_{00} * b_{00} = 1$ *and* $a_{kl} * b_{kl} = 0$ *for all* $(k,l) \in \mathbb{N}_1^2$, *so that* $\sum_{k,l} a_{kl} * b_{kl}$ *is convergent and its sum is 1.*

We shall now prove analogues of the theorems of Mertens and Abel for double series which are boundedly convergent, that is, which are convergent and their partial sums are bounded. In fact, we shall show that Mertens' result admits a converse for such double series in respect of absolute convergence. The proofs will be based on the following result for a transformation of a double sequence by a 4-fold infinite matrix. It is an analogue of the well-known Kojima-Schur Theorem given by Theorem 3.3.3 for boundedly convergent double sequences.

Let $\mathbf{a}_{mnkl} \in \mathbb{R}$ for all $m,n,k,l \in \mathbb{N}$ and consider the four-dimensional matrix $\mathbf{A} = (\mathbf{a}_{mnkl})$. We say that $\mathbf{A}$ maps a double sequence $x = (x_{kl})$ to the double sequence $\mathbf{A}x = \{(\mathbf{A}x)_{mn}\}_{m,n \in \mathbb{N}}$ defined by $(\mathbf{A}x)_{mn} := \sum_{k,l} \mathbf{a}_{mnkl} x_{kl}$ provided the double series on the right-hand side converges for each fixed $(m,n) \in \mathbb{N}^2$.

Lemma 7.2.16. [243, Lemma 3.2] *A matrix* $\mathbf{A} = (\mathbf{a}_{mnkl})$ *maps each bounded convergent double sequence to a bounded convergent double sequence if and only if the following conditions hold:*

(i) $\sup_{m,n \in \mathbb{N}} \sum_{k,l} |\mathbf{a}_{mnkl}| < \infty$,

(ii) *the limit* $\mathbf{a} = \lim_{m,n \to \infty} \sum_{k,l} \mathbf{a}_{mnkl}$ *exists,*

(iii) *the limit* $\mathbf{a}_{kl} = \lim_{m,n \to \infty} \mathbf{a}_{mnkl}$ *exists for each fixed* $(k,l) \in \mathbb{N}^2$,

(iv) $\lim_{m,n \to \infty} \sum_k |\mathbf{a}_{mnkl} - \mathbf{a}_{kl}| = 0$ *for each fixed* $l \in \mathbb{N}$,
$\lim_{m,n \to \infty} \sum_l |\mathbf{a}_{mnkl} - \mathbf{a}_{kl}| = 0$ *for each fixed* $k \in \mathbb{N}$.

In this event, the double series $\sum_{k,l} \mathbf{a}_{kl}$ is absolutely convergent and for any bounded convergent double sequence (x_{kl}), we have the limit formula

$$\lim_{m,n \to \infty} [\mathbf{A}x]_{mn} = \left(\mathbf{a} - \sum_{k,l} \mathbf{a}_{kl} \right) \lim_{m,n \to \infty} x_{mn} + \sum_{k,l} \mathbf{a}_{kl} x_{kl}.$$

See [197], especially conditions (c_1), (d_3) and (d_4) in §3, condition 20. $S.BC \to BC$ and limit formula (11.1) in §6, and a remark in §7 about the necessity of S. conditions, (See also [410, Theorem 4.1.2]).

Theorem 7.2.17. [243, Theorem 3.3] *Let $(a_{kl}) \in \Omega$. Then, the Cauchy product double series $\sum_{k,l} a_{kl} * b_{kl}$ is boundedly convergent for every boundedly convergent double series $\sum_{k,l} b_{kl}$ if and only if the double series $\sum_{k,l} a_{kl}$ is absolutely convergent. In this event, we have $\sum_{k,l} a_{kl} * b_{kl} := \left(\sum_{k,l} a_{kl}\right)\left(\sum_{k,l} b_{kl}\right)$.*

Proof. For $m, n \in \mathbb{N}$, let s'_{mn} and s''_{mn} be the $(m,n)^{th}$ partial sums of the double series $\sum_{k,l} b_{kl}$ and $\sum_{k,l} a_{kl} * b_{kl}$, respectively, i.e.,

$$s'_{mn} := \sum_{k,l=0}^{m,n} b_{kl} \quad \text{and} \quad s''_{mn} := \sum_{k,l=0}^{m,n} a_{kl} * b_{kl}.$$

Then, we have

$$\begin{aligned} s''_{mn} &:= \sum_{k,l=0}^{m,n}\left(\sum_{i,j=0}^{k,l} a_{k-i,l-j} b_{ij}\right) = \sum_{k,l=0}^{m,n} a_{m-k,n-l}\left(\sum_{i,j=0}^{k,l} b_{ij}\right) \\ &= \sum_{k,l=0}^{m,n} a_{m-k,n-l} s'_{kl} \quad \text{for} \quad (m,n) \in \mathbb{N}^2. \end{aligned}$$

Now the double series $\sum_{k,l} a_{kl} * b_{kl}$ (respectively, $\sum_{k,l} b_{kl}$) is boundedly convergent if and only if the double sequence (s''_{mn}) (respectively, (s'_{mn})) is boundedly convergent. Consider the matrix $\mathbf{A} = (\mathbf{a}_{mnkl})$ defined by

$$\mathbf{a}_{mnkl} := \begin{cases} a_{m-k,n-l} & , \quad 0 \le k \le m \text{ and } 0 \le l \le n, \\ 0 & , \quad \text{otherwise.} \end{cases}$$

Then, $[\mathbf{A}(s'_{kl})]_{mn} = s''_{mn}$ for all $(m,n) \in \mathbb{N}^2$. It is clear that the matrix $\mathbf{A}$ satisfies condition (i) of Lemma 7.2.16 if and only if

$$\sup_{m,n\in\mathbb{N}} \sum_{k,l=0}^{m,n} |a_{m-k,n-l}| = \sup_{m,n\in\mathbb{N}} \sum_{k,l=0}^{m,n} |a_{k,l}| < \infty,$$

i.e., if and only if the double series $\sum_{k,l} a_{kl}$ is absolutely convergent, and in that case, we have

$$\begin{aligned} \mathbf{a} := \lim_{m,n\to\infty} \sum_{k,l} \mathbf{a}_{mnkl} &:= \lim_{m,n\to\infty} \sum_{k,l=0}^{m,n} a_{kl} = \sum_{k,l} a_{kl}, \\ \mathbf{a}_{kl} := \lim_{m,n\to\infty} \mathbf{a}_{mnkl} &:= \lim_{m,n\to\infty} a_{mn} = 0 \text{ exists for each fixed } (k,l) \in \mathbb{N}^2, \\ \lim_{m,n\to\infty} \sum_{k} |\mathbf{a}_{mnkl} - \mathbf{a}_{kl}| &:= \lim_{n\to\infty} \sum_{k} |a_{k,n-l}| = 0 \text{ for each fixed } l \in \mathbb{N}, \\ \lim_{m,n\to\infty} \sum_{l} |\mathbf{a}_{mnkl} - \mathbf{a}_{kl}| &:= \lim_{m\to\infty} \sum_{l} |a_{m-k,l}| = 0 \text{ for each fixed } k \in \mathbb{N}, \end{aligned}$$

i.e., conditions (ii), (iii) and (iv) of Lemma 7.2.16 are automatically satisfied. Hence the desired result follows from Lemma 7.2.16 with

$$\sum_{k,l} a_{kl} * b_{kl} = \left(\sum_{k,l} a_{kl} - 0\right) \lim_{m,n\to\infty} s'_{mn} + 0 = \left(\sum_{k,l} a_{kl}\right)\left(\sum_{k,l} b_{kl}\right).$$

□

Remark 7.2.18. [243, Remark 3.4] *It is interesting to compare the above analogue of Mertens' theorem with the following result stated by Sheffer in Theorem 3 of* [362]. *Let* $\sum_{k,l} b_{kl}$ *be a convergent double series. Then, the Cauchy product double series* $\sum_{k,l} a_{kl} * b_{kl}$ *is convergent for every absolutely convergent double series* $\sum_{k,l} a_{kl}$ *if and only if the partial sums of* $\sum_{k,l} b_{kl}$ *are bounded, and in that event,* $\sum_{k,l} a_{kl} * b_{kl} = \left(\sum_{k,l} a_{kl}\right)\left(\sum_{k,l} b_{kl}\right)$.

We proceed to prove an analogue of Abel's theorem for boundedly convergent double series. Its proof is based on the following result which uses a matrix transformation considered in Lemma 7.2.16.

Lemma 7.2.19. [243, Lemma 3.5] *Let* $(a_{mn})_{m,n\in\mathbb{N}} \in \mathcal{C}_{bp}$ *and* $a = \lim_{m,n\to\infty} a_{mn}$. *If* $\widetilde{a}_{mn} = (\sum_{k,l=0}^{m,n} a_{kl})/(m+1)(n+1)$, *then* $(\widetilde{a}_{mn}) \in \mathcal{C}_{bp}$ *and* $\lim_{m,n\to\infty} \widetilde{a}_{mn} = a$. *Further, if* $(b_{mn}) \in \mathcal{C}_{bp}$ *with* $b = \lim_{m,n\to\infty} b_{mn}$, *then*

$$\lim_{m,n\to\infty} \frac{a_{mn} * b_{mn}}{(m+1)(n+1)} = ab.$$

Proof. Consider the matrix $\mathbf{C} = (\mathbf{c}_{mnkl})$ defined by

$$\mathbf{c}_{mnkl} := \begin{cases} 1/(m+1)(n+1) & , \quad 0 \le k \le m \text{ and } 0 \le l \le n, \\ 0 & , \quad \text{otherwise} \end{cases}$$

for all $m, n, k, l \in \mathbb{N}$. Then, $[\mathbf{C}(a_{kl})]_{mn} = \widetilde{a}_{mn}$ for all $(m,n) \in \mathbb{N}^2$. Also,

$$\begin{aligned}
\sup_{m,n\in\mathbb{N}} \sum_{k,l} |\mathbf{c}_{mnkl}| &= \lim_{m,n\to\infty} \sum_{k,l} \mathbf{c}_{mnkl} = \lim_{m,n\to\infty} \sum_{k,l=0}^{m,n} \mathbf{c}_{mnkl} = 1, \\
\mathbf{c}_{kl} = \lim_{m,n\to\infty} \mathbf{c}_{mnkl} &= \lim_{m,n\to\infty} \frac{1}{(m+1)(n+1)} = 0 \text{ for each fixed } (k,l) \in \mathbb{N}^2, \\
\lim_{m,n\to\infty} \sum_{k} |\mathbf{c}_{mnkl} - \mathbf{c}_{kl}| &= \lim_{n\to\infty} \frac{1}{n+1} = 0 \text{ for each fixed } l \in \mathbb{N}, \\
\lim_{m,n\to\infty} \sum_{l} |\mathbf{c}_{mnkl} - \mathbf{c}_{kl}| &= \lim_{m\to\infty} \frac{1}{m+1} = 0 \text{ for each fixed } k \in \mathbb{N},
\end{aligned}$$

that is, conditions (i)–(iv) of Lemma 7.2.16 are satisfied, and so by the limit formula, we obtain $\lim_{m,n\to\infty} \widetilde{a}_{mn} = (1-0)a + 0 = a$.

Next, let $(b_{mn}) \in \mathcal{C}_{bp}$ with $\mathbf{b} = \lim_{m,n\to\infty} b_{mn}$. Consider the matrix $\mathbf{A} = (\mathbf{a}_{mnkl})$ defined by

$$\mathbf{a}_{mnkl} := \begin{cases} a_{m-k,n-l}/(m+1)(n+1) & , \quad 0 \le k \le m \text{ and } 0 \le l \le n, \\ 0 & , \quad \text{otherwise} \end{cases}$$

for all $m, n, k, l \in \mathbb{N}$. Then, $[\mathbf{A}(b_{kl})]_{mn} = (a_{mn} * b_{mn})/(m+1)(n+1)$ for all $(m, n) \in \mathbb{N}^2$. Also, since $b = \sup\{|a_{mn}| : (m, n) \in \mathbb{N}^2\} < \infty$, we obtain

$$\begin{aligned}
\sup_{m,n\in\mathbb{N}} \sum_{k,l} |\mathbf{a}_{mnkl}| &= \sup_{m,n\in\mathbb{N}} \sum_{k,l=0}^{m,n} \mathbf{a}_{mnkl} \leq \mathbf{b}, \\
\mathbf{a} = \lim_{m,n\to\infty} \sum_{k,l} \mathbf{a}_{mnkl} &= \lim_{m,n\to\infty} \sum_{k,l=0}^{m,n} \mathbf{a}_{mnkl} = \lim_{m,n\to\infty} \widetilde{a}_{mn} = a, \\
\mathbf{a}_{kl} = \lim_{m,n\to\infty} \mathbf{a}_{mnkl} &= \lim_{m,n\to\infty} \frac{a_{mn}}{(m+1)(n+1)} = 0 \text{ for each fixed } (k, l) \in \mathbb{N}^2, \\
\lim_{m,n\to\infty} \sum_{k} |\mathbf{a}_{mnkl} - \mathbf{a}_{kl}| &\leq \lim_{n\to\infty} \frac{\mathbf{b}}{n+1} = 0 \text{ for each fixed } l \in \mathbb{N}, \\
\lim_{m,n\to\infty} \sum_{l} |\mathbf{a}_{mnkl} - \mathbf{a}_{kl}| &\leq \lim_{m\to\infty} \frac{\mathbf{b}}{m+1} = 0 \text{ for each fixed } k \in \mathbb{N},
\end{aligned}$$

i.e., conditions (i)–(iv) of Lemma 7.2.16 are satisfied, and so by the limit formula, we obtain $\lim_{m,n\to\infty}(a_{mn} * b_{mn})/(m+1)(n+1) = (a-0)b + 0 = ab$. □

A double series $\sum_{k,l} x_{kl}$ is said to be Cesàro summable if the following limit exists:

$$\lim_{m,n\to\infty} \frac{1}{(m+1)(n+1)} \sum_{k,l=0}^{m,n} s_{kl}$$

In that event, the above limit is called the *Cesàro sum of the double series*. It follows from Lemma 7.2.19 that a boundedly convergent double series is Cesàro summable, and its Cesàro sum is equal to its sum.

Theorem 7.2.20. [243, Theorem 3.6] *Let $\sum_{k,l} a_{kl}$ and $\sum_{k,l} b_{kl}$ be boundedly convergent double series with the sums A and B, respectively. Then, the Cauchy product double series $\sum_{k,l} a_{kl} * b_{kl}$ is Cesàro summable and its Cesàro sum is equal to AB. In particular, if the double series $\sum_{k,l} a_{kl} * b_{kl}$ is boundedly convergent, then its sum is equal to AB.*

Proof. We use the notation introduced in the proof of Theorem 7.2.17. We have

$$s''_{mn} := \sum_{k,l=0}^{m,n} a_{kl} * b_{kl} = a_{mn} * s'_{mn} := s_{mn} * b_{mn} \quad \text{for} \quad (m, n) \in \mathbb{N}^2.$$

Replacing a_{kl} and b_{kl} by s_{mn} and b_{mn}, we obtain

$$\sum_{m,n=0}^{p,q} s''_{mn} := \sum_{m,n=0}^{p,q} s_{mn} * b_{mn} := s_{pq} * s'_{pq} \quad \text{for} \quad (p, q) \in \mathbb{N}^2.$$

Hence, by Lemma 7.2.19, we have

$$\lim_{p,q\to\infty} \frac{1}{(p+1)(q+1)} \sum_{m,n=0}^{p,q} s''_{mn} := \lim_{p,q\to\infty} \frac{s_{pq} * s'_{pq}}{(p+1)(q+1)} := AB.$$

Thus, $\sum_{k,l} a_{kl} * b_{kl}$ is Cesàro summable and its Cesàro sum is equal to AB. The last part in the statement of the theorem follows easily. □

Remark 7.2.21. [243, Remark 3.7] *In Theorem 1 of* [102], *Cesari gives the following analogue of Abel's theorem: Let $\sum_{k,l} a_{kl}$ and $\sum_{k,l} b_{kl}$ be convergent double series such that*

$$\lim_{k+l\to\infty} a_{kl} = \lim_{k+l\to\infty} b_{kl} = 0.$$

Then, the double series $\sum_{k,l} a_{kl} * b_{kl}$ is restrictedly Cesàro summable to AB in the following sense: For any positive real numbers r, s with $r < s$,

$$\lim_{\substack{m,n\to\infty \\ nr<m<ns}} \frac{1}{(m+1)(n+1)} \sum_{k,l=0}^{m,n} s''_{kl} = AB.$$

We shall now consider a notion of convergence which is stronger than bounded convergence. Let us recall that a double sequence (a_{kl}) is said to be *regularly convergent* if it is convergent, and further, for each fixed $k \in \mathbb{N}$, the sequence given by $l \mapsto a_{kl}$ is convergent and for each fixed $l \in \mathbb{N}$, the sequence given by $k \mapsto a_{kl}$ is convergent. A double series is *regularly convergent* if the double sequence of its (rectangular) partial sums is regularly convergent, that is, the double series is convergent, and further, each corresponding row-series as well as each corresponding column-series is convergent. It is easy to see that a regularly convergent double sequence is bounded, and a regularly convergent double series is boundedly convergent. We have the following analogue of Theorem 7.2.17 for regularly convergent double series.

Theorem 7.2.22. [243, Theorem 3.8] *Let $(a_{kl}) \in \Omega$. Then, the Cauchy product double series $\sum_{k,l} a_{kl} * b_{kl}$ is regularly convergent for every regularly convergent double series $\sum_{k,l} b_{kl}$ if and only if the double series $\sum_{k,l} a_{kl}$ is absolutely convergent. In this event, we have $\sum_{k,l} a_{kl} * b_{kl} = \left(\sum_{k,l} a_{kl}\right)\left(\sum_{k,l} b_{kl}\right)$.*

Proof. An argument along the lines given in the proof of Theorem 7.2.17 yields the desired result if we note the following. A matrix $\mathbf{A} = (\mathbf{a}_{mnkl})$ maps each regularly convergent double sequence to a regularly convergent double sequence if and only if the conditions (i)–(iii) of Lemma 7.2.16 hold, and also the following condition holds:

(iv)′ the limit $\mathbf{b}_k = \lim_{m,n\to\infty} \sum_l \mathbf{a}_{mnkl}$ exists for each fixed $k \in \mathbb{N}$,
the limit $\mathbf{c}_l = \lim_{m,n\to\infty} \sum_k \mathbf{a}_{mnkl}$ exists for each fixed $l \in \mathbb{N}$,

where the convergence indicated in each of the conditions (ii), (iii) and (iv)′ is regular. In this event, the double series $\sum_{k,l} a_{kl}$ is absolutely convergent, the series $\sum_k \mathbf{b}_k$ and $\sum_l \mathbf{c}_l$ are absolutely convergent, and for any $(x_{kl}) \in \mathcal{C}_r$, we have the limit formula

$$\lim_{m,n\to\infty} [\mathbf{A}x]_{mn} := \left(\mathbf{a} + \sum_{k,l} \mathbf{a}_{kl} - \sum_k \mathbf{b}_k - \sum_l \mathbf{c}_l\right) \lim_{m,n\to\infty} x_{mn} + \sum_{kl} \mathbf{a}_{kl} x_{kl} + \\ + \sum_k \left[\left(\mathbf{b}_k - \sum_l \mathbf{a}_{kl}\right)\left(\lim_{l\to\infty} x_{kl}\right)\right] + \sum_l \left[\left(\mathbf{c}_l - \sum_k \mathbf{a}_{kl}\right)\left(\lim_{k\to\infty} x_{kl}\right)\right].$$

See [197], especially conditions (c_1), (d_1), (d_2), (d_3), (f_1), (f_2) and (f_3) in §3, condition 132. *S.RC* → *RC* and limit formula (9.1) in §6. (See also [410, Theorem 4.1.1]). In our case, with the matrix $\mathbf{A}$ as defined in the proof of Theorem 7.2.17, we have $\mathbf{a}_{kl} = \mathbf{b}_k = \mathbf{c}_l = 0$ for all $(k, l) \in \mathbb{N}^2$ and $\mathbf{a} = \sum_{k,l} a_{kl}$. □

Remark 7.2.23. [243, Remarks 3.9]

(i) *The necessity part of Theorem 7.2.22 was proved in Theorem 3 of* [102]. *The sufficiency part of Theorem 7.2.22 as well as of Theorem 7.2.17 can be strengthened as follows. If the Cauchy product double series $\sum_{k,l} a_{kl} * b_{kl}$ is boundedly convergent for every regularly convergent double series $\sum_{k,l} b_{kl}$, then the double series $\sum_{k,l} a_{kl}$ is absolutely convergent. To see this, one only has to note that condition (i) of Lemma 7.2.16 is a necessary condition for a matrix A to map each regularly convergent double sequence to a bounded convergent double sequence. (See condition (c_1) in §3, condition 134. S.RC → BC of §6, and a remark in §7 about the necessity of S. conditions).*

(ii) *In view of (i) above, it is worthwhile to observe that if $\sum_{k,l} a_{kl}$ is an absolutely convergent double series not all of whose terms are equal to zero, then there is a boundedly convergent double series $\sum_{k,l} b_{kl}$ such that the double series $\sum_{k,l} a_{kl} * b_{kl}$ is not regularly convergent. To see this, let $\sum_{k,l} a_{kl}$ be absolutely convergent and $a_{k_0,l_0} \neq 0$ for some $(k_0, l_0) \in \mathbb{N}^2$. Define $a_k = a_{kl_0}$ for $k \in \mathbb{N}$. We note that $\sum_k a_k$ is an absolutely convergent series and $a_{k_0} \neq 0$, so that there is a series $\sum_k b_k$ having bounded partial sums such that the series $\sum_k a_k * b_k$ is divergent. Now, define*

$$b_{kl} := \begin{cases} b_k & , \quad l = 0 \text{ and } k \in \mathbb{N}, \\ -b_k & , \quad l = l_0 + 1 \text{ and } k \in \mathbb{N}, \\ 0 & , \quad l \in \mathbb{N} \setminus \{0, l_0 + 1\} \text{ and } k \in \mathbb{N} \end{cases}$$

Then, we have

$$a_{k,l_0} * b_{k,l_0} := \sum_{i,j=0}^{k,l_0} a_{ij} b_{k-i,l_0-j} = \sum_{i=0}^{k} a_{i,l_0} b_{k-i,0} = a_k * b_k.$$

Hence, $\sum_k a_{k,l_0} * b_{k,l_0}$ diverges and so $\sum_{k,l} a_{kl} * b_{kl}$ is not regularly convergent.

Example 7.2.24. [243, Examples 3.10]

(i) *Let $x, y \in (-1, 0)$ with $x + y = -1$. Define*

$$a_{kl} := x^k y^l \quad \textit{and} \quad b_{kl} := \binom{k+l}{k} \quad \textit{for} \quad (k, l) \in \mathbb{N}^2.$$

Then, $\sum_{k,l} a_{kl}$ converges absolutely and $\sum_{k,l} b_{kl}$ converges regularly but not absolutely. The Cauchy product series is $\sum_{k,l} c_{kl}$, where

$$c_{kl} := \sum_{i,j=0}^{k,l} \binom{i+j}{i} x^k y^l = \left[\binom{k+l+2}{k+1} - 1\right] x^k y^l \quad \textit{for} \quad (k, l) \in \mathbb{N}^2,$$

the last equality being a consequence of Pascal's 3rd identity.

(ii) *Let* $x, y \in (-1; 1]$, $a, \widetilde{a} > 0$ *and* $b, \widetilde{b} \in (-1, 0]$. *Define*

$$a_{kl} := \binom{a}{k}\binom{\widetilde{a}}{l} x^k y^l \text{ and } b_{kl} := \binom{b}{k}\binom{\widetilde{b}}{l} x^k y^l \text{ for } (k,l) \in \mathbb{N}^2.$$

Then, $\sum_{k,l} a_{kl}$ *converges absolutely and* $\sum_{k,l} b_{kl}$ *converges regularly but not absolutely when* $x = 1$ *or* $y = 1$. *The Cauchy product series is* $\sum_{k,l} c_{kl}$, *where*

$$c_{kl} := \sum_{i,j=0}^{k,l} \binom{a}{i}\binom{b}{k-i}\binom{\widetilde{a}}{j}\binom{\widetilde{b}}{l-j} x^k y^l = \binom{a+b}{k}\binom{\widetilde{a}+\widetilde{b}}{l} x^k y^l \text{ for } (k,l) \in \mathbb{N}^2,$$

the last equality being a consequence of Vandermonde's Convolution Formula.

(iii) *Let* $p \in (0, 2]$. *Define* $a_{kl} := 1/2^{k+l}$ *and* $b_{kl} := (-1)^{k+l}/(k+l+1)^p$ *for* $(k,l) \in \mathbb{N}^2$. *Then,* $\sum_{k,l} a_{kl}$ *converges absolutely and* $\sum_{k,l} b_{kl}$ *converges regularly but not absolutely. The Cauchy product series is* $\sum_{k,l} c_{kl}$, *where*

$$c_{kl} := \sum_{i,j=0}^{k,l} \frac{(-1)^{i+j}}{2^{k-i+l-j}(i+j+1)^p} = \frac{1}{2^{k+l}} \sum_{i,j=0}^{k,l} \frac{(-2)^{i+j}}{(i+j+1)^p} \text{ for } (k,l) \in \mathbb{N}^2.$$

In each of the above cases, the Cauchy product double series $\sum_{k,l} c_{kl}$ is regularly convergent by Theorem 7.2.22.

As we have seen at the beginning of this subsection, absolute convergence of $\sum_{k,l} a_{kl}$ and convergence of $\sum_{k,l} b_{kl}$ do not guarantee convergence of the Cauchy product double series $\sum_{k,l} a_{kl} * b_{kl}$. In fact, the next result gives precise conditions on the double sequence (a_{kl}) for such convergence.

Theorem 7.2.25. [243, Theorem 3.11] *Let* $(a_{kl}) \in \Omega$. *Then, the Cauchy product double series* $\sum_{k,l} a_{kl} * b_{kl}$ *is convergent for every convergent double series* $\sum_{k,l} b_{kl}$ *if and only if the set* $\{(k,l) \in \mathbb{N}^2 : a_{kl} \neq 0\}$ *is finite.*

Proof. As in the case of Theorem 7.2.22, an argument along the lines given in the proof of Theorem 7.2.17 yields the desired result if we note the following. One of the six necessary and sufficient conditions for a matrix $\mathbf{A} = (\mathbf{a}_{mnkl})$ to map each convergent double sequence to a convergent double sequence is the following: For each fixed $k \in \mathbb{N}$, there is $p_0 \in \mathbb{N}$ such that $\mathbf{a}_{mnkl} = 0$ whenever $m, n, l \geq p_0$ and for each fixed $l \in \mathbb{N}$, there is $q_0 \in \mathbb{N}$ such that $\mathbf{a}_{mnkl} = 0$ whenever $m, n, k \geq q_0$. In our case, with the matrix $\mathbf{A}$ as defined in the proof of Theorem 7.2.17, this condition entails that the set $\{(k,l) \in \mathbb{N}^2 : a_{kl} \neq 0\}$ is finite, and then the remaining five conditions are automatically satisfied. See [197], especially conditions (a_1), (a_2), (b_1), (b_2), (d_1) and (d_3) in §3, condition 14. *S.C* $\to$ *C* in §6. (See also [410, Theorem 4.1.3]). □

Remark 7.2.26. [243, Remark 3.12] *We conclude this section by making some comments about the companion problem of determining conditions on a double sequence* $(a_{kl})_{k,l\in\mathbb{N}}$ *in order that, for every convergent/boundedly convergent/regularly convergent double series* $\sum_{k,l} b_{kl}$, *the double series* $\sum_{k,l} a_{kl}b_{kl}$ *would be convergent/boundedly* *convergent/regularly convergent, as per the assumption on* $\sum_{k,l} b_{kl}$. *For this purpose, let us say that a single sequence* (a_k) *is of bounded variation if the series* $\sum_k |a_k - a_{k+1}|$ *is convergent, and that a double sequence* (a_{kl}) *is of bounded bivariation if the double series* $\sum_{k,l} |a_{kl} - a_{k+1,l} - a_{k,l+1} + a_{k+1,l+1}|$ *is convergent. In the case of regular convergence, the necessary and sufficient conditions on* (a_{kl}) *are as follows:* (a_{kl}) *is of bounded bivariation and* (a_{k1}) *as well as* (a_{1l}) *is of bounded variation. In the case of bounded convergence, the necessary and sufficient conditions on* (a_{kl}) *are as follows:* (a_{kl}) *is of bounded bivariation,* $\lim_{k\to\infty}(a_{kl} - a_{k,l+1}) = 0$ *for each fixed* $l \in \mathbb{N}$ *and* $\lim_{l\to\infty}(a_{kl} - a_{k+1,l}) = 0$ *for each fixed* $k \in \mathbb{N}$. *In the case of convergence, the necessary and sufficient conditions on* (a_{kl}) *are as follows:* (a_{kl}) *is of bounded bivariation for each fixed* $l \in \mathbb{N}$, *there is* $k_l \in \mathbb{N}$ *such that* $a_{kl} = a_{k,l+1}$ *for all* $k \geq k_l$, *and for each fixed* $k \in \mathbb{N}$, *there is* $l_k \in \mathbb{N}$ *such that* $a_{kl} = a_{k+1,l}$ *for all* $l \geq l_k$. *The reader may refer to* [198] *for these results.*

7.3 THE DOUBLE SEQUENCE SPACE $\mathcal{L}_q$

In this section, we introduce the space $\mathcal{L}_q$ of q-summable double sequences corresponding to the space ℓ_q of q-summable single sequences which is defined by Başar and Sever [111], as follows

$$\mathcal{L}_q := \left\{ (x_{ij}) \in \Omega : \sum_{i,j} |x_{ij}|^q < \infty \right\}, (1 < q < \infty)$$

and examine some properties of the space. We give the theorem which states that $\mathcal{L}_q$ is a sequence space and is a Banach space endowed with the norm $\|\cdot\|_q$, firstly. Subsequent to giving some inclusion relations concerning the space $\mathcal{L}_q$, we establish that the α- and γ-duals of a space of double sequences are identical whenever it is solid, and $\mathcal{L}_q$ is solid if $q > 1$ and determine the $\beta(v)$-dual of the space $\mathcal{L}_q$ for $v \in \{p, bp, r\}$ which coincides with the α- and γ-duals of the space $\mathcal{L}_q$.

Theorem 7.3.1. *Let* $1 < q < \infty$. *Then, the set* $\mathcal{L}_q$ *becomes a linear space with the co-ordinatewise addition and scalar multiplication of double sequences, and* $\mathcal{L}_q$ *is a Banach space endowed with the norm*

$$\|x\|_q = \left(\sum_{i,j} |x_{ij}|^q \right)^{1/q}. \tag{7.3.1}$$

Proof. The proof of the first part of the theorem is a routine verification and so, we omit details.

Furthermore, the statement "*a sequence space* ν *is a Banach space endowed with the norm* $\|\cdot\|_\nu$ *if and only if the sequence space* $T^{-1}(\nu) = \lambda$ *is a Banach space endowed with the norm* $\|\cdot\|_\lambda$" holds by Boos [94, Corollary 6.3.41]. Therefore, the restriction of the transformation defined by (7.1.3) to the space $\mathcal{L}_q$ which is norm preserving isomorphism yields the fact that $\mathcal{L}_q = T^{-1}(\ell_q)$ is also a Banach space endowed with the norm $\|\cdot\|_q$ defined by (7.3.1) because of ℓ_q is a Banach space.

This step concludes the proof. □

Theorem 7.3.2. *Let $1 < q < s < \infty$. Then, the inclusions $\mathcal{L}_q \subset \mathcal{L}_s \subset \mathcal{C}_{r0} \subset \mathcal{M}_u$ hold.*

Proof. Let us take any $x = (x_{ij}) \in \mathcal{L}_q$. Then, $\sum_{\max\{i,j\}>n_0} |x_{ij}|^q < \varepsilon < 1$ for sufficiently large $n_0 \in \mathbb{N}$. Since $q < s$, it is obvious that $|x_{ij}|^q \geq |x_{ij}|^s$ for all $i, j \in \mathbb{N}$ such that $\max\{i, j\} > n_0$. Thus,

$$\begin{aligned}
\sum_{i,j} |x_{ij}|^s &= \sum_{i,j=0}^{n_0} |x_{ij}|^s + \sum_{\max\{i,j\}>n_0} |x_{ij}|^s \\
&\leq A + \sum_{\max\{i,j\}>n_0} |x_{ij}|^q \\
&< A + \varepsilon,
\end{aligned}$$

which leads us to the fact that $x \in \mathcal{L}_s$, where $A = \sum_{i,j=0}^{n_0} |x_{ij}|^s$. Hence, $\mathcal{L}_q \subset \mathcal{L}_s$.

Besides one can easily deduce, by means of the suitable restrictions of the isomorphism T defined by (7.1.3) and taking into account the fact that the space $\mathcal{C}_{r0}$ consists of all sequences $x = (x_{mn})$ such that $\lim_{\max\{m,n\}\to\infty} x_{mn} = 0$, from the known inclusions $\ell_s \subset c_0 \subset \ell_\infty$ for $1 < s < \infty$ that

$$T^{-1}(\ell_s) = \mathcal{L}_s \subset \mathcal{C}_{r0} = T^{-1}(c_0) \subset T^{-1}(\ell_\infty) = \mathcal{M}_u.$$

This step completes the proof. □

The *α-dual* λ^α, *$\beta(\upsilon)$-dual* $\lambda^{\beta(\upsilon)}$ with respect to the υ-convergence for $\upsilon \in \{p, bp, r\}$ and the *γ-dual* λ^γ of a double sequence space λ are respectively defined by

$$\begin{aligned}
\lambda^\alpha &:= \left\{ a = (a_{ij}) \in \Omega : \sum_{i,j} |a_{ij}x_{ij}| < \infty \text{ for all } x = (x_{ij}) \in \lambda \right\}, \\
\lambda^{\beta(\upsilon)} &:= \left\{ a = (a_{ij}) \in \Omega : \upsilon - \sum_{i,j} a_{ij}x_{ij} \text{ exists for all } x = (x_{ij}) \in \lambda \right\}, \\
\lambda^\gamma &:= \left\{ a = (a_{ij}) \in \Omega : \sup_{k,l\in\mathbb{N}} \left| \sum_{i,j=0}^{k,l} a_{ij}x_{ij} \right| < \infty \text{ for all } x = (x_{ij}) \in \lambda \right\}.
\end{aligned}$$

It is easy to see for any two spaces λ, μ of double sequences that $\mu^\alpha \subset \lambda^\alpha$ whenever $\lambda \subset \mu$ and $\lambda^\alpha \subset \lambda^\gamma$. Additionally, it is known that the inclusion $\lambda^\alpha \subset \lambda^{\beta(\upsilon)}$ holds while the inclusion $\lambda^{\beta(\upsilon)} \subset \lambda^\gamma$ does not hold, since the υ-convergence of the sequence of partial sums of a double series does not imply its boundedness.

The space λ of double sequences is said to be *solid* if and only if

$$\widetilde{\lambda} = \{x = (x_{kl}) \in \Omega : \exists x = (x_{kl}) \in \lambda \text{ such that } |u_{kl}| \leq |x_{kl}| \text{ for all } k, l \in \mathbb{N}\} \subset \lambda.$$

The space λ of double sequences is also said to be *monotone* if and only if $m_0\lambda \subset \lambda$, where m_0 is the span of the set of all sequences of zeros and ones and $m_0\lambda = \{ax =$

$(a_{ij}x_{ij}) : a = (a_{ij}) \in m_0, x = (x_{ij}) \in \lambda\}$. If λ is monotone, then $\lambda^\alpha = \lambda^{\beta(v)}$ (cf. Zeltser [410, p. 36]) and λ is monotone whenever λ is solid.

Prior to giving the theorem which asserts that the α- and γ-duals of a solid space of double sequences are identical, we quote two lemmas which are needed in proving the theorem.

Lemma 7.3.3. [204, Theorem 2, p. 279] *A positive term double series converges to its l.u.b. (that is the l.u.b. of its partial sums) if it is bounded above. Otherwise, it diverges to* $+\infty$.

Lemma 7.3.4. [57, p. 382] *A double series is absolutely convergent if and only if the set* $\left\{\sum_{i,j=0}^{m,n} |x_{ij}| : m, n \in \mathbb{N}\right\}$ *is a bounded set of real numbers.*

Now, we may give the following theorem:

Theorem 7.3.5. [77, Theorem 2.5] *If a given double sequence space* λ *is solid, then the equality* $\lambda^\alpha = \lambda^\gamma$ *holds.*

Proof. To prove the theorem, it is enough to show that the inclusion $\lambda^\gamma \subset \lambda^\alpha$ holds. Suppose that the sequence space λ is solid and take any $y = (y_{kl}) \in \lambda^\gamma$. Then, $\sup_{m,n\in\mathbb{N}} \left|\sum_{k,l=0}^{m,n} x_{kl}y_{kl}\right| < \infty$ for any $x = (x_{kl}) \in \lambda$. Now, define the sequence $z = (z_{kl})$ via the sequence $x = (x_{kl}) \in \lambda$ by $z_{kl} = x_{kl}\text{sgn}(x_{kl}y_{kl})$ for all $k, l \in \mathbb{N}$. Then, $z = (z_{kl}) \in \lambda$ since λ is solid and $|z_{kl}| \leq |x_{kl}|$ for all $k, l \in \mathbb{N}$. Therefore,

$$\begin{aligned} \sup_{m,n\in\mathbb{N}} \sum_{k,l=0}^{m,n} |x_{kl}y_{kl}| &= \sup_{m,n\in\mathbb{N}} \sum_{k,l=0}^{m,n} x_{kl}y_{kl}\text{sgn}(x_{kl}y_{kl}) \\ &= \sup_{m,n\in\mathbb{N}} \left|\sum_{k,l=0}^{m,n} y_{kl}z_{kl}\right| < \infty. \end{aligned}$$

This shows that the sequence of partial sums of the positive term double series $\sum_{k,l} |x_{kl}y_{kl}|$ is bounded which is convergent by Lemma 7.3.3. Therefore, one can see by Lemma 7.3.4 that $(x_{kl}y_{kl})_{k,l\in\mathbb{N}} \in \mathcal{L}_1$. Since $x \in \lambda$ is arbitrary, y must be in λ^α, i.e., the inclusion $\lambda^\gamma \subset \lambda^\alpha$ holds.

This step terminates the proof. □

As an easy consequence of Theorem 7.3.5, we have

Corollary 7.3.6. [77, Corollary 2.6] *If* λ *is solid, then* $\lambda^\alpha = \lambda^{\beta(v)} = \lambda^\gamma$.

One can easily observe that the double sequence space $\mathcal{L}_q$ is solid, if $q > 1$. This yields that the double sequence space $\mathcal{L}_q$ is monotone which implies the fact that the α- and the $\beta(v)$-duals of the space $\mathcal{L}_q$ are identical.

Now, we may give the theorem on the $\beta(v)$-dual of the space $\mathcal{L}_q$.

Theorem 7.3.7. [77, Theorem 2.7] *The* $\beta(v)$*-dual of the space* $\mathcal{L}_q$ *is the space* $\mathcal{L}_{q'}$*, where* $q > 1$ *and* $q^{-1} + q'^{-1} = 1$.

Proof. Let $q > 1$ and $q^{-1} + q'^{-1} = 1$. Let us take any $x = (x_{mn}) \in \mathcal{L}_{q'}$ and $y = (y_{mn}) \in \mathcal{L}_q$. Consider the inequalities

$$|x_{mn}y_{mn}| \leq \frac{|x_{mn}|^{q'}}{q'} + \frac{|y_{mn}|^q}{q} \leq |x_{mn}|^{q'} + |y_{mn}|^q$$

satisfied for all $m, n \in \mathbb{N}$. Therefore, we derive that

$$\sum_{m,n} |x_{mn}y_{mn}| \leq \sum_{m,n} |x_{mn}|^{q'} + \sum_{m,n} |y_{mn}|^q < \infty,$$

which leads us to the fact that $x \in \mathcal{L}_q^{\alpha}$, i.e., the inclusions

$$\mathcal{L}_{q'} \subset \mathcal{L}_q^{\alpha} \subset \mathcal{L}_q^{\beta(v)} \tag{7.3.2}$$

hold.

Conversely, take any $y = (y_{mn}) \in \mathcal{L}_q^{\beta(v)}$. For establishing the inclusion $\mathcal{L}_q^{\beta(v)} \subset \mathcal{L}_{q'}$, we use the analogous idea employing by Boos [94, p. 344, Theorem 7.1.11.c] for single sequences. Let us consider the linear functional f_n and the double sequence $y^{[n]}$ defined by

$$\begin{array}{rccc} f_n & : & \mathcal{L}_q & \longrightarrow & \mathbb{R} \\ & & x = (x_{kl}) & \longmapsto & f_n(x) = \sum_{k,l=0}^{n} x_{kl}y_{kl} \end{array}$$

and

$$y^{[n]} = \begin{bmatrix} y_{00} & y_{01} & y_{02} & \cdots & y_{0n} & 0 & \cdots \\ y_{10} & y_{11} & y_{12} & \cdots & y_{1n} & 0 & \cdots \\ y_{20} & y_{21} & y_{22} & \cdots & y_{2n} & 0 & \cdots \\ \vdots & \vdots & \vdots & \ddots & \vdots & \vdots & \cdots \\ y_{n0} & y_{n1} & y_{n2} & \cdots & y_{nn} & 0 & \cdots \\ 0 & 0 & 0 & \cdots & 0 & 0 & \cdots \\ \vdots & \vdots & \vdots & \cdots & \vdots & \vdots & \ddots \end{bmatrix}$$

for every $n \in \mathbb{N}$. Then, since $y^{[n]} \in \mathcal{L}_{q'}$, we obtain by Hölder's inequality that

$$|f_n(x)| \leq \sum_{k,l=0}^{n} |x_{kl}y_{kl}| = \sum_{k,l} \left| x_{kl}y_{kl}^{[n]} \right| \leq \|x\|_q \|y^{[n]}\|_{q'}$$

for each $x = (x_{kl}) \in \mathcal{L}_q$ which yields the continuity of the linear functionals f_n. Therefore, we have

$$\|f_n\| \leq \|y^{[n]}\|_{q'} \quad \text{for each } n \in \mathbb{N}. \tag{7.3.3}$$

Let us consider the sequence $x^{(n)} = \{x_{kl}^{(n)}\}_{k,l \in \mathbb{N}}$ to prove the reverse inequality, defined by

$$x_{kl}^{(n)} := \begin{cases} \frac{|y_{kl}|^{q'}}{y_{kl}} & , \text{ if } y_{kl} \neq 0 \text{ and } k, l \leq n, \\ 0 & , \text{ otherwise.} \end{cases}$$

Then, it is clear that $x^{(n)} \in \mathcal{L}_q$ and one can see that

$$\left\|x^{(n)}\right\|_q = \left(\sum_{k,l=0}^{n} |y_{kl}|^{(q'-1)q}\right)^{1/q} = \left(\sum_{k,l=0}^{n} |y_{kl}|^{q'}\right)^{1/q} = \left(\left\|y^{[n]}\right\|_{q'}\right)^{q'/q}.$$

This leads us to the consequence for all $n \in \mathbb{N}$ that

$$\frac{\left|f_n\left(x^{(n)}\right)\right|}{\|x^{(n)}\|_q} = \frac{\sum_{k,l=0}^{n} |y_{kl}|^{q'}}{\|x^{(n)}\|_q} = \left\|y^{[n]}\right\|_{q'}.$$

Hence,

$$\left\|y^{[n]}\right\|_{q'} \leq \|f_n\| \text{ for all } n \in \mathbb{N}. \tag{7.3.4}$$

Therefore, we have by (7.3.3) and (7.3.4) that

$$\|f_n\| = \left\|y^{[n]}\right\|_{q'} \quad \text{for all } n \in \mathbb{N}.$$

By applying the Banach-Steinhaus theorem, one can observe by our hypothesis that the sequence (f_n) of linear functionals converges pointwise. Since $(\mathcal{L}_q, \|\cdot\|_q)$ and $(\mathbb{C}, |\cdot|)$ are Banach spaces, the linear functional defined by

$$\begin{array}{rccl} f_y & : & \mathcal{L}_q & \longrightarrow & \mathbb{R} \\ & & x = (x_{kl}) & \longmapsto & f_y(x) = \lim_{n\to\infty} f_n(x) = \sum_{k,l} x_{kl}y_{kl} \end{array}$$

is continuous, and

$$\|f_y\| \leq \sup_{n\in\mathbb{N}} \|f_n\| = \sup_{n\in\mathbb{N}} \left\|y^{[n]}\right\|_{q'} < \infty$$

holds. Thus, we have $y \in \mathcal{L}_{q'}$, because of

$$\|f_y\| \leq \sup_{n\in\mathbb{N}} \left\|y^{[n]}\right\|_{q'} = \sup_{n\in\mathbb{N}} \left(\sum_{k,l=0}^{n} |y_{kl}|^{q'}\right)^{1/q'} = \left(\sum_{k,l} |y_{kl}|^{q'}\right)^{1/q'} < \infty.$$

That is to say that the inclusion

$$\mathcal{L}_q^{\beta(v)} \subset \mathcal{L}_{q'} \tag{7.3.5}$$

holds.

By combining the inclusions (7.3.2) and (7.3.5), the desired result immediately follows.

This completes the proof. □

As a direct consequence of Theorem 7.3.7, we have

Corollary 7.3.8. [77, Corollary 2.8] *The α-, $\beta(v)$- and γ-duals of the space $\mathcal{L}_q$ are the space $\mathcal{L}_{q'}$, where $q > 1$ and $q^{-1} + q'^{-1} = 1$.*

7.4 SOME NEW SPACES OF DOUBLE SEQUENCES

In this section, we introduce the series spaces $\mathcal{BS}$, $\mathcal{BS}(t)$ and CS_v of double sequences, and the space $\mathcal{BV}$ of double sequences of bounded bivariation defined by

$$\begin{aligned}
\mathcal{BS} &:= \left\{(x_{ij}) \in \Omega : \sup_{m,n\in\mathbb{N}} |s_{mn}| < \infty\right\}, \\
\mathcal{BS}(t) &:= \left\{(x_{ij}) \in \Omega : \sup_{m,n\in\mathbb{N}} |s_{mn}|^{t_{mn}} < \infty\right\}, \\
CS_v &:= \{(x_{ij}) \in \Omega : (s_{mn}) \in C_v\}, \\
\mathcal{BV} &:= \left\{(x_{ij}) \in \Omega : \sum_{i,j} |x_{ij} - x_{i-1,j} - x_{i,j-1} + x_{i-1,j-1}| < \infty\right\}
\end{aligned}$$

and examine their some properties, where $t = (t_{mn})$ is a sequence with positive terms and $v \in \{p, r, bp\}$.

Theorem 7.4.1. [24, Theorem 1.1] *The set $\mathcal{BS}$ becomes a linear space with the co-ordinatewise addition and scalar multiplication of double sequences and $\mathcal{BS}$ is a Banach space endowed with the norm*

$$\|x\|_{\mathcal{BS}} := \sup_{m,n\in\mathbb{N}} \left| \sum_{i,j=0}^{m,n} x_{ij} \right|, \tag{7.4.1}$$

which is norm isomorphic to the space $\mathcal{M}_u$.

Proof. The first part of the theorem is a routine verification and so, we omit details.

Now, we can show that $\mathcal{BS}$ is a Banach space endowed with the norm defined by (7.4.1). Let $(x^l)_{l\in\mathbb{N}}$ be any Cauchy sequence in the space $\mathcal{BS}$, where $x^l = \{x_{mn}^{(l)}\}_{m,n=0}^{\infty}$ for every fixed $l \in \mathbb{N}$. Then, for a given $\varepsilon > 0$ there exists a positive integer $n_0(\varepsilon)$ such that

$$\|x^l - x^r\|_{\mathcal{BS}} = \sup_{m,n\in\mathbb{N}} \left| \sum_{i,j=0}^{m,n} \left(x_{ij}^l - x_{ij}^r\right) \right| < \varepsilon$$

for all $l, r > n_0(\varepsilon)$ which yields for each fixed $m, n \in \mathbb{N}$ that

$$\left| \sum_{i,j=0}^{m,n} x_{ij}^l - \sum_{i,j=0}^{m,n} x_{ij}^r \right| < \varepsilon.$$

This means that $\left(\sum_{i,j=0}^{m,n} x_{ij}^l\right)_{l\in\mathbb{N}}$ is a Cauchy sequence of complex entries for every fixed $m, n \in \mathbb{N}$. Since $\mathbb{C}$ is complete, it converges, say

$$\lim_{l\to\infty} \sum_{i,j=0}^{m,n} x_{ij}^l = \sum_{i,j=0}^{m,n} x_{ij}. \tag{7.4.2}$$

Using these infinitely many limits, we define the sequence $\left(\sum_{i,j=0}^{m,n} x_{ij}\right)_{m,n\in\mathbb{N}}$. It is seen by (7.4.2) that

$$\lim_{l\to\infty}\left\|\sum_{i,j=0}^{m,n} x_{ij}^{l}-\sum_{i,j=0}^{m,n} x_{ij}\right\|_{\mathcal{BS}}=0. \tag{7.4.3}$$

Since $\left(\sum_{i,j=0}^{m,n} x_{ij}^{l}\right)_{m,n\in\mathbb{N}} \in \mathcal{M}_u$ for each fixed $l \in \mathbb{N}$, there exists a $K_l \in \mathbb{R}^+$ such that

$$\sup_{m,n\in\mathbb{N}}\left|\sum_{i,j=0}^{m,n} x_{ij}^{l}\right| \leq K_l.$$

Therefore, taking supremum over $m, n \in \mathbb{N}$ in the following inequality

$$\left|\sum_{i,j=0}^{m,n} x_{ij}\right| \leq \left|\sum_{i,j=0}^{m,n} x_{ij}-\sum_{i,j=0}^{m,n} x_{ij}^{l}\right|+\left|\sum_{i,j=0}^{m,n} x_{ij}^{l}\right| \leq \varepsilon + K_l$$

which shows that $x = (x_{ij}) \in \mathcal{BS}$. Since $(x^l)_{l\in\mathbb{N}}$ was an arbitrary Cauchy sequence, the space $\mathcal{BS}$ is complete.

To prove the fact "$\mathcal{BS}$ is norm isomorphic to $\mathcal{M}_u$", we should show the existence of a norm preserving linear bijection between the spaces $\mathcal{BS}$ and $\mathcal{M}_u$. Consider the transformation T defined, with the notation of (7.1.1), from $\mathcal{BS}$ to $\mathcal{M}_u$ by $x \mapsto Tx = s$. It is trivial that T is linear and $x = \theta$ whenever $Tx = \theta$ which says us that T is injective.

Let $s = (s_{mn}) \in \mathcal{M}_u$ and define the sequence $x = (x_{ij})$ via s by

$$x_{ij} = s_{ij} - s_{i,j-1} - s_{i-1,j} + s_{i-1,j-1} \text{ for all } i, j \in \mathbb{N}. \tag{7.4.4}$$

Then, since

$$\|x\|_{\mathcal{BS}} = \sup_{m,n\in\mathbb{N}}\left|\sum_{i,j=0}^{m,n} x_{ij}\right| = \sup_{m,n\in\mathbb{N}} |s_{mn}| = \|s\|_{\mathcal{M}_u} < \infty,$$

$x = (x_{ij})$ defined by (7.4.4) is in $\mathcal{BS}$ and this yields that T is surjective and is norm preserving.

This step concludes the proof. □

Theorem 7.4.2. [24, Theorem 1.2] *The inclusions $\mathcal{L}_u \subset \mathcal{BS} \subset \mathcal{M}_u$ strictly hold.*

Proof. Let us take any $x = (x_{ij}) \in \mathcal{L}_u$. Then, there exists a $K_x \in \mathbb{R}^+$ such that $\sum_{i,j} |x_{ij}| = K_x$. Therefore, we obtain by taking supremum over $m, n \in \mathbb{N}$ in the inequality that

$$\left|\sum_{i,j=0}^{m,n} x_{ij}\right| \leq \sum_{i,j=0}^{m,n} |x_{ij}|$$

$\|x\|_{\mathcal{BS}} \leq K_x$. This means that $x \in \mathcal{BS}$, i.e., $\mathcal{L}_u \subset \mathcal{BS}$.

Consider the sequence $x = (x_{mn})$ defined by

$$x_{mn} := \begin{cases} (-1)^m & , \quad n = 0, m \in \mathbb{N}, \\ 0 & , \quad \text{otherwise}. \end{cases} \tag{7.4.5}$$

It is obvious that $x \in \mathcal{BS} \setminus \mathcal{L}_u$ which says that the inclusion $\mathcal{L}_u \subset \mathcal{BS}$ strictly holds.

Let $x = (x_{ij}) \in \mathcal{BS}$. Then, since there exists a $K \in \mathbb{R}^+$ such that

$$\sup_{m,n \in \mathbb{N}} \left| \sum_{i,j=0}^{m,n} x_{ij} \right| < K$$

one can observe that

$$\begin{aligned} |x_{mn}| &= \left| \sum_{i,j=0}^{m,n} x_{ij} - \sum_{i,j=0}^{m-1,n} x_{ij} - \sum_{i,j=0}^{m,n-1} x_{ij} + \sum_{i,j=0}^{m-1,n-1} x_{ij} \right| \\ &\leq \left| \sum_{i,j=0}^{m,n} x_{ij} \right| + \left| \sum_{i,j=0}^{m-1,n} x_{ij} \right| + \left| \sum_{i,j=0}^{m,n-1} x_{ij} \right| + \left| \sum_{i,j=0}^{m-1,n-1} x_{ij} \right| . \end{aligned} \tag{7.4.6}$$

Then, we see by taking supremum over $m, n \in \mathbb{N}$ in (7.4.6) that $\|x\|_\infty \leq 4K$, i.e., $x \in \mathcal{M}_u$. Hence, the inclusion $\mathcal{BS} \subset \mathcal{M}_u$ holds. Further, since $x = (x_{ij})$ defined by $x_{ij} = 1$ for all $i, j \in \mathbb{N}$ is in $\mathcal{M}_u$ but not in $\mathcal{BS}$, the inclusion $\mathcal{BS} \subset \mathcal{M}_u$ is strict. □

Theorem 7.4.3. [24, Theorem 1.3] *The space $\mathcal{BS}$ is not monotone.*

Proof. To prove this, we must find a sequence $y \in \{0,1\}^{\mathbb{N}\times\mathbb{N}}$ such that $xy = (x_{mn}y_{mn}) \notin \mathcal{BS}$ for a sequence $x \in \mathcal{BS}$, where $\{0,1\}^{\mathbb{N}\times\mathbb{N}}$ denotes the set of all double sequences consisting of 0's and 1's.

Let us define the sequence $y = (y_{mn}) \in \{0,1\}^{\mathbb{N}\times\mathbb{N}}$ by

$$y_{mn} := \begin{cases} 1 & , \quad n = 0, m = 2k, \ k \in \mathbb{N}, \\ 0 & , \quad \text{otherwise}. \end{cases}$$

Then, the sequence $z = (z_{mn})$ obtained from the co-ordinatewise product of y by the sequence x defined by (7.4.5) that

$$z_{mn} = x_{mn}y_{mn} := \begin{cases} 1 & , \quad n = 0, m = 2k, k \in \mathbb{N}, \\ 0 & , \quad \text{otherwise}, \end{cases}$$

is not in the space $\mathcal{BS}$ which completes the proof. □

Theorem 7.4.4. [24, Theorem 1.4] *Let (t_{mn}) be any sequence in Ω with positive reals. Then, the following statements hold:*

(i) *$\mathcal{BS}(t) \subset \mathcal{BS}$ if and only if $t = \inf_{m,n\in\mathbb{N}} t_{mn} > 0$.*

(ii) $\mathcal{BS}(t) \supset \mathcal{BS}$ *if and only if* $T = \sup_{m,n\in\mathbb{N}} t_{mn} < \infty$.

(iii) $\mathcal{BS}(t) = \mathcal{BS}$ *if and only if* $0 < t \leq T < \infty$.

Proof. (i) ($\Rightarrow$) : Suppose that $\mathcal{BS}(t) \subset \mathcal{BS}$ but $t = 0$. Then, there exist the sequences $\{m(i)\}$ and $\{n(j)\}$ one of them is strictly increasing and the other one is non-decreasing such that

$$t_{m(i),n(j)} < \frac{1}{i+1} \quad \text{for all} \quad i,j \in \mathbb{N}. \tag{7.4.7}$$

Let us define the sequence $x = \sum_{i,j} ib^{m(i)n(j)}$, where

$$b_{ij}^{mn} := \begin{cases} 1 & , \quad i=m, j=n, \\ -1 & , \quad i=m+1, j=n, \\ -1 & , \quad i=m, j=n+1, \\ 1 & , \quad i=m+1, j=n+1, \\ 0 & , \quad \text{otherwise} \end{cases}$$

for all $i, j, m, n \in \mathbb{N}$. Then,

$$\sum_{i,j=0}^{m,n} x_{ij} := \begin{cases} i & , \quad m=m(i), n=n(j), \\ 0 & , \quad \text{otherwise.} \end{cases}$$

Therefore, it follows by (7.4.7) for every natural numbers i and j that

$$\left| \sum_{i,j=0}^{m,n} x_{ij} \right|^{t_{mn}} = |i|^{t_{m(i)n(j)}} < i^{1/(i+1)} < 2,$$

which shows that $x \in \mathcal{BS}(t)$. On the other hand, since

$$\sup_{m,n\in\mathbb{N}} \left| \sum_{i,j=0}^{m,n} x_{ij} \right| = \infty,$$

$x \notin \mathcal{BS}$, a contradiction. Hence, t must be positive.

($\Leftarrow$): Let $x = (x_{ij}) \in \mathcal{BS}(t)$ and $t > 0$. Then, there exists a $K_x \in \mathbb{R}^+$ such that

$$\left| \sum_{i,j=0}^{m,n} x_{ij} \right|^{t_{mn}} \leq K_x$$

for all $m, n \in \mathbb{N}$, which leads us to the consequence that

$$\left| \sum_{i,j=0}^{m,n} x_{ij} \right| \leq K_x^{1/t_{mn}} \leq \max\{1, K_x^{1/t}\}.$$

Hence, $x \in \mathcal{BS}$ as desired.

(ii) ($\Rightarrow$) : Suppose that $\mathcal{BS}(t) \supset \mathcal{BS}$ but $T = \infty$. Then, there exist the sequences $\{m(i)\}$ and $\{n(j)\}$ one of them is strictly increasing and the other one is non-decreasing such that

$$t_{m(i)n(j)} > i + j \quad \text{for all} \quad i, j \in \mathbb{N}. \tag{7.4.8}$$

Now, consider the sequence $x = (x_{ij}) := \left\{\sum_{i,j} 2b^{m(i)n(j)}\right\}$ for which

$$\sum_{i,j=0}^{m,n} x_{ij} := \begin{cases} 2 & , \quad m = m(i), n = n(j), \\ 0 & , \quad \text{otherwise.} \end{cases}$$

Hence, since $\|x\|_{\mathcal{BS}} = 2$, $x \in \mathcal{BS}$. Nevertheless, one can easily see by (7.4.8) that

$$\left|\sum_{i,j=0}^{m,n} x_{ij}\right|^{t_{mn}} = 2^{t_{m(i)n(j)}} > 2^{i+j},$$

i.e., $x \notin \mathcal{BS}(t)$ which is a contradiction. That is to say that $T < \infty$ must be hold.

($\Leftarrow$): Suppose that $x = (x_{ij}) \in \mathcal{BS}$ and $T < \infty$. Then, there exists a $K_x \in \mathbb{R}^+$ such that

$$\left|\sum_{i,j=0}^{m,n} x_{ij}\right| \leq K_x$$

for all $m, n \in \mathbb{N}$ which gives the fact that

$$\left|\sum_{i,j=0}^{m,n} x_{ij}\right|^{t_{mn}} \leq K_x^{t_{mn}} \leq \max\{1, K_x^T\}.$$

This says that $x \in \mathcal{BS}(t)$, i.e., $\mathcal{BS} \subset \mathcal{BS}(t)$.

(iii) This follows from Part (i) and Part (ii). □

Theorem 7.4.5. [24, Theorem 1.5] *The set $\mathcal{BS}(t)$ forms a linear space with the co-ordinatewise addition and scalar multiplication of double sequences if and only if $T = \sup_{m,n \in \mathbb{N}} t_{mn} < \infty$.*

Proof. ($\Rightarrow$): Let $\mathcal{BS}(t)$ be a linear space but $T = \infty$. Then, there exist the sequences $\{m(i)\}$ and $\{n(j)\}$ one of them is strictly increasing and the other one is non-decreasing such that (7.4.8) holds for all $i, j \in \mathbb{N}$. Define the sequence $x = (x_{ij}) := \left\{\sum_{i,j} b^{m(i)n(j)}/2\right\}$ for which

$$\sup_{m,n \in \mathbb{N}} \left|\sum_{i,j=0}^{m,n} x_{ij}\right|^{t_{mn}} = \sup_{m,n \in \mathbb{N}} \frac{1}{2^{t_{m(i)n(j)}}} \leq \sup_{i,j \in \mathbb{N}} \frac{1}{2^{i+j}} \leq \frac{1}{2},$$

i.e., $x \in \mathcal{BS}(t)$ but $4x \notin \mathcal{BS}(t)$ which contradicts the fact that $\mathcal{BS}(t)$ is a linear space. Hence, $T < \infty$.

($\Leftarrow$) : Let $T < \infty$, $M = \max\{1, T\}$ and $x = (x_{ij}), y = (y_{ij}) \in \mathcal{BS}(t)$. Then, there exist the positive constants K_x and K_y such that

$$\left|\sum_{i,j=0}^{m,n} x_{ij}\right|^{t_{mn}} \leq K_x \text{ and } \left|\sum_{i,j=0}^{m,n} y_{ij}\right|^{t_{mn}} \leq K_y$$

for all $m, n \in \mathbb{N}$. Since $t_{mn}/M \leq 1$ for all $m, n \in \mathbb{N}$ one can see that

$$\left|\sum_{i,j=0}^{m,n} x_{ij} + y_{ij}\right|^{t_{mn}/M} \leq \left|\sum_{i,j=0}^{m,n} x_{ij}\right|^{t_{mn}/M} + \left|\sum_{i,j=0}^{m,n} y_{ij}\right|^{t_{mn}/M},$$

which yields for all $m, n \in \mathbb{N}$ that

$$\left|\sum_{i,j=0}^{m,n} x_{ij} + y_{ij}\right|^{t_{mn}} \leq \left(K_x^{1/M} + K_y^{1/M}\right)^M.$$

Hence, $x + y \in \mathcal{BS}(t)$.

Let $\alpha \in \mathbb{C}$ and $x = (x_{ij}) \in \mathcal{BS}(t)$. Then, since the inequality

$$|\alpha|^{t_{mn}} \leq \max\left\{1, |\alpha|^M\right\}$$

holds for all $m, n \in \mathbb{N}$, it is easily obtained that

$$\left|\sum_{i,j=0}^{m,n} \alpha x_{ij}\right|^{t_{mn}} = |\alpha|^{t_{mn}} K_x \leq K_x \max\left\{1, |\alpha|^M\right\},$$

which means that $\alpha x \in \mathcal{BS}(t)$.

This completes the proof. □

Theorem 7.4.6. [24, Theorem 1.6] *Suppose that* $M = \max\{1, T\}$, $t = \inf_{m,n\in\mathbb{N}} t_{mn}$ *and* $T = \sup_{m,n\in\mathbb{N}} t_{mn}$. *Then,* $\mathcal{BS}(t)$ *is a complete paranormed space with* g *defined by*

$$g(x) := \sup_{m,n\in\mathbb{N}} \left|\sum_{i,j=0}^{m,n} x_{ij}\right|^{t_{mn}/M} \tag{7.4.9}$$

if and only if $t > 0$.

Proof. ($\Rightarrow$): Suppose that $(\mathcal{BS}(t), g)$ be a paranormed space but $t = 0$. Consider the sequence $(x^l)_{l\in\mathbb{N}} = \left(x_{ij}^l\right)_{i,j,l\in\mathbb{N}} \subset \mathcal{BS}(t)$ defined for all $l \in \mathbb{N}$ by

$$x_{ij} := \begin{cases} 1 & , \quad i = 0, j = 0, \\ 0 & , \quad \text{otherwise,} \end{cases}$$

and the sequence $\alpha = (\alpha_l) := \{1/(l+1)\}$ of scalars. Then, since $\alpha_l \to 0$ as $l \to \infty$ and $\sum_{i,j=0}^{m,n} x_{ij} = 1$ for all $m, n \in \mathbb{N}$ we obtain that

$$g(\alpha_l x^l) = \sup_{m,n\in\mathbb{N}} \left| \sum_{i,j=0}^{m,n} \alpha_l x_{ij}^l \right|^{t_{mn}/M} = \sup_{m,n\in\mathbb{N}} |\alpha_l|^{t_{mn}/M} = 1,$$

which contradicts the fact that $g(\alpha_l x^l) \to 0$, as $l \to \infty$. Hence, $t > 0$.

($\Leftarrow$): Suppose that $t > 0$ and show that the onto function g, defined by (7.4.9), is a paranorm on the space $\mathcal{BS}(t)$.

It is trivial that g is a real valued function and $g(\theta) = 0$ whenever $x = \theta \in \mathcal{BS}(t)$. It is also easy to see that $g(-x) = g(x)$ for all $x \in \mathcal{BS}(t)$ and

$$g(x+y) \leq g(x) + g(y)$$

holds for all $x, y \in \mathcal{BS}(t)$.

Let (x^l) be any sequence of the points of the space $\mathcal{BS}(t)$ such that $g(x^l - x) \to 0$ and (α_l) also be any sequence of scalars such that $\alpha_l \to \alpha$, as $l \to \infty$. Then, since there exists a positive integer K such that $|\alpha_l| \leq K$, we thus have

$$\begin{aligned} g(\alpha_l x^l - \alpha x) &= \sup_{m,n\in\mathbb{N}} \left| \sum_{i,j=0}^{m,n} (\alpha_l x_{ij}^l - \alpha x_{ij}) \right|^{t_{mn}/M} \\ &\leq \sup_{m,n\in\mathbb{N}} \left(|\alpha_l| \left| \sum_{i,j=0}^{m,n} x_{ij}^l + \sum_{i,j=0}^{m,n} x_{ij} \right| \right)^{t_{mn}/M} + \\ &+ \sup_{m,n\in\mathbb{N}} \left(|\alpha_l - \alpha| \left| \sum_{i,j=0}^{m,n} x_{ij} \right| \right)^{t_{mn}/M} \\ &\leq Kg(x^l - x) + |\alpha_l - \alpha|^{t_{mn}/M} g(x), \end{aligned}$$

which yields that

$$\lim_{l\to\infty} g(\alpha_l x^l - \alpha x) = 0.$$

Now, we establish the completeness of the space $\mathcal{BS}(t)$. Let (x^r) be any Cauchy sequence in the space $\mathcal{BS}(t)$, where $x^r = \left(x_{ij}^r\right)_{i,j\in\mathbb{N}}$ for each fixed $r \in \mathbb{N}$. Then, for a given $\varepsilon > 0$ there exists a positive integer $n_0(\varepsilon)$ such that

$$g(x^r - x^s) = \sup_{m,n\in\mathbb{N}} \left| \sum_{i,j=0}^{m,n} (x_{ij}^r - x_{ij}^s) \right|^{t_{mn}/M} < \varepsilon \tag{7.4.10}$$

for all $r, s > n_0(\varepsilon)$. Using the definition of g, we obtain for each fixed $m, n \in \mathbb{N}$ that

$$\left| \sum_{i,j=0}^{m,n} x_{ij}^r - \sum_{i,j=0}^{m,n} x_{ij}^s \right|^{t_{mn}/M} \leq \sup_{m,n\in\mathbb{N}} \left| \sum_{i,j=0}^{m,n} x_{ij}^r - \sum_{i,j=0}^{m,n} x_{ij}^s \right|^{t_{mn}/M}, \tag{7.4.11}$$

which leads us to the fact that $\left(\sum_{i,j=0}^{m,n} x_{ij}^r\right)_{r\in\mathbb{N}}$ is a Cauchy sequence of complex numbers. Since $\mathbb{C}$ is complete, it converges, say

$$\lim_{r\to\infty}\left|\sum_{i,j=0}^{m,n} x_{ij}^r - \sum_{i,j=0}^{m,n} x_{ij}\right|^{t_{mn}/M} = 0.$$

Since (x^r) is a Cauchy sequence in the linear space $\mathcal{BS}(t)$, there exists a positive number K such that $g(x^r) < K$. So, we observe by passing to limit as $r \to \infty$, in the inequality

$$g(x) = \sup_{m,n\in\mathbb{N}}\left|\sum_{i,j=0}^{m,n} x_{ij}\right|^{t_{mn}/M} \leq g(x^r) + \sup_{m,n\in\mathbb{N}}\left|\sum_{i,j=0}^{m,n} x_{ij} - \sum_{i,j=0}^{m,n} x_{ij}^r\right|^{t_{mn}/M}$$

that $g(x) < K + \varepsilon$ which shows that $x \in \mathcal{BS}(t)$.

This completes the proof. □

7.5 THE SPACES $\mathcal{CS}_p$, $\mathcal{CS}_{bp}$, $\mathcal{CS}_r$ AND $\mathcal{BV}$ OF DOUBLE SERIES

In the present section, we deal with the sets $\mathcal{CS}_p$, $\mathcal{CS}_{bp}$, $\mathcal{CS}_r$ and $\mathcal{BV}$ consisting of the double series whose sequences of partial sums are convergent in the Pringsheim's sense, bounded Pringsheim's sense, regular sense and the set of double sequences of bounded bivariation, respectively. We show that the sets $\mathcal{CS}_p$, $\mathcal{CS}_{bp}$ and $\mathcal{CS}_r$ are the complete seminormed linear spaces which are isomorphic to the spaces $\mathcal{C}_p$, $\mathcal{C}_{bp}$ and $\mathcal{C}_r$, respectively. Also, we establish that $\mathcal{BV}$ is a Banach space and is isomorphic to the space $\mathcal{L}_u$.

Theorem 7.5.1. [24, Theorem 2.1] *The set $\mathcal{CS}_p$ becomes a linear space with the co-ordinatewise addition and scalar multiplication of double sequences which is linearly isomorphic to the space $\mathcal{C}_p$, and $\mathcal{CS}_p$ is a complete seminormed space with the seminorm*

$$\|x\|_\infty = \lim_{n\to\infty}\left(\sup_{k,l\geq n}\left|\sum_{i,j=0}^{k,l} x_{ij}\right|\right).$$

Proof. Because of this is easily obtained in the similar way used in proving Theorem 7.4.1, we omit the detailed proof. □

Theorem 7.5.2. [24, Theorem 2.2] *Neither of the spaces $\mathcal{BS}$ and $\mathcal{CS}_p$ includes the other one.*

Proof. For this, we must show the existence at least one sequence in the sets $\mathcal{BS}\setminus\mathcal{CS}_p$ and $\mathcal{CS}_p \setminus \mathcal{BS}$. Consider the sequences $x = (x_{mn})$ and $x' = (x'_{mn})$ defined by

$$x_{mn} := \begin{cases} 1 & , \quad m = 0,\ n \in \mathbb{N}, \\ -1 & , \quad m = 1,\ n \in \mathbb{N}, \\ 0 & , \quad \text{otherwise} \end{cases} \quad \text{and} \quad x'_{mn} := (-1)^{m+n}$$

for all $m, n \in \mathbb{N}$. Then, the sequences $s = (s_{mn})$ and $s' = (s'_{mn})$ of partial sums of double series $\sum_{m,n} x_{mn}$ and $\sum_{m,n} x'_{mn}$ are given by

$$s_{mn} := \begin{cases} n & , \quad m = 0 \text{ and } n \in \mathbb{N}, \\ 0 & , \quad \text{otherwise} \end{cases} \quad \text{and} \quad s'_{mn} := \begin{cases} 1 & , \quad m, n = 2k, \ k \in \mathbb{N}, \\ 0 & , \quad \text{otherwise,} \end{cases}$$

for all $m, n \in \mathbb{N}$, respectively. In this situation, it is easy to see that

$$\sup_{m,n\in\mathbb{N}} \left| \sum_{i,j=0}^{m,n} x_{ij} \right| = \infty,$$

while $(s_{mn}) \in \mathcal{C}_p$. That is to say that $x \in \mathcal{CS}_p \setminus \mathcal{BS}$. Since (s'_{mn}) is in $\mathcal{M}_u$ but not in $\mathcal{C}_p$, it is trivial that $x' \in \mathcal{BS} \setminus \mathcal{CS}_p$. Hence, the sequence spaces $\mathcal{BS}$ and $\mathcal{CS}_p$ overlap but neither contains the other.

This completes the proof. □

Theorem 7.5.3. [24, Theorem 2.3] *$\mathcal{CS}_p$ is the subspace of $\mathcal{C}_p$.*

Proof. Let us take $x = (x_{ij}) \in \mathcal{CS}_p$. Then, for a given $\varepsilon > 0$, there exists an $n_x(\varepsilon) \in \mathbb{N}$ such that $\left|\sum_{i,j=0}^{k,m} x_{ij} - l\right| < \varepsilon/4$ for all $k, m > n_x(\varepsilon)$. Then,

$$\begin{aligned} |x_{km}| &= \left| \sum_{i,j=0}^{k,m} x_{ij} - \sum_{i,j=0}^{k-1,m} x_{ij} - \sum_{i,j=0}^{k,m-1} x_{ij} - \sum_{i,j=0}^{k-1,m-1} x_{ij} \right| \\ &\leq \left| \sum_{i,j=0}^{k,m} x_{ij} - l \right| + \left| \sum_{i,j=0}^{k-1,m} x_{ij} - l \right| + \left| \sum_{i,j=0}^{k,m-1} x_{ij} - l \right| + \left| \sum_{i,j=0}^{k-1,m-1} x_{ij} - l \right| \\ &< \frac{\varepsilon}{4} + \frac{\varepsilon}{4} + \frac{\varepsilon}{4} + \frac{\varepsilon}{4} = \varepsilon \end{aligned}$$

for sufficiently large k, m; which means that $p - \lim x_{ij} = 0$. Hence, $x \in \mathcal{C}_p$ that is to say that $\mathcal{CS}_p \subset \mathcal{C}_p$ holds, as expected. □

To avoid the repetition of the similar statements, we give the following theorems without proof, related to the spaces $\mathcal{CS}_{bp}$ and $\mathcal{CS}_r$ consisting of the series whose sequences of partial sums are in $\mathcal{C}_{bp}$ and $\mathcal{C}_r$, respectively, since they may be proved in the similar way used in proving Theorem 7.5.1 and Theorem 7.5.2.

Theorem 7.5.4. [24, Theorem 2.4] *The sets $\mathcal{CS}_{bp}$ and $\mathcal{CS}_r$ become linear spaces with the co-ordinatewise addition and scalar multiplication of double sequences which are norm isomorphic to the spaces $\mathcal{C}_{bp}$ and $\mathcal{C}_r$, respectively.*

Theorem 7.5.5. [24, Theorem 2.5] *The linear spaces $\mathcal{CS}_{bp}$ and $\mathcal{CS}_r$ are Banach spaces endowed with the norm $\|x\|_\infty = \sup_{k,l\in\mathbb{N}} \left|\sum_{i,j=0}^{k,l} x_{ij}\right|$.*

Theorem 7.5.6. [24, Theorem 2.6] *The space $\mathcal{CS}_{bp}$ includes the space $\mathcal{CS}_r$.*

We continue by giving some knowledge concerning with the set $\mathcal{BV}$ of double sequences of bounded bivariation.

Theorem 7.5.7. [24, Theorem 2.7] *The set $\mathcal{BV}$ becomes a linear space with the coordinatewise addition and scalar multiplication of double sequences which is norm isomorphic to the space $\mathcal{L}_u$ of absolutely convergent double series.*

Proof. This is similar to the proof of Theorem 7.4.1 with T defined from $\mathcal{BV}$ to $\mathcal{L}_u$ by $x \mapsto Tx = (x_{ij} - x_{i-1,j} - x_{i,j-1} + x_{i-1,j-1})$ instead of T defined from $\mathcal{BS}$ to $\mathcal{M}_u$ by $x \mapsto Tx = s$. So, we can leave the detail to the reader. □

Because of the isomorphism T, defined in the proof of Theorem 7.5.7, is onto we have:

Theorem 7.5.8. [24, Theorem 2.8] *The linear space $\mathcal{BV}$ is a Banach space endowed with the norm defined by*

$$\|x\|_{\mathcal{BV}} = \sum_{i,j} |x_{ij} - x_{i-1,j} - x_{i,j-1} + x_{i-1,j-1}|. \tag{7.5.1}$$

Theorem 7.5.9. [24, Theorem 2.9] *Let $\vartheta \in \{p, bp, r\}$. Then, the inclusions $\mathcal{BV} \subset \mathcal{C}_\vartheta$ and $\mathcal{BV} \subset \mathcal{M}_u$ strictly hold.*

Proof. Let $x = (x_{ij}) \in \mathcal{BV}$. Then, since

$$\sum_{i,j} |x_{ij} - x_{i-1,j} - x_{i,j-1} + x_{i-1,j-1}| < \infty,$$

the series $\sum_{i,j}(x_{ij} - x_{i-1,j} - x_{i,j-1} + x_{i-1,j-1})$ converges. In this situation, since the sequence x can be written in the form

$$x_{mn} = \sum_{i,j=0}^{m,n} (x_{ij} - x_{i-1,j} - x_{i,j-1} + x_{i-1,j-1}) \text{ for all } m, n \in \mathbb{N}, \tag{7.5.2}$$

it is easily seen that $x \in \mathcal{C}_\vartheta$, i.e., the inclusion $\mathcal{BV} \subset \mathcal{C}_\vartheta$ holds.

To prove the inclusion $\mathcal{BV} \subset \mathcal{M}_u$, consider the inequality derived by (7.5.2) for any $x = (x_{ij}) \in \mathcal{BV}$ that

$$\begin{aligned} \sup_{m,n\in\mathbb{N}} |x_{mn}| &= \sup_{m,n\in\mathbb{N}} \left| \sum_{i,j=0}^{m,n} (x_{ij} - x_{i-1,j} - x_{i,j-1} + x_{i-1,j-1}) \right| \\ &\leq \sup_{m,n\in\mathbb{N}} \sum_{i,j=0}^{m,n} |x_{ij} - x_{i-1,j} - x_{i,j-1} + x_{i-1,j-1}| < \infty. \end{aligned}$$

This shows that $x \in \mathcal{M}_u$, i.e., the inclusion $\mathcal{BV} \subset \mathcal{M}_u$ also holds.

Additionally, since the sequence $x = (x_{mn})$ defined by

$$x_{mn} := \begin{cases} 2/(n+2) & , \quad m = 0 \text{ and } n \text{ is even}, \\ 0 & , \quad \text{otherwise}, \end{cases}$$

for all $m, n \in \mathbb{N}$ is in both $\mathcal{C}_\vartheta$ and $\mathcal{M}_u$ but not in $\mathcal{BV}$, the inclusions $\mathcal{BV} \subset \mathcal{C}_\vartheta$ and $\mathcal{BV} \subset \mathcal{M}_u$ are strict.

This completes the proof. □

Theorem 7.5.10. [24, Theorem 2.10] *The space $\mathcal{BV}$ is not monotone.*

Proof. This is easily obtained by proceeding as in the proof of Theorem 7.4.3 with the sequences $x = (x_{mn}) \in \mathcal{BV}$ and $y = (y_{mn}) \in \{0,1\}^{\mathbb{N}\times\mathbb{N}}$ defined by

$$x_{mn} := \begin{cases} 1 & , \quad m = 0, \ n \in \mathbb{N}, \\ 0 & , \quad m \in \mathbb{N}_1 \end{cases} \quad \text{and} \quad y_{mn} := \begin{cases} 1 & , \quad m+n \text{ is odd}, \\ 0 & , \quad \text{otherwise} \end{cases}$$

whose co-ordinatewise product $z = (z_{mn})$ is

$$z_{mn} := \begin{cases} 1 & , \quad m = 0 \text{ and } n \text{ is even}, \\ 0 & , \quad \text{otherwise}, \end{cases}$$

which is not in $\mathcal{BV}$. This shows that the space $\mathcal{BV}$ is not monotone. □

7.6 THE α- AND β-DUALS OF THE SPACES OF DOUBLE SERIES

In this section, we determine the α- and β-duals of some spaces of double series. Although the α-duals of the spaces of double series are unique, their β-duals may be more than one with respect to υ-convergence.

Theorem 7.6.1. [24, Theorem 3.1] *The α-dual of the spaces $\mathcal{BS}$, $\mathcal{BV}$ and $\mathcal{CS}_{bp}$ is the space $\mathcal{L}_u$.*

Proof. Let $\lambda \in \{\mathcal{BS}, \mathcal{CS}_{bp}, \mathcal{BV}\}$. Since $\lambda \subset \mathcal{M}_u$ and $\mathcal{M}_u^\alpha = \mathcal{L}_u$, the validity of the inclusion

$$\mathcal{L}_u \subset \lambda^\alpha \tag{7.6.1}$$

is clear.

Conversely, suppose that $a = (a_{ij}) \in \mathcal{BS}^\alpha$ but $a = (a_{ij}) \notin \mathcal{L}_u$. Then, one can easily see in the special case $x = (x_{ij}) = \{(-1)^{i+j}\} \in \mathcal{BS}$ that

$$\sum_{i,j} |a_{ij}x_{ij}| = \sum_{i,j} |a_{ij}| = \infty.$$

This means that $a \notin \mathcal{BS}^\alpha$ which contradicts the hypothesis. Hence, $a = (a_{ij})$ must be in $\mathcal{L}_u$ that is to say that the inclusion

$$\mathcal{BS}^\alpha \subset \mathcal{L}_u \tag{7.6.2}$$

must be hold. By combining the inclusions (7.6.1) and (7.6.2), the desired result immediately follows.

In the similar way, the inclusions $\mathcal{BV}^\alpha \subset \mathcal{L}_u$ and $\mathcal{CS}_{bp}^\alpha \subset \mathcal{L}_u$ easily hold.

This completes the proof. □

Now, we may give the β-duals of the spaces of double series with respect to the υ-convergence for $\upsilon \in \{p, bp, r\}$ using the technique in [65] and [66] for the single sequences.

Lemma 7.6.2. [413, p. 86] *The matrix $A = (a_{mnkl})$ is $\mathcal{C}_\vartheta$-conservative for $\vartheta = r$ if and only if*

$$\sup_{m,n\in\mathbb{N}} \sum_{k,l} |a_{mnkl}| < \infty, \tag{7.6.3}$$

$$\vartheta - \lim_{m,n\to\infty} a_{mnkl} = \alpha_{kl} \text{ for each } (k,l) \in \mathbb{N}^2, \tag{7.6.4}$$

$$\vartheta - \lim_{m,n\to\infty} \sum_{k} a_{mnkl_0} = u^{l_0} \text{ for a fixed } l_0 \in \mathbb{N}, \tag{7.6.5}$$

$$\vartheta - \lim_{m,n\to\infty} \sum_{l} a_{mnk_0l} = y_{k_0} \text{ for a fixed } k_0 \in \mathbb{N},$$

$$\vartheta - \lim_{m,n\to\infty} \sum_{k,l} a_{mnkl} = v \text{ exists.} \tag{7.6.6}$$

Theorem 7.6.3. [24, Theorem 3.3] *The $\beta(r)$-dual of the space $\mathcal{CS}_r$ is the space $\mathcal{BV}$.*

Proof. Suppose that $x = (x_{mn}) \in \mathcal{CS}_r$. Then, the double sequence $s = (s_{mn})$, defined by (7.1.1), is in the space $\mathcal{C}_r$, by Theorem 7.5.4. Let us determine the necessary and sufficient condition in order to the series

$$\sum_{i,j} a_{ij}x_{ij} \tag{7.6.7}$$

to be r-convergent for a sequence $a = (a_{ij}) \in \Omega$.

We obtain by applying the Abel generalized transformation for double sequences to the m, n^{th} partial sums of the series in (7.6.7) that

$$\begin{aligned} z_{mn} &= \sum_{i,j=0}^{m,n} a_{ij}x_{ij} \\ &= \sum_{i,j=0}^{m-1,n-1} s_{ij}\Delta_{11}a_{ij} + \sum_{i=0}^{m-1} s_{in}\Delta_{10}a_{in} + \sum_{j=0}^{n-1} s_{mj}\Delta_{01}a_{mj} + s_{mn}a_{mn} \end{aligned} \tag{7.6.8}$$

for all $m, n \in \mathbb{N}$. (7.6.8) may be rewritten by the matrix representation as follows:

$$z_{mn} = \sum_{i,j=0}^{m,n} b_{mnij}s_{ij} = (Bs)_{mn} \tag{7.6.9}$$

for all $m, n \in \mathbb{N}$, where the four-dimensional matrix $B = (b_{mnij})$ is defined by

$$b_{mnij} := \begin{cases} \Delta_{11}a_{ij} & , \quad 0 < i \leq m-1,\ 0 < j \leq n-1, \\ \Delta_{10}a_{in} & , \quad 0 < i \leq m-1,\ j = n, \\ \Delta_{01}a_{mj} & , \quad i = m,\ 0 < j \leq n-1, \\ a_{mn} & , \quad i = m,\ j = n, \\ 0 & , \quad \text{otherwise} \end{cases} \tag{7.6.10}$$

for all $i, j, m, n \in \mathbb{N}$. We therefore read from the equality (7.6.9) that $ax = (a_{mn}x_{mn}) \in \mathcal{CS}_r$ whenever $x = (x_{mn}) \in \mathcal{CS}_r$ if and only if $z = (z_{mn}) \in \mathcal{C}_r$ whenever $s = (s_{mn}) \in \mathcal{C}_r$ which leads us to the fact that $B = (b_{mnij})$, defined by (7.6.10), is in the class $(\mathcal{C}_r : \mathcal{C}_r)$. Thus, we see from Lemma 7.6.2 for the matrix B, defined above, that the conditions

$$\begin{aligned}
&\sup_{m,n\in\mathbb{N}} \sum_{k,l=0}^{m,n} |b_{mnkl}| = \\
&= \sup_{m,n\in\mathbb{N}} \left\{ \sum_{k,l=0}^{m-1,n-1} |\Delta_{11}a_{kl}| + \sum_{k=0}^{m-1} |\Delta_{10}a_{kn}| + \sum_{l=0}^{n-1} |\Delta_{01}a_{ml}| + |a_{mn}| \right\} < \infty, \\
&r - \lim_{m,n\to\infty} b_{mnkl} = \Delta_{11}a_{kl} \text{ for each } (k,l) \in \mathbb{N}^2, \\
&r - \lim_{m,n\to\infty} \sum_k b_{mnkl_0} = \Delta_{01}a_{1l_0} \text{ for a fixed } l_0 \in \mathbb{N}, \\
&r - \lim_{m,n\to\infty} \sum_l b_{mnk_0l} = \Delta_{10}a_{k_01} \text{ for a fixed } k_0 \in \mathbb{N}, \\
&r - \lim_{m,n\to\infty} \sum_{k,l} b_{mnkl} = a_{00}
\end{aligned}$$

hold. Therefore, we derive from the conditions (7.6.3)–(7.6.6) that

$$\sum_{k,l} |\Delta_{11}a_{kl}| < \infty, \tag{7.6.11}$$

$$\sup_{n\in\mathbb{N}} \sum_k |\Delta_{10}a_{kn}| < \infty, \tag{7.6.12}$$

$$\sup_{m\in\mathbb{N}} \sum_l |\Delta_{01}a_{ml}| < \infty. \tag{7.6.13}$$

Further, since the equality

$$\sum_{k=0}^{m} |b_{mnk0}| = \sum_{k=0}^{m-1} |\Delta_{10}a_{k0}| + |a_{m0}|$$

holds, the supremum over $m \in \mathbb{N}$ of this expression gives that

$$(a_{k0})_{k\in\mathbb{N}} \in bv \tag{7.6.14}$$

and similarly for $m = 0$, the equality

$$\sum_{l=0}^{n} |b_{mn0l}| = \sum_{l=0}^{n-1} |\Delta_{01}a_{0l}| + |a_{0n}|$$

holds which also gives by taking supremum over $n \in \mathbb{N}$ that

$$(a_{0l})_{l\in\mathbb{N}} \in bv. \tag{7.6.15}$$

Besides, by using the equality

$$\Delta_{10}a_{kn} = \Delta_{10}a_{k0} - \sum_{l=0}^{n-1} \Delta_{11}a_{kl}$$

we see that

$$\begin{aligned}\sum_k |\Delta_{10}a_{kn}| &= \sum_k \left|\Delta_{10}a_{k0} - \sum_{l=0}^{n-1}\Delta_{11}a_{kl}\right| \\ &\leq \sum_k |\Delta_{10}a_{k0}| + \sum_k \left|\sum_{l=0}^{n-1}\Delta_{11}a_{kl}\right|.\end{aligned} \tag{7.6.16}$$

Since, it is seen by (7.6.14) that $\sum_k |\Delta_{10}a_{k0}| < \infty$ and (7.6.14) implies (7.6.12) when (7.6.11) considered in the equality (7.6.16). Similarly, by employing the equality

$$\Delta_{01}a_{ml} = \Delta_{01}a_{0l} - \sum_{k=0}^{m-1}\Delta_{11}a_{kl}$$

one can see from here that

$$\begin{aligned}\sum_l |\Delta_{01}a_{ml}| &= \sum_l \left|\Delta_{01}a_{0l} - \sum_{k=0}^{m-1}\Delta_{11}a_{kl}\right| \\ &\leq \sum_l |\Delta_{01}a_{0l}| + \sum_l \left|\sum_{k=0}^{m-1}\Delta_{11}a_{kl}\right|\end{aligned}$$

by (7.6.15) and (7.6.11) that (7.6.13) holds. Therefore, it is deduced that the conditions on the sequence $a = (a_{ij}) \in \Omega$ in order for the series in (7.6.7) to be r-convergent are

$$\sum_{k,l} |\Delta_{11}a_{kl}| < \infty \quad \text{and} \quad (a_{k0})_{k\in\mathbb{N}},\ (a_{0l})_{l\in\mathbb{N}} \in bv.$$

This shows that $\mathcal{CS}_r^{\beta(r)} = \mathcal{BV}$, which completes the proof. □

Now, we may give our theorem exhibiting the $\beta(\vartheta)$-dual of the series space $\mathcal{CS}_\eta$ in the case $\eta, \vartheta \in \{p, bp, r\}$, without proof.

Theorem 7.6.4. [24, Theorem 3.4] *Let $\eta, \vartheta \in \{p, bp, r\}$. Then, the $\beta(\vartheta)$-dual of the space $\mathcal{CS}_\eta$ is the set*

$$\{a = (a_{ij}) \in \Omega : B = (b_{mnij}) \in (\mathcal{C}_\eta : \mathcal{C}_\vartheta)\},$$

where $B = (b_{mnij})$ is defined by (7.6.10).

As a consequence of Theorem 7.6.4, we have:

$$\mathcal{CS}_r^{\beta(p)} := \mathcal{CS}_r^{\beta(r)} = \mathcal{BV},$$

$$\mathcal{CS}_{bp}^{\beta(bp)} := \left\{(a_{kl}) \in \Omega : \sum_{k,l} |\Delta_{11}a_{kl}| < \infty,\ (\Delta_{10}a_{in})_{n\in\mathbb{N}}, (\Delta_{10}a_{mj})_{m\in\mathbb{N}} \in c_0;\ (i, j \in \mathbb{N})\right\},$$

$$\mathcal{CS}_{bp}^{\beta(p)} := \left\{(a_{kl}) \in \Omega : \sum_{k,l} |\Delta_{11}a_{kl}| < \infty,\ (\Delta_{10}a_{kl})_{l\in\mathbb{N}} \in c_0;\ (k \in \mathbb{N})\right\}.$$

7.7 CHARACTERIZATION OF SOME CLASSES OF FOUR-DIMENSIONAL MATRICES

In this section, we characterize some matrix mappings from some double series spaces to the double sequence spaces $\mathcal{C}_p$ and $\mathcal{C}_r$. Although the theorem characterizing the class $(\mathcal{CS}_{bp} : \mathcal{C}_p)$ is stated and proved, the necessary and sufficient conditions on a four-dimensional matrix belonging to the classes $(\mathcal{CS}_r : \mathcal{C}_p)$, $(\mathcal{CS}_r : \mathcal{C}_r)$ and $(\mathcal{CS}_p : \mathcal{C}_p; p)$ also given without proof.

Theorem 7.7.1. [24, Theorem 4.1] $A = (a_{mnkl}) \in (\mathcal{CS}_{bp} : \mathcal{C}_p)$ *if and only if*

$$\sup_{m,n\in\mathbb{N}} \sum_{k,l} \left|\Delta_{11}^{kl} a_{mnkl}\right| < \infty,$$

$$\lim_{l\to\infty} \Delta_{10}^{kl} a_{mnkl} = 0 \quad \text{for each } k, m, n \in \mathbb{N},$$

$$\lim_{k\to\infty} \Delta_{01}^{kl} a_{mnkl} = 0 \quad \text{for each } l, m, n \in \mathbb{N},$$

$$bp - \lim_{m,n\to\infty} \Delta_{11}^{kl} a_{mnkl} = a_{kl} \quad \text{for each } (k,l) \in \mathbb{N}^2,$$

$$bp - \lim_{m,n\to\infty} \sum_{l} \left|\Delta_{11}^{kl} a_{mnkl}\right| = \sum_{k} |a_{kl}|.$$

Proof. Let us take any $x = (x_{ij}) \in \mathcal{CS}_{bp}$ and define the sequence $y = (y_{kl})$ by

$$y_{kl} := \sum_{i,j=0}^{k,l} x_{ij} \quad \text{for all} \quad (k,l) \in \mathbb{N}^2.$$

Then, $y = (y_{kl}) \in \mathcal{C}_{bp}$ by Theorem 7.5.4. By applying the double Abel transformation to the s,t^{th} partial sums $(Ax)_{mn}^{[s,t]}$ of the series $\sum_{k,l} a_{mnkl} x_{kl}$ for each $m, n \in \mathbb{N}$, we derive that

$$\begin{aligned} (Ax)_{mn}^{[s,t]} &= \sum_{k,l=0}^{s,t} a_{mnkl} x_{kl} \qquad (7.7.1) \\ &= \sum_{k,l=0}^{s-1,t-1} \Delta_{11}^{kl} a_{mnkl} y_{kl} + \sum_{k=0}^{s-1} \Delta_{10}^{kt} a_{mnkt} y_{kt} + \\ &\quad + \sum_{l=0}^{t-1} \Delta_{01}^{sl} a_{mnsl} y_{sl} + a_{mnst} y_{st}. \end{aligned}$$

Define $B_{mn} = (b_{stkl}^{mn})$ by

$$b_{stkl}^{mn} := \begin{cases} \Delta_{11}^{kl} a_{mnkl} &, \quad k \leq s-1, \ l \leq t-1, \\ \Delta_{10}^{kl} a_{mnkl} &, \quad k = s, \ l \leq t-1, \\ \Delta_{01}^{kl} a_{mnkl} &, \quad k \leq s-1, \ l = t, \\ a_{mnst} &, \quad k = s, \ l = t, \\ 0 &, \quad \text{otherwise} \end{cases} \qquad (7.7.2)$$

for all $s, t, k, l \in \mathbb{N}$. Then, the equality (7.7.1) can be rewritten as in the form

$$(Ax)_{mn}^{[s,t]} := (B_{mn} y)_{[s,t]}. \qquad (7.7.3)$$

Therefore, one can see by (7.7.3) that the convergence of the partial sums $(Ax)_{mn}^{[s,t]}$ in the Pringsheim's sense for all $m, n \in \mathbb{N}$ and for all $x \in \mathcal{CS}_{bp}$ implies the fact that $B_{mn} \in (\mathcal{C}_{bp} : \mathcal{C}_{bp})$ and hence, the conditions

$$\sum_{k,l} \left|\Delta_{11}^{kl} a_{mnkl}\right| < \infty, \ \lim_{l\to\infty} \Delta_{10}^{kl} a_{mnkl} = 0 \ \text{ for each } \ k \in \mathbb{N} \tag{7.7.4}$$
$$\text{and } \lim_{k\to\infty} \Delta_{01}^{kl} a_{mnkl} = 0 \ \text{ for each } \ l \in \mathbb{N}$$

must be satisfied for every fixed $m, n \in \mathbb{N}$. In this situation,

$$p - \lim_{s,t\to\infty} b_{stkl}^{mn} = \Delta_{11}^{kl} a_{mnkl} \ \text{ and } \ p - \lim_{s,t\to\infty} (Ax)_{mn}^{[s,t]} = p - (\lim By)_{mn}$$

hold. Thus, we derive from the two-sided implication "Ax is in $\mathcal{C}_p$ whenever $x \in \mathcal{CS}_{bp}$ if and only if $B = (\Delta_{11}^{kl} a_{mnkl}) \in (\mathcal{C}_{bp} : \mathcal{C}_p)$" that

$$\sup_{m,n\in\mathbb{N}} \sum_{k,l} \left|\Delta_{11}^{kl} a_{mnkl}\right| < \infty, \tag{7.7.5}$$

$$bp - \lim_{m,n\to\infty} \Delta_{11}^{kl} a_{mnkl} = a_{kl} \ \text{ and } \ bp - \lim_{m,n\to\infty} \sum_{l} \left|\Delta_{11}^{kl} a_{mnkl}\right| = \sum_{k} |a_{kl}|. \tag{7.7.6}$$

Now, it is deduced by the conditions (7.7.4)–(7.7.6) that $A = (a_{mnkl}) \in (\mathcal{CS}_{bp} : \mathcal{C}_p)$ if and only if the following conditions

$$\sup_{m,n\in\mathbb{N}} \sum_{k,l} \left|\Delta_{11}^{kl} a_{mnkl}\right| < \infty,$$
$$\lim_{l\to\infty} \Delta_{10}^{kl} a_{mnkl} = 0 \ \text{ for each } \ k, m, n \in \mathbb{N},$$
$$\lim_{k\to\infty} \Delta_{01}^{kl} a_{mnkl} = 0 \ \text{ for each } \ l, m, n \in \mathbb{N},$$
$$bp - \lim_{m,n\to\infty} \Delta_{11}^{kl} a_{m} = a_{kl} \ \text{ for all } \ k, l \in \mathbb{N},$$
$$bp - \lim_{m,n\to\infty} \sum_{l} \left|\Delta_{11}^{kl} a_{mnkl}\right| = \sum_{k} |a_{kl}|$$

hold.

This step concludes the proof. □

Theorem 7.7.2. [24, Theorem 4.2] *$A = (a_{mnkl}) \in (\mathcal{CS}_r : \mathcal{C}_p)$ if and only if*

$\sup_{m,n\in\mathbb{N}} \sum_{k,l} \left|\Delta_{11}^{kl} a_{mnkl}\right| < \infty,$
$(a_{mnk0})_{k\in\mathbb{N}}, \ (a_{mn0l})_{l\in\mathbb{N}} \in bv$ for each $(m, n) \in \mathbb{N}^2$,
$p - \lim \sum_{l} \left|\Delta_{11}^{kl} a_{mnkl}\right| = \sum_{k} \left|\Delta_{11}^{kl} a_{mnkl}\right|,$
There exists $L \in \mathbb{N}$ such that $\Delta_{11}^{kl} a_{mnkl} = 0$ for each $k \in \mathbb{N}$ whenever $m, n, l > L$,
There exists $K \in \mathbb{N}$ such that $\Delta_{11}^{kl} a_{mnkl} = 0$ for each $l \in \mathbb{N}$ whenever $m, n, k > K$.

Theorem 7.7.3. [24, Theorem 4.3] $A = (a_{mnij}) \in (\mathcal{CS}_r : \mathcal{C}_r)$ *if and only if*

$$\sup_{m,n\in\mathbb{N}} \sum_{k,l} \left|\Delta_{11}^{kl} a_{mnkl}\right| < \infty,$$

$$(a_{mnk0})_{k\in\mathbb{N}}, (a_{mn0l})_{l\in\mathbb{N}} \in bv \text{ for each } (m,n) \in \mathbb{N}^2,$$

$$r - \lim_{m,n\to\infty} \Delta_{11}^{kl} a_{mnkl} = a_{kl} \text{ for each } (k,l) \in \mathbb{N}^2,$$

$$r - \lim_{m,n\to\infty} \sum_{k} \Delta_{11}^{kl} a_{mnkl_0} = u^{l_0} \text{ for a fixed } l_0 \in \mathbb{N},$$

$$r - \lim_{m,n\to\infty} \sum_{l} \Delta_{11}^{kl} a_{mnk_0l} = y_{k_0} \text{ for a fixed } k_0 \in \mathbb{N},$$

$$r - \lim_{m,n\to\infty} \sum_{k,l} \Delta_{11}^{kl} a_{mnkl} = v.$$

Theorem 7.7.4. [24, Theorem 4.4] $A = (a_{mnij}) \in (\mathcal{CS}_p : \mathcal{C}_p; p)$ *if and only if*

$$p - \lim_{m,n\to\infty} \Delta_{11}^{kl} a_{mnkl} = 0 \text{ for each } k,l \in \mathbb{N},$$

$$p - \lim_{m,n\to\infty} \sum_{k,l} \Delta_{11}^{kl} a_{mnkl} = 1,$$

$$p - \lim_{m,n\to\infty} \sum_{k} \left|\Delta_{11}^{kl} a_{mnkl}\right| = 0 \text{ for each } l \in \mathbb{N},$$

$$p - \lim_{m,n\to\infty} \sum_{l} \left|\Delta_{11}^{kl} a_{mnkl}\right| = 0 \text{ for each } k \in \mathbb{N},$$

$$p - \lim_{m,n\to\infty} \sum_{k,l} \left|\Delta_{11}^{kl} a_{mnkl}\right| = v,$$

$$\sup_{B\in\mathbb{N}} \sum_{k,l>B} \left|\Delta_{11}^{kl} a_{mnkl}\right| < \infty.$$

7.8 BINOMIAL SPACES OF DOUBLE SEQUENCES

In this section, following Demiriz and Erdem [143], we essentially deal with the spaces $\mathcal{B}_\infty^{r,s}$, $\mathcal{B}^{r,s}$, $\mathcal{B}_{bp}^{r,s}$ and $\mathcal{B}_{reg}^{r,s}$ of double sequences generated by the domain of four-dimensional binomial matrix $B^{(r,s)}$ in the spaces $\mathcal{M}_u$, $\mathcal{C}_p$, $\mathcal{C}_{bp}$ and $\mathcal{C}_r$, respectively.

Let us remember the definition of four-dimensional triangle matrix. For all $k,l,i,j \in \mathbb{N}$, we say that $D = (d_{klij})$ is a *triangular matrix* if $d_{klij} = 0$ for $i > k$ or $j > l$ or both [1]. By following Adams [1], we also say that a triangular matrix $D = (d_{klij})$ is called a *triangle* if $d_{klkl} \neq 0$ for all $k,l \in \mathbb{N}$. Referring to Cooke [111, Remark (a), p. 22], one can conclude that every triangle matrix has an unique inverse which is also a triangle.

Now, we deal with on the matrix mapping. Let us consider the double sequence spaces Ψ and Λ, and the four-dimensional complex infinite matrix $D = (d_{klij})$. If for every $x = (x_{ij}) \in \Psi$, $(Du)_{kl} = \vartheta - \sum_{i,j} d_{klij} u_{ij}$ exists and is in Λ, then it is said that D is a matrix mapping from Ψ into Λ and is written as $D : \Psi \to \Lambda$.

Let $(\Psi : \Lambda) = \{D = (d_{klij}) | D : \Psi \to \Lambda\}$. Here, $D \in (\Psi : \Lambda)$ iff $D_{kl} \in \Psi^{\beta(\vartheta)}$, where $D_{kl} = (d_{klij})_{i,j\in\mathbb{N}}$ for each fixed $k,l \in \mathbb{N}$.

The domain $\Psi_D^{(\vartheta)}$ of D in a double sequence space Ψ consists of whose D-transforms are in Ψ, that is,

$$\Psi_D^{(\vartheta)} := \left\{ x = (x_{ij}) \in \Omega : Dx = \left(\vartheta - \sum_{i,j} d_{klij} x_{ij} \right)_{k,l \in \mathbb{N}} \text{ exists and is in } \Psi \right\} \quad (7.8.1)$$

Let r, s and $r+s$ are non-zero real numbers. We define the four-dimensional binomial matrix $B^{(r,s)} = (b_{klij}^{r,s})$ of orders r and s by

$$b_{klij}^{r,s} := \begin{cases} \frac{1}{(r+s)^{k+l}} \binom{k}{i} \binom{l}{j} s^{k+j-i} r^{l+i-j} & , \quad 0 \leq i \leq k \ , \ 0 \leq j \leq l, \\ 0 & , \quad \text{otherwise}, \end{cases} \quad (7.8.2)$$

for all $k, l, i, j \in \mathbb{N}$. Thus, the $B^{(r,s)}$-transform of $x = (x_{ij}) \in \Omega$ is given by

$$y_{kl} := \left(B^{(r,s)} x \right)_{kl} = \sum_{i,j}^{k,l} \frac{1}{(r+s)^{k+l}} \binom{k}{i} \binom{l}{j} s^{k+j-i} r^{l+i-j} u_{ij}. \quad (7.8.3)$$

It will be assume unless stated otherwise that the double sequences (u_{ij}) and (y_{ij}) are as in the equality (7.8.3) and r, s and $r+s$ are non-zero real numbers. We would like touch on a point, while $r+s=1$, it is obtained four-dimensional Euler matrix $E(r,s) = (e_{klij}^{r,s})$ from the four-dimensional binomial matrix. So, $B^{(r,s)}$ is more general then $E(r,s)$. Consider that the four-dimensional unit matrix $I = (\delta_{klij})$ defined by

$$\delta_{klij} := \begin{cases} 1 & , \quad (k,l) = (i,j), \\ 0 & , \quad \text{otherwise}. \end{cases}$$

for all $k, l, i, j \in \mathbb{N}$. The following equality

$$\delta_{klij} = \sum_{m,n} b_{klmn}^{r,s} c_{mnij}^{r,s},$$

leads to the inverse $\{B^{(r,s)}\}^{-1} = C^{(r,s)} = (c_{klij}^{r,s})$ given by

$$c_{klij}^{r,s} := \begin{cases} (-1)^{k+l-(i+j)} \binom{k}{i} \binom{l}{j} \binom{k}{i} \binom{l}{j} s^{k-l-i} r^{l-k-j} (r+s)^{i+j} & , \quad 0 \leq i \leq k, \ 0 \leq j \leq l, \\ 0 & , \quad \text{otherwise}, \end{cases}$$

for all $k, l, i, j \in \mathbb{N}$. Now, we define the spaces $\mathcal{B}_\infty^{r,s}$, $\mathcal{B}^{r,s}$, $\mathcal{B}_{bp}^{r,s}$ and $\mathcal{B}_{reg}^{r,s}$ by the domain of four-dimensional binomial matrix $B^{(r,s)}$ in the spaces $\mathcal{M}_u$, $\mathcal{C}_p$, $\mathcal{C}_{bp}$ and $\mathcal{C}_r$, respectively, that is,

$$
\begin{aligned}
\mathcal{B}_{\infty}^{r,s} &:= \left\{x=(x_{ij})\in\Omega : \sup_{k,l\in\mathbb{N}}\left|(B^{(r,s)}x)_{kl}\right|<\infty\right\},\\
\mathcal{B}^{r,s} &:= \left\{x=(x_{ij})\in\Omega : \exists L\in\mathbb{C}\ni p-\lim_{k,l\to\infty}\left|(B^{(r,s)}x)_{kl}-L\right|=0\right\},\\
\mathcal{B}_{bp}^{r,s} &:= \left\{x=(x_{ij})\in\Omega : B^{(r,s)}x\in\mathcal{C}_{bp}\right\},\\
\mathcal{B}_{reg}^{r,s} &:= \left\{x=(x_{ij})\in\Omega : B^{(r,s)}x\in\mathcal{C}_{r}\right\}.
\end{aligned}
$$

The spaces $\mathcal{B}_{\infty}^{r,s}$, $\mathcal{B}^{r,s}$, $\mathcal{B}_{bp}^{r,s}$ and $\mathcal{B}_{reg}^{r,s}$ can be rewritten with the notation of (7.8.1) as $\mathcal{B}_{\infty}^{r,s} := (\mathcal{M}_u)_{B^{(r,s)}}$, $\mathcal{B}^{r,s} := (\mathcal{C}_p)_{B^{(r,s)}}$, $\mathcal{B}_{bp}^{r,s} := (\mathcal{C}_{bp})_{B^{(r,s)}}$ and $\mathcal{B}_{reg}^{r,s} := (\mathcal{C}_r)_{B^{(r,s)}}$, respectively.

Definition 7.8.1 ([197, 335]). *$D=(d_{klij})$ is entitled RH-regular if D transform all bounded p-convergent sequences into a p-convergent sequence with the same p-limit.*

Lemma 7.8.2 ([197, 335]). *A triangle $D=(d_{klij})$ is RH-regular iff*

$$
\begin{aligned}
RH_1 &: p-\lim_{k,l\to\infty} d_{klij}=0 \quad \text{for each} \quad i,j\in\mathbb{N},\\
RH_2 &: p-\lim_{k,l\to\infty}\sum_{i,j} d_{klij}=1,\\
RH_3 &: p-\lim_{k,l\to\infty}\sum_{i}|d_{klij}|=0 \quad \text{for each} \quad j\in\mathbb{N},\\
RH_4 &: p-\lim_{k,l\to\infty}\sum_{j}|d_{klij}|=0 \quad \text{for each} \quad i\in\mathbb{N},\\
RH_5 &: \sum_{i,j>\mu}|d_{klij}|<\eta, \quad \text{where } \mu \text{ and } \eta \text{ finite positive integers.}
\end{aligned}
$$

It is immediate that the four-dimensional matrix $B^{(r,s)}$ given by (7.8.2) is *RH*-regular whenever $rs>0$. In the rest of this section, we assume that $rs>0$.

Let X denote any of the spaces $\mathcal{M}_u$, $\mathcal{C}_p$, $\mathcal{C}_{bp}$ and $\mathcal{C}_r$ and the sequence $y=(y_{kl})$ be the $B^{(r,s)}$-transform of the sequence $x=(x_{ij})$. Consider the mapping T defined by

$$
\begin{aligned}
T \ : \ X_{B^{(r,s)}} &\longrightarrow X\\
x &\longmapsto Tx=y=B^{(r,s)}x.
\end{aligned}
$$

Now, one can derive by a straightforward calculation for $y\in X$ that

$$
\begin{aligned}
x_{kl} &:= \{(B^{(r,s)})^{-1}y\}_{kl}\\
&= \sum_{i,j=0}^{k,l}(-1)^{k+l-(i+j)}\binom{k}{i}\binom{l}{j}s^{k-l-i}r^{l-k-j}(r+s)^{i+j}y_{ij} \qquad (7.8.4)
\end{aligned}
$$

for every $k, l \in \mathbb{N}$. Since the mapping T has the matrix representation $B^{(r,s)}$ defined by (7.8.2) which is triangle, we therefore conclude that the mapping T is a norm preserving linear bijection. Hence, the space $X_{B^{(r,s)}}$ is linearly isomorphic to the space X, that is, $\mathcal{B}_{\infty}^{r,s} \cong \mathcal{M}_u$, $\mathcal{B}^{r,s} \cong \mathcal{C}_p$, $\mathcal{B}_{bp}^{r,s} \cong \mathcal{C}_{bp}$ and $\mathcal{B}_{reg}^{r,s} \cong \mathcal{C}_r$.

Theorem 7.8.3. *The set $\mathcal{B}^{r,s}$ is linearly isomorphic to the space $\mathcal{C}_p$ and is a complete semi-normed space with the semi-norm*

$$\|x\|_{\mathcal{B}^{r,s}} = \lim_{i\to\infty}\left[\sup_{k,l\geq i}\left|(B^{(r,s)}x)_{kl}\right|\right].$$

Proof. This is easily obtained by the usual way, so we omit details. □

Now, we give some inclusion relations.

Theorem 7.8.4. *The inclusion $\mathcal{M}_u \subset \mathcal{B}_{\infty}^{r,s}$ strictly holds.*

Proof. Suppose that $x = (x_{ij}) \in \mathcal{M}_u$ is an arbitrary double sequence. At that time, $\|x\|_\infty = \sup_{i,j\in\mathbb{N}} |x_{ij}| \leq \eta$, where $\eta \in \mathbb{R}^+$. Thus, it can be immediately seen from the inequality

$$\|x\|_{\mathcal{B}_{\infty}^{r,s}} = \sup_{k,l\in\mathbb{N}}\left|\frac{1}{(r+s)^{k+l}}\sum_{i,j=0}^{k,l}\binom{k}{i}\binom{l}{j}s^{k+j-i}r^{l+i-j}x_{ij}\right| \leq \|x\|_\infty$$

that the inclusion $\mathcal{M}_u \subset \mathcal{B}_{\infty}^{r,s}$ holds.

To show that the inclusion is strict, let us consider the sequence $x = (x_{kl}) = \left((-s-r)^{k+l}/(r^k s^l)\right)_{k,l\in\mathbb{N}}$. Then, $x = (x_{kl}) \notin \mathcal{M}_u$ but $B^{(r,s)}x = ((-1)^{k+l}r^k s^l/(r+s)^{k+l}) \in \mathcal{M}_u$, that is $x = (x_{kl}) \in \mathcal{B}_{\infty}^{r,s}$. Hence, $x \in \mathcal{B}_{\infty}^{r,s} \setminus \mathcal{M}_u$, as desired. □

Theorem 7.8.5. $\mathcal{C}_{bp} \subset \mathcal{B}^{r,s}$.

Proof. Consider the sequence $x = (x_{ij}) \in \mathcal{C}_{bp}$ with $p - \lim_{i,j\to\infty} u_{ij} = L$. Since the four-dimensional binomial matrix is RH-regular, then $p - \lim_{k,l\to\infty} y_{kl} = L$, where $y = (y_{kl}) = B^{(r,s)}x$. Hence, we see that $\mathcal{C}_{bp} \subset \mathcal{B}^{r,s}$, as desired. □

7.8.1 Dual Spaces of the Binomial Spaces

Throughout this subsection, we write the notations $\{\mathcal{B}_{\infty}^{r,s}\}^{\kappa_1}$, $\left\{\mathcal{B}_{bp}^{r,s}\right\}^{\kappa_2}$, $\{\mathcal{B}^{r,s}\}^{\beta(\vartheta)}$ and $\left\{\mathcal{B}_{reg}^{r,s}\right\}^{\beta(\vartheta)}$, where $\kappa_1 \in \{\alpha, \beta(bp), \beta(p), \gamma\}$ and $\kappa_2 \in \{\beta(\vartheta), \gamma\}$.

Theorem 7.8.6. $\{\mathcal{B}_{\infty}^{r,s}\}^{\alpha} = \mathcal{L}_u$.

Proof. Let us take $t = (t_{kl}) \in \mathcal{L}_u$ and $x = (x_{kl}) \in \mathcal{B}_{\infty}^{r,s}$. Then, $y = (y_{kl}) \in \mathcal{M}_u$ by (7.8.3) and there is $\eta \in \mathbb{R}^+$ such that $\sup_{k,l\in\mathbb{N}} |y_{kl}| < \eta$. Therefore, it is easy to see that

$$\begin{aligned}
\sum_{k,l} |t_{kl}x_{kl}| &= \sum_{k,l} |t_{kl}| \left| \sum_{i,j=0}^{k,l} (-1)^{k+l-(i+j)} \binom{k}{i}\binom{l}{j} s^{k-l-i} r^{l-k-j} (r+s)^{i+j} y_{ij} \right| \\
&\leq \sum_{k,l} |t_{kl}| \left| \frac{1}{r^k s^l} \sum_{i,j=0}^{k,l} \binom{k}{i}\binom{l}{j} (-s)^{k-i}(r+s)^i(-r)^{l-j}(r+s)^j \right| |y_{ij}| \\
&\leq \eta \sum_{k,l} |t_{kl}| \left| \frac{1}{r^k s^l} \sum_{i=0}^{k} \binom{k}{i} (-s)^{k-i}(r+s)^i \sum_{j=0}^{l} \binom{l}{j} (-r)^{l-j}(r+s)^j \right| \\
&= \eta \sum_{k,l} |t_{kl}|
\end{aligned}$$

that $t = (t_{kl}) \in \{\mathcal{B}_{\infty}^{r,s}\}^{\alpha}$. Hence, the inclusion $\mathcal{L}_u \subset \{\mathcal{B}_{\infty}^{r,s}\}^{\alpha}$ holds.

Conversely, let us assume the sequence $t = (t_{kl}) \in \{\mathcal{B}_{\infty}^{r,s}\}^{\alpha} - \mathcal{L}_u$. So, $\sum_{k,l} |t_{kl}x_{kl}| < \infty$ for every $x = (x_{kl}) \in \mathcal{B}_{\infty}^{r,s}$. If we consider $e \in \mathcal{B}_{\infty}^{r,s}$, then it is clear that $te = t \notin \mathcal{L}_u$, that is $t \notin \{\mathcal{B}_{\infty}^{r,s}\}^{\alpha}$, a contradiction. Thus, it is seen that $\{\mathcal{B}_{\infty}^{r,s}\}^{\alpha} \subset \mathcal{L}_u$. This completes the proof. □

Now, we give some lemmas which are needed here and after, (see [197, 413, 414]).

Lemma 7.8.7. *Suppose that $D = (d_{klij})$ be a four-dimensional infinite matrix. Then, $D = (d_{klij}) \in (\mathcal{C}_{bp} : \mathcal{C}_{\vartheta})$ if and only if following conditions hold:*

$$\sup_{k,l\in\mathbb{N}} \sum_{i,j} |d_{klij}| < \infty, \tag{7.8.5}$$

$$\exists d_{ij} \in \mathbb{C} \ni \vartheta - \lim_{k,l\to\infty} d_{klij} = d_{ij} \text{ for all } i,j \in \mathbb{N}, \tag{7.8.6}$$

$$\exists L \in \mathbb{C} \ni \vartheta - \lim_{k,l\to\infty} \sum_{i,j} d_{klij} = L \text{ exists}, \tag{7.8.7}$$

$$\exists i_0 \in \mathbb{N} \ni \vartheta - \lim_{k,l\to\infty} \sum_{j} |d_{kli_0j} - d_{i_0j}| = 0, \tag{7.8.8}$$

$$\exists j_0 \in \mathbb{N} \ni \vartheta - \lim_{k,l\to\infty} \sum_{i} |d_{klij_0} - d_{ij_0}| = 0. \tag{7.8.9}$$

In the case of (7.8.9), $d = (d_{ij}) \in \mathcal{L}_u$ and

$$\vartheta - \lim_{k,l\to\infty} (Dx)_{kl} = \sum_{i,j} d_{ij}x_{ij} + \left(L - \sum_{i,j} d_{ij} \right) bp - \lim_{k,l\to\infty} x_{kl}$$

satisfies for $x \in \mathcal{C}_{bp}$.

Lemma 7.8.8. *Suppose that $D = (d_{klij})$ be a four-dimensional infinite matrix. Then, $D = (d_{klij}) \in (\mathcal{C}_p : \mathcal{C}_{\vartheta})$ if and only if (7.8.5)–(7.8.7) and followings are provided:*

$$\forall i \in \mathbb{N}\ \exists j_0 \in \mathbb{N} \ni d_{klij} = 0 \text{ for every } j > j_0 \text{ and } k,l \in \mathbb{N}, \tag{7.8.10}$$

$$\forall j \in \mathbb{N}\ \exists i_0 \in \mathbb{N} \ni d_{klij} = 0 \text{ for every } i > i_0 \text{ and } k,l \in \mathbb{N}. \tag{7.8.11}$$

In the case of (7.8.11), $\exists i_0, j_0 \in \mathbb{N}$ such that $d = (d_{ij}) \in \mathcal{L}_u$ and $(d_{ij_0})_{i\in\mathbb{N}}, (d_{i_0j})_{j\in\mathbb{N}} \in \wp$, where $\wp$ represents the space of every finitely sequences which are non-equivalent zero and

$$\vartheta - \lim_{k,l\to\infty} (Dx)_{kl} = \sum_{i,j} d_{ij}x_{ij} + \sum_{i} \left(L - \sum_{i,j} d_{ij} \right) p - \lim_{k,l\to\infty} x_{kl}$$

satisfies for $x \in \mathcal{C}_p$.

Lemma 7.8.9. *Suppose that $D = (d_{klij})$ be a four-dimensional infinite matrix. Then, $D = (d_{klij}) \in (\mathcal{C}_r : \mathcal{C}_\vartheta)$ if and only if (7.8.5)–(7.8.7) and followings are provided:*

$$\exists j_0 \in \mathbb{N} \ni \vartheta - \lim_{k,l\to\infty} \sum_{i} d_{klij_0} = x_{j_0}, \tag{7.8.12}$$

$$\exists i_0 \in \mathbb{N} \ni \vartheta - \lim_{k,l\to\infty} \sum_{j} d_{kli_0j} = y_{i_0}. \tag{7.8.13}$$

In the case of (7.8.13), $d = (d_{ij}) \in \mathcal{L}_u$ and $(x_j), (y_i) \in \ell_1$, and

$$\begin{aligned}\vartheta - \lim_{k,l\to\infty} (Dx)_{kl} &= \sum_{i,j} d_{ij}x_{ij} + \sum_{i} \left(y_i - \sum_{j} d_{ij} \right) x_i + \sum_{j} \left(x_j - \sum_{i} d_{ij} \right) u_j \\ &\quad + \left(L + \sum_{i,j} d_{ij} - \sum_{i} y_i - \sum_{j} x_j \right) r - \lim_{k,l\to\infty} x_{kl}\end{aligned}$$

satisfies for $x \in \mathcal{C}_r$.

Lemma 7.8.10. [387] *Suppose that $D = (d_{klij})$ be a four-dimensional infinite matrix. Then, $D = (d_{klij}) \in (\mathcal{C}_{bp} : \mathcal{M}_u)$ if and only if the condition (7.8.5) holds.*

Lemma 7.8.11. [115] *Suppose that $D = (d_{klij})$ be a four-dimensional infinite matrix. Then, $D = (d_{klij}) \in (\mathcal{M}_u : \mathcal{C}_{bp})$ if and only if the conditions (7.8.5), (7.8.6) hold and*

$$\exists d_{ij} \in \mathbb{C} \ni bp - \lim_{k,l\to\infty} \sum_{ij} |d_{klij} - d_{ij}| = 0. \tag{7.8.14}$$

$$bp - \lim_{k,l\to\infty} \sum_{j=0}^{l} d_{klij} \quad \text{exists for each} \quad i \in \mathbb{N}. \tag{7.8.15}$$

$$bp - \lim_{k,l\to\infty} \sum_{i=0}^{k} d_{klij} \quad \text{exists for each} \quad j \in \mathbb{N}. \tag{7.8.16}$$

$$\sum_{i,j} |d_{klij}| \quad \text{converges.} \tag{7.8.17}$$

Lemma 7.8.12. [402] *Suppose that $D = (d_{klij})$ be a four-dimensional infinite matrix. Then, $D = (d_{klij}) \in (\mathcal{M}_u : \mathcal{M}_u)$ if and only if the condition (7.8.5) holds.*

Lemma 7.8.13. [403] *Suppose that $D = (d_{klij})$ be a four-dimensional infinite matrix. Then, $D = (d_{klij}) \in (\mathcal{M}_u : \mathcal{C}_p)$ if and only if the conditions (7.8.6), (7.8.10) and (7.8.11) hold.*

Lemma 7.8.14. [404] *Suppose that $D = (d_{klij})$ be a four-dimensional infinite matrix. Then, the following statements hold:*

(i) *If $0 < p \leq 1$, then $D \in (\mathcal{L}_p : \mathcal{M}_u)$ if and only if*

$$\sup_{k,l,i,j \in \mathbb{N}} |d_{klij}| < \infty. \tag{7.8.18}$$

(ii) *If $1 < p < \infty$, then $D \in (\mathcal{L}_p : \mathcal{M}_u)$ if and only if*

$$\sup_{k,l \in \mathbb{N}} \sum_{i,j} |d_{klij}|^{p'} < \infty. \tag{7.8.19}$$

Lemma 7.8.15. [404] *Suppose that $D = (d_{klij})$ be a four-dimensional infinite matrix. Then, the following statements hold:*

(i) *If $0 < p \leq 1$, then $D \in (\mathcal{L}_p : \mathcal{C}_{bp})$ if and only if the conditions (7.8.6) and (7.8.18) hold with $\vartheta = bp$.*

(ii) *If $1 < p < \infty$, then $D \in (\mathcal{L}_p : \mathcal{C}_{bp})$ if and only if the conditions (7.8.6) and (7.8.19) hold with $\vartheta = bp$.*

Let us define the sets $w_1 - w_{13}$ by

$$w_1 := \left\{ t = (t_{ij}) \in \Omega : \sup_{k,l \in \mathbb{N}} \sum_{i,j} |\chi(k,l,i,j,m,n)| < \infty \right\},$$

$$w_2 := \left\{ t = (t_{ij}) \in \Omega : \exists d_{ij} \in \mathbb{C} \ni \vartheta - \lim_{k,l \to \infty} \chi(k,l,i,j,m,n) = d_{ij} \right\},$$

$$w_3 := \left\{ t = (t_{ij}) \in \Omega : \exists L \in \mathbb{C} \ni \vartheta - \lim_{k,l \to \infty} \sum_{i,j} \chi(k,l,i,j,m,n) = L \quad \text{exists} \right\},$$

$$w_4 := \left\{ t = (t_{ij}) \in \Omega : \exists j_0 \in \mathbb{N} \ni \vartheta - \lim_{k,l \to \infty} \sum_{i} |\chi(k,l,i,j_0,m,n) - d_{ij_0}| = 0 \right\},$$

$$w_5 := \left\{ t = (t_{ij}) \in \Omega : \exists i_0 \in \mathbb{N} \ni \vartheta - \lim_{k,l \to \infty} \sum_{j} |\chi(k,l,i_0,j,m,n) - d_{i_0 j}| = 0 \right\},$$

$$w_6 := \left\{ t = (t_{ij}) \in \Omega : \forall i \in \mathbb{N}, \exists j_0 \in \mathbb{N} \ni \chi(k,l,i,j,m,n) = 0, \forall j > j_0, \forall k,l \in \mathbb{N} \right\},$$

$$w_7 := \left\{ t = (t_{ij}) \in \Omega : \forall j \in \mathbb{N}, \exists i_0 \in \mathbb{N} \ni \chi(k,l,i,j,m,n) = 0, \forall i > i_0, \forall k,l \in \mathbb{N} \right\},$$

$$w_8 := \left\{t=(t_{ij})\in\Omega : \exists j_0\in\mathbb{N}\ni \vartheta-\lim_{k,l\to\infty}\sum_i \chi(k,l,i,j_0,m,n)=d_{j_0}\right\},$$

$$w_9 := \left\{t=(t_{ij})\in\Omega : \exists i_0\in\mathbb{N}\ni \vartheta-\lim_{k,l\to\infty}\sum_j \chi(k,l,i_0,j,m,n)=d_{i_0}\right\},$$

$$w_{10} := \left\{t=(t_{ij})\in\Omega : \exists d_{ij}\in\mathbb{C}\ni bp-\lim_{k,l\to\infty}\sum_{i,j} |\chi(k,l,i,j,m,n)-d_{ij}|=0\right\},$$

$$w_{11} := \left\{t=(t_{ij})\in\Omega : \forall i\in\mathbb{N}\ni bp-\lim_{k,l\to\infty}\sum_{j=0}^{l} \chi(k,l,i,j,m,n)\quad \text{exists}\right\},$$

$$w_{12} := \left\{t=(t_{ij})\in\Omega : \forall j\in\mathbb{N}\ni bp-\lim_{k,l\to\infty}\sum_{i=0}^{k} \chi(k,l,i,j,m,n)\quad \text{exists}\right\},$$

$$w_{13} := \left\{t=(t_{ij})\in\Omega : \sum_{i,j} |\chi(k,l,i,j,m,n)|\quad \text{converges}\right\},$$

where

$$\chi(k,l,i,j,m,n)=\sum_{m=i}^{k}\sum_{n=j}^{l}(-1)^{m+n-(i+j)}\binom{m}{i}\binom{n}{j}s^{m-n-i}r^{n-m-j}(r+s)^{i+j}t_{mn}.$$

Theorem 7.8.16. *The following statements are satisfied:*

(i) $\left\{\mathcal{B}^{r,s}_{bp}\right\}^{\beta(\vartheta)} := \bigcap_{k=1}^{5} w_k.$

(ii) $\{\mathcal{B}^{r,s}\}^{\beta(\vartheta)} := \bigcap_{k=1}^{3} w_k \cap w_6 \cap w_7.$

(iii) $\left\{\mathcal{B}^{r,s}_{reg}\right\}^{\beta(\vartheta)} := \bigcap_{k=1}^{3} w_k \cap w_8 \cap w_9.$

(iv) $\{\mathcal{B}^{r,s}_{\infty}\}^{\beta(bp)} := w_1 \cap w_2 \bigcap_{k=10}^{13} w_k.$

(v) $\{\mathcal{B}^{r,s}_{\infty}\}^{\beta(p)} := w_2 \cap w_6 \cap w_7.$

(vi) $\{\mathcal{B}^{r,s}_{\infty}\}^{\gamma} := w_1.$

(vii) $\left\{\mathcal{B}^{r,s}_{bp}\right\}^{\gamma} := w_1.$

Proof. (i) Suppose that $t=(t_{kl})\in\Omega$ and $x=(x_{kl})\in\mathcal{B}^{r,s}_{bp}$. Then, we can conclude from (7.8.3) that $y=(y_{kl})\in\mathcal{C}_{bp}$. Now, let us define the four-dimensional matrix $O^{r,s}=\left(o^{r,s}_{klij}\right)$ by

$$o^{r,s}_{klij} := \begin{cases} \chi(k,l,i,j,m,n) & , \quad 0\le i\le k, 0\le j\le l, \\ 0 & , \quad \text{otherwise}, \end{cases}$$

for all $k, l, i, j \in \mathbb{N}$. Therefore, we obtain by the relation (7.8.4) that

$$\begin{aligned} z_{kl} &= \sum_{i,j=0}^{k,l} t_{ij}x_{ij} \qquad (7.8.20)\\ &= \sum_{i,j=0}^{k,l} t_{ij}\left\{\sum_{m,n=0}^{i,j}(-1)^{i+j-(m+n)}\binom{i}{m}\binom{j}{n}s^{i-j-m}r^{j-i-n}(r+s)^{m+n}y_{mn}\right\}\\ &= \sum_{i,j=0}^{k,l}\left\{\sum_{m=i}^{k}\sum_{n=j}^{l}(-1)^{m+n-(i+j)}\binom{m}{i}\binom{n}{j}s^{m-n-i}r^{n-m-j}(r+s)^{i+j}t_{mn}\right\}y_{ij}\\ &= (O^{r,s}y)_{kl} \end{aligned}$$

for every $k, l \in \mathbb{N}$. Therefore, we conclude from (7.8.20) that $tx = (t_{kl}x_{kl}) \in \mathcal{CS}_{\vartheta}$ whenever $x = (x_{kl}) \in \mathcal{B}_{bp}^{r,s}$ iff $z = (z_{kl}) \in \mathcal{C}_{\vartheta}$ whenever $y = (y_{kl}) \in \mathcal{C}_{bp}$. This implies that $t = (t_{kl}) \in \left(\mathcal{B}_{bp}^{r,s}\right)^{\beta(\vartheta)}$ if and only if $O^{r,s} \in (\mathcal{C}_{bp} : \mathcal{C}_{\vartheta})$ and the proof is completed in view of Lemma 7.8.7.

The other parts of the theorem can be proved analogously by using Lemmas 7.8.8, 7.8.9, 7.8.11, 7.8.13, 7.8.12 and 7.8.10, respectively. So, we omit details. □

7.8.2 Characterizations of Some Matrix Classes

Theorem 7.8.17. *Assume that $H = (h_{klij})$ be defined via $D = (d_{klab})$, as follows:*

$$h_{klij} := \sum_{a=i}^{\infty}\sum_{b=j}^{\infty}(-1)^{a+b-(i+j)}\binom{a}{i}\binom{b}{j}s^{a-b-i}r^{b-a-j}(r+s)^{i+j}d_{klab} \qquad (7.8.21)$$

for all $i, j, k, l \in \mathbb{N}$. Then, $D \in (\Psi_{B(r,s)} : \Lambda)$ if and only if $D_{kl} \in [\Psi_{B(r,s)}]^{\beta(\vartheta)}$ for every $k, l \in \mathbb{N}$ and $H \in (\Psi : \Lambda)$, where Ψ and $\Lambda \in \{\mathcal{M}_u, \mathcal{C}_p, \mathcal{C}_{bp}, \mathcal{C}_r\}$.

Proof. Let $D \in (\Psi_{B(r,s)} : \Lambda)$. Then, Du exists and is in Λ for every $u \in \Psi_{B(r,s)}$, which implies that the fact that $D_{kl} \in [\Psi_{B(r,s)}]^{\beta(\vartheta)}$ for every $k, l \in \mathbb{N}$. So, by bearing in mind (7.8.4), the relation

$$\begin{aligned} &\sum_{i,j}^{k,l} d_{klij}x_{ij} = \qquad (7.8.22)\\ &= \sum_{i,j}^{k,l} d_{klij}\sum_{a,b=0}^{i,j}(-1)^{i+j-(a+b)}\binom{i}{a}\binom{j}{b}s^{i-j-a}r^{j-i-b}(r+s)^{a+b}y_{ab}\\ &= \sum_{i,j}^{k,l}\left[\sum_{a,b=i,j}^{k,l}(-1)^{a+b-(i+j)}\binom{a}{i}\binom{b}{j}s^{a-b-i}r^{b-a-j}(r+s)^{i+j}d_{klab}\right]y_{ij} \end{aligned}$$

is obtained for every $k, l \in \mathbb{N}$. Then, by taking ϑ-limit on (7.8.22) as $k, l \to \infty$, we have $Dx = Hy$. Therefore, we obtain that $Hy \in \Lambda$ whenever $y \in \Psi$, i.e., $H \in (\Psi, \Lambda)$.

For the sufficiency part, assume the sequence $D_{kl} \in [\Psi_{B(r,s)}]^{\beta(\vartheta)}$ for every $k, l \in \mathbb{N}$,

$H \in (\Psi : \Lambda)$ and $x = (x_{ij}) \in \Psi_{B^{(r,s)}}$ such that $y = B^{(r,s)}x$. Then, Dx exists and therefore, rectangular partial sums for $\sum_{i,j} d_{klij}x_{ij}$ obtained as

$$\begin{aligned} (Dx)_{kl}^{[m,n]} &= \sum_{i,j}^{m,n} d_{klij}x_{ij} \qquad (7.8.23)\\ &= \sum_{i,j}^{m,n}\left[\sum_{a,b=i,j}^{m,n}(-1)^{a+b-(i+j)}\binom{a}{i}\binom{b}{j}s^{a-b-i}r^{b-a-j}(r+s)^{i+j}d_{klab}\right]y_{ij} \end{aligned}$$

for every $k, l, m, n \in \mathbb{N}$. By taking ϑ-limit in (7.8.23) as $m, n \to \infty$, it can be easily obtain from the following equality

$$\sum_{i,j} d_{klij}x_{ij} = \sum_{i,j} h_{klij}y_{ij}$$

for every $k, l \in \mathbb{N}$ that $Dx = Hy$ which leads to fact that $D \in (\Psi_{B^{(r,s)}} : \Lambda)$. □

Corollary 7.8.18. *Suppose that $D = (d_{klij})$ be a four-dimensional matrix. The following statements are satisfied:*

(i) *$D \in (\mathcal{B}^{r,s} : \mathcal{C}_\vartheta)$ if and only if the conditions (7.8.5)–(7.8.7), (7.8.10) and (7.8.11) are satisfied with h_{klij} instead of d_{klij}.*

(ii) *$D \in \left(\mathcal{B}_{bp}^{r,s} : \mathcal{C}_\vartheta\right)$ if and only if the conditions (7.8.5)–(7.8.9) are satisfied with h_{klij} instead of d_{klij}.*

(iii) *$D \in \left(\mathcal{B}_{bp}^{r,s} : \mathcal{M}_u\right)$ if and only if the condition (7.8.5) is satisfied with h_{klij} instead of d_{klij}.*

(iv) *$D \in \left(\mathcal{B}_{reg}^{r,s} : \mathcal{C}_\vartheta\right)$ if and only if the conditions (7.8.5)–(7.8.7), (7.8.12) and (7.8.13) are satisfied with h_{klij} instead of d_{klij}.*

(v) *$D \in (\mathcal{B}_\infty^{r,s} : \mathcal{C}_{bp})$ if and only if the conditions (7.8.5), (7.8.6), (7.8.14)–(7.8.17) are satisfied with h_{klij} instead of d_{klij}.*

(vi) *$D \in (\mathcal{B}_\infty^{r,s} : \mathcal{C}_p)$ if and only if the conditions (7.8.6), (7.8.10) and (7.8.11) are satisfied with h_{klij} instead of d_{klij}.*

Lemma 7.8.19. [404] *Let $B = (b_{klij})$ be a triangle matrix. Then, $D = (d_{klij}) \in (\Psi : \Lambda_B)$ if and only if $BD \in (\Psi : \Lambda)$.*

Now, let us define the four-dimensional matrix $G = (g_{klij})$ by

$$g_{klij} := \sum_{m,n=0}^{k,l} b_{klmn}^{r,s} d_{mnij}$$

for all $k, l, i, j \in \mathbb{N}$ and give the following corollary:

Corollary 7.8.20. *Suppose that $D = (d_{klij})$ be a four-dimensional matrix. Then, the following statements hold:*

(i) $D \in (\mathcal{C}_p : (\mathcal{C}_\vartheta)_{B^{(r,s)}})$ *if and only if the conditions (7.8.5)–(7.8.7), (7.8.10) and (7.8.11) are satisfied with g_{klij} instead of d_{klij}.*

(ii) $D \in (\mathcal{C}_{bp} : (\mathcal{C}_\vartheta)_{B^{(r,s)}})$ *if and only if the conditions (7.8.5)–(7.8.9) are satisfied with g_{klij} instead of d_{klij}.*

(iii) $D \in (\mathcal{C}_r : (\mathcal{C}_\vartheta)_{B^{(r,s)}})$ *if and only if the conditions (7.8.5)–(7.8.7), (7.8.12) and (7.8.13) are satisfied with g_{klij} instead of d_{klij}.*

(iv) $D \in \left(\mathcal{L}_p : \mathcal{B}^{r,s}_{bp}\right)$ *if and only if the conditions (7.8.6) and (7.8.18) are satisfied for $0 < p \leq 1$ and $\vartheta = bp$ with g_{klij} instead of d_{klij}.*

(v) $D \in \left(\mathcal{L}_p : \mathcal{B}^{r,s}_{bp}\right)$ *if and only if the conditions (7.8.6) and (7.8.19) are satisfied for $1 < p < \infty$ and $\vartheta = bp$ with g_{klij} instead of d_{klij}.*

(vi) $D \in (\mathcal{L}_p : \mathcal{B}^{r,s}_{\infty})$ *if and only if the condition (7.8.18) is satisfied for $0 < p \leq 1$ with g_{klij} instead of d_{klij}.*

(vii) $D \in (\mathcal{L}_p : \mathcal{B}^{r,s}_{\infty})$ *if and only if the condition (7.8.19) is satisfied for $1 < p < \infty$ with g_{klij} instead of d_{klij}.*

(viii) $D \in \left(\mathcal{M}_u : \mathcal{B}^{r,s}_{bp}\right)$ *if and only if the conditions (7.8.5),(7.8.6), (7.8.14)–(7.8.17) are satisfied with g_{klij} instead of d_{klij}.*

(ix) $D \in (\mathcal{M}_u : \mathcal{B}^{r,s})$ *if and only if the conditions (7.8.6), (7.8.10) and (7.8.11) are satisfied with g_{klij} instead of d_{klij}.*

(x) $D \in (\mathcal{C}_{bp} : \mathcal{B}^{r,s}_{\infty})$ *if and only if the condition (7.8.5) is satisfied with g_{klij} instead of d_{klij}.*

7.9 CONCLUSION

Each story on single sequences has been usually experienced over double sequences.

Constructing a new set of single or double sequence spaces and relating them to determine the location of those spaces between the other sequence spaces, characterizing matrix transformations on one of these spaces into another one, are among the qualified problems. In this section, we introduced some double sequence spaces by using the domains of four-dimensional binomial matrix on some classical double sequence spaces. Moreover, it is investigated their some properties and inclusion relations, computed duals and characterized some matrix classes. We conclude that the results obtained from the matrix $B^{(r,s)}$ is more general and extensive than the existent results of the authors [377, 406]. We expect that the results given in this section might be a reference for further studies in this field.

CHAPTER 8

Sequences of Fuzzy Numbers

This chapter is devoted to the sequences of fuzzy numbers. After presenting the fundamental facts concerning convergent sequences of fuzzy numbers, some results on statistical convergence of sequences of fuzzy numbers and related results are given. Also, the α-, β- and γ-duals of the classical sets $\ell_\infty(F)$, $c(F)$, $c_0(F)$ and $\ell_p(F)$ of all bounded, convergent, null and absolutely p-summable sequences of fuzzy numbers are determined and the classes $(\mu(F) : \ell_\infty(F))$, $(c_0(F) : c(F))$, $(c_0(F) : c_0(F))$, $(c(F) : c(F); p)$, $(\ell_p(F) : c(F))$, $(\ell_p(F) : c_0(F))$ and $(\ell_\infty(F) : c_0(F))$ of infinite matrices of fuzzy numbers are characterized, where $\mu \in \{\ell_\infty, c, c_0, \ell_p\}$. Finally, the quasilinearity of the classical sets of sequences of fuzzy numbers is investigated.

Keywords: Fuzzy number, level set, fuzzy valued sequences, statistical convergence of fuzzy sequences, bounded set from above and below, infinite matrix of fuzzy numbers, convergence and boundedness of a fuzzy sequence, limit superior and limit inferior and core of a sequence of fuzzy numbers, statistical monotonic and bounded sequences of fuzzy numbers, statistical cluster point and limit point of a sequence of fuzzy numbers, statistical limit superior and limit inferior of a sequence of fuzzy numbers, α-, β- and γ-duals of the classical sets $\ell_\infty(F)$, $c(F)$, $c_0(F)$ and $\ell_p(F)$ of all bounded, convergent, null and absolutely p-summable sequences of fuzzy numbers, characterization of matrix transformations between the sets of sequences of fuzzy numbers, quasilinearity of the classical sets of sequences of fuzzy numbers, sets of sequences of fuzzy numbers defined by a modulus.

8.1 INTRODUCTION

Zadeh introduced the concepts of fuzzy sets and fuzzy set operations, in his significant article [408]. Subsequently, several authors discussed various aspects of the theory and applications of fuzzy sets such as fuzzy topological spaces, similarity relations and fuzzy orderings, fuzzy measures of fuzzy events, fuzzy mathematical programming.

In the literature, there are different definitions of a fuzzy number characterized by a function. We will base ourselves on the definition given by Chang and Zadeh [103], which is used widely. Similarly, we encounter various metrics defined on the set

DOI: 10.1201/9781003294153-8

of fuzzy numbers. In addition, we see that a fuzzy number valued metric can also be defined on any set which is not necessarily the set of fuzzy numbers (see [210]). In this chapter, we will use the *supremum metric* D, defined by (8.2.2) whose domain is the set of fuzzy numbers and range is the set of non-negative real numbers. Under the metric D, it is proved by Theorem 8.3.4 and Theorem 8.3.5 that the convergence and the statistical convergence of a sequence of fuzzy numbers are equivalent to the uniform convergence and uniform statistical convergence of the sequences of functions which are obtained by α-cuts of fuzzy numbers in the fuzzy number sequence. Therefore, the supremum metric is also called the *uniform metric*.

Of course, it is not possible to consider here all the documents related to sequences of fuzzy numbers. So, we will consider only the fundamental studies related to the convergence and statistical convergence of sequences of fuzzy numbers, and concerning with the classical sets of sequences of fuzzy numbers and their quasilinearity. Besides these studies, there are some others which are related to the difference [36, 35, 128, 161, 164]/ lacunary [192, 234, 235, 305]/ double [33, 348, 357, 355]/ statistically Cauchy [350, 352] sequences of fuzzy numbers; and those related to the levelwise [209, 169, 400]/ levelwise statistical [53, 54, 372]/ almost [122, 353]/ ideal [306] / λ-statistical [349, 351, 354, 388] convergence of a sequence of fuzzy numbers; and the ones related to the spaces of sequences of fuzzy numbers [302, 351]/ related to the classical sets of sequences of fuzzy numbers [379]/ related to the quasilinearity of the classical sets of sequences of fuzzy numbers [381].

8.2 CONVERGENCE OF A SEQUENCE OF FUZZY NUMBERS

First, we recall some of the basic notations and definitions in the theory of fuzzy numbers and we refer to [149, 151, 152, 214, 274, 302, 401] for more details.

A *fuzzy number* is a fuzzy set on the real axis, i.e., a mapping $u : \mathbb{R} \to [0, 1]$ which satisfies the following four conditions:

(i) u is normal, i.e., there exists an $\alpha_0 \in \mathbb{R}$ such that $u(\alpha_0) = 1$.

(ii) u is fuzzy convex, i.e., $u[\alpha s + (1-\alpha)t] \geq \min\{u(s), u(t)\}$ for all $s, t \in \mathbb{R}$ and for all $\alpha \in [0, 1]$.

(iii) u is upper semi-continuous.

(iv) The set $[u]_0 = \overline{\{t \in \mathbb{R} : u(t) > 0\}}$ is compact, (cf. Chang and Zadeh [103]), where $\overline{\{t \in \mathbb{R} : u(t) > 0\}}$ denotes the closure of the set $\{t \in \mathbb{R} : u(t) > 0\}$ in the usual topology of $\mathbb{R}$.

We denote *the set of all fuzzy numbers on* $\mathbb{R}$ by $L(\mathbb{R})$ and called it as *the space of fuzzy numbers*. α*-level set* $[u]_\alpha$ of $u \in L(\mathbb{R})$ is defined by

$$[u]_\alpha := \begin{cases} \{t \in \mathbb{R} : u(t) \geq \alpha\} & , \quad 0 < \alpha \leq 1, \\ \overline{\{t \in \mathbb{R} : u(t) > \alpha\}} & , \quad \alpha = 0. \end{cases}$$

The set $[u]_\alpha$ is closed, bounded and non-empty interval for each $\alpha \in [0,1]$ which is defined by $[u]_\alpha = [u^-(\alpha), u^+(\alpha)]$. $\mathbb{R}$ can be embedded in $L(\mathbb{R})$, since each $r \in \mathbb{R}$ can be regarded as a fuzzy number $\overline{r}$ defined by $\overline{r}(t) := \begin{cases} 1 & , \quad t = r, \\ 0 & , \quad t \neq r. \end{cases}$

Representation Theorem. [185, Theorem 1.1] *Let* $[u]_\alpha = [u^-(\alpha), u^+(\alpha)]$ *for* $u \in L(\mathbb{R})$ *and for each* $\alpha \in [0,1]$. *Then, the following statements hold:*

(i) u^- *is a bounded and non-decreasing left continuous function on* $(0,1]$.

(ii) u^+ *is a bounded and non-increasing left continuous function on* $(0,1]$.

(iii) *The functions* u^- *and* u^+ *are right continuous at the point* $\alpha = 0$.

(iv) $u^-(1) \leq u^+(1)$.

Conversely, if the pair of functions u^- *and* u^+ *satisfies the conditions (i)–(iv), then there exists a unique* $u \in L(\mathbb{R})$ *such that* $[u]_\alpha = [u^-(\alpha), u^+(\alpha)]$ *for each* $\alpha \in [0,1]$. *The fuzzy number* u *corresponding to the pair of functions* u^- *and* u^+ *is defined by* $u : \mathbb{R} \to [0,1]$, $u(x) := \sup\{t \in [0,1] : u^-(t) \leq x \leq u^+(t)\}$.

Given two fuzzy numbers $u, v \in L(\mathbb{R})$, we define

(i) their algebraic sum $u \mp v$ [214, p. 14] by

$$\begin{aligned} u + v = w \quad &\Leftrightarrow \quad [w]_\alpha = [u]_\alpha + [v]_\alpha \\ &\Leftrightarrow \quad w^-(\alpha) = u^-(\alpha) + v^-(\alpha) \ \text{ and } w^+(\alpha) = u^+(\alpha) + v^+(\alpha), \end{aligned}$$

$$\begin{aligned} u - v = w \quad &\Leftrightarrow \quad [w]_\alpha = [u]_\alpha - [v]_\alpha \\ &\Leftrightarrow \quad w^-(\alpha) = u^-(\alpha) - v^+(\alpha) \ \text{ and } \ w^+(\alpha) = u^+(\alpha) - v^-(\alpha). \end{aligned}$$

(ii) their product $u \cdot v$ [214, p. 21] by $uv = w \Leftrightarrow [w]_\alpha = [u]_\alpha [v]_\alpha$, where it is immediate that

$$\begin{aligned} w^-(\alpha) &= \min\{u^-(\alpha)v^-(\alpha), u^-(\alpha)v^+(\alpha), u^+(\alpha)v^-(\alpha), u^+(\alpha)v^+(\alpha)\}, \\ w^+(\alpha) &= \max\{u^-(\alpha)v^-(\alpha), u^-(\alpha)v^+(\alpha), u^+(\alpha)v^-(\alpha), u^+(\alpha)v^+(\alpha)\} \end{aligned}$$

for all $\alpha \in [0,1]$.

(iii) scalar multiplication with $\lambda \in \mathbb{R}$ of u [214, p. 25] by $[\lambda u]_\alpha = \lambda [u]_\alpha$.

If $u \in L(\mathbb{R})$ and $0 \notin \{\alpha \in \mathbb{R} : u(\alpha) > 0\}$, then we have by Kaufmann and Gupta [214, p. 31] that

$$\left(\frac{1}{u}\right)^-(\alpha) = \frac{1}{u^+(\alpha)} \quad \text{and} \quad \left(\frac{1}{u}\right)^+(\alpha) = \frac{1}{u^-(\alpha)}.$$

Let W be the set of all closed bounded intervals A of real numbers with endpoints $\underline{A}$ and $\overline{A}$, i.e., $A = [\underline{A}, \overline{A}]$. Define the relation d on W by

$$d(A, B) := \max\{|\underline{A} - \underline{B}|, |\overline{A} - \overline{B}|\}. \tag{8.2.1}$$

Then, it can easily be observed that d is a metric on W (cf. Diamond and Kloeden [148]) and (W, d) is a complete metric space, (cf. Nanda [302]).

Now, we define the metric D on $L(\mathbb{R})$ by means of the *Hausdorff metric* d by

$$\begin{aligned} D(u, v) &:= \sup_{\alpha\in[0,1]} d([u]_\alpha, [v]_\alpha) \\ &= \sup_{\alpha\in[0,1]} \max\{|u^-(\alpha) - v^-(\alpha)|, |u^+(\alpha) - v^+(\alpha)|\}. \end{aligned} \tag{8.2.2}$$

The partial ordering relation on $L(\mathbb{R})$ is defined, as follows:

$$u \preceq v \Leftrightarrow [u]_\alpha \preceq [v]_\alpha \Leftrightarrow u^-(\alpha) \leq v^-(\alpha) \text{ and } u^+(\alpha) \leq v^+(\alpha) \text{ for all } \alpha \in [0, 1].$$

This leads us to the consequence that $u \prec v$ if $u \preceq v$ and there exists $\alpha_0 \in [0, 1]$ such that $u^-(\alpha_0) < v^-(\alpha_0)$ or $u^+(\alpha_0) < v^+(\alpha_0)$ [195, Definition 2.4]. It is trivial that $u + \overline{a}$, $u - \overline{a}$, $\overline{a}u \in L(\mathbb{R})$. If $\overline{a} \succ \overline{0}$, then we have $u - \overline{a} \prec u \prec u + \overline{a}$ and $D(u, u + \overline{a}) = D(u, u - \overline{a}) = \overline{a}$ [49].

Two fuzzy numbers u and v are said to be *incomparable* if neither $u \preceq v$ nor $v \preceq u$ holds. In this case, we use the notation $u \not\sim v$ [49, p. 977].

Definition 8.2.1. [210, p. 216] *$u \in L(\mathbb{R})$ is a non-negative fuzzy number if and only if $u(\alpha) = 0$ for all $\alpha < 0$. It is immediate that $u \succeq \overline{0}$ if u is a non-negative fuzzy number.*

One can see that

$$D(u, \overline{0}) = \sup_{\alpha\in[0,1]} \max\{|u^-(\alpha)|, |u^+(\alpha)|\} = \max\{|u^-(0)|, |u^+(0)|\}. \tag{8.2.3}$$

Now, we may give:

Proposition 8.2.2. [379, Proposition 2.2] *Let $u, v, w, z \in L(\mathbb{R})$ and $\alpha \in \mathbb{R}$. Then, the following statements hold:*

(i) *$(L(\mathbb{R}), D)$ is a complete metric space,* (cf. Puri and Ralescu [321, Theorem 2.1]).

(ii) $D(\alpha u, \alpha v) = |\alpha| D(u, v)$.

(iii) $D(u + v, w + v) = D(u, w)$.

(iv) $D(u + v, w + z) \leq D(u, w) + D(v, z)$.

(v) $|D(u, \overline{0}) - D(v, \overline{0})| \leq D(u, v) \leq D(u, \overline{0}) + D(v, \overline{0})$.

Definition 8.2.3. (cf. Tripathy and Nanda [386]) *The absolute value $|u|$ of a fuzzy number u is defined by*

$$|u|(\alpha) := \begin{cases} \max\{u(\alpha), u(-\alpha)\} & , \quad \alpha \geq 0, \\ 0 & , \quad \alpha < 0. \end{cases}$$

α-level set $[|u|]_\alpha$ of the absolute value of $u \in L(\mathbb{R})$ is in the form $[|u|]_\alpha = [|u|^-(\alpha), |u|^+(\alpha)]$, where

$$\begin{aligned} |u|^-(\alpha) &= \max\{0, u^-(\alpha), -u^+(\alpha)\}, \\ |u|^+(\alpha) &= \max\{|u^-(\alpha)|, |u^+(\alpha)|\}. \end{aligned}$$

Proposition 8.2.4. [188] *Let $u, v, m \in L(\mathbb{R})$ with $m \succeq \overline{0}$ and $\alpha \in \mathbb{R}$. Then, the following statements hold:*

(i) $|u| := \begin{cases} u & , \quad u \succeq \overline{0}, \\ -u & , \quad u \prec \overline{0}. \end{cases}$

(ii) $|u+v| \preceq |u| + |v|$.

(iii) $|\alpha u| = |\alpha||u|$.

(iv) $|u| = \overline{0}$ *if and only if* $u = \overline{0}$.

(v) $|u| \preceq m$ *if and only if* $-m \preceq u \preceq m$.

We give two lemmas related with some required inequalities, below.

Lemma 8.2.5. [379, Lemma 2.5] *The absolute value $|uv|$ of the product uv of $u, v \in L(\mathbb{R})$ satisfies the inequalities*

$$\begin{aligned} & |uv|^-(\alpha) \leq |uv|^+(\alpha) \leq \\ & \leq \max\{|u^-(\alpha)||v^-(\alpha)|, |u^-(\alpha)||v^+(\alpha)|, |u^+(\alpha)||v^-(\alpha)|, |u^+(\alpha)||v^+(\alpha)|\}. \end{aligned} \tag{8.2.4}$$

Proof. For the α-level set $[|uv|]_\alpha$ of $|uv|$, we have

$$\begin{aligned} |uv|^-(\alpha) &= \max\{0, (uv)^-(\alpha), -(uv)^+(\alpha)\}, \\ |uv|^+(\alpha) &= \max\{|(uv)^-(\alpha)|, |(uv)^+(\alpha)|\}. \end{aligned}$$

Therefore, since

$$\begin{aligned} |(uv)^-(\alpha)| &= |\min\{u^-(\alpha)v^-(\alpha), u^-(\alpha)v^+(\alpha), u^+(\alpha)v^-(\alpha), u^+(\alpha)v^+(\alpha)\}| \\ &\leq \max\{|u^-(\alpha)||v^-(\alpha)|, |u^-(\alpha)||v^+(\alpha)|, |u^+(\alpha)||v^-(\alpha)|, |u^+(\alpha)||v^+(\alpha)|\} \end{aligned}$$

and

$$\begin{aligned} |(uv)^+(\alpha)| &= |\max\{u^-(\alpha)v^-(\alpha), u^-(\alpha)v^+(\alpha), u^+(\alpha)v^-(\alpha), u^+(\alpha)v^+(\alpha)\}| \\ &\leq \max\{|u^-(\alpha)||v^-(\alpha)|, |u^-(\alpha)||v^+(\alpha)|, |u^+(\alpha)||v^-(\alpha)|, |u^+(\alpha)||v^+(\alpha)|\}, \end{aligned}$$

one can deduce that (8.2.4) holds. □

Lemma 8.2.6. [379, Lemma 2.6] *The following statements hold:*

(i) $D(uv, \overline{0}) \leq D(u, \overline{0})D(v, \overline{0})$ *for all* $u, v \in L(\mathbb{R})$.

(ii) *If* $u_k \to u$, *as* $k \to \infty$ *then* $D(u_k, \overline{0}) \to D(u, \overline{0})$, *as* $k \to \infty$; *where* $(u_k) \in \omega(F)$.

Proof. (i) It is trivial that the inequalities $|u^-(\alpha)| \leq D(u, \overline{0})$ and $|u^+(\alpha)| \leq D(u, \overline{0})$ hold for $\alpha \in [0,1]$. By considering these facts, one can see that

$$\begin{aligned} D(uv, \overline{0}) &= \sup_{\alpha \in [0,1]} \max\{|(uv)^-(\alpha)|, |(uv)^+(\alpha)|\} \\ &\leq \sup_{\alpha \in [0,1]} \max\{|u^-(\alpha)||v^-(\alpha)|, |u^-(\alpha)||v^+(\alpha)|, |u^+(\alpha)||v^-(\alpha)|, |u^+(\alpha)||v^+(\alpha)|\} \\ &\leq \sup_{\alpha \in [0,1]} \max\{D(u, \overline{0})|v^-(\alpha)|, D(u, \overline{0})|v^+(\alpha)|, D(u, \overline{0})|v^-(\alpha)|, D(u, \overline{0})|v^+(\alpha)|\} \\ &= D(u, \overline{0}) \sup_{\alpha \in [0,1]} \max\{|v^-(\alpha)|, |v^+(\alpha)|\} \\ &= D(u, \overline{0})D(v, \overline{0}), \end{aligned}$$

which completes the proof of Part (i).

(ii) This is trivial by using the fact given by Part (v) of Proposition 8.2.2. □

A subset E of $L(\mathbb{R})$ is said to be *bounded from above* if there exists a fuzzy number u, called an *upper bound* of E, such that $v \preceq u$ for every $v \in E$. u is called the *least upper bound* or *supremum* of E if u is an upper bound and $u \preceq u'$ for all upper bounds u'. Similarly, a subset E of $L(\mathbb{R})$ is said to be *bounded from below* if there exists a fuzzy number v, called a *lower bound* of E, such that $v \preceq u$ for every $u \in E$. v is called the *greatest lower bound* or *infimum* of E if v is a lower bound and $v' \preceq v$ for all upper bounds v'. E is said to be *bounded* if it is both bounded from above and below, (cf. Nanda [302]). Equivalently, we can also state this definition using the supremum metric on the set of fuzzy numbers, as follows: The set E is called *bounded* provided that there exists a real number M such that $D(v,\overline{0}) \leq M$ for each $v \in E$.

The notions of "sup" and "inf" have been defined only for bounded sets of fuzzy numbers. An important fact, proved by Wu and Wu [401, Theorem 2.1], states that if the set $E \subset L(\mathbb{R})$ is bounded, then its supremum and infimum exist. The following theorem essentially comes from [401]. It is noteworthy that the following theorem given by Fang and Huang [169, Theorem 3.1] which presents the expressions for endpoints of level sets of the supremum and infimum is concise and convenient for use.

Theorem 8.2.7. [169, Theorem 3.1] *Let A be a non-empty subset of the set $L(\mathbb{R})$. If A has a lower bound, then its infimum $\mu \in L(\mathbb{R})$ must exist and has the following expressions:*

$$\begin{aligned}
\mu &:= \inf A = \bigcup_{\lambda\in(0,1]} \lambda\left[\inf_{\gamma\in A}\gamma^-(\lambda), \inf_{\gamma\in A}\gamma^+(\lambda)\right],\\
\mu^-(\lambda) &:= \sup_{r<\lambda}\inf_{\gamma\in A}\gamma^-(r), \quad \mu^+(\lambda) = \inf_{\gamma\in A}\gamma^+(\lambda),\\
\mu^-(0) &:= \inf_{\gamma\in A}\gamma^-(0), \quad \mu^+(0) = \sup_{\lambda>0}\inf_{\gamma\in A}\gamma^+(\lambda)
\end{aligned}$$

for each $\lambda \in (0,1]$. Dually, if A has an upper bound, then its supremum $\nu \in L(\mathbb{R})$ must exist and has the following expressions:

$$\begin{aligned}
\nu &:= \sup A = \bigcup_{\lambda\in(0,1]} \lambda\left[\sup_{\gamma\in A}\gamma^-(\lambda), \sup_{\gamma\in A}\gamma^+(\lambda)\right],\\
\nu^-(\lambda) &:= \sup_{\gamma\in A}\gamma^-(\lambda), \quad \nu^+(\lambda) = \inf_{r<\lambda}\sup_{\gamma\in A}\gamma^+(r),\\
\nu^-(0) &:= \inf_{\lambda>0}\sup_{\gamma\in A}\gamma^-(\lambda), \quad \nu^+(0) = \sup_{\gamma\in A}\gamma^+(0)
\end{aligned}$$

for each $\lambda \in (0,1]$.

Definition 8.2.8. [274, Definition 3.1] *A sequence $u = (u_k)$ of fuzzy numbers is a function u from the set $\mathbb{N}_1$ into the set $L(\mathbb{R})$. The fuzzy number u_k denotes the value of the function at $k \in \mathbb{N}_1$ and is called as the kth term of the sequence.*

By $\omega(F)$, we denote the set of all sequences of fuzzy numbers.

Definition 8.2.9. [379, Definition 2.8] *An infinite matrix $A = (a_{nk})$ of fuzzy numbers is a double sequence of fuzzy numbers defined by a function A from the set $\mathbb{N}_1 \times \mathbb{N}_1$ into the set $L(\mathbb{R})$. The fuzzy number a_{nk} denotes the value of the function at $(n,k) \in \mathbb{N}_1^2$ and is called as the entry of the matrix which stands on the nth row and k^{th} column.*

Definition 8.2.10. [274, Definition 3.2] *A sequence $(u_k) \in \omega(F)$ is called convergent with limit $u \in L(\mathbb{R})$, if and only if for every $\varepsilon > 0$ there exists an $n_0 = n_0(\varepsilon) \in \mathbb{N}_1$ such that $D(u_k, u) < \varepsilon$ for all $k \geq n_0$.*

If the sequence $(u_k) \in \omega(F)$ converges to a fuzzy number u, then by the definition of D the sequences of functions $\{u_k^-(\alpha)\}$ and $\{u_k^+(\alpha)\}$ are uniformly convergent to $u^-(\alpha)$ and $u^+(\alpha)$ in $[0,1]$, respectively. Indeed, by combining the definition of D and the fact $\lim_{k\to\infty} D(u_k, u) = 0$ one can observe that

$$\lim_{k\to\infty} \sup_{\alpha\in[0,1]} \max\{|u_k^-(\alpha) - u^-(\alpha)|, |u_k^+(\alpha) - u^+(\alpha)|\} = 0.$$

Therefore, we have

$$\lim_{k\to\infty} \sup_{\alpha\in[0,1]} |u_k^-(\alpha) - u^-(\alpha)| = 0 \quad \text{and} \quad \lim_{k\to\infty} \sup_{\alpha\in[0,1]} |u_k^+(\alpha) - u^+(\alpha)| = 0,$$

which yield that the sequences of functions $\{u_k^-(\alpha)\}$ and $\{u_k^+(\alpha)\}$ are uniformly convergent to $u^-(\alpha)$ and $u^+(\alpha)$ in $[0,1]$, and conversely.

By $c(F)$ and $c_0(F)$, we denote the set of all convergent and null sequences of fuzzy numbers.

Now, we give the following main theorem concerning with the convergent sequences of fuzzy numbers which can easily be proved by using the definition of the supremum metric:

Theorem 8.2.11. [274, Theorem 4.1] *Let $(u_k), (v_k) \in \omega(F)$ with $u_k \to a$, $v_k \to b$, as $k \to \infty$. Then, the following statements hold:*

(i) *$u_k + v_k \to a + b$, as $k \to \infty$.*

(ii) *$u_k - v_k \to a - b$, as $k \to \infty$.*

(iii) *$u_k v_k \to ab$, as $k \to \infty$.*

(iv) *$u_k / v_k \to a/b$, as $k \to \infty$; where $0 \notin [v_k]_0$ for all $k \in \mathbb{N}$ and $0 \notin [b]_0$.*

Example 8.2.12. *Consider the sequence $x = (x_k) \in \omega(F)$ of fuzzy numbers* (cf. [274, Example]) *given by*

$$x_k(t) := \begin{cases} \frac{k}{2k-1} t & , \ t \in \left[0, \frac{2k-1}{k}\right), \\ 1 & , \ t \in \left[\frac{2k-1}{k}, \frac{2k+1}{k}\right), \\ -\frac{k}{2k-1}(t-4) & , \ t \in \left[\frac{2k+1}{k}, 4\right], \\ 0 & , \ \text{otherwise}. \end{cases} \tag{8.2.5}$$

One can see that $\lim_{k\to\infty} x_k = a$, *where* $a(t) := \begin{cases} t/2 & , \quad 0 \le t < 2, \\ (-t+4)/2 & , \quad 2 \le t \le 4, \\ 0 & , \quad \text{otherwise.} \end{cases}$ *To show this, take* $\varepsilon > 0$. *We wish to establish that there exists a number* k_0 *such that* $D(x_k, a) < \varepsilon$ *whenever* $k > k_0$. *We have* $D(x_k, a) = 1/k$. *So, we must show that there exists a number* k_0 *such that* $D(x_k, a) = 1/k < \varepsilon$ *whenever* $k > k_0$. *If we let* $k_0 = 1/\varepsilon$, *then* $k > k_0$ *implies that* $1/k < 1/k_0 = \varepsilon$. *This means that* $\lim_{k\to\infty} x_k = a$.

A neighborhood of a fuzzy number a with radius ε is the set of all fuzzy numbers μ in $L(\mathbb{R})$ such that $D(a, \mu) < \varepsilon$ and is denoted by $K(a, \varepsilon)$. It is clear that a fuzzy number a is the *limit* of a sequence $(x_k) \in \omega(F)$ iff every neighborhood of a contains infinitely many terms of the sequence in it and finitely many terms of the sequence out of it [274, Theorem 3.1].

Definition 8.2.13. [379, Definition 2.11] *A sequence* $(u_k) \in \omega(F)$ *is called bounded if and only if the set* $\{u_k : k \in \mathbb{N}_1\}$ *of fuzzy numbers consisting of the terms of the sequence* (u_k) *is a bounded set* [274, Definition 3.3]. *That is to say that a sequence* $(u_k) \in \omega(F)$ *is said to be bounded if and only if there exist two fuzzy numbers* m *and* M *such that* $m \preceq u_k \preceq M$ *for all* $k \in \mathbb{N}_1$ *which is equivalent to the uniform boundedness of the functions* u_k^- *and* u_k^+ *on* $[0, 1]$. *This means that* $m^-(\alpha) \le u_k^-(\alpha) \le M^-(\alpha)$ *and* $m^+(\alpha) \le u_k^+(\alpha) \le M^+(\alpha)$ *for* $\alpha \in [0, 1]$. *This leads us to the fact that the boundedness of a sequence* $(u_k) \in \omega(F)$ *is equivalent to the fact that*

$$\sup_{k\in\mathbb{N}} D(u_k, \overline{0}) = \sup_{k\in\mathbb{N}} \sup_{\alpha\in[0,1]} \max\{|u_k^-(\alpha)|, |u_k^+(\alpha)|\} < \infty.$$

By $\ell_\infty(F)$, *we denote the set of all bounded sequences of fuzzy numbers.*

Theorem 8.2.14. [274, Theorem 3.4] *Any convergent sequence* $(x_k) \in \omega(F)$ *of fuzzy numbers is bounded.*

Proof. Let a be the limit of the sequence (x_k). Then, there exists a number k_0 such that $D(x_k, a) < 1$ whenever $k > k_0$. Let $\rho = \max\{D(x_1, a), D(x_2, a), \ldots, D(x_{k_0}, a), 1\}$. Then, $D(x_k, a) \le \rho$ for any $k \in \mathbb{N}_1$. Since D is a metric on $L(\mathbb{R})$, we observe by triangle inequality that

$$D(x_k, \overline{0}) \le D(x_k, a) + D(a, \overline{0}) \le \rho + D(a, \overline{0}) = K$$

for all $k, m \in \mathbb{N}_1$. This completes the proof of the theorem. □

The proof of Theorem 8.2.14 reminds us the proofs in the analysis textbooks. Since the set of fuzzy numbers is a metric space under the metric D, many of the proofs that do not require and are analogous to the proofs given in the classical analysis. Therefore, the readers can easily prove many of the theorems which will be presented but not proved here, by basing themselves on the proofs given in the classical analysis.

A fuzzy number γ is called a *limit point* of the sequence $x = (x_k)$, if it has a subsequence converging to γ. We denote by L_x the set of all limit points of the sequence $x = (x_k) \in \omega(F)$, [48, Definition 2.2].

Theorem 8.2.15. [274, Theorem 3.2] *If a sequence* $(x_k) \in \omega(F)$ *is convergent, then it has only one limit.*

Theorem 8.2.16. [274, Theorem 3.3] *If there exists a* k_0 *such that* $x_k \preceq y_k \preceq z_k$ *for all* $k > k_0$ *and if* $\lim_{k\to\infty} x_k = \mu = \lim_{k\to\infty} z_k$*, then the sequence* $(y_k) \in \omega(F)$ *also converges to the fuzzy number* μ*.*

Let (k_n) be an increasing sequence of natural numbers. Then, the sequence (x_{k_n}) is called a *subsequence* of $(x_k) \in \omega(F)$, [274, Definition 3.4].

Theorem 8.2.17. [274, Theorem 3.6] *If the sequence* $(x_k) \in \omega(F)$ *converges to a fuzzy number, then any subsequence of* (x_k) *converges to the same fuzzy number.*

A sequence $(x_k) \in \omega(F)$ of fuzzy numbers is said to be a *Cauchy sequence* if for every $\varepsilon > 0$, there exists a $k_0 = k_0(\varepsilon) \in \mathbb{N}_1$ such that $D(x_k, x_m) < \varepsilon$ for all $k, m > k_0$ [302, Definition 2.2]. It is straightforward to see that every convergent sequence of fuzzy numbers is a Cauchy sequence.

A sequence $(x_k) \in \omega(F)$ is said to be *monotone increasing* provided that $x_k \preceq x_{k+1}$ for each $k \in \mathbb{N}_1$. A *monotone decreasing* sequence can be defined, similarly. Either monotone increasing or monotone decreasing sequence can be simply called as *monotonic*, [195, Definition 4.3].

Example 8.2.18. [50, Example] *Define the sequence* $x = (x_k) \in \omega(F)$ *of fuzzy numbers by*

$$x_k(t) := \begin{cases} t+1 & , \quad -1 \leq t < 0, \\ -t+1 & , \quad 0 \leq t < \frac{1}{2}, \\ \frac{1}{2} & , \quad \frac{1}{2} \leq t < 1, \\ -\frac{1}{5k}(t-1) + \frac{1}{2} & , \quad 1 \leq t < \frac{15k-2}{2(5k-1)}, \\ -t+2 & , \quad \frac{15k-2}{2(5k-1)} \leq t \leq 2, \\ 0 & , \quad \text{otherwise.} \end{cases}$$

The sequence $x = (x_k) \in \omega(F)$ *is monotone increasing and bounded from above but the limit of this sequence does not exist.*

A *closed interval* in $L(\mathbb{R})$ is defined as

$$[\mu, \nu] := \{\gamma \in L(\mathbb{R}) : \mu \preceq \gamma \preceq \nu\}, \tag{8.2.6}$$

where $\mu, \nu \in L(\mathbb{R})$; [195, Definition 4.1]. It is clear that the limit points of a sequence $x = (x_k) \in \omega(F)$ whose elements are in the interval $[\mu, \nu]$, also belong to the interval $[\mu, \nu]$.

A set E of fuzzy numbers is called *convex* if $\alpha u_1 + (1-\alpha) u_2 \in E$ for all $\alpha \in [0,1]$ and $u_1, u_2 \in E$ [334, p. 10]. It is clear that each interval $[\mu, \nu]$ is a convex closed set.

Consider a subset $E \subset L(\mathbb{R})$. A closed interval $[\mu', \nu']$ is called a *cover* of the set E if $E \subset [\mu', \nu']$. The set $[\mu, \nu]$ is called the *least closed interval covering the set E* if $[\mu, \nu] \subset [\mu', \nu']$ for all intervals $[\mu', \nu']$ covering E, (cf. Aytar [55]).

Throughout this section, we will write BSFN to abbreviate a bounded sequence of fuzzy numbers.

8.2.1 The Limit Superior and Limit Inferior of a Sequence of Fuzzy Numbers

In this subsection, we define the concepts of limit inferior and limit superior of a BSFN in a similar way as in the case of real numbers, and investigate their properties. We present two different definitions for limit inferior and limit superior which are equivalent for real numbers but are different for BSFNs. We observe that some properties established for sequences with real terms are not valid for BSFNs. Indeed, there exists a BSFN such that its limit superior is not the greatest limit point of this sequence.

Definition 8.2.19. [55, Definition 1] *The limit inferior and limit superior of $(x_k) \in \ell_\infty(F)$ are defined by*

$$\inf_{k\geq n}\liminf_{k\to\infty} x_k := \lim_{n\to\infty}\inf_{k\geq n} x_k \text{ and } \limsup_{k\to\infty} x_k := \lim_{n\to\infty}\sup_{k\geq n} x_k.$$

This definition assumes the convergence of the sequences $\{\inf_{k\geq n} x_k\}$ and $\{\sup_{k\geq n} x_k\}$. As was seen by Example 8.2.26, below, we should note that the limit inferior or the limit superior of a BSFN may not exist.

The definitions of these concepts are the same as in [195, Definition 4.4] with the difference that the limit is considered with respect to the metric used in [274].

Lemma 8.2.20. [55, Lemma 1] *Assume that (x_k) and (y_k) are convergent sequences of fuzzy numbers. If there is a number $k_0 \in \mathbb{N}_1$ such that $x_k \preceq y_k$ for all $k > k_0$, then $\lim_{k\to\infty} x_k \preceq \lim_{k\to\infty} y_k$.*

Proof. Let $(x_k), (y_k) \in c(F)$ with $x_k \to \mu$ and $y_k \to \nu$, as $k \to \infty$. The condition $x_k \preceq y_k$ for all $k > k_0$ means for each $\alpha \in [0,1]$ that $x_k^-(\alpha) \leq y_k^-(\alpha)$ and $x_k^+(\alpha) \leq y_k^+(\alpha)$ for all $k > k_0$. Taking limit as $k \to \infty$, we obtain for each $\alpha \in [0,1]$ that the inequalities $\mu^-(\alpha) \leq \nu^-(\alpha)$ and $\mu^+(\alpha) \leq \nu^+(\alpha)$ hold. This means that $\mu \preceq \nu$. □

Applying Lemma 8.2.20, the following two theorems can easily be proved:

Theorem 8.2.21. [55, Theorem 2] *Let $x = (x_k)$ be a BSFN. If $\liminf_{k\to\infty} x_k$ and $\limsup_{k\to\infty} x_k$ exist, then $\liminf_{k\to\infty} x_k \preceq \limsup_{k\to\infty} x_k$.*

Theorem 8.2.22. [55, Theorem 3] *Let $x = (x_k), y = (y_k) \in \ell_\infty(F)$ and assume that there is a number $n_0 \in \mathbb{N}$ such that $x_k \preceq y_k$ for all $k > n_0$. If $\liminf_{k\to\infty} x_k$, $\limsup_{k\to\infty} x_k$, $\liminf_{k\to\infty} y_k$ and $\limsup_{k\to\infty} y_k$ exist, then $\liminf_{k\to\infty} x_k \preceq \liminf_{k\to\infty} y_k$ and $\limsup_{k\to\infty} x_k \preceq \limsup_{k\to\infty} y_k$.*

Now, we prove the following lemma:

Lemma 8.2.23. [55, Lemma 4] *Let $x = (x_k) \in \ell_\infty(F)$ and $\nu = \liminf_{k\to\infty} x_k$, and $\mu = \limsup_{k\to\infty} x_k$ exist. Then, the sets*

$$M_1(\varepsilon) := \{k \in \mathbb{N}_1 : x_k \prec \nu - \overline{\varepsilon}\} \text{ and } M_2(\varepsilon) := \{k \in \mathbb{N}_1 : x_k \succ \mu + \overline{\varepsilon}\}$$

are finite for every $\varepsilon > 0$.

Proof. We prove that the set $M_1(\varepsilon)$ has only a finite number of elements for every $\varepsilon > 0$.

Suppose to the contrary that there exists $\varepsilon > 0$ such that $M_1(\varepsilon)$ has an infinite number of elements. By definition of "inf", we have $\inf_{k\geq n} x_k \preceq x_n$ for all $n \in \mathbb{N}_1$. Then, $\inf_{k\geq n} x_k \preceq x_n \prec \nu - \overline{\varepsilon}$ for all $n \in M_1(\varepsilon)$. We obtain $D\left(\inf_{k\geq n} x_k, \nu\right) > \varepsilon$ for all $n \in M_1(\varepsilon)$. Since the set $M_1(\varepsilon)$ has an infinite number of elements, the last inequality contradicts the fact that $\lim_{n\to\infty} \inf_{k\geq n} x_k = \nu$.

This completes the proof. □

Now, we give an alternative approach to define the notions of limit superior and limit inferior. Given a sequence $x = (x_k) \in \omega(F)$, we define the following sets:

$$
\begin{aligned}
A_x &:= \{\mu \in L(\mathbb{R}) : \{k \in \mathbb{N}_1 : x_k \prec \mu\} \in \mathcal{P}_I(\mathbb{N}_1)\}, \\
\overline{A}_x &:= \{\mu \in L(\mathbb{R}) : \{k \in \mathbb{N}_1 : x_k \succ \mu\} \in \mathcal{P}_\Delta(\mathbb{N}_1)\}, \\
B_x &:= \{\mu \in L(\mathbb{R}) : \{k \in \mathbb{N}_1 : x_k \succ \mu\} \in \mathcal{P}_I(\mathbb{N}_1)\}, \\
\overline{B}_x &:= \{\mu \in L(\mathbb{R}) : \{k \in \mathbb{N}_1 : x_k \prec \mu\} \in \mathcal{P}_\Delta(\mathbb{N}_1)\}, \text{ (cf. Aytar [55]), where} \\
\mathcal{P}_I(\mathbb{N}_1) &:= \{A \in \mathcal{P}(\mathbb{N}_1) : A \text{ has infinite number of elements}\}, \\
\mathcal{P}_\Delta(\mathbb{N}_1) &:= \{A \in \mathcal{P}(\mathbb{N}_1) : \mathbb{N}_1 \backslash A \text{ has finite number of elements}\}.
\end{aligned}
$$

It is clear that if the sequence $x = (x_k) \in \omega(F)$ is bounded, then the sets $A_x, \overline{A}_x, B_x$ and $\overline{B}_x$ are non-empty. It is also clear that the sets A_x and $\overline{B}_x$ have lower bounds, and the sets $\overline{A}_x$ and B_x have upper bounds. Therefore, by Theorem 8.2.7, we conclude that $\inf A_x$, $\sup \overline{A}_x$, $\sup B_x$ and $\inf \overline{B}_x$ exist. We show in Theorem 8.2.25 below, that the relations $\inf A_x = \sup \overline{A}_x$ and $\sup B_x = \inf \overline{B}_x$ hold, (cf. Aytar [55]).

Therefore, we give the following definition:

Definition 8.2.24. [55, Definition 2] *The notions Liminf and Limsup are defined for a given $x = (x_k) \in \ell_\infty(F)$ by*

$$
\begin{aligned}
\mathop{Liminf}_{k\to\infty} x_k &:= \inf A_x = \sup \overline{A}_x, \\
\mathop{Limsup}_{k\to\infty} x_k &:= \sup B_x = \inf \overline{B}_x.
\end{aligned}
$$

Theorem 8.2.25. [55, Theorem 5] *If $x = (x_k) \in \ell_\infty(F)$, then $\inf A_x = \sup \overline{A}_x$ and $\sup B_x = \inf \overline{B}_x$.*

Proof. We prove only the first equality. Denote $\nu = \inf A_x$ and $\mu = \sup \overline{A}_x$. By definition of the sets A_x and $\overline{A}_x$, we have $\nu \preceq \widetilde{\nu}$ for all $\widetilde{\nu} \in A_x$, and $\mu \succeq \widetilde{\mu}$ for all $\widetilde{\mu} \in \overline{A}_x$.

$\widetilde{\nu} \in A_x$ means that the set $\{k \in \mathbb{N}_1 : x_k \prec \widetilde{\nu}\}$ is infinite. On the other hand, from $\widetilde{\mu} \in \overline{A}_x$ it follows that the set $\mathbb{N}_1 \setminus \{k \in \mathbb{N}_1 : x_k \succ \widetilde{\mu}\}$ is finite; that is, there is a number $K_{\widetilde{\mu}} < \infty$ such that $x_k \succ \widetilde{\mu}$ for all $k \geq K_{\widetilde{\mu}}$. Thus, for every $\widetilde{\nu} \in A_x$ and $\widetilde{\mu} \in \overline{A}_x$, there is a number $k' \geq K_{\widetilde{\mu}}$ such that $\widetilde{\mu} \prec x_{k'} \prec \widetilde{\nu}$. This implies that

$$
\widetilde{\mu} \prec \widetilde{\nu} \text{ for all } \widetilde{\nu} \in A_x, \ \widetilde{\mu} \in \overline{A}_x. \tag{8.2.7}
$$

From (8.2.7), it follows that $\widetilde{\mu}$ is a lower bound of the set A_x. Then, we have by definition of infimum that $\widetilde{\mu} \preceq \nu = \inf A_x$. This inequality holds for all $\widetilde{\mu} \in \overline{A}_x$. Then, we get by definition of supremum that

$$\mu \preceq \nu. \tag{8.2.8}$$

Now, we show that the case $\mu \prec \nu$ is impossible.

To the contrary, assume that $\mu \prec \nu$. This means that, there is a number $\alpha \in [0,1]$ such that $\mu^-(\alpha) < \nu^-(\alpha)$ or $\mu^+(\alpha) < \nu^+(\alpha)$. For the sake of definiteness, we only consider the case

$$\mu^-(\alpha) < \nu^-(\alpha) \tag{8.2.9}$$

and show that it leads a contradiction.

Denote $b = \nu(\mu^-(\alpha))$. It is clear that $b < \alpha$ (b may be zero). Moreover, for all $\lambda \in (b, \alpha]$ the inequality $\mu^-(\lambda) < \nu^-(\lambda)$ holds. Since the functions $\mu(x)$ and $\nu(x)$ are upper semi-continuous, there is a point (z, β) such that $z \in (\mu^-(\alpha), \nu^-(\alpha))$, $\beta \in (b, \alpha)$ and

$$\mu^-(\lambda) < z, \ \nu^-(\lambda) < z \ \text{ for all } \ \lambda \in [\beta, \alpha]. \tag{8.2.10}$$

We define the numbers $\gamma_1, \gamma_2 \in L(\mathbb{R})$ by

$$\gamma_1(t) := \begin{cases} 0 & , \quad t < \underline{t}^0, \\ \beta & , \quad t \in [\underline{t}^0, z), \\ 1 & , \quad t = z, \\ 0 & , \quad t > z, \end{cases} \quad \text{and} \quad \gamma_2(t) := \begin{cases} 0 & , \quad t < z, \\ \beta & , \quad t \in [z, \overline{t}^0), \\ 1 & , \quad t = \overline{t}^0, \\ 0 & , \quad t > \overline{t}^0, \end{cases}$$

where the numbers $\underline{t}^0 = \liminf x_k^-(0) - 1$ and $\overline{t}^0 = \limsup x_k^+(0) + 1$ are finite.

It is not difficult to observe by (8.2.10) that

$$\mu^-(\beta) \geq \liminf_{k\to\infty} x_k^-(\beta) \geq \liminf_{k\to\infty} x_k^-(0) > \underline{t}^0 = \gamma_1^-(\beta), \quad \mu^-(\alpha) < z = \gamma_1^-(\alpha)$$

and

$$\nu^-(b) \leq \mu^-(\alpha) < z = \gamma_2^-(b), \quad \nu^-(\beta) > z = \gamma_2^-(\beta).$$

This means that

$$\mu \not\preceq \gamma_1 \ \text{ and } \ \nu \not\preceq \gamma_2. \tag{8.2.11}$$

Now, we define the following sets

$$C_1 := \{k \in \mathbb{N}_1 : x_k^-(\lambda) \leq z \ \text{ for some } \lambda \in (\beta, \alpha]\},$$
$$C_2 := \{k \in \mathbb{N}_1 : x_k^-(\lambda) \geq z \ \text{ for some } \lambda \in (\beta, \alpha]\}.$$

Clearly, we have

$$C_1 \cup C_2 = \mathbb{N}_1. \tag{8.2.12}$$

First we assume that the set C_1 contains an infinite number of elements. Taking into account the structure of the fuzzy number γ_2 and the real number $\overline{t}^0$, we obtain that $x_k \prec \gamma_2$ for all $k \in C_1 \backslash K_1$, where $K_1 := \left\{k \in \mathbb{N}_1 : x_k^+(\lambda) > \overline{t}^0 \text{ for some } \lambda \in [0,1]\right\}$. It is clear that the set K_1 is finite. Then, we conclude that the number of $k \in \mathbb{N}_1$, for which $x_k \prec \gamma_2$ holds, is infinite. This means that $\gamma_2 \in A_x$ and therefore, from the definition of $\inf A_x$ we obtain $\gamma_2 \succeq \nu = \inf A_x$. This contradicts (8.2.11), that is, $\nu \not\prec \gamma_2$.

Hence, we have shown that the set C_1 is finite. In this case, from (8.2.12), it follows that the set C_2 is infinite. Taking into account the structure of the fuzzy number γ_1 and the real number $\underline{t}^0$, we obtain $x_k \succ \gamma_1$ for all $k \in C_2 \backslash (C_1 \cup K_2)$, where $K_2 := \{k \in \mathbb{N}_1 : x_k^-(\lambda) < \underline{t}^0\}$ for some $\lambda \in [0, \beta]$. It is obvious that the set K_2 is finite, hence we conclude that the set $C_2 \backslash (C_1 \cup K_2)$ is infinite. This implies that $\gamma_1 \in \overline{A}_x$. Thus, $\gamma_1 \preceq \mu = \sup \overline{A}_x$. This contradicts (8.2.11), that is, $\mu \not\prec \gamma_1$.

This completes the proof. □

Now, we give an example which shows the difference of Definition 8.2.19 and Definition 8.2.24.

Example 8.2.26. [55, Example 3.1] *Consider the sequences $x = (x_k)$, $\gamma = (\gamma_k)$ and $\nu = (\nu_k)$ of fuzzy numbers together with the fuzzy number μ defined by*

$$x_k(t) := \begin{cases} t+1 & , \quad -1 \le t \le 0, \\ -t+1 & , \quad 0 < t \le \frac{5k-2}{2(5k-1)}, \\ -\frac{1}{5k}(t-1)+\frac{1}{2} & , \quad \frac{5k-2}{2(5k-1)} < t \le \frac{15k-2}{2(5k-1)}, \\ -t+2 & , \quad \frac{15k-2}{2(5k-1)} < t \le 2, \\ 0 & , \quad \text{otherwise}, \end{cases}$$

$$\mu(t) := \begin{cases} t+1 & , \quad -1 \le t \le 0, \\ -t+1 & , \quad 0 < t \le \frac{1}{2}, \\ \frac{1}{2} & , \quad \frac{1}{2} < t \le \frac{3}{2}, \\ -t+2 & , \quad \frac{3}{2} < t \le 2, \\ 0 & , \quad \text{otherwise}, \end{cases}$$

$$\gamma_k(t) := \begin{cases} t+1 & , \quad -1 \le t \le 0, \\ -t+1 & , \quad 0 < t \le \frac{5k-2}{2(5k-1)}, \\ -\frac{1}{5k}(t-1)+\frac{1}{2} & , \quad \frac{5k-2}{2(5k-1)} < t \le 1, \\ \frac{1}{2} & , \quad 1 < t \le \frac{3}{2}, \\ -t+2 & , \quad \frac{3}{2} < t \le 2, \\ 0 & , \quad \text{otherwise}, \end{cases}$$

$$\nu_k(t) := \begin{cases} t+1 & , \quad -1 \le t \le 0, \\ -t+1 & , \quad 0 < t \le \frac{1}{2}, \\ \frac{1}{2} & , \quad \frac{1}{2} < t \le 1, \\ -\frac{1}{5k}(t-1)+\frac{1}{2} & , \quad 1 < t \le \frac{15k-2}{2(5k-1)}, \\ -t+2 & , \quad \frac{15k-2}{2(5k-1)} < t \le 2, \\ 0 & , \quad \text{otherwise}. \end{cases}$$

Then, it is not difficult to see that $\inf A_x = \inf_{k \in \mathbb{N}_1} \gamma_k$.

We obtain by using Theorem 8.2.7 that $\inf_{k\in\mathbb{N}_1}\gamma_k = \mu$. *Therefore, by Definition 8.2.24, we have* $\operatorname*{Liminf}_{k\to\infty} x_k = \mu$. *Similarly, we observe that* $\sup A_x = \sup_{k\in\mathbb{N}_1}\nu_k$. *Moreover,* $\sup_{k\in\mathbb{N}_1}\nu_k = \mu$. *Then, by Definition 8.2.24, we have* $\operatorname*{Limsup}_{k\to\infty} x_k = \mu$.

On the other hand, $\liminf_{k\to\infty} x_k$ *(and similarly* $\limsup_{k\to\infty} x_k$*) does not exist, since the sequence*

$$\inf_{k\geq n} x_k(t) := \begin{cases} t+1 & , & -1\leq t<0, \\ -t+1 & , & 0\leq t<\frac{1}{2}, \\ \frac{1}{2} & , & \frac{1}{2}\leq t<1, \\ -\frac{t-1}{5n}+\frac{1}{2} & , & 1\leq t<\frac{15n-2}{2(5n-1)}, \\ -t+2 & , & \frac{15n-2}{2(5n-1)}\leq t\leq 2, \\ 0 & , & \text{otherwise} \end{cases}$$

does not converge to any fuzzy number. We note that $D(\inf_{k\geq n} x_k, \mu)\geq \frac{1}{2}$, *since*

$$\left|\left(\inf_{k\geq n} x_k\right)^+\left(\frac{1}{2}\right)-\mu^+\left(\frac{1}{2}\right)\right| = \left|1-\frac{3}{2}\right| = \frac{1}{2} \text{ for all } n\in\mathbb{N}_1.$$

Example 8.2.26 forms the basis for the following results:

1. The limit of the sequence $(x_k)\in\omega(F)$ may not exist while $\operatorname*{Liminf}_{k\to\infty} x_k = \operatorname*{Limsup}_{k\to\infty} x_k$.
2. $\operatorname*{Liminf}_{k\to\infty} x_k$ and $\operatorname*{Limsup}_{k\to\infty} x_k$ exist for each BSFN but $\liminf_{k\to\infty} x_k$ and $\limsup_{k\to\infty} x_k$ may not exist.
3. For some BSFN (x_k), there may not exist any subsequence that converges to $\operatorname*{Liminf}_{k\to\infty} x_k$ and $\operatorname*{Limsup}_{k\to\infty} x_k$.
4. The set of limit points L_x of a BSFN (x_k) may be empty.

Remark 8.2.27. *It is clear that if* $\liminf_{k\to\infty} x_k$ *exists, then Definitions 8.2.19 and 8.2.24 are equivalent.*

8.2.2 The Core of a Sequence of Fuzzy Numbers

As was seen in Chapter 6, the notion of the core of a sequence of complex numbers was introduced by Knopp [224], (see [111, p. 137]). Below, we extend this notion to the sequences of fuzzy numbers.

Consider a sequence $x=(x_k)\in\omega(F)$. Let R_n be the least closed interval of $L(\mathbb{R})$ containing the set $\{x_n, x_{n+1}, x_{n+2}, \ldots\}$. Then, obviously we have

$$R_1\supseteq R_2\supseteq R_3\supseteq\cdots.$$

Definition 8.2.28. [55, Definition 3] *The intersection* $core\{x\}=\bigcap_{n=1}^{\infty} R_n$ *is called the core of the bounded sequence* $x=(x_k)$ *of fuzzy numbers.*

Proposition 8.2.29. [55, Proposition 6] *Let* $x = (x_k)$ *be a BSFN. If* $\liminf\limits_{k\to\infty} x_k$ *and* $\limsup\limits_{k\to\infty} x_k$ *exist, then* $core\{x\} = \left[\liminf\limits_{k\to\infty} x_k, \limsup\limits_{k\to\infty} x_k\right]$.

Proof. By Definition 8.2.28, we have $\text{core}\{x\} = \bigcap\limits_{n=1}^{\infty} R_n$, where $R_n = [A_n, B_n]$ is the least interval containing the set $\{x_n, x_{n+1}, x_{n+2}, \ldots\}$:

$$\{x_n, x_{n+1}, x_{n+2}, \ldots\} \subset R_n = [A_n, B_n] \quad \text{for all} \quad n \in \mathbb{N}_1. \tag{8.2.13}$$

First, we show the inclusion

$$\left[\liminf_{k\to\infty} x_k, \limsup_{k\to\infty} x_k\right] \subset core\{x\} \tag{8.2.14}$$

holds. Take any $\gamma \in \left[\liminf\limits_{k\to\infty} x_k, \limsup\limits_{k\to\infty} x_k\right]$ and any number $n_0 \in \mathbb{N}_1$.

From (8.2.13), we have

$$A_{n_0} \preceq x_k \preceq B_{n_0} \quad \text{for all} \quad k \geq n_0. \tag{8.2.15}$$

From the left-hand side of this inequality, it follows that A_{n_0} is a lower bound of the set $\{x_{n_0}, x_{n_0+1}, x_{n_0+2}, \ldots\}$. This means that $A_{n_0} \preceq \inf_{k\geq n_0} x_k$. For all $i \in \mathbb{N}$, write $\xi_{n_0+i} = \inf_{k\geq n_0+i} x_k$. Since $\{x_{n_0}, x_{n_0+1}, \ldots\} \supseteq \{x_{n_0+1}, x_{n_0+2}, \ldots\} \supseteq \cdots$, we have $\xi_{n_0} \preceq \xi_{n_0+1} \preceq \cdots$. Thus, $A_{n_0} \preceq \xi_{n_0+i}$ for all $i \in \mathbb{N}$. Then, from Lemma 8.2.20 it follows that

$$A_{n_0} \preceq \lim_{i\to\infty} \xi_{n_0+i} = \lim_{i\to\infty} \inf_{k\geq n_0+i} x_k = \lim_{n\to\infty} \inf_{k\geq n} x_k = \liminf_{k\to\infty} x_k.$$

We obtain by taking into account $\gamma \in \left[\liminf\limits_{k\to\infty} x_k, \limsup\limits_{k\to\infty} x_k\right]$ that $A_{n_0} \preceq \gamma$.

Similarly, considering the right-hand side of the inequality, (8.2.15) we can prove that $B_{n_0} \succeq \limsup\limits_{k\to\infty} x_k \succeq \gamma$. Hence, we have $\gamma \in [A_{n_0}, B_{n_0}] = R_{n_0}$. As $n_0 \in \mathbb{N}_1$ is an arbitrary number, we obtain $\gamma \in \bigcap\limits_{n=1}^{\infty} R_n = core\{x\}$. Therefore, the inclusion (8.2.14) holds.

Now, we show

$$core\{x\} \subset \left[\liminf_{k\to\infty} x_k, \limsup_{k\to\infty} x_k\right]. \tag{8.2.16}$$

Take any $\gamma \in core\{x\}$. It is clear that

$$\gamma \in [A_n, B_n] \quad \text{for all} \quad n \in \mathbb{N}_1. \tag{8.2.17}$$

Denote $\xi_n = \inf\limits_{k\geq n} x_k$ and $\eta_n = \sup\limits_{k\geq n} x_k$ for all $n \in \mathbb{N}_1$. Clearly, $\xi_n \preceq x_k \leq \eta_n$ for all $k \geq n$; that is,

$$\{x_n, x_{n+1}, x_{n+2}, \ldots\} \subset [\xi_n, \eta_n] \quad \text{for all} \quad n \in \mathbb{N}_1. \tag{8.2.18}$$

On the other hand, by Definition 8.2.28, $R_n = [A_n, B_n]$ is the least interval containing the set $\{x_n, x_{n+1}, x_{n+2}, \ldots\}$. This means that $[A_n, B_n] \subset [\xi_n, \eta_n]$ for all $n \in \mathbb{N}_1$. Then, from (8.2.17), it follows that $\gamma \in [\xi_n, \eta_n]$; that is, $\xi_n \preceq \gamma \preceq \eta_n$ for all $n \in \mathbb{N}_1$. Applying Lemma 8.2.20, we obtain

$$\lim_{n\to\infty} \xi_n \preceq \gamma \preceq \lim_{n\to\infty} \eta_n \quad \text{or} \quad \gamma \in \left[\liminf_{k\to\infty} x_k, \limsup_{k\to\infty} x_k\right].$$

Therefore, the inclusion (8.2.16) holds.

Consequently, the inclusions (8.2.14) and (8.2.16) give the required result. □

Theorem 8.2.30. [55, Theorem 9] *Let $x = (x_k) \in \ell_\infty(F)$ such that $core\{x\} \neq \emptyset$. Then, $L_x \subset core\{x\}$.*

Proof. Let $\gamma \in L_x$ and $core\{x\} = \bigcap_{n\in\mathbb{N}_1} [A_n, B_n]$. We have to show that

$$\gamma \in [A_n, B_n] \quad \text{for all} \quad n \in \mathbb{N}_1. \tag{8.2.19}$$

Take any $n \in \mathbb{N}_1$. By definition of $core\{x\}$, we have

$$A_n \preceq x_k \preceq B_n \quad \text{for all} \quad k \geq n. \tag{8.2.20}$$

Since γ is a limit point of the sequence $(x_k) \in \omega(F)$, there is a subsequence (x_{k_m}) that converges to γ. We observe from (8.2.20) that $A_n \preceq x_{k_m} \preceq B_n$ for all $k_m \geq n, m \in \mathbb{N}$. Applying Lemma 8.2.20, we obtain $A_n \preceq \gamma \preceq B_n$; that is, (8.2.19) holds. □

Theorem 8.2.31. [55, Theorem 8] *Suppose that $\liminf_{k\to\infty} x_k$ and $\limsup_{k\to\infty} x_k$ exist for $x = (x_k) \in \ell_\infty(F)$. Then, $L_x \subset \left[\liminf_{k\to\infty} x_k, \limsup_{k\to\infty} x_k\right]$.*

Proof. This directly follows from Proposition 8.2.29 and Theorem 8.2.30. □

Since $\mathbb{R}$ can be embedded into $L(\mathbb{R})$, the following conditions which do not hold for sequences of real numbers are not also valid for sequences of fuzzy numbers:

(i) The converse of the inclusion in Theorem 8.2.30 does not hold, in general.

(ii) Two BSFNs may have the same core while the sets of their limit points are different.

Although, some properties obtained for sequences of fuzzy numbers are also valid for sequences of real numbers, some properties may be different as was seen in the following example:

Example 8.2.32. [55, Example 4.1] *The following statements hold:*

(a) *Let the sequence $x = (x_k) \in \omega(F)$ and the fuzzy number μ be given as in Example 8.2.26. Define the sequences $z = (z_k)$ and $y = (y_k)$ by*

$$z_k := \begin{cases} x_k & , \quad k \text{ is odd number,} \\ \nu & , \quad \text{otherwise} \end{cases} \quad \text{and } y_k := \nu \text{ for all } k \in \mathbb{N}_1,$$

where $\nu(t) := \begin{cases} t-6 & , \ t \in [6,7], \\ -t+8 & , \ t \in (7,8], \\ 0 & , \ \text{otherwise.} \end{cases}$ *Then, one can see that* $L_z = L_y = \{\nu\}$, $core\{z\} = [\mu, \nu]$ *and* $core\{y\} = \{\nu\}$. *That is, two BSFNs may have the same limit set while their cores are different.*

(b) *For the sequence* $x = (x_k) \in \omega(F)$ *in Example 8.2.26, we obtain that* $core\{x\} = \{\mu\}$ *but* $\lim_{k\to\infty} x_k$ *does not exist.*

(c) *For the bounded sequence* $z = (z_k)$ *defined in Part (a), we have* $L_z = \{\nu\}$. *But the sequence* $z = (z_k)$ *does not converge to* ν.

(d) *Let the sequence* $x = (x_k) \in \omega(F)$ *and the fuzzy number* μ *be given as in Example 8.2.26. Define the sequence* $s = \{s_k\}$ *by*

$$s_k := \begin{cases} x_k & , \ k \text{ is odd number,} \\ \mu & , \ \text{otherwise.} \end{cases}$$

Then, although $L_s = core\{s\} = \{\mu\}$, *the limit of* $s = (s_k)$ *does not exist.*

Now, we can give a sufficient condition for a BSFN to be convergent.

Theorem 8.2.33. [55] *Assume that* $x = (x_k) \in \ell_\infty(F)$ *with*

$$core\{x\} = \left[\liminf_{k\to\infty} x_k, \limsup_{k\to\infty} x_k\right] = \{\mu\}.$$

If there is a number $\varepsilon^0 > 0$ *such that for each* $\varepsilon \in (0, \varepsilon^0)$, *the sets*

$$K_1(\varepsilon) := \{k \in \mathbb{N}_1 : x_k \not\succ \mu - \overline{\varepsilon}\} \text{ and } K_2(\varepsilon) := \{k \in \mathbb{N}_1 : x_k \not\prec \mu + \overline{\varepsilon}\}$$

are finite, then $\lim_{k\to\infty} x_k = \mu$.

Proof. Take any $\varepsilon > 0$. Let $N_1 := \max\{n : n \in K_1(\varepsilon)\}$ and $N_2 := \max\{n : n \in K_2(\varepsilon)\}$. By Lemma 8.2.23, the sets

$$M_1(\varepsilon) := \{k \in \mathbb{N}_1 : x_k \preceq \mu - \overline{\varepsilon}\} \quad \text{and} \quad M_2(\varepsilon) := \{k \in \mathbb{N}_1 : x_k \succeq \mu + \overline{\varepsilon}\}$$

are finite. Denote the largest elements of the sets $M_1(\varepsilon)$ and $M_2(\varepsilon)$ by N_3 and N_4, respectively. Since

$$K_1(\varepsilon) \cup M_1(\varepsilon) \cup \{k \in \mathbb{N}_1 : x_k \succ \mu - \overline{\varepsilon}\} = \mathbb{N}_1$$

we have

$$x_k \succ \mu - \overline{\varepsilon} \text{ for all } k > \max\{N_1, N_3\}.$$

Similarly, we obtain

$$x_k \prec \mu + \overline{\varepsilon} \text{ for all } k > \max\{N_2, N_4\}.$$

Hence, we have

$$\mu - \overline{\varepsilon} \prec x_k \prec \mu + \overline{\varepsilon} \text{ for all } k > \max\{N_1, N_2, N_3, N_4\}.$$

This means that $D(x_k, \mu) < \varepsilon$; that is, $\lim_{k\to\infty} x_k = \mu$. □

If $x = (x_k) \in \omega(F)$ is a sequence of fuzzy numbers, then the set core$\{x\}$ which is the intersection of the smallest closed and convex sets that contain the tail terms of the sequence $x = (x_k) \in \omega(F)$ coincides with the interval $[\liminf_{k\to\infty} x_k, \limsup_{k\to\infty} x_k]$. This is the main result of this subsection. Since the core and the singleton which contains the limit of the sequence coincide for convergent sequences, the concept of core does not provide additional information for such sequences. If the sequence is not convergent, then we can interpret certain properties of the sequence via the concept of core. For instance, we can determine the behaviors of the extreme limit points and the other limit points of the sequence. As another example, consider the sequence $x = (x_k) \in \omega(F)$ in Example 8.2.26 which is not convergent but almost all terms of the sequence cluster about $core\{x\}$ geometrically. Therefore, the core of a sequence of fuzzy numbers is important, especially for a sequence that is not convergent, (cf. Aytar [55]).

8.3 STATISTICAL CONVERGENCE OF A SEQUENCE OF FUZZY NUMBERS

Statistical convergence is a generalization of the usual notion of convergence that parallels to the usual theory of convergence. If K is a subset of positive integers $\mathbb{N}_1$, then K_n denotes the set $\{k \in K : k \leq n\}$. The natural density of K is given by $\delta(K) = \lim_{n\to\infty} |K_n|/n$, where $|K_n|$ denotes the number of elements in K_n. The natural density may not be exist for some sets K. But, the upper density $\overline{\delta}$ always exists for each set K identified as $\overline{\delta}(K) = \limsup_{n\to\infty} |K_n|/n$. Moreover, the natural density of K is different from zero which means $\overline{\delta}(K) > 0$. Clearly, finite subsets have natural density zero and $\delta(K^c) = 1 - \delta(K)$; where $K^c = \mathbb{N}_1 \backslash K$, i.e., the complement of K. If $K_1 \subseteq K_2$, then $\delta(K_1) \leq \delta(K_2)$; [173].

Statistical convergence of a sequence of fuzzy numbers was introduced by Nuray and Savaş [307, Definition 2.1], as follows: A sequence of fuzzy numbers $x = (x_k)$ is *statistically convergent* to $a \in L(\mathbb{R})$ provided that for every $\varepsilon > 0$, the set $\{k \in \mathbb{N}_1 : D(x_k, a) \geq \varepsilon\}$ has natural density zero; in this case we write $st-\lim_{k\to\infty} x_k = a$.

Theorem 8.3.1. [307, Theorem 2.2] *If $x = (x_k)$ is a sequence of fuzzy numbers for which there is a convergent sequence $y = (y_k)$ such that $x_k = y_k$ for almost all k, i.e., $\delta(\{k \in \mathbb{N}_1 : x_k \neq y_k\}) = 0$, then the sequence $(x_k) \in \omega(F)$ is statistically convergent.*

Clearly, if a sequence of fuzzy numbers is convergent to a, then for each $\varepsilon > 0$, there remain only finite number of elements outside of each ε-neighborhood of a. Since the natural density of a finite set is zero, a convergent sequence must be statistically convergent. As was seen by the following example, the converse of this claim does not hold, in general:

Example 8.3.2. [48, Example 2.1] *Define the sequence $x = (x_k)$ of fuzzy numbers by*

$$x_k(t) := \begin{cases} \left.\begin{array}{ll} 0 & , \; t \in (-\infty, k-1) \cup (k+1, \infty) \\ t-(k-1) & , \; t \in [k-1, k] \\ -t+(k+1) & , \; t \in (k, k+1] \end{array}\right\} & , \quad k = n^2 \text{ for } n \in \mathbb{N}_1 \\ \mu(t) & , \quad \text{otherwise,} \end{cases}$$

where

$$\mu(t) := \begin{cases} 0 & , \; t \in (-\infty, 0) \cup (2, \infty), \\ t & , \; t \in [0, 1], \\ -t+2 & , \; t \in (1, 2]. \end{cases}$$

Then, we have for every $\varepsilon \in (0, 1)$

$$K = \{k \in \mathbb{N}_1 : D(x_k, \mu) \geq \varepsilon\} = \{4, 9, 16, 25, \ldots\}.$$

Since $\delta(\{K\}) = 0$, $x = (x_k) \in \omega(F)$ is statistically convergent to μ while it does not converge to μ.

A sequence of fuzzy numbers $x = (x_k)$ is said to be a *statistically Cauchy* if for every $\varepsilon > 0$ there exists a positive number $n_0 = n_0(\varepsilon)$ such that

$$\delta\left(\{k \in \mathbb{N}_1 : D\left(x_k, x_{n_0}\right) \geq \varepsilon\}\right) = 0, \text{ [307, Definition 2.2].}$$

It is clear that every statistically convergent sequence of fuzzy numbers is statistically Cauchy.

Theorem 8.3.3. [233, Theorem 3.3] *If a sequence $x = (x_k)$ of fuzzy numbers is statistically Cauchy, then there is a convergent sequence $y = (y_k)$ such that $\delta\left(\{k \in \mathbb{N}_1 : x_k \neq y_k\}\right) = 0$.*

Combining Theorems 8.3.1 and 8.3.3, one can conclude that if a sequence of fuzzy numbers is statistically Cauchy, then it is statistically convergent.

8.3.1 Statistical Convergence of a Sequence of Fuzzy Numbers and the Statistical Convergence of the Corresponding Sequence of α-Cuts

In this subsection, we prove that the concepts of statistical convergence of a sequence of fuzzy numbers with respect to the supremum metric and the uniform statistical convergence of the corresponding sequences of endpoints of α-cuts are equivalent to each other.

Theorem 8.3.4. [53, Theorem 2.1] *Let $(x_k) \in \omega(F)$. Then, (x_k) is statistically convergent to a fuzzy number a if and only if the sequences $\{x_k^-(\alpha)\}$ and $\{x_k^+(\alpha)\}$ of real valued functions converge statistically uniformly on $[0, 1]$ to $a^-(\alpha)$ and $a^+(\alpha)$, respectively.*

Proof. Suppose that $st-\lim_{k\to\infty} x_k = a$. By definition, we have

$$\delta\left(\{k \in \mathbb{N}_1 : D(x_k, a) \geq \varepsilon\}\right) = 0$$

for every $\varepsilon > 0$. Then, for every $\varepsilon > 0$, there exists a set $K \subset \mathbb{N}_1$ which has natural density 1 and $n_0 = n_0(\varepsilon) \in \mathbb{N}_1$ such that

$$D(x_k, a) = \sup_{\alpha\in[0,1]} \max\left\{\left|x_k^-(\alpha) - a^-(\alpha)\right|, \left|x_k^+(\alpha) - a^+(\alpha)\right|\right\} < \varepsilon$$

for all $k \geq n_0$ and $k \in K$. This implies that

$$\max\left\{\left|x_k^-(\alpha) - a^-(\alpha)\right|, \left|x_k^+(\alpha) - a^+(\alpha)\right|\right\} < \varepsilon,$$

hence the result follows for $\alpha \in [0, 1]$.

To prove the converse implication, fix $\varepsilon > 0$. Then, there exists $K_1 \subset \mathbb{N}_1$ with $\delta(K_1) = 1$, there exists $n_1 = n_1(\varepsilon)$ such that $\left|x_k^-(\alpha) - a^-(\alpha)\right| < \varepsilon$ for all $k \in K_1$ such that $k \geq n_1$ and for each $\alpha \in [0, 1]$. Similarly, there exists $K_2 \subset \mathbb{N}_1$ with $\delta(K_2) = 1$, there exists $n_2 = n_2(\varepsilon)$ such that $\left|x_k^+(\alpha) - a^+(\alpha)\right| < \varepsilon$ for all $k \in K_2$ such that $k \geq n_2$ and for each $\alpha \in [0, 1]$. Let $n_0 = \max\{n_1, n_2\}$ and $K = K_1 \cap K_2$. Then, $\delta(K) = 1$ since $\delta(K_1) = 1$ and $\delta(K_2) = 1$. Thus, for every $\varepsilon > 0$ there exists $K \subset \mathbb{N}_1$ with $\delta(K) = 1$ such that

$$\sup_{\alpha\in[0,1]} \max\left\{\left|x_k^-(\alpha) - a^-(\alpha)\right|, \left|x_k^+(\alpha) - a^+(\alpha)\right|\right\} = D(x_k, a) < \varepsilon$$

for all $k \in K$ such that $k \geq n_0$. Since $\{k \in K : D(x_k, a) \geq \varepsilon\} = \emptyset$ and $\delta(K^c) = 0$, we have $\delta\left(\{k \in \mathbb{N}_1 : D(x_k, a) \geq \varepsilon\}\right) = 0$ for every $\varepsilon > 0$. □

Theorem 8.3.5. [53, Theorem 2.2] *Let $(x_k) \in \omega(F)$. Then, (x_k) is a statistically Cauchy sequence if and only if the sequences $\{x_k^-(\alpha)\}$ and $\{x_k^+(\alpha)\}$ of real valued functions are statistically uniform Cauchy on $[0, 1]$.*

Proof. Since the proof is similar to that of Theorem 8.3.4, we omit details. □

8.3.2 Statistically Monotonic and Statistically Bounded Sequences of Fuzzy Numbers

A sequence $x = (x_k) \in \omega(F)$ is said to be *statistically monotone increasing* if there exists a subset $K = \{k_1 < k_2 < \cdots\}$ of $\mathbb{N}_1$ such that $\delta(K) = 1$ and $x_{k_n} \preceq x_{k_{n+1}}$ for all $n \in \mathbb{N}_1$. A *statistically monotone decreasing* sequence can be defined, similarly. Either statistically monotone increasing or statistically monotone decreasing sequence can be simply called as *statistically monotonic*, (cf. Aytar [50, Definition 2.1]).

If the sequence $x = (x_k) \in \omega(F)$ is statistically monotonic, then the set $E = \{x_k : k \in \mathbb{N}_1\}$ is almost linearly ordered. In addition, it is clear that if a sequence is monotonic, then it is statistically monotonic. The converse of this claim does not hold in general, which can be seen by the following example:

Example 8.3.6. [50, Example 2.2] *Define the sequence* $x = (x_k) \in \omega(F)$ *by*

$$x_k(t) := \begin{cases} x_{k-2}(t) & , \quad 1 \neq k = n^2 \ (n \in \mathbb{N}_1) \\ \left.\begin{array}{lll} 0 & , & t \in \left(-\infty, k - \frac{5}{2}\right) \cup \left(k - \frac{3}{2}, \infty\right) \\ 2\left(t - k + \frac{5}{2}\right) & , & t \in \left[k - \frac{5}{2}, k - 2\right] \\ -2\left(t - k + \frac{3}{2}\right) & , & t \in \left(k - 2, k - \frac{3}{2}\right] \end{array}\right\} & , \quad 2 \neq k = n^2 + 1 \ (n \in \mathbb{N}_1) \\ \left.\begin{array}{lll} 0 & , & t \in (-\infty, k-1) \cup (k+1, \infty) \\ t - (k-1) & , & t \in [k-1, k] \\ -t + (k+1) & , & t \in (k, k+1] \end{array}\right\} & , \quad \text{otherwise.} \end{cases}$$

Let $K := \{1, 2, 3, 6, 7, 8, 11, 12, 13, 14, 15, 18, 19, \ldots\}$. *Then,* $\delta(K) = 1$ *and* $x_{k_n} \preceq x_{k_{n+1}}$ *for all* $n \in \mathbb{N}_1$. *Hence, we say that* $x = (x_k) \in \omega(F)$ *is statistically monotone increasing sequence but is not monotone increasing.*

One of the properties of a statistically monotone increasing sequence is that the index set of the decreasing or incomparable elements of that sequence has density zero. This property may be given more precisely by the following proposition:

Proposition 8.3.7. [50, Proposition 2.3] *Let* $x = (x_k) \in \omega(F)$. *Then, the following statements hold:*

(a) *If* $x = (x_k)$ *is a statistically monotone increasing sequence, then we have*

$$\delta(\{k \in \mathbb{N}_1 : x_k \npreceq x_{k+1}\}) = 0. \tag{8.3.1}$$

(b) *Similarly, if* $x = (x_k)$ *is a statistically monotone decreasing sequence, then we have* $\delta(\{k \in \mathbb{N}_1 : x_k \nsucceq x_{k+1}\}) = 0$.

Proof. (a) Let $x = (x_k) \in \omega(F)$ be a statistically monotone increasing sequence. Then, there exists a subset $K = \{k_1 < k_2 < \cdots\}$ of $\mathbb{N}_1$ such that $\delta(K) = 1$ and (x_{k_n}) is monotone increasing. Since $\{k \in \mathbb{N}_1 : x_k \preceq x_{k+1}\} \supseteq \{k \in K : x_k \preceq x_{k+1}\}$ and $\delta(\{k \in K : x_k \preceq x_{k+1}\}) = 1$, we have $\delta(\{k \in \mathbb{N}_1 : x_k \preceq x_{k+1}\}) = 1$. Hence, $\delta(\{k \in \mathbb{N}_1 : x_k \npreceq x_{k+1}\}) = 0$.

The proof of Part (b) is similar to that of Part (a). □

The following example shows that the implication in Part (a) of Proposition 8.3.7 cannot be reversed:

Example 8.3.8. [50, Example 2.4] *Define* $x = (x_k) \in \omega(F)$ *by*

$$x_k(t) := \begin{cases} \left.\begin{array}{lll} 0 & , & t \in (-\infty, 0) \cup (2, \infty) \\ t & , & t \in [0, 1] \\ 2 - t & , & t \in (1, 2] \end{array}\right\} & , \quad (2n-1)^2 \leq k < (2n)^2; n \in \mathbb{N}_1, \\ \left.\begin{array}{lll} 0 & , & t \in (-\infty, 1) \cup (3, \infty) \\ t - 1 & , & t \in [1, 2] \\ 3 - t & , & t \in (2, 3] \end{array}\right\} & , \quad \text{otherwise.} \end{cases}$$

Thus, it is clear that

$$\begin{aligned} \delta(\{k \in \mathbb{N}_1 : x_k \npreceq x_{k+1}\}) &= \delta(\{k \in \mathbb{N}_1 : x_k \succ x_{k+1}\} \cup \{k \in \mathbb{N}_1 : x_k \nsim x_{k+1}\}) \\ &= \delta(\{9, 25, 49, 81, \ldots\}) = 0. \end{aligned}$$

But, there does not exist a subset $K = \{k_1 < k_2 < \cdots\}$ *of* $\mathbb{N}_1$ *such that* $\delta(K) = 1$ *and* (x_{k_n}) *is monotone increasing.*

As a conclusion of this example, we remark that a sequence which satisfies the condition (8.3.1) may not be statistically monotone increasing.

Now, we give the decomposition theorem for a statistically monotonic sequence of fuzzy numbers. It is important to remark that we have to impose an additional condition on such a sequence, since $L(\mathbb{R})$ does not has a group structure.

Theorem 8.3.9. [50, Theorem 2.5] *Let* $x = (x_k) \in \omega(F)$ *be a sequence such that there exists an index set* $K = \{k_1 < k_2 < \cdots\}$ *with* $\delta(K) = 1$ *and* $x^+_{k_n}(0) \leq x^-_{k_{n+1}}(0)$ *for all* $n \in \mathbb{N}_1$*. Then, we can write* $x = y + z$*, where* $y = (y_k)$ *is monotone increasing and* $\delta(\{k \in \mathbb{N}_1 : z_k \neq \overline{0}\}) = 0$.

We note that the hypothesis of this theorem guarantees that the sequence $x = (x_k) \in \omega(F)$ *is statistically monotone increasing.*

Proof. Define $q_k := \max\{n \in K : n \leq k\}$ and $p_k := \min\{n \in K : n \geq k\}$ for all $k \in \mathbb{N}_1$. Now, we construct the sequences $y = (y_k)$ and $z = (z_k)$ as follows:

$$y_k := \begin{cases} \overline{\left(x^-_{k_1}(0)\right)} & , \quad k < k_1, \\ x_k & , \quad k \in K, \\ \overline{\left(\frac{x^+_{q_k}(0) + (x^-_{p_k}(0))}{2}\right)} & , \quad \text{otherwise} \end{cases}$$

and

$$z_k := \begin{cases} x_k - \overline{\left(x^-_{k_1}(0)\right)} & , \quad k < k_1, \\ \overline{0} & , \quad k \in K, \\ x_k - \overline{\left(\frac{x^+_{q_k}(0) + x^-_{p_k}(0)}{2}\right)} & , \quad \text{otherwise} \end{cases}$$

for all $k \in \mathbb{N}_1$. Clearly, we have $x_k = y_k + z_k$ for all $k \in \mathbb{N}_1$. Since $x^+_{k_n}(0) \leq x^-_{k_{n+1}}(0)$, we have

$$\overline{\left(x^-_{k_1}(0)\right)} \preceq x_{k_n} \preceq \overline{\left(\frac{x^+_{k_n}(0) + x^-_{k_{n+1}}(0)}{2}\right)} \preceq x_{k_{n+1}}$$

for every $n \in \mathbb{N}_1$. It follows from the definitions of (y_k) and (z_k) that (y_k) is monotone increasing and

$$\delta(\{k \in \mathbb{N}_1 : z_k \neq \overline{0}\}) = \delta(\{k \in \mathbb{N}_1 : k < k_1\} \cup \{k \in \mathbb{N}_1 : k \notin K\}) = 0.$$

□

The analogue of Theorem 8.3.9 can also be given for statistically monotone decreasing sequences.

In the following definition, we define the statistical boundedness of a sequence of fuzzy numbers similar to the real case. Meanwhile, we must take all incomparable elements of $L(\mathbb{R})$ into account because the set $L(\mathbb{R})$ is a partially ordered set.

Definition 8.3.10. [50, Definition 2.6] *A sequence $x = (x_k) \in \omega(F)$ is said to be statistically bounded above if there exists a fuzzy number μ such that*

$$\delta\left(\{k \in \mathbb{N}_1 : x_k \succ \mu\} \cup \{k \in \mathbb{N}_1 : x_k \not\sim \mu\}\right) = 0.$$

Statistical boundedness below can be defined, similarly. If the sequence $x = (x_k) \in \omega(F)$ is both statistically bounded above and below, then it is called statistically bounded.

We can also state this definition, as follows: A sequence $x = (x_k) \in \omega(F)$ is called *statistically bounded*, if there exists a real number $M > 0$ such that the set $\{k \in \mathbb{N}_1 : D(x_k, \overline{0}) > M\}$ has natural density zero [50].

It is easy to see that if a sequence is bounded, it is also statistically bounded. In general, the converse of this claim does not hold, as can be seen in the following example:

Example 8.3.11. [50, Example 2.7] *Let $x = (x_k) \in \omega(F)$ defined by*

$$x_k(t) := \begin{cases} \left.\begin{array}{ll} 0 & ,\ t \in (-\infty, k-1) \cup (k+1, \infty) \\ t-(k-1) & ,\ t \in [k-1, k] \\ -t+(k+1) & ,\ t \in (k, k+1] \end{array}\right\} & ,\ k \text{ is an even square}, \\ \left.\begin{array}{ll} 0 & ,\ t \in (-\infty, -k-1) \cup (-k+1, \infty) \\ t+k+1 & ,\ t \in [-k-1, -k] \\ -t-k+1 & ,\ t \in (-k, -k+1] \end{array}\right\} & ,\ k \text{ is an odd square}, \\ \mu_1(t) & ,\ k \text{ is an odd nonsquare}, \\ \mu_2(t) & ,\ k \text{ is an even nonsquare}, \end{cases}$$

where

$$\begin{aligned} \mu_1(t) &:= \begin{cases} 0 & ,\ t \in (-\infty, 0) \cup (2, \infty), \\ t & ,\ \alpha \in [0, 1], \\ -t+2 & ,\ t \in (1, 2], \end{cases} \\ \mu_2(t) &:= \begin{cases} 0 & ,\ t \in (-\infty, 3) \cup (5, \infty), \\ t-3 & ,\ t \in [3, 4], \\ -t+5 & ,\ t \in (4, 5]. \end{cases} \end{aligned}$$

Since

$$\begin{aligned} \delta\left(\{k \in \mathbb{N}_1 : x_k \prec \mu_1\} \cup \{k \in \mathbb{N}_1 : x_k \not\sim \mu_1\}\right) &= \delta\left(\{1, 9, 25, \ldots\} \cup \emptyset\right) = 0, \\ \delta\left(\{k \in \mathbb{N}_1 : x_k \succ \mu_2\} \cup \{k \in \mathbb{N}_1 : x_k \not\sim \mu_2\}\right) &= \delta\left(\{16, 36, 64, \ldots\} \cup \emptyset\right) = 0, \end{aligned}$$

the sequence $x = (x_k)$ is statistically bounded. However, the sequence $x = (x_k)$ is unbounded, since there does not exist a fuzzy number ν such that $x_k \preceq \nu$ for all $k \in \mathbb{N}_1$.

The following second decomposition theorem shows that a statistically bounded sequence of fuzzy numbers can be decomposed into two parts consisting of a statistical null sequence and a bounded sequence.

Theorem 8.3.12. [50, Theorem 2.8] *If $x = (x_k) \in \omega(F)$ is a statistically bounded sequence, then we can write $x = y + z$; where $\delta(\{k \in \mathbb{N}_1 : z_k \neq \overline{0}\}) = 0$ and $y = (y_k)$ is bounded.*

Proof. Let $x = (x_k) \in \omega(F)$ be a statistically bounded sequence. For $M > 0$ large enough, the set $K = \{k \in \mathbb{N}_1 : D(x_k, \overline{0}) > M\}$ satisfies $\delta(K) = 0$. Define the sequences $y = (y_k) \in \omega(F)$ and $z = (z_k) \in \omega(F)$ by

$$y_k := \begin{cases} x_k & , \quad k \in K^c, \\ \overline{0} & , \quad \text{otherwise} \end{cases} \quad \text{and} \quad z_k := \begin{cases} x_k & , \quad k \in K, \\ \overline{0} & , \quad \text{otherwise.} \end{cases}$$

Clearly, $y = (y_k) \in \omega(F)$ is bounded and $\delta(\{k \in \mathbb{N}_1 : z_k \neq \overline{0}\}) = 0$. Therefore, from the construction of $y = (y_k) \in \omega(F)$ and $z = (z_k) \in \omega(F)$ it follows that $x_k = y_k + z_k$ for all $k \in \mathbb{N}_1$. □

To demonstrate the usefulness of Theorem 8.3.12, we provide the following example:

Example 8.3.13. [50, Example 2.9] *Let the sequence $x = (x_k) \in \omega(F)$ be given as in Example 8.2.26. Define $u = (u_k) \in \omega(F)$ by*

$$u_k(t) := \begin{cases} \left.\begin{array}{ll} 0 & , \ t \in (-\infty, k-1) \cup (k+1, \infty) \\ t - (k-1) & , \ t \in [k-1, k] \\ -t + (k+1) & , \ \text{otherwise} \end{array}\right\} & , \ k = n^2 \text{ for } n \in \mathbb{N}_2, \\ x_k(t) & , \ \text{otherwise.} \end{cases}$$

It is clear that $u = (u_k) \in \omega(F)$ is statistically bounded. Hence, we can write $u = y + z$; where the sequences $y = (y_k) \in \omega(F)$ and $z = (z_k) \in \omega(F)$ are given by

$$y_k(t) := \begin{cases} \overline{0} & , \quad k = n^2 \text{ for } n \in \mathbb{N}_2, \\ x_k(t) & , \quad \text{otherwise,} \end{cases}$$

$$z_k(t) := \begin{cases} \left.\begin{array}{ll} 0 & , \ t \in (-\infty, k-1) \cup (k+1, \infty) \\ t - (k-1) & , \ t \in [k-1, k] \\ -t + (k+1) & , \ \text{otherwise} \end{array}\right\} & , \ k = n^2 \text{ for } n \in \mathbb{N}_2, \\ \overline{0} & , \ \text{otherwise.} \end{cases}$$

Here $y = (y_k)$ is bounded and $z = (z_k)$ is a statistically null sequence, i.e.,

$$\delta(\{k \in \mathbb{N}_1 : z_k \neq \overline{0}\}) = 0.$$

Theorem 8.3.14. [50, Lemma 2.10] *A sequence $x = (x_k) \in \omega(F)$ is statistically bounded if and only if there exists a subset $K = \{k_1 < k_2 < \cdots\}$ of $\mathbb{N}_1$ such that $\delta(K) = 1$ and (x_{k_n}) is bounded.*

The proof is clear from Definition 8.3.10 as well as from Theorem 8.3.12.

As can be seen from Example 8.2.18, since the classical monotone convergence theorem is not valid for sequences of fuzzy numbers, the statistical analogue of this theorem is not also valid for sequences of fuzzy numbers.

8.3.3 Statistical Cluster Points and Statistical Limit Points of a Sequence of Fuzzy Numbers

Aytar [48] introduced the concepts of statistical limit and statistical cluster points of a sequence of fuzzy numbers, based on the definitions given by Fridy [175] for sequences of real numbers. Later, he obtained the relations between the sets of ordinary limit points, statistical limit points and statistical cluster points of a sequence of fuzzy numbers.

Definition 8.3.15. [48, Definition 2.4] *If $(x_{k(j)})$ is a subsequence of $x = (x_k) \in \omega(F)$ and $K = \{k(j) \in \mathbb{N}_1 : j \in \mathbb{N}\}$, then we abbreviate $(x_{k(j)})$ by $\{x\}_K$. In case $\delta(\{K\}) = 0$, the subsequence $(x_{k(j)})$ is called a subsequence of density zero or thin subsequence. In addition, $\{x\}_K$ is a nonthin subsequence of $x = (x_k) \in \omega(F)$ if K does not has density zero.*

Definition 8.3.16. [48, Definition 3.1] *A fuzzy number ν is called a statistical limit point of the sequence of fuzzy numbers $x = (x_k)$ provided that there is a nonthin subsequence of $x = (x_k) \in \omega(F)$ that converges to ν. We denote the set of all statistical limit points of the sequence $x = (x_k) \in \omega(F)$ by Λ_x.*

Definition 8.3.17. [48, Definition 3.2] *A fuzzy number μ is called a statistical cluster point of the sequence of fuzzy numbers $x = (x_k)$ provided that $\overline{\delta}(\{k \in \mathbb{N}_1 : D(x_k, \mu) < \varepsilon\}) > 0$ for every $\varepsilon > 0$. We denote the set of all statistical cluster points of the sequence $x = (x_k) \in \omega(F)$ by Γ_x.*

Theorem 8.3.18. [48, Theorem 3.1] *If $x = (x_k)$, $y = (y_k) \in \omega(F)$ such that $\delta(\{k \in \mathbb{N}_1 : x_k \neq y_k\}) = 0$, then $\Lambda_x = \Lambda_y$ and $\Gamma_x = \Gamma_y$.*

Theorem 8.3.19. [48, Theorem 3.2] *For any sequence $x = (x_k) \in \omega(F)$, we have $\Lambda_x \subseteq \Gamma_x$.*

Theorem 8.3.20. [48, Theorem 3.3] *For any sequence $x = (x_k) \in \omega(F)$, we have $\Gamma_x \subseteq L_x$.*

Proof. Assume that $\mu \in \Gamma_x$, that is, $\overline{\delta}(\{k \in \mathbb{N}_1 : D(x_k, \mu) < \varepsilon\}) > 0$ for every $\varepsilon > 0$. We set $\{x\}_K$ a nonthin subsequence of $x = (x_k) \in \omega(F)$ such that $K = \left\{k(j) \in \mathbb{N}_1 : D(x_{k(j)}, \mu) < \varepsilon\right\}$ for every $\varepsilon > 0$ and $\overline{\delta}(\{K\}) > 0$. Since there are infinitely many elements in K, we have $\mu \in L_x$. □

The converse of Theorem 8.3.20 does not hold in general, as can be seen by the following example:

Example 8.3.21. [48, Example 3.1] *Define the sequence $x = (x_k) \in \omega(F)$ by*

$$x_k(t) := \begin{cases} \mu_1(t) & , \quad \text{if } k = n^2 \text{ for } n \in \mathbb{N}, \\ \mu_2(t) & , \quad \text{otherwise,} \end{cases}$$

where

$$\mu_1(t) := \begin{cases} 0 & , \ t \in (-\infty, 0) \cup (2, \infty), \\ t & , \ t \in [0,1], \\ -t+2 & , \ t \in (1,2], \end{cases}$$

$$\mu_2(t) := \begin{cases} 0 & , \ t \in (-\infty, 3) \cup (5, \infty), \\ t-3 & , \ t \in [3,4], \\ -t+5 & , \ t \in (4,5]. \end{cases}$$

Then, $L_x = \{\mu_1, \mu_2\}$ *for every* $\varepsilon > 0$*, but* $\Gamma_x = \{\mu_2\}$.

Theorem 8.3.22. [48, Theorem 3.4] *If* $a = st - \lim_{k\to\infty} x_k$ *for* $x = (x_k) \in \omega(F)$*, then* $\Lambda_x = \Gamma_x = \{a\}$.

Proof. First, we show that $\Lambda_x = \{a\}$. We assume that $\Lambda_x = \{a, b\}$ such that $D(a,b) > 2\varepsilon$ for every $\varepsilon > 0$. In this case, there exist nonthin subsequences $(x_{k(j)})$ and $(x_{l(i)})$ of $x = (x_k) \in \omega(F)$ that converge to a and b, respectively. Since $(x_{l(i)})$ converges to b, then $\left\{l(i) \in \mathbb{N}_1 : D(x_{l(i)}, b) \geq \varepsilon\right\}$ is a finite set, for every $\varepsilon > 0$.

$$\{l(i) \in \mathbb{N}_1 : i \in \mathbb{N}_1\} := \left\{l(i) \in \mathbb{N}_1 : D\left(x_{l(i)}, b\right) < \varepsilon\right\} \cup \left\{l(i) \in \mathbb{N}_1 : D\left(x_{l(i)}, b\right) \geq \varepsilon\right\},$$

which implies that

$$\begin{aligned} \overline{\delta}(\{l(i) \in \mathbb{N}_1 : i \in \mathbb{N}_1\}) &= \overline{\delta}\left(\left\{l(i) \in \mathbb{N}_1 : D\left(x_{l(i)}, b\right) < \varepsilon\right\}\right) + \\ &\quad + \overline{\delta}\left(\left\{l(i) \in \mathbb{N}_1 : D\left(x_{l(i)}, b\right) \geq \varepsilon\right\}\right). \end{aligned}$$

Hence,

$$\overline{\delta}\left(\left\{l(i) \in \mathbb{N}_1 : D\left(x_{l(i)}, b\right) < \varepsilon\right\}\right) > 0. \tag{8.3.2}$$

Since $a = st - \lim_{k\to\infty} x_k$, we have

$$\delta(\{k \in \mathbb{N}_1 : D(x_k, a) \geq \varepsilon\}) = 0 \tag{8.3.3}$$

for every $\varepsilon > 0$. Therefore, we can write

$$\overline{\delta}\left(\{k \in \mathbb{N}_1 : D(x_k, a) < \varepsilon\}\right) > 0.$$

For every $0 < 2\varepsilon < D(a,b)$, we have

$$\left\{l(i) \in \mathbb{N}_1 : D\left(x_{l(i)}, b\right) < \varepsilon\right\} \cap \{k \in \mathbb{N}_1 : D(x_k, a) < \varepsilon\} = \emptyset$$

hence

$$\left\{l(i) \in \mathbb{N}_1 : D\left(x_{l(i)}, b\right) < \varepsilon\right\} \subseteq \{k \in \mathbb{N}_1 : D(x_k, a) \geq \varepsilon\}.$$

Therefore,

$$\overline{\delta}\left(\left\{l(i) \in \mathbb{N}_1 : D\left(x_{l(i)}, b\right) < \varepsilon\right\}\right) \leq \overline{\delta}(\{k \in \mathbb{N}_1 : D(x_k, a) \geq \varepsilon\}) = 0,$$

which contrary (8.3.2). Hence, $\Lambda_x = \{a\}$.

Now, we assume that $\Gamma_x = \{a, z_0\}$ such that $D(a, z_0) > 2\varepsilon$ for some $\varepsilon > 0$. Then,

$$\overline{\delta}(\{k \in \mathbb{N}_1 : D(x_k, z_0) < \varepsilon\}) > 0. \tag{8.3.4}$$

Since

$$\{k \in \mathbb{N}_1 : D(x_k, a) < \varepsilon\} \cap \{k \in \mathbb{N}_1 : D(x_k, z_0) < \varepsilon\} = \emptyset$$

for every $0 < 2\varepsilon < D(a, z_0)$, we have

$$\{k \in \mathbb{N}_1 : D(x_k, a) \geq \varepsilon\} \supseteq \{k \in \mathbb{N}_1 : D(x_k, z_0) < \varepsilon\}.$$

Therefore,

$$\overline{\delta}(\{k \in \mathbb{N}_1 : D(x_k, a) \geq \varepsilon\}) \geq \overline{\delta}(\{k \in \mathbb{N}_1 : D(x_k, z_0) < \varepsilon\}). \tag{8.3.5}$$

From (8.3.4), the right-hand side of (8.3.5) is greater than zero and from (8.3.3), the left-hand side of (8.3.5) equals to zero, a contradiction. Hence, $\Gamma_x = \{a\}$. □

8.3.4 The Statistical Limit Inferior and the Statistical Limit Superior of a Statistically Bounded Sequence of Fuzzy Numbers

In this subsection, we introduce the concepts of statistical limit superior and statistical limit inferior for statistically bounded sequences of fuzzy numbers. Given a sequence $x = (x_k)$ of fuzzy numbers. Following Aytar et al. [49], let us define

$$\begin{aligned}
A'_x &:= \{\mu \in L(\mathbb{R}) : \delta(\{k \in \mathbb{N}_1 : x_k \prec \mu\}) \neq 0\}, \\
\overline{A'_x} &:= \{\mu \in L(\mathbb{R}) : \delta(\{k \in \mathbb{N}_1 : x_k \succ \mu\}) = 1\}, \\
B'_x &:= \{\mu \in L(\mathbb{R}) : \delta(\{k \in \mathbb{N}_1 : x_k \succ \mu\}) \neq 0\}, \\
\overline{B'_x} &:= \{\mu \in L(\mathbb{R}) : \delta(\{k \in \mathbb{N}_1 : x_k \prec \mu\}) = 1\}.
\end{aligned}$$

The sets $\overline{A'_x}$ and $\overline{B'_x}$ are the sets of statistical lower bounds and statistical upper bounds, respectively. It is clear that if the sequence $x = (x_k) \in \omega(F)$ is statistically bounded, then these sets are non-empty. It is also clear that the sets A'_x and $\overline{B'_x}$ have lower bounds, and the sets $\overline{A'_x}$ and B'_x have upper bounds. Therefore, by the existence theorem of supremum and infimum (Theorem 8.2.7), we conclude that $\inf A'_x$, $\sup \overline{A'_x}$, $\sup B'_x$, and $\inf \overline{B'_x}$ exist, (cf. Aytar et al. [49]).

Theorem 8.3.23. [49, Theorem 1] *If the sequence* $x = (x_k) \in \omega(F)$ *is statistically bounded, then* $\inf A'_x = \sup \overline{A'_x}$ *and* $\sup B'_x = \inf \overline{B'_x}$.

This theorem can be proved in the similar way used in the proof of Theorem 8.2.25 by using the properties of natural density δ.

Definition 8.3.24. [49, Definition 2] *If $x = (x_k)$ is a statistically bounded sequence of fuzzy numbers, then the statistical limit inferior of $x = (x_k)$ is given by*

$$st - \liminf_{k\to\infty} x_k = \inf A'_x.$$

Also, the statistical limit superior of $x = (x_k)$ is given by

$$st - \limsup_{k\to\infty} x_k = \sup B'_x.$$

By Theorem 8.3.23, we get $st - \liminf_{k\to\infty} x_k = \sup \overline{A'_x}$ and $st - \limsup_{k\to\infty} x_k = \inf \overline{B'_x}$. The following simple example may serve to illustrate the concepts just defined, above:

Example 8.3.25. [49, Example 1] *Define the sequence $x = (x_k) \in \omega(F)$ by*

$$x_k(t) := \begin{cases} \left.\begin{array}{lll} 0 & , & t \in (-\infty, k-1) \cup (k+1, \infty) \\ t-(k-1) & , & t \in [k-1, k] \\ -t+(k+1) & , & t \in (k, k+1] \end{array}\right\} & , \ k = n^2 \text{ for } n \in \mathbb{N}_1, \\ \mu_1(t) & , \ k \text{ is an odd nonsquare}, \\ \mu_2(t) & , \ \text{otherwise}, \end{cases}$$

where

$$\mu_1(t) := \begin{cases} 0 & , \ t \in (-\infty, 0) \cup (2, \infty), \\ t & , \ \alpha \in [0, 1], \\ -t+2 & , \ \text{otherwise}, \end{cases}$$

$$\mu_2(t) := \begin{cases} 0 & , \ t \in (-\infty, 3) \cup (5, \infty), \\ t-3 & , \ t \in [3, 4], \\ -t+5 & , \ \text{otherwise}. \end{cases}$$

Let the sequence $y = (y_k) \in \omega(F)$ be given by

$$y_k(t) := \begin{cases} \mu_1(t) & , \ k \text{ is an even nonsquare}, \\ \mu_2(t) & , \ k \text{ is an odd square}, \\ \overline{0} & , \ k \text{ is an odd nonsquare}, \\ x_k(t) & , \ \text{otherwise}. \end{cases}$$

Note that although $y = (y_k) \in \omega(F)$ is unbounded from above, it is statistically bounded since the set of squares has density zero. Thus, $st - \limsup_{k\to\infty} y_k = \sup B'_y = \mu_1$. Similarly, $st - \liminf_{k\to\infty} y_k = \inf A'_y = \overline{0}$.

Theorem 8.3.26. [49, Theorem 2] *Let $x = (x_k)$ be a statistically bounded sequence of fuzzy numbers. If $\nu = st - \liminf_{k\to\infty} x_k$, then*

$$\delta\left(\{k \in \mathbb{N}_1 : x_k \prec \nu - \overline{\varepsilon}\}\right) = 0 \tag{8.3.6}$$

and

$$\delta\left(\{k \in \mathbb{N}_1 : x_k \prec \nu + \overline{\varepsilon}\} \cup \{k \in \mathbb{N}_1 : x_k \not\succ \nu + \overline{\varepsilon}\}\right) \neq 0 \tag{8.3.7}$$

for every $\varepsilon > 0$.

Proof. To contrary, we assume that there exists $\varepsilon > 0$ such that $\delta(\{k \in \mathbb{N}_1 : x_k \prec \nu - \overline{\varepsilon}\}) \neq 0$. This means that $\nu - \overline{\varepsilon} \in A'_x$. Then, we get $\nu \preceq \nu - \overline{\varepsilon}$ which is a contradiction.

Now, let us show that the inequality (8.3.7) holds. Assume that it is not true, that is, there exists $\varepsilon > 0$ such that

$$\delta(\{k \in \mathbb{N}_1 : x_k \prec \nu + \overline{\varepsilon}\}) = 0 \quad \text{and} \quad \delta(\{k \in \mathbb{N}_1 : x_k \not\sim \nu + \overline{\varepsilon}\}) = 0. \tag{8.3.8}$$

For each $k \in \mathbb{N}_1$, only the following three cases are possible: $x_k \prec \nu + \overline{\varepsilon}$, $x_k \not\sim \nu + \overline{\varepsilon}$ and $x_k \succeq \nu + \overline{\varepsilon}$. Then,

$$\{k \in \mathbb{N}_1 : x_k \prec \nu + \overline{\varepsilon}\} \cup \{k \in \mathbb{N}_1 : x_k \not\sim \nu + \overline{\varepsilon}\} \cup \{k \in \mathbb{N}_1 : x_k \succeq \nu + \overline{\varepsilon}\} = \mathbb{N}_1.$$

Thus, from (8.3.8), we have

$$\delta(\{k \in \mathbb{N}_1 : x_k \succeq \nu + \overline{\varepsilon}\}) = 1.$$

This means that $\nu + \overline{\varepsilon} \in \overline{A'_x}$. Hence, we can write $\nu + \overline{\varepsilon} \preceq \sup \overline{A'_x} = \inf A'_x = \nu$, which is a contradiction. □

Remark 8.3.27. [49, Remark 1] *We note that the converse of Theorem 8.3.26 holds for sequences of real numbers* (see [176, Theorem 1']), *but it may not be true for sequences of fuzzy numbers as can be seen in the following example:*

Define the sequence $x = (x_k) \in \omega(F)$ *by*

$$x_k(t) := \begin{cases} \mu_1(t) & , \ k \text{ is an odd number}, \\ \mu_2(t) & , \ \text{otherwise}, \end{cases}$$

where

$$\mu_1(t) := \begin{cases} 0 & , \ t \in (-\infty, 1) \cup (6, \infty), \\ (t-1)/4 & , \ t \in [1, 5], \\ 6 - t & , \ \text{otherwise}, \end{cases}$$

$$\mu_2(t) := \begin{cases} 0 & , \ t \in (-\infty, 2) \cup (4, \infty), \\ t - 2 & , \ t \in [2, 3], \\ 4 - t & , \ \text{otherwise}. \end{cases}$$

Although for each $\varepsilon > 0$, *the conditions* (8.3.6) *and* (8.3.7) *are satisfied for the number* μ_1, *we have* $\mu_1 \neq st - \liminf_{k\to\infty} x_k$, *where*

$$st - \liminf_{k\to\infty} x_k := \begin{cases} 0 & , \ t \in (-\infty, 1) \cup (4, \infty), \\ (t-1)/4 & , \ t \in [1, 7/3], \\ t - 2 & , \ t \in (7/3, 3], \\ 4 - t & , \ \text{otherwise}. \end{cases}$$

The dual statement of Theorem 8.3.26 for $st - \limsup_{k\to\infty} x_k$ may be given, as follows:

Theorem 8.3.28. [49, Theorem 3] *Let* $x = (x_k) \in \omega(F)$ *be a statistically bounded sequence. If* $\mu = st - \limsup_{k\to\infty} x_k$, *then*

$$\begin{aligned} &\delta\left(\{k \in \mathbb{N}_1 : x_k \succ \mu + \overline{\varepsilon}\}\right) = 0, \\ &\delta\left(\{k \in \mathbb{N}_1 : x_k \succ \mu - \overline{\varepsilon}\} \cup \{k \in \mathbb{N}_1 : x_k \not\sim \mu - \overline{\varepsilon}\}\right) \neq 0 \end{aligned} \tag{8.3.9}$$

for every $\varepsilon > 0$.

This can be proved by the similar way as was used in proving Theorem 8.3.26. In addition, similar to Remark 8.3.27, the converse of Theorem 8.3.28 does not hold, in general.

Now, we have the following assertion:

Theorem 8.3.29. [49, Theorem 4] $st - \liminf_{k\to\infty} x_k \preceq st - \limsup_{k\to\infty} x_k$ *for any statistically bounded sequence* $x = (x_k) \in \omega(F)$.

Proof. We have $st - \liminf_{k\to\infty} x_k = \sup \overline{A'_x}$ and $st - \limsup_{k\to\infty} x_k = \sup B'_x$. By definition of the sets $\overline{A'_x}$ and B'_x, we see that $\overline{A'_x} \subset B'_x$. Hence, we get $\sup \overline{A'_x} \leq \sup B'_x$ which completes the proof. □

Theorem 8.3.30. [49, Theorem 5] *Assume that the sequence* $x = (x_k) \in \omega(F)$ *is statistically convergent to* a. *Then,* $st - \liminf_{k\to\infty} x_k = st - \limsup_{k\to\infty} x_k = a$.

Proof. Take any $\varepsilon > 0$. Since the sequence $x = (x_k) \in \omega(F)$ is statistically convergent to a, we have

$$\delta(\{k \in \mathbb{N}_1 : D(x_k, a) \geq \varepsilon\}) = 0;$$

$$\delta\left(\{k \in \mathbb{N}_1 : D(x_k, a) < \varepsilon\}\right) = 1.$$

In other words, $D(x_k, a) < \varepsilon$ for almost all k, i.e.,

$$\sup_{\alpha\in[0,1]} \left|x_k^-(\alpha) - a^-(\alpha)\right| < \varepsilon \quad \text{and} \quad \sup_{\alpha\in[0,1]} \left|x_k^+(\alpha) - a^+(\alpha)\right| < \varepsilon.$$

Then, for almost all k, the inequality $a - \overline{\varepsilon} \prec x_k \prec a + \overline{\varepsilon}$ holds. Therefore, we have:

(1) $\delta(\{k \in \mathbb{N}_1 : x_k \prec a + \overline{\varepsilon}\}) = 1$. This means that $a + \overline{\varepsilon} \in \overline{B'_x}$. Then,

$$st - \limsup_{k\to\infty} x_k = \inf \overline{B'_x} = \mu \preceq a + \overline{\varepsilon}.$$

(2) $\delta(\{k \in \mathbb{N}_1 : x_k \succ a - \overline{\varepsilon}\}) = 1$. This means that $a - \overline{\varepsilon} \in \overline{A'_x}$. Then,

$$st - \liminf_{k\to\infty} x_k = \sup \overline{A'_x} = \nu \succeq a - \overline{\varepsilon}.$$

We obtain from these inequalities and Theorem 8.3.29 that

$$a - \overline{\varepsilon} \preceq \nu \preceq \mu \preceq a + \overline{\varepsilon}.$$

Since $\varepsilon > 0$ is an arbitrary number, we get $\nu = \mu = a$. □

The following example shows that the converse of Theorem 8.3.30 does not hold, in general:

Example 8.3.31. [49, Example 2] *Let $x = (x_k) \in \omega(F)$ and $\mu \in L(\mathbb{R})$ be given as in Example 8.2.26. Then, we conclude that $st - \limsup_{k\to\infty} x_k = st - \liminf_{k\to\infty} x_k = \mu$. On the other hand, since $\left|x_k^+\left(\frac{1}{2}\right) - \mu^+\left(\frac{1}{2}\right)\right| = \left|1 - \frac{1}{2}\right| = \frac{1}{2}$ for all $k \in \mathbb{N}_1$, we have $D(x_k, \mu) \geq \frac{1}{2}$ for all $k \in \mathbb{N}_1$. Hence, we conclude that*

$$\delta\left(\left\{k \in \mathbb{N}_1 : D(x_k, \mu) \geq \frac{1}{2}\right\}\right) = 1.$$

Consequently, $st - \lim_{k\to\infty} x_k$ does not exist.

Before giving a sufficient condition for a sequence of fuzzy numbers to be statistically convergent, we state and prove the following lemma:

Lemma 8.3.32. [49, Lemma 6] *ν and μ be given two fuzzy numbers, and let $\varepsilon > 0$. Then, the following statements are equivalent:*

(i) $D(\nu, \mu) \leq \varepsilon$.

(ii) $\mu - \overline{\varepsilon} \preceq \nu \preceq \mu + \overline{\varepsilon}$.

Proof. Since

$$D(\nu, \mu) = \sup_{\alpha\in[0,1]} \max\left\{|\nu^-(\alpha) - \mu^-(\alpha)|, |\nu^+(\alpha) - \mu^+(\alpha)|\right\},$$

the inequality (i) is equivalent to

$$\left|\nu^-(\alpha) - \mu^-(\alpha)\right| \leq \varepsilon \text{ and } \left|\nu^+(\alpha) - \mu^+(\alpha)\right| \leq \varepsilon \text{ for all } \alpha \in [0,1],$$

which means that the following inequalities hold for all $\alpha \in [0, 1]$:

$$\begin{aligned} \nu^-(\alpha) &\leq \mu^-(\alpha) + \varepsilon, \quad \nu^+(\alpha) \leq \mu^+(\alpha) + \varepsilon, \\ \nu^+(\alpha) &\geq \mu^+(\alpha) - \varepsilon, \quad \nu^+(\alpha) \geq \mu^+(\alpha) - \varepsilon. \end{aligned}$$

In other words, $\nu \preceq \mu + \overline{\varepsilon}$ and $\nu \succeq \mu - \overline{\varepsilon}$. □

Theorem 8.3.33. [49, Theorem 7] *Assume that*

$$st - \limsup_{k\to\infty} x_k = st - \liminf_{k\to\infty} x_k = \mu$$

and there is a number $\varepsilon^0 > 0$ such that for each $\varepsilon \in (0, \varepsilon^0)$ the sets $\{k \in \mathbb{N}_1 : x_k \not\preceq \mu + \overline{\varepsilon}\}$ and $\{k \in \mathbb{N}_1 : x_k \not\succeq \mu - \overline{\varepsilon}\}$ have natural densities zero. Then, we have $st - \lim_{k\to\infty} x_k = \mu$.

Proof. Take any number $\varepsilon \in (0, \varepsilon^0)$. Since $st - \liminf_{k\to\infty} x_k = \mu$, by Theorem 8.3.26 we have

$$\delta(\{k \in \mathbb{N}_1 : x_k \prec \mu - \overline{\varepsilon}\}) = 0 \tag{8.3.10}$$

for all $\varepsilon > 0$. Similarly, since $st - \limsup_{k\to\infty} x_k = \mu$, by Theorem 8.3.28 we obtain

$$\delta(\{k \in \mathbb{N}_1 : x_k > \mu + \overline{\varepsilon}\}) = 0 \tag{8.3.11}$$

for every $\varepsilon > 0$.

From $\delta(\{k \in \mathbb{N}_1 : x_k \not\succ \mu - \overline{\varepsilon}\}) = 0$, $\delta(\{k \in \mathbb{N}_1 : x_k \not\prec \mu + \overline{\varepsilon}\}) = 0$ and the equalities (8.3.10), (8.3.11), we conclude that $\delta(K_1(\varepsilon)) = 1$ and $\delta(K_2(\varepsilon)) = 1$; where $K_1(\varepsilon) = \{k \in \mathbb{N}_1 : x_k \succeq \mu - \overline{\varepsilon}\}$ and $K_2(\varepsilon) = \{k \in \mathbb{N}_1 : x_k \preceq \mu + \overline{\varepsilon}\}$.

Clearly,

$$K_1(\varepsilon) \cap K_2(\varepsilon) = \{k \in \mathbb{N}_1 : \mu - \overline{\varepsilon} \preceq x_k \preceq \mu + \overline{\varepsilon}\}.$$

By Lemma 8.3.32, we obtain

$$K_1(\varepsilon) \cap K_2(\varepsilon) = \{k \in \mathbb{N}_1 : D(x_k, \mu) \leq \varepsilon\}.$$

Therefore, $\delta(\{k \in \mathbb{N}_1 : D(x_k, \mu) \leq \varepsilon\}) = 1$. In this case, $\delta(\{k \in \mathbb{N}_1 : D(x_k, \mu) > \varepsilon\}) = 0$. Since $\varepsilon > 0$ is an arbitrary number, we conclude that $st - \lim_{k\to\infty} x_k = \mu$. □

Theorem 8.3.34. [49, Theorem 8] *If $x = (x_k)$ and $y = (y_k)$ are statistically bounded sequences of fuzzy numbers such that $\delta(\{k \in \mathbb{N}_1 : x_k \neq y_k\}) = 0$ then, the following statements hold:*

(i) $st - \limsup_{k\to\infty} x_k = st - \limsup_{k\to\infty} y_k$.

(ii) $st - \liminf_{k\to\infty} x_k = st - \liminf_{k\to\infty} y_k$.

Proof. We prove only Part (i). Since $\delta(\{k \in \mathbb{N}_1 : x_k \neq y_k\}) = 0$, we have

$$\{\mu \in L(\mathbb{R}) : \delta(\{k \in \mathbb{N}_1 : x_k \succ \mu\}) \neq 0\} = \{\mu \in L(\mathbb{R}) : \delta(\{k \in \mathbb{N}_1 : y_k \succ \mu\}) \neq 0\},$$

i.e., $B'_x = B'_y$. Then, we have $\sup B'_x = \sup B'_y$, as was desired. □

8.3.5 Further Results

The results given in this subsection are also valid if we replace the δ-natural density with any density which satisfies the axiomatic properties of the density function given by Freedman and Sember [173]. For instance, we say that a set $K \subset \mathbb{N}_1$ has C-density if $\delta_C(K) = \lim_{n\to\infty} \sum_{k\in K} c_{nk}$ exists, where $C = (c_{nk})$ is a non-negative regular matrix. The sequence $x = (x_k) \in \omega(F)$ is C-statistically convergent to a provided that for every $\varepsilon > 0$ the set $K(\varepsilon) = \{k \in \mathbb{N}_1 : D(x_k, a) \geq \varepsilon\}$ has C-density zero. In the case $C = C_1$, the Cesàro mean of order 1, this reduces to the usual definition of statistical convergence. In this subsection, if we replace the density δ with the δ_C density, then the results will be invariant. By replacing the concept of δ-density in this way, we can say that the results of this subsection are also valid for more general types of convergence.

8.3.6 Relation Between Statistical Cluster Points and Statistical Extreme Limit Points

Fridy and Orhan [176] showed that the statistical limit inferior or statistical limit superior of a sequence of real numbers is a statistical cluster point of the sequence. But, this result may not hold for a sequence of fuzzy numbers in general, as can be seen in the following example:

Example 8.3.35. [52, Example 1] *Define the sequence* $x = (x_k) \in \omega(F)$ *by*

$$x_k(t) := \begin{cases} \mu_1(t) & , \quad k \text{ is an odd number}, \\ \mu_2(t) & , \quad \text{otherwise}, \end{cases}$$

where

$$\mu_1(t) \quad := \quad \begin{cases} 0 & , \quad t \in (-\infty, 1) \cup (6, \infty), \\ (t-1)/4 & , \quad t \in [1, 5], \\ 6 - t & , \quad \text{otherwise}, \end{cases}$$

$$\mu_2(t) \quad := \quad \begin{cases} 0 & , \quad t \in (-\infty, 2) \cup (4, \infty), \\ t - 2 & , \quad t \in [2, 3], \\ 4 - t & , \quad \text{otherwise}. \end{cases}$$

It follows that

$$st - \liminf_{k\to\infty} x_k := \begin{cases} 0 & , \quad t \in (-\infty, 1) \cup (4, \infty), \\ (t-1)/4 & , \quad t \in [1, 7/3], \\ t - 2 & , \quad t \in (7/3, 3], \\ 4 - t & , \quad \text{otherwise} \end{cases}$$

and $\Gamma_x = \{\mu_1, \mu_2\}$*, i.e.,* $st - \liminf_{k\to\infty} x_k \notin \Gamma_x$.

Example 8.3.35 shows that the statistical limit inferior and statistical limit superior of a sequence of fuzzy numbers may not be a statistical cluster point of the sequence. The following two theorems give the sufficient conditions in order that the statistical limit inferior or statistical limit superior can be a statistical cluster point.

Theorem 8.3.36. [52, Theorem 3] *Let* $x = (x_k) \in \omega(F)$ *be statistically bounded from above. Assume that* $st - \limsup_{k\to\infty} x_k = \mu$ *and there is a number* $\varepsilon^0 > 0$ *such that for each* $\varepsilon \in (0, \varepsilon^0)$*, the sets* $\{k \in \mathbb{N}_1 : x_k \not\succ \mu + \overline{\varepsilon}\}$ *and* $\{k \in \mathbb{N}_1 : x_k \not\succ \mu - \overline{\varepsilon}\}$ *have natural densities zero. Then,* $\mu \in \Gamma_x$.

Proof. Take any number $\varepsilon \in (0, \varepsilon^0)$. By hypothesis and Theorem 8.3.28, we have

$$\delta\left(\{k \in \mathbb{N}_1 : x_k \succ \mu - \overline{\varepsilon}\}\right) \neq 0. \tag{8.3.12}$$

By Theorem 8.3.28, we also have $\delta\left(\{k \in \mathbb{N}_1 : x_k \succ \mu + \overline{\varepsilon}\}\right) = 0$. Now, we get

$$\delta\left(\{k \in \mathbb{N}_1 : x_k \preceq \mu + \overline{\varepsilon}\}\right) = 1, \tag{8.3.13}$$

since $\delta\left(\{k \in \mathbb{N}_1 : x_k \not\prec \mu + \overline{\varepsilon}\}\right) = 0$. From the equalities (8.3.12) and (8.3.13), we have

$$\delta\left(\{k \in \mathbb{N}_1 : \mu - \overline{\varepsilon} \prec x_k \preceq \mu + \overline{\varepsilon}\}\right) \neq 0.$$

Thus, we can write

$$\delta\left(\{k \in \mathbb{N}_1 : D\left(x_k, \mu\right) \leq \varepsilon\}\right) \neq 0.$$

Since the last inequality holds for all $\varepsilon \in (0, \varepsilon^0)$, we have

$$\delta\left(\{k \in \mathbb{N}_1 : D\left(x_k, \mu\right) < \varepsilon\}\right) \neq 0.$$

Consequently, we have $\mu \in \Gamma_x$. □

The dual statement of Theorem 8.3.36 for $st - \liminf_{k\to\infty} x_k$ may be presented, as follows;

Theorem 8.3.37. [52, Theorem 4] *Let $x = (x_k) \in \omega(F)$ be statistically bounded from below. Assume that $st - \liminf_{k\to\infty} x_k = \nu$ and there is a number $\varepsilon^0 > 0$ such that for each $\varepsilon \in (0, \varepsilon^0)$, the sets $\{k \in \mathbb{N}_1 : x_k \not\prec \nu + \overline{\varepsilon}\}$ and $\{k \in \mathbb{N}_1 : x_k \not\prec \nu - \overline{\varepsilon}\}$ have natural densities zero. Then, $\nu \in \Gamma_x$.*

Now, we prove the next lemma which is needed for the proof of Theorem 8.3.39, below:

Lemma 8.3.38. [52, Lemma 5] *If $\mu \preceq \nu + \overline{\varepsilon}$ for every $\varepsilon > 0$, then $\mu \preceq \nu$.*

Proof. To the contrary, assume that $\mu \npreceq \nu$. Then, there exists $\alpha_0 \in [0, 1]$ such that $\mu^-(\alpha_0) > \nu^-(\alpha_0)$ or $\mu^+(\alpha_0) > \nu^+(\alpha_0)$. For the sake of definiteness, we will consider the case $\mu^-(\alpha_0) > \nu^-(\alpha_0)$. It is clear that there exists $\varepsilon > 0$ such that $\mu^-(\alpha_0) > \nu^-(\alpha_0) + \varepsilon$. Then, we can write $\mu^-(\alpha_0) > \nu^-(\alpha_0) + \overline{\varepsilon}^-(\alpha_0)$. So, we have $\mu \npreceq \nu + \overline{\varepsilon}$, which contradicts the hypothesis of the theorem. □

The following theorem states that all the statistical cluster points of a sequence of fuzzy numbers are among the statistical extreme limit points.

Theorem 8.3.39. [52, Theorem 6] *Let $x = (x_k) \in \omega(F)$ be statistically bounded. If $\gamma \in \Gamma_x$, then $st - \liminf_{k\to\infty} x_k \preceq \gamma \preceq st - \limsup_{k\to\infty} x_k$.*

Proof. We prove that the inequality $\gamma \preceq st - \limsup_{k\to\infty} x_k$ holds. The other inequality can be obtained, similarly.

Take any fuzzy number $\gamma \in \Gamma_x$. Then, we have

$$\delta\left(\{k \in \mathbb{N}_1 : D\left(x_k, \gamma\right) < \varepsilon\}\right) \neq 0$$

for every $\varepsilon > 0$. Hence, we can write

$$\delta\left(\{k \in \mathbb{N}_1 : \gamma - \overline{\varepsilon} \prec x_k \prec \gamma + \overline{\varepsilon}\}\right) \neq 0 \tag{8.3.14}$$

for every $\varepsilon > 0$. Since we have

$$\{k \in \mathbb{N}_1 : \gamma - \bar{\varepsilon} \prec x_k \prec \gamma + \bar{\varepsilon}\} \subseteq \{k \in \mathbb{N}_1 : \gamma - \bar{\varepsilon} \prec x_k\}$$

from the relation (8.3.14), we get $\delta\left(\{k \in \mathbb{N}_1 : \gamma - \bar{\varepsilon} \prec x_k\}\right) \neq 0$ for every $\varepsilon > 0$. By definition of B'_x, it is clear that $\gamma - \bar{\varepsilon} \in B'_x$. Now, by definition of supremum we have $\gamma - \bar{\varepsilon} \preceq \sup B'_x$ for every $\varepsilon > 0$. Consequently, by Lemma 8.3.38, we can write $\gamma \preceq \sup B'_x = st - \limsup_{k\to\infty} x_k$. □

Combining Theorem 8.3.37 and Theorem 8.3.39, we derive the next corollary which gives the condition under which the statistical limit infimum point of a statistically bounded sequence becomes the least statistical cluster point of the sequence.

Corollary 8.3.40. [52, Corollary 7] *Let $x = (x_k) \in \omega(F)$ be statistically bounded. Assume that $st - \liminf_{k\to\infty} x_k = \nu$ and there is a number $\varepsilon^0 > 0$ such that for each $\varepsilon \in (0, \varepsilon^0)$, the sets $\{k \in \mathbb{N}_1 : x_k \not\succ \nu + \bar{\varepsilon}\}$ and $\{k \in \mathbb{N}_1 : x_k \not\succ \nu - \bar{\varepsilon}\}$ have natural densities zero. Then, $st - \liminf_{k\to\infty} x_k$ is the least cluster point of the sequence $x = (x_k) \in \omega(F)$.*

Under the conditions of Theorem 8.3.36, we similarly say that the statistical limit supremum of any statistically bounded sequence $x = (x_k) \in \omega(F)$ is the greatest statistical cluster point of the sequence.

Kostyrko et al. [229] showed that the set $\mathcal{I}(\Gamma_x)$ is closed in any metric space M for all sequence $x = (x_k)_{k\in\mathbb{N}_1}$ of elements of M, where $\mathcal{I}$ is an admissible ideal and $\mathcal{I}(\Gamma_x)$ is the set of all $\mathcal{I}$-cluster points of x. It is known that $(L(\mathbb{R}), D)$ is a complete metric space and the family of the sets which have natural densities zero is an admissible ideal (see [229, Example 3.1.a]). Hence, we can say from Kostyrko's theorem that the set of statistical cluster points of a sequence of fuzzy numbers is closed with respect to the topology which is generated by the supremum metric D. This can be stated, as follows:

Theorem 8.3.41. [52, Theorem 8] *For any sequence $x = (x_k) \in \omega(F)$, Γ_x is closed.*

We know that for a statistically bounded sequence $x = (x_k)$ of real numbers, Γ_x is a non-empty set (see [175, Corollary]). But, this result may not hold for a statistically bounded sequence of fuzzy numbers in general, as can be seen in the following example:

Example 8.3.42. [52, Example 2] *Let $x = (x_k) \in \omega(F)$ and $\mu \in L(\mathbb{R})$ be given as in Example 8.2.26. By definition of statistical limit superior and statistical limit inferior, we conclude that $st - \limsup x_k = st - \liminf_{k\to\infty} x_k = \mu$. On the other hand, since*

$$\left|(x_k)^+\left(\frac{1}{2}\right) - \mu^+\left(\frac{1}{2}\right)\right| = \left|1 - \frac{1}{2}\right| = \frac{1}{2}$$

for all $k \in \mathbb{N}_1$, we have $D(x_k, \mu) \geq \frac{1}{2}$ for all $k \in \mathbb{N}_1$. Then, we get

$$\delta\left(\left\{k \in \mathbb{N}_1 : D(x_k, \mu) \geq \frac{1}{2}\right\}\right) = 1,$$

hence $\mu \notin \Gamma_x$. *It is clear that* $\Gamma_x = \emptyset$ *and there does not exist a statistically convergent subsequence of* $x = (x_k) \in \omega(F)$.

By Example 8.3.42, we can say that a statistically bounded sequence of fuzzy numbers may not have a statistically convergent subsequence. Now, as the consequences of Theorems 8.3.36 and 8.3.37, we present the next theorem giving the sufficient condition in order that a statistically bounded sequence of fuzzy numbers may have a statistically convergent subsequence.

Theorem 8.3.43. *Let* $x = (x_k) \in \omega(F)$. *Then, the following statements hold:*

(i) [52, Theorem 9] *Assume that* $st - \limsup_{k\to\infty} x_k = \mu$ *and there is a number* $\varepsilon^0 > 0$ *such that for each* $\varepsilon \in (0, \varepsilon^0)$, *the sets* $\{k \in \mathbb{N}_1 : x_k \not\preceq \mu + \overline{\varepsilon}\}$ *and* $\{k \in \mathbb{N}_1 : x_k \not\succeq \mu - \overline{\varepsilon}\}$ *have natural densities zero. If* $x = (x_k)$ *is statistically bounded from above, then there exists a subsequence of* $x = (x_k)$ *which statistically converges to* μ.

(ii) [52, Theorem 10] *Assume that* $st - \liminf_{k\to\infty} x_k = \nu$ *and there is a number* $\varepsilon^0 > 0$ *such that for each* $\varepsilon \in (0, \varepsilon^0)$, *the sets* $\{k \in \mathbb{N}_1 : x_k \not\preceq \nu + \overline{\varepsilon}\}$ *and* $\{k \in \mathbb{N}_1 : x_k \not\succeq \nu - \overline{\varepsilon}\}$ *have natural densities zero. If* $x = (x_k)$ *is statistically bounded from below, then there exists a subsequence of* $x = (x_k)$ *which statistically converges to* ν.

8.4 THE CLASSICAL SETS OF SEQUENCES OF FUZZY NUMBERS

In this section, we study the classical sets $\ell_\infty(F)$, $c(F)$, $c_0(F)$ and $\ell_p(F)$ consisting of the bounded, convergent, null and absolutely p-summable sequences of fuzzy numbers, i.e.,

$$\begin{aligned}
\ell_\infty(F) &:= \left\{(u_k) \in \omega(F) : \sup_{k\in\mathbb{N}_1} D(u_k, \overline{0}) < \infty\right\}, \\
c(F) &:= \left\{(u_k) \in \omega(F) : \exists l \in L(\mathbb{R}) \ni \lim_{k\to\infty} D(u_k, l) = 0\right\}, \\
c_0(F) &:= \left\{(u_k) \in \omega(F) : \lim_{k\to\infty} D(u_k, \overline{0}) = 0\right\}, \\
\ell_p(F) &:= \left\{(u_k) \in \omega(F) : \sum_k [D(u_k, \overline{0})]^p < \infty\right\}.
\end{aligned}$$

Throughout this section, we also suppose that $1 \le p < \infty$ with $p^{-1} + q^{-1} = 1$. The classical sets are considered in more details of next subsections. In [302], it was shown that $c(F)$ and $\ell_\infty(F)$ are complete metric spaces with the Hausdorff metric D_∞ defined by

$$D_\infty(u, v) := \sup_{k\in\mathbb{N}} D(u_k, v_k),$$

where $u = (u_k)$, $v = (v_k)$ are the elements of the sets $c(F)$ or $\ell_\infty(F)$. Of course, $c_0(F)$ is also a complete metric space with respect to the Hausdorff metric D_∞. Besides this,

Nanda [302] introduced and proved that the set $\ell_p(F)$ is a complete metric space with the Hausdorff metric D_p defined by

$$D_p(u, v) := \left\{ \sum_k [D(u_k, v_k)]^p \right\}^{1/p},$$

where $u = (u_k)$ and $v = (v_k)$ are the points of $\ell_p(F)$. Further, Nuray and Savaş [307] have recently showed that the set $\ell(p, F)$ of sequences of fuzzy numbers

$$\ell(p, F) := \left\{ (u_k) \in \omega(F) : \sum_k [D(u_k, \overline{0})]^{p_k} < \infty \right\}$$

is a complete metric space with the metric h defined by

$$h(u, v) := \left\{ \sum_k [D(u_k, v_k)]^{p_k} \right\}^{1/M},$$

where $p = (p_k)$ is a bounded sequence of strictly positive real numbers and $M = \max\{1, \sup_{k\in\mathbb{N}} p_k\}$, and $u = (u_k)$, $v = (v_k)$ are the points of the set $\ell(p, F)$. Mursaleen and Başarır [290] have recently introduced some new sets of sequences of fuzzy numbers generated by a non-negative regular matrix A some of which reduced to the Maddox spaces $\ell_\infty(p, F)$, $c(p, F)$, $c_0(p, F)$ and $\ell(p, F)$ of sequences of fuzzy numbers for the special cases of that matrix A. Altın et al. [36] have recently defined the concepts of lacunary statistical convergence and lacunary strongly convergence of generalized difference sequences of fuzzy numbers. They have also given some relations concerning with these concepts and showed that lacunary Δ^m-statistical convergence and lacunary strongly $\Delta^m_{(p)}$-convergence are equivalent for Δ^m-bounded sequences of fuzzy numbers. Çolak et al. [127] and Çolak et al. [128] have recently worked on the difference sequences and the generalized difference sequences of fuzzy numbers. Quite recently; Talo and Başar [379] have extended the main results of Başar and Altay [66] to the fuzzy numbers. Also, Talo and Başar [381] have recently studied the normed quasilinearity of the classical sets $\ell_\infty(F)$, $c(F)$, $c_0(F)$ and $\ell_p(F)$ of sequences of fuzzy numbers and derived some related results. Finally, Talo and Başar [380] have introduced the sets $\ell_\infty(F, f)$, $c(F, f)$, $c_0(F, f)$ and $\ell_p(F, f, s)$ of sequences of fuzzy numbers defined by a modulus function f, and given some topological properties of the sets together with some inclusion relations.

The main purpose of the present section is to study the corresponding sets $\ell_\infty(F)$, $c(F)$, $c_0(F)$ and $\ell_p(F)$ of sequences of fuzzy numbers to the classical spaces ℓ_∞, c, c_0 and ℓ_p of sequences with real or complex entries. Following Talo and Başar [379], we essentially proceed with some classes of matrix transformations between the classical sets of sequences of fuzzy numbers. To do this, since we need the β-dual of the sets $\ell_\infty(F)$, $c(F)$, $c_0(F)$ and $\ell_p(F)$ we find them together with their α- and γ-duals. Furthermore, we emphasize the solidness of the sets $\ell_\infty(F)$, $c_0(F)$ and $\ell_p(F)$.

The rest of this section is organized, as follows:

In Subsection 8.4.1, we give some required definitions and consequences related with the sequences and series of fuzzy numbers. Additionally, we present an example on the convergence of series of fuzzy numbers and also report the most relevant, and recent literature in Subsection 8.4.1. This subsection terminates with the condensation of the results on the sum of the series of the fuzzy sets given by M. Stojaković and Z. Stojaković in [369]. Subsection 8.4.2 is devoted to the calculation of the α-, β- and γ-duals of the classical sets $\ell_\infty(F)$, $c(F)$, $c_0(F)$ and $\ell_p(F)$. It is also established in Subsection 8.4.2 that the classical sets $\ell_\infty(F)$, $c_0(F)$ and $\ell_p(F)$ are solid. In Subsection 8.4.3, we characterize the classes $(\mu(F) : \ell_\infty(F))$, $(c_0(F) : c(F))$, $(c_0(F) : c_0(F))$, $(c(F) : c(F), p)$, $(\ell_p(F) : c(F))$, $(\ell_p(F) : c_0(F))$ and $(\ell_\infty(F) : c_0(F))$ of infinite matrices of fuzzy numbers, where $\mu \in \{\ell_\infty, c, c_0, \ell_p\}$. Furthermore, we construct four examples concerning the matrix transformations of sequences of fuzzy numbers.

8.4.1 Preliminaries, Background and Notations

Following Matloka [274], we give some definitions concerning the sequences of fuzzy numbers, below, which are needed.

Now, prior to stating and proving the lemma concerning the sum of a convergent series of fuzzy numbers we give the following definition (cf. Kim and Ghil [218]):

Definition 8.4.1. [379, Definition 2.12] *Let $(u_k) \in \omega(F)$. Then, the expression $u_0 + u_1 + u_2 + \cdots + u_k + \cdots$, in short $\sum_k u_k$, is called a series of fuzzy numbers. Denote $s_n = \sum_{k=0}^{n} u_k$ for all $n \in \mathbb{N}$, if the sequence (s_n) converges to a fuzzy number u, then we say that the series $\sum_k u_k$ of fuzzy numbers converges to u and write $\sum_k u_k = u$ which implies that*

$$\lim_{n\to\infty} \sum_{k=0}^{n} u_k^-(\alpha) = u^-(\alpha) \text{ and } \lim_{n\to\infty} \sum_{k=0}^{n} u_k^+(\alpha) = u^+(\alpha),$$

uniformly in $\alpha \in [0,1]$. Conversely, if the fuzzy numbers

$$u_k = \{(u_k^-(\alpha), u_k^+(\alpha)) : \alpha \in [0,1]\}, \ \sum_k u_k^-(\alpha) = u^-(\alpha) \text{ and } \sum_k u_k^+(\alpha) = u^+(\alpha)$$

converge uniformly in α, then $u = \{(u^-(\alpha), u^+(\alpha)) : \alpha \in [0,1]\}$ defines a fuzzy number such that $u = \sum_k u_k$.

We say otherwise the series of fuzzy numbers diverges. Additionally, if the sequence (s_n) is bounded, then we say that the series $\sum_k u_k$ of fuzzy numbers is bounded. By $cs(F)$ and $bs(F)$, we denote the sets of all convergent and bounded series of fuzzy numbers, respectively.

Lemma 8.4.2. [379, Lemma 2.13] *If $\sum_k u_k^-(\alpha) = u^-(\alpha)$ and $\sum_k u_k^+(\alpha) = u^+(\alpha)$ converge uniformly in α for the fuzzy numbers $u_k = \{(u_k^-(\alpha), u_k^+(\alpha)) : \alpha \in [0,1]\}$, then $u = \{(u^-(\alpha), u^+(\alpha)) : \alpha \in [0,1]\}$ defines a fuzzy number such that $u = \sum_k u_k$.*

Proof. To prove the lemma, we must show that the pair of functions u^- and u^+ satisfies the conditions of Representation Theorem. For this, we prove that u^- is a bounded, non-decreasing, left continuous function on $(0,1]$ and right continuous at the point $\alpha = 0$. u_k^-'s are the bounded, non-decreasing, left continuous functions on $(0,1]$ and right continuous at the point $\alpha = 0$ for each $k \in \mathbb{N}$.

(i) Let $\alpha_1 < \alpha_2$. Then, $u_k^-(\alpha_1) \leq u_k^-(\alpha_2)$ for each $k \in \mathbb{N}$. Therefore, we have $\sum_k u_k^-(\alpha_1) \leq \sum_k u_k^-(\alpha_2)$ which yields the fact that $u^-(\alpha_1) \leq u^-(\alpha_2)$. Hence, u^- is non-decreasing.

(ii) By taking into account the uniform convergence in α of $\lim_{\alpha\to\alpha_0^-} u_k^-(\alpha) = u_k^-(\alpha_0)$ for each $k \in \mathbb{N}$, $\sum_k u_k^-(\alpha) = u^-(\alpha)$, we obtain for $\alpha_0 \in (0,1]$ that

$$\begin{aligned} \lim_{\alpha\to\alpha_0^-} u^-(\alpha) &= \lim_{\alpha\to\alpha_0^-} \sum_k u_k^-(\alpha) \\ &= \sum_k \lim_{\alpha\to\alpha_0^-} u_k^-(\alpha) = \sum_k u_k^-(\alpha_0) = u^-(\alpha_0), \end{aligned} \tag{8.4.1}$$

which shows that the function u^- is left continuous on $(0,1]$.

(iii) By using the uniform convergence in α in the expressions $\lim_{\alpha\to 0^+} u_k^-(\alpha) = u_k^-(0)$ for each $k \in \mathbb{N}$ and $\sum_k u_k^-(\alpha) = u^-(\alpha)$, we see that

$$\begin{aligned} \lim_{\alpha\to 0^+} u^-(\alpha) &= \lim_{\alpha\to 0^+} \sum_k u_k^-(\alpha) \\ &= \sum_k \lim_{\alpha\to 0^+} u_k^-(\alpha) = \sum_k u_k^-(0) = u^-(0). \end{aligned} \tag{8.4.2}$$

This means that u^- is a right continuous function at the point $\alpha = 0$.

(iv) There exists $M_k > 0$ such that $|u_k^-(\alpha)| \leq M_k$ for all $\alpha \in [0,1]$ and for all $k \in \mathbb{N}$. Since the series $\sum_k u_k^-(\alpha) = u^-(\alpha)$ converges uniformly in α there exists $n_0 \in \mathbb{N}$ for all $\varepsilon > 0$ such that $\left|\sum_{k=n+1}^{\infty} u_k^-(\alpha)\right| < \varepsilon$ for all $n \geq n_0$ and for $\alpha \in [0,1]$. Therefore, we have

$$\begin{aligned} |u^-(\alpha)| = \left|\sum_k u_k^-(\alpha)\right| &= \left|\sum_{k=0}^{n} u_k^-(\alpha) + \sum_{k=n+1}^{\infty} u_k^-(\alpha)\right| \\ &\leq \sum_{k=0}^{n} |u_k^-(\alpha)| + \left|\sum_{k=n+1}^{\infty} u_k^-(\alpha)\right| \\ &\leq \sum_{k=0}^{n} M_k + \varepsilon \leq K_\varepsilon. \end{aligned}$$

This leads us to the fact that u^- is a bounded function.

Since one can establish in the similar way that u^+ is a bounded, non-increasing, left continuous function on $(0,1]$, and right continuous at the point $\alpha = 0$, we omit details.

Finally, we show that $u^-(1) \leq u^+(1)$. Since $u_k^-(1) \leq u_k^+(1)$ for each $k \in \mathbb{N}$, we derive that $\sum_k u_k^-(1) \leq \sum_k u_k^+(1)$ which shows that $u^-(1) \leq u^+(1)$.

Therefore, it is deduced that $[u]_\alpha = [u^-(\alpha), u^+(\alpha)]$ defines a fuzzy number. Finally, we show that $\sum_k u_k = u$. Since the series of functions $\sum_k u_k^-(\alpha)$ and $\sum_k u_k^+(\alpha)$

converge uniformly in α to $u^-(\alpha)$ and $u^+(\alpha)$, respectively, for all $\varepsilon > 0$ there exists $n_0 \in \mathbb{N}$ such that

$$\begin{aligned} D\left(\sum_{k=0}^{n} u_k, u\right) &= \sup_{\alpha\in[0,1]} \max\left\{\left|\sum_{k=0}^{n} u_k^-(\alpha) - u^-(\alpha)\right|, \left|\sum_{k=0}^{n} u_k^+(\alpha) - u^+(\alpha)\right|\right\} \\ &\leq \max\left\{\sup_{\alpha\in[0,1]}\left|\sum_{k=0}^{n} u_k^-(\alpha) - u^-(\alpha)\right|, \sup_{\alpha\in[0,1]}\left|\sum_{k=0}^{n} u_k^+(\alpha) - u^+(\alpha)\right|\right\} < \varepsilon \end{aligned}$$

for all $n \geq n_0$, the sequence $\left(\sum_{k=0}^{n} u_k\right)$ converges to the fuzzy number u, i.e., $\sum_k u_k = u$.

This step completes the proof. □

Lemma 8.4.2 provides that uniform convergence does not necessary in order for defining a fuzzy number by the series $\sum_k u_k^-(\alpha) = u^-(\alpha)$ and $\sum_k u_k^+(\alpha) = u^+(\alpha)$. This can be seen from the criteria related with the level convergence of a sequence of fuzzy numbers given by Fang and Huang [169], as follows:

Definition 8.4.3. *Let $\{f_k\}$ be a sequence of functions defined on $[a,b]$ and $\alpha_0 \in (a,b]$. Then, $\{f_k\}$ is said to be eventually equi-left-continuous at α_0 if for any $\varepsilon > 0$ there exist $n_0 \in \mathbb{N}$ and $\delta > 0$ such that $|f_k(\alpha) - f_k(\alpha_0)| < \varepsilon$ whenever $\alpha \in (\alpha_0 - \delta, \alpha_0]$ and $k \geq n_0$. Similarly, eventually equi-right-continuity at $\alpha_0 \in [a,b)$ of $\{f_k\}$ can be defined.*

Theorem 8.4.4. [379, Theorem 2.15] *Let (u_k) be a sequence of fuzzy numbers such that $u_k^-(\alpha) \to \lambda(\alpha)$ and $u_k^+(\alpha) \to \mu(\alpha)$ for each $\alpha \in [0,1]$, as $k \to \infty$. Then, the pair of functions λ and μ determines a fuzzy number if and only if the sequences of functions $\{u_k^-(\alpha)\}$ and $\{u_k^+(\alpha)\}$ are eventually equi-left-continuous at each $\alpha \in (0,1]$ and eventually equi-right-continuous at $\alpha = 0$.*

Thus, it is deduced that the series $\sum_k u_k^-(\alpha) = u^-(\alpha)$ and $\sum_k u_k^+(\alpha) = u^+(\alpha)$ define a fuzzy number if the sequences

$$\{s_k^-(\alpha)\} := \left\{\sum_{k=0}^{n} u_k^-(\alpha)\right\} \text{ and } \{s_k^+(\alpha)\} := \left\{\sum_{k=0}^{n} u_k^+(\alpha)\right\}$$

satisfy the conditions of Theorem 8.4.4. Of course, this is a weaker condition than the uniform convergence.

Example 8.4.5. [379, Example 2.16] *As an example for convergent series, consider the series $\sum_k u_k$ with*

$$u_k(t) := \begin{cases} 1-(k+1)^2 t &, \quad 0 \leq t \leq \frac{1}{(k+1)^2}, \\ 0 &, \quad \text{otherwise,} \end{cases}$$

for all $k \in \mathbb{N}$. It is trivial that $u_k^-(\alpha) = 0$ and $u_k^+(\alpha) = (1-\alpha)/(k+1)^2$ for every $\alpha \in [0,1]$. Therefore, we see that

$$\sum_k u_k^-(\alpha) = 0 \text{ and } \sum_k u_k^+(\alpha) = \sum_k \frac{1-\alpha}{(k+1)^2} = \frac{(1-\alpha)\pi^2}{6}.$$

Then, it is conclude that $\sum_k u_k = u$*, where*

$$u(t) := \begin{cases} 1 - \frac{6}{\pi^2}t & , \quad 0 \leq t \leq \frac{\pi^2}{6}, \\ 0 & , \quad \text{otherwise.} \end{cases}$$

Now, we may give two well-known consequences concerning the series of functions:

Lemma 8.4.6. *Let for the series of functions* $\sum_k u_k(t)$ *and* $\sum_k v_k(t)$*, there exists an* $n_0 \in \mathbb{N}$ *such that* $|u_k(t)| \leq v_k(t)$ *for all* $k \geq n_0$ *and for every* $t \in [a, b]$ *with* $u_k : [a, b] \to \mathbb{R}$ *and* $v_k : [a, b] \to \mathbb{R}$*. If the series* $\sum_k v_k(t)$ *converges uniformly in* $[a, b]$*, then the series* $\sum_k |u_k(t)|$ *and* $\sum_k u_k(t)$ *are uniformly convergent in* $[a, b]$*.*

Weierstrass M test. *Let* $u_k : [a, b] \to \mathbb{R}$ *be given. If there exists an* $M_k \geq 0$ *such that* $|u_k(t)| \leq M_k$ *for all* $k \in \mathbb{N}$ *and the series* $\sum_k M_k$ *converges, then the series* $\sum_k u_k(t)$ *is uniformly and absolutely convergent in* $[a, b]$*.*

Now, we can give the following theorem concerning the convergent series of fuzzy numbers:

Theorem 8.4.7. [379, Theorem 2.18] *If* $\sum_k u_k$ *and* $\sum_k v_k$ *converge, then the inequality* $D(\sum_k u_k, \sum_k v_k) \leq \sum_k D(u_k, v_k)$ *holds.*

Proof. Let $\sum_k u_k = u$ and $\sum_k v_k = v$, i.e., $\lim_{n\to\infty} D(\sum_{k=0}^n u_k, u) = 0$ and $\lim_{n\to\infty} D(\sum_{k=0}^n v_k, v) = 0$. It is obvious that there is no problem in the case $\sum_k D(u_k, v_k) = \infty$.

Suppose that $\sum_k D(u_k, v_k) < \infty$. Then, by using the properties of the metric D given by Proposition 8.2.2, we derive that

$$\begin{aligned} D(u, v) &= D\left(u + \sum_{k=0}^n u_k + \sum_{k=0}^n v_k, v + \sum_{k=0}^n u_k + \sum_{k=0}^n v_k\right) \\ &\leq D\left(\sum_{k=0}^n u_k, u\right) + D\left(\sum_{k=0}^n v_k, v\right) + D\left(\sum_{k=0}^n u_k, \sum_{k=0}^n v_k\right) \\ &\leq D\left(\sum_{k=0}^n u_k, u\right) + D\left(\sum_{k=0}^n v_k, v\right) + \sum_{k=0}^n D(u_k, v_k), \end{aligned}$$

which leads us by passing to limit as $n \to \infty$ that $D(u, v) = D(\sum_k u_k, \sum_k v_k) \leq \sum_k D(u_k, v_k)$. □

We conclude this subsection by summarizing the concerned results given by M. Stojaković and Z. Stojaković in [369].

Let $(\mathfrak{X}, \|\cdot\|)$ be a real Banach space and X, Y be the subsets of $\mathfrak{X}$. Then, the Hausdorff distance h is defined by

$$h(X, Y) := \max\left\{\sup_{t\in X}\inf_{y\in Y}\|x - y\|, \sup_{y\in Y}\inf_{t\in X}\|x - y\|\right\}.$$

Denote $h(X, \{0\}) = \sup_{t\in X} \|x\|$ by $|X|$. The mapping $u : \mathfrak{X} \to [0,1]$ is a fuzzy set on $\mathfrak{X}$. By $F(\mathfrak{X})$, let us denote the set of fuzzy sets such that α-level sets $[u]_\alpha = \{t \in \mathfrak{X} : u(t) \geq \alpha\}$ are non-empty for every $\alpha \in (0,1]$.

If $u_k \in F(\mathfrak{X})$ for $k \in \mathbb{N}$, then it is defined that

$$\left(\sum_k u_k\right)(x) := \sup\left\{\inf\{u_k(x_k)\}_{k\in\mathbb{N}},\ x = \sum_k x_k\right\},\ (x \in \mathfrak{X}). \tag{8.4.3}$$

Theorem 8.4.8. [379, Theorem 2.19] *If α-level sets of u_k are compact for all $k \in \mathbb{N}$ and for all $\alpha \in (0,1]$, and if $\sum_k |[u_k]_\alpha| < \infty$ for all $\alpha \in (0,1]$, then*

$$\left[\sum_k u_k\right]_\alpha = \sum_k [u_k]_\alpha \text{ for every } \alpha \in (0,1]. \tag{8.4.4}$$

Theorem 8.4.9. [379, Theorem 2.20] *If α-level sets of u_k are upper semicontinuous fuzzy sets with compact supports for all $k \in \mathbb{N}$ and if $\sum_k |supp\ u_k| < \infty$, then (8.4.4) holds, where supp u_k denotes the set of real numbers t such that $u_k(t) > 0$.*

Theorem 8.4.10. [379, Theorem 2.21] *If $u_k : \mathbb{R} \to [0,1]$ are fuzzy sets for $k \in \mathbb{N}$ such that α-level sets $[u_k]_\alpha$ are non-empty, bounded and closed for every $\alpha \in (0,1]$, then the α-level sets of sum are non-empty, bounded and closed, and the following statements hold:*

(i) $[\sum_{k=0}^n u_k]_\alpha = \sum_{k=0}^n [u_k]_\alpha$.

(ii) *If $\sum_k |[u_k]_\alpha| < \infty$ for every $\alpha \in (0,1]$, then (8.4.4) holds.*

Theorem 8.4.11. [379, Theorem 2.22] *If α-level sets of u_k are compact and convex for all $k \in \mathbb{N}$ and for every $\alpha \in (0,1]$, and if $\sum_k |[u_k]_\alpha| < \infty$, then the α-level set of $\sum_k u_k$ is also compact and convex for every $\alpha \in (0,1]$.*

If $u \in L(\mathbb{R})$, then since $[u]_\alpha$ is compact and convex for $\alpha \in (0,1]$ one can see that Theorem 8.4.10 and Theorem 8.4.11 are satisfied. Additionally, the sets $supp\ u_k = \overline{\{t \in R : u_k(t) > 0\}}$ are compact and u_k's are the upper semicontinuous functions. Thus, Theorem 8.4.10 is also satisfied. Therefore, it can conclude that $\sum_k |[u_k]_t| \leq \sum_k |supp\ u_k|$. Thus, we have:

Corollary 8.4.12. [379, Corollary 2.23] *Let $u_k \in L(\mathbb{R})$ for all $k \in \mathbb{N}$. If $\sum_k |supp\ u_k| < \infty$, then (8.4.4) holds.*

Proof. Let $u_k \in L(\mathbb{R})$ for all $k \in \mathbb{N}$. Then, since we use the real field it is obtained from here that

$$|supp\ u_k| = \sup_{x\in[u_k]_0} |x| = \sup_{x\in[u_k^-(0),u_k^+(0)]} |x| = \max\left\{\left|u_k^-(0)\right|, \left|u_k^+(0)\right|\right\} = D(u_k, \overline{0}).$$

This leads us to the consequence that if $u = (u_k) \in \ell_1(F)$, then (8.4.4) holds. Additionally, since $[u]_\alpha = [u^-(\alpha), u^+(\alpha)]$ for $\alpha \in [0,1]$ when $u \in L(\mathbb{R})$, it follows that

$$\sum_k [u_k]_\alpha = \left[\sum_k u_k^-(\alpha), \sum_k u_k^+(\alpha)\right].$$

Since $|u_k^-(\alpha)| \leq D(u_k, \overline{0})$ and $|u_k^+(\alpha)| \leq D(u_k, \overline{0})$ for all $k \in \mathbb{N}$, Weierstrass criterion yields the uniform convergence of the series $\sum_k u_k^-(\alpha)$ and $\sum_k u_k^+(\alpha)$.

Finally, by using the fact that convergence is uniform one can see for the case $\alpha = 0$ that

$$\left[\sum_k u_k\right]_0 = \lim_{\alpha\to 0^+} \left[\sum_k u_k\right]_\alpha = \lim_{\alpha\to 0^+} \sum_k [u_k]_\alpha = \sum_k \lim_{\alpha\to 0^+} [u_k]_\alpha = \sum_k [u_k]_0.$$

This step completes the proof. □

M. Stojaković and Z. Stojaković [369] used the concept of convergence of series of fuzzy numbers defined by Zadeh's Extension Principle. In this section, we emphasize the fact that $u = \sum_k u_k$ if and only if $\sum_k u_k^-(\alpha) = u^-(\alpha)$ and $\sum_k u_k^+(\alpha) = u^+(\alpha)$, uniformly in α. Therefore, in the case $\sum_k D(u_k, \overline{0}) < \infty$ our definition coincides with the convergence of a series given by (8.4.3). However, it is possible to define a fuzzy number by using (8.4.3) even if the condition $\sum_k D(u_k, \overline{0}) < \infty$ is not satisfied. Then, the sums of a convergent series may be different in the sense of our definition and in the sense of M. Stojaković and Z. Stojaković [369]. Nevertheless, there is no need to elaborate on the condition of uniform convergence given in Theorem 8.4.4 for defining a fuzzy number by the series $\sum_k [u_k]_\alpha = \left[\sum_k u_k^-(\alpha), \sum_k u_k^+(\alpha)\right]$. Even in finding the sum of two fuzzy numbers, we used the sum of α-level sets instead of Zadeh's Extension Principle. Of course, calculating the sum of a convergent series of fuzzy numbers via Zadeh's Extension Principle is very difficult. So, after calculating the sum of the series of α-level sets in evaluating the response of the question: *Is the sum a fuzzy number?*, it becomes easier and much more applicable than the usage of the Zadeh's Extension Principle.

If we consider the convergent series $\sum_k u_k$ in the sense of the present section, then

$$\begin{aligned}\sum_k u_k = u \quad &\Leftrightarrow \quad \lim_{n\to\infty} D\left(\sum_{k=0}^{n} u_k, u\right) = 0 \\ &\Leftrightarrow \quad \lim_{n\to\infty} \sup_{\alpha\in[0,1]} d\left(\sum_{k=0}^{n} [u_k]_\alpha, [u]_\alpha\right) = 0,\end{aligned}$$

where $\mathfrak{X}$ is a Banach space, $u_k \in F(\mathfrak{X})$ for all $k \in \mathbb{N}$ and d is the Hausdorff metric. This is equivalent to the fact that $\sum_k [u_k]_\alpha = [u]_\alpha$ uniformly in $\alpha \in [0,1]$.

8.4.2 Determination of Duals of the Classical Sets of Sequences of Fuzzy Numbers

Firstly, we define the α-, β- and γ-duals of a set $\mu(F) \subset \omega(F)$ which are respectively denoted by $\{\mu(F)\}^\alpha$, $\{\mu(F)\}^\beta$ and $\{\mu(F)\}^\gamma$, as follows:

$$\begin{aligned}\{\mu(F)\}^\alpha &:= \{(u_k) \in \omega(F) : (u_k v_k) \in \ell_1(F) \text{ for all } (v_k) \in \mu(F)\}, \\ \{\mu(F)\}^\beta &:= \{(u_k) \in \omega(F) : (u_k v_k) \in cs(F) \text{ for all } (v_k) \in \mu(F)\}, \\ \{\mu(F)\}^\gamma &:= \{(u_k) \in \omega(F) : (u_k v_k) \in bs(F) \text{ for all } (v_k) \in \mu(F)\}.\end{aligned}$$

Now, following Talo and Başar [379], we can give the results concerning the α-, β- and γ-duals of the classical sets $\ell_\infty(F)$, $c(F)$, $c_0(F)$ and $\ell_p(F)$ of sequences of fuzzy numbers.

We begin with a lemma which yields a useful result on the absolute convergence of the series of fuzzy numbers.

Lemma 8.4.13. [379, Lemma 3.1] *Let d denotes the set of all absolutely convergent series of fuzzy numbers, i.e.,*

$$d := \left\{ (u_k) \in \omega(F) : \sum_k |u_k| < \infty \right\}.$$

Then, the set d is identical to the set $\ell_1(F)$.

Proof. Let $(u_k) \in \ell_1(F)$. Then, since $\sum_k D(u_k, \overline{0}) < \infty$ and

$$\max\left\{\left|u_k^-(\alpha)\right|, \left|u_k^+(\alpha)\right|\right\} \leq D(u_k, \overline{0})$$

for $\alpha \in [0,1]$, we can see that

$$\sum_k \max\left\{\left|u_k^-(\alpha)\right|, \left|u_k^+(\alpha)\right|\right\} \leq \sum_k D(u_k, \overline{0}) < \infty,$$

which leads us to the consequence $(u_k) \in d$. Hence, $\ell_1(F) \subset d$.

Conversely, let $(u_k) \in d$. Therefore, $\sum_k \max\left\{\left|u_k^-(\alpha)\right|, \left|u_k^+(\alpha)\right|\right\}$ converges for $\alpha \in [0,1]$. This gives for $\alpha = 0$ that

$$\sum_k D(u_k, \overline{0}) = \sum_k \max\left\{\left|u_k^-(0)\right|, \left|u_k^+(0)\right|\right\} < \infty,$$

which means that $(u_k) \in \ell_1(F)$, as desired. Hence, $d \subset \ell_1(F)$.

This step completes the proof. □

Theorem 8.4.14. [379, Theorem 3.2] *The α-dual of the sets $c_0(F)$, $c(F)$ and $\ell_\infty(F)$ of sequences of fuzzy numbers is the set $\ell_1(F)$.*

Proof. Since the proof is similar for the sets $c_0(F)$ and $c(F)$, we give the proof only for the set $\ell_\infty(F)$.

Let $(u_k) \in \ell_\infty(F)$. Then, there exists a $K > 0$ such that $D(u_k, \overline{0}) \leq K$ for all $k \in \mathbb{N}$. Therefore, we derive by using the fact given Part (i) of Lemma 8.2.6 that

$$\sum_k D(u_k v_k, \overline{0}) \leq \sum_k D(u_k, \overline{0}) D(v_k, \overline{0}) \leq K \sum_k D(v_k, \overline{0}).$$

If we take $(v_k) \in \ell_1(F)$, then we have $\sum_k D(u_k v_k, \overline{0}) < \infty$ which gives that

$$\ell_1(F) \subseteq \{\ell_\infty(F)\}^\alpha. \tag{8.4.5}$$

Conversely, let $(v_k) \in \{\ell_\infty(F)\}^\alpha$. If we consider $(u_k) = (\overline{1}) \in \ell_\infty(F)$, then the series $\sum_k D(u_k v_k, \overline{0}) = \sum_k D(v_k, \overline{0})$ converges, that is to say that $(v_k) \in \ell_1(F)$. Therefore, we have

$$\{\ell_\infty(F)\}^\alpha \subseteq \ell_1(F). \tag{8.4.6}$$

Thus, the desired result follows by combining the inclusion relations (8.4.5) and (8.4.6). □

Theorem 8.4.15. [379, Theorem 3.5] *The β-dual of the sets $c_0(F)$, $c(F)$ and $\ell_\infty(F)$ of sequences of fuzzy numbers is the set $\ell_1(F)$.*

Proof. Since the proof is similar for the sets $c_0(F)$ and $c(F)$, we give the proof only for the set $\ell_\infty(F)$.

Let $(u_k) \in \ell_\infty(F)$ and $(v_k) \in \omega(F)$. Then, there exists a $K > 0$ such that $D(u_k, \overline{0}) \leq K$ for all $k \in \mathbb{N}$. Since

$$\begin{aligned} |(u_k v_k)^-(\alpha)| &\leq D(u_k v_k, \overline{0}) \leq D(u_k, \overline{0}) D(v_k, \overline{0}) \leq K \cdot D(v_k, \overline{0}) \\ |(u_k v_k)^+(\alpha)| &\leq D(u_k v_k, \overline{0}) \leq D(u_k, \overline{0}) D(v_k, \overline{0}) \leq K \cdot D(v_k, \overline{0}), \end{aligned}$$

Weierstrass M test yields that $\sum_k (u_k v_k)^-(\alpha)$ and $\sum_k (u_k v_k)^+(\alpha)$ converge uniformly and hence $\sum_k u_k v_k$ converges whenever $\sum_k D(v_k, \overline{0})$ converges. Therefore, we have

$$\ell_1(F) \subseteq \{\ell_\infty(F)\}^\beta. \tag{8.4.7}$$

Conversely, suppose that $(v_k) \in \{\ell_\infty(F)\}^\beta$. Then, the series $\sum_k u_k v_k$ converges for all $(u_k) \in \ell_\infty(F)$. This also holds for the sequence (u_k) of fuzzy numbers defined by $u_k = \chi_{[-1,1]}$ for all $k \in \mathbb{N}$. Then, since $u_k^-(\alpha) = -1$ and $u_k^+(\alpha) = 1$ for $\alpha \in [0, 1]$, the series

$$\begin{aligned} \sum_k (u_k v_k)^+(\alpha) &= \sum_k \max\{u_k^-(\alpha)v_k^-(\alpha), u_k^-(\alpha)v_k^+(\alpha), u_k^+(\alpha)v_k^-(\alpha), u_k^+(\alpha)v_k^+(\alpha)\} \\ &= \sum_k \max\{-v_k^-(\alpha), -v_k^+(\alpha), v_k^-(\alpha), v_k^+(\alpha)\} \\ &= \sum_k \max\{|v_k^-(\alpha)|, |v_k^+(\alpha)|\} \end{aligned}$$

converges uniformly which gives that $(v_k) \in \ell_1(F)$. Hence,

$$\{\ell_\infty(F)\}^\beta \subseteq \ell_1(F). \tag{8.4.8}$$

Thus, the desired result is obtained by combining the inclusions (8.4.7) and (8.4.8). □

Theorem 8.4.16. [381, Theorems 4.1 and 4.2] *The β- and α-duals of the set $\ell_1(F)$ are the set $\ell_\infty(F)$.*

Proof. (i) As a first step, we show that β-dual of the set $\ell_1(F)$ is the set $\ell_\infty(F)$.

Suppose that $(u_k) \in \ell_1(F)$ and $(v_k) \in \omega(F)$. Then, $\sum_k \|u_k\| < \infty$ and the inequalities

$$\begin{aligned} |(u_k v_k)^-(\alpha)| &\leq \|u_k v_k\| \leq \|u_k\|\|v_k\|, \\ |(u_k v_k)^+(\alpha)| &\leq \|u_k v_k\| \leq \|u_k\|\|v_k\| \end{aligned}$$

hold for each $\alpha \in [0,1]$. If $(v_k) \in \ell_\infty(F)$, then since $\|v\|_\infty = \sup_{k\in\mathbb{N}} \|v_k\| < \infty$ one can immediately see that

$$\sum_k \|u_k\|\|v_k\| \leq \|v\|_\infty \sum_k \|u_k\| < \infty.$$

Since the series $\sum_k (u_k v_k)^-(\alpha)$ and $\sum_k (u_k v_k)^+(\alpha)$ are uniformly convergent by Weierstrass M test, the series $\sum_k u_k v_k$ is also convergent. Hence, the inclusion

$$\ell_\infty(F) \subseteq \{\ell_1(F)\}^\beta \tag{8.4.9}$$

holds.

Conversely, let us take any point $v = (v_k)$ in the set $\{\ell_1(F)\}^\beta$. Then, the series $\sum_k u_k v_k$ is convergent for each $(u_k) \in \ell_1(F)$. Define the sequence (f_n) of bounded quasilinear operators on $\ell_1(F)$ by

$$f_n(u) := \sum_{k=0}^{n} u_k v_k$$

for all $n \in \mathbb{N}$. One can observe by using Proposition 8.5.8 and Proposition 8.5.14 at this stage that the operator f defined by

$$f(u) := \lim_{n\to\infty} f_n(u) = \sum_k u_k v_k$$

is a bounded quasilinear operator on the space $\ell_1(F)$. Consider the sequence $\{u_k^{(n)}\}_{k=0}^\infty \in \ell_1(F)$ defined by

$$u_k^{(n)} := \begin{cases} \overline{1} \quad , & n = k, \\ \overline{0} \quad , & n \neq k, \end{cases} \tag{8.4.10}$$

for every fixed $n \in \mathbb{N}$. Therefore, by the boundedness of f we see from the fact

$$\left\|\sum_k u_k^{(n)} v_k\right\| \leq \|f\| \left\|u_k^{(n)}\right\|_1$$

that $\|(v_k)\| \leq \|f\|$. This yields that $(v_k) \in \ell_\infty(F)$. That is to say that the inclusion

$$\{\ell_1(F)\}^\beta \subseteq \ell_\infty(F) \tag{8.4.11}$$

holds. Now, the desired result follows by combining the inclusions (8.4.9) and (8.4.11).

(ii) As a second step, we show that α-dual of the set $\ell_1(F)$ is also the set $\ell_\infty(F)$.

Suppose that $(u_k) \in \ell_1(F)$ and $(v_k) \in \omega(F)$. Then, $\sum_k \|u_k\| < \infty$. If $(v_k) \in \ell_\infty(F)$, then since $\|v\|_\infty = \sup_{k\in\mathbb{N}} \|v_k\| < \infty$ one can immediately see that

$$\sum_k \|u_k v_k\| \leq \sum_k \|u_k\|\|v_k\| \leq \|v\|_\infty \sum_k \|u_k\| < \infty.$$

Hence, the inclusion

$$\ell_\infty(F) \subseteq \{\ell_1(F)\}^\alpha \tag{8.4.12}$$

holds.

Conversely, since the inclusion $\{\ell_1(F)\}^\alpha \subseteq \{\ell_1(F)\}^\beta = \ell_\infty(F)$ holds we derive together with (8.4.12) that the α-dual of the space $\ell_1(F)$ is the space $\ell_\infty(F)$. □

Now, we may state the definition of the concept of perfectness of a set of sequences of fuzzy numbers and give our easy result concerning the perfectness of the spaces $\ell_\infty(F)$ and $\ell_1(F)$.

Definition 8.4.17. [381] *A set $\mu(F) \subset \omega(F)$ is said to be perfect if $\{[\mu(F)]^\alpha\}^\alpha = \mu(F)$.*

Proposition 8.4.18. [381, Proposition 4.4] *The sets $\ell_\infty(F)$ and $\ell_1(F)$ are perfect.*

Proof. One can immediately see by combining the consequences given by Theorem 8.4.14 and the second part of Theorem 8.4.16 that

$$\begin{aligned} \{[\ell_\infty(F)]^\alpha\}^\alpha &= \{\ell_1(F)\}^\alpha = \ell_\infty(F), \\ \{[\ell_1(F)]^\alpha\}^\alpha &= \{\ell_\infty(F)\}^\alpha = \ell_1(F), \end{aligned}$$

as desired. □

Now, following Sarma [345] we state the fuzzy analogues of the concept of solidity of a set of sequences of complex terms and next give our result concerning the solidity of the sets $\ell_\infty(F)$, $c_0(F)$ and $\ell_p(F)$.

Definition 8.4.19. [379, Definition 3.3] *A set $\mu(F) \subset \omega(F)$ is said to be solid if $(v_k) \in \mu(F)$ whenever $D(v_k, \overline{0}) \leq D(u_k, \overline{0})$ for all $k \in \mathbb{N}$, for some $(u_k) \in \mu(F)$. Therefore, one can conclude that the α- and γ-duals of a set of sequences of fuzzy numbers are identical if it is solid.*

Theorem 8.4.20. [379, Theorem 3.4] *The sets $\ell_\infty(F)$, $c_0(F)$ and $\ell_p(F)$ are solid.*

Proof. Let $\mu(F)$ denotes the any of the sets $\ell_\infty(F)$, $c_0(F)$ and $\ell_p(F)$, and suppose that

$$D(v_k, \overline{0}) \leq D(u_k, \overline{0}) \tag{8.4.13}$$

holds for some $(u_k) \in \mu(F)$. Therefore, one can easily see by (8.4.13) that

$$\sup_{k\in\mathbb{N}} D(v_k, \overline{0}) \leq \sup_{k\in\mathbb{N}} D(u_k, \overline{0}) < \infty, \tag{8.4.14}$$

$$\lim_{k\to\infty} D(v_k, \overline{0}) \leq \lim_{k\to\infty} D(u_k, \overline{0}) = 0, \tag{8.4.15}$$

$$\sum_k [D(v_k, \overline{0})]^p \leq \sum_k [D(u_k, \overline{0})]^p < \infty. \tag{8.4.16}$$

The inequalities (8.4.14), (8.4.15) and (8.4.16) yield the desired consequence that $(v_k) \in \mu(F)$.

This completes the proof. □

Theorem 8.4.21. [379, Theorem 3.6] *The α- and β-duals of the set $\ell_p(F)$ of sequences of fuzzy numbers are the set $\ell_q(F)$, where $1 < p < \infty$.*

Proof. We give the proof only for β-dual. Since the implication "$(u_k) \in \ell_p(F)$ if and only if the series $\sum_k D(u_k, \overline{0})^p$ converges, i.e., $\{D(u_k, \overline{0})\} \in \ell_p$" holds, $\{D(v_k, \overline{0})\} \in \ell_q$ whenever $(v_k) \in \ell_q(F)$. Because of the well-known result $\ell_p^\beta = \ell_q$, the series $\sum_k D(u_k, \overline{0}) D(v_k, \overline{0})$ converges. Therefore, since

$$\begin{aligned} |(u_k v_k)^-(\alpha)| &\leq D(u_k v_k, \overline{0}) \leq D(u_k, \overline{0}) D(v_k, \overline{0}), \\ |(u_k v_k)^+(\alpha)| &\leq D(u_k v_k, \overline{0}) \leq D(u_k, \overline{0}) D(v_k, \overline{0}), \end{aligned}$$

Weierstrass M test implies that the series $\sum_k (u_k v_k)^-(\alpha)$ and $\sum_k (u_k v_k)^+(\alpha)$ converge uniformly, hence the series $\sum_k u_k v_k$ also converges. Thus, we have

$$\ell_q(F) \subseteq \{\ell_p(F)\}^\beta. \tag{8.4.17}$$

Conversely, let $(v_k) \in \{\ell_p(F)\}^\beta$. Then, the series $\sum_k u_k v_k$ converges for all $(u_k) \in \ell_p(F)$ which yields that the series $\sum_k (u_k v_k)^+(\alpha)$ converges uniformly in $\alpha \in [0, 1]$. Define the sequence $\{u_k^{(n)}\} \in \ell_p(F)$ by

$$u_k^{(n)} := \begin{cases} \chi_{[-(D(v_k, \overline{0}))^{q-1}, (D(v_k, \overline{0}))^{q-1}]} &, \quad k \leq n, \\ \overline{0} &, \quad \text{otherwise}, \end{cases}$$

for all $k, n \in \mathbb{N}$. Therefore, $[u_k^{(n)}]_\alpha = [-(D(v_k, \overline{0}))^{q-1}, (D(v_k, \overline{0}))^{q-1}]$ for $k \leq n$. Thus, one can observe that

$$\begin{aligned} &\left(u_k^{(n)} v_k\right)^+(\alpha) = \\ &= \max\{-[D(v_k, \overline{0})]^{q-1} v_k^-(\alpha), -[D(v_k, \overline{0})]^{q-1} v_k^+(\alpha), [D(v_k, \overline{0})]^{q-1} v_k^-(\alpha), [D(v_k, \overline{0})]^{q-1} v_k^+(\alpha)\} \\ &= [D(v_k, \overline{0})]^{q-1} \max\{-v_k^-(\alpha), -v_k^+(\alpha), v_k^-(\alpha), v_k^+(\alpha)\} \\ &= [D(v_k, \overline{0})]^{q-1} \max\{|v_k^-(\alpha)|, |v_k^+(\alpha)|\}, \end{aligned}$$

which gives for $\alpha = 0$ that

$$\begin{aligned} \left(u_k^{(n)} v_k\right)^+(0) &= [D(v_k, \overline{0})]^{q-1} \max\{|v_k^-(0)|, |v_k^+(0)|\} \\ &= [D(v_k, \overline{0})]^{q-1} D(v_k, \overline{0}) = [D(v_k, \overline{0})]^q. \end{aligned}$$

This leads us to the consequence that $\sum_{k=0}^n [D(v_k, \overline{0})]^q = \sum_k [u_k^{(n)} v_k]^+(0)$ for all $n \in \mathbb{N}$, which means that $v = (v_k) \in \ell_q(F)$. Hence, we have

$$\{\ell_p(F)\}^\beta \subseteq \ell_q(F). \tag{8.4.18}$$

The desired result is obtained by combining the inclusions (8.4.17) and (8.4.18). □

By combining Theorem 8.4.20 with Theorems 8.4.14 and 8.4.21, we derive:

Corollary 8.4.22. [379, Corollary 3.7] *The γ-dual of the sets $\ell_\infty(F)$, $c_0(F)$ and $\ell_p(F)$ of sequences of fuzzy numbers are the sets $\ell_1(F)$ and $\ell_q(F)$, respectively.*

8.4.3 Matrix Transformations Between Some Sets of Sequences of Fuzzy Numbers

In this subsection, we emphasize matrix transformations between the sets of sequences of fuzzy numbers.

Let $\mu_1(F), \mu_2(F) \subset \omega(F)$ and $A = (a_{nk})$ be an infinite matrix of fuzzy numbers. Then, we say that A defines a matrix transformation from $\mu_1(F)$ into $\mu_2(F)$, and denote it by writing $A : \mu_1(F) \to \mu_2(F)$, if for all sequences $u = (u_k) \in \mu_1(F)$ the sequence $Au = \{(Au)_n\}$, the A-transform of u, exists and is in $\mu_2(F)$; where

$$(Au)_n := \sum_k a_{nk}u_k \quad \text{for all} \quad n \in \mathbb{N}. \tag{8.4.19}$$

By $(\mu_1(F) : \mu_2(F))$, we denote the class of all matrices A such that $A : \mu_1(F) \to \mu_2(F)$. Thus, $A \in (\mu_1(F) : \mu_2(F))$ if and only if the series on the right-hand side of (8.4.19) converges for each $n \in \mathbb{N}$ and every $u \in \mu_1(F)$, and we have $Au = \{(Au)_n\}_{n\in\mathbb{N}} \in \mu_2(F)$ for all $u \in \mu_1(F)$. A sequence u is said to be A-summable to α if Au converges to α which is called as the A-limit of u. Also, by $(c(F) : c(F); p)$, we mean the class of all regular matrices A such that $A : c(F) \to c(F)$ with the A-limit of u equals to the limit of u, for all $u \in c(F)$.

In this subsection, we characterize the classes $(\mu(F) : \ell_\infty(F))$, $(c_0(F) : c(F))$, $(c_0(F) : c_0(F))$, $(c(F) : c(F); p)$, $(\ell_p(F) : c(F))$, $(\ell_p(F) : c_0(F))$ and $(\ell_\infty(F) : c_0(F))$ of infinite matrices of fuzzy numbers, where $\mu \in \{\ell_\infty, c, c_0, \ell_p\}$.

Now, we can begin with the following basic theorem:

Basic Theorem. [379] *Let $A = (a_{nk})$ be an infinite matrix of fuzzy numbers. Then, the following statements hold:*

(i) $A \in (\ell_\infty(F) : \ell_\infty(F))$ *if and only if*

$$M := \sup_{n\in\mathbb{N}} \sum_k D(a_{nk}, \overline{0}) < \infty. \tag{8.4.20}$$

(ii) $A \in (c(F) : \ell_\infty(F))$ *if and only if (8.4.20) holds.*

(iii) $A \in (c_0(F) : \ell_\infty(F))$ *if and only if (8.4.20) holds.*

(iv) $A \in (\ell_p(F) : \ell_\infty(F))$ *if and only if*

$$C := \sup_{n\in\mathbb{N}} \sum_k [D(a_{nk}, \overline{0})]^q < \infty. \tag{8.4.21}$$

Proof. Since the proof can also be obtained in the similar way for the other parts, to avoid undue repetition in the statements, we prove only Part (i).

(i) Let $A = (a_{nk}) \in (\ell_\infty(F) : \ell_\infty(F))$ and $u = (u_k) \in \ell_\infty(F)$. Then, the series $\sum_k a_{nk}u_k$ converges for each fixed $n \in \mathbb{N}$, since Au exists. Hence, $A_n \in \{\ell_\infty(F)\}^\beta$ for all $n \in \mathbb{N}$. Define the sequence $u = (u_k) \in \ell_\infty(F)$ by $u_k = \chi_{[-1,1]}$ for all $k \in \mathbb{N}$. Then, $Au \in \ell_\infty(F)$ which yields for all $n \in \mathbb{N}$ and for $\alpha \in [0,1]$ that

$$\begin{aligned}\sum_k (a_{nk}u_k)^+(\alpha) &= \sum_k \max\{a_{nk}^-(\alpha)u_k^-(\alpha), a_{nk}^-(\alpha)u_k^+(\alpha), a_{nk}^+(\alpha)u_k^-(\alpha), a_{nk}^+(\alpha)u_k^+(\alpha)\} \\ &= \sum_k \max\{-a_{nk}^-(\alpha), -a_{nk}^+(\alpha), a_{nk}^-(\alpha), a_{nk}^+(\alpha)\} \\ &= \sum_k \max\{|a_{nk}^-(\alpha)|, |a_{nk}^+(\alpha)|\}.\end{aligned}$$

In the special case $\alpha = 0$, the sequence

$$\left\{\sum_k D(a_{nk}, \overline{0})\right\}_{n\in\mathbb{N}} = \left\{\sum_k \max\{|a_{nk}^-(0)|, |a_{nk}^+(0)|\}\right\}_{n\in\mathbb{N}}$$

is bounded, which means that (8.4.20) holds.

Conversely suppose that (8.4.20) holds and $u = (u_k) \in \ell_\infty(F)$. Then, since $A_n \in \{\ell_\infty(F)\}^\beta$ for each $n \in \mathbb{N}$, Au exists. Therefore, one can observe by using Lemma 8.2.6 together with the condition (8.4.20) that

$$\begin{aligned}D_\infty(Au, \theta) = \sup_{n\in\mathbb{N}} D((Au)_n, \overline{0}) &= \sup_{n\in\mathbb{N}} D\left(\sum_k a_{nk}u_k, \overline{0}\right) \\ &\leq \sup_{n\in\mathbb{N}} \sum_k D(a_{nk}u_k, \overline{0}) \\ &\leq \sup_{n\in\mathbb{N}} \sum_k D(a_{nk}, \overline{0}) D(u_k, \overline{0}) \\ &\leq \sup_{k\in\mathbb{N}} D(u_k, \overline{0}) \sup_{n\in\mathbb{N}} \sum_k D(a_{nk}, \overline{0}) < \infty,\end{aligned}$$

i.e., $Au \in \ell_\infty(F)$.

This step concludes the proof. □

Now, we give an example about an infinite matrix in the class $(\ell_\infty(F) : \ell_\infty(F))$.

Example 8.4.23. [384] *Define the matrix* $A = (a_{nk})$ *by*

$$a_{nk} := \begin{cases} u_k & , \quad k = n, \\ \overline{0} & , \quad k \neq n \end{cases}$$

with

$$u_k(t) := \begin{cases} 1 + (k+1)t & , \quad -\frac{1}{k+1} \leq t < 0, \\ 1 - (k+1)t & , \quad 0 \leq t \leq \frac{1}{k+1}, \\ 0 & , \quad \text{otherwise} \end{cases}$$

for all $k, n \in \mathbb{N}$. Then, since

$$D(a_{nk}, \overline{0}) := \begin{cases} \frac{1}{n+1} & , \quad k = n, \\ 0 & , \quad k \neq n, \end{cases}$$

we have

$$\sup_{n \in \mathbb{N}} \sum_k D(a_{nk}, \overline{0}) = \sup_{n \in \mathbb{N}} \frac{1}{n+1} < \infty.$$

This shows by Part (i) of Basic Theorem that $A = (a_{nk}) \in (\ell_\infty(F) : \ell_\infty(F))$.

Theorem 8.4.24. [379, Theorem 4.2] *$A = (a_{nk}) \in (c_0(F) : c(F))$ if and only if (8.4.20) holds and there exists $(\alpha_k) \in \omega(F)$ such that*

$$\lim_{n \to \infty} D(a_{nk}, \alpha_k) = 0 \text{ for each } k \in \mathbb{N}. \tag{8.4.22}$$

If $A = (a_{nk}) \in (c_0(F) : c(F))$, then $(\alpha_k) \in \ell_1(F)$ and $\sum_k a_{nk} u_k \to \sum_k \alpha_k u_k$, as $n \to \infty$.

Proof. Suppose that (8.4.20) and (8.4.22) hold. Then, there exists an $n_K \in \mathbb{N}$ for $K \in \mathbb{N}$ and $\varepsilon > 0$ such that

$$\sum_{k=0}^{K} D(a_{nk}, \alpha_k) < \varepsilon$$

for all $n \geq n_K$. Since,

$$\begin{aligned} \sum_{k=0}^{K} D(\alpha_k, \overline{0}) &= \sum_{k=0}^{K} D(a_{nk} + \alpha_k, a_{nk}) \\ &\leq \sum_{k=0}^{K} D(a_{nk}, \alpha_k) + \sum_{k=0}^{K} D(a_{nk}, \overline{0}) \\ &\leq \varepsilon + M \end{aligned}$$

for $n \geq n_K$ by (8.4.22) one can see that $(\alpha_k) \in \ell_1(F)$ and $\sum_k D(\alpha_k, \overline{0}) \leq M$.

Let $u = (u_k) \in c_0(F)$. Then, one can choose a $k_0 \in \mathbb{N}$ for $\varepsilon > 0$ such that

$$D(u_k, \overline{0}) < \frac{\varepsilon}{2(2M+1)}$$

for each fixed $k \geq k_0$. Additionally, since $a_{nk} \to \alpha_k$, as $n \to \infty$ by (8.4.22), we have $a_{nk} u_k \to \alpha_k u_k$, as $n \to \infty$ for each fixed $k \in \mathbb{N}$. That is to say that

$$\lim_{n \to \infty} D(a_{nk} u_k, \alpha_k u_k) = 0.$$

Hence, there exists an $N = N(k_0) \in \mathbb{N}$ such that

$$\sum_{k=0}^{k_0} D(a_{nk} u_k, \alpha_k u_k) < \frac{\varepsilon}{2}$$

for all $n \geq N$. Thus, since

$$\begin{aligned} D\left(\sum_k a_{nk}u_k, \sum_k \alpha_k u_k\right) &\leq \sum_k D(a_{nk}u_k, \alpha_k u_k) \\ &= \sum_{k=0}^{k_0} D(a_{nk}u_k, \alpha_k u_k) + \sum_{k=k_0+1}^{\infty} D(a_{nk}u_k, \alpha_k u_k) \\ &\leq \frac{\varepsilon}{2} + \sum_{k=k_0+1}^{\infty} [D(a_{nk}u_k, \overline{0}) + D(\alpha_k u_k, \overline{0})] \\ &\leq \frac{\varepsilon}{2} + \sum_{k=k_0+1}^{\infty} D(a_{nk}, \overline{0})D(u_k, \overline{0}) + \sum_{k=k_0+1}^{\infty} D(\alpha_k, \overline{0})D(u_k, \overline{0}) \\ &\leq \frac{\varepsilon}{2} + \frac{\varepsilon}{2(2M+1)}\left[\sum_{k=k_0+1}^{\infty} D(a_{nk}, \overline{0}) + \sum_{k=k_0+1}^{\infty} D(\alpha_k, \overline{0})\right] \\ &\leq \frac{\varepsilon}{2} + \frac{\varepsilon}{2(2M+1)}(M + M + 1) = \varepsilon \text{ for all } n \geq N, \end{aligned}$$

the series $\sum_k a_{nk}u_k$ are convergent for each $n \in \mathbb{N}$ and $D(\sum_k a_{nk}u_k, \sum_k \alpha_k u_k) \to 0$, as $n \to \infty$. This means that $Au \in c(F)$.

Conversely, let $A = (a_{nk}) \in (c_0(F) : c(F))$ and $u = (u_k) \in c_0(F)$. Then, since Au exists and the inclusion $(c_0(F) : c(F)) \subset (c_0(F) : \ell_\infty(F))$ holds, the necessity of (8.4.20) is trivial by Part (iii) of Basic Theorem.

Now, consider the sequence $u^{(n)} = \{u_k^{(n)}\} \in c_0(F)$ defined by (8.4.10). Hence, $Au^{(n)} = (a_{nk})_{n=0}^{\infty} \in c(F)$ for each fixed $k \in \mathbb{N}$, i.e., the condition (8.4.22) is also necessary.

Thus, the proof is completed. □

As an easy consequence of Theorem 8.4.24, we have

Corollary 8.4.25. [379, Corollary 4.3] *$A = (a_{nk}) \in (c_0(F) : c_0(F))$ if and only if (8.4.20) holds and (8.4.22) also holds with $\alpha_k = \overline{0}$ for all $k \in \mathbb{N}$.*

Now, we give an example on the construction of an infinite matrix belonging to the class $(c_0(F) : c_0(F))$.

Example 8.4.26. [379, Example 4.4] *Consider the matrix $A = (a_{nk})$ defined by*

$$a_{nk} := \begin{cases} u_n & , \quad k \leq n, \\ \overline{0} & , \quad k > n, \end{cases}$$

where

$$u_k(t) := \begin{cases} (k+1)t - 1 & , \quad \frac{1}{k+1} \leq t < \frac{2}{k+1}, \\ 3 - (k+1)t & , \quad \frac{2}{k+1} < t \leq \frac{3}{k+1}, \\ 0 & , \quad \text{otherwise}. \end{cases}$$

Then, since

$$D(a_{nk}, \overline{0}) := \begin{cases} \frac{3}{n+1} & , \quad k \leq n, \\ 0 & , \quad k > n, \end{cases}$$

we have

$$\sup_{n\in\mathbb{N}}\sum_k D(a_{nk},\overline{0}) = \sup_{n\in\mathbb{N}}\sum_{k=0}^{n}\frac{3}{n+1} < \infty.$$

Additionally, since $u_k^-(\alpha) = (\alpha+1)/(k+1)$, $u_k^+(\alpha) = (3-\alpha)/(k+1)$ *and* $u_k \to \overline{0}$, *as* $k \to \infty$, *and* $a_{nk} \to \overline{0}$, *as* $n \to \infty$, *for each* $k \in \mathbb{N}$ *and each* $\alpha \in [0,1]$. *Hence, because of the matrix* A *satisfies the conditions of Corollary 8.4.25,* $A \in (c_0(F) : c_0(F))$.

Example 8.4.27. [379, Example 4.5] *Let us consider the sequence* $\{v_k(t)\}$ *defined by Rojes-Medar and Roman-Flores* [336], *as follows:*

$$v_k(t) := \begin{cases} 1-(k+1)t &, \quad 0 \le t \le \frac{1}{k+1}, \\ 0 &, \quad \text{otherwise} \end{cases}$$

for all $k \in \mathbb{N}$ *and try to obtain the A-transform of this sequence. Then, it is trivial that* $v_k \to \overline{0}$, *as* $k \to \infty$, *since* $v_k^-(\alpha) = 0$ *and* $v_k^+(\alpha) = (1-\alpha)/(k+1)$ *for* $\alpha \in [0,1]$. *Therefore, one can see that* $\sum_k (a_{nk}v_k)^-(\alpha) = 0$ *and*

$$\sum_k (a_{nk}v_k)^+(\alpha) = \sum_{k=0}^{n}(u_n v_k)^+(\alpha) = \frac{(3-\alpha)(1-\alpha)}{n+1}\sum_{k=0}^{n}\frac{1}{k+1},$$

which yields that $[(3-\alpha)(1-\alpha)/(n+1)]\sum_{k=0}^{n} 1/(k+1) \to 0$, *as* $n \to \infty$. *Thus,* $\sum_k a_{nk}v_k \to \overline{0}$, *as* $n \to \infty$. *Let us suppose that* $Av = u$. *Therefore, it is clear that* $u_n^-(\alpha) = 0$ *for every* $\alpha \in [0,1]$. *Then,* $u_n^+(\alpha) = (3-\alpha)(1-\alpha)b_n$ *with* $b_n = 1/(n+1)\sum_{k=0}^{n} 1/(k+1)$ *for all* $n \in \mathbb{N}$. *Thus, we obtain the A-transform of the sequence* (v_k) *as*

$$(Av)_n(t) := \begin{cases} 2-\sqrt{\frac{t}{b_n}+1} &, \quad 0 \le t \le 3b_n, \\ 0 &, \quad \text{otherwise.} \end{cases}$$

Theorem 8.4.28. [379, Theorem 4.6] *Let* $a_{nk} \succeq \overline{0}$ *for all* $k, n \in \mathbb{N}$. *Then,* $A = (a_{nk}) \in (c(F) : c(F); p)$ *if and only if (8.4.20) holds and*

$$\lim_{n\to\infty} a_{nk} = \overline{0} \text{ for each } k \in \mathbb{N}, \tag{8.4.23}$$

$$\lim_{n\to\infty}\sum_k a_{nk} = \overline{1}. \tag{8.4.24}$$

Proof. Suppose that (8.4.20), (8.4.23) and (8.4.24) hold, and $(u_k) \in c(F)$. Then, since Au exists the series $\sum_k a_{nk}u_k$ converges for each fixed $n \in \mathbb{N}$. Hence, $A_n \in \{c(F)\}^\beta$ for all $n \in \mathbb{N}$. It is obvious that (8.4.20) holds if and only if

$$\sup_{n\in\mathbb{N}}\sum_k \sup_{\alpha\in[0,1]} |a_{nk}^-(\alpha)| < \infty,$$

$$\sup_{n\in\mathbb{N}}\sum_k \sup_{\alpha\in[0,1]} |a_{nk}^+(\alpha)| < \infty,$$

(8.4.23) holds if and only if

$$\begin{aligned}\lim_{n\to\infty}\sup_{\alpha\in[0,1]}|a^-_{nk}(\alpha)| &= 0,\\ \lim_{n\to\infty}\sup_{\alpha\in[0,1]}|a^+_{nk}(\alpha)| &= 0\end{aligned}$$

and (8.4.24) holds if and only if

$$\begin{aligned}\lim_{n\to\infty}\sup_{\alpha\in[0,1]}\left|\sum_k a^-_{nk}(\alpha)-1\right| &= 0,\\ \lim_{n\to\infty}\sup_{\alpha\in[0,1]}\left|\sum_k a^+_{nk}(\alpha)-1\right| &= 0.\end{aligned}$$

By taking into account the above conditions, we see that the matrices $A^-(\alpha) = [a^-_{nk}(\alpha)]_{k,n\in\mathbb{N}}$ and $A^+(\alpha) = [a^+_{nk}(\alpha)]_{k,n\in\mathbb{N}}$ are regular for $\alpha \in [0,1]$. Now, suppose that $u_k \to u$, as $k \to \infty$. Then, $u^-_k(\alpha) \to u^-(\alpha)$, as $k \to \infty$ and $u^+_k(\alpha) \to u^+(\alpha)$, as $k \to \infty$, uniformly in α's. Since the matrices $A^-(\alpha)$ and $A^+(\alpha)$ are regular, we have uniformly that

$$\begin{aligned}\lim_{n\to\infty}\textstyle\sum_k a^-_{nk}(\alpha)u^-_k(\alpha) &= u^-(\alpha), & \lim_{n\to\infty}\textstyle\sum_k a^+_{nk}(\alpha)u^-_k(\alpha) &= u^-(\alpha),\\ \lim_{n\to\infty}\textstyle\sum_k a^-_{nk}(\alpha)u^+_k(\alpha) &= u^+(\alpha), & \lim_{n\to\infty}\textstyle\sum_k a^+_{nk}(\alpha)u^+_k(\alpha) &= u^+(\alpha)\end{aligned}$$

for $\alpha \in [0,1]$. Indeed since

$$\begin{aligned}&\left|\sum_k a^-_{nk}(\alpha)u^-_k(\alpha)-u^-(\alpha)\right| =\\ &= \left|\sum_k a^-_{nk}(\alpha)u^-_k(\alpha)-u^-(\alpha)\sum_k a^-_{nk}(\alpha)+u^-(\alpha)\sum_k a^-_{nk}(\alpha)-u^-(\alpha)\right|\\ &\le \left|\sum_k a^-_{nk}(\alpha)u^-_k(\alpha)-u^-(\alpha)\sum_k a^-_{nk}(\alpha)\right|+\left|u^-(\alpha)\sum_k a^-_{nk}(\alpha)-u^-(\alpha)\right|\\ &\le \sum_k |a^-_{nk}(\alpha)||u^-_k(\alpha)-u^-(\alpha)|+|u^-(\alpha)|\left|\sum_k a^-_{nk}(\alpha)-1\right|\\ &\le \sum_k \sup_{\alpha\in[0,1]}|a^-_{nk}(\alpha)|\sup_{\alpha\in[0,1]}|u^-_k(\alpha)-u^-(\alpha)|+\sup_{\alpha\in[0,1]}|u^-(\alpha)|\sup_{\alpha\in[0,1]}\left|\sum_k a^-_{nk}(\alpha)-1\right|,\end{aligned}$$

we have $\sup_{\alpha\in[0,1]}\left|\sum_k a^-_{nk}(\alpha)u^-_k(\alpha)-u^-(\alpha)\right| \to 0$, as $n \to \infty$. Since $a_{nk} \succeq \overline{0}$ for all $k, n \in \mathbb{N}$, $u^-_k(\alpha) \le u^+_k(\alpha)$ for $\alpha \in [0,1]$, which implies that

$$\begin{aligned}a^-_{nk}(\alpha)u^-_k(\alpha) &\le a^-_{nk}(\alpha)u^+_k(\alpha)\\ a^+_{nk}(\alpha)u^-_k(\alpha) &\le a^+_{nk}(\alpha)u^+_k(\alpha).\end{aligned}$$

This leads us to the consequence that

$$\begin{aligned}(a_{nk}u_k)^-(\alpha) &= \min\{a_{nk}^-(\alpha)u_k^-(\alpha), a_{nk}^-(\alpha)u_k^+(\alpha), a_{nk}^+(\alpha)u_k^-(\alpha), a_{nk}^+(\alpha)u_k^+(\alpha)\}\\ &= \min\{a_{nk}^-(\alpha)u_k^-(\alpha), a_{nk}^+(\alpha)u_k^-(\alpha)\}\\ (a_{nk}u_k)^+(\alpha) &= \max\{a_{nk}^-(\alpha)u_k^-(\alpha), a_{nk}^-(\alpha)u_k^+(\alpha), a_{nk}^+(\alpha)u_k^-(\alpha), a_{nk}^+(\alpha)u_k^+(\alpha)\}\\ &= \max\{a_{nk}^-(\alpha)u_k^+(\alpha), a_{nk}^+(\alpha)u_k^+(\alpha)\}.\end{aligned}$$

Consequently,

$$\lim_{n\to\infty}\sum_k(a_{nk}u_k)^-(\alpha) = \lim_{n\to\infty}\sum_k\min\{a_{nk}^-(\alpha)u_k^-(\alpha), a_{nk}^+(\alpha)u_k^-(\alpha)\} = u^-(\alpha),$$
$$\lim_{n\to\infty}\sum_k(a_{nk}u_k)^+(\alpha) = \lim_{n\to\infty}\sum_k\max\{a_{nk}^-(\alpha)u_k^+(\alpha), a_{nk}^+(\alpha)u_k^+(\alpha)\} = u^+(\alpha),$$

uniformly in α. Hence, $\sum_k a_{nk}u_k \to u$, as $n\to\infty$, as expected.

Conversely, let $A=(a_{nk})\in(c(F):c(F);p)$ and $u=(u_k)\in c(F)$. Then, since Au exists and the inclusion $(c(F):c(F);p)\subset(c(F):\ell_\infty(F))$ holds, the necessity of (8.4.20) is trivial by Part (ii) of Basic Theorem.

Consider the sequences $u^{(n)}=\{u_k^{(n)}\}$ defined by (8.4.10) and $z=(z_k)=(\overline{1})$ in the set $c(F)$. Then, the necessity of (8.4.23) and (8.4.24) are immediate, since $Au^{(n)}$, $Az\in c(F)$.

This step conclude the proof. □

We give an example of an infinite matrix in the class $(c(F):c(F);p)$, below:

Example 8.4.29. [379, Example 4.7] *Consider the matrix $A=(a_{nk})$ defined by*

$$a_{nk} := \begin{cases} u_k & , \quad k=n,\\ \overline{0} & , \quad k\neq n,\end{cases}$$

where

$$u_k(t) := \begin{cases} (k+1)(t-1) & , \quad 1\le t<1+\frac{1}{k+1},\\ 2-(k+1)(t-1) & , \quad 1+\frac{1}{k+1}\le t\le 1+\frac{2}{k+1},\\ 0 & , \quad \text{otherwise}.\end{cases}$$

Then, since

$$D(a_{nk},\overline{0}) := \begin{cases} 1+\frac{2}{n+1} & , \quad k=n,\\ 0 & , \quad k\neq n,\end{cases}$$

$\sup_{n\in\mathbb{N}}\sum_k D(a_{nk},\overline{0}) = \sup_{n\in\mathbb{N}}\{1+2/(n+1)\}<\infty$. *Additionally, since* $u_k^-(\alpha)=1+\alpha/(k+1)$, $u_k^+(\alpha)=1+(2-\alpha)/(k+1)$ *and* $u_k\to\overline{1}$, *as* $k\to\infty$, $a_{nk}\to\overline{0}$, *as* $n\to\infty$ *for each* $k\in\mathbb{N}$. $\lim_{n\to\infty}\sum_k a_{nk} = \lim_{n\to\infty}u_k=\overline{1}$. *Hence, A is a regular matrix since it satisfies the conditions of Theorem 8.4.28.*

Now, we have:

Theorem 8.4.30. [379, Theorem 4.8] $A = (a_{nk}) \in (\ell_p(F) : c(F))$ *if and only if (8.4.21) and (8.4.22) hold.*

Proof. Let $A = (a_{nk}) \in (\ell_p(F) : c(F))$ and $u = (u_k) \in \ell_p(F)$. Then, since Au exists and the inclusion $(\ell_p(F) : c(F)) \subset (\ell_p(F) : \ell_\infty(F))$ holds, the necessity of (8.4.21) is trivial by Part (iv) of Basic Theorem.

Consider the sequences $u^{(n)} = \{u_k^{(n)}\}$ in the set $\ell_p(F)$ defined by (8.4.10). Then, since $Au^{(n)} \in c(F)$, the necessity of (8.4.22) is clear.

Conversely suppose that the conditions (8.4.21) and (8.4.22) hold, and $u = (u_k) \in \ell_p(F) \subset c_0(F)$. Then, since Au exists and $A_n \in \{\ell_p(F)\}^\beta$ for each $n \in \mathbb{N}$ by taking into account the validity of (8.4.21) one can see by Hölder's inequality that

$$\begin{aligned}\sum_{k=0}^{m} D(\alpha_k u_k, \overline{0}) &= \lim_{n\to\infty} \sum_{k=0}^{m} D(a_{nk}u_k, \overline{0}) \\ &\leq \left\{\sum_k [D(u_k, \overline{0})]^p\right\}^{1/p} \left\{\sup_{n\in\mathbb{N}} \sum_k [D(a_{nk}, \overline{0}]^q\right\}^{1/q} < \infty\end{aligned}$$

for all $m \in \mathbb{N}$, which says us that $(\alpha_k) \in \ell_q(F)$. For simplicity in notation, here and after we write B instead of $\left\{\sum_k [D(\alpha_k, \overline{0})]^q\right\}^{1/q}$. Since $u = (u_k) \in \ell_p(F)$, one can choose a $k_0 \in \mathbb{N}$ for $\varepsilon > 0$ such that

$$\sum_{k=k_0+1}^{\infty} [D(u_k, \overline{0})]^p < \frac{\varepsilon^p}{2^p[C^{1/q} + B]^p}$$

for each fixed $k \geq k_0$. Additionally, since $a_{nk} \to \alpha_k$, as $n \to \infty$ by (8.4.22), we have $a_{nk}u_k \to \alpha_k u_k$, as $n \to \infty$ for each fixed $k \in \mathbb{N}$. That is to say that

$$\lim_{n\to\infty} D(a_{nk}u_k, \alpha_k u_k) = 0.$$

Hence, there exists an $N = N(k_0) \in \mathbb{N}$ such that

$$\sum_{k=0}^{k_0} D(a_{nk}u_k, \alpha_k u_k) < \frac{\varepsilon}{2}$$

for all $n \geq N$. Thus, since

$$\begin{aligned}
D\left(\sum_k a_{nk}u_k, \sum_k \alpha_k u_k\right) &\leq \sum_k D(a_{nk}u_k, \alpha_k u_k) \\
&= \sum_{k=0}^{k_0} D(a_{nk}u_k, \alpha_k u_k) + \sum_{k=k_0+1}^{\infty} D(a_{nk}u_k, \alpha_k u_k) \\
&\leq \frac{\varepsilon}{2} + \sum_{k=k_0+1}^{\infty} [D(a_{nk}u_k, \overline{0}) + D(\alpha_k u_k, \overline{0})] \\
&\leq \frac{\varepsilon}{2} + \sum_{k=k_0+1}^{\infty} D(a_{nk}, \overline{0})D(u_k, \overline{0}) + \sum_{k=k_0+1}^{\infty} D(\alpha_k, \overline{0})D(u_k, \overline{0}) \\
&\leq \frac{\varepsilon}{2} + \left\{\sum_{k=k_0+1}^{\infty} [D(u_k, \overline{0})]^p\right\}^{1/p} \left\{\sum_{k=k_0+1}^{\infty} [D(a_{nk}, \overline{0})]^q\right\}^{1/q} \\
&\quad + \left\{\sum_{k=k_0+1}^{\infty} [D(u_k, \overline{0})]^p\right\}^{1/p} \left\{\sum_{k=k_0+1}^{\infty} [D(\alpha_k, \overline{0}]^q\right\}^{1/q} \\
&\leq \frac{\varepsilon}{2} + \frac{\varepsilon}{2(C^{1/q} + B)}(C^{1/q} + B) = \varepsilon \text{ for all } n \geq N,
\end{aligned}$$

the series $\sum_k a_{nk}u_k$ are convergent for each $n \in \mathbb{N}$ and $D\left(\sum_k a_{nk}u_k, \sum_k \alpha_k u_k\right) \to 0$, as $n \to \infty$. This means that $Au \in c(F)$.

This step terminates the proof. □

Finally, we have:

Corollary 8.4.31. [379, Corollary 4.9] $A = (a_{nk}) \in (\ell_p(F) : c_0(F))$ *if and only if (8.4.21) holds and (8.4.22) also holds with* $\alpha_k = \overline{0}$ *for all* $k \in \mathbb{N}$.

Theorem 8.4.32. [379, Theorem 4.10] $A = (a_{nk}) \in (\ell_\infty(F) : c_0(F))$ *if and only if*

$$\lim_{n\to\infty} \sum_k D(a_{nk}, \overline{0}) = 0. \tag{8.4.25}$$

Proof. Let $A = (a_{nk}) \in (\ell_\infty(F) : c_0(F))$ and $u = (u_k) \in \ell_\infty(F)$. Then, the series $\sum_k a_{nk}u_k$ converges for each fixed $n \in \mathbb{N}$, since Au exists. Hence, $A_n \in \{\ell_\infty(F)\}^\beta$ for all $n \in \mathbb{N}$. Define the sequence $u = (u_k) \in \ell_\infty(F)$ by $u_k = \chi_{[-1,1]}$ for all $k \in \mathbb{N}$. Then, $Au \in c_0(F)$ which yields for all $n \in \mathbb{N}$ and for $\alpha \in [0, 1]$ that

$$\begin{aligned}
\sum_k (a_{nk}u_k)^+(\alpha) &= \sum_k \max\left\{a^-_{nk}(\alpha)u^-_k(\alpha), a^-_{nk}(\alpha)u^+_k(\alpha), a^+_{nk}(\alpha)u^-_k(\alpha), a^+_{nk}(\alpha)u^+_k(\alpha)\right\} \\
&= \sum_k \max\left\{ - a^-_{nk}(\alpha), -a^+_{nk}(\alpha), a^-_{nk}(\alpha), a^+_{nk}(\alpha)\right\} \\
&= \sum_k \max\left\{|a^-_{nk}(\alpha)|, |a^+_{nk}(\alpha)|\right\}.
\end{aligned}$$

In the special case $\alpha = 0$, we have

$$\lim_{n\to\infty} \sum_k D(a_{nk}, \overline{0}) = \lim_{n\to\infty} \sum_k \max\left\{|a^-_{nk}(0)|, |a^+_{nk}(0)|\right\} = 0.$$

Conversely suppose that (8.4.25) holds and $u = (u_k) \in \ell_\infty(F)$. Then, since $A_n \in \{\ell_\infty(F)\}^\beta$ for each $n \in \mathbb{N}$, Au exists. Therefore, one can observe by using Part (ii) of Lemma 8.2.6 together with the condition (8.4.25) that

$$\begin{aligned}\lim_{n\to\infty} D((Au)_n, \overline{0}) &= \lim_{n\to\infty} D\left(\sum_k a_{nk}u_k, \overline{0}\right)\\ &\leq \lim_{n\to\infty} \sum_k D(a_{nk}u_k, \overline{0})\\ &\leq \lim_{n\to\infty} \sum_k D(a_{nk}, \overline{0})D(u_k, \overline{0})\\ &\leq \sup_{k\in\mathbb{N}} D(u_k, \overline{0}) \lim_{n\to\infty} \sum_k D(a_{nk}, \overline{0}) = 0,\end{aligned}$$

which means that $Au \in c_0(F)$, as desired.

This step completes the proof. □

Theorem 8.4.33. [381, Theorem 4.5] *$A \in (\ell_1(F) : \ell_p(F))$ if and only if*

(i) $M := \sup_{k\in\mathbb{N}} \sum_n \|a_{nk}\|^p < \infty, \ (1 \leq p < \infty)$.

(ii) $\sup_{k,n\in\mathbb{N}} \|a_{nk}\| < \infty, \ (p = \infty)$.

Here and after, the norm $\|\cdot\|$ on the quasilinear space $L(\mathbb{R})$ is defined by (8.5.1).

Proof. Since the proofs of Part (i) and Part (ii) are similar, we give the proof only for Part (i).

(i) Suppose that $(u_k) \in \ell_1(F)$. Then, by taking into account the fact that the order of summation can be reversed by the absolute convergence of the series $\sum_n \sum_k a_{nk}u_k$ we obtain from Minkowski's inequality that

$$\begin{aligned}\left(\sum_n \left\|\sum_k a_{nk}u_k\right\|^p\right)^{1/p} &\leq \left[\sum_n \left(\sum_k \|a_{nk}u_k\|\right)^p\right]^{1/p}\\ &\leq \sum_k \left(\sum_n \|a_{nk}u_k\|^p\right)^{1/p}\\ &\leq \sum_k \|u_k\| \left(\sum_n \|a_{nk}\|^p\right)^{1/p}\\ &\leq M^{1/p}\|u\|_1 < \infty.\end{aligned}$$

This means that $A \in (\ell_1(F) : \ell_p(F))$.

Conversely, let us suppose that $A \in (\ell_1(F) : \ell_p(F))$. Then, since

$$\sum_i \|(Au)_i\|^p < \infty$$

for all $u \in \ell_1(F)$, the series $\sum_k a_{ik}u_k$ converges for all $u \in \ell_1(F)$ and for each $i \in \mathbb{N}$. Hence, $\sup_{k\in\mathbb{N}} \|a_{ik}\| < \infty$ by Theorem 8.4.16. Define the operators q_n's on the space $\ell_1(F)$ by

$$q_n(u) := \left[\sum_{i=0}^{n} \|(Au)_i\|^p\right]^{1/p} \quad \text{for all} \quad n \in \mathbb{N}.$$

The bounded quasilinearity of A_i's on the space $\ell_1(F)$ implies the bounded quasilinearity of q_n's on the space $\ell_1(F)$. Therefore, there exists a number $H > 0$ by Proposition 8.5.14 such that

$$\sup_{n\in\mathbb{N}} q_n(u) = \sum_i \|(Au)_i\|^p \leq H\|u\|_1. \tag{8.4.26}$$

Thus, (8.4.26) gives for the sequence $\{u_k^{(n)}\}$ defined by (8.4.10) that $\sum_{i=0}^{\infty} \|a_{ik}\|^p \leq H$, as desired.

This step completes the proof. □

Now, we prove that a matrix is in the class $(\ell_p(F) : \ell_p(F))$ if it is both in the classes $(\ell_\infty(F) : \ell_\infty(F))$ and $(\ell_1(F) : \ell_1(F))$ with $1 < p < \infty$:

Theorem 8.4.34. [381, Theorem 4.6] *Let $1 < p < \infty$. If $A \in (\ell_\infty(F) : \ell_\infty(F)) \cap (\ell_1(F) : \ell_1(F))$, then $A \in (\ell_p(F) : \ell_p(F))$.*

Proof. Suppose that $A = (a_{nk}) \in (\ell_\infty(F) : \ell_\infty(F)) \cap (\ell_1(F) : \ell_1(F))$ and $1 < p < \infty$. Then, Part (i) of Basic Theorem and Theorem 8.4.33 together yield that

$$M := \sup_{n\in\mathbb{N}} \sum_k \|a_{nk}\| < \infty, \; N := \sup_{k\in\mathbb{N}} \sum_n \|a_{nk}\| < \infty.$$

Therefore, we derive for $(u_k) \in \ell_p(F)$ that

$$\begin{aligned} \left\|\sum_k a_{nk}u_k\right\| &\leq \sum_k \|a_{nk}u_k\| \\ &\leq \sum_k \|a_{nk}\|^{1/p}\|a_{nk}\|^{1/q}\|u_k\| \\ &\leq \left(\sum_k \|a_{nk}\|\|u_k\|^p\right)^{1/p} \left(\sum_k \|a_{nk}\|\right)^{1/q} \end{aligned}$$

which leads us to the consequence that

$$\left\|\sum_k a_{nk}u_k\right\|^p \leq \left(\sum_k \|a_{nk}\|\|u_k\|^p\right) \left(\sum_k \|a_{nk}\|\right)^{p/q}.$$

Thus, since

$$\begin{aligned} \sum_n \|(Au)_n\|^p = \sum_n \left\|\sum_k a_{nk}u_k\right\|^p &\leq \sum_n \sum_k \|a_{nk}\|\|u_k\|^p M^{p/q} \\ &\leq M^{p/q} \sum_k \|u_k\|^p \sum_n \|a_{nk}\| \\ &\leq M^{p/q} N \sum_k \|u_k\|^p < \infty, \end{aligned}$$

$Au \in \ell_p(F)$, as desired. □

8.5 QUASILINEARITY OF THE CLASSICAL SETS OF SEQUENCES OF FUZZY NUMBERS

In the present section, we prove that the classical sets $\ell_\infty(F)$, $c(F)$, $c_0(F)$ and $\ell_p(F)$ of sequences of fuzzy numbers are normed quasilinear spaces. Besides this, we show that $\ell_\infty(F)$ and $c(F)$ are normed quasialgebras and an operator defined by an infinite matrix belonging to the class $(\ell_\infty(F) : \ell_\infty(F))$ is bounded and quasilinear.

In [39], Aseev introduced the concepts of quasilinear space and quasilinear operator which enable us to consider both linear and nonlinear spaces of subsets and multivalued mappings from a single point of view. Following Aseev [39], Rojas-Medar et al. have recently extended some results of linear functional analysis to the fuzzy context, in [337]. Their work has motivated Talo and Başar to study the quasilinearity of the classical sets of sequences of fuzzy numbers, in [381]. The main emphasis of this section is to study the quasilinearity of the classical sets $\ell_\infty(F)$, $c(F)$, $c_0(F)$ and $\ell_p(F)$ of sequences of fuzzy numbers and to prove that $\ell_\infty(F)$ and $c(F)$ are normed quasialgebras and an operator defined by an infinite matrix belonging to the class $(\ell_\infty(F) : \ell_\infty(F))$ is bounded and quasilinear.

Following Aseev [39], we begin with defining the concepts of quasilinear space, normed quasilinear space and quasilinear operator:

Definition 8.5.1. *A set X is called a quasilinear space if a partial order relation $\preceq$, an algebraic sum operation $(+)$ and an operation of multiplication by real numbers $(\cdot)$ are defined on it and satisfy the following properties for any elements $x, y, z, v \in X$ and any real numbers α, β:*

(q.1) $x \preceq x$.

(q.2) $x \preceq y,\ y \preceq z \Rightarrow x \preceq z$.

(q.3) $x \preceq y,\ y \preceq x \Rightarrow x = y$.

(q.4) $x + y = y + x$.

(q.5) $x + (y + z) = (x + y) + z$.

(q.6) *There exists an element $\theta \in X$, called neutral element, such that $x + \theta = x$.*

(q.7) $\alpha \cdot (\beta \cdot x) = (\alpha \cdot \beta) \cdot x$.

(q.8) $\alpha \cdot (x + y) = \alpha \cdot x + \alpha \cdot y$.

(q.9) $1 \cdot x = x$.

(q.10) $0 \cdot x = \theta$.

(q.11) $(\alpha + \beta) \cdot x \preceq \alpha \cdot x + \beta \cdot x$.

(q.12) $x \preceq y$ *and* $z \preceq v \Rightarrow x + z \preceq y + v$.

(q.13) $x \preceq y \Rightarrow \alpha \cdot x \preceq \alpha \cdot y$.

This quasilinear space will be denoted by $(X, \preceq, +, \cdot)$.

An element $x' \in X$ is called an *inverse* of $x \in X$, if $x + x' = \theta$. Obviously, if an element x has an inverse x', then it is unique. If any element x in the quasilinear space X has an inverse element $x' \in X$, then the partial order on X is determined by equality and consequently X is a linear space with scalars in $\mathbb{R}$.

Definition 8.5.2. *Let X be a quasilinear space. A real function $\|\cdot\|_X : X \to \mathbb{R}$ is called a norm if the following conditions hold for any $x, y \in X$ and any $\alpha \in \mathbb{R}$:*

(n.1) $\|x\|_X > 0$ *if* $x \neq \theta$.

(n.2) $\|x + y\|_X \leq \|x\|_X + \|y\|_X$.

(n.3) $\|\alpha \cdot x\|_X = |\alpha| \cdot \|x\|_X$.

(n.4) *If* $x \preceq y$, *then* $\|x\|_X \leq \|y\|_X$.

(n.5) *If for any* $\varepsilon > 0$ *there exists an element* $x_\varepsilon \in X$ *such that* $x \preceq y + x_\varepsilon$ *and* $\|x_\varepsilon\|_X \leq \varepsilon$ *then* $x \preceq y$.
A quasilinear space X with a norm defined on it, is called a normed quasilinear space.

If any $x \in X$ has an inverse element $x' \in X$, then the concept of a normed quasilinear space coincides with the concept of a real normed linear space.

The Hausdorff metric H on a normed quasilinear space X is defined by

$$H(x, y) := \inf \left\{ r \geq 0 : x \preceq y + a_1^r,\ y \preceq x + a_2^r,\ a_i^r \in X,\ \|a_i^r\| \leq r,\ i = 1, 2 \right\}.$$

Lemma 8.5.3. [381, Lemma 2.3] *The operations of algebraic sum and multiplication by real numbers are continuous with respect to the Hausdorff metric. The norm is a continuous function with respect to the Hausdorff metric.*

Lemma 8.5.4. [381, Lemma 2.4] *The following statements hold:*

(i) *Suppose that* $x_k \to x$ *and* $y_k \to y$, *as* $k \to \infty$ *and that* $x_k \preceq y_k$ *for any positive integer* k. *Then,* $x \preceq y$.

(ii) *Suppose that* $x_k \to x$ *and* $z_k \to x$, *as* $k \to \infty$. *If* $x_k \preceq y_k \preceq z_k$ *for any positive integer* k, *then* $y_k \to x$, *as* $k \to \infty$.

(iii) *If* $x_k + y_k \to x$ *and* $y_k \to \theta$, *as* $k \to \infty$, *then* $x_k \to x$, *as* $k \to \infty$.

The set W of all closed bounded intervals is a quasilinear space with the algebraic operations, above. Define the function $\|\cdot\|$ on W by

$$\|A\| := \max \left\{ |\underline{A}|, |\overline{A}| \right\}.$$

Then, W is a normed quasilinear space with this norm. The Hausdorff metric d obtained from this norm, on W is defined by (8.2.1). Now, we define the quasilinearity of an operator.

Definition 8.5.5. [381, Definition 2.5] *Let X and Y be quasilinear spaces. If the operator $T : X \to Y$ satisfies the following conditions:*

(o.1) $T(\alpha \cdot x) = \alpha \cdot Tx$

(o.2) $T(x + y) \preceq Tx + Ty$

(o.3) *If $x \preceq y$, then $Tx \preceq Ty$*

for all $x, y \in X$ and for all $\alpha \in \mathbb{R}$, then T is called a quasilinear operator. By $L(X : Y)$, we denote the set of all quasilinear operators from X to Y. Let $T_1, T_2 \in L(X : Y)$. Then, we write $T_1 \preceq T_2$ whenever $T_1x \preceq T_2x$ for all $x \in X$. It is easy to see that $L(X : Y)$ is a quasilinear space with the usual algebraic operations addition and scalar multiplication of operators.

A quasilinear operator $T : X \to Y$ is said to be bounded if and only if there exists a $K > 0$ such that $\|Tx\|_Y \leq K\|x\|_X$ for all $x \in X$. By $B(X : Y)$, we denote the set of all bounded quasilinear operators from X to Y. It is natural that the order on the set $B(X : Y)$ can be defined by the similar way used in Definition 8.5.5, above. It is not hard to show that $B(X : Y)$ is a normed quasilinear space with the usual algebraic operations addition and scalar multiplication of operators, normed by

$$\|T\| := \sup_{\|x\|_X = 1} \|Tx\|_Y.$$

Now, we give four known propositions concerned some properties of quasilinear operators of which the final one is analogous to the Banach-Steinhaus theorem.

Proposition 8.5.6. [381, Proposition 2.6] *Let $T \in L(X : Y)$. Then, the following statements hold:*

(i) *$T\theta = \theta$.*

(ii) *T is bounded if and only if T is continuous on $\theta \in X$.*

(iii) *If T is continuous at $\theta \in X$, then T is uniformly continuous on X.*

Proposition 8.5.7. [381, Proposition 2.7] *The following statements hold:*

(a) *Let $T \in B(X : Y)$. Then, the Lipschitz condition $\|Tx\|_Y \leq \|T\|_{B(X:Y)}\|x\|_X$ holds for all $x \in X$.*

(b) *The composition $T_2 \circ T_1$ is in the quasilinear space $B(X : Z)$ if $T_1 \in B(X : Y)$ and $T_2 \in B(Y : Z)$.*

Proposition 8.5.8. [381, Proposition 2.8] *If the sequence $\{T_k\} \subset B(X : Y)$ is convergent for each $x \in X$, then the operator T defined by $Tx := \lim_{k\to\infty} T_kx$ is quasilinear.*

The operator T does not need to be bounded. But, if the space X is complete with respect to the metric defined on it, then $T \in B(X : Y)$.

Proposition 8.5.9. [381, Proposition 2.9] *Suppose that X is a complete normed quasilinear space and Y a normed quasilinear space. Assume that the sequence $\{T_k\}$ of elements of the space $B(X:Y)$ is bounded at each point $x \in X$. Then, the sequence $\{\|T_k\|\}$ of norms is also bounded.*

It is known that $L(\mathbb{R})$ is a quasilinear space with the partial ordering relation and the algebraic operations, defined above, (see Rojas-Medar et al. [337]). The unit element with respect to the operation $(+)$ of the space $L(\mathbb{R})$ is $\overline{0}$. We define the norm on $L(\mathbb{R})$ by

$$\|u\| := \sup_{\alpha\in[0,1]} \max\left\{|u^-(\alpha)|, |u^+(\alpha)|\right\} = \max\left\{|u^-(0)|, |u^+(0)|\right\}. \tag{8.5.1}$$

Thus, $L(\mathbb{R})$ is a normed quasilinear space endowed with the norm defined by (8.5.1). The Hausdorff metric D obtained from the norm given by (8.5.1) is defined by

$$\begin{aligned} D(u,v) &= \sup_{\alpha\in[0,1]} d([u]_\alpha,[v]_\alpha) \qquad (8.5.2)\\ &= \sup_{\alpha\in[0,1]} \max\left\{|u^-(\alpha)-v^-(\alpha)|, |u^+(\alpha)-v^+(\alpha)|\right\}. \end{aligned}$$

It is immediate that $\|\cdot\| = D(\cdot,\overline{0})$.

8.5.1 The Quasilinearity of the Classical Sets of Sequences of Fuzzy Numbers

In this subsection, our purpose is to give the basic theorem on the quasilinearity of the classical sets of sequences of fuzzy numbers together with some other results.

The addition $(+)$ and scalar multiplication $(\cdot)$ are defined on $\omega(F)$ by

$$(u_k) + (v_k) = (u_k + v_k) \text{ and } \alpha(u_k) = (\alpha u_k),$$

as usual, for (u_k), $(v_k) \in \omega(F)$ and $\alpha \in \mathbb{R}$. $\theta = (\overline{0})$ is the unit element of the space $\omega(F)$ with respect to addition.

Let us define the partial ordering relation $\preceq$ on $\omega(F)$, as follows: Let $u = (u_k)$, $v = (v_k) \in \omega(F)$. Then,

$$u \preceq v \Leftrightarrow u_k \subseteq v_k \text{ for all } k \in \mathbb{N}. \tag{8.5.3}$$

It is a routine verification that $\omega(F)$ is a quasilinear space with the partial ordering relation defined by (8.5.3) and the usual algebraic operations addition and scalar multiplication.

Now, we give the main theorem:

Main Theorem. [381] *The following statements hold:*

(i) *Let $\mu(F)$ denotes any of the sets $\ell_\infty(F)$, $c(F)$ and $c_0(F)$. Then, the set $\mu(F)$ is a quasilinear space with the partial ordering relation $\preceq$ defined by (8.5.3) and usual algebraic operations addition, scalar multiplication. Additionally, the set $\mu(F)$ is a normed quasilinear space endowed with the norm $\|\cdot\|_\infty$, defined by $\|u\|_\infty := \sup_{k\in\mathbb{N}} \|u_k\|$, where $u = (u_k) \in \mu(F)$.*

(ii) *The set $\ell_p(F)$ is a quasilinear space with the partial ordering relation defined by (8.5.3) and usual algebraic operations addition, scalar multiplication. Additionally, the set $\ell_p(F)$ is a normed quasilinear space endowed with the norm $\|\cdot\|_p$, defined by*

$$\|u\|_p := \left(\sum_k \|u_k\|^p\right)^{1/p}, \quad \textit{where} \quad u = (u_k) \in \ell_p(F). \tag{8.5.4}$$

Proof. Since the proofs are similar for the sets $c_0(F)$, $c(F)$, $\ell_\infty(F)$ and $\ell_p(F)$, we give the proof only for the set $\ell_p(F)$.

(ii) Firstly, we establish that the set $\ell_p(F)$ is a quasilinear space. Since the inclusion $\ell_p(F) \subseteq \omega(F)$ holds, the conditions (q.1)–(q.13) are directly satisfied. So, it is sufficient to show that the set $\ell_p(F)$ is closed under the co-ordinatewise operations addition and scalar multiplication.

Let $u = (u_k)$, $v = (v_k) \in \ell_p(F)$ and $\alpha \in \mathbb{R}$. Then, since

$$\begin{aligned} \left(\sum_k \|u_k + v_k\|^p\right)^{1/p} &\leq \left[\sum_k (\|u_k\| + \|v_k\|)^p\right]^{1/p} \\ &\leq \left(\sum_k \|u_k\|^p\right)^{1/p} + \left(\sum_k \|v_k\|^p\right)^{1/p} < \infty \end{aligned} \tag{8.5.5}$$

and

$$\sum_k \|\alpha u_k\|^p = |\alpha|^p \sum_k \|u_k\|^p < \infty, \tag{8.5.6}$$

$u + v \in \ell_p(F)$ and $\alpha u \in \ell_p(F)$. It is trivial that $\theta = (\overline{0}) \in \ell_p(F)$. Hence, $\ell_p(F)$ is a quasilinear space.

Finally, we should show that the relation (8.5.4) satisfies the conditions (n.1)–(n.5) on the set $\ell_p(F)$.

(a) Let $u \in \ell_p(F) \setminus \{\theta\}$. Then, there exists a $k \in \mathbb{N}$ such that $u_k \neq \overline{0}$, i.e., $\|u_k\| > 0$. Thus, one can easily see that $\|u\|_p = (\sum_k \|u_k\|^p)^{1/p} \geq \|u_k\|^{1/p} > 0$, i.e., (n.1) holds.

(b) (8.5.5) and (8.5.6) give the conditions (n.2) and (n.3).

(c) Let $u \preceq v$. Then, $u_k \subseteq v_k$ for all $k \in \mathbb{N}$ which gives us $\|u_k\| \leq \|v_k\|$. Therefore, it is immediate that $\|u\|_p \leq \|v\|_p$, i.e., (n.4) holds.

(d) Let us suppose that for an $\varepsilon > 0$, there is an u_ε in the space $\ell_p(F)$ such that $u \preceq v + u_\varepsilon$ and $\|u_\varepsilon\|_p = \left(\sum_k \|u_\varepsilon^k\|^p\right)^{1/p} \leq \varepsilon$. Then, $u_k \subseteq v_k + u_\varepsilon^k$ for all $k \in \mathbb{N}$ and $\|u_\varepsilon^k\| \leq \varepsilon$. Therefore, we have $u_k \subseteq v_k$ for each $k \in \mathbb{N}$. This means that $u \preceq v$, i.e., (n.5) holds.

This step completes the proof of Part (ii). □

Now, we may give the definition of the concepts of quasialgebra and normed quasialgebra for the quasilinear spaces.

Definition 8.5.10. [381, Definition 3.1] *Let $(X, \preceq, +, \cdot)$ be a quasilinear space. Then, the set X is called an quasialgebra if the following conditions hold for the operation $(\star)$ defined on X:*

(c.1) $(\alpha \cdot x) \star y = \alpha \cdot (x \star y)$.

(c.2) $x \star (y \star z) = (x \star y) \star z$.

(c.3) $x \star (y + z) \preceq (x \star y) + (x \star z)$ *and* $(y + z) \star x \preceq (y \star x) + (z \star x)$

for any $x, y, z \in X$ and any $\alpha \in \mathbb{R}$.

Definition 8.5.11. [381, Definition 3.2] *Let the normed quasilinear space $(X, \|\cdot\|)$ be a quasialgebra with the well–defined operation $(\cdot)$ on X. If $\|x \cdot y\| \leq \|x\|\|y\|$ and $\|1\| = 1$ for all x, $y \in X$, then the space X is called as a normed quasialgebra.*

It is trivial that $L(\mathbb{R})$ is a normed quasialgebra and the normed quasilinear space $\omega(F)$ is also a quasialgebra with

$$u \cdot v = (u_k) \cdot (v_k) = (u_k \cdot v_k);\ u = (u_k), v = (v_k) \in \omega(F).$$

The unit element with respect to the multiplication is $1 = (\overline{1})$.

Now, we may give:

Theorem 8.5.12. [381, Theorem 3.4] *The normed quasilinear spaces $\ell_\infty(F)$ and $c(F)$ are normed quasialgebras with $\|\cdot\|_\infty$.*

The spaces $c_0(F)$ and $\ell_p(F)$ do not have a unit element with respect to the multiplication.

Now, we give the theorem on the boundedness and the quasilinearity of an operator defined by an infinite matrix in the class $(\ell_\infty(F) : \ell_\infty(F))$.

Theorem 8.5.13. [381, Theorem 3.6] *The operator defined by an infinite matrix $A = (a_{nk})$ from $\ell_\infty(F)$ to $\ell_\infty(F)$ is bounded and quasilinear.*

Proof. A-transform of a sequence $u \in \ell_\infty(F)$ is the sequence $Au = \{(Au)_n\}$ defined by (8.4.19). Let us show that this transformation satisfies the conditions (o.1)–(o.3). Suppose that $u = (u_k)$, $v = (v_k) \in \ell_\infty(F)$ and $\alpha \in \mathbb{R}$.

(i) Since

$$\{A(\alpha u)\}_n = \sum_k a_{nk} \cdot (\alpha u_k) = \alpha \sum_k a_{nk} \cdot u_k = \alpha (Au)_n$$

for all $n \in \mathbb{N}$, it is obvious that $A(\alpha u) = \alpha Au$, i.e., (o.1) is satisfied.

(ii) Because of

$$\begin{aligned} & a_{nk} \cdot (u_k + v_k) \subseteq a_{nk} \cdot u_k + a_{nk} \cdot v_k \\ \Rightarrow\ & \sum_{k=0}^{m} a_{nk} \cdot (u_k + v_k) \subseteq \sum_{k=0}^{m} a_{nk} \cdot u_k + \sum_{k=0}^{m} a_{nk} \cdot v_k \\ \Rightarrow\ & \lim_{m\to\infty} \sum_{k=0}^{m} a_{nk} \cdot (u_k + v_k) \subseteq \lim_{m\to\infty} \sum_{k=0}^{m} a_{nk} \cdot u_k + \lim_{m\to\infty} \sum_{k=0}^{m} a_{nk} \cdot v_k \\ \Rightarrow\ & \sum_k a_{nk} \cdot (u_k + v_k) \subseteq \sum_k a_{nk} \cdot u_k + \sum_k a_{nk} \cdot v_k \\ \Rightarrow\ & \{A(u+v)\}_n \subseteq (Au)_n + (Av)_n, \end{aligned}$$

one can observe that $A(u + v) \preceq (Au + Av)$. That is, (o.2) is also satisfied.

(iii) Suppose that $u \preceq v$. Then, $u_k \subseteq v_k$ and hence $a_{nk} \cdot u_k \subseteq a_{nk} \cdot v_k$ for all $k \in \mathbb{N}$ and for each fixed $n \in \mathbb{N}$. Therefore, one can easily establish by analogy to Part (ii) that $Au \preceq Av$, i.e., (o.3) holds.

Finally, we prove that the quasilinear operator A is bounded. By considering the fact $\sup_{n\in\mathbb{N}} \|(Au)_n\| < \infty$ by Part (i) of Basic Theorem for all $u \in \ell_\infty(F)$ one can see by Part (i) of Lemma 2.6 of Talo and Başar [381] that

$$\begin{aligned}
\sup_{n\in\mathbb{N}} \|(Au)_n\| &= \sup_{n\in\mathbb{N}} \left\| \sum_k a_{nk} \cdot u_k \right\| \\
&\leq \sup_{n\in\mathbb{N}} \sum_k \|a_{nk} \cdot u_k\| \\
&\leq \sup_{n\in\mathbb{N}} \sum_k \|a_{nk}\| \sup_{k\in\mathbb{N}} \|u_k\| \\
&= \|u\|_\infty \left(\sup_{n\in\mathbb{N}} \sum_k \|a_{nk}\| \right).
\end{aligned}$$

Thus, we conclude by taking supremum over $\|u\|_\infty = 1$ that $\|A\| \leq \sup_{n\in\mathbb{N}} \sum_k \|a_{nk}\|$ what we wished to prove. □

Proposition 8.5.14. [381, Proposition 3.7] *Let $\mu_i(F)$ denotes any of the spaces $\ell_\infty(F)$, $c(F)$, $c_0(F)$ and $\ell_p(F)$ for $i = 1, 2$. Suppose that $(\mu_1(F), D)$ is a complete metric space and $\{T_n\} \subset L(\mu_1(F) : \mu_2(F))$ such that $\sup_{n\in\mathbb{N}} \|T_n x\|_{\mu_2(F)} < \infty$ for all $x \in \mu_1(F)$. Then, there exists a number $C > 0$ such that $\|T_n x\|_{\mu_2(F)} \leq C\|x\|_{\mu_1(F)}$ for each $n \in \mathbb{N}$ and for all $x \in \mu_1(F)$.*

Proof. Since $\mu_1(F)$ is one of the quasilinear spaces $\ell_\infty(F)$, $c(F)$, $c_0(F)$ and $\ell_p(F)$, the pointwise boundedness of the sequence $\{T_n\}$ of operators between these spaces implies its uniform boundedness, as was desired. □

8.6 CERTAIN SETS OF SEQUENCES OF FUZZY NUMBERS DEFINED BY A MODULUS

In this final section, following Talo and Başar [380], we introduce the sets $\ell_\infty(F, f)$, $c(F, f)$, $c_0(F, f)$ and $\ell_p(F, f, s)$ of sequences of fuzzy numbers derived from the classical sets $\ell_\infty(F)$, $c(F)$, $c_0(F)$ and $\ell_p(F)$ of sequences of fuzzy numbers by a modulus function f.

In the present section, we essentially deal with the metric spaces $\ell_\infty(F, f)$, $c(F, f)$, $c_0(F, f)$ and $\ell_p(F, f, s)$ of sequences of fuzzy numbers defined by a modulus function f which are the generalization of the metric spaces $\ell_\infty(F)$, $c(F)$, $c_0(F)$ and $\ell_p(F)$ of sequences of fuzzy numbers. Additionally, we state and prove some inclusion theorems related to those sets. Finally, we establish that the sets $\ell_\infty(F, f)$, $c_0(F, f)$ and $\ell_p(F, f, s)$ are solid as a consequence of the fact that the sets $\ell_\infty(F)$, $c_0(F)$ and $\ell_p(F)$ are solid.

8.6.1 The Spaces of Sequences of Fuzzy Numbers Defined by a Modulus Function

Let f be a modulus function. We introduce the sets $\ell_\infty(F,f)$, $c(F,f)$, $c_0(F,f)$ and $\ell_p(F,f,s)$ of sequences of fuzzy numbers defined by f, as follows;

$$\begin{aligned}
\ell_\infty(F,f) &:= \left\{x=(x_k)\in\omega(F) : \sup_{k\in\mathbb{N}} f[D(x_k,\overline{0})]<\infty\right\},\\
c(F,f) &:= \left\{x=(x_k)\in\omega(F) : \lim_{k\to\infty} f[D(x_k,l)]=0 \text{ for some } l\in L(\mathbb{R})\right\},\\
c_0(F,f) &:= \left\{x=(x_k)\in\omega(F) : \lim_{k\to\infty} f[D(x_k,\overline{0})]=0\right\},\\
\ell_p(F,f,s) &:= \left\{x=(x_k)\in\omega(F) : \sum_k \frac{\{f[D(x_k,\overline{0})]\}^p}{k^s}<\infty\right\},\ (s\geq 0).
\end{aligned}$$

Now, we may begin with the following theorem which is essential in this section:

Theorem 8.6.1. [380, Theorem 2.1] *The sets $\ell_\infty(F,f)$, $c(F,f)$, $c_0(F,f)$ and $\ell_p(F,f,s)$ of sequences of fuzzy numbers defined by a modulus function f are closed under the co-ordinatewise addition and scalar multiplication of sequences.*

Proof. Since it is not hard to show that the sets $\ell_\infty(F,f)$, $c(F,f)$, $c_0(F,f)$ and $\ell_p(F,f,s)$ are closed with respect to the co-ordinatewise addition and scalar multiplication of sequences, we leave the detailed proof to the reader. □

Theorem 8.6.2. [380, Theorem 2.2] *The sets $\ell_\infty(F,f)$, $c(F,f)$, $c_0(F,f)$ and $\ell_p(F,f,s)$ of sequences of fuzzy numbers defined by a modulus function f are complete metric spaces with respect to the metrics $\overline{D}_\infty$ and $\overline{D}_p$ defined by*

$$\begin{aligned}
\overline{D}_\infty(x,y) &:= \sup_{k\in\mathbb{N}} f[D(x_k,y_k)],\\
\overline{D}_p(x,y) &:= \left\{\sum_k \frac{\{f[D(x_k,y_k)]\}^p}{k^s}\right\}^{1/p},
\end{aligned}$$

respectively; where $x=(x_k)$, $y=(y_k)$ are the elements of the sets $\ell_\infty(F,f)$, $c(F,f)$, $c_0(F,f)$ or $\ell_p(F,f,s)$.

Proof. Since the proof is analogue for the spaces $\ell_\infty(F,f)$, $c(F,f)$ and $\ell_p(F,f,s)$, we consider only the space $c_0(F,f)$. One can easily establish that $\overline{D}_\infty$ defines a metric on $c_0(F,f)$ which is a routine verification, so we omit details.

It remains to prove the completeness of the space $c_0(F,f)$. Let $\{x^i\}$ be any Cauchy sequence in the space $c_0(F,f)$, where $x^i=\left\{x_0^{(i)},x_1^{(i)},x_2^{(i)},\ldots\right\}$. Then, for a given $\varepsilon>0$ there exists a natural number $n_0(\varepsilon)$ such that

$$\overline{D}_\infty(x^i,x^j)=\sup_{k\in\mathbb{N}} f\left[D\left(x_k^{(i)},x_k^{(j)}\right)\right]<\varepsilon \tag{8.6.1}$$

for all $i,j\geq n_0(\varepsilon)$. We obtain for each fixed $k\in\mathbb{N}$ from (8.6.1) that

$$f\left[D\left(x_k^{(i)},x_k^{(j)}\right)\right]<\varepsilon \tag{8.6.2}$$

for all $i, j \geq n_0(\varepsilon)$. (8.6.2) means that

$$\lim_{i,j\to\infty} f\left[D\left(x_k^{(i)}, x_k^{(j)}\right)\right] = 0. \tag{8.6.3}$$

Since f is continuous, we have from (8.6.3) that

$$f\left[\lim_{i,j\to\infty} D\left(x_k^{(i)}, x_k^{(j)}\right)\right] = 0. \tag{8.6.4}$$

Therefore, since f is a modulus function one can derive by (8.6.4) that

$$\lim_{i,j\to\infty} D\left(x_k^{(i)}, x_k^{(j)}\right) = 0, \tag{8.6.5}$$

which means that $\left\{x_k^{(i)}\right\}$ is a Cauchy sequence in $L(\mathbb{R})$ for each fixed $k \in \mathbb{N}$. Since $L(\mathbb{R})$ is complete, it converges, say $x_k^{(i)} \to x_k$, as $i \to \infty$ for each fixed $k \in \mathbb{N}$. Using these infinitely many limits, we define the sequence $x = (x_0, x_1, x_2, \ldots)$. Let us pass to limit firstly as $j \to \infty$ and next take supremum over $k \in \mathbb{N}$ in (8.6.2) to obtain $\overline{D}_\infty(x^i, x) \leq \varepsilon/2$. Since $x^i = \left\{x_k^{(i)}\right\} \in c_0(F, f)$ for each $i \in \mathbb{N}$, there exists $k_0(\varepsilon) \in \mathbb{N}$ such that $f\left[D\left(x_k^{(i)}, \overline{0}\right)\right] \leq \varepsilon/2$ for all $k \geq k_0(\varepsilon)$ and for each fixed $i \in \mathbb{N}$. Therefore, since

$$f\left[D(x_k, \overline{0})\right] < f\left[D\left(x_k^{(i)}, x_k\right)\right] + f\left[D\left(x_k^{(i)}, \overline{0}\right)\right]$$

holds by triangle inequality for all $i, k \in \mathbb{N}$ for fixed $i \geq n_0(\varepsilon)$, we have

$$f\left[D(x_k, \overline{0})\right] \leq \varepsilon \text{ for all } k \geq k_0(\varepsilon).$$

This shows that $x = (x_k) \in c_0(F, f)$. Since $\{x^i\}$ was an arbitrary Cauchy sequence, the space $c_0(F, f)$ is complete.

This step concludes the proof. □

Theorem 8.6.3. [380, Theorem 2.3] *Let $\lambda(F, f)$ denotes the any of the spaces $\ell_\infty(F, f)$, $c(F, f)$, $c_0(F, f)$ and f_1, f_2 be two modulus functions. Then, the following inclusion relations hold:*

(a) $\lambda(F, f_1) \cap \lambda(F, f_2) \subseteq \lambda(F, f_1 + f_2)$.

(b) $\lambda(F, f_1) \subseteq \lambda(F, f_2 \circ f_1)$.

(c) $\lambda(F, f_2) \subseteq \lambda(F, f_1)$, *if $f_1(t) \leq f_2(t)$ for all $t \in [0, \infty)$.*

Proof. (a) Let $x = (x_k) \in \lambda(F, f_1) \cap \lambda(F, f_2)$. Since

$$(f_1 + f_2)[D(x_k, \overline{0})] = f_1[D(x_k, \overline{0})] + f_2[D(x_k, \overline{0})], \tag{8.6.6}$$

one can see by passing to limit as $k \to \infty$ or taking supremum over $k \in \mathbb{N}$ from (8.6.6) that $x \in \lambda(F, f_1 + f_2)$, where $\lambda \in \{\ell_\infty, c, c_0\}$.

(b) Let $x = (x_k) \in \lambda(F, f_1)$. Since f_2 is continuous, there exists an $\eta > 0$ such that $f_2(\eta) < \varepsilon$ for all $\varepsilon > 0$. Since $x = (x_k) \in \lambda(F, f_1)$, there exists an $k_0 \in \mathbb{N}$ such that $f_1[D(x_k, \overline{0})] < \eta$ for all $k \geq k_0$. Therefore, one can derive by applying f_2 that $f_2\{f_1[D(x_k, \overline{0})]\} < f_2(\eta) < \varepsilon$, i.e., $x \in \lambda(F, f_2 \circ f_1)$.

(c) Since $f_1(t) \leq f_2(t)$ for all $t \in [0, \infty)$, we have $f_1[D(x_k, \overline{0})] \leq f_2[D(x_k, \overline{0})]$. This leads us to the consequence that $x \in \lambda(F, f_2)$ which implies that $x \in \lambda(F, f_1)$, as expected. □

Lemma 8.6.4. (cf. [257]) *Let f_1 and f_2 be two modulus functions, and $0 < \delta < 1$. If $f_1(t) > \delta$, then*

$$(f_2 \circ f_1)(t) \leq \frac{2f_2(1)}{\delta} f_1(t) \tag{8.6.7}$$

holds for all $t \in [0, \infty)$.

Theorem 8.6.5. [380, Theorem 2.5] *f_1 and f_2 be two modulus functions. Then, the following inclusion relations hold:*

(a) $\ell_p(F, f_1, s) \cap \ell_p(F, f_2, s) \subseteq \ell_p(F, f_1 + f_2, s)$.

(b) *If* $s > 1$, $\ell_p(F, f_1, s) \subseteq \ell_p(F, f_2 \circ f_1, s)$.

(c) *If* $\limsup_{t \to \infty} [f_1(t)/f_2(t)] < \infty$, $\ell_p(F, f_2, s) \subseteq \ell_p(F, f_1, s)$.

(d) *If* $s_1 \leq s_2$, $\ell_p(F, f_1, s_1) \subseteq \ell_p(F, f_1, s_2)$.

Proof. (a) Let $x = (x_k) \in \ell_p(F, f_1, s) \cap \ell_p(F, f_2, s)$. Then, one can see that the inequality

$$\{(f_1 + f_2)[D(x_k, \overline{0})]\}^p \leq 2^{p-1} \{[f_1[D(x_k, \overline{0})]]^p + [f_2[D(x_k, \overline{0})]]^p\}$$

holds which yields us by taking summation over $k \in \mathbb{N}$ that $x = (x_k) \in \ell_p(F, f_1 + f_2, s)$, as desired.

(b) Since f_2 is continuous from the right at 0, there is δ with $0 < \delta < 1$ such that $f_2(t) < \varepsilon$ for all $\varepsilon > 0$ whenever $0 \leq t \leq \delta$. Define the sets N_1 and N_2 by

$$\begin{aligned} N_1 &:= \{k \in N : f_1[D(x_k, \overline{0})] \leq \delta\}, \\ N_2 &:= \{k \in N : f_1[D(x_k, \overline{0})] > \delta\}. \end{aligned}$$

Then, we obtain from Lemma 8.6.4 for $f_1[D(x_k, \overline{0})] > \delta$ that

$$(f_2 \circ f_1)[D(x_k, \overline{0})] \leq \frac{2f_2(1)}{\delta} f_1[D(x_k, \overline{0})]. \tag{8.6.8}$$

Therefore, we derive for $x = (x_k) \in \ell_p(F, f_1, s)$ with $s > 1$ that

$$\begin{aligned} \sum_k \frac{\{(f_2 \circ f_1)[D(x_k, \overline{0})]\}^p}{k^s} &= \sum_{k \in N_1} \frac{\{(f_2 \circ f_1)[D(x_k, \overline{0})]\}^p}{k^s} + \sum_{k \in N_2} \frac{\{(f_2 \circ f_1)[D(x_k, \overline{0})]\}^p}{k^s} \\ &\leq \sum_{k \in N_1} \frac{\varepsilon^p}{k^s} + \sum_{k \in N_2} \frac{\left\{\frac{2f_2(1)}{\delta} f_1[D(x_k, \overline{0})]\right\}^p}{k^s} \\ &= \varepsilon^p \sum_{k \in N_1} \frac{1}{k^s} + \left[\frac{2f_2(1)}{\delta}\right]^p \sum_{k \in N_2} \frac{\{f_1[D(x_k, \overline{0})]\}^p}{k^s} < \infty. \end{aligned}$$

Hence, $x = (x_k) \in \ell_p(F, f_2 \circ f_1, s)$.

(c) Suppose that $\limsup_{t\to\infty} [f_1(t)/f_2(t)] < \infty$. Then, there is a number $M > 0$ such that $[f_1(t)/f_2(t)] \leq M$ for all $t \in [0, \infty)$. Since $D(x_k, \overline{0}) \geq 0$ for all $k \in \mathbb{N}$ and for all $x = (x_k) \in \ell_p(F, f_2, s)$, we have $f_1[D(x_k, \overline{0})] \leq M f_2[D(x_k, \overline{0})]$ which leads us

$$\sum_k \frac{\{f_1[D(x_k, \overline{0})]\}^p}{k^s} \leq \sum_k \frac{\{M f_2[D(x_k, \overline{0})]\}^p}{k^s} = M^p \sum_k \frac{\{f_2[D(x_k, \overline{0})]\}^p}{k^s} < \infty.$$

Thus, $x = (x_k) \in \ell_p(F, f_1, s)$, as desired.

(d) Let $s_1 \leq s_2$. Since $0 < k^{-1} \leq 1$ for all $k \in \mathbb{N}_1$, it is immediate that $k^{-s_2} \leq k^{-s_1}$. Then, one can see that

$$\sum_k \frac{1}{k^{s_2}} \{f_1[D(x_k, \overline{0})]\}^p \leq \sum_k \frac{1}{k^{s_1}} \{f_1[D(x_k, \overline{0})]\}^p < \infty$$

holds, for all $x = (x_k) \in \ell_p(F, f_1, s_1)$. This means that $x = (x_k) \in \ell_p(F, f_1, s_2)$, which completes the proof. □

Corollary 8.6.6. [380, Corollary 2.6] *Define the sets* $\ell_p(F, s)$ *and* $\ell_p(F, f)$ *by*

$$\ell_p(F, s) := \left\{ x = (x_k) \in \omega(F) : \sum_k \frac{1}{k^s} [D(x_k, \overline{0})]^p < \infty \right\}; \ (s \geq 0),$$

$$\ell_p(F, f) := \left\{ x = (x_k) \in \omega(F) : \sum_k \{f[D(x_k, \overline{0})]\}^p < \infty \right\}.$$

Then, the following inclusions hold:

(a) *If* $s > 1$, $\ell_p(F, s) \subseteq \ell_p(F, f, s)$.

(b) $\ell_p(F, f) \subseteq \ell_p(F, f, s)$.

Proof. (a) This follows from Part (b) of Theorem 8.6.5 with $f_1(t) = t$ and $f_2 = f$.

(b) This is immediate from Part (d) of Theorem 8.6.5 by taking $s_1 = 0$, $s_2 = s$ and $f_1 = f$. □

Now, we can give the next theorem.

Theorem 8.6.7. [380, Theorem 2.7] *Let* $s > 1$. *Then, the following relations hold:*

(a) $\ell_\infty(F) \subseteq \ell_p(F, f, s)$.

(b) *If* f *is bounded, then* $\ell_p(F, f, s) = \omega(F)$.

Proof. (a) Let $x = (x_k) \in \ell_\infty(F)$. Then, there is a number $M > 0$ such that $D(x_k, \overline{0}) \leq M$ for all $k \in \mathbb{N}$. Since f is continuous and increasing, there is a number $N > 0$ such that $f[D(x_k, \overline{0})] \leq f(M) \leq N$. Therefore, we get for $s > 1$ that

$$\sum_k \frac{\{f[D(x_k, \overline{0})]\}^p}{k^s} \leq N^p \sum_k \frac{1}{k^s} < \infty.$$

Hence, $x = (x_k) \in \ell_p(F, f, s)$.

(b) Suppose that f is bounded. Then, one can find a number $N > 0$ such that $f(t) \leq N$ for all $t \in [0, \infty)$. Thus, for $x = (x_k) \in \omega(F)$ we have

$$\sum_k \frac{\{f[D(x_k, \overline{0})]\}^p}{k^s} \leq N^p \sum_k \frac{1}{k^s} < \infty,$$

which says that the inclusion $\omega(F) \subseteq \ell_p(F, f, s)$ holds. Since the inclusion $\ell_p(F, f, s) \subseteq \omega(F)$ also holds, we conclude that $\omega(F) = \ell_p(F, f, s)$, as desired. □

Prior to giving the final theorem of this subsection on the solidity of the spaces $\ell_\infty(F, f)$, $c_0(F, f)$ and $\ell_p(F, f, s)$, we state and prove the lemma concerning the solidity of the spaces $\ell_\infty(F)$, $c_0(F)$ and $\ell_p(F)$.

Lemma 8.6.8. [380, Lemma 2.8] *The spaces $\ell_\infty(F)$, $c_0(F)$ and $\ell_p(F)$ are solid.*

Proof. Let $\lambda(F)$ denotes the any of the spaces $\ell_\infty(F)$, $c_0(F)$ and $\ell_p(F)$. Suppose that

$$D(y_k, \overline{0}) \leq D(x_k, \overline{0}) \tag{8.6.9}$$

holds, for some $x = (x_k) \in \lambda(F)$. Therefore, one can easily see by (8.6.9) that

$$\begin{aligned}
\sup_{k \in \mathbb{N}} D(y_k, \overline{0}) &\leq \sup_{k \in \mathbb{N}} D(x_k, \overline{0}) < \infty, \\
\lim_{k \to \infty} D(y_k, \overline{0}) &\leq \lim_{k \to \infty} D(x_k, \overline{0}) = 0, \\
\sum_k [D(y_k, \overline{0})]^p &\leq \sum_k [D(x_k, \overline{0})]^p < \infty.
\end{aligned}$$

The above inequalities yield the desired consequence that $y = (y_k) \in \lambda(F)$.
This completes the proof of the lemma. □

Theorem 8.6.9. [380, Theorem 2.9] *The spaces $\ell_\infty(F, f)$, $c_0(F, f)$ and $\ell_p(F, f, s)$ are solid.*

Proof. This is immediate by Lemma 8.6.8, since the modulus function f is increasing. □

8.7 CONCLUSION

Nanda introduced the classical sets $\ell_\infty(F)$, $c(F)$ and $\ell_p(F)$ of sequences of fuzzy numbers, and proved that $(\ell_\infty(F), D_\infty)$, $(c(F), D_\infty)$ and $(\ell_p(F), D_p)$ are complete metric spaces, in [302]. Quite recently Talo and Başar have essentially dealt with the determination of the α-, β- and γ-duals of the classical sets of sequences of fuzzy numbers and the characterization of the classes of infinite matrices of fuzzy numbers transforming one of the classical sets to the another one, (see [379]).

The table on the characterizations of the matrix transformations between certain spaces of sequences with real or complex terms were given in the famous article by Stieglitz and Tietz [368]. To prepare the fuzzy analogues of this table, Talo and Başar have characterized the classes $(\mu(F) : \ell_\infty(F))$, $(c_0(F) : c(F))$, $(c_0(F) : c_0(F))$,

$(c(F) : c(F); p)$, $(\ell_p(F) : c(F))$, $(\ell_p(F) : c_0(F))$ and $(\ell_1(F) : \ell_p(F))$ of infinite matrices of fuzzy numbers, as a beginning; where $\mu \in \{\ell_\infty, c, c_0, \ell_p\}$, (see [379, 381]).

Of course, during the completion of the table of matrix transformations from the set $\mu_1(F)$ to the set $\mu_2(F)$, there are several open problems depending on the choice of $\mu_1(F)$ and $\mu_2(F)$. It is obvious that even the characterization of the class $(c(F) : c(F); p)$ of regular matrices remains open, without the restriction in Theorem 8.4.28 that the infinite matrix to be non-negative. Başar and Altay [66] have determined the α-, β- and γ-duals of some new spaces of sequences with real or complex terms by using the characterization of related matrix classes, which was a new development of the matrix transformations. As a new approach, Altay and Başar have derived recently some topological properties of certain spaces of sequences with real or complex terms from the characterization of the related matrix classes, (see [29]). It is natural that both of these techniques can be applied to the sets of sequences of fuzzy numbers. Then, the table of characterizations of the matrix classes between certain sets of sequences of fuzzy numbers is needed. Indeed, if we study the new set $\mu(F)_A$ obtained by the domain of an infinite matrix A of fuzzy numbers in a set $\mu(F)$ then it is necessary to know the characterization of the matrix transformations from $\mu(F)$ to $\ell_1(F)$, to $c(F)$ and to $\ell_\infty(F)$ for calculating the α-, β- and γ-duals of the set $\mu(F)_A$.

CHAPTER 9

Absolute Summability

In this chapter, we focus on the absolute summability which is an expansion of the concept of convergence and is a generalization of the concept of absolute convergence. We consider absolute summability and investigate some properties of absolute summability methods. Absolute summability is discussed regarding inclusion theorems and summability factors theorems. We deal with the general absolute inclusion theorem involving a pair of triangles and, as corollaries, inclusion theorems for special classes of triangles.

Keywords: Absolute Summability, Infinite Series, Cesáro Matrix, Matrix Transformations, Double Sequences.

9.1 BACKGROUND, PRELIMINARIES AND NOTATIONS

Throughout this chapter, by (s_k) we denote the sequence of partial sums of the infinite series $\sum_n a_n$.

Definition 9.1.1. [94, p. 142] *A real sequence $x = (x_k)$ is called totally monotone if $\Delta^n x_k \geq 0$ for all $n, k \in \mathbb{N}$. We call a complex sequence $x = (x_k) = (u_k + iv_k)$ is totally monotone if the real sequences (u_k) and (v_k) are totally monotone.*

Theorem 9.1.2. [94, p. 144] *The following statements hold:*

(i) *A real sequence $p = (p_n)$ is totally monotone if and only if there exists a non-decreasing function $\Phi : [0, 1] \to \mathbb{R}$ with*

$$p_n := \int_0^1 t_n d\Phi(t) \quad \text{for all} \quad n \in \mathbb{N}.$$

(ii) *A sequence $p = (p_n) \in \omega$ is representable as the difference of two totally monotone sequences if and only if p is a moment sequence, that is $\exists \Phi \in BV([0, 1])$ such that*

$$p_n := \int_0^1 t_n d\Phi(t) \quad \text{for every} \quad n \in \mathbb{N}.$$

In both cases, we say that Φ generates p.

DOI: 10.1201/9781003294153-9

9.2 ABSOLUTE SUMMABILITY OF SEQUENCES AND SERIES

Following Volkov [391], we would like to present the beginning concepts of the absolute summability of sequences and series. A special type of summability of series and sequences, which differs from ordinary summability in that additional restrictions are imposed. For matrix summability methods the requirement is that the series and sequences obtained as a result of the transformation corresponding to the given summability method must be absolutely convergent.

Let a sequence-to-sequence summability method A be defined by means of the matrix $(a_{nk})_{n,k\in\mathbb{N}}$. If a sequence $x = (x_k)$ is A-summable to the limit l and its A-transform is of bounded variation, i.e.

$$\lim_{n\to\infty} (Ax)_n = l,$$

$$\sum_{n=0}^{\infty} |(Ax)_n - (Ax)_{n-1}| < \infty \tag{9.2.1}$$

then we say that *the sequence is absolutely A-summable* or *$|A|$-summable to the same value.*

By replacing the sequences $x = (x_k)$ and $s = (s_k)$, it is obtained the absolute summability of the series $\sum_{n=0}^{\infty} a_n$ by the method A to the sum l. The condition (9.2.1) is an additional requirement which makes absolute summability different from ordinary summability.

Let a series-to-sequence summability method B be defined by means of the matrix $(b_{nk})_{n,k\in\mathbb{N}}$. If a series $\sum_{n=0}^{\infty} a_n$ is B-summable to the sum l and its B-transform is of bounded variation, i.e.

$$\lim_{n\to\infty} \sum_{k=0}^{\infty} b_{nk}a_k = l,$$

$$\sum_{n=0}^{\infty} |(Ba)_n - (Ba)_{n-1}| < \infty$$

then we say that *the series is absolutely B-summable* or *$|B|$-summable to the same sum.*

Absolute summability is defined in a similar manner for the series-to-series and sequence-to-series methods. Indeed, if the method A corresponds to the identity transformation of a series-to-series or sequence-to-series summability method A which is defined by means of the matrix $(a_{nk})_{n,k\in\mathbb{N}}$ then, absolute summability of a series is the same as absolute convergence.

The supplementary requirements are suitably modified for non-matrix summability methods. Thus, in the case of the Abel summability method such a requirement is that the function $f(x) = \sum_{k=0}^{\infty} a_k x^k$ is of bounded variation on the interval $0 \le x < 1$. In the case of integral summability methods, absolute summability is distinguished by the requirement of absolute convergence of the corresponding integrals. Thus, for the Borel summability method the integral

$$\int_0^{\infty} e^{-x} \sum_{k=0}^{\infty} \frac{a_k x^k}{k!} dx$$

must be absolutely convergent.

A summability method is said to preserve absolute convergence of a series if it absolutely sums each absolutely convergent series. If each such series is summable by this method to a sum equal to the sum to which it converges, then the method is called absolutely regular. For instance, the Cesàro summability method $C^{(m)}$ of order m is absolutely regular with $m \geq 0$. The Abel method is absolutely regular. Knopp-Lorentz theorem gives the following necessary and sufficient conditions for the absolute regularity of a summability method given by a series-to-series transformation defined by means of a matrix (b_{nk}):

$$\sum_{n=0}^{\infty} |b_{nk}| \leq M,$$
$$\sum_{n=0}^{\infty} b_{nk} = 1 \quad \text{for each} \quad k \in \mathbb{N}.$$

A generalization of absolute summability is absolute summability of order p, where $p \geq 1$. In this case, the condition

$$\sum_{n=0}^{\infty} |(As)_n - (As)_{n-1}|^p < \infty$$

is the supplementary restriction by which absolute summability of order p is distinguished from ordinary summability.

The concept of absolute summability was introduced by Borel for one of his methods, but was stated differently from its modern formulation: Absolute summability was distinguished by imposing the condition

$$\int_0^{\infty} e^{-x} \left| \sum_{k=0}^{\infty} \frac{a_{k+p}x^k}{k!} \right| dx < \infty$$

for each $p \in \mathbb{N}$. Absolute summability was initially used in the study of the summability of power series outside the disc of convergence. In view of the problems involved in the multiplication of summable series, absolute summability was defined and studied by the Cesàro summability method $|C|_k$. The general definition of absolute summability is more recent, and has found extensive application in studies on the summation of Fourier series.

Now, we can give the absolute summability of order p concerning with certain summability methods defined by triangle matrices.

The absolute summability $\left|C^{(m)}\right|$ of the Cesàro summability method $C^{(m)}$ of order m of a series was defined Fekete [171] for the case m is an integer, and in the general case by Kogbetliantz [227].

A sequence (s_n) is said to be of bounded variation (i.e., belongs to the space bv) if $\sum_n |(\Delta s)_n| < \infty$. Thus, to say that a series is absolutely summable by a matrix A is equivalent to saying that the A-transform of the sequence of partial sums is in bv. The necessary and sufficient conditions characterized the class $(bv : bv)$ are known (see Mears [275], Stieglitz and Tietz [368]).

Definition 9.2.1. *We have the following statements:*

(i) *If the C_1-transform of the sequence s is in the space bv, i.e.,*

$$\sum_{n=1}^{\infty} |(C_1 s)_n - (C_1 s)_{n-1}| < \infty$$

then, the series $\sum_n a_n$ is called absolute Cesáro summable of order 1 or shortly $|C_1|$-summable.

(ii) *If the $C^{(m)}$-transform of the sequence s is in the space bv, i.e.,*

$$\sum_{n=1}^{\infty} \left|\left(C^{(m)} s\right)_n - \left(C^{(m)} s\right)_{n-1}\right| < \infty$$

then, the series $\sum_n a_n$ is called $|C^{(m)}|$-summable.

The definition of absolute Nörlund summability was given by Mears [275].

Definition 9.2.2. *If the N^t-transform of the sequence s is in the space bv, i.e.,*

$$\sum_{n=1}^{\infty} |(N^t s)_n - (N^t s)_{n-1}| < \infty$$

then, the series $\sum_n a_n$ is called absolute (N^t)-summable or shortly $|N^t|$-summable.

Absolute Riesz summability, written as $\left|\bar{N}^t\right|$ or $|R^t|$, was defined by Sunuochi [371] as follows:

Definition 9.2.3. *If the R^t-transform of the sequence s is in the space bv, i.e.,*

$$\sum_{n=1}^{\infty} |(R^t s)_n - (R^t s)_{n-1}| < \infty$$

then, the series $\sum_n a_n$ is called absolute Riesz summable or shortly $|R^t|$-summable.

Definition 9.2.4. (Flett [172]) *Let $k \geq 1$. If*

$$\sum_{n=1}^{\infty} n^{k-1} \left|\left(C^{(m)} s\right)_n - \left(C^{(m)} s\right)_{n-1}\right|^k < \infty \tag{9.2.2}$$

then, the series $\sum_n a_n$ is called $|C^{(m)}|_k$-summable.

Let $x = (na_n)$. The $C^{(m)}$-transform of the sequence x with $m > -1$ is given by

$$\left(C^{(m)} x\right)_n := \frac{1}{A_n^m} \sum_{k=1}^{n} A_{n-k}^{m-1} k a_k$$

and is also defined by Kogbetliantz [227], as follows;

$$\left(C^{(m)} x\right)_n := n\left[\left(C^{(m)} s\right)_n - \left(C^{(m)} s\right)_{n-1}\right].$$

Hence, the condition (9.2.2) is equivalent to the condition

$$\sum_{n=1}^{\infty} \frac{1}{n} \left| \left(C^{(m)}x\right)_n \right|^k < \infty. \tag{9.2.3}$$

If we put $k = 1$ *in (9.2.3), then we obtain* $|C^{(m)}|$ *summability method given by Definition 9.2.1.*

Let $k \geq 1$. *If*

$$\sum_{k=1}^{\infty} n^{k-1} \left| (C_1 s)_n - (C_1 s)_{n-1} \right|^k < \infty$$

then, the series $\sum_n a_n$ *is called* $|C_1|_k$*-summable. If we put* $k = 1$*, then we obtain* $|C_1|$ *summability method given by Definition 9.2.1.*

Definition 9.2.5. (Borwein and Cass [98]) *Let* $k \geq 1$. *If*

$$\sum_{n=1}^{\infty} n^{k-1} \left| (N^t s)_n - (N^t s)_{n-1} \right|^k < \infty$$

then, the series $\sum_n a_n$ *is called* $|N^t|_k$*-summable. If we put* $k = 1$ *then we obtain* $|N^t|$ *summability method given by Definition 9.2.2.*

For the kth *order absolute Nörlund summability* the following different definition is also given by Umar and Khan [390]:

Definition 9.2.6. *Let* $k \geq 1$. *If*

$$\sum_{n=1}^{\infty} \left(\frac{T_n}{t_n} \right)^{k-1} \left| (N^t s)_n - (N^t s)_{n-1} \right|^k < \infty$$

then, the series $\sum_n a_n$ *is called* $|N^t|_k$*-summable. If we put* $k = 1$*, then we obtain* $|N^t|$ *summability method given by Definition 9.2.2.*

Definition 9.2.7. (Bor, [97]) *Let* $k \geq 1$. *If*

$$\sum_{n=1}^{\infty} n^{k-1} \left| (R^t s)_n - (R^t s)_{n-1} \right|^k < \infty$$

then, the series $\sum_n a_n$ *is called* $|R^t|_k$ *summable. If we put* $k = 1$*, then we obtain* $|R^t| = \left|\overline{N}^t\right|$ *summability method given by Definition 9.2.3.*

Definition 9.2.8. (cf. Miller [382] and Rhoades [329]) *Let* $T = (t_{nk})$ *be an arbitrary triangle matrix. Therefore, an appropriate extension of (9.2.2) for the matrix* T *can be given by*

$$\sum_{n=1}^{\infty} n^{k-1} \left| \Delta (Ts)_{n-1} \right|^k < \infty. \tag{9.2.4}$$

Then, the series $\sum_n a_n$ *is said to be summable* $|T|_k$ *with* $k \geq 1$ *if (9.2.4) is satisfied.*

However, Umar and Khan [390] defined $|N^t|_k$ summability by using

$$\sum_{n=1}^{\infty}\left(\frac{T_n}{t_n}\right)^{k-1}\left|(N^t s)_n-(N^t s)_{n-1}\right|^k<\infty$$

and Bor [96] defined $\left|\bar{N}^t\right|_k$ summability by using

$$\sum_{n=1}^{\infty}\left(\frac{T_n}{t_n}\right)^{k-1}\left|(R^t s)_n-(R^t s)_{n-1}\right|^k<\infty$$

instead of (9.2.4). Apparently the authors have interpreted n in (9.2.2) to represent the reciprocal of the nth diagonal term of the matrices N^t and $\bar{N}^t$, respectively. Unfortunately, this interpretation cannot be correct, (see Rhoades [329]). If it was true, then since the nth diagonal entry of $C^{(m)}$ is $O(n^\alpha)$, (9.2.2) would take the form

$$\sum_{n=1}^{\infty}(n^\alpha)^{k-1}|(C^{(m)}s)_{n-1}^{\alpha}-(C^{(m)}s)_{n}^{\alpha}|^k<\infty.$$

However, Flett stays with (9.2.2).

In 1970, by using Definition 9.2.4, Das [139] introduced the concept of absolute k^{th} power conservativeness.

Definition 9.2.9. *Let A be a sequence-to-sequence transformation mapping the sequence (s_n) into $\{(As)_n\}$. If $\{(As)_n\}$ converges absolutely whenever (s_n) converges absolutely, then A is called absolutely conservative. If the absolute convergence of (s_n) implies absolute convergence of $\{(As)_n\}$ to the same limit, then A is called absolutely regular. For some given $k\geq 1$ if*

$$\sum_{n=1}^{\infty}n^{k-1}\left|s_n-s_{n-1}\right|^k<\infty \tag{9.2.5}$$

implies

$$\sum_{n=1}^{\infty}n^{k-1}\left|(As)_n-(As)_{n-1}\right|^k<\infty \tag{9.2.6}$$

then, A is called absolutely k^{th} power conservative.

For $k=1$, condition (9.2.5) guarantees the convergence of (s_n). Note that when $k>1$, (9.2.5) does not necessarily imply the convergence of (s_n). For example, take

$$s_n=\sum_{v=1}^{n}\frac{1}{v\log(v+1)}$$

for all $n\in\mathbb{N}$. Then, (9.2.5) holds but, (s_n) does not converge. Thus, since the limit of (s_n) need not exist, the concept of absolute k^{th} power regularity cannot be introduced for $k>1$.

9.3 INCLUSION THEOREMS

We associate two infinite matrices $\breve{T} = (\breve{t}_{nk})$ and $\widehat{T} = (\widehat{t}_{nk})$ with the matrix $T = (t_{nk})$, as follows:

$$\breve{t}_{nk} := \sum_{j=k}^{n} t_{nj} \quad \text{for all} \quad n, k \in \mathbb{N} \tag{9.3.1}$$

$$\widehat{t}_{nk} := \breve{t}_{nk} - \breve{t}_{n-1,k} \quad \text{for all} \quad n, k \in \mathbb{N}_1, \tag{9.3.2}$$

where $\widehat{t}_{00} = \breve{t}_{00} = t_{00}$. When the factorable matrix $G(u, v)$ replaces by the matrix T, the matrices $\breve{A} = (\breve{a}_{nk})$ and $\widehat{A} = (\widehat{a}_{nk})$ can be expressed as was same as the matrices $\breve{T}$ and $\widehat{T}$. That is to say that the matrices $\breve{A} = (\breve{a}_{nk})$ and $\widehat{A} = (\widehat{a}_{nk})$ are defined by

$$\breve{a}_{nk} := u_n \sum_{j=k}^{n} v_j \quad \text{for all} \quad n, k \in \mathbb{N},$$

$$\widehat{a}_{nk} := \breve{a}_{nk} - \breve{a}_{n-1,k} \quad \text{for all} \quad n, k \in \mathbb{N}_1.$$

If $T = (t_{nv})$ is a lower triangular matrix, then $\breve{T}$ and $\widehat{T}$ are also lower triangular matrices. If T is a triangle, then $\widehat{t}_{nn} = \breve{t}_{nn} = t_{nn} \neq 0$ for all $n \in \mathbb{N}$, hence $\breve{T}$ and $\widehat{T}$ are triangles. Throughout this chapter, we suppose that the matrices $\breve{T}$ and $\widehat{T}$ (the matrices $\breve{A}$ and $\widehat{A}$) are connected to the matrix T (to the matrix A) by the relations (9.3.1) and (9.3.2), respectively.

It is immediate that

$$(Ts)_n := \sum_{k=0}^{n} t_{nk} s_k = \sum_{k=0}^{n} t_{nk} \sum_{i=0}^{k} a_i = \sum_{i=0}^{n} a_i \sum_{k=i}^{n} t_{nk} = \sum_{i=0}^{n} \breve{t}_{ni} a_i,$$

$$Y_n := (Ts)_n - (Ts)_{n-1} = \sum_{k=0}^{n} \widehat{t}_{nk} a_k \tag{9.3.3}$$

for all $n \in \mathbb{N}$. Since $\widehat{T}$ is a triangle, it has a unique two-sided inverse $\widehat{T}^{-1} = (u_{nk})$, (cf. Cooke [111, (2.2, II); p. 24]). Then, we can solve (9.3.3) for a_n and thus obtain

$$a_n := \sum_{k=0}^{n} u_{nk} Y_k \quad \text{for all} \quad n \in \mathbb{N}. \tag{9.3.4}$$

Rhoades and Savaş [332] gave the following absolute inclusion theorem involving a pair of triangles which contains inclusion theorems for the special classes of triangles:

Theorem 9.3.1. [332, Theorem 1] *Let $\widehat{B}^{-1} = (v_{nk})$ and $A = (a_{nk})$, $B = (b_{nk})$ be two triangles satisfying the following conditions:*

(i) $\dfrac{|a_{nn}|}{|b_{nn}|} = O(1)$.

(ii) $\left|\dfrac{b_{n+1,n} - b_{nn}}{b_{nn} b_{n+1,n+1}}\right| = O(1)$.

(iii) $\sum_{i=0}^{n-1} |\Delta_i \widehat{a}_{ni}| = O(|a_{nn}|)$.

(iv) $\sum_{n=i+1}^{\infty} (n|a_{nn}|)^{k-1} |\Delta_i \widehat{a}_{ni}| = O(i^{k-1}|a_{ii}|^k)$.

(v) $\sum_{i=0}^{n-1} |a_{ii}||\widehat{a}_{n,i+1}| = O(|a_{nn}|)$.

(vi) $\sum_{n=i+1}^{\infty} (n|a_{nn}|)^{k-1} |\widehat{a}_{n,i+1}| = O((i|a_{ii}|)^{k-1})$.

(vii) $\sum_{j=0}^{n-2} |a_{nj}| \left| \sum_{i=2}^{n} \widehat{a}_{ni} v_{ir} \right| = O(|a_{nn}|)$.

(viii) $\sum_{n=j+2}^{\infty} (n|a_{nn}|)^{k-1} \left| \sum_{i=j+2}^{n} \widehat{a}_{ni} v_{ij} \right| = O((j|a_{jj}|)^{k-1})$,

If $\sum_n a_n$ *is summable* $|B|_k$*, then it is summable* $|A|_k$*, where* $k \geq 1$.

Proof. One can easily see for all $n \in \mathbb{N}$ that

$$X_n := (Bs)_n - (Bs)_{n-1} = \sum_{i=0}^{n} \widehat{b}_{ni} a_i. \tag{9.3.5}$$

Since $\widehat{B}^{-1}$ is a triangle, we can solve (9.3.5) for a_n and thus we derive that

$$a_n := \sum_{i=0}^{n} v_{ni} X_i. \tag{9.3.6}$$

$$\begin{aligned} Y_n &= \sum_{k=0}^{n} \widehat{a}_{nk} \sum_{i=0}^{k} v_{ki} X_i \qquad (9.3.7) \\ &= \sum_{k=0}^{n} \widehat{a}_{nk} \Big(v_{kk} X_k + v_{k,k-1} X_{k-1} + \sum_{i=0}^{k-2} v_{ki} X_i \Big) \\ &= \frac{a_{nn} X_n}{b_{nn}} + \sum_{k=0}^{n-1} \frac{\Delta_k \widehat{a}_{nk} X_k}{b_{kk}} + \sum_{k=0}^{n-1} \widehat{a}_{n,k+1} \Big(\frac{b_{kk} - b_{k+1,k}}{b_k b_{k+1,k+1}} \Big) X_k + \sum_{k=2}^{n} \widehat{a}_{nk} \sum_{i=0}^{k-2} v_{ki} X_i \\ &= T_{n,1} + T_{n,2} + T_{n,3} + T_{n,4}, \quad \text{say.} \end{aligned}$$

By Minkowski's inequality, it is sufficient to prove that

$$\sum_{n=1}^{\infty} n^{k-1} |T_{n,i}|^k < \infty, \quad i = 1, 2, 3, 4.$$

Using the condition (i), we have

$$\sum_{n=1}^{\infty} n^{k-1}|T_{n,1}|^k = \sum_{n=1}^{\infty} n^{k-1}\Big|\frac{a_{nn}X_n}{b_{nn}}\Big|^k = O(1)\sum_{n=1}^{\infty} n^{k-1}|X_n|^k = O(1),$$

since $\sum_n a_n$ is summable $|B|_k$.

Using the condition (i), Hölder's inequality, the conditions (iii) and (iv), we see that

$$\begin{aligned}
\sum_{n=1}^{\infty} n^{k-1}|T_{n,2}|^k &\le \sum_{n=1}^{\infty} n^{k-1}\Big(\sum_{i=0}^{n-1}|b_{ii}|^{-1}|\Delta_i\widehat{a}_{ni}||X_i|\Big)^k \\
&= O(1)\sum_{n=1}^{\infty} n^{k-1}\Big(\sum_{i=0}^{n-1}|a_{ii}|^{-k}|\Delta_i\widehat{a}_{ni}||X_i|^k\Big)\Big(\sum_{i=0}^{n-1}|a_{ii}||\Delta_i\widehat{a}_{ni}|\Big)^{k-1} \\
&= O(1)\sum_{i=0}^{\infty}|a_{ii}|^{-k}|X_i|^k\sum_{n=i+1}^{\infty}(n|a_{nn}|)^{k-1}|\Delta_i\widehat{a}_{ni}| \\
&= O(1)\sum_{i=1}^{\infty} i^{k-1}|X_i|^k = O(1).
\end{aligned}$$

Using the conditions (ii), (v) and (vi), one can see that

$$\begin{aligned}
\sum_{n=1}^{\infty} n^{k-1}|T_{n,3}|^k &= \sum_{n=1}^{\infty} n^{k-1}\Big|\sum_{i=0}^{n-1}\widehat{a}_{n,i+1}\Big(\frac{b_{ii}-b_{i+1,i}}{b_{\nu\nu}b_{i+1,i+1}}\Big)X_i\Big|^k \\
&= O(1)\sum_{n=1}^{\infty} n^{k-1}\Big(\sum_{i=0}^{n-1}|\widehat{a}_{n,i+1}||X_i|\Big)^k \\
&= O(1)\sum_{n=1}^{\infty} n^{k-1}\Big(\sum_{i=0}^{n-1}|a_{ii}|^{1-k}|\widehat{a}_{n,i+1}||X_i|^k\Big)\Big(\sum_{i=0}^{n-1}|a_{ii}||\widehat{a}_{n,i+1}|\Big)^{k-1} \\
&= O(1)\sum_{n=1}^{\infty}(n|a_{nn}|)^{k-1}\sum_{i=0}^{n-1}|a_{ii}|^{1-k}|\widehat{a}_{n,i+1}||X_i|^k \\
&= O(1)\sum_{i=0}^{\infty} i^{k-1}|X_i|^k = O(1).
\end{aligned}$$

From the conditions (vii) and (viii), we derive that

$$\begin{aligned}\sum_{n=1}^{\infty} n^{k-1}|T_{n,4}|^k &= \sum_{n=1}^{\infty} n^{k-1}\Big|\sum_{i=2}^{n}\widehat{a}_{ni}\sum_{j=0}^{i-2}X_j\Big|^k \\ &\le \sum_{n=1}^{\infty} n^{k-1}\sum_{j=0}^{n-2}|a_{rr}|^{1-k}|X_j|^k\Big|\sum_{i=j+2}^{n}\widehat{a}_{ni}v_{ij}\Big|\Big(\sum_{j=0}^{n-2}|a_{rr}|\Big|\sum_{i=j+2}^{n}\widehat{a}_{ni}v_{ij}\Big|\Big)^{k-1} \\ &= O(1)\sum_{n=1}^{\infty}\sum_{j=0}^{n-2}|a_{rr}|^{1-k}|X_j|^k\Big|\sum_{i=j+2}^{n}\widehat{a}_{ni}v_{ij}\Big|(n|a_{nn}|)^{k-1} \\ &= O(1)\sum_{j=0}^{\infty}|a_{jj}|^{1-k}|X_j|^k(j|a_{jj}|)^{k-1} \\ &= O(1)\sum_{j=1}^{\infty}j^{k-1}|X_j|^k = O(1).\end{aligned}$$

□

Theorem 9.3.1 represents that two arbitrary normal matrices have been used in an absolute inclusion theorem and also represents one of the most general such inclusion theorems that one can expect to obtain.

Corollary 9.3.2. [332, Corollary 1] *Let A and B be triangles, B factorable, satisfying the conditions (i)–(vi) of Theorem 9.3.1. Then, if $\sum_n a_n$ is summable $|B|_k$, it is summable $|A|_k$ with $k \ge 1$.*

It is well-known that, if B is a weighted mean matrix, then $\widehat{B}$ has the entries

$$\widehat{b}_{nk} := \frac{T_{k-1}t_n}{T_nT_{n-1}} \quad \text{for all} \quad n \in \mathbb{N}.$$

Hence, $\widehat{B}$ is factorable.

Corollary 9.3.3. [332, Corollary 2] *Let (t_n) be a positive sequence, A be a triangle, satisfying*

(i) $\dfrac{T_n}{t_n}|a_{nn}| = O(1)$

and the conditions (iii)–(vi) of Theorem 9.3.1. If $\sum_n a_n$ is summable $|\overline{N}^t|_k$, then it is summable $|A|_k$ with $k \ge 1$.

Proof. With $B = \overline{N}^t$, the condition (i) of Theorem 9.3.1 reduces to the condition (i) of Corollary 9.3.3

$$\Big|\frac{b_{n+1,n}-b_{nn}}{b_{nn}b_{n+1,n+1}}\Big| = \Big|\frac{t_n/T_{n+1}-t_n/T_n}{t_nt_{n+1}/T_nT_{n+1}}\Big| = \frac{t_nt_{n+1}}{t_nt_{n+1}} = 1$$

and the condition (ii) of Theorem 9.3.1 is automatically satisfied. □

9.4 SUMMABILITY FACTORS THEOREMS

If a sequence (λ_n) has the property that the series $\sum_n a_n\lambda_n$ is summable by a method Y whenever $\sum_n a_n$ is summable by a method X, then λ_n is called a *summability factor of type* (X, Y), and we write $\lambda_n \in (X, Y)$.

Lemma 9.4.1. [255, p. 227] *Let* $1 < p < \infty$. *Then, an infinite matrix* $T = (t_{nk}) \in (\ell_1 : \ell_p)$ *if and only if*

$$\sup_{k\in\mathbb{N}} \sum_{n=0}^{\infty} |t_{nk}|^p < \infty.$$

Theorem 9.4.2. [331, Theorem 2.1] *Let* $1 < k < \infty$ *and* $T = (t_{nj})$ *be a lower triangular matrix with positive entries and row sums* 1. *Then,* $\lambda \in (|\overline{N}^t|, |T|_k)$, *i.e.,* $\sum_n a_n$ *is summable* $|\overline{N}^t|$, *then* $\sum_n a_n\lambda_n$ *is summable* $|T|_k$ *if and only if*

(i) $|t_{jj}\lambda_j| \dfrac{T_j}{t_j} = O\left(j^{1/k-1}\right)$,

(ii) $\left(\displaystyle\sum_{n=j+1}^{\infty} n^{k-1} \left|\Delta_j(\widehat{t}_{nj}\lambda_j)\right|^k\right)^{1/k} = O\left(\dfrac{t_j}{T_j}\right)$,

(iii) $\left(\displaystyle\sum_{n=j+1}^{\infty} n^{k-1} \left|\widehat{t}_{n,j+1}\lambda_{j+1}\right|^k\right)^{1/k} = O(1)$.

Proof. Let (x_n) be the $\overline{N}^t$-transform of the sequence $s = (s_j)$. Therefore, by definition we have

$$x_n := \frac{1}{T_n}\sum_{j=0}^{n} t_j s_j = \frac{1}{T_n}\sum_{j=0}^{n}(T_n - T_{j-1})a_j.$$

Hence,

$$X_n := x_n - x_{n-1} = \frac{t_n}{T_nT_{n-1}}\sum_{j=1}^{n} T_{j-1}a_j \tag{9.4.1}$$

for all $n \in \mathbb{N}_1$. Let y_n be the T-transform of the sequence $s = (s_j)$. Then, we see by definition that

$$y_n := \sum_{i=0}^{n} t_{ni}s_i = \sum_{i=0}^{n} t_{ni}\sum_{j=0}^{i}\lambda_j a_j = \sum_{j=0}^{n}\lambda_j a_j\sum_{i=j}^{n} t_{ni} = \sum_{j=0}^{n}\breve{t}_{nj}\lambda_j a_j.$$

Thus, we derive for $n \in \mathbb{N}_1$ that

$$Y_n := y_n - y_{n-1} = \sum_{j=0}^{n}(\breve{t}_{nj} - \breve{t}_{n-1,j})\lambda_j a_j = \sum_{j=0}^{n}\widehat{t}_{nj}\lambda_j a_j. \tag{9.4.2}$$

By the hypothesis of the theorem,

$$\sum_{n=1}^{\infty} n^{k-1} |Y_n|^k < \infty \tag{9.4.3}$$

whenever

$$\sum_{n=1}^{\infty} |X_n| < \infty. \tag{9.4.4}$$

For $k \geq 1$, we define the sets $\mathcal{B}$ and $\mathcal{C}$ as follows:

$$\begin{aligned} \mathcal{B} &:= \left\{(a_i) \in \omega : \sum a_i \text{ is summable } |\overline{N}^t|\right\} \\ \mathcal{C} &:= \left\{(a_i) \in \omega : \sum a_i\lambda_i \text{ is summable } |T|_k\right\} \end{aligned}$$

which are Banach spaces endowed with the norms defined by

$$\|X\|_{\mathcal{B}} := \sum |X_n|, \quad \|Y\|_{\mathcal{C}} := \{|Y_0|^k + \sum n^{k-1}|Y_n|^k\}^{1/k}. \tag{9.4.5}$$

Since $\sum_n a_n$ is summable by $|\overline{N}^t|$ implies that $\sum_n a_n\lambda_n$ is summable by $|T|_k$, by the Banach-Steinhaus theorem, there exists a positive constant M such that

$$\|Y\| \leq M \|X\| \tag{9.4.6}$$

for all sequences satisfying (9.4.4).

Applying (9.4.1) and (9.4.2) to the sequence (a_j) given by

$$a_j := \begin{cases} (-1)^{n-j} e_j & , \quad n \leq j \leq n+1, \\ 0 & , \quad \text{otherwise} \end{cases},$$

where e_j is the jth coordinate sequence, we obtain

$$\begin{aligned} X_n &:= \begin{cases} 0 & , \quad n < j, \\ \frac{t_j}{T_j} & , \quad n = j \\ \frac{-t_j t_n}{T_n T_{n-1}} & , \quad n > j \end{cases} \\ Y_n &:= \begin{cases} 0 & , \quad n < j \\ \widehat{t}_{jj}\lambda_j & , \quad n = j, \\ \Delta_j(\widehat{t}_{nj}\lambda_j) & , \quad n > j \end{cases} \end{aligned}$$

By (9.4.5), it follows that $\|X\| = 2t_j/T_j$ and

$$\|Y\| = \left[v^{k-1} |t_{vj}\lambda_j|^k + \sum_{n=j+1}^{\infty} n^{k-1} |\Delta_j(\widehat{t}_{nj}\lambda_j)|^k \right]^{1/k}.$$

Thus, it follows from (9.4.6) that

$$v^{k-1} |t_{vj}\lambda_j|^k + \sum_{n=j+1}^{\infty} n^{k-1} |\Delta_j(\widehat{t}_{nj}\lambda_j)|^k \leq (2M)^k \left(\frac{t_j}{T_j}\right)^k.$$

Since this inequality holds for every $j \in \mathbb{N}_1$, we have

$$v^{k-1}\left|t_{vj}\lambda_\nu\right|^k + \sum_{n=v+1}^{\infty} n^{k-1}|\Delta_j(\widehat{t}_{nj}\lambda_j)|^k = O\left[\left(\frac{t_j}{T_j}\right)^k\right].$$

The above inequality holds iff each term on the left-hand side is $O\left((t_j/T_j)^k\right)$. By taking the first term, it is immediate that

$$\frac{T_j}{t_j}\left|t_{jj}\lambda_j\right| = O\left(j^{1/k-1}\right),$$

i.e., the condition (i) is necessary. Taking the second term, we obtain

$$\left[\sum_{n=j+1}^{\infty} n^{k-1}|\Delta_j(\widehat{t}_{nj}\lambda_j)|^k\right]^{1/k} = O\left(\frac{t_j}{T_j}\right),$$

which is the condition (ii).

To prove the necessity of (iii), we again apply (9.4.1) and (9.4.2) to the sequence (a_j) with $a_j = e_{j+1}$. Then, we obtain

$$X_n := \begin{cases} \frac{T_j t_n}{T_n T_{n-1}} & , \quad n \geq j+1, \\ 0 & , \quad n < j+1 \end{cases} \quad \text{and} \quad Y_n := \begin{cases} \widehat{t}_{n,j+1}\lambda_{j+1} & , \quad n \geq j+1, \\ 0 & , \quad n < j+1 \end{cases}.$$

Using (9.4.5), we obtain $\|X\| = 1$, and

$$\|Y\| = \left(\sum_{n=j+1}^{\infty} n^{k-1}|\widehat{t}_{n,j+1}\lambda_{j+1}|^k\right)^{1/k}.$$

From (9.4.6), it follows that

$$\left(\sum_{n=j+1}^{\infty} n^{k-1}|\widehat{t}_{n,j+1}\lambda_{j+1}|^k\right)^{1/k} = O(1),$$

which gives the necessity of (iii).

To prove the sufficiency of the conditions, from (9.4.1) we have

$$\frac{T_n T_{n-1} X_n}{t_n} = \sum_{j=1}^{n} T_{j-1} a_j,$$

$$\frac{T_{n-1} T_{n-2} X_{n-1}}{t_{n-1}} = \sum_{j=1}^{n-1} T_{j-1} a_j,$$

$$\frac{T_n T_{n-1} X_n}{t_n} - \frac{T_{n-1} T_{n-2} X_{n-1}}{t_{n-1}} = T_{n-1} a_n.$$

Thus, we have

$$a_n = \frac{T_n X_n}{t_n} - \frac{T_{n-2} X_{n-1}}{t_{n-1}}. \tag{9.4.7}$$

Then, one can see by substituting (9.4.7) in (9.4.2) that

$$\begin{aligned}
Y_n &= \sum_{j=0}^{n} \widehat{t}_{nj}\lambda_j a_j = \widehat{t}_{n0}\lambda_0 a_0 + \sum_{j=1}^{n} \widehat{t}_{nj}\lambda_j \left(\frac{T_j X_j}{t_j} - \frac{T_{j-2}X_{j-1}}{t_{j-1}}\right) \\
&= \widehat{t}_{n0}\lambda_0 a_0 + t_{nn}\lambda_n \frac{T_n X_n}{t_n} + \sum_{j=1}^{n-1} (\widehat{t}_{nj}\lambda_j T_j - \widehat{t}_{n,j+1}\lambda_{j+1}T_{j-1})\frac{X_j}{t_j} \\
&= t_{nn}\lambda_n \frac{T_n}{t_n} X_n + \sum_{j=1}^{n-1} \left(\widehat{t}_{nj}\lambda_j T_j - \widehat{t}_{n,j+1}\lambda_{j+1}T_{j-1}\right)\frac{X_j}{t_j} \\
&= t_{nn}\lambda_n \frac{T_n}{t_n} X_n + \sum_{j=1}^{n-1} \left[\frac{T_j}{t_j}\Delta_j(\widehat{t}_{nj}\lambda_j) + \widehat{t}_{n,j+1}\lambda_{j+1}\right] X_j.
\end{aligned}$$

Write $Y_n^* = n^{1-1/k}Y_n$. Then,

$$Y_n^* = n^{1-1/k}\left\{t_{nn}\lambda_n \frac{T_n}{t_n} X_n + \sum_{j=1}^{n-1} \left[\frac{T_j}{t_j}\Delta_j(\widehat{t}_{nj}\lambda_j) + \widehat{t}_{n,j+1}\lambda_{j+1}\right] X_j\right\}.$$

Hence, we can write $Y_n^* = \sum\limits_{j=1}^{n} a_{nj}X_j$, where

$$a_{nj} := \begin{cases} n^{1-1/k}\left[\frac{T_j}{t_j}\Delta_j(\widehat{t}_{nj}\lambda_j) + \widehat{t}_{n,j+1}\lambda_{j+1}\right] & , \quad 1 \le j \le n-1, \\ n^{1-1/k}\frac{T_n}{t_n}t_{nn}\lambda_n & , \quad n = j, \\ 0 & , \quad n > j. \end{cases}$$

Then, the series $\sum a_j\lambda_j$ is summable $|T|_k$ with $k \ge 1$ whenever $\sum a_j$ is summable $|\overline{N}^t|$, is equivalent to $\sum |Y_n^*|^k < \infty$ whenever $\sum |X_n| < \infty$, or, equivalently,

$$\sup_{j\in\mathbb{N}} \sum_n |a_{nj}|^k < \infty \tag{9.4.8}$$

by Lemma 9.4.1. From the definition of T, it follows that

$$\sum_{n=j}^{\infty} |a_{nj}|^k = j^{k-1}\left(\frac{T_n}{t_n}|t_{nn}\lambda_n|\right)^k + \sum_{n=j+1}^{\infty} n^{k-1}\left|\frac{T_j}{t_j}\Delta_j(\widehat{t}_{nj}\lambda_j) + \widehat{t}_{n,j+1}\lambda_{j+1}\right|^k.$$

Therefore, the conditions (i)–(iii) and Minkowski's inequality imply that

$$\sum_{n=j}^{\infty} |a_{nj}|^k = O(1)$$

as $j \to \infty$. This completes the proof. □

9.4.1 An Application of Quasi Power Increasing Sequences

By applying the concept of a quasi β-power increasing sequence, Şevli and Leindler [376] obtained the sufficient conditions for $\sum_n a_n\lambda_n$ to be absolutely summable of order k by a triangular matrix method.

A positive sequence (b_n) is said to be an *almost increasing sequence*, if there exist an increasing sequence (c_n) and positive constants A, B such that $Ac_n \le b_n \le Bc_n$, (see [17]). Obviously, every increasing sequence is almost increasing. However, the converse need not be true as can be seen by taking $b_n = ne^{(-1)^n}$, (see [273]).

A positive sequence $\gamma = (\gamma_n)$ is said to be a *quasi β-power increasing sequence* if there exists a constant $K = K(\beta, \gamma) \ge 1$ such that

$$Kn^{\beta}\gamma_n \ge m^{\beta}\gamma_m \tag{9.4.9}$$

holds for all $n \ge m \ge 1$. It should be noted that almost every increasing sequence is a quasi β-power increasing sequence for any non-negative β, but the converse need not be true as can be seen by taking an example, say $\gamma_n = n^{-\beta}$ for $\beta > 0$ (see [240]). If (9.4.9) stays with $\beta = 0$, then γ is simply called a quasi-increasing sequence. In [241], Leindler showed that the class of quasi-increasing sequences is equivalent to the class of almost increasing sequences. It is clear that if (γ_n) is quasi β-power increasing, then $(n^{\beta}\gamma_n)$ is quasi-increasing. Sulaiman [370] generalized this definition by considering $f = (f_n) := \{n^{\beta}(\log n)^{\mu}\}$, $\mu > 0$, $0 < \beta < 1$ instead of n^{β}. A positive sequence $\gamma = (\gamma_n)$ is said to be a *quasi f-power increasing sequence*, if there exists a constant $K = K(\gamma, f) \ge 1$ such that $Kf_n\gamma_n \ge f_m\gamma_m$ holds for all $n \ge m \ge 1$.

Prior to stating the next theorems, we designate $A = (a_{nj})$ belonging to the class $\mathcal{A}$ if the following conditions hold:

$$A \text{ is lower triangular,} \tag{9.4.10}$$

$$a_{nj} \ge 0 \quad \text{for all} \quad n, j \in \mathbb{N}, \tag{9.4.11}$$

$$a_{n-1,j} \ge a_{nj} \quad \text{for all} \quad n \ge j+1, \tag{9.4.12}$$

$$\breve{a}_{n0} = 1 \quad \text{for all } n \in \mathbb{N}. \tag{9.4.13}$$

Rhoades and Savaş [333] obtained the sufficient conditions for $\sum_n a_n\lambda_n$ to be summable $|A|_k$ with $k \ge 1$, as follows:

Theorem 9.4.3. [333, Theorem 1] *Let $A \in \mathcal{A}$ satisfying*

$$na_{nn} = O(1), \quad n \to \infty, \tag{9.4.14}$$

and (λ_n) be a real sequence. If (X_n) is an almost increasing sequence such that

$$|\lambda_m| X_m = O(1), \quad m \to \infty, \tag{9.4.15}$$

$$\sum_{n=1}^{m} nX_n \left|\Delta^2\lambda_n\right| = O(1), \quad m \to \infty, \tag{9.4.16}$$

$$\sum_{n=1}^{m} \frac{1}{n} |t_n|^k = O(X_m), \quad m \to \infty, \tag{9.4.17}$$

then the series $\sum_n a_n\lambda_n$ is summable $|A|_k$ with $k \ge 1$.

Savaş [356] obtained a further generalization of Theorem 9.4.3 under weaker conditions by using a quasi β-power increasing sequence instead of an almost increasing sequence, as follows:

Theorem 9.4.4. *Let $A \in \mathcal{A}$ satisfying the condition (9.4.14), and $(\lambda_n) \in bv_0$. If (X_n) is a quasi β-power increasing sequence for some $0 < \beta < 1$ such that the conditions (9.4.15)–(9.4.17) of Theorem 3.3 are satisfied, then the series $\sum_n a_n\lambda_n$ is summable $|A|_k$, $k \geq 1$.*

Theorem 9.4.5. [376, Theorem 2.1] *Let $A \in \mathcal{A}$ satisfying the condition (9.4.14) and (λ_n) be a sequence of real numbers satisfying*

$$\sum_{n=1}^{m} \lambda_n = o(m), \quad m \to \infty \tag{9.4.18}$$

$$\sum_{n=1}^{m} |\Delta\lambda_n| = o(m), \quad m \to \infty. \tag{9.4.19}$$

If (X_n) is a quasi f-increasing sequence and the conditions (9.4.17) and

$$\sum_{n=1}^{\infty} nX_n(\beta, \mu) \, |\Delta \, |\Delta\lambda_n|| < \infty \tag{9.4.20}$$

are satisfied, then the series $\sum_n a_n\lambda_n$ is summable $|A|_k$ with $k \geq 1$, where $(f_n) := \{n^\beta (\log n)^\mu\}$ and $X_n(\beta, \mu) = \max\{n^\beta(\log n)^\mu X_n, \ \log n\}$, $\mu \geq 0$ and $0 \leq \beta < 1$.

We need the following lemmas for proving Theorem 9.4.5.

Lemma 9.4.6. [329, p. 86] *Let $A \in \mathcal{A}$. Then, we have*

$$\sum_{j=1}^{n-1} |\Delta_n(a_{n-1,j})| \leq a_{nn}, \tag{9.4.21}$$

$$\sum_{n=j+1}^{m+1} |\Delta_n(a_{n-1,j})| \leq a_{jj}. \tag{9.4.22}$$

Lemma 9.4.7. [329, p. 86] *Let $A \in \mathcal{A}$. Then,*

$$\widehat{a}_{n,j+1} \leq a_{nn} \text{ for } n \geq j+1 \tag{9.4.23}$$

$$\sum_{n=j+1}^{m+1} \widehat{a}_{n,j+1} \leq 1, \quad j \in \mathbb{N}. \tag{9.4.24}$$

Lemma 9.4.8. [240, Lemma 3.2] *Let (φ_n) be a sequence of real numbers and define Φ_n and Ψ_n for all $n \in \mathbb{N}_1$ by*

$$\Phi_n := \sum_{k=1}^{n} \varphi_k \text{ and } \Psi_n := \sum_{k=n}^{\infty} |\Delta\varphi_k| .$$

If $\Phi_n = o(n)$, then there exists a natural number N such that $|\varphi_n| \leq 2\Psi_n$ for all $n \geq N$.

Lemma 9.4.9. [376, Lemma 3.4] *Let (X_n) be a quasi f-increasing sequence, where*

$$(f_n) := \left\{n^\beta(\log n)^\mu\right\},\ \mu \geq 0,\ 0 \leq \beta < 1.$$

If the conditions (9.4.18)–(9.4.20) of Theorem 9.4.4 are satisfied, then

$$\lim_{n\to\infty} \lambda_n = 0, \tag{9.4.25}$$

$$|\lambda_n|\, X_n = O(1),\ n \to \infty, \tag{9.4.26}$$

$$\sum_{n=1}^{\infty} X_n\, |\Delta\lambda_n| < \infty. \tag{9.4.27}$$

Proof of Theorem 9.4.5. Let (y_n) denote the A-transform of the sequence of partial sums of the series $\sum_n a_n\lambda_n$. Then, by definition, we have

$$y_n = \sum_{i=0}^{n} a_{ni}s_i = \sum_{i=0}^{n} a_{ni}\sum_{j=0}^{i}\lambda_j a_j = \sum_{j=0}^{n}\lambda_j a_j\sum_{i=j}^{n} a_{ni} = \sum_{j=0}^{n}\breve{a}_{nj}\lambda_j a_j.$$

Then, for $n \geq 1$, we have

$$Y_n = y_n - y_{n-1} = \sum_{j=0}^{n}(\breve{a}_{nj} - \breve{a}_{n-1,j})\lambda_j a_j = \sum_{j=0}^{n}\widehat{a}_{nj}\lambda_j a_j. \tag{9.4.28}$$

Applying Abel's transformation to the right-hand side of (9.4.28), we may write

$$\begin{aligned} Y_n &= \sum_{j=1}^{n}\left(\frac{\widehat{a}_{nj}\lambda_j}{j}\right) ja_j = \sum_{j=1}^{n-1}\Delta_j\left(\frac{\widehat{a}_{nj}\lambda_j}{j}\right)\sum_{r=1}^{v} ra_r + \frac{\widehat{a}_{nn}\lambda_n}{n}\sum_{j=1}^{n} ja_j \\ &= \sum_{j=1}^{n-1}\Delta_j\left(\frac{\widehat{a}_{nj}\lambda_j}{j}\right)(j+1)t_j + \frac{\widehat{a}_{nn}\lambda_n}{n}(n+1)\,t_n. \end{aligned}$$

Since

$$\Delta_j\left(\frac{\widehat{a}_{nj}\lambda_j}{j}\right) = \frac{-\lambda_j}{j}\Delta_n(a_{n-1,j}) + \frac{1}{j}\Delta\lambda_j\widehat{a}_{n,j+1} + \Delta\left(\frac{1}{j}\right)\lambda_{j+1}\widehat{a}_{n,j+1},$$

we have

$$\begin{aligned} Y_n &= \frac{(n+1)a_{nn}t_n\lambda_n}{n} - \sum_{j=1}^{n-1}\Delta_n(a_{n-1,j})\lambda_j t_j\frac{j+1}{j} \\ &\quad + \sum_{j=1}^{n-1}\widehat{a}_{n,j+1}\Delta\lambda_j t_j\frac{j+1}{j} + \sum_{j=1}^{n-1}\widehat{a}_{n,j+1}\lambda_{j+1}t_j\frac{1}{j} \\ &= Y_{n,1} + Y_{n,2} + Y_{n,3} + Y_{n,4}, \text{ say.} \end{aligned}$$

Since

$$|Y_{n,1} + Y_{n,2} + Y_{n,3} + Y_{n,4}|^k \leq 4^k\left(|Y_{n,1}|^k + |Y_{n,2}|^k + |Y_{n,3}|^k + |Y_{n,4}|^k\right),$$

to complete the proof, it is sufficient to show that

$$\sum_{n=1}^{\infty} n^{k-1} |Y_{n,r}|^k < \infty \quad \text{for} \quad r = 1, 2, 3, 4. \tag{9.4.29}$$

Since (λ_n) is bounded by (9.4.25) of Lemma 9.4.9 and using (9.4.14), we have

$$\begin{aligned} I_1 &= \sum_{n=1}^{m} n^{k-1} |Y_{n,1}|^k = \sum_{n=1}^{m} n^{k-1} \left| \frac{(n+1)a_{nn} t_n \lambda_n}{n} \right|^k \\ &= O(1) \sum_{n=1}^{m} (n a_{nn})^{k-1} a_{nn} |t_n|^k |\lambda_n|^{k-1} |\lambda_n| \\ &= O(1) \sum_{n=1}^{m} a_{nn} |\lambda_n| |t_n|^k . \end{aligned}$$

Using summation by parts, properties (9.4.26) and (9.4.27) of Lemma 9.4.9 and (9.4.17) yield

$$\begin{aligned} I_1 &= O(1) \sum_{n=1}^{m-1} |\Delta \lambda_n| \sum_{j=1}^{n} a_{vv} |t_j|^k + O(1) |\lambda_m| \sum_{j=1}^{m} a_{jj} |t_j|^k \\ &= O(1) \sum_{n=1}^{m-1} |\Delta \lambda_n| \sum_{j=1}^{n} \frac{1}{j} |t_j|^k + O(1) |\lambda_m| \sum_{j=1}^{m} \frac{1}{j} |t_j|^k \\ &= O(1) \sum_{n=1}^{m-1} |\Delta \lambda_n| X_n + O(1) |\lambda_m| X_m = O(1), \quad \text{as } m \to \infty. \end{aligned}$$

Applying Hölder's inequality, we get

$$\begin{aligned} I_2 &= \sum_{n=2}^{m+1} n^{k-1} |Y_{n,2}|^k = O(1) \sum_{n=2}^{m+1} n^{k-1} \left(\sum_{v=1}^{n-1} |\Delta_n (a_{n-1,v})| |t_v| |\lambda_v| \right)^k \\ &= O(1) \sum_{n=2}^{m+1} n^{k-1} \sum_{v=1}^{n-1} |\Delta_n (a_{n-1,v})| |t_v|^k |\lambda_v|^k \times \left[\sum_{v=1}^{n-1} |\Delta_n (a_{n-1,v})| \right]^{k-1} . \end{aligned}$$

Using boundedness of (λ_n) and the properties of Lemma 9.4.6,

$$\begin{aligned} I_2 &= O(1) \left(\sum_{n=2}^{m+1} (n a_{nn} \right)^{k-1} \sum_{j=1}^{n-1} |\Delta_n (a_{n-1,j})| |t_j|^k |\lambda_j|^{k-1} |\lambda_j| \\ &= O(1) \sum_{j=1}^{m} |\lambda_j| |t_j|^k \sum_{n=j+1}^{m+1} |\Delta_n (a_{n-1,j})| \\ &= O(1) \sum_{j=1}^{m} a_{jj} |\lambda_j| |t_j|^k = O(1), \quad \text{as } m \to \infty, \end{aligned}$$

as in the proof of I_1.

Using Hölder's inequality,

$$\begin{aligned} I_3 &= \sum_{n=2}^{m+1} n^{k-1} |Y_{n,3}|^k = O(1) \sum_{n=2}^{m+1} n^{k-1} \left(\sum_{j=1}^{n-1} \widehat{a}_{n,j+1} |\Delta\lambda_j| |t_j| \right)^k \\ &= O(1) \sum_{n=2}^{m+1} n^{k-1} \sum_{j=1}^{n-1} \widehat{a}_{n,j+1} |\Delta\lambda_j| |t_j|^k \times \left(\sum_{j=1}^{n-1} \widehat{a}_{n,j+1} |\Delta\lambda_j| \right)^{k-1}. \end{aligned}$$

By Lemma 9.4.8, the condition (9.4.19) implies that

$$\sum_{n=1}^{\infty} |\Delta\lambda_n| \le 2 \sum_{n=1}^{\infty} \sum_{k=n}^{\infty} |\Delta(\Delta\lambda_k)| = 2 \sum_{k=1}^{\infty} k |\Delta(\Delta\lambda_k)|$$

holds. Thus, (9.4.20) implies that $\sum_{n=1}^{\infty} |\Delta\lambda_n|$ converges. Therefore, there exists a positive constant M such that $\sum_{n=1}^{\infty} |\Delta\lambda_n| \le M$ and from (9.4.23), we obtain

$$\sum_{j=1}^{n-1} \widehat{a}_{n,j+1} |\Delta\lambda_j| \le a_{nn} \sum_{j=1}^{n-1} |\Delta\lambda_j| \le M a_{nn}.$$

Using the last inequality, from (9.4.14) and (9.4.24) we have

$$\begin{aligned} I_3 &= O(1) \sum_{n=2}^{m+1} (n a_{nn})^{k-1} \sum_{j=1}^{n-1} \widehat{a}_{n,j+1} |\Delta\lambda_j| |t_j|^k \\ &= O(1) \sum_{j=1}^{m} |\Delta\lambda_j| |t_j|^k \sum_{n=j+1}^{m+1} \widehat{a}_{n,j+1} \\ &= O(1) \sum_{j=1}^{m} j |\Delta\lambda_j| \frac{|t_j|^k}{j}. \end{aligned}$$

Using Abel transformation and (9.4.17),

$$I_3 = O(1) \sum_{j=1}^{m-1} |\Delta(j |\Delta\lambda_j|)| \left(\sum_{r=1}^{j} \frac{1}{r} |t_r|^k \right) + O(1) m |\Delta\lambda_m| \sum_{j=1}^{m} \frac{1}{j} |t_j|^k.$$

Since

$$\Delta(j |\Delta\lambda_j|) = j |\Delta\lambda_j| - (j+1) |\Delta\lambda_{j+1}| = j\Delta |\Delta\lambda_j| - \Delta\lambda_{j+1},$$

we have

$$I_3 = O(1) \sum_{j=1}^{m-1} j X_j |\Delta(\Delta\lambda_j)| + O(1) \sum_{j=1}^{m-1} X_{j+1} |\Delta\lambda_{j+1}| + O(1) m X_m |\Delta\lambda_m|.$$

Here, the first term is bounded by (9.4.20) and the second one by (9.4.27). Utilizing the quasi monotonicity of $\{n^{\beta}(\log n)^{\mu} X_n\}$ and the fact that $\{n^{1-\beta}(\log n)^{-\mu}\}$ is increasing, from Lemma 9.4.8, condition (9.4.19) implies that

$$
\begin{aligned}
mX_m\left|\Delta\lambda_m\right| &= O(1)mX_m m^{\beta}(\log m)^{\mu} m^{-\beta}(\log m)^{-\mu}\sum_{k=m}^{\infty}\left|\Delta\left(\Delta\lambda_k\right)\right| \\
&= O(1)m^{1-\beta}(\log m)^{-\mu}\sum_{k=m}^{\infty}k^{\beta}(\log k)^{\mu}X_k\left|\Delta\left(\Delta\lambda_k\right)\right| \\
&= O(1)\sum_{k=m}^{\infty}k^{1-\beta}(\log k)^{-\mu}k^{\beta}(\log k)^{\mu}X_k\left|\Delta\left(\Delta\lambda_k\right)\right| \\
&= O(1)\sum_{k=m}^{\infty}kX_k\left|\Delta\left(\Delta\lambda_k\right)\right|,
\end{aligned}
$$

and the third term is bounded by (9.4.20). Therefore $I_3 = O(1)$, as $m \to \infty$.

Finally, again using Hölder's inequality,

$$
\begin{aligned}
I_4 &= \sum_{n=2}^{m+1} n^{k-1}\left|Y_{n,4}\right|^k \leq \sum_{n=2}^{m+1} n^{k-1}\left(\sum_{j=1}^{n-1}\widehat{a}_{n,j+1}\left|\lambda_{j+1}\right|\left|t_j\right|\frac{1}{j}\right)^k \\
&= O(1)\sum_{n=2}^{m+1} n^{k-1}\sum_{j=1}^{n-1}\widehat{a}_{n,j+1}\left|\lambda_{j+1}\right|\left|t_j\right|^k\frac{1}{j}\left(\sum_{j=1}^{n-1}\widehat{a}_{n,j+1}\frac{1}{j}\left|\lambda_{j+1}\right|\right)^{k-1}.
\end{aligned}
$$

By Lemma 9.4.7, the conditions (9.4.18) and (9.4.19) imply that

$$
\left|\lambda_n\right| = O(1)\sum_{k=n}^{\infty}k\left|\Delta\left(\Delta\lambda_k\right)\right|.
$$

Thus, from (9.4.20),

$$
\begin{aligned}
\sum_{j=1}^{\infty}\frac{1}{j}\left|\lambda_j\right| &= O(1)\sum_{j=1}^{\infty}\frac{1}{j}\sum_{k=j}^{\infty}k\left|\Delta\left(\Delta\lambda_k\right)\right| \\
&= O(1)\sum_{k=1}^{\infty}k\log k\left|\Delta\left(\Delta\lambda_k\right)\right| = O(1).
\end{aligned}
$$

Therefore, from (9.4.23), there exists a positive constant K such that

$$
\sum_{j=1}^{n-1}\widehat{a}_{n,j+1}\frac{1}{j}\left|\lambda_{j+1}\right| \leq a_{nn}\sum_{j=1}^{n-1}\frac{1}{j}\left|\lambda_j\right| \leq Ka_{nn}.
$$

In view of the last inequality and using (9.4.14) and (9.4.24)

$$
\begin{aligned}
I_4 &= O(1)\sum_{n=2}^{m+1}(na_{nn})^{k-1}\sum_{j=1}^{n-1}\widehat{a}_{n,j+1}\left|\lambda_{j+1}\right|\frac{\left|t_j\right|^k}{j} \\
&= O(1)\sum_{j=1}^{m}\frac{\left|\lambda_{j+1}\right|}{j}\left|t_j\right|^k\sum_{n=j+1}^{m+1}\widehat{a}_{n,j+1} \\
&= O(1)\sum_{j=1}^{m}\left|\lambda_{j+1}\right|\frac{\left|t_j\right|^k}{j}.
\end{aligned}
$$

Hence,

$$\begin{aligned} I_4 &= O(1)\sum_{j=1}^{m-1}|\Delta\lambda_{j+1}|\sum_{r=1}^{j+1}\frac{1}{r}|t_r|^k + O(1)|\lambda_{m+1}|\sum_{j=1}^{m+1}\frac{1}{j}|t_j|^k \\ &= O(1)\sum_{j=1}^{m-1}X_{j+1}|\Delta\lambda_{j+1}| + O(1)X_{m+1}|\lambda_{m+1}| \\ &= O(1) \text{ as } m\to\infty, \end{aligned}$$

by virtue of (9.4.17) and the properties (9.4.26) and (9.4.27) of Lemma 9.4.9.

So, we obtain (9.4.29). This completes the proof. □

Theorem 9.4.5 includes the following theorem with the special case $\mu = 0$.

Theorem 9.4.10. [376, Theorem 2.2] *Let $A \in \mathcal{A}$ satisfying the condition (9.4.14) and (λ_n) be a sequence of real numbers satisfying the conditions (9.4.18) and (9.4.19). If (X_n) is a quasi β-power increasing sequence for some $0 \leq \beta < 1$ and the condition (9.4.17), and*

$$\sum_{n=1}^{\infty} nX_n(\beta)|\Delta(\Delta\lambda_n)| < \infty \tag{9.4.30}$$

are satisfied, where $X_n(\beta) = \max\{n^\beta X_n, \log n\}$, then the series $\sum_n a_n\lambda_n$ is summable $|A|_k$ with $k \geq 1$.

Corollary 9.4.11. [376, Corollary 5.1] *Let (t_n) be a positive sequence satisfying*

$$nt_n = O(T_n), \quad \text{as} \quad n \to \infty, \tag{9.4.31}$$

and (X_n) be a quasi f-increasing sequence, where $(f_n) := \{n^\beta(\log n)^\mu\}$, $\mu \geq 0$, $0 \leq \beta < 1$. Then, under the conditions (9.4.17)–(9.4.20), $\sum_n a_n\lambda_n$ is summable $|R^t|_k$ with $k \geq 1$.

Corollary 9.4.12. [376, Corollary 5.2] *Let (t_n) is a positive sequence satisfying (9.4.31) and (X_n) be a quasi β-power increasing sequence for some $0 \leq \beta < 1$. Then, under the conditions (9.4.17)–(9.4.19) and (9.4.30), $\sum_n a_n\lambda_n$ is summable $|R^t|_k$ with $k \geq 1$.*

Bibliography

[1] C.R. Adams, *On non-factorable transformations of double sequences*, Proc. Natl. Acad. Sci. USA. **19** (5) (1933), 564–567.

[2] R.P. Agnew, *The behavior of bound and oscillations of sequences of functions under regular transformations*, Trans. Amer. Math. Soc. **32** (1930), 669–708.

[3] R.P. Agnew, *On ranges of inconsistency of regular transformations and allied topics*, Ann. Math. **32** (2) (1931), 715–722.

[4] R.P. Agnew, *Cores of complex sequences and of their transforms*, Amer. J. Math. **61** (1939), 178–186.

[5] R.P. Agnew, *On cores of bounded divergent complex sequences and of their transforms by square matrices*, Revista Ci. Lima **47** (1945), 87–103.

[6] R.P. Agnew, *Abel transforms and partial sums of Tauberian series*, Ann. Math. **50** (1949), 110–117.

[7] Z.U. Ahmad, *On difference sequence spaces*, Analysis and Applications (Ujjain, 1999), 1–18, Narosa, New Delhi, 2002.

[8] Z.U. Ahmad, Mursaleen, *Köthe-Toeplitz duals of some new sequence spaces and their matrix maps*, Publ. Inst. Math. (Beograd) (N.S.) **42** (1987), 57–61.

[9] Z.U. Ahmad, Mursaleen, A.H.A. Bataineh, *On some generalized spaces of difference sequences,* Southeast Asian Bull. Math. **29** (4) (2005), 635–649.

[10] Z.U. Ahmad, Mursaleen, Q.A. Khan, *Invariant means and some matrix transformations*, Indian J. Pure Appl. Math. **25** (3) (1994), 353–359.

[11] Z.U. Ahmad, Mursaleen, Q.A. Khan, *Generalized almost convergence and Knopp's core theorem*, Math. Slovaca **46** (1996), 245–253.

[12] A.M. Akhmedov, F. Başar, *On spectrum of the Cesaro operator*, Proc. Inst. Math. Mech. Natl. Acad. Sci. Azerb. **19** (2003), 3–8.

[13] A.M. Akhmedov, F. Başar, *On the fine spectrum of the Cesàro operator in* c_0, Math. J. Ibaraki Univ. **36** (2004), 25–32.

[14] A.M. Akhmedov, F. Başar, *On the fine spectra of the difference operator* Δ *over the sequence space* ℓ_p, $(1 \leq p < \infty)$, Demonstratio Math. **39** (3) (2006), 585–595.

[15] A.M. Akhmedov, F. Başar, *The fine spectra of the difference operator* Δ *over the sequence space* bv_p, $(1 \leq p < \infty)$, Acta Math. Sin. Eng. Ser. **23** (10) (2007), 1757–1768.

[16] A.M. Akhmedov, F. Başar, *The fine spectra of the Cesàro operator* C_1 *over the sequence space* bv_p, $(1 \leq p < \infty)$, Math. J. Okayama Univ. **50** (2008), 135–147.

[17] S. Alijancic, D. Arendelovic, *O-regularly varying functions*, Publ. Inst. Math. (Beograd) (N.S.) **22** (36) (1977), 5–22.

[18] H.S. Allen, *T-transformations which leave the core of every bounded sequence invariant*, J. London Math. Soc. **19** (1944), 42–46.

[19] R.S. Alsaedi, A.H.A. Bataineh, *On a generalized difference sequence spaces defined by a sequence of Orlicz functions*, Int. Math. Forum **2** (1-4) (2007), 167–177.

[20] B. Altay, *On the space of p-summable difference sequences of order m*, $(1 \leq p < \infty)$, Stud. Sci. Math. Hungar. **43** (4) (2006), 387–402.

[21] B. Altay, F. Başar, *On the paranormed Riesz sequence spaces of non-absolute type*, Southeast Asian Bull. Math. **26** (5) (2002), 701–715.

[22] B. Altay, F. Başar, *On the fine spectrum of the difference operator* Δ *on* c_0 *and* c, Inform. Sci. **168** (2004), 217–224.

[23] B. Altay, F. Başar, *Some Euler sequence spaces of non-absolute type*, Ukrainian Math. J. **57** (1) (2005), 1–17.

[24] B. Altay, F. Başar, *Some new spaces of double sequences*, J. Math. Anal. Appl. **309** (1) (2005), 70–90.

[25] B. Altay, F. Başar, *On the fine spectrum of the generalized difference operator* $B(r,s)$ *over the sequence spaces* c_0 *and* c, Int. J. Math. Math. Sci. **18** (2005), 3005–3013.

[26] B. Altay, F. Başar, *Some paranormed Riesz sequence spaces of non-absolute type*, Southeast Asian Bull. Math. **30** (5) (2006), 591–608.

[27] B. Altay, F. Başar, *Some paranormed sequence spaces of non-absolute type derived by weighted mean*, J. Math. Anal. Appl. **319** (2) (2006), 494–508.

[28] B. Altay, F. Başar, *Generalization of the sequence space* $\ell(p)$ *derived by weighted mean*, J. Math. Anal. Appl. **330** (1) (2007), 174–185.

[29] B. Altay, F. Başar, *Certain topological properties and duals of the matrix domain of a triangle matrix in a sequence space*, J. Math. Anal. Appl. **336** (1) (2007), 632–645.

[30] B. Altay, F. Başar, *The matrix domain and the fine spectrum of the difference operator* Δ *on the sequence space* ℓ_p, $(0 < p < 1)$, Commun. Math. Anal. **2** (2) (2007), 1–11.

[31] B. Altay, F. Başar, M. Mursaleen, *On the Euler sequence spaces which include the spaces* ℓ_p *and* ℓ_∞ *I*, Inform. Sci. **176** (10) (2006), 1450–1462.

[32] B. Altay, M. Karakuş, *On the spectrum and fine spectrum of the Zweier matrix as an operator on some sequence spaces*, Thai J. Math. **3** (2005), 153–162.

[33] Y. Altın, *A note on lacunary statistically convergent double sequences of fuzzy numbers*, Commun. Korean Math. Soc. **23** (2) (2008), 179–185.

[34] Y. Altın, M. Et, *Generalized difference sequence spaces defined by a modulus function in a locally convex space*, Soochow J. Math. **31** (2) (2005), 233–243.

[35] Y. Altın, M. Et, M. Başarır, *On some generalized difference sequences of fuzzy numbers*, Kuwait J. Sci. Eng. **34** (1) (2007), 1–14.

[36] Y. Altın, M. Et, R. Çolak, *Lacunary statistical and lacunary strongly convergence of generalized difference sequences of fuzzy numbers*, Comput. Math. Appl. **52** (2006), 1011–1020.

[37] H. Altınok, M. Et, Y. Altın, *Strongly almost summable difference sequences*, Vietnam J. Math. **34** (3) (2006), 331–339.

[38] J. Appell, E. Pascale, A. Vignoli, *Nonlinear Spectral Theory*, Walter de Gruyter, Berlin · New York, 2004.

[39] S.M. Aseev, *Quasilinear operators and their application in the theory of multivalued mappings*, Proc. Steklov Inst. Math. **2** (1986), 23–52.

[40] Ç. Asma, R. Çolak, *On the Köthe-Toeplitz duals of some generalized sets of difference sequences*, Demonstratio Math. **33** (2000), 797–803.

[41] Ç. Asma, M. Et, *On v-invariant sequence spaces*, Soochow J. Math. **24** (4) (1998), 305–308.

[42] C. Aydın, F. Başar, *On the new sequence spaces which include the spaces* c_0 *and* c, Hokkaido Math. J. **33** (2) (2004), 383–398.

[43] C. Aydın, F. Başar, *Some new paranormed sequence spaces*, Inform. Sci. **160** (1-4) (2004), 27–40.

[44] C. Aydın, F. Başar, *Some new difference sequence spaces*, Appl. Math. Comput. **157** (3) (2004), 677–693.

[45] C. Aydın, F. Başar, *Some new sequence spaces which include the spaces* ℓ_p *and* ℓ_∞, Demonstratio Math. **38** (3) (2005), 641–656.

[46] C. Aydın, F. Başar, *Some generalizations of the sequence space a_p^r*, Iran. J. Sci. Technol. Trans. A: Sci. **30** (A2) (2006), 175–190.

[47] C. Aydın, F. Başar, *Some topological and geometric properties of the domain of the generalized difference matrix $B(r,s)$ in the sequence space $\ell(p)$*, Thai J. Math. **12** (1) (2014), 113–132.

[48] S. Aytar, *Statistical limit points of sequences of fuzzy numbers*, Inform. Sci. **165** (2004), 129–138.

[49] S. Aytar, M. Mammadov, S. Pehlivan, *Statistical limit inferior and limit superior for sequences of fuzzy numbers*, Fuzzy Sets and Systems **157** (7) (2006), 976–985.

[50] S. Aytar, S. Pehlivan, *Statistically monotonic and statistically bounded sequences of fuzzy numbers*, Inform. Sci. **176** (2006), 734–744.

[51] S. Aytar, S. Pehlivan, *Addendum to "Statistically monotonic and statistically bounded sequences of fuzzy numbers [Inform. Sci. **176** (2006), 734–744]"*, Inform. Sci. **177** (2) (2007), 655–656.

[52] S. Aytar, S. Pehlivan, *Statistical cluster and extreme limit points of sequences of fuzzy numbers*, Inform. Sci. **177** (16) (2007), 3290–3296.

[53] S. Aytar, S. Pehlivan, *Statistical convergence of sequences of fuzzy numbers and sequences of α-cuts*, Int. J. Gen. Syst. **37** (2) (2008), 231–237.

[54] S. Aytar, S. Pehlivan, *Levelwise statistical cluster and limit points of a sequence of fuzzy numbers*, Int. J. Gen. Syst. **37** (5) (2008), 573–584.

[55] S. Aytar, S. Pehlivan, M. Mammadov, *The core of a sequence of fuzzy numbers*, Fuzzy Sets and Systems **159** (24) (2008), 3369–3379.

[56] S.B. Baron, *Derivation of tests for the convergence of double numerical series*, Tartu Riikl. Ülik. Toimetised, **55** (1958), 9–19 (in Russian).

[57] R.G. Bartle, *The Elements of Real Analysis*, John Wiley & Sons Inc. New York, 1964.

[58] F. Başar, *Strongly-conservative sequence-to-series matrix transformations*, Erc. Üni. Fen Bil. Derg. **5** (12) (1989), 888–893.

[59] F. Başar, *f-conservative matrix sequences*, Tamkang J. Math. **22** (2) (1991), 205–212.

[60] F. Başar, *Infinite matrices and almost boundedness*, Boll. Unione Mat. Ital. A(7) **6** (1992), 395–402.

[61] F. Başar, *A note on the triangle limitation methods*, Fırat Üniv. Fen & Müh. Bil. Dergisi **5** (1) (1993), 113–117.

[62] F. Başar, *Infinite matrices and Cesàro sequence spaces of non-absolute type*, Math. J. Ibaraki Univ. **31** (1999), 1–12.

[63] F. Başar, *Matrix transformations between certain sequence spaces of* X_p *and* ℓ_p, Soochow J. Math. **26** (2) (2000), 191–204.

[64] F. Başar, *Dual summability medhods with a new approach*, *Modern Methods in Analysis and Its Applications*, pp. 56–67 Anamaya Publishers, New Delhi, 2010.

[65] F. Başar, B. Altay, *Matrix mappings on the space* $bs(p)$ *and its* α-, β- *and* γ-*duals*, Aligarh Bull. Math. **21** (1) (2001), 79–91.

[66] F. Başar, B. Altay, *On the space of sequences of* p-*bounded variation and related matrix mappings*, Ukrain. Mat. Zh. **55** (1) (2003), 108–118, reprinted in Ukrainian Math. J. **55** (1) (2003), 136–147.

[67] F. Başar, B. Altay, M. Mursaleen, *Some generalizations of the space* bv_p *of* p-*bounded variation sequences*, Nonlinear Anal. **68** (2) (2008), 273–287.

[68] F. Başar, E. Malkowsky, B. Altay, *Matrix transformations on the matrix domains of triangles in the spaces of strongly* C_1-*summable and bounded sequences*, Publ. Math. Debrecen **73** (1-2) (2008), 193–213.

[69] F. Başar, R. Çolak, *On the strongly regular dual summability methods of the second kind*, Comm. Fac. Sci. Univ. Ank. Ser. *A1* Math. Statist. **35** (1-2) (1986), 15–18.

[70] F. Başar, R. Çolak, *Almost-conservative matrix transformations*, Turk. J. Math. **13** (3) (1989), 91–100.

[71] F. Başar, N. Durna, M. Yıldırım, *Subdivisions of the spectra for generalized difference operator over certain sequence spaces*, Thai J. Math. **9** (2) (2011), 261–271.

[72] F. Başar, N. Durna, M. Yıldırım, *Subdivisions of the spectra for the triple band matrix over certain sequence spaces*, Gen. Math. Notes **4** (1) (2011), 35–48.

[73] F. Başar, H. Dutta, *Summable Spaces and Their Duals, Matrix Transformations and Geometric Properties*, Monographs and Research Notes in Mathematics, CRC Press/Taylor & Francis Group, Boca Raton · London · New York, 2020. ISBN: 978-0-8153-5177-1.

[74] F. Başar, M. Kirişçi, *Almost convergence and generalized difference matrix*, Comput. Math. Appl. **61** (3) (2011), 602–611.

[75] F. Başar, E. Öztürk, *On a generalization of the dual summability methods*, Comm. Fac. Sci. Univ. Ank. Ser. *A1* Math. Statist. **31** (10) (1982), 89–94.

[76] F. Başar, H. Roopaei, *On the factorable spaces of absolutely p-summable, null, convergent and bounded sequences*, Math. Slovaca **71** (6)(2021), 1375–1400.

[77] F. Başar, Y. Sever, *The space $\mathcal{L}_p$ of double sequences*, Math. J. Okayama Univ. **51** (2009), 149–157.

[78] F. Başar, İ. Solak, *Almost-coercive matrix transformations*, Rend. Mat. Appl. (7) **11** (2) (1991), 249–256.

[79] M. Başarır, *On some new sequence spaces and related matrix transformations*, Indian J. Pure Appl. Math. **26** (10) (1995), 1003–1010.

[80] M. Başarır, *On the generalized Riesz B-difference sequence spaces*, Filomat **24** (4) (2010), 35–52.

[81] M. Başarır, S. Altundağ, *Some difference sequence spaces defined by a sequence of ϕ-functions*, Rend. Circ. Mat. Palermo **57** (2) (2008), 149–160.

[82] M. Başarır, S. Altundağ, *Some generalized difference sequence spaces defined by a sequence of Orlicz functions*, Fasc. Math. **41** (2009), 17–30.

[83] M. Başarır, F. Başar, E.E. Kara, *On the spaces of Fibonacci difference absolutely p-summable, null and convergent sequences*, Sarajevo J. Math. **12 (25)** (2) (2016), 167–182.

[84] M. Başarır, M. Kayıkçı, *On the generalized B^m-Riesz sequence space and β-property*, J. Inequal. Appl. 2009 (2009), Article ID 385029, 18 pp.

[85] M. Başarır, M. Öztürk, *On the Riesz difference sequence space*, Rend. Circ. Mat. Palermo **(2) 57** (2008), 377–389.

[86] Ç.A. Bektaş, *On some new generalized sequence spaces*, J. Math. Anal. Appl. **277** (2003), 681–688.

[87] Ç.A. Bektaş, *Absolute and ordinary Köthe-Toeplitz duals of some sets of difference sequences*, Hokkaido Math. J. **33** (3) (2004), 207–214.

[88] Ç.A. Bektaş, R. Çolak, *On some generalized difference sequence spaces*, Thai J. Math. **3** (1) (2005), 83–98.

[89] Ç.A. Bektaş, M. Et, *The dual spaces of the sets of difference sequences of order m*, J. Inequal. Pure Appl. Math. **7** (3) (2006), Article 101, 10 pp.

[90] Ç.A. Bektaş, M. Et, R. Çolak, *Generalized difference sequence spaces and their dual spaces*, J. Math. Anal. Appl. **292** (2) (2004), 423–432.

[91] V.K. Bhardwaj, I. Bala, *On some lacunary generalized difference sequence spaces defined by a modulus function in a locally convex space*, Southeast Asian Bull. Math. **33** (4) (2009), 665–683.

[92] O. Biermann, *Über unendliche Doppelreihen und unendliche Doppelproducte*, Monats. Math. Phys. **8** (1897), 115–137.

[93] H. Bilgiç, H. Furkan, *On the fine spectrum of the generalized difference operator $B(r, s)$ over the sequence spaces ℓ_p and bv_p* $(1 < p < \infty)$, Nonlinear Anal. **68** (3) (2008), 499–506.

[94] J. Boos, *Classical and Modern Methods in Summability*, Oxford University Press Inc., New York, 2000.

[95] J. Boos, T. Leiger, K. Zeller, *Consistency theory for SM-methods*, Acta Math. Hungar. **76** (1-2) (1997), 109–142.

[96] H. Bor, *On $\left|\bar{N}, p_n\right|_k$ summability factors of infinite series*, Tamkang J. Math. **16** (1985), 13–20.

[97] H. Bor, *On the relative strength of two absolute summability methods*, Proc. Amer. Math. Soc. **113** (1991), 1009–1012.

[98] D. Borwein, F.P. Cass, *Strong Nörlund summability*, Math. Z. **103** (1968), 94–111.

[99] R.C. Buck, *Generalized asymptotic density*, Amer. J. Math. **75** (1953), 335–346.

[100] R.C. Buck, *Advanced Calculus*, 3rd edn. McGraw-Hill, New York, 1978.

[101] J.P. Cartlidge, *Weighted Mean Matrices as Operators on ℓ^p*, Ph.D. Dissertation, Indiana University, 1978.

[102] L. Cesari, *Sulla moltiplicazione delle serie doppie*, Ann. Scuola Norm. Sup. Pisa, **12** (1947), 189–204.

[103] S.S.L. Chang, L.A. Zadeh, *On fuzzy mapping and control*, IEEE Trans. Systems Man Cybernet **2** (1972), 30–34.

[104] B. Choudhary, *An extension of Knopp's core theorem*, J. Math. Anal. Appl. **132** (1988), 226–233.

[105] B. Choudhary, S.K. Mishra, *On Köthe-Toeplitz duals of certain sequence spaces and their matrix transformations*, Indian J. Pure Appl. Math. **24** (5) (1993), 291–301.

[106] B. Choudhary, S.K. Mishra, *A note on certain sequence spaces*, J. Anal. **1** (1993), 139–148.

[107] B. Choudhary, S. Nanda, *Functional Analysis with Applications*, John Wiley & Sons Inc. New Delhi, 1989.

[108] J.S. Connor, *The statistical and strong p-Cesàro convergence of sequences*, Analysis **8** (1-2) (1988), 47–63.

[109] J. Connor, *On strong matrix summability with respect to a modulus and statistical convergence*, Canad. Math. Bull. **32** (1989), 194–198.

[110] J. Connor, J.A. Fridy, C. Orhan, *Core equality results for sequences*, J. Math. Anal. Appl. **321** (2006), 515–523.

[111] R.G. Cooke, *Infinite Matrices and Sequence Spaces*, Macmillan and Co. Limited, London, 1950.

[112] C. Coşkun, *The spectra and fine spectra for p-Cesàro operators*, Turkish J. Math. **21** (1997), 207–212.

[113] H. Coşkun, F. Başar, *On the characterization of certain matrix classes*, Boll. Unione Mat. Ital. A(7) **8** (2) (1994), 215–222.

[114] C. Çakan, *Lacunary strong σ-convergence and σ-core*, Southeast Asian Bull. Math. **28** (2004), 203–212.

[115] C. Çakan, B. Altay, M. Mursaleen, *The σ-convergence and σ-core of double sequences*, Appl. Math. Lett. **19** (2006), 387-399.

[116] C. Çakan, C. Aydın, *Some inequalities related to a new sequence space which include c*, Thai J. Math. **2** (1) (2004), 115–122.

[117] C. Çakan, C. Aydın, *On the K_r-core of complex sequences and the absolute equivalence of summability matrices*, Math. Comput. Appl. **9** (3) (2004), 409–416.

[118] C. Çakan, H. Çoşkun, *A class of I-conservative matrices*, Int. J. Math. Math. Sci. **21** (2005), 3443–3452.

[119] C. Çakan, H. Çoşkun, *On the A-statistical core of sequences of complex numbers*, Acta Comm. Univ. Tartu Math. **7** (2003), 31–36.

[120] C. Çakan, H. Çoşkun, *A class of matrices mapping σ-core into A-statistical core*, Tamsui Oxford J. Math. Sci. **20** (1) (2004), 17–25.

[121] C. Çakan, Z. Kasap, *$F_{\mathcal{B}}$-convergence and a generalization of Knopp's core*, Georgian Math. J. **17** (3) (2010), 459–467.

[122] C. Çakan, M. Şengönül, *The ordinary and almost convergence of sequences of fuzzy numbers*, Thai J. Math. **1** (2) (2003), 67–73.

[123] R. Çolak, *On some generalized sequence spaces*, Comm. Fac. Sci. Univ. Ankara Ser. *A1* Math. Statist. **38** (1-2) (1989), 35–46.

[124] R. Çolak, *On invariant sequence spaces*, Erciyes Univ. J. Sci. **5** (1-2) (1989), 881–887.

[125] R. Çolak, *Lacunary strong convergence of difference sequences with respect to a modulus function*, Filomat **17** (2003), 9–14.

[126] R. Çolak, *Statistical convergence of order α*, *Modern Methods in Analysis and Its Applications*, pp. 121–129, Anamaya Publishers, New Delhi, 2010.

[127] R. Çolak, Y. Altın, Mursaalen, *On some sets of difference sequences of fuzzy numbers*, Soft Comput. **15** (4) (2010), 787–793.

[128] R. Çolak, H. Altınok, M. Et, *Generalized difference sequences of fuzzy numbers*, Chaos, Solitons Fractals **40** (3) (2009), 1106–1117.

[129] R. Çolak, Ç.A. Bektaş, *λ-statistical convergence of order α*, Acta Math. Sci. **31B** (3) (2011), 953–959.

[130] R. Çolak, Ö. Çakar, *Banach limits and related matrix transformations*, Stud. Sci. Math. Hungar. **24** (1989), 429–436.

[131] R. Çolak, M. Et, *On some generalized difference sequence spaces and related matrix transformations*, Hokkaido Math. J. **26** (3) (1997), 483–492.

[132] R. Çolak, M. Et, *On some difference sequence sets and their topological properties*, Bull. Malays. Math. Sci. Soc. (2) **28** (2) (2005), 125–130.

[133] R. Çolak, M. Et, E. Malkowsky, *Some Topics of Sequence Spaces*, Lecture Notes in Mathematics, pp. 1–63 Fırat Univ. Elâzığ, Turkey, 2004, ISBN: 975-394-038-6.

[134] H. Çoşkun, C. Çakan, *σ-core theorems for real bounded sequences*, Far East J. Math. Sci. **2** (1) (2000), 79–86.

[135] H. Çoşkun, C. Çakan, *Infinite matrices and σ-core*, Demonstratio Math. **34** (2001), 825–830.

[136] H. Çoşkun, C. Çakan, *On some new inequalities related to $F_{\mathcal{B}}$-convergence*, Tamsui Oxford J. Math. Sci. **19** (2) (2003), 131–140.

[137] H. Çoşkun, C. Çakan, *A class of statistical and σ-conservative matrices*, Czechoslovak Math. J. **55** (3) (2005), 791–801.

[138] H. Çoşkun, C. Çakan, Mursaleen, *On the statistical and σ-cores*, Studia Math. **154** (1) (2003), 29–35.

[139] G. Das, *A Tauberian theorem for absolute summability*, Proc. Cambridge. Philos. Soc. **67** (1970), 321–326.

[140] G. Das, *Banach and other limits*, J. London Math. Soc. **7** (1973), 501–507.

[141] G. Das, *Sublinear functionals and a class of conservative matrices*, Bull. Inst. Math. Acad. Sinica **15** (1987), 89–106.

[142] N.A. Davydov, *Inclusion of kernels of sequences given with the help of regular matrices*, Ukrainskii Math. Z. **21** (4) (1979), 421–425.

[143] S. Demiriz, S. Erdem, *Domain of binomial matrix in some spaces of double sequences*, Punjab Univ. J. Math. (Lahore) **52** (11) (2020), 65–79.

[144] S.L. Devi, *Banach limits and infinite matrices*, J. London Math. Soc. **12** (2) (1976), 397–401.

[145] K. Demirci, *A-statistical core of a sequence*, Demonstratio Math. **33** (2000), 343–353.

[146] K. Demirci, *$\mathcal{I}$-limit superior and limit inferior*, Math. Commun. **6** (2001), 165–172.

[147] K. Demirci, S. Yardımcı, *σ-core and $\mathcal{I}$-core of bounded sequences*, J. Math. Anal. Appl. **290** (2004), 414–422.

[148] P. Diamond, P. Kloeden, *Metric spaces of fuzzy sets*, Fuzzy Sets and Systems **35** (1990), 241–249.

[149] P. Diamond, P. Kloeden, *Metric Spaces of Fuzzy Sets: Theory and Applications*, World Scientific Publishing Co., Singapore, 1994.

[150] J. Diestel, *Sequence and Series in Banach Spaces*, Graduate Texts in Mathematics, **92**, Springer-Verlag, 1984.

[151] D. Dubois, H. Prade, *Operation on fuzzy numbers*, Internat J. System Sci. **9** (1978), 613–626.

[152] D. Dubois, H. Prade, *Fuzzy Sets and Systems*, Academic Press, New York, 1980.

[153] O. Duman, *The effects of matrix summability methods on bounds of function sequences*, Appl. Math. Lett. **17** (2004), 1221–1229.

[154] J.P. Duran, *Infinite matrices and almost convergence*, Math. Z. **128** (1972), 75–83.

[155] M. Et, *On some difference sequence spaces*, Turkish J. Math. **17** (1993), 18–24.

[156] M. Et, *On some generalized Cesàro difference sequence spaces*, İstanbul Üniv. Fen Fak. Mat. Derg. **55-56** (1996-1997), 221–229.

[157] M. Et, *On some topological properties of generalized difference sequence spaces*, Int. J. Math. Math. Sci. **24** (11) (2000), 785–791.

[158] M. Et, *Strongly almost summable difference sequences of order m defined by a modulus*, Studia Sci. Math. Hungar. **40** (4) (2003), 463–476.

[159] M. Et, *Spaces of Cesàro difference sequences of order r defined by a modulus function in a locally convex space*, Taiwanese J. Math. **10** (4) (2006), 865–879.

[160] M. Et, Y. Altın, H. Altınok, *On some generalized difference sequence spaces defined by a modulus function*, Filomat **17** (2003), 23–33.

[161] M. Et, Y. Altın, H. Altınok, *On almost statistical convergence of generalized difference sequences of fuzzy numbers*, Math. Model. Anal. **10** (4) (2005), 345–352.

[162] M. Et, Y. Altın, B. Choudhary, B.C. Tripathy, *On some classes of sequences defined by sequences of Orlicz functions*, Math. Inequal. Appl. **9** (2) (2006), 335–342.

[163] M. Et, H. Altınok, Y. Altın, *On some generalized sequence spaces*, Appl. Math. Comput. **154** (1) (2004), 167–173.

[164] M. Et, H. Altınok, R. Çolak, *On λ-statistical convergence of difference sequences of fuzzy numbers*, Inform. Sci. **176** (15) (2006), 2268–2278.

[165] M. Et, M. Başarır, *On some new generalized difference sequence spaces*, Period. Math. Hungar. **35** (3) (1997), 169–175.

[166] M. Et, R. Çolak, *On some generalized difference sequence spaces*, Soochow J. Math. **21** (4) (1995), 377–386.

[167] M. Et, A. Esi, *On Köthe-Toeplitz duals of generalized difference sequence spaces*, Bull. Malays. Math. Sci. Soc. (2) **23** (1) (2000), 25–32.

[168] M. Et, F. Nuray, *Δ^m-statistical convergence*, Indian J. Pure Appl. Math. **32** (6) (2001), 961–969.

[169] J. Fang, H. Huang, *On the level convergence of sequence of fuzzy numbers*, Fuzzy Sets and Systems **147** (2004), 417–435.

[170] H. Fast, *Sur la convergence statistique*, Colloq. Math. **2** (1951), 241–244.

[171] M. Fekete, *Zur theorie der divergenten reihen*, Math. ès Termezs Èrtesitö (Budapest) **29** (1911), 719–726.

[172] T.M. Flett, *On an extension of absolute summability and some theorems of Littlewood and Paley*, Proc. London Math. Soc. **7** (1957), 113–141.

[173] A.R. Freedman, J.J. Sember, *Densities and summability*, Pacific J. Math. **95** (1981), 293–305.

[174] J.A. Fridy, *On statistical convergence*, Analysis **5** (4) (1985), 301–313.

[175] J.A. Fridy, *Statistical limit points*, Proc. Amer. Math. Soc. **4** (1993), 1187–1192.

[176] J.A. Fridy, C. Orhan, *Statistical limit superior and limit inferior*, Proc. Amer. Math. Soc. **125** (12) (1997), 3625–3631.

[177] J.A. Fridy, C. Orhan, *Statistical core theorems*, J. Math. Anal. Appl. **208** (1997), 520–527.

[178] H. Furkan, H. Bilgiç, F. Başar, *On the fine spectrum of the operator $B(r, s, t)$ over the sequence spaces ℓ_p and bv_p, $(1 < p < \infty)$*, Comput. Math. Appl. **60** (7) (2010), 2141–2152.

[179] J. Garcia-Falset, *Stability and fixed points for nonexpansive mappings*, Houston J. Math. **20** (1994), 495–505.

[180] J. Garcia-Falset, *The fixed point property in Banach spaces with NUS-property*, J. Math. Anal. Appl. **215** (2) (1997), 532–542.

[181] A.K. Gaur, M. Mursaleen, *Difference sequence spaces*, Internat. J. Math. Math. Sci. **21** (4) (1998), 701–706.

[182] A.K. Gaur, M. Mursaleen, *Difference sequence spaces defined by a sequence of moduli*, Demonstratio Math. **31** (2) (1998), 275–278.

[183] S.R. Ghorpade, B.V. Limaye, *A Course in Calculus and Real Analysis*, Springer, New York, 2006.

[184] C. Gnanaseelan, P.D. Srivastava, *The α-, β- and γ-duals of some generalised difference sequence spaces*, Indian J. Math. **38** (2) (1996), 111–120.

[185] R. Goetschel, W. Voxman, *Elementary fuzzy calculus*, Fuzzy Sets and Systems **18** (1986), 31–43.

[186] C. Goffman, G.M. Petersen, *Consistent limitation methods*, Proc. Amer. Math. Soc. **7** (1956), 367–369.

[187] S. Goldberg, *Unbounded Lineer Operators*, Dover Publications Inc., New York, 1985.

[188] Z. Gong, C. Wu, *Bounded variation, absolute continuity and absolute integrability for fuzzy-number-valued functions*, Fuzzy Sets and Systems **129** (2002), 83–94.

[189] M. Gonzàlez, *The fine spectrum of the Cesàro operator in ℓ_p $(1 < p < \infty)$*, Arch. Math. **44** (1985), 355–358.

[190] A. Gökhan, R. Çolak, *The double sequence spaces $c_2^P(p)$ and $c_2^{PB}(p)$*, Appl. Math. Comput. **157** (2) (2004), 491–501.

[191] A. Gökhan, R. Çolak, *Double sequence space $\ell_2^\infty(p)$*, Appl. Math. Comput. **160** (1) (2005), 147–153.

[192] A. Gökhan, M. Et, Mursaleen, *Almost lacunary statistical and strongly almost lacunary convergence of sequences of fuzzy numbers*, Math. Comput. Modelling **49** (3-4) (2009), 548–555.

[193] K.-G. Grosse-Erdmann, *Matrix transformations between the sequence spaces of Maddox*, J. Math. Anal. Appl. **180** (1993), 223–238.

[194] K.-G. Grosse-Erdmann, *On ℓ^1-invariant sequence spaces*, J. Math. Anal. Appl. **262** (2001), 112–132.

[195] Z. Guangquan, *Fuzzy continuous function and its properties*, Fuzzy Sets and Systems **43** (1991), 159–171.

[196] V.I. Gurarii, *On differential properties of the convexity moduli of Banach spaces*, Mat. Issled. **2** (1969), 141–148.

[197] H.J. Hamilton, *Transformations of multiple sequences*, Duke Math. J. **2** (1936), 29–60.

[198] H.J. Hamilton, *On transformations of double series*, Bull. Amer. Math. Soc. **42** (1936), 275–283.

[199] G.H. Hardy, J.E. Littlewood, *Theorems concerning the summability of series by Borel's exponential method*, Rend. Circ. Mat. Palermo **41** (1916), 36–53.

[200] G.H. Hardy, J.E. Littlewood, G. Polya, *Inequalities*, Cambridge University Press, 1952.

[201] W.A. Hurwitz, *Some properties of methods of evaluation of divergent sequences*, Proc. London Math. Soc. **2** (26) (1927), 231–248.

[202] M. Imaninezhad, M.R. Miri, *The dual space of the sequence space bv_p ($1 \leq p < \infty$)*, Acta Math. Univ. Comen. **79** (1) (2010), 143–149.

[203] G.L. Isacs, *A class of regular matrices*, Proc. Amer. Math. Soc. **58** (1976), 211–214.

[204] V.G. Iyer, *Mathematical Analysis*, Tata McGraw-Hill Publishig Company Ltd. New Delhi, 1985.

[205] C. Jardas, N. Sarapa, *On the summability of pairs of sequences*, Glasnik Math. III. Ser. **26** (46) (1-2) (1991), 67–78.

[206] A.M. Jarrah, E. Malkowsky, *BK spaces, bases and linear operators*, Rend. Circ. Mat. Palermo (2) **52** (1990), 177–191.

[207] A.M. Jarrah, E. Malkowsky, *Ordinary, absolute and strong summability and matrix transformations*, Filomat **17** (2003), 59–78.

[208] U. Kadak, F. Başar, *Alternating and binomial series of fuzzy numbers with the level sets*, Contemp. Anal. Appl. Math. **3** (2) (2015), 310–328.

[209] O. Kaleva, *On the convergence of fuzzy sets*, Fuzzy Sets and Systems **17** (1985), 53–65.

[210] O. Kaleva, S. Seikkala, *On fuzzy metric spaces*, Fuzzy Sets and Systems **12** (1984), 215–229.

[211] P.K. Kamthan, M. Gupta, *Sequence Spaces and Series*, Marcel Dekker Inc., New York and Basel, 1981.

[212] E.E. Kara, M. Başarır, *On compact operators and some Euler $B(m)$-difference sequence spaces*, J. Math. Anal. Appl. **379** (2011), 499–511.

[213] E.E. Kara, M. Öztürk, M. Başarır, *Some topological and geometric properties of generalized Euler sequence spaces*, Math. Slovaca **60** (3) (2010), 385–398.

[214] A. Kaufmann, M.M. Gupta, *Introduction to Fuzzy Arithmetic*, Van Nostrand Reinhold, New York, 1984.

[215] V.A. Khan, Q.M.D. Lohani, *On quasi almost lacunary strong convergence difference sequence spaces defined by a sequence of moduli*, Mat. Vesnik **60** (2) (2008), 95–100.

[216] F.M. Khan, M.F. Rahman, *Matrix transformations on Cesàro sequence spaces of nonabsolute type*, J. Analysis **4** (1996), 97–101.

[217] H. Kızmaz, *On certain sequence spaces*, Canad. Math. Bull. **24** (2) (1981), 169–176.

[218] Y.K. Kim, B.M. Ghil, *Integrals of fuzzy-number-valued functions*, Fuzzy Sets and Systems **86** (1997), 213–222.

[219] J.P. King, *Almost summable sequences*, Proc. Amer. Math. Soc. **17** (1966), 1219–1225.

[220] M. Kirişçi, F. Başar, *Some new sequence spaces derived by the domain of generalized difference matrix*, Comput. Math. Appl. **60** (5) (2010), 1299–1309.

[221] H. Knaust, *Orlicz sequence spaces of Banach-Saks type*, Arch. Math. **59** (1992), 562–565.

[222] K. Knopp, *Über das Eulersche summerungsverfahren I*, Math. Z. **15** (1922), 226–253.

[223] K. Knopp, *Über das Eulersche summerungsverfahren II*, Math. Z. **16** (1923), 125–156.

[224] K. Knopp, *Zur theorie der limitierungsverfahren (Erste Mitteilung)*, Math. Z. **31** (1930), 97–127.

[225] K. Knopp, *Infinite Sequences and Series*, Dover Publications Inc., New York, 1956.

[226] K. Knopp, *Theory and Application of Infinite Series*, Dover Publications Inc., New York, 1990.

[227] E. Kogbetliantz, *Sur les séries absolument sommables par la méthode des moyennes arithmétiques*, Bull. Sci. Math. **49** (1925), 234–256.

[228] E. Kolk, *Inclusion relations between the statistical convergence and strong summability*, Acta Comment. Univ. Tartu. Math. **2** (1998), 39–54.

[229] P. Kostyrko, T. Salát, W. Wilcsynski, *$\mathcal{I}$-convergence*, Real Anal. Exchange **26** (2) (2001), 669–685.

[230] M.A. Krasnoselkii, Y.B. Rutickii, *Convex Functions and Orlicz Spaces*, Groningen, Netherlands, 1961.

[231] E. Kreyszig, *Introductory Functional Analysis with Applications*, John Wiley & Sons Inc., New York · Chichester · Brisbane · Toronto, 1978.

[232] B. Kuttner, *On dual summability methods*, Proc. Camb. Phil. Soc. **71** (1972), 67–73.

[233] J.S. Kwon, *On statistical and p-Cesaro convergence of fuzzy numbers*, Korean J. Comput. Appl. Math. **7** (2000), 195–203.

[234] J.S. Kwon, S.H. Sung, *On lacunary statistical and p-Cesàro summability of fuzzy numbers*, J. Fuzzy Math. **9** (3) (2001), 603–610.

[235] J.S. Kwon, H.T. Shim, *Remark on lacunary statistical convergence of fuzzy numbers*, Fuzzy Sets and Syst. **123** (1) (2001), 85–88.

[236] C.G. Lascarides, *A study of certain sequence spaces of Maddox and a generalization of a theorem of Iyer*, Pacific J. Math. **38** (1971), 487–500.

[237] C.G. Lascarides, I.J. Maddox, *Matrix transformations between some classes of sequences*, Proc. Camb. Phil. Soc. **68** (1970), 99–104.

[238] G. Leibowitz, *The Cesàro operators and their generalizations: Examples in infinite dimensional linear analysis*, Amer. Math. Monthly **80** (1973), 654–661.

[239] G. Leibowitz, *Rhaly matrices*, J. Math. Anal. Appl. **128** (1987), 272–286.

[240] L. Leindler, *A note on the absolute Riesz summability factors*, J. Inequal. Pure Appl. Math. **6** (2005) Art. 96.

[241] L. Leindler, *A recent note on the absolute Riesz summability factors*, J. Inequal. Pure Appl. Math. **7** (2006) Art. 44.

[242] J. Lie, J.A. Fridy, *Matrix transformations of statistical cores of complex sequences*, Analysis **20** (2000), 15–34.

[243] B.V. Limaye, M. Zeltser, *On the Pringsheim convergence of double series*, Proc. Est. Acad. Sci. Phys. Math. **58** (2) (2009), 108–121.

[244] K. Lindberg, *On subspaces of Orlicz sequence spaces*, Studia Math. **45** (1973), 119–146.

[245] J. Lindenstrauss, L. Tzafriri, *On Orlicz sequence spaces*, Israel J. Math. **10** (1971), 379–390.

[246] G.G. Lorentz, *Über Limitierungsverfahren die von einem Stieltjes-Integral abhängen*, Acta Math. **79** (1947), 255–272.

[247] G.G. Lorentz, *A contribution to the theory of divergent sequences*, Acta Math. **80** (1948), 167–190.

[248] G.G. Lorentz, K. Zeller, *Summation of sequences and summation of series*, Proc. Amer. Math. Soc. **15** (1964), 743–746.

[249] I.J. Maddox, *On theorems of Steinhaus type*, J. London Math. Soc. **42** (1967), 239–244.

[250] I.J. Maddox, *Spaces of strongly summable sequences*, Quart. J. Math. Oxford (2) **18** (1967), 345–355.

[251] I.J. Maddox, *Paranormed sequence spaces generated by infinite matrices*, Proc. Camb. Phil. Soc. **64** (1968), 335–340.

[252] I.J. Maddox, *On Kuttner's theorem*, J. London Math. Soc. **43** (1968), 285–290.

[253] I.J. Maddox, *Some properties of paranormed sequence spaces*, London J. Math. Soc. (2) **1** (1969), 316–322.

[254] I.J. Maddox, *Continuous and Köthe-Toeplitz duals of certain sequence spaces*, Proc. Camb. Phil. Soc. **65** (1969), 431–435.

[255] I.J. Maddox, *Elements of Functional Analysis*, Cambridge University Press, Cambridge · London · New York · Melbourne, 1970.

[256] I.J. Maddox, *Some analogues of Knopp's core theorem*, Internat. J. Math. Math. Sci. **2** (1979), 605–614.

[257] I.J. Maddox, *Sequence spaces defined by a modulus*, Math. Proc. Cambridge Philos. Soc. **100** (1986), 161–166.

[258] I.J. Maddox, *Elements of Functional Analysis*, 2nd edn. The University Press, Cambridge, 1988.

[259] B. de Malafosse, *Properties of some sets of sequences and application to the spaces of bounded difference sequences of order μ*, Hokkaido Math. J. **31** (2002), 283–299.

[260] E. Malkowsky, *Sequence Spaces, Matrix Transformations and Compact Operators*, Lecture Notes, in preparation.

[261] E. Malkowsky, *Absolute and ordinary Köthe-Toeplitz duals of some sets of sequences and matrix transformations*, Publ. Inst. Math. (Beograd) (N.S.) **46** (60) (1989), 97–103.

[262] E. Malkowsky, *A note on the Köthe-Toeplitz duals of generalized sets of bounded and convergent difference sequences*, J. Anal. **4** (1996), 81–91.

[263] E. Malkowsky, *Recent results in the theory of matrix transformations in sequence spaces*, Mat. Vesnik **49** (1997), 187–196.

[264] E. Malkowsky, *Linear operators between some matrix domains*, Rend. Circ. Mat. Palermo (2) **68** (2002), 641–655.

[265] E. Malkowsky, Mursaleen, *Some matrix transformations between the difference sequence spaces* $\Delta c_0(p)$, $\Delta c(p)$ and $\Delta \ell_\infty(p)$, Filomat **15** (2001), 353–363.

[266] E. Malkowsky, Mursaleen, Qamaruddin, *Generalized sets of difference sequences, their duals and matrix transformations*, Sequence Spaces and Applications, pp. 68–83, Narosa, New Delhi, 1999.

[267] E. Malkowsky, Mursaleen, S. Suantai, *The dual spaces of sets of difference sequences of order* m *and matrix transformations*, Acta Math. Sin. Eng. Ser. **23** (3) (2007), 521–532.

[268] E. Malkowsky, S.D. Parashar, *Matrix transformations in space of bounded and convergent difference sequences of order* m, Analysis **17** (1997), 87–97.

[269] E. Malkowsky, V. Rakočević, *An Introduction into the Theory of Sequence Spaces and Measures of Noncompactness*, Zbornik Radova, Matematički Institut SANU, Belgrade, **9** (17) (2000), 143–234.

[270] E. Malkowsky, V. Rakočević, *On matrix domains of triangle*, Appl. Math. Comput. **189** (2) (2007), 1146–1163.

[271] E. Malkowsky, V. Rakočević, S. Živković-Zlatanović, *Matrix transformations between the sequence spaces* bv^p *and certain BK spaces*, Bul. Acad. Serbe des Sciences et des Arts, Sciences Mathématique **27** (2002), 33–46.

[272] E. Malkowsky, E. Savaş, *Matrix transformations between sequence spaces of generalized weighted means*, Appl. Math. Comput. **147** (2004), 333–345.

[273] S.M. Mazhar, *Absolute summability factors of infinite series*, Kyungpook Math. J. **39** (1999), 67–73.

[274] M. Matloka, *Sequences of fuzzy numbers*, BUSEFAL **28** (1986), 28–37.

[275] F.M. Mears, *Some multiplication theorems for the Nörlund mean*, Bull. Amer. Math. Soc. **41** (1935), 875–880.

[276] M.R. Mehdi, *Summability factors for generalized absolute summability*, Proc. London Math. Soc. **10** (1960), 180–200.

[277] H.I. Miller, C. Orhan, *On almost convergent and staistically convergent subsequences*, Acta Math. Hungar. **93** (2001), 135–151.

[278] S.K. Mishra, *Matrix maps involving certain sequence spaces*, Indian J. Pure Appl. Math. **24** (2) (1993), 125–132.

[279] S.L. Mishra, B. Satapathy, N. Rath, *Invariant means and σ-core*, J. Indian Math. Soc. **60** (1994), 151–158.

[280] F. Móricz, *Extensions of the spaces c and c_0 from single to double sequences*, Acta Math. Hungar. **57** (1-2) (1991), 129–136.

[281] Mursaleen, *Matrix transformations and generalized almost convergence II*, Comm. Fac. Sci. Univ. Ank. Ser. *A1* Math. Statist. **27** (1978), 41–51.

[282] Mursaleen, *On some new invariant matrix methods of summability*, Quart. J. Math. Oxford Ser. 2 **34** (1983), 77–86.

[283] Mursaleen, *Generalized spaces of difference sequences*, J. Math. Anal. Appl. **203** (3) (1996), 738–745.

[284] Mursaleen, *Almost strongly regular matrices and a core theorem for double sequences*, J. Math. Anal. Appl. **293** (2) (2004), 523–531.

[285] M. Mursaleen, *Elements of Metric Spaces*, Anamaya Publishers, New Delhi, 2005.

[286] M. Mursaleen, *Almost convergence and some related methods*, in: M. Mursaleen (Ed.), *Modern Methods of Analysis and Its Applications*, pp. 1–10, Anamaya Publishers, New Delhi, 2010.

[287] M. Mursaleen, *Applied Summability Methods*, Springer Briefs, 2014.

[288] M. Mursaleen, F. Başar, *Sequence Spaces: Topics in Modern Summability Theory*, Series: Mathematics and Its Applications, CRC Press/Taylor & Francis Group, Boca Raton · London · New York, 2020.

[289] Mursaleen, F. Başar, B. Altay, *On the Euler sequence spaces which include the spaces ℓ_p and ℓ_∞ II*, Nonlinear Anal. **65** (3) (2006), 707–717.

[290] Mursaleen, M. Başarır, *On some new sequence spaces of fuzzy numbers*, Indian J. Pure Appl. Math. **34** (9) (2003), 1351–1357.

[291] Mursaleen, A.K. Gaur, A.H. Saifi, *Some new sequence spaces and their duals and matrix transformations*, Bull. Calcutta Math. Soc. **88** (3) (1996), 207–212.

[292] Mursaleen, O.H.H. Edely, *On some statistical core theorems*, Analysis **22** (2002), 265–276.

[293] Mursaleen, O.H.H. Edely, *Statistical convergence of double sequences*, J. Math. Anal. Appl. **288** (1) (2003), 223–231.

[294] Mursaleen, O.H.H. Edely, *Almost convergence and a core theorem for double sequences*, J. Math. Anal. Appl. **293** (2) (2004), 532–540.

[295] Mursaleen, O.H.H. Edely, *Generalized statistical concergence*, Inform. Sci. **162** (3-4) (2004), 287–294.

[296] Mursaleen, A.K. Gaur, T.A. Ghishti, *On some new sequence spaces of invariant means*, Acta Math. Hungar. **75** (3) (1997), 209–214.

[297] Mursaleen, M.A. Khan, Qamaruddin, *Difference sequence spaces defined by Orlicz functions*, Demonstratio Math. **32** (1) (1999), 145–150.

[298] H. Nakano, *Modulared sequence spaces*, Proc. Japan Acad. **27** (2) (1951), 508–512.

[299] H. Nakano, *Concave modulars*, J. Math. Soc. Japan **5** (1953), 29–49.

[300] S. Nanda, *Matrix Transformations and Sequence Spaces*, Internal report, I.C.T.P. (1983), IC/83/55.

[301] S. Nanda, *Matrix transformations and almost boundedness*, Glas. Mat. 14 (34) (1979), 99–107.

[302] S. Nanda, *On sequences of fuzzy numbers*, Fuzzy Sets and Systems **33** (1989), 123–126.

[303] P.-N. Ng, *Matrix transformations on Cesàro sequence spaces of non-absolute type*, Tamkang J. Math. **10** (2) (1979), 215–221.

[304] P.-N. Ng, P.-Y. Lee, *Cesàro sequence spaces of non-absolute type*, Comment. Math. Prace Mat. **20** (2) (1978), 429–433.

[305] F. Nuray, *Lacunary statistical convergence of sequences of fuzzy numbers*, Fuzzy Sets and Systems **99** (1998), 353–355.

[306] F. Nuray, *$\mathcal{I}$-convergence of sequences of fuzzy numbers*, New Math. Nat. Comput. **4** (2) (2008), 231–236.

[307] F. Nuray, E. Savaş, *Statistical convergence of sequences of fuzzy numbers*, Math. Slovaca **45** (3) (1995), 269–273.

[308] J.T. Okutoyi, *On the spectrum of C_1 as an operator on bv*, Commun. Fac. Sci. Univ. Ank. Ser. *A1* Math. Statist. **41** (1992), 197–207.

[309] C. Orhan, *Sublinear functionals and Knopp's core theorem*, Internat. J. Math. Math. Sci. **13** (1990), 461–468.

[310] C. Orhan, E. Öztürk, *On f-regular dual summability methods*, Bull. Inst. Math. Acad. Sinica **14** (1) (1986), 99–104.

[311] W. Orlicz, *Linear Functional Analysis* (Transl. from the Chinese by Lee Peng Yee, Addendum by W. Congxin), World Scientific Publishing Co., Singapore, 1992.

[312] E. Öztürk, *On dual summability methods*, Comm. Fac. Sci. Univ. Ank. Ser. *A1* Math. Statist. **25** (1) (1976), 1–19.

[313] E. Öztürk, *Iraksak Seriler Teorisi (Theory of Divergent Series)*, unpublished lecture notes, (in Turkish).

[314] R.F. Patterson, *Invariant core theorems for double sequences*, Southeast Asian Bull. Math. **24** (3) (2000), 421–429.

[315] R.F. Patterson, *Comparison theorems for four dimensional regular matrices*, Southeast Asian Bull. Math. **26** (2) (2002), 299–305.

[316] G.M. Petersen, *Regular Matrix Transformations*, McGraw-Hill Publishing Company Limited, London · New York · Toronto · Sidney, 1966.

[317] A. Peyerimhoff, *Lectures on Summability*, Lecture Notes in Mathematics, Springer-Verlag, Berlin · Heidelberg · New York, 1969.

[318] H. Polat, F. Başar, *Some Euler spaces of difference sequences of order m*, Acta Math. Sci. **27B** (2) (2007), 254–266.

[319] R.E. Powell, S.M. Shah, *Summability Theory and Its Applications*, Van Nostrand Reinhold Company, London · New York · Cincinnati · Toronto · Melbourne, 1972.

[320] A. Pringsheim, *Elementare Theorie der unendliche Doppel-reihen*, Sitsungsber. Akad. Wiss. München, **27** (1897), 101–153.

[321] M.L. Puri, D.A. Ralescu, *Differentials for fuzzy functions*, J. Math. Anal. Appl. **91** (1983), 552–558.

[322] Qamaruddin, A.H. Saifi, *Generalized difference sequence spaces defined by a sequence of Orlicz functions*, Southeast Asian Bull. Math. **29** (6) (2005), 1125–1130.

[323] R. Raimi, *Invariant means and invariant matrix methods of summability*, Duke Math. J. **30** (1963), 81–94.

[324] D. Rath, B.K. Tripathy, *On some classes of regular matrices*, Duke Math. J. **84** (1981), 530-533.

[325] J.B. Reade, *On the spectrum of the Cesaro operator*, Bull. Lond. Math. Soc. **17** (1985), 263–267.

[326] B.E. Rhoades, *Some properties of totally coregular matrices*, Illinois J. Math. **4** (1960), 518–525.

[327] B.E. Rhoades, *The fine spectra for weighted mean operators*, Pacific J. Math. **104** (1) (1983), 219–230.

[328] B.E. Rhoades, *The spectrum of weighted mean operators*, Canad. Math. Bull. **30** (4) (1987), 446–449.

[329] B.E. Rhoades, *Inclusion theorems for absolute matrix summability methods*, J. Math. Anal. Appl. **238** (1999), 82–90.

[330] B.E. Rhoades, *Some sequence spaces which include* c_0 *and* c, Hokkaido Math. J. **35** (2006), 587–599.

[331] B.E. Rhoades, E. Savaş, *A characterization of absolute summability factors*, Taiwanese J. Math. **8** (2004), 453–465.

[332] B.E. Rhoades, E. Savaş, *General inclusion theorems for absolute summability of order* $k \geq 1$, Math. Inequal. Appl. **8** (2005), 505–520.

[333] B.E. Rhoades, E. Savaş, *On* $|A|_k$ *summability factors*, Acta Math. Hungar. **112** (2006), 15–23.

[334] R. Rockafellar, R. Tyrrell, *Convex Analysis*, Princeton University Press, New Jersey, 1970.

[335] G.M. Robison, *Divergent double sequences and series,* Trans. Amer. Math. Soc. **28** (1) (1926), 50–73.

[336] M. Rojes-Medar, H. Roman-Flores, *On the equivalence of convergences of fuzzy sets*, Fuzzy Sets and Systems **80** (1996), 217–224.

[337] M.A. Rojas-Medar, M.D. Jiménez-Gamero, Y. Chalco-Cano, A.J. Viera-Brandão, *Fuzzy quasilinear spaces and applications*, Fuzzy Sets and Systems **152** (2005), 173–190.

[338] H. Roopaei, D. Foroutannia, M. İlkhan, E.E. Kara, *Cesàro spaces and norm of operators on these matrix domains*, Mediterr. J. Math. **17** (4) (2020), 121, https://doi.org/10.1007/s00009-020-01557-9.

[339] H. Roopaei, F. Başar, *On the spaces of Cesàro absolutely p-summable, null and convergent sequences*, Math. Meth. Appl. Sci. **44** (5) (2021), 3670–3685.

[340] W.H. Ruckle, *FK-spaces in which the sequence of coordinate vectors is bounded*, Canad. J. Math. **25** (1973), 973–978.

[341] T. Šalát, *On statistically convergent sequences of real numbers*, Math. Slovaca **30** (2) (1980), 139–150.

[342] L. Sánchez, A. Ullán, *Some properties of Gurarii's modulus of continuity*, Arch. Math. **71** (1998), 399–406.

[343] M.A. Sarıgöl, *On difference sequence spaces*, J. Karadeniz Tech. Univ. Fac. Arts Sci. Ser. Math.-Phys. **10** (1987), 63–71.

[344] M.A. Sarıgöl, *Necessary and sufficient conditions for the equivalence of the summability methods* $\left|\bar{N}, p_n\right|_k$ *and* $|C,1|_k$, Indian J. Pure Appl. Math. **22** (1991), 483–489.

[345] B. Sarma, *On a class of sequences of fuzzy numbers defined by modulus function*, Internat. J. Sci. Technol. **2** (1) (2007), 25–28.

[346] E. Savaş, *σ-regular dual summability methods*, Indian J. Pure Appl. Math. **18** (1987), 1011–1015.

[347] E. Savaş, *Matrix transformation of X_p into Cs*, J. Math. Punjab Univ. **24** (1991), 59–66.

[348] E. Savaş, *A note on double sequences of fuzzy numbers*, Turkish J. Math. **20** (2) (1996), 175–178.

[349] E. Savaş, *On strongly λ-summable sequences of fuzzy numbers*, Inform. Sci. **125** (2000), 181–186.

[350] E. Savaş, *A note on sequence of fuzzy numbers*, Inform. Sci. **124** (1-4) (2000), 297–300.

[351] E. Savaş, *Spaces of double sequences of fuzzy numbers*, Math. Comput. Appl. **5** (1) (2000), 35–40.

[352] E. Savaş, *On statistically convergent sequences of fuzzy numbers*, Inform. Sci. **137** (2001), 272–282.

[353] E. Savaş, *On almost convergent sequences of fuzzy numbers*, New Math. Nat. Comput. **2** (2) (2006), 123–130.

[354] E. Savaş, *On asymptotically λ-statistical equivalent sequences of fuzzy numbers*, New Math. Nat. Comput. **3** (3) (2007), 301–306.

[355] E. Savaş, *On lacunary statistically convergent double sequences of fuzzy numbers*, Appl. Math. Lett. **21** (2) (2008), 134–141.

[356] E. Savaş, *On quasi β-power increasing sequences*, Nonlinear Anal. **70** (2009), 1459–1464.

[357] E. Savaş, Mursaleen, *On statistically convergent double sequences of fuzzy numbers*, Inform. Sci. **162** (3-4) (2004), 183–192.

[358] P. Schaefer, *Core theorems for coregular matrices*, Illinois J. Math. **9** (1965), 207–211.

[359] P. Schaefer, *Infinite matrices and invariant means*, Proc. Amer. Math. Soc. **36** (1972), 104–110.

[360] I.J. Schoenberg, *The integrability of certain functions and related summability methods*, Amer. Math. Monthly **66** (1959), 361–375.

[361] A.A. Shcherbakov, *Kernels of sequences of complex numbers and their regular transformations*, Math. Notes **22** (1977), 948–953.

[362] I.M. Sheffer, *Note on multiply-infinite series*, Bull. Amer. Math. Soc. **52** (1946), 1036–1041.

[363] J.A. Sıddıqi, *Infinite matrices summing every almost periodic sequences*, Pac. J. Math. **39** (1) (1971), 235–251.

[364] S. Simons, *The sequence spaces $\ell(p_v)$ and $m(p_v)$*, Proc. London Math. Soc. (3) **15** (1965), 422–436.

[365] S. Simons, *Banach limits, infinite matrices and sublinear functionals*, J. Math. Anal. Appl. **26** (1969), 640–655.

[366] P.D. Srivastava, S. Kumar, *Fine spectrum of the generalized difference operator Δ_ν on sequence space ℓ_1*, Thai J. Math. **8** (2) (2010), 7–19.

[367] M. Stieglitz, *Eine verallgenmeinerung des begriffs festkonvergenz*, Math. Japon. **18** (1973), 53–70.

[368] M. Stieglitz, H. Tietz, *Matrix transformationen von folgenräumen eine ergebnisübersicht*, Math. Z. **154** (1977), 1–16.

[369] M. Stojaković, Z. Stojaković, *Addition and series of fuzzy sets*, Fuzzy Sets and Systems **83** (1996), 341–346.

[370] W.T. Sulaiman, *Extension on absolute summability factors of infinite series*, J. Math. Anal. Appl. **322** (2006), 1224–1230.

[371] G. Sunouchi, *Notes on Fourier analysis. XVIII. Absolute summability of a series with constant terms*, Tôhoku. Math. **1** (2) (1949), 57–65.

[372] C. Şençimen, *On the level statistical convergence of a sequence of fuzzy numbers*, Int. J. Pure Appl. Math. **28** (2) (2006), 257–264.

[373] M. Şengönül, F. Başar, *Some matrix transformations into the Cesàro sequence spaces of non-absolute type*, Math. J. Ibaraki Univ. **31** (1999), 13–20.

[374] M. Şengönül, F. Başar, *Some new Cesàro sequence spaces of non-absolute type which include the spaces c_0 and c*, Soochow J. Math. **31** (1) (2005), 107–119.

[375] H. Şevli, *Mutlak Toplanabilme Metotları*, Yüzüncü Yıl University, 2005, unpublished PhD thesis, (in Turkish).

[376] H. Şevli, L. Leindler, *On the absolute summability factors of inifinite series involving quasi-power-increasing sequences*, Comput. Math. Appl. **57** (2009), 702–709.

[377] G. Talebi, *Operator norms of four-dimensional Hausdorff matrices on the double Euler sequence spaces*, Linear Multilinear Algebra **65** (11) (2017), 2257–2267.

[378] Ö. Talo, F. Başar, *On the space $bv_p(F)$ of sequences of p-bounded variation of fuzzy numbers*, Acta Math. Sin. Eng. Ser. **24** (7) (2008), 1205–1212.

[379] Ö. Talo, F. Başar, *Determination of the duals of classical sets of sequences of fuzzy numbers and related matrix transformations*, Comput. Math. Appl. **58** (2009), 717–733.

[380] Ö. Talo, F. Başar, *Certain spaces of sequences of fuzzy numbers defined by a modulus function*, Demonstratio Math. **43** (1) (2010), 139–149.

[381] Ö. Talo, F. Başar, *Quasilinearity of the classical sets of sequences of the fuzzy numbers and some applications*, Taiwanese J. Math. **14** (5) (2010), 1799–1819.

[382] N. Tanovic-Miller, *On strong summability*, Glas. Mat. Ser. III **14** (1) (1979), 87–97.

[383] B.C. Tripathy, *Matrix transformations between some classes of sequences*, J. Math. Anal. Appl. **206** (2) (1997), 448–450.

[384] B.C. Tripathy, A.J. Dutta, *On fuzzy real-valued double sequence space ${}_2\ell^p_F$*, Math. Comput. Modelling **46** (9-10) (2007), 1294–1299.

[385] B.C. Tripathy, M. Et, Y. Altın, *Generalized difference sequence spaces defined by Orlicz function in a locally convex space*, J. Anal. Appl. **1** (3) (2003), 175–192.

[386] B.K. Tripathy, S. Nanda, *Absolute value of fuzzy real numbers and fuzzy sequence spaces*, J. Fuzzy Math. **8** (4) (2000), 883–892.

[387] O. Tuğ, *Four-dimensional generalized difference matrix and some double sequence spaces*, J. Inequal. Appl., 149 (2017). https://doi.org/10.1186/s13660-017-1423-y,

[388] A. Tuncer, F.B. Benli, *λ-statistical limit points of the sequences of fuzzy numbers*, Inform. Sci. **177** (16) (2007), 3297–3304.

[389] A. Türkmenoğlu [A. Gökhan], *Bazı Çift İndisli Dizi Uzayları*, Fırat Üniv. Fen Bil. Enst. 1993, unpublished PhD thesis, (in Turkish).

[390] S. Umar, H.H. Khan, *On $|N_p, \gamma, \alpha|_k$ summability of infinite series*, Indian J. Pure Appl. Math. **8** (1977), 752–757.

[391] I.I. Volkov (originator), *Absolute summability*, Encyclopedia of Mathematics, URL: http://encyclopediaofmath.org/index.php?title=Absolute_summability&oldid=44716.

[392] N.N. Vorob'ev, *Theory of Series*, 5th edn. Nauka, Moscow, 1986 (in Russian).

[393] C.-S. Wang, *On Nörlund sequence spaces*, Tamkang J. Math. **9** (1978), 269–274.

[394] R.B. Wenger, *The fine spectra of Hölder summability operators*, Indian J. Pure Appl. Math. **6** (1975), 695–712.

[395] A. Wilansky, *Functional Analysis*, Blaisdell Publishing Company, New York · Toronto · London, 1964.

[396] A. Wilansky, *Summability through Functional Analysis*, Volume 85, North-Holland Mathematics Studies, Amsterdam · New York · Oxford, 1984.

[397] A. Wilansky, K. Zeller, *The inverse matrix in summability*, J. London Math. Soc. **32** (1957), 397–408.

[398] A. Wilansky, K. Zeller, *Banach algebra and summability*, Illionis J. Math. **2** (1958), 378–385.

[399] J.M. Whittaker, *The absolute summability of a series*, Proc. Edinburgh Math. Soc. **2** (1930), 1–5.

[400] C. Wu, G. Wang, *Convergence of sequences of fuzzy numbers and fixed point theorems for increasing fuzzy mappings and application*, Fuzzy Sets and Systems **130** (2002), 383–390.

[401] C. Wu, C. Wu, *The supremum and infimum of the set of fuzzy numbers and its application*, J. Math. Anal. Appl. **210** (1997), 499–511.

[402] M. Yeşilkayagil, F. Başar, *Mercerian theorem for four dimensional matrices*, Commun. Fac. Sci. Univ. Ankara, Ser. *A1* Math. Stat. **65** (1) (2016), 147–155.

[403] M. Yeşilkayagil, F. Başar, *On the characterization of some classes of four dimensional matrices and Steinhaus type theorems*, Kragujevac J. Math. **40** (1) (2016), 35–45.

[404] M. Yeşilkayagil, F. Başar, *Domain of Riesz mean in the space* $\mathcal{L}_s$, Filomat **31** (4) (2017), 925–940.

[405] M. Yeşilkayagil, F. Başar, *Domain of Riesz mean in some spaces of double sequences*, Indag. Math. (N.S.) **29** (3) (2018), 1009–1029.

[406] M. Yeşilkayagil, F. Başar, *Domain of Euler mean in the space of absolutely* p*-summable double sequences with* $0 < p < 1$, Anal. Theory Appl. **34** (3) (2018), 241–252.

[407] M. Yıldırım, *On the spectrum and fine spectrum of the compact Rhally operators*, Indian J. Pure Appl. Math. **27** (8) (1996), 779–784.

[408] L.A. Zadeh, *Fuzzy sets*, Information and Control **8** (1965), 338–353.

[409] M. Zeltser, *Matrix transformations of double sequences*, Acta Comment. Univ. Tartu. Math. **4** (2000), 39–51.

[410] M. Zeltser, *Investigation of Double Sequence Spaces by Soft and Hard Analitical Methods*, Diss. Math. Univ. Tartu. 25, Tartu University Press, Tartu, 2001.

[411] M. Zeltser, *On conservative and coercive SM-methods*, Proc. Est. Acad. Sci. Phys. Math. **50** (2) (2001), 76–85.

[412] M. Zeltser, *Weak sequential completeness of β-duals of double sequence spaces*, Anal. Math. **27** (3) (2001), 223–238.

[413] M. Zeltser, *On conservative matrix methods for double sequence spaces*, Acta Math. Hungar. **95** (3) (2002), 225–242.

[414] M. Zeltser, M. Mursaleen, S.A. Mohiuddine, *On almost conservative matrix mathods for double sequence spaces*, Publ. Math. Debrecen **75** (2009), 387–399.

Index